D0965648

THE AMIDE LINKAGE

THE AMIDE LINKAGE

SELECTED STRUCTURAL ASPECTS IN CHEMISTRY, BIOCHEMISTRY, AND MATERIALS SCIENCE

Editors

ARTHUR GREENBERG
CURT M. BRENEMAN
JOEL F. LIEBMAN

A JOHN WILEY & SONS, INC., PUBLICATION

New York / Chichester / Weinheim / Brisbane / Singapore / Toronto

This book is printed on acid-free paper. ♾

Published simultaneously in Canada.

Library of Congress Cataloging-in-Publication Data:

The amide linkage : selected structural aspects in chemistry,
biochemistry, and materials science / editors, Arthur Greenberg,
Curt M. Breneman, and Joel F. Liebman.
p. cm.
Includes index.
ISBN 0-471-35893-2 (alk. paper)
1. Amides. I. Greenberg, Arthur. II. Breneman, Curt M.
III. Liebman, Joel F.
QD305.A7A35 2000 99-15235
547'.042–dc21

Printed in the United States of America.

10 9 8 7 6 5 4 3 2 1

DEDICATION AND DESIDERATA

Amour	Chimie	Joie
Sustenance	Reading	Friendship
Generosity	Writing	Creativity

CONTENTS

PREFACE

The amide linkage is a common and very important functional group which contributes to the special properties of peptides, proteins, beta-lactam antibiotics, and numerous synthetic polymers. The discovery of its rigid planarity and propensity for hydrogen bonding sped Linus Pauling's discovery of the alpha-helix and now offers a template for the design of exotic new materials. The present book presents many diverse aspects of the structure, bonding, and dynamics of amides and their influences on chemical, physical, biological, and material properties. It emphasizes a structural, energetics, and bonding perspective with a view toward understanding physical and chemical properties. The approach is also pedagogical in nature—its eighteen chapters provide background and context for the non-expert before going into depth in a particular area. This book is meant to be complementary to Zabicky's classic *The Chemistry of Amides* in the Patai series published in 1970, in that it is broader in scope and places less emphasis on the synthesis and reactions of amides. The Zabicky book continues to provide a great deal of useful background information for readers of the present book.

The amide linkage is undergoing a renaissance of interest at the start of the twenty-first century. When amide planarity and rigidity was discovered by Pauling and co-workers in the 1930s, this important attribute was rationalized by resonance theory—one of the widely accepted canons of modern chemistry. However, theories and "accepted knowledge" are challenged every generation and the first two chapters, by Breneman and then Wiberg, apply molecular orbital (MO) theory, density functional theory (DFT), and Bader's Atoms in Molecules (AIM) approach to achieve new levels of understanding of the structures and dynamics of amides. Chapter 3, by Greenberg, uses the more traditional resonance canon to understand the effects of distortion on the ligand properties (e.g., the site of protonation) of amides. Chapter 4 by Brown details the very subtle physical organic chemistry investigations of the minute details of acid- and base-catalyzed hydrolysis of lactams (cyclic amides). Although amides and lactams are normally stable to hydrolysis around neutral pH, he describes the kinetics and mechanistic details for highly distorted lactams that react under these conditions. Liebman, Afeefy, and Slayden (Chapter 5) summarize the thermochemical data for amides. As valuable as this information is for practical problems in engineering and for understanding structure and bonding, there is surprisingly little published data. Chapter 6 (Hoffman) and Chapter 7 (Bose,

Manhas, Banik, and Srirajan) treat selected aspects of the stereoselective synthesis of alpha-lactams and beta-lactams. Chapter 7 also provides very interesting historical perspective on the role of serendipity in the discovery of penicillin and related antibiotics. In Chapter 8, Yamada explores the structural, spectroscopic, and chemical properties of acyclic amides twisted by virtue of steric repulsion. Rademacher (Chapter 9) explores the use of UV photoelectron spectroscopy (PES) as a means of understanding the nature of chemical bonding in planar and distorted amides and lactams.

Starting with Chapter 10 (Palmore and MacDonald), the emphasis switches towards larger amide-containing molecules. Palmore and MacDonald use hydrogen bonding, so vital to the structures of alpha-, beta- and other protein structures, to design new materials in which the sum of numerous weak-moderate H-bonds acting in concert more than equals the strength of attachment due to covalent bonds. Boyd's Chapter 11 nicely describes the role of the ab initio MO theory in the exploration of beta-lactam antibiotics. It provides additional perspective to this interesting history and leads into Chapter 12 (Bohacek and Shakespeare) which describes the design of new enzyme inhibitors as potential pharmaceutical agents. In Chapter 13, Perczel and Csizmadia explore the current limits of computing full conformational energy surfaces (Ramachandran plots) for peptides and proteins at accessible ab initio levels. Cassady (Chapter 14) explores the interactions of peptides and proteins with ions in order to elucidate both proton affinity and structural information. The idea of obtaining mass spectra of proteins seemed to be heretical until recently. In Chapter 15, Maitra and Nowick consider the interactions between beta-sheet proteins in order to gain important insights into molecular recognition. Chapter 16, by Spatola and Romanovskis, describes the generation of cyclic peptide libraries for use in the exploding field of combinatorial chemistry. Chapter 17 (Cieplak) and Chapter 18 (Kallenbach, Bell, and Spek) explore some of the myriad complexities in understanding the subtleties of protein folding that is so vital to our understanding of the biosynthesis and functioning of proteins including enzymes.

We hope that readers will find this book valuable, and will be encouraged to seek as yet unknown applications and understanding for the ubiquitous amide group.

ARTHUR GREENBERG
Charlotte, North Carolina

CURT M. BRENEMAN
Troy, New York

JOEL F. LIEBMAN
Baltimore, Maryland

CONTRIBUTORS

DR. HUSSEIN Y. AFEEFY, Department of Chemistry and Biochemistry, University of Maryland Baltimore County, Baltimore, MD 21250 (USA).

DR. B.K. BANIK, Department of Experimental Molecular Pathology, M.D. Anderson Cancer Center, Houston, TX 777030 (USA).

DR. ANTHONY J. BELL, JR., Department of Chemistry, New York University, 29 Washington Place, New York, NY 10003 (USA).

DR. REGINE S. BOHACEK, Ariad Pharmaceuticals, Inc., 26 Landsdowne Street, Cambridge, MA 02139-4234 (USA).

DR. AJAY K. BOSE, Chemistry and Chemical Biology Department, Stevens Institute of Technology, Castle Point on the Hudson, Hoboken, NJ 07030 (USA).

DR. DONALD B. BOYD, Department of Chemistry, Indiana University—Purdue University, Indianapolis, IN 46202-3274 (USA).

DR. CURT M. BRENEMAN, Department of Chemistry, Cogswell Laboratory 319A, Rensselaer Polytechnic Institute, Troy, NY 12180 (USA).

DR. R.S. BROWN, Department of Chemistry, Queen's University, Kingston, Ontario, 5KL 1B8 (Canada).

DR. CAROLYN J. CASSADY, Department of Chemistry and Biochemistry, University of Alabama, Box 870336, Tuscaloosa, AL 35487-0336 (USA).

DR. ANDRZEJ S. CIEPLAK, Department of Chemistry, Bilkent University, 06533 Bilkent, Ankara (Turkey).

DR. IMRE G. CSIZMADIA, Department of Chemistry, University of Toronto, Toronto Ontario, M5S 1A1 (Canada).

DR. ARTHUR GREENBERG, Department of Chemistry, University of North Carolina at Charlotte, Charlotte, NC 28223 (USA).

DR. ROBERT V. HOFFMAN, Department of Chemistry and Biochemistry, New Mexico State University, Las Cruces, NM 88003-0001 (USA).

Dr. Neville R. Kallenbach, Department of Chemistry, New York University, 29 Washington Place, New York, NY 10003 (USA).

Dr. Joel F. Liebman, Department of Chemistry and Biochemistry, University of Maryland Baltimore County, Baltimore, MD 21250 (USA).

Dr. John C. MacDonald, Department of Chemistry, Northern Arizona University, Flagstaff, AZ 86011 (USA).

Dr. Santanu Maitra, Department of Chemistry, University of California, Irvine, Irvine, CA 92697-2025 (USA).

Dr. Maghar Singh Manhas, Chemistry and Chemical Biology Department, Stevens Institute of Technology, Castle Point on the Hudson, Hoboken, NJ 07030 (USA).

Dr. Martin Martinov, Department of Chemistry, Cogswell Laboratory 319A, Rensselaer Polytechnic Institute, Troy, NY 12180 (USA).

Dr. James S. Nowick, Department of Chemistry, University of California, Irvine, Irvine, CA 92697-2025 (USA).

Dr. G. Tayhas R. Palmore, Department of Chemistry, University of California, Davis, Davis, CA 95616 (USA).

Dr. Andras Perczel, Department of Chemistry, L. Eotvos University, 112 Budapest P.O.B. 32, H-1518 (Hungary).

Dr. Paul Rademacher, Institut für Organische Chemie, Universität GH Essen, Universitatsstrasse 5-7, D-45117 Essen (Germany).

Dr. Peteris Romanovskis, Department of Chemistry, University of Louisville, Louisville, KY 40292 (USA).

Dr. William C. Shakespeare, Ariad Pharmaceuticals, Inc., 26 Landsdowne Street, Cambridge, MA 02139-4234 (USA).

Dr. Suzanne W. Slayden, Department of Chemistry, George Mason University, Fairfax, VA 22030 (USA).

Dr. Arno F. Spatola, Department of Chemistry, University of Louisville, Louisville, KY 40292 (USA).

Dr. Erik J. Spek, Department of Chemistry, New York University, 29 Washington Place, New York, NY 10003 (USA).

Dr. V. Srirajan, Department of Chemistry, Temple University, Philadelphia, PA 19122 (USA).

Dr. Kenneth B. Wiberg, Department of Chemistry, Yale University, New Haven, CT 06520-8107 (USA).

Dr. Shinj Yamada, Department of Chemistry, Faculty of Science, Ochanomizu University, Bunkyo-ku Tokyo 112-8610 (Japan).

THE AMIDE LINKAGE

CHAPTER 1

THE ELECTRON DENSITY DISTRIBUTION OF AMIDES AND RELATED COMPOUNDS

CURT M. BRENEMAN and MARTIN MARTINOV
Department of Chemistry, Rensselaer Polytechnic Institute

1. INTRODUCTION

The peculiar characteristics and ubiquitous nature of amide bonds have stimulated a great deal of research concerning the electronic effects responsible for their conformational preferences and unusual stabilities. The following chapter contains an overview of several investigations into the phenomenon of amide stability and an examination of several different interpretations made by workers in this field. Since one of the traditional explanations for the special nature of amide linkages involves aspects of the qualitative resonance theory of allylic compounds, the first section of the work is devoted to a general review of this traditional model. More recent alternative views of amide stability and reactivity offered by us and by other groups are also included in the chapter. Since delocalization of charge density is a core issue in both resonance theory and AIM (atoms in molecules) analysis, this review includes a discussion of the nature of atomic charges in general. Finally, certain characteristics of the electron density distribution responsible for the chemical properties of several amide-like compounds such as enaminonitriles, fluorosulfonamides and sulfonamides are compared with those of amides. The comparison yields a striking similarity of electronic effects among this set of loosely related compounds.

The Amide Linkage: Selected Structural Aspects in Chemistry, Biochemistry, and Materials Science,
Edited by Arthur Greenberg, Curt M. Breneman, and Joel F. Liebman
ISBN 0-471-35893-2

2. RESONANCE AND AMIDES

2.1. Introduction to Amide Resonance

The special properties of amides are of great importance in determining the conformations, electrostatics, and geometries of peptides, proteins, and related compounds. Consequently, a number of theoretical and experimental studies have been devoted to amides, especially on their planarity and the high rotational barrier about the C–N bond.[1–16] Many of the experimentally observed properties are traditionally explained by invoking the following resonance formalism:

$$HC(=O)NH_2 \longleftrightarrow HC(-O^-)=NH_2^+$$

The assumption of a partially double C–N bond readily explains the nature of the rotation barrier, since this kind of stabilizing interaction would be lost when the amino group is rotated out of planarity. The resonance model also accounts for the amino group planarity in amides since interaction of its lone pair of electrons with the adjacent carbonyl group is required for resonance stabilization. The resonance model also provides a convenient explanation for the slow addition rates of nucleophiles to amide carbonyls, since such reactions would necessitate loss of resonance energy. Similarly, the observed thermodynamic stability of amides may be attributed to resonance interactions. The magnitude of this effect is exemplified by the following isodesmic reaction (kcal mol^{-1}):[17]

$$HC(=O)CH_3 + N(CH_3)_3 \longrightarrow HC(=O)N(CH_3)_2 + CH_3{-}CH_3$$

$$\Delta H_f = -39.37 \qquad \Delta H_f = -5.67 \qquad \Delta H_f = -45.8 \qquad \Delta H_f = -20.24$$

$$\Delta\Delta H = -21.8\ \text{kcal mole}^{-1}$$

The energy change observed in this example is close to that of the rotational barrier of dimethylformamide, suggesting that a hypothetical isodesmic reaction involving the 90° rotated structure would be close to thermoneutral.

Even though the resonance model is successful in explaining several experimental observations, some of the limitations of resonance theory are exposed when amides are examined by appropriate computational means. Some of the results obtained by ab initio calculations are shown in Table 1.1.[14]

The data in Table 1.1 show that the C–N bond increased in length by 0.08 Å on a 90° rotation, while the C=O carbonyl bond length decreased by only 0.01 Å. If the resonance model were truly descriptive, one might have expected a

TABLE 1.1. Structural Parameters of Formamide Conformers[a]

	Planar			90° Structure (**A**)		270° Structure (**B**)	
Parameter	HF	MP2	Obs	HF	MP2	HF	MP2
Energy (6–31G*)	−168.93070	−169.40538		−168.90569	−169.37878	−168.90114	−169.37462
ΔE (kcal mol^{-1})	0.0	0.0		15.69	16.69	18.55	19.30
Energy (6–31G**)	−168.94048			−168.91510		−168.91072	
ΔE (kcal mol^{-1})	0.0			15.98		18.67	
$r_{C=O}$	1.1927	1.2238	1.2	1.1832	1.2169	1.1789	1.2124
r_{CN}	1.3489	1.3606	1.3	1.4273	1.4421	1.423	1.4368
r_{CH}	1.0910	1.1046	1.0	1.0876	1.1004	1.0943	1.1090
r_{NH}	0.9957[b]	1.0107[b]	1.0	1.0055	1.0225	1.0046	1.0220
	0.9929[c]	1.0084[c]	1.0				
N–C–O	124.95	124.73	124.0	125.05	125.38	123.27	122.83
H–C–N	112.66	112.37	112.0	113.48	113.06	116.39	116.71
C–N–H	119.33[b]	118.95[b]	118.0	108.48	107.24	106.78	108.35
	121.79[c]	121.83[c]	121.0				
τ[d]	0.0	0.0	0.0	57.06	55.74	121.71	123.27

[a] Total energies are in Hartrees, bond lengths in Å, and bond angles in degrees. Geometries were determined using the HF/6–31G* method except where noted.
[b] Hydrogen eclipsed with carbonyl.
[c] Hydrogen eclipsed with aldehyde hydrogen.
[d] O=C–N–H torsional angle.

Planar A B

more equal change in bond lengths. Such observations, combined with the importance of the amide functionality, have justified numerous computational and theoretical studies on the subject. Several of the more recent investigations have challenged the widely accepted resonance explanation and pointed out some of its limitations. The additional questions to be addressed in the remainder of this chapter are:

1. Does the nitrogen of an amide actually have a partial positive charge in the planar ground state of the molecule?
2. Does the amide oxygen absorb the donated π-density of the amide nitrogen in an allylic fashion?
3. Is the carbonyl carbon a bystander or a participant in "resonance" behavior?
4. What do we mean by atomic charges?
5. Do infrared stretching frequencies constitute adequate indicators of bond order?

3. ALLYLIC RESONANCE AND CHARGE DENSITY DELOCALIZATION

3.1. Introduction to Allylic Resonance

Much work has been done for the purpose of understanding the added stability afforded to allylic cations, radicals and anions by virtue of their ability to delocalize charge or spin density.[18] Accompanying this extra stability are rotational barriers about the partial π-bonds involved in the allylic backbone. As a natural consequence of the coexistence of these two observable phenomena, the chemistry community has generally accepted the concept of allylic resonance between canonical forms describing each Kekule structure. At issue, however, is the extent to which the empirical evidence can be adequately explained on the basis of symbolic resonance structures alone.

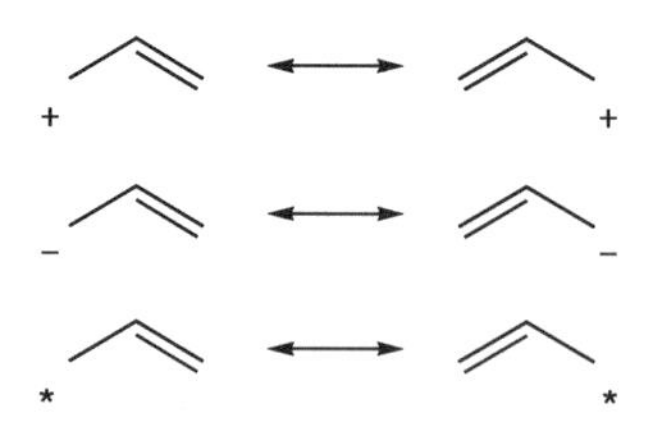

As a result of this controversy, resonance interactions in allylic systems have come under theoretical scrutiny. The early predominance of Huckel molecular orbital theory and valence bond theory as semiquantitative ways of examining π-system interactions has helped to reinforce the popular concept of resonance interactions. Of special interest are amide-type allylic systems by virtue of their charge neutrality, high C–N bond rotational barrier (18–22 kcal mol^{-1}) and importance in protein chemistry. It is interesting to note that when computational methods such as AM1 and MOPAC became commonplace, it became necessary to add a special additive molecular-mechanics correction for calculations performed on molecules containing amide linkages. Without this special correction, these semiempirical quantum mechanical methods failed to fully account for the geometric preferences and added stability of amides. Good ab initio calculations have been able to quantitatively reproduce experimental results, but by themselves cannot provide answers concerning the electronic nature of resonance interactions. Why? In order to gain perspective into this issue, the reader must take a critical look at the way computational results have been applied in support of the widespread qualitative concepts of resonance.

During the early stages of frontier molecular orbital (FMO) theory development, the concept of allylic resonance was examined using the available theoretical models. Methods such as Huckel molecular orbital theory were not capable of reproducing the observed resonance energies quantitatively, but were able to provide an easy qualitative explanation for the effect: favorable $\pi-\pi$ interactions resulting in electron delocalization. This expected result comes from an exercise in circular logic, since the Huckel method attributes *all* electronic effects to various $\pi-\pi$ interactions. Consequently, Huckel calculations were not particularly useful for the analysis of resonance interactions.

More advanced semiempirical quantum mechanical methods (CNDO/2, INDO, MNDO, MINDO/3, NDDO, AM1, PM3, and others) account for mixed $\sigma-\pi$ interactions in a semiquantitative way by using a Hamiltonian of parameterized electronic interactions within a self-consistent field of many electrons. Since such semiempirical LCAO-SCF calculations provide a wealth of orbital coefficients and bond orders as part of their output stream, it was quite tempting to use such information to perform FMO analyses of resonance phenomena without appreciating the inherent limitations of the method. The advent of widely available ab initio packages incorporating many basis set choices (such as the Gaussian series of programs)[19] enabled more research groups to discover that most of the popular charge density analysis methods are dependent upon basis set selection. Two fundamental questions had to be addressed in order to continue:

1. Besides conformational preferences, geometry changes, and added stability, what other observable phenomena are related to "resonance"?
2. Is it possible to reliably detect and analyze any such observables using available experimental or theoretical methods?

In answer to the first question, the pedagogical model for allylic resonance may be used as a guide. Within this model, the extra stability, bond length (or at least bond order) changes and a significant rotational barrier attributed to resonance interactions involves the influence of at least one additional charge-delocalized canonical form of the structure. In the case of symmetrical allyl ions and radicals, both symmetric forms would be of equal importance in determining the electronic character of the molecule. In the special cases of heteroatom-containing allylic systems, the second canonical form is often zwitterionic and contributes only a small fraction of its character to the molecule due to the relatively high energy of charge-separated electronic configurations. Amide linkages are members of this special class of hetero-allyl systems.

Since bond orders correlate with bond lengths for a given pair of atoms, it is reasonable to expect that allylic donation of the π-type lone-pair electron density should result in nearly equal changes in both C–N and C=O bond lengths. The figure below illustrates this type of hetero-allylic resonance.

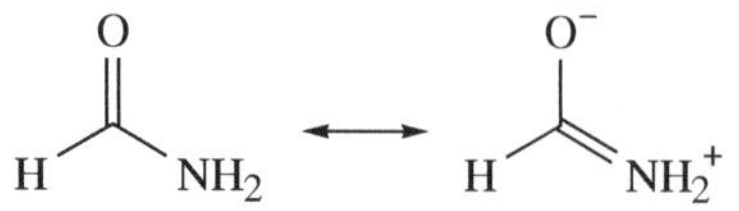

O O$^-$

H NH_2 ⟷ H NH_2^+

The contribution of the zwitterionic canonical form on the right would therefore be expected to shorten the C–N bond and elongate the C=O bond by nearly equivalent amounts. This effect should be removed when the amide NH_2 group is rotated out of conjugation with the carbonyl π system.

The data in Table 1.1 show that this expectation is not met.[14] While the C–N and C=O bond lengths are observed to change in appropriate directions, they do not change in equivalent amounts. Both HF/6–31G* and MP2/6–31G* optimized geometries of the planar and both rotated forms are consistent in this regard. The significance of this observation has been discussed in the literature by Wiberg and Breneman.[14]

Another potential indicator of bond order may be found in the harmonic force constants of the carbonyl group and C–N bond in each rotational isomer of formamide. Table 1.2 shows the HF/6–31G* vibrational force constants for the planar structure as well as the two rotated structures.

The data in Table 1.2 show that the C–N stretching force constant is reduced in the rotated structures (A and B) by approximately 20%, and that the C=O force constant is increased by only 7% (A) or 9% (B). These observations are consistent with the computed geometry changes, but are not in quantitative agreement with a simple resonance model. Since this data is derived from a computational model instead of experiment, the quality of the theoretical method must be evaluated. In this case, it has been established that the HF/6–31G* theoretical model gives good geometries with a systematic shortening of

TABLE 1.2. Some Significant Force Constants in Formamide Conformers[a]

Force Constant	Planar	90° Structure (**A**)	270° Structure (**B**)
C–N Stretch	8.22	6.15	6.20
C=O Stretch	15.80	16.95	17.30
C=O/C–N Interaction	1.59	1.01	1.00

[a]Force constants are given in mdyne/Å.

bond lengths by 1%.[20] This results in an increase in calculated force constants and introduces the need to scale the associated vibrational frequencies 0.88 below 2000 cm^{-1} and 0.91 above 2000 cm^{-1}. While the scaled frequencies differ from experimental values by around 25 cm^{-1}, it has been established that trends are well reproduced using this theoretical model and scaling scheme.[21–23] The experimental and computed vibrational frequencies of formamide are shown in Table 1.3.

The data in Table 1.3 shows that most of the vibrational frequencies for planar formamide compare well with the experimental results. The frequency of the low energy out-of-plane wagging mode of the amino group is underestimated by a large amount, however, due to the harmonic approximation used in the frequency analysis. Other work has shown that this wagging mode has a large quartic component in its vibrational potential well, and good results would not be expected when using a harmonic approximation.

When taken together, the thermochemical, geometric, and vibrational data for formamide indicate that the effects attributed to amide resonance are observable in both theoretical and experimental studies without invoking specific models of charge density behavior. The data also shows that while the simple resonance model provides a qualitative explanation of the observed effects, it also suggests that a more detailed examination of the flow of charge and energy is required to understand hetero-allylic resonance at a fundamental level. The next section describes the motivations and theories which drove the development of one particularly effective type of charge density analysis.

TABLE 1.3. Calculated and Experimental Vibrational Frequencies of Formamide (cm^{-1})

Experimental[a]	Planar[b]	90° Structure (**A**)[b]	270° Structure (**B**)[b]
289	99	−444	−398
565	544	565	595
602	593	850	833
1030	1021	904	836
1059	1041	1068	1029
1255	1213	1216	1213
1378	1380	1363	1384
1572	1575	1573	1583
1734[c]	1760[c]	1788[c]	1816[c]
2852	2923	2962	2880
3451	3492	3362	3368
3545	3614	3433	3440

[a] Data from Evans, J. C. *J. Chem. Phys.* **1954**, *22*, 1228 and King, S.T. *J. Phys. Chem.* **1971**, *75*, 405.

[b] Corrected by multiplying all values under 2000 cm^{-1} by 0.88, and all those over 2000 cm^{-1} by 0.91.

[c] Carbonyl stretching mode.

4. ATOMIC CHARGE MODELS

When ab initio quantum methods became generally available, the analysis of resonance began in earnest. Early ab initio work produced a variety of method-dependent results. When reasonably large basis sets and good theoretical models were used in addition to full geometry optimization, it was found that resonance and aromaticity phenomena could be accurately reproduced. The ability to reproduce experimental observations quantitatively using ab initio models was a necessary first step towards building a full conceptual understanding of resonance effects. By itself, these computations were not sufficient for interpreting such effects without an effective way to study the computed wavefunctions. When charge density analysis methods such as PROAIM[24] were first applied to the problem, there emerged an alternative way of thinking about the nature of the resonance effect. This evolution is described below.

Traditional ab initio calculations have proven to be useful for estimating the relative energies and molecular geometries of both known and unknown compounds. However, when such calculations are used by themselves, they often fail to provide insights about the ultimate atomic-level molecular mechanisms responsible for the observed phenomena. The main reason for this limitation stems from the fact that the quantum mechanics of open systems can only be used for complete molecular systems—the postulates of traditional quantum mechanics do not contain any definition of atoms in molecules. Therefore, any direct decomposition of the orbital representation into atomic components will constitute a partitioning of the Hilbert-space representation of the wavefunction and will thus be explicitly dependent on the basis set and, as long as it contains a reference to the basis functions, it will be ambiguous. This often hinders any attempt to connect the results of a sophisticated ab initio calculations directly with a set of clear and intuitive chemical concepts[25] and the need for better interpretative tools has been long recognized. An outstanding effort to this end is the popular topological theory of atoms in molecules (AIM).[26] This theory characterizes the gradient field of the molecular electron density and relates it to chemical properties of atoms in molecules. A remarkable feature of this theory is that by adding a single postulate—one that identifies atoms as spatial regions defined by the so-called zero-flux surfaces—to the set of postulates of quantum mechanics, it became possible to provide a rigorous definition of any atomic property. Next we present a brief overview of the AIM formalism and discuss its applications to the study of the properties of amides.

5. ATOMS-IN-MOLECULES (AIM) THEORY

The Born–Oppenheimer electron density, $\rho(\boldsymbol{r})$, is a real, scalar, Cartesian function that is finite everywhere and infinitely differentiable over all space except at the positions of the nuclei. Associated with this scalar field is a uniquely defined vector field, $\nabla\rho(\boldsymbol{r})$ for each nuclear configuration. The topological properties of this vector field are used to partition the Cartesian molecular space into rigorous and well-defined atomic components. A gradient path through a point $\boldsymbol{r}_n$ is a trajectory in this vector field and is defined by Eq. (1.1)

$$g_n = \left\{ \boldsymbol{r}(s) \left| \frac{\mathrm{d}\boldsymbol{r}(s)}{\mathrm{d}s} = \nabla\rho(\boldsymbol{r}(s)) \right. \right\}, \tag{1.1}$$

where s is a path parameter that approaches $-\infty$ as the path approaches the path origin, $+\infty$ near the path terminus, and is equal to $\mathbf{0}$ at $\boldsymbol{r}_n$, $r(\mathbf{0}) = \boldsymbol{r}_n$. A gradient path at each point is orthogonal to the corresponding electron density isosurface containing this point. Families of electron density isosurfaces have found numerous applications in various problems such as molecular shape and similarity analysis[27] and in the generation of molecular descriptors.[28] It is often helpful to use the relation between gradient paths and isosurfaces for connection between AIM and other fields, such as QSAR and QSPR.[28] Also, it is possible to use this orthogonality condition to construct convenient local coordinate systems for the calculation of observable molecular properties.[29] Every gradient path of the $\nabla\rho(\boldsymbol{r})$ field originates either at infinity or at a critical point of the same field, and terminates at another critical point (Fig. 1.1). A critical point, $\boldsymbol{c}$, is a point for which the gradient vanishes, i.e. when $\nabla\rho(\boldsymbol{c}) = \mathbf{0}$.

A convenient classification of stable critical points uses the eigenvalues of the gradient field Hessian matrix as given by Eq. (1.2)

$$h_{ij} = \left(\frac{\partial^2 \rho(\boldsymbol{r})}{\partial r_i \partial r_j} \right)_{\boldsymbol{r}=\boldsymbol{c}} \tag{1.2}$$

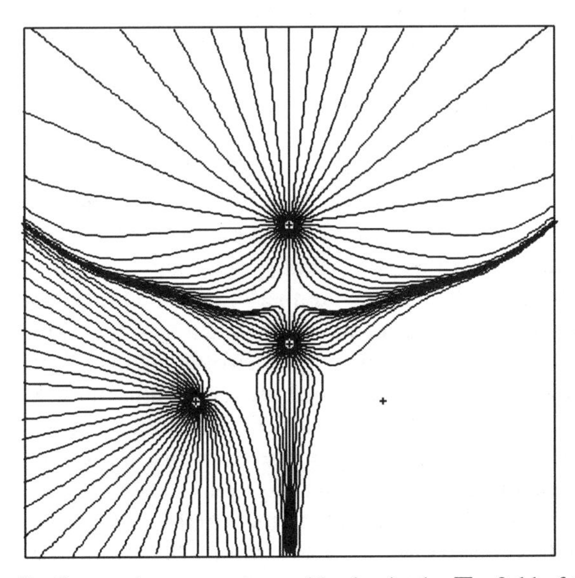

FIGURE 1.1. Gradient paths, attractors, and basins in the $\nabla\rho$ field of formaldehyde.

The rank of the critical point is equal to the number of non-zero Hessian eigenvalues and the signature is the difference between the number of the positive and the number of negative eigenvalues. Thus, four types of critical points can be identified. The local maxima in the scalar field of $\rho(\boldsymbol{r})$, which are also called attractors, are labeled as $(3,-3)$-type points since they possess a rank 3 and a signature -3. The local minima, also called for cage points chemical reasons, are of type $(3,-3)$. The remaining types of critical points, $(3,-1)$ and $(3,1)$, are called bond and ring points respectively, and represent saddle points in the scalar field of the electron density. It should be noted that the Hessian eigenvalues of most attractors can be defined only in a topological sense due to the nonexisting derivative of $\rho(\boldsymbol{r})$ at the position of the nuclei, which constitute the vast majority of attractors. The identification of the types of critical points in the gradient field suffices for the definition of important entities such as atoms in molecules, atomic surfaces, bond paths, and molecular graphs. An atom is defined as the union of the attractor and the space spanned by the gradient paths terminating at this attractor (this space is called atomic basin). The existence of non-nuclear attractors has been shown in some peculiar chemical systems,[30,31] but their incorporation in AIM presents no difficulty. The boundary of the atomic basin is called atomic surface. The atomic surface is a zero-flux surface, S,

$$S(\boldsymbol{r}) = \{\boldsymbol{r}|\rho(\boldsymbol{r}) \cdot \boldsymbol{n}(\boldsymbol{r}) = \boldsymbol{0}\}, \tag{1.3}$$

which does not contain an attractor. In Eq. (1.3), $\boldsymbol{n}(\boldsymbol{r})$ denotes a vector normal to the surface S. The common boundary of two atomic basins is called interatomic surface and necessarily contains a $(3,-1)$ point at which all the gradient paths of the interatomic surface terminate. Also, the positive eigenvalue of this $(3,-1)$ point is associated with two gradient paths starting at this point and ending respectively at the two attractors at whose basins the interatomic surface delineates. For this reason, the critical point is called a *bond point* and its union with the two gradient path it originates is called a *bond path*. More generally, they are called an *interaction* bond and path respectively, since their existence sometimes signifies the presence a weak nonbonded interaction. One can proceed further to define the set of all bond paths in the system as a molecular graph and use basic graph theory to identify rings and cages and show that their respective numbers are equal to the number of (3,1) and (3,3) points. The above definitions provide a valuable theory of chemical structure, however, the most remarkable connection between the properties of the field of $\nabla\rho(\boldsymbol{r})$ and quantum chemistry is hidden in the definition of atoms in molecules. Since atoms are regions of Cartesian space bound by a zero-flux surface, it is possible to use Schwinger's quantum action principle[32] to show that the so-defined atoms are valid independent quantum-mechanical systems. Therefore, the fundamental relations of quantum mechanics such as the virial and Ehrenfest theorems are satisfied for them. This allows for the rigorous definition and calculation of atomic properties in molecular systems. Indeed, let a one-electron operator P

correspond to a particular property P. Then, if $\Gamma(\boldsymbol{x},\boldsymbol{x}')$ is the reduced first order density matrix and $\boldsymbol{x}=\{\boldsymbol{s},\boldsymbol{r}\}$ is a four-dimensional spin-spatial coordinate, its expectation value is given by Eq. (1.4)

$$\langle \mathsf{P} \rangle = \int [(\mathsf{P}\Gamma(\boldsymbol{x},\boldsymbol{x}'))]|_{\boldsymbol{x}=\boldsymbol{x}'}\mathrm{d}\boldsymbol{x} \tag{1.4}$$

It is also equal to the sum of the expectation values for all basins present in the molecular system,

$$\langle \mathsf{P} \rangle = \sum \langle \mathsf{P} \rangle_{\mathrm{A}}, \tag{1.5}$$

where the integrations has been performed over the respective basins,

$$\langle \mathsf{P} \rangle_{\mathrm{A}} = \int_{\Omega(\mathrm{A})} [(\mathsf{P}\,\Gamma(\boldsymbol{x},\boldsymbol{x}'))]|_{\boldsymbol{x}=\boldsymbol{x}'}\mathrm{d}\boldsymbol{x} \tag{1.6}$$

Here $\Omega(\mathrm{A})$ is the basin corresponding to the attractor A. The nuclei of the overwhelming majority of molecules correspond uniquely to attractors in the field of $\nabla\rho(\boldsymbol{r})$. A similar one-to-one relationship exists between atoms and basins and it can be exploited to calculate atomic properties by straightforward application of the above equation. Indeed, this has been done for a large variety of properties, such as bond orders,[33,34] atomic Fukui indices,[35] electronegativities[36] and energy components,[37] and molecular similarities,[38] to name only a few. This approach has also served as a starting point for algorithms reconstructing molecular properties from atomic components.[6,39–43] Even though this is a powerful formalism that yields a variety of atomic properties, it is probably most frequently used to evaluate atomic charges, Q_{A},

$$Q_{\mathrm{A}} = Z_{\mathrm{A}} - \int \rho(\boldsymbol{r})\mathrm{d}\boldsymbol{r}, \tag{1.7}$$

where Z_{A} is the respective nuclear charge. Of course, when using this definition of atomic charges, one should bear in mind that the corresponding atomic basins are highly nonspherical, so the resulting atomic dipole moments can often be substantial. Indeed, it was the ability to rigorously define and compute atomic charges that inspired the early theoretical studies of the properties of amides and the validity of the associated resonance picture. However, tools now exist which enable even more detailed investigations of atomic and molecular properties, such as the atomic distribution of electrostatic potential, electronic kinetic energy density, local average ionization potential, and other properties of electron density.

6. A CHARGE-DENSITY ANALYSIS OF AMIDE RESONANCE

Once an appropriate set of charge-density analysis tools were available, Wiberg and co-workers[12,14,15] performed a detailed analysis of amide resonance. The aim of their study was to understand the way lone pair/carbonyl resonance affected the electronic structure of formamide. This investigation involved both PROAIM and more traditional methods of atomic charge determination (Mulliken,[25] CHELPG[2]). The purpose of this multi-faceted approach was to discover whether these vastly different methods of charge analysis could produce consistent or complementary results.

In this study, the rotational pathway from Structure A was followed as it traveled along the rotational isomerization reaction coordinate from the *syn* rotational transition state (A) to the planar ground state structure. A total of nine intermediate geometries were examined using all three population analysis methods. The rotational pathway and all electronic properties were computed at the HF/6–31G* level of theory.[2] The results presented in Figs. 1.2–1.4 illustrate the changes in electron population of the atoms of formamide using the planar form as a reference. In the interest of clarity, the amino group atomic populations were combined.

In order to interpret the data in Figs. 1.2–1.4, it is instructive to consider the results that might be expected from the viewpoint of the resonance theory.[5,9,11] Within the guidelines of this paradigm, one might expect that the carbonyl

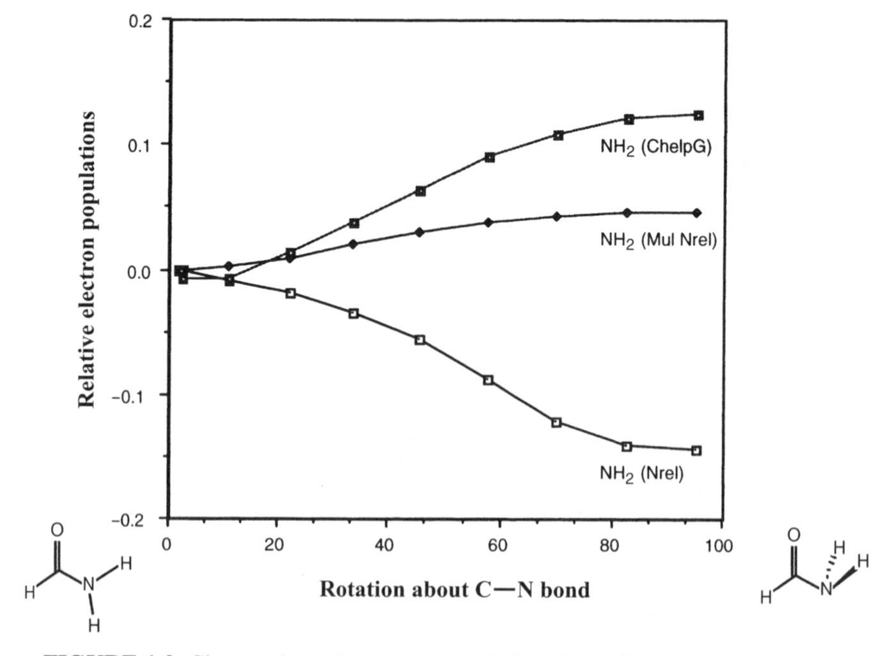

FIGURE 1.2. Changes in amino group population along the reaction coordinate.

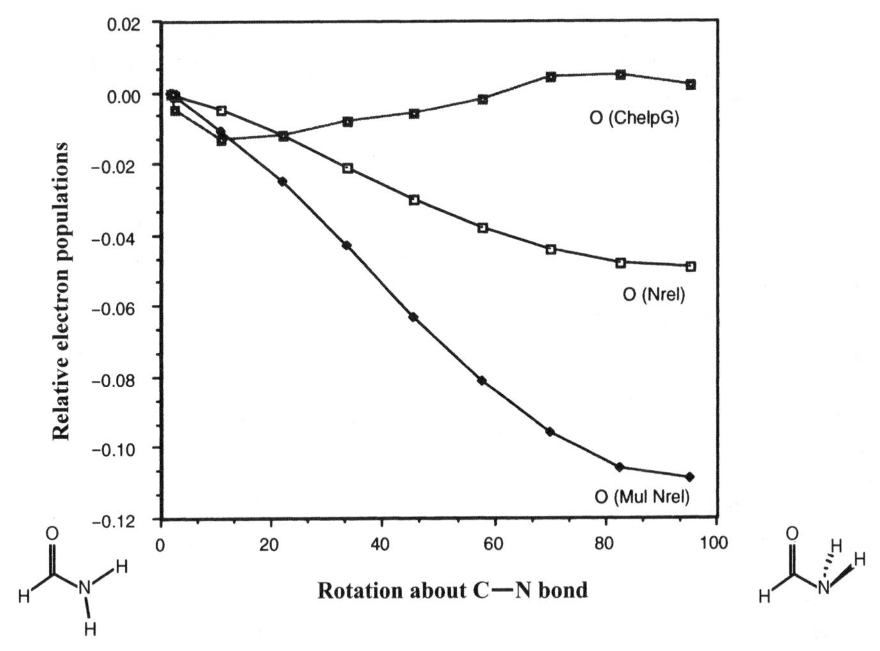

FIGURE 1.3. Changes in carbonyl oxygen population along the reaction coordinate.

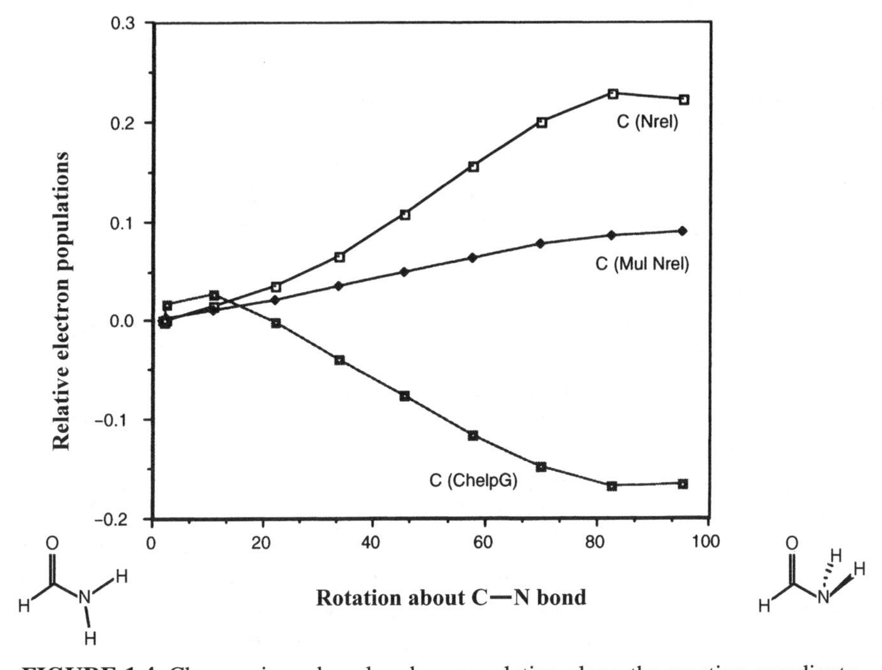

FIGURE 1.4. Changes in carbonyl carbon population along the reaction coordinate.

carbon would be unaffected by rotational isomerism, since it does not bear a charge in either of the canonical resonance forms of formamide shown earlier. In contrast, the nitrogen atom should gain electron population at the expense of the carbonyl oxygen when the nitrogen lonepair is rotated out of conjugation. As shown in the figures above, the CHELPG, Mulliken and AIM (Nrel) populations change in quite different ways.

Upon examination of Fig. 1.2, it becomes clear that both the CHELPG and Mulliken amino group electron populations rise when the C–N bond is rotated from planarity. The AIM populations however, are observed to follow the opposite trend. Figure 1.3 shows a different pattern for the carbonyl oxygen populations. In this case, the electrostatic potential-derived CHELPG populations do not show much change during the rotation, while both Mulliken and AIM populations change in a small but regular fashion. The data illustrated in Fig. 1.4 shows that an unexpectedly large charge redistribution occurs on the carbonyl carbon. It is interesting to note that the CHELPG carbonyl carbon populations are quite different than both Mulliken and AIM results. While each method produces conflicting predictions in all cases, none of these three methods provide results consistent with a simple resonance model.

A careful analysis of each atomic population method reveals the reasons for their varied results. The Mulliken population analysis method has been available to computational chemists for many years. The major strength of this technique is its speed—Mulliken charges are routinely computed as the last step of most semiempirical and ab initio computations. Mulliken atomic charges are known to be basis-set dependent, and are derived from atomic orbital and overlap populations represented in an associated density matrix.[25] This quick but approximate method splits overlap density equally between bonded atoms, which leads to an underestimation of electron density polarization. For these reasons, the Mulliken method can be said to represent the electron density distribution from the viewpoint of atomic orbitals.

CHELPG is one of several available methods that use a molecular wavefunction to produce electrostatic potential-derived charges.[2] Such charges are derived by constructing an array of points around a molecule, and then sampling the electrostatic potential field at each point. The resulting data is then used in a fitting algorithm to derive the values of a set of atom-centered monopoles capable of reproducing the electrostatic potential field. Such potential-derived charges may be viewed as indicators of molecular polarity from an external perspective. By nature, monopole "atomic charges" imply spherically symmetric electron density distributions in the neighborhood of each atom. For this reason, potential-derived charges should not be directly compared to AIM atomic populations.

As described in the earlier theory section, the AIM method utilizes a molecular electron density partitioning technique based upon the gradient field of the electron density distribution. Graphical analysis shows that the shapes of most AIM atoms are far from spherical.[29] The shape of a typical sp^2-hybridized carbon atom is shown in Fig. 1.5.

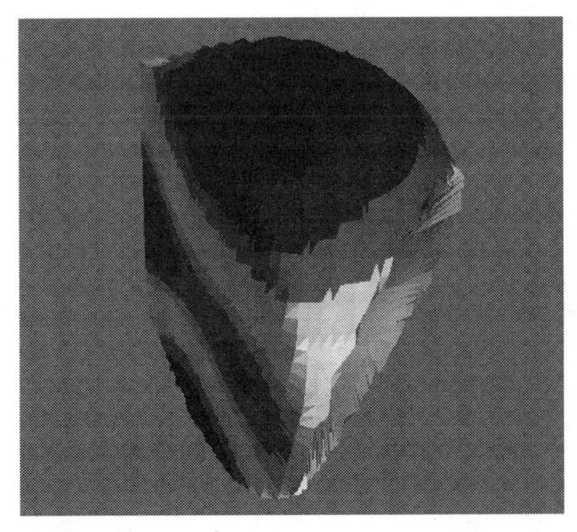

FIGURE 1.5. Property-encoded sp^2 carbon atom (AIM basin with electron density isosurface cutoff at $0.002\,e\,au^{-3}$).

The nonspherical nature of these atomic representations is only one feature which sets them apart from other methods of electron density analysis. As a result of the ability of AIM technology to integrate numerous properties of the electron density within each atomic basin, additional information about each atom may be obtained. Integration of the electronic kinetic energy density within each atomic basin reveals the contribution of each atom to the total molecular electronic energy. These changes in energy are helpful in understanding the flow of charge indicated in the AIM (Nrel) data (Figs. 1.2–1.4).

Figure 1.6 illustrates changes in the relative energy of each atom in formamide during rotation about the C–N bond. Significantly, the data in this graph show that only the amide nitrogen and the carbonyl carbon undergo large changes during this conformational isomerism. The energy of the carbonyl oxygen is almost completely unchanged as a result of the rotation. This feature demonstrates an important electronic effect responsible for the barrier to rotation of this simple amide: electron density *and its associated kinetic energy density* is shown to move between atoms during the isomerization. The data in Figs. 1.2, 1.4 and 1.6 clearly show that the major electronic changes associated with loss of conjugation is an exchange of charge and energy *between the carbonyl carbon and the amide nitrogen.* A more detailed discussion of this effect may be found in the recent literature.

An interpretation of these results by Wiberg and co-workers[2] concluded that the CHELPG potential-derived charges represent an external view of the formamide molecule, while the AIM results were considered to be an internal view of the electron density distribution. While CHELPG populations were not considered representative of internal charge and energy flow, they were found to correlate with chemical intuition concerning the reactivity of the formamide molecule. When lone pair donation is possible, the carbonyl oxygen atom

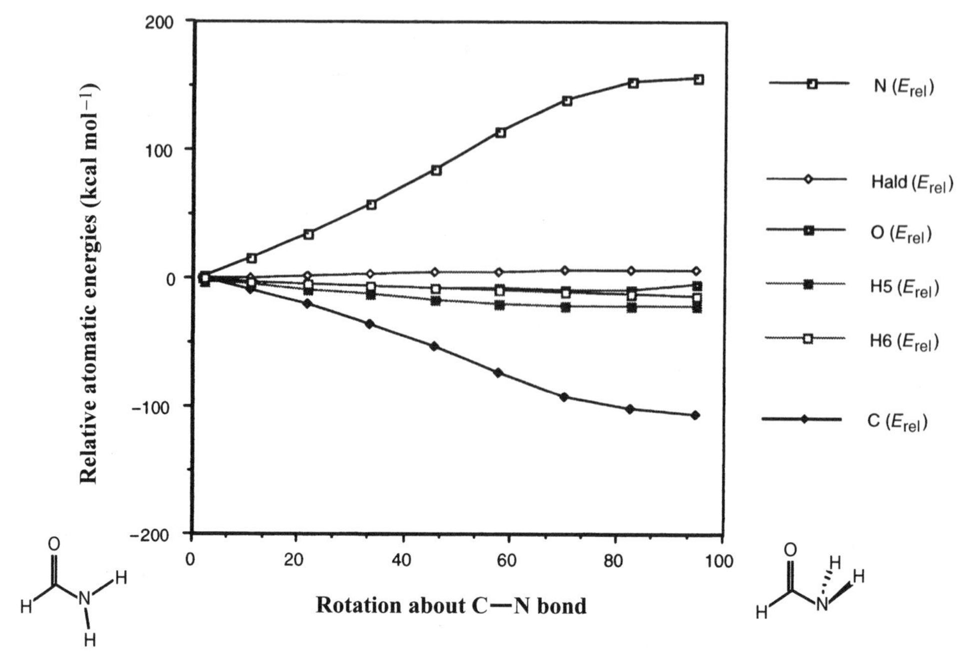

FIGURE 1.6. Changes in atomic energy upon C–N bond rotation.

appears to be surrounded by slightly more negative electrostatic potential than in the rotated conformer. This observation is consistent with the enhanced basicity of the amide carbonyl oxygen. The CHELPG populations also show that the amide nitrogen exerts less negative electrostatic potential in the planar form than in the rotated form. This much larger change is consistent with the large difference in basicity between an amide nitrogen and amine nitrogen. Though not at all consistent with the resonance model, the CHELPG populations show that the carbonyl carbon is far more positive in the rotated, unconjugated structure than in planar formamide. This result is in complete agreement with the accelerated hydrolysis rates of twisted amides.

The AIM populations tell a different story. From an internal viewpoint, the redistribution of charge and kinetic energy is responsible for the rotational barrier. Wiberg has shown that it is possible to separately analyze the σ and π electron density changes within the AIM methodology.[18,44] Using this technique, it was shown that an increase in π density was, in fact, observed on the carbonyl oxygen relative to the rotated forms. No such comparison is possible for the rotating amino group, since its lone pair loses Cs symmetry when rotated from planarity. It was also shown that the rotated forms resist planarization of the amino group by several kcal mol^{-1}. This is indicative of the VSEPR repulsion of the electrons in the lone pair with the N–C and N–H bonds. If some of the lone-pair density can be donated to an adjacent π bond, such as a carbonyl group, the resistance to planarization is reduced and rehybridization may take place. Once the amide amino group becomes sp^2 hybridized, it becomes significantly more electronegative and withdraws much more σ density from the adjacent carbonyl carbon than was donated to the carbonyl π system. This is the source of the apparent discripancy between the CHELPG and AIM results.

There is a large energetic advantage to the amino group driving the rehybridization. This effect is clearly shown in Fig. 1.6. Note that in the planar conformer of formamide, the carbonyl oxygen is functioning only as a π electron sink which facilitates the amino group rehybridization.

Greenberg and co-workers[45–47] made some interesting observations concerning the effects of deformation on experimental core ionization potentials and theoretical electron density distributions of amides. In their work, changes in the 1s core ionization energies for the pertinent carbon, nitrogen and oxygen atoms were determined for several deformed amide linkages. In planar amides, Greenberg found that the nitrogen core potentials were higher than in deformed conformations. Likewise, the oxygen core potentials were found to be lower in planar amides. The prevailing interpretation of core ionization potential data suggests that higher values indicate a loss of valence electron density, while lower potentials result from valence electron density enrichment. For the purpose of comparison with AIM results, a closer look at the important physical effects is required. An increase in core ionization potential in atoms with reduced local electron density depends on the assumption that atomic electron density is spherically arranged around each nucleus, thereby coming under its influence. While this is largely true for core density, it is not true for valence electron density. As shown earlier, AIM atomic shapes are often highly nonspherical. Such anisotropic atomic basins result in atomic ownership of valence electron density considerably removed from nuclear positions. Although electronic forces from this density contribute to the overall character of an atom, changes in the electron density gradient vector field resulting in the motion of interatomic surfaces may not exert large changes in electron repulsion in the neighborhood of the core electrons. Consequently, AIM atomic populations would not be expected to correlate with core ionization potentials. The flow of charge and electronic kinetic energy density described by AIM analysis provides a different view of electron density redistribution.

If such π-donation/σ-withdrawal schemes are of general importance in amides, it was reasonable to expect that similar effects were operative in other chemical systems too. The next section addresses the rotational barrier and conformational preferences of the sulfonamide linkage.

7. A CHARGE-DENSITY ANALYSIS OF SULFONAMIDE BONDING

The sulfonamide linkage is also an important component of many molecules, and is also known to exhibit a significant barrier to S–N rotation. The underlying reasons for this barrier have been postulated by Catalan.[48] In this work, it was stated that the factors controlling sulfonamide rotational barriers were different from those involved in amide resonance. The results of another study on sulfonamide bonding was reported by Schleyer and co-workers.[49] In the study, Schleyer examined whether d-orbital participation involves dsp^3 or d^2sp^3 hybridization, or if "negative hyperconjugation" is a better description of the bonding situation. From the results of that work, Schleyer was able to conclude that the "negative hyperconjugation" model provided a better description of hypervalent bonding. An alternative explanation of sulfonamide stability and conformational preferences was recently published in which the origins of sulfonamide and fluorosulfonamide rotational barriers were re-examined by Breneman and co-workers using the AIM technique which had proven successful in the analysis of formamide resonance.[3,14]

In this work, an exhaustive examination of the electron density distributions around the sulfur- and amino groups of sulfonamide and fluorosulfonamide was performed. As in the case of formamide, the geometries and wavefunctions were determined at regular increments of rotation about each S–N bond, as defined by the H(6)–N(5)–S(2)–H(1)(F1) dihedral angle. Each minimum energy conformation was also included in the examination as well as both nitrogen inversion and N–H rotational transition states.

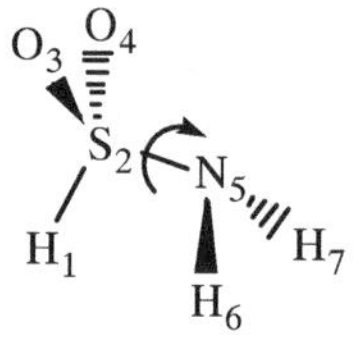

O_3 O_4 S_2 N_5 H_1 H_7 H_6

Sulfonamide

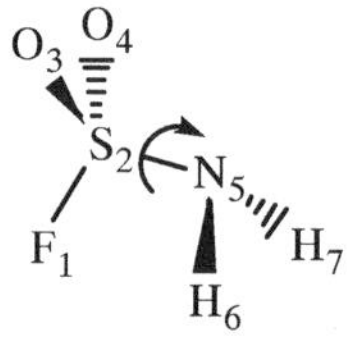

O_3 O_4 S_2 N_5 F_1 H_7 H_6

Fluorosulfonamide

The computed results showed that the mechanism of conformer stabilization in sulfonamides is very similar to that reported earlier for amides, in which stabilization of the amide nitrogen was found to occur through σ withdrawal of electron density from an adjacent polarized atom. In the case of an amide, this atom is the carbonyl carbon. When it was first proposed, this interpretation was controversial. Catalan and co-workers stated that the hybridization of the amino nitrogen in sulfonamide has little effect on the total energy of the sulfonamide system. In an effort to gain more understanding of the sulfonamide system, ab initio calculations and electron density distribution analyses were performed on both sulfonamide and fluorosulfonamide. The introduction of fluorine into the molecule allowed the normal electron populations of the sulfur and oxygens to be altered. The introduction of an electronegative fluorine substituent placed more electron demand on the sulfur, and served to enhance the observed charge density reorganization. Prior to evaluating the interatomic charge flow data, the relative energies of each conformer were examined and the geometric stationary points were characterized. The rotational potentials were found to be quite similar in shape for both sulfonamide and fluorosulfonamide. The important difference between the two curves is that the geometries at which the molecules reach their global and relative minima are reversed. The relative minimum of sulfonamide occurs at an H–N–S–H dihedral angle 285° while the global minimum occurs at 115°. In fluorosulfonamide, the relative minimum occurs at the corresponding H–N–S–F dihedral of 110° while the global minimum occurs at 291°. This means that sulfonamide is most stable when its nitrogen lone pair is *syn* to the S–H bond. The substitution of a fluorine atom for the hydrogen in fluorosulfonamide causes the global and relative minima to reverse in energy ordering. In this case, the molecule is most stable when the nitrogen lone pair is *trans* to the S–F bond. It is also important to note that the minima in both sulfonamide and fluorosulfonamide are very close to their respective amino nitrogen inversion transition states, indicating a propensity for planarization of the nitrogen.

Upon examination of AIM atomic integration results, an obvious intramolecular charge shift trend was observed. The data indicated that a charge shift was taking place between the sulfur and the amino group at geometries corresponding to both minima in fluorosulfonamide and the relative minimum in sulfonamide. At these geometries, the amino group was observed to gain electron population, while the sulfur population was depleted by a similar amount. The other atoms within the molecules exhibited little change when compared to the sulfur and nitrogen.

The relative energy by atom showed that the kinetic energy flow closely followed the charge flow from atom to atom during S–N bond rotation. The charge and energy transfer interactions were observed mainly between the sulfur and amino groups, with the amino group reaching a minimum and the sulfur reaching a maximum in energy. The maximum charge and energy transfer was observed near the minimum energy structures of both molecules. This was to be expected, because motion of electrons from a less electronegative atom (S) to a more electronegative atom (N) is favorable for the whole molecule as the amino group approaches planarity.

When the relative sulfur and amino group energies were analyzed with respect to their relative electron populations, the best fit least-squares slopes showed that the amino group was 1.08 and 1.16 times more electronegative than the sulfur atom in sulfonamide and fluorosulfonamide, respectively ($R^2 > 0.999$ in each case). Therefore, for a given amount of electron density transferred, the amino group would gain 8–16% more stability than what was lost by the sulfur.

This electronegativity difference suggests that the rotationally induced changes in the amino group and sulfur energies in sulfonamide and fluorosulfonamide should be examined in greater detail. The amino groups became more stable at geometries where both sulfonamide and fluorosulfonamide are most stable. In addition, the sulfur energy varies in an opposite manner at the same conformations. Thus, there is evidence that sulfur destabilization occurs in unison with the stabilization of the amino group. If relative electronegativity changes are affecting the stabilization of the amino group and the destabilization of the sulfur, then the bond connecting them is the most likely place to look for changes.

The S–N bond reaches its shortest length when both sulfonamide and fluorosulfonamide are in their most stable geometries. This bond length contraction implies that the bond is increasing in s character as the amino group becomes sp^2 hybridized. It is also important to note that for all conformers, the N–S bond length in fluorosulfonamide is shorter than in sulfonamide.

The amino group pyramidalization decreases and approaches planarity at the most stable geometries of both sulfonamide and fluorosulfonamide. At these geometries, (1) the amino group gains electron population while the sulfur atom loses electron population; (2) the amino group becomes more stable while the sulfur atom becomes less stable; (3) the S–N bond shortens, and (4) the amino group is most planar. These observations point to the conclusion that nitrogen rehybridization to a more nearly sp^2 structure increases its electronegativity and gains overall stability for the molecule.

Further studies on the amino nitrogens revealed additional important information. The local minima in the Laplacian ($\nabla^2\rho$) and electrostatic potential were located in the "lone pair region" of space of the nitrogen by using a gradient search procedure.[19] The nitrogen lone pair Laplacian and electrostatic potential minima were found to approach a maximum value (least negative) at geometries where both sulfonamide and fluorosulfonamide are most stable.[3] The positions of these maxima implied that nitrogen lone pairs were least basic at those geometries. These results also show that the nitrogen lone pair is more "p"-like and very slightly delocalized in the minimum energy structures. One interpretation of these observations is that the delocalization of a small amount of π electron density allows the more electronegative amino group to withdraw a large amount of σ electron density from the less electronegative sulfur, thereby increasing the amino group's overall electron population. This results in rehybridization of the amino nitrogen from sp^3 to nearly sp^2, thus increasing the s character of the S–N bond. The amino group, therefore, becomes more planar and consequently more stable. The sulfur, having lost electron population, would be destabilized to a lesser degree, causing the molecule to become more stable overall.

This leaves but one question: Where did the small amount of delocalized nitrogen lone pair density go? The nitrogen lone pair could have possibly been partially donated into the S–F bond. This hypothesis is supported by the fact that in the global minimum structure, the S–F bond is favorably oriented with respect to the nitrogen lone pair to allow delocalization of the lone pair into the S–F antibonding orbital. This donation of nonbonded electron density would account for the observed increase of the S–F bond length. In sulfonamide, however, the corresponding S–H bond-elongation effect was not observed to such a large extent.

Sulfonamide was found to be most stable when its nitrogen lone pair is oriented *syn* to the S–H bond but *anti* to the vector sum of the sulfonamide S–O bonds. It was therefore postulated that the global minimum in sulfonamide may be controlled by the best electron sink available for the amino lone pair electrons. Since fluorine is more electronegative than oxygen, the fluorine appears to be acting as a more efficient electron sink in fluorosulfonamide than the two oxygens are in sulfonamide. The effect of this electron sink was also revealed in the barrier to rotation in fluorosulfonamide.

Fluorine substitution in fluorosulfonamide caused the nitrogen to undergo greater changes in hybridization and energy than observed in sulfonamide. The cause of this phenomenon was also apparent in the computed S–N bond lengths. The fact that the S–N bond lengths were shorter for all conformers when compared to sulfonamide suggested that the more electronegative nitrogen underwent a greater degree of rehybridization in fluorosulfonamide than in similar geometries for sulfonamide. In sulfonamide, the two S–O bonds were found to be less effective electron acceptors since the energy of their σ^* nonbonding orbitals are higher than those of the S–O and S–F bonds in fluorosulfonamide. Consequently, donation of electron density into the S–F bond in fluorosulfonamide is more favorable than in sulfonamide, thus allowing its nitrogen to rehybridize more readily.

For this new model of sulfonamide stability to be rational, it must be able to reconcile Catalan's recent results[48] which state that amino nitrogen hybridization does not appear to be important to the total energy of the sulfonamide system. In Catalan's work, the idea was tested by examining the extreme examples of an sp^2 hybridized nitrogen versus an sp^3 hybridized nitrogen during rotation about the S–N bond of sulfonamide. We must conclude that Catalan inadvertently biased the results of his analysis by not allowing the nitrogen atoms to change hybridization during rotation.

To address these points further, the nitrogen rehybridization scenario proposed in this work was further tested by comparing the values of ρ and $\nabla^2\rho$ at the S–N bond critical points in each conformer. Covalent bond orders were also calculated using the BONDER method.[50,51] On the basis of this data, it was possible to assess the degree and type of bonding between each pair of atoms. The results of the critical point electron density analysis show an increase of electron density (ρ) at the S–N bond critical points for both sulfonamide and fluorosulfonamide when the torsional angles approached those of the minimum energy geometries. These increases in ρ indicate an increase in the total bond order of the S–N bond at those minimum energy geometries. This observation is consistent with the hypothesis that donation of a small amount of nonbonded electron density from the nitrogen lone pair into the antibonding orbital of the S–O, S–F or S–H bonds is required to enable a comparatively larger withdrawal of sigma density from the sulfur by the amino group. The withdrawal of this sigma density along the S–N bond would be expected to increase the s character of the bond resulting in a decrease in its bond length.

The value of $\nabla^2\rho$ at a bond critical point may be interpreted as describing the relative nature (ionic or covalent) of a bond. The Laplacian data showed that for both sulfonamide and fluorosulfonamide, $\nabla^2\rho$ at the C–N bond critical points reaches its most positive value when nitrogen planarization and rehybridization take place. This data implies that the covalent S–N bond order decreases and the bond becomes more ionic at these geometries. This observation further supports the hypothesis that electron withdrawal along the S–N bond takes place during nitrogen rehybridization.

The BONDER results were consistent with the Laplacian results, and reveal that the S–N covalent bond orders dramatically decrease at the most stable geometries of both sulfonamide and fluorosulfonamide. This implies that less electron density is being shared between the two nuclei at these geometries and is being held more tightly by the sulfur and the nitrogen. In the cases of sulfonamide and fluorosulfonamide, it is the more electronegative amino group which is becoming the anionic partner of the ionic portion of the bond, and is therefore acting as an electron sink. This sigma polarization was manifested by increases in the electron density (ρ) at the S–N bond critical points and by the more ionic S–N bond. The increased ionic character of this bond was established by comparing the ρ and $\nabla^2\rho$ results with BONDER covalent bond orders. The observed relationship between the *total* bond order and the value of ρ at the bond critical point allowed separation of the total bonding scheme into a combination of ionic and covalent parts.

Thus, the results of ρ, $\nabla^2\rho$, and BONDER analysis all support a central hypothesis: The rehybridization of sulfonamide amino groups is facilitated by the donation of a small amount of lone pair electron density into the σ^* orbital of another bond, resulting in a hybridization-driven increase in amino group electronegativity. This change causes a large flux of σ electron density towards the nitrogen. This flow of charge and electronic kinetic energy density results in the stabilization of the entire molecule, since the more electronegative amino group is stabilized at the expense of sulfur stabilization.

8. CONCLUSIONS

The redistribution of electron density and electronic kinetic energy between the sulfur and nitrogen atoms in sulfonamides were found to be the factors controlling conformational stability. The results showed that in sulfonamide, at least one minimum was strongly dependent upon the hybridization of the amino nitrogen atom, where as in fluorosulfonamide, both minima were clearly hybridization controlled. These results lead to the first conclusion: *The total stabilization of both molecules is directly related to the redistribution of charge density between the sulfur and amino nitrogens in a manner analogous to formamide stabilization.*

The S–N bonds were found to be shortest at those geometries where both sulfonamide and fluorosulfonamide were most stable. This observation implied that in both molecules, the S–N bond increased in s character in the most stable conformers.

The positions and magnitudes of electrostatic and Laplacian ($\nabla^2\rho$) minima in the vicinity of the nitrogen "lone pair" region, as well as the amino group pyramidalization altitudes were found to be reliable indicators of nitrogen hybridization. The fluorosulfonamide results demonstrated that donation of a small amount of electron density from a nitrogen lone pair into the antibonding orbital of an adjacent S–F bond allowed a much larger withdrawal of σ electron density from the sulfur atom.

All the results presented in this work indicate that nitrogen rehybridization controls the torsional preferences in sulfonamide and fluorosulfonamide. The concept of *dynamic relative electronegativity* was found to rationalize the geometry of one minimum of sulfonamide and both minima of fluorosulfonamide. Within this interpretation, individual atomic stabilities are rearranged to serve the "greater good" of the total molecule. In this case, providing the more electronegative amino group with more electron density transferred from the less electronegative sulfur atom allows the nitrogen to rehybridize and stabilize the entire molecule.

The fact that fluorosulfonamide has a greater rotation barrier than sulfonamide was found to be a direct result of the electronic consequences of fluorine substitution. *The fluorine was found to act as an excellent lone pair electron sink*, thereby allowing the amino nitrogen to rehybridize to a larger degree than is possible in sulfonamide. These results show that at all geometries, the S–N bonds in fluorosulfonamide are consistently shorter than those in sulfonamide.

Through the implementation of electron density partitioning methods, the seemingly dissimilar phenomena of amide and sulfonamide stability was explained by one unifying mechanism. In both cases, it was clear that the added stability and conformational preferences of each system were controlled by a redistribution of electron density and its associated kinetic energy.

9. A CHARGE-DENSITY ANALYSIS OF ENAMINONITRILE BONDING

Another moiety closely related to the amide linkage is the enaminonitrile system. Enaminonitrile monomers may be polymerized to produce materials with unusual physical and electronic properties. The resulting polymers contain backbone chains with restricted rotational properties characteristic of a nylon or other polyamide but have much higher dielelectric constants. During a previous study, several model compounds containing one or more of the enaminonitrile "push–pull olefin" linkanges were examined by variable temperature ^{1}H NMR spectroscopy.[52] The dynamic NMR data showed barriers to rotation about the C–N bonds of 15–17 kcal mol^{-1}. It has been demonstrated that the C–N internal rotational barriers of enaminonitriles may be successfully reproduced by quantum mechanical calculations. Therefore, ab initio calculations at the HF/6–31G**//HF/6–31G* level of theory were performed on Compound **1** and its two rotational transition state structures.[42] The calculated rotational barriers (15.6 and 15.4 kcal) were found to agree well with the NMR data. Since the interatomic flow of charge and electronic kinetic energy density have been shown to control the rotational preferences in amides and sulfonamides, the electronic structure of **1** was investigated to see if the same effects were operative in these systems.

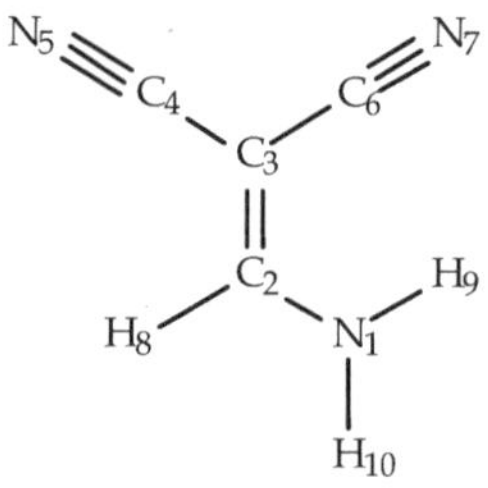

Planar Enaminonitrile

1

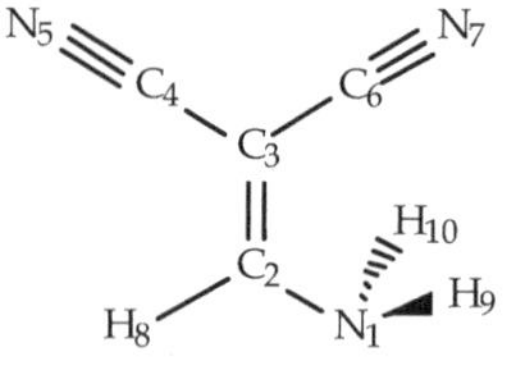

syn Enaminonitrile TS

$\mathbf{1}_{syn}$

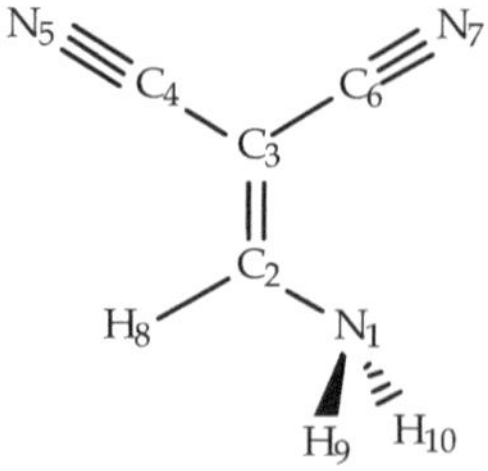

anti Enaminonitrile TS

$\mathbf{1}_{anti}$

As in the study of amide resonance, it was important to begin by defining the phenomenon to be explained. In the case of Compound **1**, the barrier to rotation is accompanied by geometric changes occurring in concert with charge density redistribution. The most pronounced changes are in the C–N bond length and in the pyramidalization of the amino group. This geometric effect can be seen in the C(2)–N(1)–H(9) and C(2)–N(1)–H(10) bond angles. As shown in Table 1.4, the N(1)–C(2) bond length elongates 0.0816 Å (Å) in going to the *syn* transition

TABLE 1.4. Geometric Changes in Enaminonitrile 1 Upon C–N Rotation (6–31G/6–31G*)**

Parameter	Planar Minimum	*syn* Transition State	*anti* Transition State
Distances			
N(1)–C(2)	1.3309	1.4125	1.4080
C(2)–C(3)	1.3552	1.3326	1.3289
C(3)–C(4)	1.4325	1.4427	1.4434
C(4)–N(5)	1.1377	1.1353	1.1345
C(3)–C(6)	1.4311	1.4422	1.4441
C(6)–N(7)	1.1373	1.1351	1.1351
C(2)–H(8)	1.0742	1.0756	1.0793
N(1)–H(9)	0.9967	1.0030	1.0030
N(1)–H(10)	0.9928	1.0030	1.0030
Angles			
N(1)–C(2)–C(3)	126.13	125.71	121.71
C(2)–C(3)–C(4)	120.48	121.39	122.66
C(2)–C(3)–C(6)	120.08	120.98	120.14
C(3)–C(2)–H(8)	118.44	118.35	118.00
C(2)–N(1)–H(9)	121.39	111.68	111.09
C(2)–N(1)–H(10)	120.89	111.68	111.09
Dihedrals			
H(9)–N(1)–C(2)–C(3)	0.0	60.32	120.07
H(10)–N(1)–C(2)–C(3)	180.0	−60.32	−120.07

state and 0.0771 Å in the *anti* transition state. This geometric change is consistent with the 0.0790 Å C–N bond length change calculated at the same level of theory for formamide undergoing rotational isomerism.

The next most pronounced geometric change is in the C(2)–C(3) olefin bond length. This bond is observed to shorten by 0.0226 and 0.0263 Å at the two transition states, respectively. The corresponding C=O bond shortening in formamide was calculated to be 0.0100 Å, which is appreciably less than the other bond length changes in the molecule. The C(3)–C(4) and C(3)–C(6) bonds were found to elongate by 0.011 Å in the transition states, while the cyano group triple bonds shortened by about 20% of that value. This geometric data suggests that rehybridization of the amino group occurs during C–N bond rotation in a manner similar to that of amides and sulfonamides, but that the vinylidenecyanide carbonyl analog responds in a way somewhat different than a standard carbonyl group. Some canonical resonance forms for the planar form of **1** are shown in the scheme below.

In contrast to the formamide and sulfonamide cases, it is possible to rationalize the calculated results of enaminonitrile rotational isomerism using conventional thought: When interactions with the enamine nitrogen lone pair are lost due to C–N bond rotation, the C(2)–N(1) bond should lengthen, the C(2)–C(3) bond should shorten, the C(3)–C(4) and C(3)–C(6) bonds should lengthen and the C(4)–N(5) and C(6)–N(7) bonds should shorten to some extent.

While traditional π-density delocalization was found to play a more important part in the stabilization of enaminonitriles, it was nevertheless important to examine the flow of charge and energy in both internal (PROAIM) and external (Electrostatic, CHELPG) perspectives. The external viewpoint of charge distribution is well represented by the molecular electrostatic potential field, so electrostatic potential-derived atomic point charges calculated by the CHELPG method were used to provide a good measure of this view. The results of the CHELPG calculations are listed in Table 1.5. Similar to the results for formamide, the amino nitrogen appears significantly more negative in the transition state structures than in the planar ground state. This can be interpreted in two ways: Either the nitrogen is truly partially positively charged in the ground state or the transition state sp^3 hybridized nitrogens are simply more basic than the planar sp^2 nitrogen in the ground state.

The "atoms in molecules" PROAIM results shown in Table 1.6 favor the latter explanation. Carbon (2) is calculated to be more positive in the transition

TABLE 1.5. CHELPG Monopole Charges of Enaminonitrile 1 (6–31G/6–31G*)**

Atom	Planar	*syn* TS	*anti* TS
N(1)	−0.8565	−0.9155	−0.9387
C(2)	0.3485	0.4316	0.3804
C(3)	−0.4686	−0.3913	−0.2590
C(4)	0.6026	0.5598	0.5453
N(5)	−0.5628	−0.5172	−0.5123
C(6)	0.4673	0.4561	0.3971
N(7)	−0.5177	−0.4737	−0.4590
H(8)	0.1457	0.0846	0.0528
H(9)	0.4167	0.3825	0.3967
H(10)	0.4247	0.3830	0.3967
Total	−0.0001	−0.0001	0.0000

states than in the ground state, while C(3) is considerably more negative in the ground state than in either of the rotational transition states. The cyano group nitrogens were also calculated to be more negative in the planar ground state than either transition state. The cyano group carbons show a similar, but opposite effect. All of these data appeared to be consistent with the traditional resonance model with the exception of the PROAIM integrated atomic population data, and the somewhat unusual electrostatic behavior of C(2).

The integrated atomic population data shown in Table 1.6 and plotted in Fig. 1.7 provides an "internal" viewpoint of the flow of charge density during enaminonitrile conformational isomerism. In going from the planar form to either of the rotational transition states, the amino nitrogen was observed to actually *lose* electron population in a manner similar to that observed for formamide! The greatest gains in electron population in the transition states were realized by C (2) and 3, as well as the two amino hydrogens H(9) and H(10). Very modest changes were found for the other atoms in the molecule.

Because the real goal of the investigation was to pinpoint the energetic effects of charge density redistribution, it was important to separately examine the electronic kinetic energy changes for each atom in the molecule. Since the virially corrected electron kinetic energy of an atom is equal to the negative of the total energy of that atom, changes in the overall molecular energy could be subdivided among the atoms. Figure 1.8 shows changes in electron kinetic energy for each atom in **1** going from the planar ground state to both rotational transition states. Note that while this figure is somewhat correlated with the charge data in Fig. 1.7, different atoms realize different amounts of stabilization from the same increment of charge density. Once again, it is quite clear that the major energetic changes taking place involve the amino nitrogen and the olefin carbon atoms. The remainder of the energetic changes nearly cancel out when taken together. Figure 1.9 illustrates the origin of the rotational barrier in enaminonitrile **1**. Here, the energy changes associated with the amino group

TABLE 1.6. Integrated Atomic Properties of Enaminonitrile 1 (6–31G/6–31G*)**

Atom	n	L	$T=-E$	$1-(V/T)$
Rotomer: Planar; E(HF) = −316.5485; E_{rel} = 0.0 kcal				
N(1)	8.480	0.0004	55.1001	1.00178026
C(2)	5.223	−0.0083	37.3181	
C(3)	5.685	0.0012	37.5667	
C(4)	4.832	0.0005	37.1116	
N(5)	8.501	−0.0001	55.1759	
C(6)	4.810	0.0002	37.0931	
N(7)	8.500	−0.0001	55.1744	
H(8)	0.953	0.0001	0.6200	
H(9)	0.503	0.0001	0.4071	
H(10)	0.516	0.0001	0.4176	
Total	48.003		315.9846	316.5471
Rotomer: *syn* TS; E(HF) = −316.5237; E_{rel} = 15.6 kcal				
N(1)	8.210	−0.0001	54.8293	1.00173297
C(2)	5.380	−0.0043	37.4498	
C(3)	5.774	−0.0035	37.6576	
C(4)	4.815	0.0003	37.1020	
N(5)	8.465	−0.0001	55.1680	
C(6)	4.803	0.0004	37.0932	
N(7)	8.460	−0.0001	55.1647	
H(8)	0.958	0.0001	0.6227	
H(9)	0.569	0.0001	0.4443	
H(10)	0.569	0.0001	0.4443	
Total	48.003		315.9759	316.5235
Rotomer: *anti* TS; E(HF) = −316.5240; E_{rel} = 15.4 kcal				
N(1)	8.210	−0.0004	54.8410	1.00171870
C(2)	5.374	−0.0037	37.4416	
C(3)	5.762	−0.0004	37.6538	
C(4)	4.779	0.0003	37.0783	
N(5)	8.458	−0.0001	55.1666	
C(6)	4.811	−0.0001	37.0981	
N(7)	8.465	−0.0001	55.1673	
H(8)	0.985	0.0001	0.6317	
H(9)	0.581	0.0001	0.4512	
H(10)	0.581	0.0001	0.4512	
Total	48.006		315.9808	316.5239

moiety are combined and plotted as "NH_2 E_{rel}," while the energy changes principally involving the olefinic carbons are plotted as "E_{Res}." The changes in molecular electronic kinetic energy are plotted as "Total energy." Note that the barrier to rotation in this compound comes from the ability of the planar amino

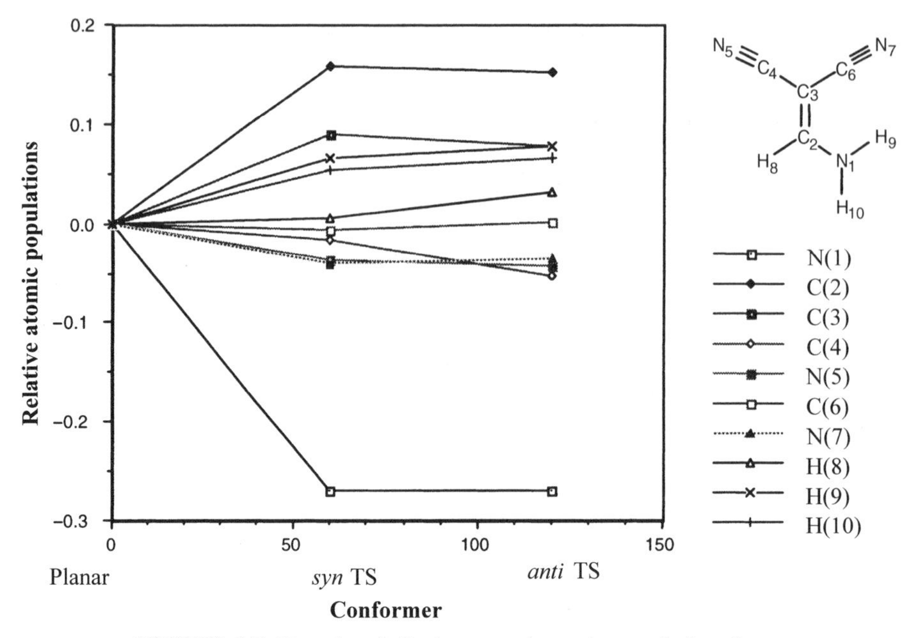

FIGURE 1.7. Enaminonitrile integrated atomic population data.

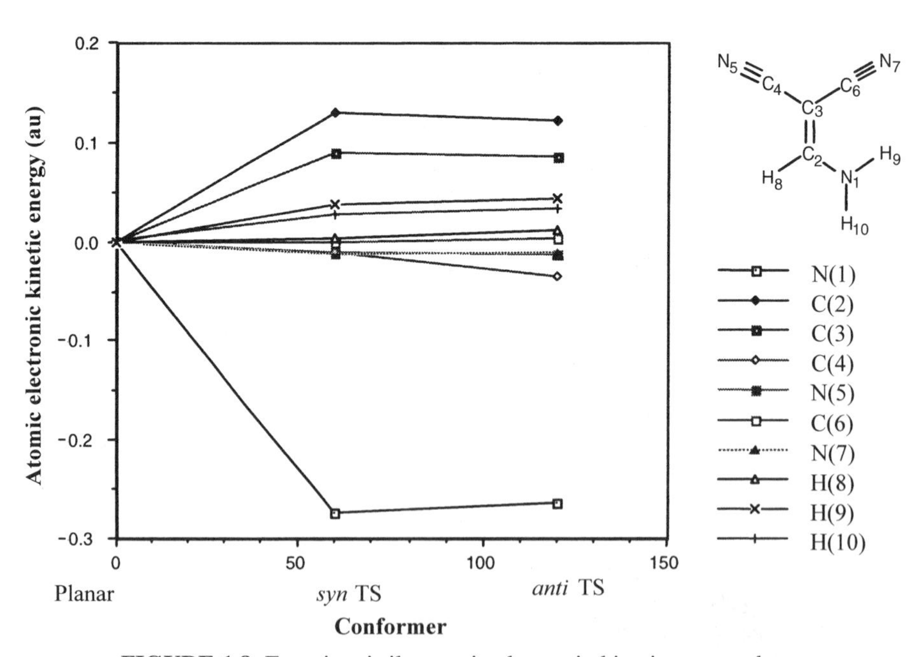

FIGURE 1.8. Enaminonitrile atomic electronic kinetic energy data.

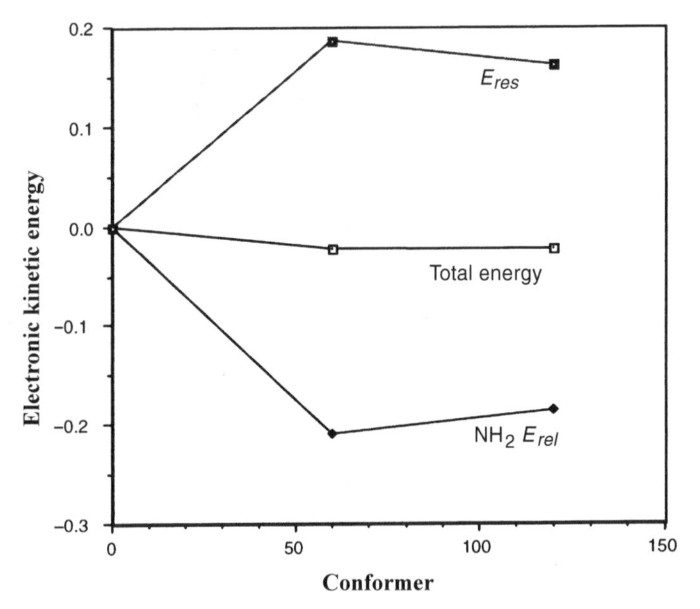

FIGURE 1.9. Enaminonitrile molecular fragment electronic kinetic energy.

group to stabilize an increment of electron density more than the remainder of the molecule. When a small portion of that charge is shifted from the amino group to the rest of the molecule, the entire system becomes destabilized. As in formamide, this destabilization arises in the rotated conformers because the amino group "lone pair" of electrons cannot properly align with the vinylidenecyanide π system. Thus, the nitrogen is not able to donate the small amount of π density required to make rehybridization to sp^2 energetically feasible. This small amount of π accounts for the observed electrostatic (CHELPG) effects in formamide, sulfonamide and in **1**, but the σ back-donation from the adjacent sp^2 carbon atom to the sp^2 amino group in planar **1** is truly the source of its "resonance" stabilization energy.

10. A SUMMARY OF AMIDE AND AMIDE-RELATED BONDING

One of the most important aspects of the recent investigations into the electronic structure of amides and related compounds is the emergence of a unified concept of molecular stability. These studies have also shown that molecular electron density distributions appear to have different properties when examined from "internal" or "external" points of view. In the end, it is clear that AIM populations should not be directly compared with "atomic charges" obtained from other methods. The roots of this discord may be traced to the underlying assumptions upon which different charge density analysis methods are based. The AIM approach produces results indicative of the complex way by which the forces of electron–nuclear attraction and electron–electron repulsion govern the characteristics of a molecular electron density distribution. The interatomic surfaces defined within the AIM method move in response to geometric or electronic perturbations. Such surface motion is intimately coupled with the electron density itself. Any charge density analysis based upon fixed atomic boundaries is bound to give questionable results. The highly anisotropic shapes of AIM atomic basins demand that any use of basin populations as atom-centered monopoles must be supplemented by the electric multipole components of each atomic basin.

The CHELPG method (or other molecular electrostatic potential-derived charge technique) produces results indicative of the electrostatic potential field surrounding a molecule. Since this potential consists of a balance between the electron repulsion forces and nuclear attraction forces, it reveals portions of a molecule where more or less electron density is available for interaction. For this reason, the electrostatic potential distribution often reflects chemical intuition concerning reactivity and functional group behavior. In the amide study, the electrostatic potential field was found to be indicative of π density delocalization, but did not adequately reflect the much larger σ withdrawal associated with rehybridization. Thus, while not perfect, the CHELPG charges changed in a way that was more consistent with the empirical predictions of resonance theory.

The best choice of an electron density analysis technique will largely depend upon the kind of questions being posed during an investigation. If the internal mechanisms of molecular stabilization are to be studied, AIM analysis would be the method of choice. On the other hand, long-range interaction potentials would be best interpreted using an electrostatic potential-related method.

In each of the three amide-related studies described in this work, the mechanism of molecular stability was most effectively elucidated using the AIM method. The planar ground state geometry of each type of molecule was stabilized by a favorable flow of charge density and its associated electronic kinetic energy density. For this stabilizing effect to take place, a small amount of π electron density had to be displaced from the lone pair of each nitrogen. In each case, π donation was followed by a much larger amount of sigma withdrawal. The delocalization pathway of the π density is well described by allylic "resonance" structures, but no popular qualitative model exists yet for the important σ density redistribution.

REFERENCES

1. Bennet, A. J.; Wang, Q. P.; Slebocka-Tilk, H.; Somayaji, V.; Brown, R. S.; Santarsiero, B. D. *J. Am. Chem. Soc.* 1990, **112**, 6383.
2. Breneman, C. M.; Wiberg, K. B. *J. Comp. Chem.* 1990, **11**(3), 361.
3. Breneman, C. M.; Weber, L. W. *Can. J. Chem.* 1996, **74**, 1271.
4. Brown, R. D.; Godfrey, P. D.; Kleibomer, B. *J. Molec. Struct.* 1987, **124**, 34.
5. Drakenberg, T.; Forsen, S. *J. Phys. Chem.* 1970, **74**, 1.

6. Faerman, C. H.; Price, S. L. *J. Am. Chem. Soc.* 1990, **112**(12), 4915–4926.
7. Greenberg, A. In *Structure and Reactivity*; Greenberg A.; Liebman, J. Eds.; VCH Publishers: New York, 1988.
8. Hirota, E.; Sugisaki, R.; Nielsen, C. J.; Sorensen, G. *J. Molec. Spectro.* 1974, **49**, 251.
9. Kamei, H. *Bull. Chem. Soc. Japan* 1968, **41**, 2269.
10. Laidig, K. E.; Bader, R. F. W. *J. Am. Chem. Soc.* 1991, **113**, 6312.
11. Sunner, B.; Piette, L. H.; Scheider, W.G. *Can. J. Chem.* 1960, **38**, 681.
12. Wiberg, K. B.; Laidig, K. E. *J. Am. Chem. Soc.* 1987, **109**, 5935.
13. Wiberg, K. B.; Breneman, C. M.; Laidig, K. E.; Rosenberg, R. *Pure Appl. Chem.* 1989, **61**, 635.
14. Wiberg, K. B.; Breneman, C. M. *J. Am. Chem. Soc.* 1992, **114**, 831.
15. Wiberg, K. B.; Rablen, P. R.; Rush, D. J.; Keith, T. A. *J. Am. Chem. Soc.* 1995, **117**, 4261.
16. Wiberg, K. B.; Rablen, P. R. *J. Am. Chem. Soc.* 1995, **117**, 2201.
17. Cox, J. D.; Pilcher, G. *Thermochemistry of Organic and Organometallic Compounds*; Academic Press: New York, 1970.
18. Wiberg, K. B.; Breneman, C. M.; LePage, T. J. *J. Am. Chem. Soc.* 1990, **112**, 61.
19. Frisch, M. J.; Trucks, G. W.; Schlegel, H. B.; Gill, P. M. W.; Johnson, B. G.; Robb, M. A.; Cheeseman, J. R.; Keith, T.; Petersson, G. A.; Montgomery, J. A.; Raghavachari, K.; Al-Laham, M. A.; *Gaussian 94*; Gaussian, Inc.: Pittsburgh, 1995.
20. Sellers, H.; Pulay, P.; Boggs, J. E. *J. Am. Chem. Soc.* 1985, **107**, 6487.
21. Wiberg, K. B.; Walters, V. A.; Wong, K. N.; Colson, S. D. *J. Phys. Chem.* 1984, **88**, 6067.
22. Wiberg, K. B.; Walters, V.; Colson, S. D. *J. Phys. Chem.* 1984, **88**, 4723.
23. Wiberg, K. B.; Dempsey, R. C.; Wendolowsky, J. J. *J. Phys. Chem.* 1984, **88**, 5596.
24. Bader, R. F. W. *Accounts Chem. Res.* 1985, **18**, 9.
25. Mulliken, R. S. *J. Chem. Phys.* 1965, **43** S2).
26. Bader, R. F. W. *Atoms in Molecules: A Quantum Theory*; Oxford: Oxford Press, 1990.
27. Mezey, P. G. *Shape in Chemistry*; VCH: New York, 1993.
28. Breneman, C. M.; Martinov, M. *The Use of Electrostatic Potential Fields in QSAR and QSPR*, in *Molecular Electrostatic Potential: Concept and Applications*; Murray, J. S.; Sen, K.; Ed.; Elsevier: Amsterdam, 1996, p. 143–179.
29. Martinov, M.; Breneman, C. M. *SURFPROP*, 1993.
30. Gatti, C.; Fantucci, P.; Pacchioni, G. *Theoretica Chemica Acta* 1987, **72**, 433.
31. Lammertsma, K.; Leszzynski, J. *J. Phys. Chem.* 1990, **94**, 5533.
32. Schwinger, J. *Phys. Rev.* 1951, **82**, 1914.
33. Bader, R. F. W.; Slee, T. S.; Cremer, D.; Kraka, E. *J. Am. Chem. Soc.* 1983, **105**, 5061.
34. Cioslowski, J.; Mixon, S.; *J. Am. Chem. Soc.* 1991, **113**, 4142.
35. Cioslowski, J.; Martinov, M. Mixon, S. T. *J. Phys. Chem.* 1993, **97**, 10948.
36. Cioslowski, J.; Martinov, M. *J. Phys. Chem.* 1996, **100**, 6156.
37. Martinov, M.; Cioslowski, J. *J. Molec. Phys.* 1995, **85**, 121.
38. Cioslowski, J.; Stefanov, B.; Constans, P. *J. Comp. Chem.* 1996, **17**, 1352.

39. Breneman, C. M.; Weber, L. W. Transferable atom equivalents. Assembling accurate electrostatic potential fields for large molecules from ab-initio and PROAIMS results on model systems; In *The Application of Charge Density Research to Chemistry and Drug Design*; Jeffrey G. A.; Piniella, J. F. Eds.; Plenum Press, 1991.
40. Breneman, C. M.; Thompson, T. Modeling the hydrogen bond with transferable atom equivalents; In *Modeling the Hydrogen Bond*; Smith, D. Ed.; ACS Symposium Series: Washington, D.C., 1993, pp. 152–174.
41. Breneman, C. M.; Thompson, T. R.; Rhem, M.; Dung, M.; *Comput. and Chem.* 1995, **19**(3), 161.
42. Breneman, C. M.; Moore, J. A. *Structural Chemistry* 1997, **8**, 13.
43. Breneman, C. M.; Rhem, M. *J. Comp. Chem.* 1997, **18**, 182.
44. Wiberg, K. B.; Rablen, P. R.; *J. Am. Chem. Soc.* 1993, **115**, 9234.
45. Greenberg, A.; Moore, D. T.; Dubois, T. D. *J. Am. Chem. Soc.* 1996, **118**, 8658.
46. Greenberg, A.; Moore, D. T.; *J. Molec. Struct.* 1997, **413**, 477.
47. Greenberg, A.; Hsing, H. J.; Liebman, J. F. *Theochem. J. Molec. Struct.* 1995, **338**, 83–100.
48. Elguero, J.; Goya, P.; Rozas, I.; Catalan, J.; De Paz, J. L. *Theochem* 1989, **184**, 115.
49. Reed, A. E.; Schleyer, P. v. R. *J. Am. Chem. Soc.* 1990, **112**, 1434.
50. Cioslowski, J. *BONDER*, SCRI: Tallahassee, 1990.
51. Wiberg, K. B.; Hadad, C. M.; Rablen, P. R.; Cioslowski, J. *J. Am. Chem. Soc.* 1992, **114**, 8644.
52. Moore, J. A.; Mehta, P. G.; Breneman, C. M. *Struct. Chem.* 1997, **8**, 21.

CHAPTER 2

ORIGIN OF THE AMIDE ROTATIONAL BARRIER

KENNETH B. WIBERG
Department of Chemistry, Yale University

1. INTRODUCTION

The rotational barrier about the amide C–N bond is a major factor in determining the secondary structure of peptides. As a result, it has received considerable study, both experimental and theoretical. The origin of the barrier has been controversial, and the purpose of this essay is to present a consistent picture of the interactions that occur in the amide bond.

It seems appropriate to first examine the experimental observations that bear on the question:

1. Amides generally prefer to be planar or near-planar.
2. The barrier to rotation about the C–N bond is of the order of 16–20 $kcal\,mol^{-1}$.
3. The barrier is increased on going to polar solvents.
4. Amides are less basic than amines and are protonated at oxygen rather than nitrogen.
5. Thioamides have somewhat larger rotational barriers than amides.

The structures of many amides have been determined via electron diffraction and X-ray crystallography, and they are generally planar or near-planar at the amide nitrogen.[1] However, significant distortions from the planar structure are not uncommon. Even acetamide has been found to be nonplanar by both X-ray crystallography[2] and high level ab-initio calculations.[3,4] Amides have a low barrier for the out-of-plane wagging motion of the amide group. The energy

The Amide Linkage: Selected Structural Aspects in Chemistry, Biochemistry, and Materials Science, Edited by Arthur Greenberg, Curt M. Breneman, and Joel F. Liebman
ISBN 0-471-35893-2

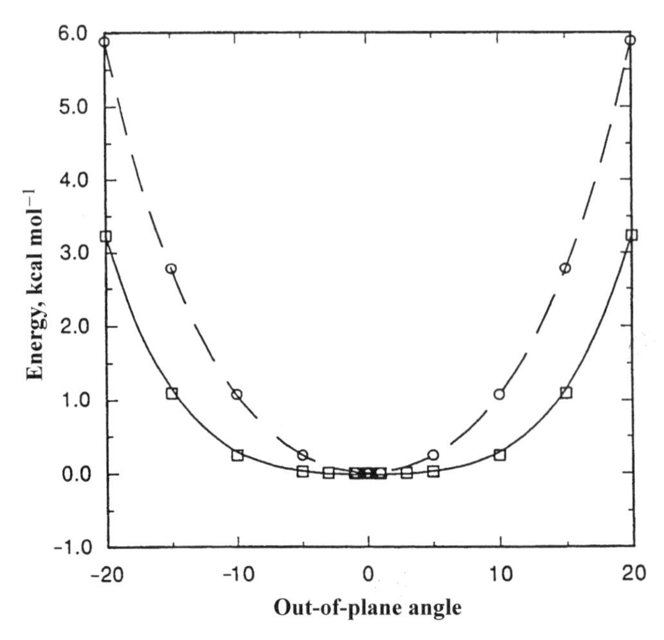

FIGURE 2.1. Out-of-plane NR_2 distortion energies for *N,N*-dimethylformamide (open squares) and *N,N*-dimethylthioformamide (open circles).

calculated as a function of out-of-plane angle for *N,N*-dimethylformamide is shown in Fig. 2.1, and it can be seen that fairly large distortions from planarity are possible without much of an increase in energy.[5] This contributes to the flexibility of the peptide backbone.

It should be remembered that amines are pyramidal, and that the planar form is a transition state for inversion at nitrogen. An inversion barrier of 5.4 kcal mol^{-1} was found for ammonia.[6,7] In ammonia and most simple amines, the H–N–H bond angle is significantly smaller than tetrahedral (106.7° for NH_3),[8] and the bond path angle (the angle between the N–H orbitals at the nitrogen) is even smaller (105.5°).[9] The pyramidal structure and the small H–N–H angle places the lone pair in an orbital with high s character. This is desirable because the lone pair electrons are associated with a only one center and are thus not as strongly bound as the electrons in the N–H bonds. The bonding can be increased (and the energy decreased) by placing the electrons in an orbital with a high s character since s electrons are more strongly bound than p electrons.[10]

The planar or near planar amide nitrogen requires that the nitrogen adopt a hybridization near sp^2, placing the lone pair electrons in a p orbital. The higher energy of this arrangement is compensated by the interaction of these electrons with the adjacent electron-deficient carbon, and such an interaction is not possible with electrons in an s orbital because of the difference in symmetry. Thus, the out-of-plane distortion at the amide nitrogen represents a tension between a pyramidal geometry which leads to local stabilization of the lone pair, and a planar geometry which leads to stabilization of the lone pair by the adjacent carbon.

It is possible to prepare amides that are structurally prevented from achieving a planar geometry, and in extreme cases, such as the 2-quinuclidones (**A**), the amide resembles the transition state for amide rotation. The properties of these amides are markedly different than the normal amides. They have enhanced basicity, and protonation now occurs at nitrogen rather than at oxygen.[11–12]

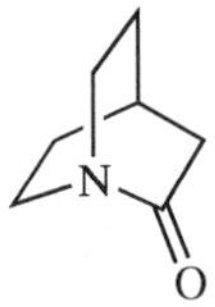

A

Rotation about the C–N bond of an amide may be studied experimentally via variable temperature NMR spectroscopy. In *N,N*-dimethylformamide, for example, the two methyl groups have different chemical shifts, but when rotation occurs, they become equivalent.[13] The change in spectrum with temperature may be converted into rate constants for rotation, and from these data one may obtain the activation enthalpy that corresponds to the barrier height.

The rotational barrier for dimethylformamide has been determined in the gas phase and is 19.7 ± 0.3 kcal mol^{-1},[14,15] and that for *N,N*-dimethylacetamide is 15.8 ± 1.1 kcal mol^{-1}.[13] Both of these barriers increase on going to solutions.[11,12] Although many chemists think of hydrocarbon solvents such as cyclohexane as almost equivalent to the gas phase, this is not true for polar solutes. The change in dielectric constant on going from the gas phase ($\varepsilon = 1$) to cyclohexane ($\varepsilon = 2$) leads to significant stabilization for these molecules. The reaction field model of solvent effects[16,17,18] suggests that cyclohexane will give about 40% of the maximum effect of a very polar aprotic solvent such as acetonitrile or dimethylsulfoxide, and this has been experimentally verified.[19,20]

The experimentally determined change in barrier with medium for the above amides is given in Table 2.1.[21] The fact that the barrier increases with increasing polarity of the solvent, indicates that the transition state for rotation has a lower dipole moment than that of the ground state, as has been confirmed by calculations.

Amines are, of course, readily protonated at nitrogen, although it is interesting to note that the charge is distributed over the groups attached to nitrogen, and that the nitrogen in an ammonium ion has a negative charge because of its greater electronegativity than either hydrogen or carbon.[22] Amides are much less basic than amines, and when they are protonated, the proton normally goes to oxygen rather than nitrogen.[10] This shows that the amide nitrogen is strongly coupled to the carbonyl group.

TABLE 2.1. Effect of Solvents on the Rotational Barrier for DMF and DMA

		$\Delta G^{\ddagger}$ kcal mol^{-1}	
Solvent	ε	DMF	DMA
Gas phase	1.00	19.3	15.3
Cyclohexane	2.02	19.7	16.4
Carbon tetrachloride	2.23	20.1	16.9
Benzene	2.27	20.2	17.3
Di-n-butyl ether	3.06	20.0	16.7
Acetone	20.7	20.5	17.5
Acetonitrile	36.7	20.6	17.8
Methanol	32.7	21.3	18.7
Water	78.4	22.0	19.1

Since the rotational barrier of amides is associated with the carbonyl group, it is interesting to note that thioamides have somewhat larger barriers,[23,24] and also a stiffer out-of-plane bending mode (Fig. 2.1).[5] The electrical character of the C=S and C=O bonds is quite different. The significance of this difference is examined below.

The structures of some amides that are structurally prevented from adopting the normal amide geometry have been studied experimentally (Table 2.2) and theoretically.[10] In these cases, the C–N bond is significantly lengthened, whereas the C–O bond is relatively unaffected. This has also been seen in theoretical calculations for amides and their rotational transition state structures.

TABLE 2.2. Bond Lengths in Amides and Lactams (Ref. 11, 12)

Compound	Source	r(NC)	r(C=O)	τ
Formamide, planar	Calc	1.349	1.193	0.0
B	X-ray	1.374	1.201	20.8
C	X-ray	1.380	1.213	16.3
D	X-ray	1.401	1.216	35.6
E	X-ray	1.413	1.225	37.8
	X-ray	1.419	1.233	41.2
A	Calc	1.433	1.183	90.0
Formamide, TS	Calc	1.423	1.179	90.0

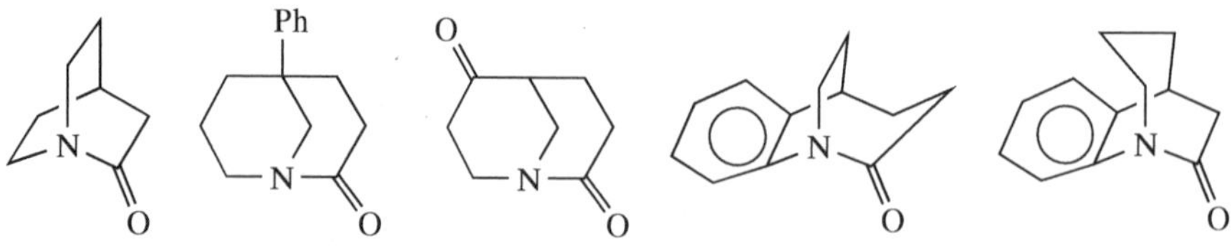

A B C D E

There is a limit to how much information about amides may be derived from just experimental studies. One of the major developments of the past decade has been the increased integration of experiment and theory to the benefit of both. The major theoretical tool is ab initio MO theory,[25] and the principal observations concerning amides that may be derived from these calculations are:

1. On rotation from the planar form to the rotational transition state, the C–O bond in formamide shortens only slightly, (0.01 Å) whereas the C–N bond length increases considerably (0.08 Å).[26] The observed barrier height is reproduced by the calculations.[27]
2. The electron population at oxygen changes only slightly on rotation from the planar form to the rotational transition state.[26,27]
3. The effect of polar aprotic solvents on the barrier is related to the decrease in dipole moment in going to the rotational transition state. The experimental data for these solvents are well reproduced by reaction field calculations of solvent effects.[21] These calculations do not correctly take into account specific solvent-solute interactions such as hydrogen bonding. The effect of protic solvents such as water and methanol may be reproduced by Monte Carlo free energy perturbation methods.[28,29]
4. The electron population at sulfur in the thioamides changes considerably more on rotation than the oxygen in amides.[5,30]

The theoretical calculations by themselves only provide the energy, structure, dipole moment, and vibrational frequencies of the compound or transition state being studied. In order to gain more information, it is useful to examine the electron density distribution in the amides and their rotational transition states. Thus, the wavefunctions that are obtained from the calculations may be analyzed by one of several methods.

The simplest method makes use of the Mulliken population analysis that assigns to each atom the electrons associated with the orbitals used for that atom, and splits the electrons in the overlap regions equally between the two atoms involved.[31] This method is strongly basis set dependent, and when diffuse functions are used, they may actually "belong" to a different atom than the one they were originally placed in. In the case of isobutene with the 6–311 + G* basis set, the Mulliken analysis for the central carbon leads to a charge of − 1 e.[32] This is clearly unphysical since there is little difference in electronegativity between the carbons, and any difference that may exist would be expected to give only a small charge, and of the opposite sign. These problems with this analysis have been recognized by many workers.[33,34]

The Weinhold–Reed Natural Population Analysis[35,36] is a great improvement over the Mulliken scheme. Here, the molecular orbitals that are derived from the ab initio calculation are localized into orthogonal bond orbitals for C–C, C–O, C–H, and other bonds, and lone pairs. Such a localization leaves small amounts of electron density unaccounted, and this density is assigned largely to unoccupied orbitals in such a way that they are orthogonal to each other and to the largely filled bond orbitals. This allows a population analysis that is almost basis set independent.

A third approach is Bader's theory of atoms in molecules.[37,38] Here, the orbitals are not used, rather the electron density is examined in a topological fashion. Between any pair of atoms joined by a chemical bond there is a point having the lowest electron density. This is known as the bond critical point, and is minimum in electron density only along the bond. In the directions perpendicular to the bond, it is a point of maximum electron density. Starting at the bond critical point, one traces rays of maximum rate of decrease of electron density, leading to a surface separating the two atoms. Across this surface, the change in electron density with change in distance is zero, and thus it is known as a zero-flux surface. These surfaces serve to separate atoms into distinct volumes, and the electron population for each atom may be obtained by integrating the electron density with the appropriate volume element.

All these methods have both advantages and disadvantages. The NPA method enforces a relatively constant "size" for each atom, whereas the topology of the charge distribution leads one to think that the "size" is dependent on the environment and hybridization of a given atom. The AIM method leads to large changes in atom "size" with changes in relative electronegativity. Although these changes can be justified theoretically, they sometimes lead to pictures of charge distributions that are not intuitive.[39,40,41] This results in part because the charge distribution is just the first term of an expansion, and atomic dipoles and higher moments at each atom also need to be taken into account.

In the case of amides, all methods agree that there is little charge transfer to the carbonyl oxygen. The NPA procedure finds the dominant process to be π-electron transfer from nitrogen to carbon, whereas the AIM method finds σ-electron transfer from carbon to nitrogen, caused by rehybridization at nitrogen, to be about as important as the π-electron transfer but in the opposite direction.

One way in which to avoid problems associated with electron populations is to directly examine changes in electron density distributions. It is possible to calculate three-dimensional arrays of electron density ("cubes") from the calculated wave functions. If that for the planar amide is subtracted from that of the rotated amide and a 3D plot of the difference density is prepared, one can "see" how the electron density has changed on rotation.[5] This is shown for formamide, formamidine and thioformamide in Fig. 2.2. The dashed contours indicate regions where electron density is lost on rotation away from the planar form, and solid contours indicate regions where electron density is increased. It can be seen that there is a π electron component which increases as the electronegativity of the terminal doubly bonded atom decreases, and that there is a σ electron shift in the opposite direction.

It is possible to obtain more quantitative information by integrating the electron density differences, and this can be done separately for the σ and π electrons. With an amide, the π-electron donation to the oxygen in the planar form is 0.088 e, but at the same time there is 0.031 e σ-electron loss, leading to a net increase of 0.057 e. The corresponding values for the thioamide are 0.162 π-

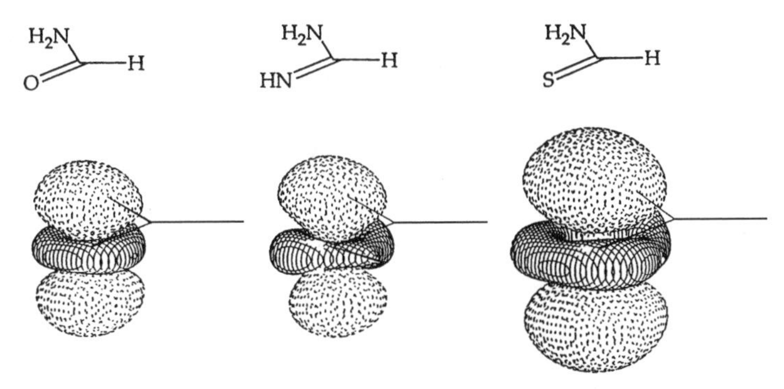

FIGURE 2.2. Electron density difference plots for the rotation of formamide, formamidine and thioformamide. The plots show the change in electron density at the =X atom on going from the planar to the rotated form. The dashed contours indicate regions in which the electron density is decreased, and the solid contours indicate regions in which it is increased. The amount of π charge transfer depends on the electronegativity of =X.

electron gain and 0.046 σ-electron loss leading to a net gain of 0.116 e.[5] Thus, electron transfer from N to S is twice as large as from N to O. With the thioamides, the C–N bond was also examined. Here it is found that π-electron density is transferred from N to *both* C and S.[5]

The σ-electron back donation from O and S is at least in part a reflection of the increased electronegativity of the planar amide nitrogen as compared to the pyramidal nitrogen in the rotational transition state.

With these results in hand, it is now possible to examine the models that have been proposed to explain the high C–N rotational barrier in amides. The amide rotational barrier has generally been assumed to arise from the contribution of the dipolar resonance structure to the stabilization of the planar form.[42] Rotation about the C–N bond would bring the lone pair of nitrogen out of conjugation with the carbon group, and the resonance stabilization will be lost.

$$R{-}C({=}O){-}NH_2 \longleftrightarrow R{-}C(O^-){=}NH_2^+$$

Several observations indicate that this formulation is too simple.

First, it suggests that there should be significant charge transfer from nitrogen to oxygen in the planar amide. One might also expect similar but opposite changes in geometry for the C–O and C–N bonds. Neither expectation is realized. The change in the C–N bond length is *much* greater than the change in the C=O length on rotation, and only a small amount of electron density is transferred from N to O.

Second, the formulation has difficulty in accounting for the larger rotational barrier and greater charge transfer to sulfur with the thioamides. Whereas oxygen is much more electronegative than carbon and should be well able to bear a negative charge, sulfur has about the same electronegativity as carbon, and so it should be more difficult to transfer charge to it.

The recognition that oxygen is much more electronegative than carbon, is an important observation in improving the resonance formulation. All carbonyl groups are strongly polarized in the sense C^{+}–O^{-}. This is one reason why carbonyl groups are so readily attacked by nucleophiles. Thus, a more appropriate resonance formulation will make use of the following three structures

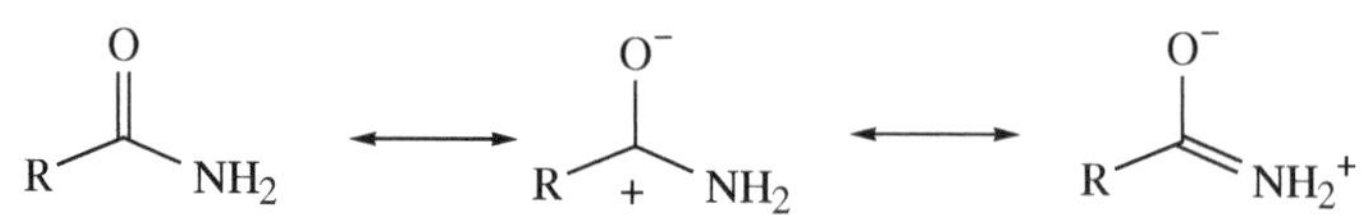

O, R, NH_2 **1** ⟷ O^-, R + NH_2 **2** ⟷ O^-, R, NH_2^+ **3**

Because of the importance of structure **2**, the main charge transfer is between nitrogen and carbon rather than between nitrogen and oxygen. The oxygen in the carbonyl group already bears a large negative charge, therefore there is little to be gained by increasing its charge. In the case of the thioamides, the middle structure has little importance, and charge transfer from nitrogen now leads to a larger change at sulfur. Thus, the interactions in the thioamides are closer to the "classical" resonance formulation than the amides.*

The π-electron interactions in both the amides and thioamides are better described by the frontier MO (FMO) approach.[43–44] Here, the amides may be considered to be formed by joining a planar amino group to a C=O or C=S group of formaldehyde or thioformaldehyde respectively. The π orbital of the carbonyl group has its largest coefficient at oxygen, and as a result the empty π^* orbital will have its largest coefficient at carbon (Fig. 2.3). Thus, the nitrogen *p*-type lone pair in its donor interaction with the empty π^* orbital will transfer charge mainly to carbon. On the other hand, with the C=S group, there is little difference in electronegativity and the C=S π-orbital will have similar coefficients at the two atoms. As a result, the π^* orbital will also have similar coefficients at the two atoms (but with opposite signs), and the interaction with the nitrogen lone pair will lead to charge transfer to *both* C and S. Charge transfer to S is facilitated by its small initial charge, as well as its high polarizability.

The logical extension of this argument would be that the charge transfer would be even greater in the selenium analogs, as has been proved by calculations.[45] It is interesting to note that calculations also found the selenoamide to prefer the "enol" form in contrast to the amides and thioamides.[46]

Although it is not obvious that it should be the case, the properties of cyclopropenone and cyclopropenethione are related to amides and thioamides. With the cyclopropene derivatives, there is π-electron transfer from the double

*For a contrary view, see Laidig, K. E.; Cameron, L. M. *J. Am. Chem. Soc.* 1996, **118**, 1737.

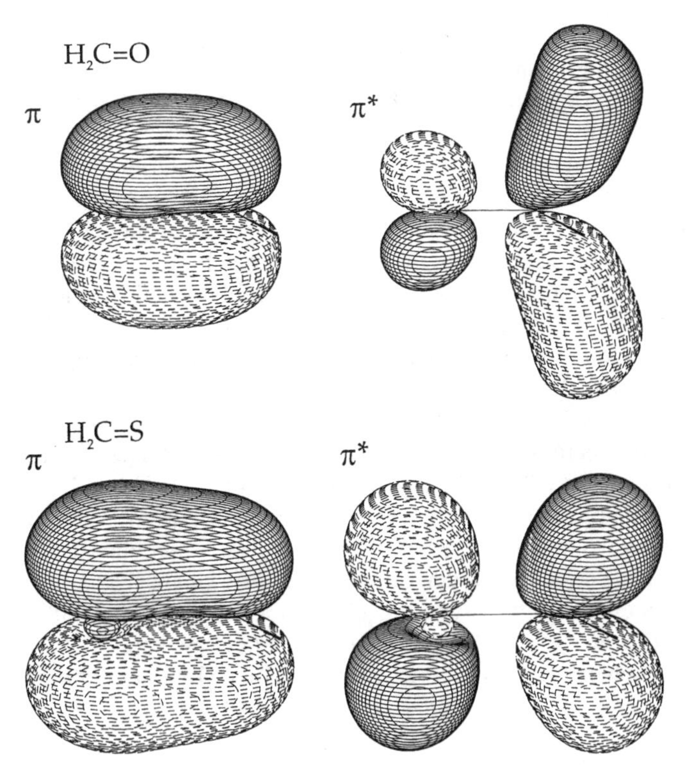

FIGURE 2.3. The π and π^* orbitals of formaldehyde and thioformaldehyde with the methylene groups at the right side. The contours correspond to an electron density of 0.05 e au^{-1}. It should be noted that the atomic orbitals at oxygen are more contracted than those of carbon, and so although the contours appear similar for the π orbital of formaldehyde, the oxygen has the larger coefficients.

bond to the C=O or C=S groups, and it is larger with C=S than C=O.[47] The driving force in this case appears to be the high electron density in the relatively short C=C bond that can be relieved by donating electron density to the C=O or C=S groups. Similarly, with the amides, it is reasonable to say that the driving force for the amide interaction is the electron repulsion in the lone pair orbital that can be relieved by donating π-density to the C=O or C=S groups.

There does not appear to be any real controversy about the nature of the π-electron interactions in amides and the thioamides. The recent valence bond study by Lauvergnat and Hiberty found that the amide resonance structures **2** and **3** given above were the important ones.[48] They also concluded that the π resonance component accounted for about one-half of the barrier for formamide and two-thirds of the barrier for thioformamide. Glendenning and Hrabal examined amides using the NPA/NRT method.[45] It is difficult to relate their results to the above since they chose to represent structures **1** and **2** above as a single structure. They evaluated the importance of this combined structure and that of structure **3**, and found that the degree of charge transfer to the =Y group increased on going from O to S and Se.

However, this is far from the whole picture. When rotation about the C–N bond occurs, there is a marked change in the geometry about nitrogen. It is usually close to planar in conjugated form, but becomes pyramidal in the rotational transition state.

As noted above, amines normally adopt a pyramidal geometry in order to place the lone pair electrons in an orbital that has high s-character. Thus, the H–N–H bond in ammonia is 106°, considerably smaller than the 109.5° characteristic of sp^3 orbitals. As a result, the bonds to nitrogen in the transition state for amide rotation (which resembles an amine nitrogen) are formed using orbitals with more than 75% p character in contrast to 67% in the ground state structure. Orbitals have decreased electronegativity when the p-character is increased. The decrease in electronegativity on rotation leads to a shift of σ electrons toward the carbon during rotation about the C–N bond.

The importance of σ interactions in stabilizing amides may be seen in a plot of the bond dissociation energies (BDE) of acetyl-X compounds against those of the corresponding methyl-X derivatives (Fig. 2.4).[49] The BDE of the methyl derivatives increase with increasing electronegativity of X (Table 2.3). The electronegative element will withdraw electron density from the methyl group, leading to a dipolar bond. This bond will be stabilized by a coulombic interaction

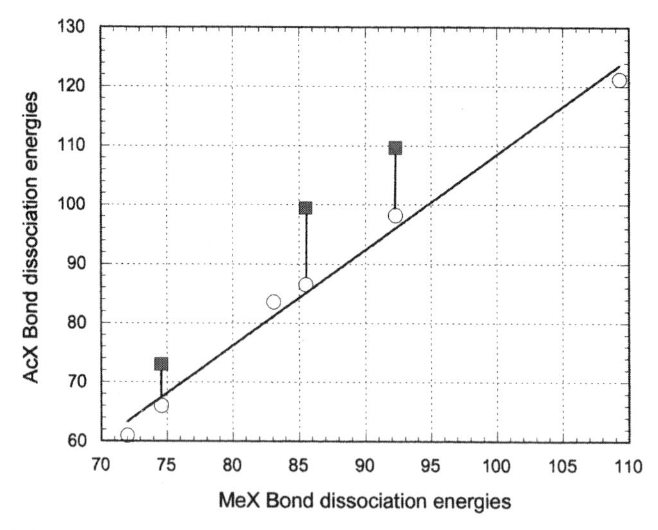

FIGURE 2.4. Correlation between the bond dissociation energies (BDE) of acetyl derivatives with the corresponding methyl derivatives. The substituents (X) are from left to right: PH_2, SH, Cl, NH_2, OH, and F. With X = SH, NH_2 and OH, the solid squares give the observed AcX BDE's. They are corrected for the π interaction (the rotational barrier) to give the corresponding open circles. The PH_2, Cl and F groups do not give π interactions with the carbonyl group. The slope of the line is 1.61.

TABLE 2.3. Bond Dissociation Energies of Methyl Derivatives

Compound	BDE (kcal/mol, 25 °C)
CH_3–CH_3	89.4 ± 0.3[a]
CH_3–NH_2	85.7 ± 1.5
CH_3–OH	92.2 ± 0.2
CH_3–F	111[b]
CH_3–SiH_3	89.6 ± 2.1
CH_3–PH_2	71[b]
CH_3–SH	74.4 ± 0.8
CH_3–Cl	83.3 ± 0.3

[a] The decrease in BDE from ethane to methylamine results from the change in hybridization.
[b] Calculated value.

between the methyl group and the substituent. In the second row, the largest BDE is found with X=SiH_3. Here, the electronegativities are reversed, with carbon being the more electronegative atom. This again leads to a dipolar bond, and to an increased BDE.

With the acetyl derivatives, there are two points in the figure for X=SH, NH_2 and OH. The closed circle represents the BDE of the rotated substituent, minimizing the π interaction. These, along with the other points, fall on a straight line with a slope of 1.6.*

The greater sensitivity of the acetyl group to the electronegativity of the substituent results from the dipolar character of the carbonyl group where the greater electronegativity of oxygen results in both σ and π electron shifts from carbon to oxygen. The dipolar character of the C=O group leads to its high energetic stability and also to its reactivity toward nucleophiles. If an electronegative substituent is attached to the carbon, it will further withdraw electron density via the σ-system (just as it did with the methyl derivatives). The dipolar character will strengthen the C–X bond. But, the increased positive charge at carbon will also strengthen the dipolar C=O bond. The strengthening of the two bonds in the acetyl derivatives leads to the high sensitivity towards the electronegativity of the substituents.

2. CONCLUSIONS

Amides are stabilized both by π donation and σ withdrawal of electron density from nitrogen to carbon. The amide oxygen has a large negative charge,

* It is interesting to note that whereas SiH_3 forms a relatively strong bond to a methyl group, it leads to destabilization of an acetyl group (ref. 50). As a result of its low electronegativity with respect to the hydrogens, the Si in the SiH_3 group bears a positive charge. The attachment of this positive charge to the positively charge carbon of the carbonyl group leads to destabilization. This is also found with other groups such as NO_2 and CN that have bonding atoms bearing positive charges.

characteristic of all carbonyl groups, and is little involved with the interaction that leads to the barrier to rotation about the C–N bond.

The classical valence bond "resonance" formulation is a relatively crude tool, and is not satisfactory for explaining the π interactions in amides, thioamides, and many other systems. Here, the FMO model is far more satisfactory in that it takes into account the changes in the atoms involved that are due to differences in electronegativity and other intramolecular effects. It provides a clear explanation of the lack of oxygen participation in the amide rotational barrier, and the greater participation of sulfur in the thioamides.

However, even the FMO model neglects the important changes that frequently occur in the σ system that is associated with the π system, and with the coulombic interactions. For example, a shift in π electrons in one direction will generally lead to a shift in σ electrons in the opposite direction. Although σ and π electrons in planar systems are orthogonal and cannot interact directly, they still repel each other and tend to move so as to minimize electron repulsion.* These interactions often have large effects on the energies of bonds and other properties of molecules. This is not just a problem associated with amides, but is a general problem in the application of the "resonance" concept.

REFERENCES

1. Dunitz, J. D.; Winkler, F. K. *Acta Crystallogr.* 1975, **B51**, 251.
2. Jeffrey, G. A.; Ruble, J. R.; McMullen, R. K.; DeFrees, D. J.; Pople, J. A. *Acta Crystallogr.* 1980, **B36**, 2292.
3. Wong, M. W.; Wiberg, K. B. *J. Phys. Chem.* 1992, **96**, 668. References to earlier ab initio calculations for amides are given therein.
4. Fogarasi, G.; Szalay, P. G. *J. Phys. Chem.* 1997, **101**, 1400 reported that MP2 calculations found formamide to be non-planar, but higher level CCSD(T)/PVTZ calculations found to to be planar.
5. Wiberg, K. B.; Rablen, P. R. *J. Am. Chem. Soc.* 1995, **117**, 2201.
6. Herzberg, G. *Molecular Spectra and Molecular Structure, Vol II*, Van Nostrand Reinhold, New York 1945.
7. Rush, D. J.; Wiberg, K. B. *J. Phys. Chem.* 1997, **101**, 3143.
8. Helminger, P.; DeLucia, F. C.; Gordy, W. *J. Mol. Spectrosc.* 1971, **39**, 94.
9. Wiberg, K. B. *Acct. Chem. Res.* 1996, **29**, 229.
10. Coulson, C. A. *Valence*, Oxford University Press, London, 1952.
11. Greenberg, A.; Venzanzi, C. A. *J. Am. Chem. Soc.* 1993, **115**, 6951.
12. Greenberg, A.; Moore, D. T.; Dubois, T. D. *J. Am. Chem. Soc.* 1996, **118**, 8658.
13. Drakenberg, T.; Dahlquist, K. S.; Forsen, S. *J. Phys. Chem.* 1972, **76**, 2178.
14. Ross, B. D.; True, N. S. *J. Am. Chem. Soc.* 1984, **88**, 2451.

* There are many examples of σ/π polarization: Cf. Wiberg, K. B.; Ochterski, J.; Streitwieser, A. *J. Am. Chem. Soc.* 1996, **118**, 8291 and ref. 22.

15. Ross, B. D.; True, N. S.; Matsen, G. B. *J. Phys. Chem.* 1984, **88**, 2675. The gas phase barrier for formamide has also been measured and was found to be Ea = 16.6 kcal mol^{-1} (Taha, A. N.; Crawford, S. M. N.; True, N. S. *J. Am. Chem. Soc.* 1998, **120**, 1934).
16. Onsager, L. *J. Am. Chem. Soc.* 1936, 58, 1486.
17. Kirkwood, J. *J. Chem. Phys.* 1934, **2**, 351.
18. Born, M. *Z. Phys.* 1920, **1**, 45.
19. Wiberg, K. B.; Keith, T. A.; Frisch, M. J.; Murcko, M. *J. Phys. Chem.* 1995, **99**, 9072.
20. Foresman, J. B.; Keith, T. A.; Wiberg, K. B.; Snoonian, J.; Frisch, M. J. *J. Phys. Chem.* 1996, **100**, 16098.
21. Wiberg, K. B.; Rablen, P. R.; Rush, D. J.; Keith, T. A. *J. Am. Chem. Soc.* 1995, **117**, 4261.
22. Wiberg, K. B.; Schleyer, P. v. R.; Streitwieser, A. *Can. J. Chem.* 1996, **74**, 892.
23. Sandstrom, J. *J. Phys. Chem.* 1967, **71**, 2318.
24. Stewart, W. E.; Siddall, T. H. *Chem. Rev.* 1970, **70**, 517.
25. Hehre, W. J.; Radom, L.; Schleyer, P. v. R.; Pople, J. A. *Ab Initio Molecular Orbital Theory*, Wiley, NY 1986.
26. Wiberg, K. B.; Laidig, K. E. *J. Am. Chem. Soc.* 1987, **109**, 5935.
27. Wiberg, K.B.; Breneman, C. M. *J. Am. Chem. Soc.* 1992, **114**, 831.
28. Duffy, E. M.; Severance, D. L.; Jorgensen, W. L. *J. Am. Chem. Soc.* 1992, **114**, 7535.
29. Gao, J. *J. Am. Chem. Soc.* 1993, **115**, 2930.
30. Ou, M.-C.; Tsai, M.-S.; Chu, S.-Y. *J. Mol. Struct.* 1994, **310**, 247.
31. Mulliken, R. S. *J. Chem. Phys.* 1962, **36**, 3428.
32. Wiberg, K. B.; Rablen, P. R. *J. Comput. Chem.* 1993, **14**, 1504.
33. Mulliken, R. S.; Politzer, P. *J. Chem. Phys.* 1971, **55**, 5135.
34. Grier, D. D.; Streitwieser, A. *J. Am. Chem. Soc.* 1982, **104**, 3556.
35. Reed, A. E.; Weinstock, R. B.; Weinhold, F. *J. Chem. Phys.* 1985, **83**, 735.
36. Reed, A. E.; Weinhold, F.; Curtiss, L. A. *Chem. Rev.* 1988, **88**, 899.
37. Bader, R. F. W. *Atoms in Molecules. A Quantum Theory*, Clarendon Press, Oxford, UK 1990.
38. Bader, R. F. W. *Acc. Chem. Res.* 1985, **18**, 9.
39. Perrin, C. L. *J. Am. Chem. Soc.* 1991, **113**, 2865.
40. Gatti, C.; Fantucci, P. *J. Phys. Chem.* 1993, **97**, 11677.
41. Laidig, K. E. *J. Am. Chem. Soc.* 1992, **114**, 7912.
42. Pauling, L. *The Nature of the Chemical Bond*, Cornell Univ. Press, Ithaca, NY.
43. Fukui, K.; Fujimoto, H. *Bull. Chem. Soc. Jap.* 1968, **41**, 1989.
44. Fukui, K.; Fujimoto, H. *Bull. Chem. Soc. Jap.* 1969, **42**, 3399.
45. Glendenning, E. D.; Hrabal, J. A. II, *J. Am. Chem. Soc.* 1997, **119**, 9478.
46. Lesczcynski, J.; Kwaitkowski, J. S.; Lesczcynska, D. *J. Am. Chem. Soc.* 1992, **114**, 10089.
47. Wiberg, K. B.; Marquez, M. *J. Am. Chem. Soc.* 1998, **120**, 2932.
48. Lauvergnat, D.; Hiberty, P. C. *J. Am. Chem. Soc.* 1997, **119**, 9478.
49. Wiberg, K. B.; Hadad, C. M.; Rablen, P. R.; Cioslowski, J. *J. Am. Chem. Soc.* 1992, **114**, 8644.

CHAPTER 3

THE AMIDE LINKAGE AS A LIGAND: ITS PROPERTIES AND THE ROLE OF DISTORTION

ARTHUR GREENBERG
Department of Chemistry, University of North Carolina at Charlotte

1. INTRODUCTION

The amide linkage[1] plays a central role in biochemistry.[2] It forms part of the backbone of proteins and various peptides, its functionality provides a scaffold for intermolecular association and recognition through hydrogen bonding, and it is the fundamental structural unit in the family of β-lactam antibiotics. Investigations into the dynamics of protein folding are forcing reexamination of the properties of the amide linkage.[3] The formal N–CO single bond possesses a relatively high rotational barrier ($\Delta G^{\ddagger} \approx 20\,\text{kcal mol}^{-1}$)[4,5] and this has been commonly explained through two major resonance contributors (**1A** and **1B**). The significant N=C double-bond character in the amide linkage is consistent with the coplanarity or near coplanarity of the six atoms that directly comprise it.[6] Indeed, it was this discovery that led Pauling to the solution of the α-helix "merely" by model building.[7]

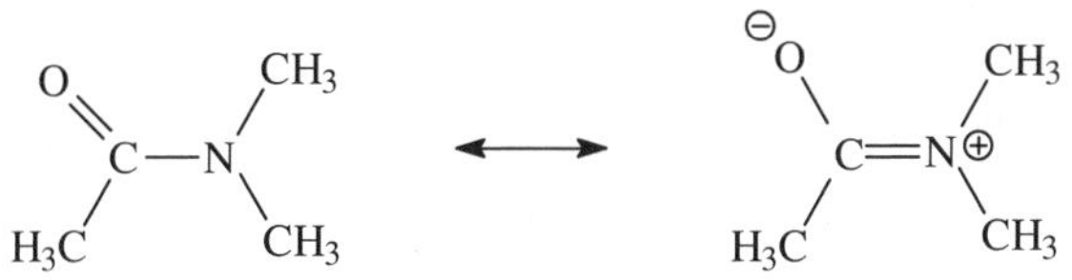

O=C(CH$_3$)–N(CH$_3$)$_2$ ⟷ $^{\ominus}$O–C(CH$_3$)=N$^{\oplus}$(CH$_3$)$_2$

1A **1B**

The present chapter will examine the amide linkage as a ligand. We will thus consider the sites of protonation, alkylation, hydrogen bonding and metallation

The Amide Linkage: Selected Structural Aspects in Chemistry, Biochemistry, and Materials Science,
Edited by Arthur Greenberg, Curt M. Breneman, and Joel F. Liebman
ISBN 0-471-35893-2

of planar as well as distorted amides. Over twenty years ago, Mock[8] made an excellent point that binding of a protein by an enzyme could significantly distort the amide linkage, decrease resonance stabilization and increase its reactivity. Distortion of the amide linkage is known to change the functionality of the amide linkage and therefore its ligation properties. Normal, unstrained amides or lactams such as *N*-methylpyrrolidinone (**2**) protonate at oxygen and, indeed, X-ray analysis of the crystalline hydrochloride of *N,N*-dimethylacetamide (**3H$^+$·Cl$^-$**) clearly shows this feature.[9] In contrast, the distorted lactam 6,6,7,7-tetramethyl-1-azabicyclo[2.2.2]octan-2-one (**4**) protonates at nitrogen.[10–13]

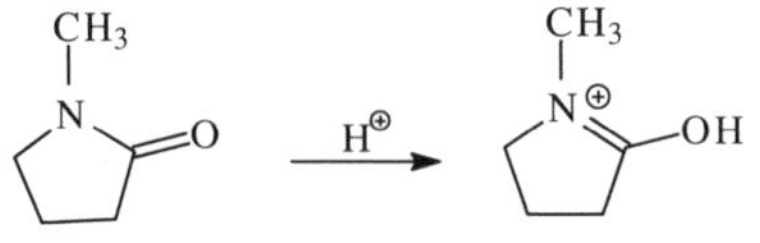

CH$_3$ N O $\xrightarrow{H^{\oplus}}$ CH$_3$ N$^{\oplus}$ OH

2 **2H$^+$**

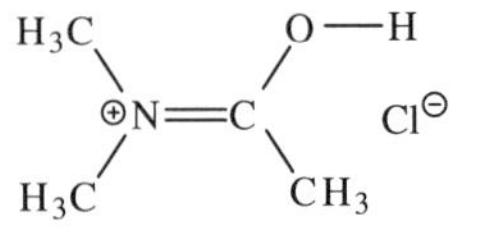

H$_3$C, H$_3$C $^{\oplus}$N=C(O—H)(CH$_3$) Cl$^{\ominus}$

3H$^+$·Cl$^-$

H$_3$C, CH$_3$, H$_3$C, CH$_3$, N, O $\xrightarrow{H^{\oplus}}$ H$_3$C, CH$_3$, H$_3$C, CH$_3$, N$^{\oplus}$, H, O

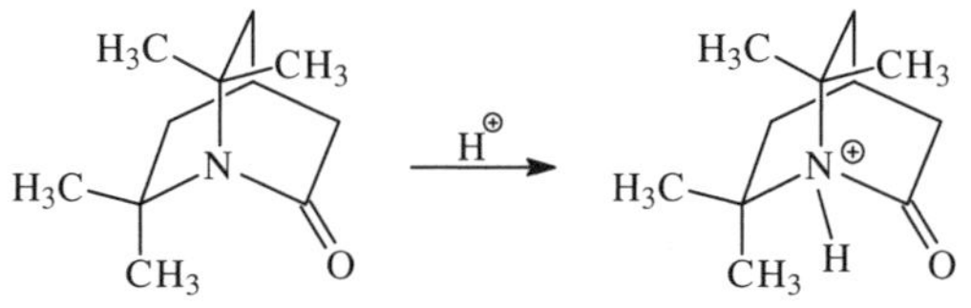

4 **4H$^+$**

In order to better understand the ligation properties of stable and distorted amides linkages, it is best to start by reviewing our understanding of the character of this structural moiety.

2. DESCRIPTION OF THE AMIDE LINKAGE

2.1. Resonance Contributors and Structural Parameters

Although the resonance picture typified by structures **1A** and **1B** has been strongly entrenched for over sixty years, the model was pointedly questioned starting about twelve years ago by Wiberg and Laidig,[14] who initially employed the Atoms-In-Molecules (AIM) approach of Bader.[15] The work has evolved as it was debated in the literature[16–20] and Wiberg[21] has recently outlined his view of structure and bonding in the amide linkage. This reexamination of the seemingly familiar amide linkage has been very beneficial. For example, Wiberg and his co-workers calculated that severe distortion of the amide linkage through pyramidalization at nitrogen (χ_N) and/or severe torsion (τ) significantly lengthens C–N but barely shortens C=O.[14,16,17] Their calculations are consistent with experimental observations and calculations on distorted bridgehead lactams such as **4**.[22–29] The effects of distortion on bridgehead lactams are mirrored in their spectroscopic properties which parallel those of acyclic amides distorted through steric effects.[30,31] Recently, 1-aza-2-adamantanone **5**, a stable crystalline compound termed "the most twisted amide" has been studied using X-ray crystallography.[32,33] The N–C and C=O bond lengths are 1.475 (11) Å and 1.196 (5) Å respectively. These values are estimated to be ca 0.12–0.13 Å longer and 0.03–0.04 Å shorter than the corresponding values in an unstrained lactam.[32]

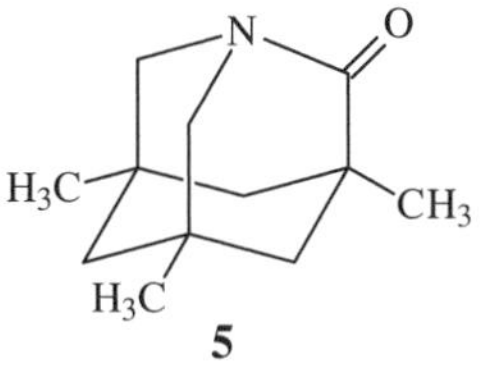

N O H_3C CH_3 H_3C

5

To a first approximation, the comparative insensitivity of C=O bond length to severe distortion is counter-intuitive since roughly equal contribution of **1A** and **1B** might be expected to give rise to similar changes (in C–N and C=O bond lengths) as a function of distortion. The work of Wiberg and co-workers[14,16,17] has also helped to focus attention on the role of the local geometry at nitrogen on its electronegativity and has prompted chemists to reexamine the role of the σ framework in understanding the structures of amides.

There is now general agreement that the source of the preferred planarity of the amide linkage is donation of the nitrogen lone pair to the carbonyl carbon. Although this is explicit in resonance structure **1B**, Wiberg and co-workers describe a slightly more complex resonance picture that involves a third major resonance contributor (**1C**) in which the extreme polarization of the carbonyl group reflects the high electronegativity of oxygen.[21] One might rationalize the small changes in the C–O bond length as a function of distortion by invoking major contributions from **1B** and **1C** in the planar structure and a disappearance of **1B** as distortion increases. In this view, the CO bond in a planar amide is essentially a single bond shortened by electrostatic attraction between carbon and oxygen and is little changed by distortion.[21] In contrast, the C–N bond significantly lengthens upon distortion clearly reflecting a reduced contribution from **1B**. In this view, the higher rotational barrier in thioamides is explicable in terms of only two major resonance contributors (**6A** and **6B**) with a higher proportional contribution from **6B** than in the amides. Since sulfur has an electronegativity roughly equal to that of carbon, one view is that resonance structure **6C** makes little contribution.

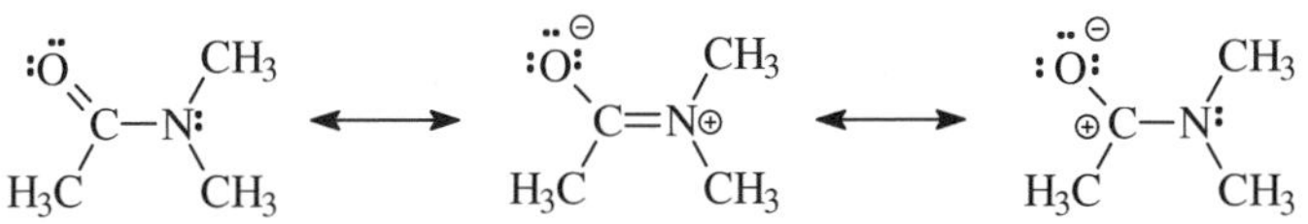

:Ö=C(–H_3C)–N̈(CH_3)$_2$ ⟷ :Ö:$^{\ominus}$–C(H_3C)=N$^{\oplus}$(CH_3)$_2$ ⟷ :Ö:$^{\ominus}$–$^{\oplus}$C(H_3C)–N̈(CH_3)$_2$

1A **1B** **1C**

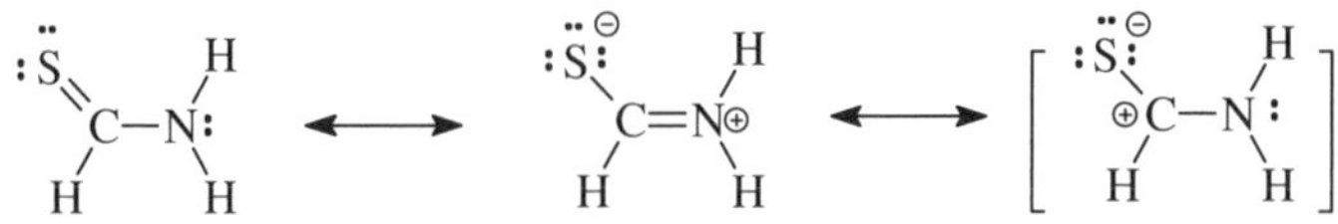

(negligible)

6A **6B** **6C**

Two recent studies have examined the rotational barrier in amides and provided a resonance picture, albeit more complex than the prevailing simple (**1A**↔**1B**) view, to account for at least a large fraction of it. Lauvergnat and Hiberty[34] suggest that amides may be represented by resonance structures such as **1A**, **1B**, and **1C**, but conclude that resonance accounts for only about half of the amide rotational barrier and about two-thirds of the thioamide rotational barrier (we will revisit their analysis below). Glendening and Hrabel[35] analyze the rotational barrier of amides and conclude that both amides and thioamides are about 60% **1A**-type and about 30% **1B**-type resonance contributors.

Despite the utility of the **1A**, **1B**, **1C** picture, Wiberg terms the resonance formulation "a relatively crude tool" and favors a frontier orbital approach.[16,21] In amides, the N lone pair is shown to donate to the carbonyl π^* orbital whose coefficient is largely on carbon due to the electronegativity difference between C and O which requires a large coefficient on oxygen in the carbonyl π orbital.[21] In contrast, the electronegativities of C and S are similar, their coefficients (albeit 2π and 3π) are more similar and the N lone pair is effectively donated to both C and S. Although this reasoning is also consistent with little contribution from **6C**, some of its subtleties do not lend themselves to simplified resonance formulations.

Our own studies in this area have supported the resonance picture. Admittedly, the resonance picture is somewhat crude. It typically employs formal charges, themselves overly simplistic electron accounting. Depictions such as **1A** and **1B** implicitly suggest that a (coincidental) 1:1 mixture would be intermediate between C–N/C=N and C=O/C–O, ignoring differences in stretching constants of these bonds. Depictions such as **1A**, **1B**, and **1C** are not very illuminating about the σ framework. Cooperativity between the bonding in σ- and π-frameworks (e.g. "push–pull") would be expected to be a stabilizing influence.

Nevertheless, chemists' comfort with canonical Lewis-type structures makes resonance theory useful as a model. The rather different conclusions about the weighting of resonance contributors in the Lauvergnat and Hiberty study,[34] on one hand, and the Glendening and Hrabel study,[35] on the other, suggests caution over too fine an interpretation of canonical contributions. However, decisions on dominant orbital interactions and the relative stabilizations due to σ- and π-orbital contributions are also hard to quantitate except perhaps a posteriori.

We have correlated stabilities of bridgehead lactams with bridgehead olefins—an explicit expression of the contribution of resonance contributors analogous to

1B.[29] More specifically we observed from a 6–31G* ab initio MO study that bridgehead lactams showed trends in strain energies, structure and spectroscopic properties which grouped together according to whether these compounds were analogues of *trans*-cyclohexene, *trans*-cycloheptene, *trans*-cyclooctene or *trans*-cyclononene (see Scheme 1).[29] It is well to recognize, however, that a perfectly orthogonal amide linkage drawn conventionally obeys the octet rule while a perfectly orthogonal alkene linkage does not.

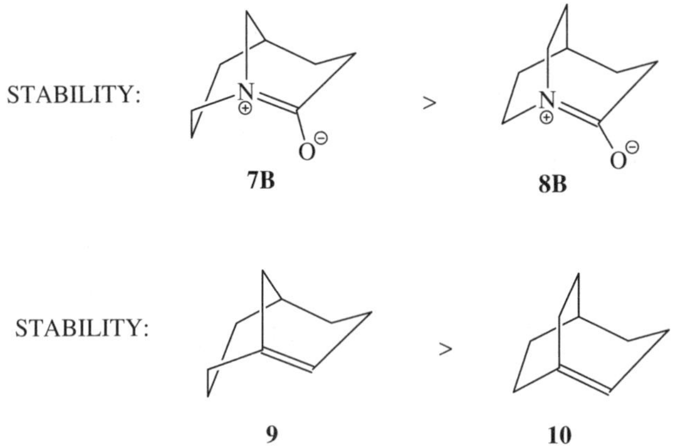

STABILITY: **7B** > **8B**

STABILITY: **9** > **10**

Scheme 1

Although we also find that the C=O bond length changes little as a function of distortion of the amide linkage,[29] the change agrees with a model that increases the contribution of **1A** as distortion increases (i.e. shorter C=O bond length). Moreover, as the amide linkage distorts, the frequency of ν_{CO} increases presumably due to the increased contribution of **1A**.[25–31]

2.2. Resonance Contributors and Atomic Charge

The relative contributions of resonance structures **1A**, **1B**, and **1C** suggest that a measure of atomic charge might be useful in understanding the relative importance of these contributors. Wiberg and Breneman[17] and, more recently, Wiberg[21] have summarized some of the factors and uncertainties pertaining to Mulliken charges, Bader charges derived from the AIM approach as well as Weinhold pure resonance charges.

One point that needs to be clarified is that formal charges are just that—“formal.” Thus, both theoretical and experimental evidence indicates that the nitrogen atom in ammonium ion (**11**) is not even positive (as its +1 formal charge would imply) but negative due to the four electropositive hydrogens attached to it.[36,37] Similarly, although silatranes (**12**) do maintain a weak but real bond between silicon and nitrogen, the metalloid silicon is surely not negative and the nitrogen is almost certainly negative and not positive.[38] Thus, the rather obvious point that has been emphasized is that the nitrogen in amides is not

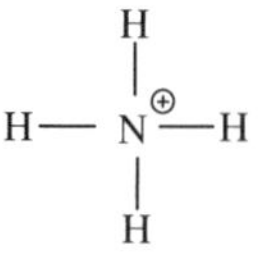

11 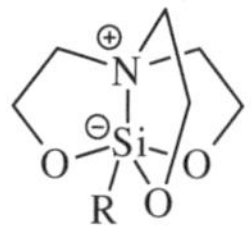 **12**

really positive.[39] The more useful point is to see the nitrogens in planar amides, richer in dipolar resonance contributors such as **1B**, as more positive (i.e., less negative) than twisted amides which have a reduced contribution from **1B**.

We have explored the use of core orbital (1s) energies as probes of charge on atoms. This measure, albeit an indirect measure of electron density on an atom, is free of the assumptions of electron apportionment. Experimental measurements of core electron binding energies using X-ray photoelectron spectroscopy (XPS, also known as ESCA) have long been accepted as measures of relative atomic charge.[40] An earlier study on solid-state aminimides employed N_{1s} core ionization energies to roughly apportion resonance contributors and correlate them with IR frequencies.[41]

It is interesting to compare the core ionization energies for nitrogen in amines and amides and for oxygen in ketones and amides.[42,43] These data are shown for *N,N*-dimethylacetamide in Scheme 2 (see Ref. 42 for discussion of some

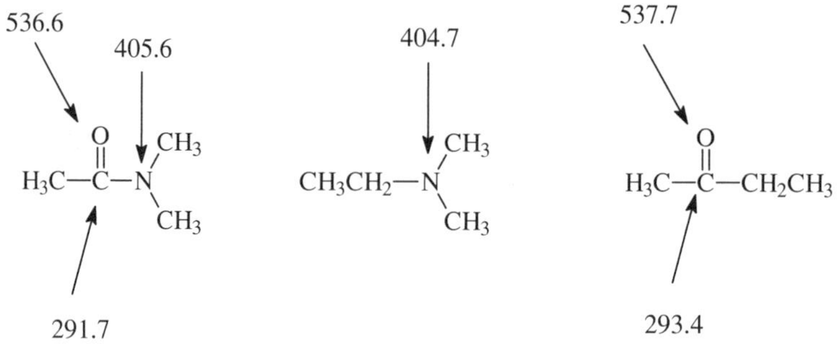

13 **14** **15**

Core Ionization Energies (eV)

Scheme 2

estimated numbers). The fact that the ionization of the N_{1s} electron in *N,N*-dimethylacetamide requires 0.90 eV *more* energy than in *N,N*-dimethylethanamine while the O_{1s} core electron requires 1.1 eV *less* energy is certainly consistent with the contribution of canonical structures **1B** and **1C**. Admittedly, the more electronegative carbonyl neighbor should increase the core ionization energy of the N_{1s} orbital even if π-donation by the nitrogen were to be ignored. However, replacement of electronegative nitrogen by carbon should decrease the ionization energy of O_{1s} rather than increase it in the ketone if simple electronegativity arguments alone are invoked. The fact that the carbonyl C_{1s} ionization energy in the amide is ca 1.7 eV less than in the ketone also contradicts electronegativity arguments and supports the concept[21] that a resonance contributor such as **15B** is important in ketones while the corresponding **1C** in amides is countered by **1B**. Again, these arguments are highly qualitative.

$H_3C{-}C({=}O){-}CH_2CH_3 \longleftrightarrow H_3C{-}C^{\oplus}(O^{\ominus}){-}CH_2CH_3$

15A **15B**

Computational studies allow one to look at model species not experimentally accessible under normal circumstances. Our own 6–31G* ab initio calculations of core ionization energies, assuming the validity of Koopmans' theorem, have provided interesting insights. For example, Scheme 3 indicates that 90° twisting of the amide linkage (which includes pyramidalization at nitrogen) decreases N_{1s} by 0.6 eV (formamide) or 0.9 eV (*N,N*-dimethylacetamide), increases O_{1s} by 1.1–1.2 eV and increases the carbonyl C_{1s} by 0.2–0.3 eV. These data are fully consistent with a picture including resonance contributors **1A**, **1B**, and **1C** in the

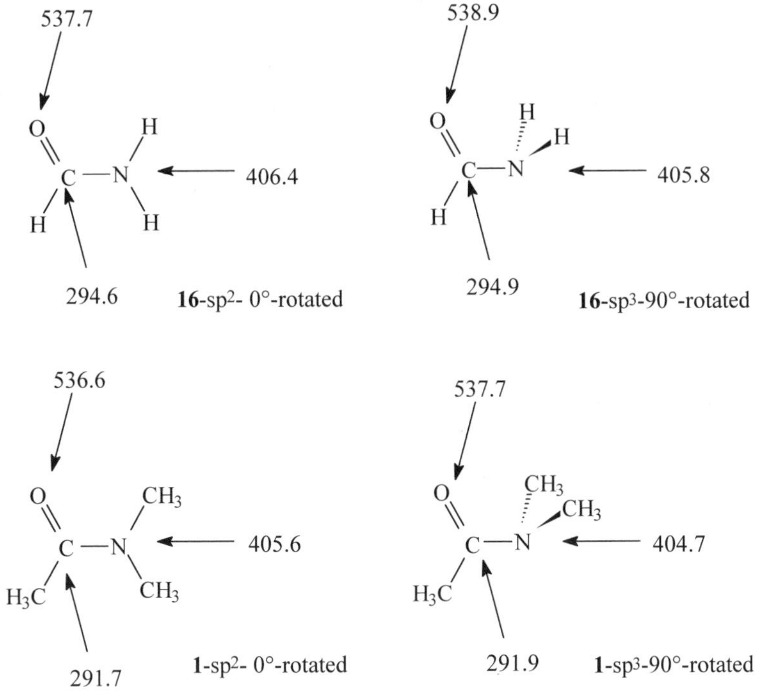

537.7 | 406.4 | 294.6 — **16**-sp2- 0°-rotated

538.9 | 405.8 | 294.9 — **16**-sp3-90°-rotated

536.6 | 405.6 | 291.7 — **1**-sp2- 0°-rotated

537.7 | 404.7 | 291.9 — **1**-sp3-90°-rotated

Calculated Core Ionization Energies (eV)

Scheme 3

planar structure and enhanced contribution of **1A** and **1C** with the loss of **1B** in the twisted structure.

Computational studies of the bridgehead lactams have also provided support for this model. In Table 3.1 we list computed N_{1s}, O_{1s} and carbonyl C_{1s} core ionization energies for the bridgehead lactams **17–25**. Listed in this table are calculated resonance energies for these systems relative to *N,N*-dimethylacetamide which is taken as 20.2 kcal mol^{-1} (these values were obtained at the 6–31G* level[29] and are employed here). A higher level of computation[44] yields a value of 18 kcal mol^{-1} for *N,N*-dimethylacetamide. The results strongly support the view that a purely planar lactam linkage has appreciable contributions from canonical structures **1A**, **1B**, and **1C** while distortion decreases the contribution of **1B** with concommitant increasing contributions from **1A** and **1C**. Of particular interest is the observation that the calculated core ionization energies and the resonance energies group according to the *trans*-cycloalkene analogy (bridgehead lactams and alkenes—see Scheme 1). This is supportive of the utility of the resonance model as dipolar structures such as **1B** support the alkene analogy for amides.

TABLE 3.1. Computed Core Ionization Energies[a] for Bridgehead Bicyclic Lactams and Computed Resonance Energies[b] (see text)

Lactam	O_{1s} (eV)	N_{1s} (eV)	C_{1s} (eV)	Resonance Energy (kcal mol^{-1})
trans-Cyclohexene Analogue				
2.2.2 (**17**)	537.35	404.75	291.01	0.0
trans-Cycloheptene Analogues				
3.**2**.2 (**18**)	536.88	404.83	290.86	6.6
3.2.2 (**19**)	536.83	404.88	290.86	6.1
3.3.**2** (**20**)	536.86	404.75	290.83	5.9
trans-Cyclooctene Analogues				
3.3.1 (**21**)	536.67 (exp)	405.07 (exp)	290.81 (exp)	11.5
3.3.2 (**22**)	536.41	405.09	290.75	13.7
3.3.3 (**23**)	536.41	404.98	290.78	15.2
trans-Cyclononene Analogues				
4.3.3 (**24**)	536.26	405.11	290.68	22.6
4.**3**.3 (**25**)	536.23	405.17	290.70	26.4
Model Compound				
1-Methylpyrrolidone (**2**)	536.36	405.40	290.78	21.5

[a] See reference 29.
[b] The 6–31G*-calculated resonance energy of *N,N*-dimethylacetamide is taken as 20.2 kcal mol^{-1} according to these estimates. The experimental value is 18.0 kcal mol^{-1} using the methyl capping method.[29]

It is also interesting to note in Table 3.1 that the bicyclic bridgehead lactam **25** is predicted to have about 6 kcal mol^{-1} higher resonance stabilization than *N,N*-dimethylacetamide. We have termed **25** a hyperstable lactam in direct analogy to Schleyer and co-workers concept of hyperstable bridgehead alkenes.[45,46] We further note that addition reactions are the characteristic alkene reactions and that hyperstable olefins will be both thermodynamically stabilized *and* unreactive. In contrast, since the characteristic amide reaction is nucleophilic acyl substitution, we anticipate that **25** will be thermodynamically "normal" but kinetically stabilized.[29]

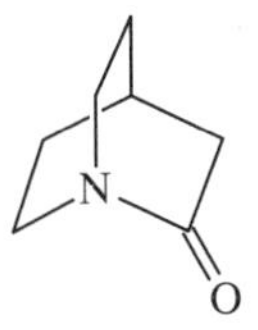

17 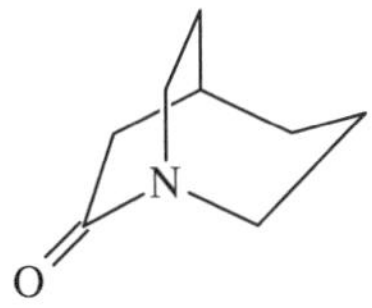 **18** 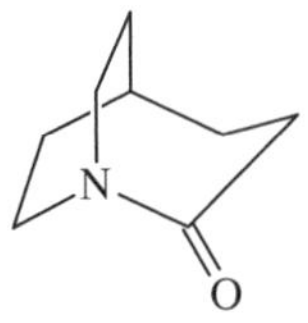 **19**

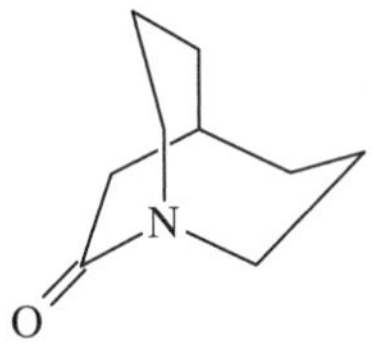

20 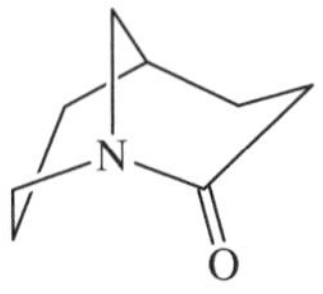 **21** 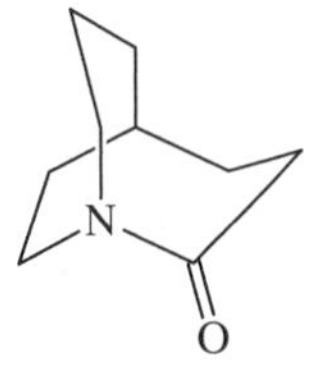 **22**

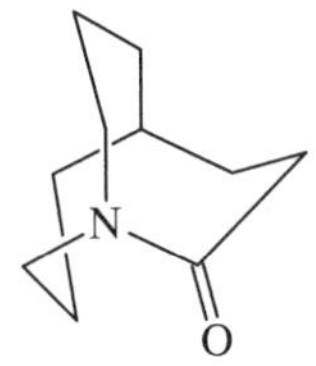

23 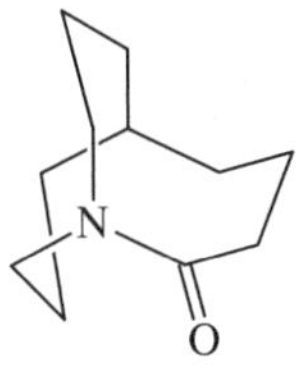 **24** 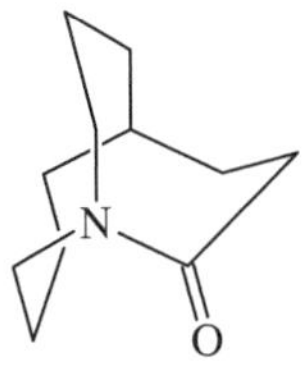 **25**

2.3. Amide Rotational Barrier

The amide rotational barrier has been recently discussed by Wiberg[21] and the analysis of the barrier in the gas-phase and in polar and nonpolar solvents has been analyzed incisively by Wiberg et al.[44] Quite recently, the gas-phase rotational barrier of formamide was determined to be 16.6 (0.3) kcal mol^{-1}.[47] The rotational barrier for amides increases as the solvent is changed from nonpolar cyclohexane to acetone to water ($k_{cyclohexane}/k_{water} \cong 60$) reflecting preferential stabilization of the more polar ground state (e.g., due to enhanced contribution from **1B**) in water.[44,48] We also note that analyses of the co-variation in solid-state structural parameters as a function of distortion of amides have also been published.[49–52]

We present here two analyses of aspects of the rotational barrier. In Fig. 3.1, we present a pictorial analysis of the formamide rotational barrier by Lauvergnat and Hiberty.[34] In the first of four "panels" in Fig. 3.1 (i.e., Fig. 3.1(A)) we note a 27.3 kcal mol^{-1} calculated energy difference between the optimized, fully-delocalized (global minimum) structure of formamide and the same geometry wherein delocalization of the nitrogen lone pair is computationally "turned off". This is termed the vertical delocalization energy (DE_v). Optimization of the N lone pair-localized structure lowers its energy by 3.1 kcal mol^{-1}. The resulting energy difference, 24.2 kcal mol^{-1}, is termed as the adiabatic delocalization energy (DE_{ad}) and is probably a more valid measure of the interaction. In

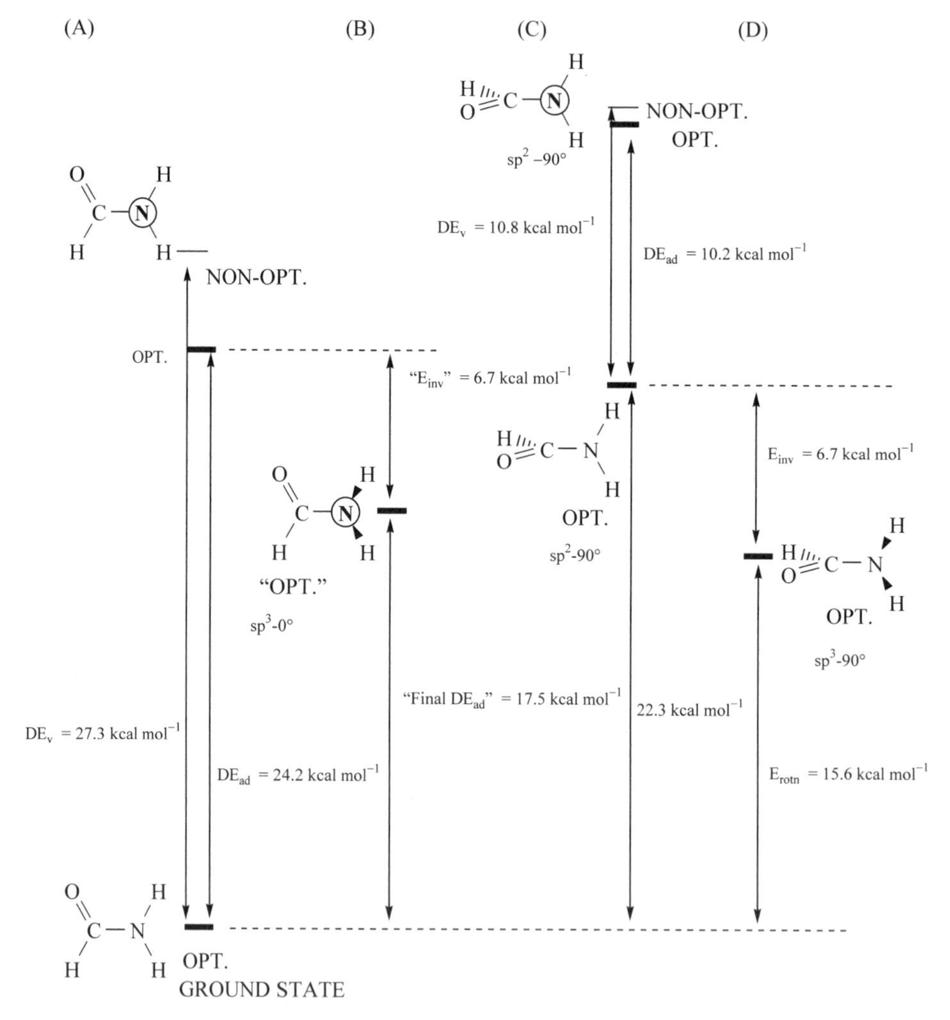

FIGURE 3.1. Dissection of components of delocalization and resonance energy of formamide following the methods of Lauvergnat and Hiberty[33] (see text). A boldface nitrogen in a circle lacks p–π interaction.

Fig. 3.1(B) there is a 6.7 kcal mol^{-1} correction for the N inversion barrier. This is the number obtained for the N inversion barrier in Fig. 3.1(D). Thus, a kind of pure "N switched-off" delocalization energy which corrects for the inversion barrier is calculated at 17.5 kcal mol^{-1} and is termed the "final" DE_{ac}. In Fig. 3.1(C) we note that a 90°-rotated formamide, fixing planar geometry at N, fully delocalized and fully optimized (except for planarity at N) and termed OPTsp^2-90° is 22.3 kcal mol^{-1} higher in energy than the global minimum. If one allows this structure to relax completely, then, not surprisingly, the nitrogen pyramidalizes. This structure (termed OPT sp^3-90°) is 15.6 kcal mol^{-1} higher than the global minimum and represents the C–N rotational barrier. This number is in very good agreement with the gas-phase experimental and computational (gas-phase) numbers cited by Wiberg et al.[44] as well as Taha et al.[48] The 6.7 kcal mol^{-1} inversion energy (E_{inv}) [see Fig. 3.1(C) and (D)], in very good agreement with typical amine inversion barriers, is considered "normal" and thus transferable for Fig. 1(B). Finally, "localization" of the N lone pair in Fig. 3.1(C), as in Fig. 3.1(A), starting with the OPT sp^2-90° structure, gives a value 10.8 kcal mol^{-1} higher in energy ($=DE_v$). Relaxation of this structure yields the OPT sp^2-90° structure which is 10.2 kcal mol^{-1} higher in energy than OPT sp^2-90°. This is considered to be the adiabatic delocalization energy (DE_{ad}) for the sp^2-90° structure. It is not conventional to imagine such a large delocalization for a structure where the N lone pair is orthogonal to the attached carbonyl π system. The 10.2 kcal mol^{-1} delocalization is attributed primarily to hyperconjugation. Lauvergnat and Hiberty[34] then subtract this 10.2 kcal mol^{-1} from the "final" DE_{ad} (17.5 kcal mol^{-1}) and compare the resulting 7.3 kcal mol^{-1} to the C–N rotational barrier (15.6 kcal mol^{-1}) to conclude that delocalization of the N lone pair into the carbonyl group in the planar structure (i.e., resonance stabilization) accounts for about half of the amide rotational barrier. A similar analysis of thioamides indicates that resonance stabilization accounts for about two-thirds of the rotational barrier.[34] We find this study insightful and compelling. However, we remain troubled by the simple subtraction of 10.2 kcal from 17.5 kcal since we feel that the two types of N lone pair delocalization are different.

For another analysis, albeit a simpler although more operational one, we turn to Fig. 3.2[53] (this is an improved version of Scheme 4 in Ref. 29). The numbers employed for *N,N*-dimethylactamide, its conformers and model compounds are derived from Wiberg et al.[44] The values for the 1-azabicyclo[2.2.2]octan-2-one molecule (**17**) are derived from our 6–31G* study.[29] We assumed an N-inversion barrier of 6 kcal mol^{-1}. The actual rotational barrier energy is calculated at ca 16 kcal mol^{-1} (versus 15.6 kcal in Fig. 3.1). The rigid rotation to form OPT sp^2-90° is calculated at 22 kcal mol^{-1} (versus 22.3 kcal in Fig. 3.1). This total loss of resonance energy, with all amide linkage atoms connected and unconnected for the N-inversion barrier is very similar to the 21 kcal mol^{-1} calculated for **17** in comparing the calculated value with the hypothetical **17** having full resonance energy—both nitrogens are pyramidal and, hence, no correction has been made for the inversion barrier.[29] Finally, Liebman's "methyl-capping" resonance

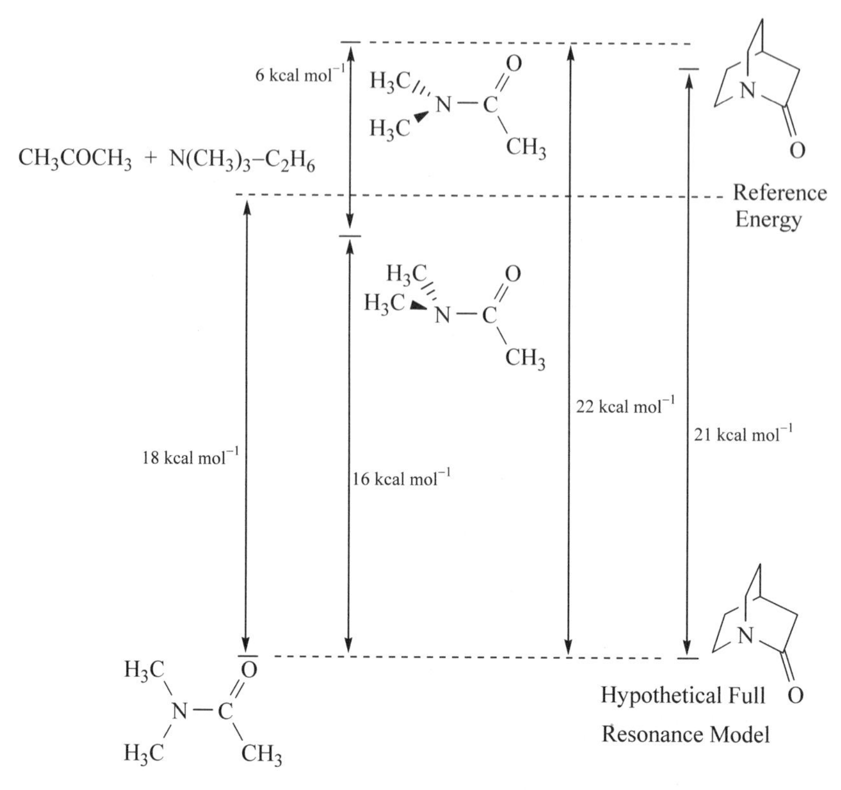

FIGURE 3.2. Relationship between "methyl capping" resonance energy and rotational barrier of *N,N*-dimethylacetamide and resonance destabilization in 1-azabicyclo[2.2.2]octan-2-one 9see text).

energy[54] (amide versus isolated ketone and amine molecules) is 18 kcal mol^{-1}. It is tempting to try and compare this with the "final" DE_{ad} = 17.5 kcal mol^{-1} in Fig. 3.1(B). However, while N lone pair delocalization is absent in both model systems and the nitrogens are pyramidal in both model systems, the model in Fig. 3.1(B) retains all σ connections while the model in Fig. 3.2 does not. The ca 2 kcal mol^{-1} stabilization in Fig. 3.2, derived from forming the amide linkage with pyramidal nitrogen which is orthogonal to the carbonyl π linkage, may be thought of as a composite of the newly formed σ-linkages and perhaps some of the hyperconjugation-derived stabilization in OPT sp^3-90° (presumably smaller than in OPT sp^2-90°).

3. PROTONATION OF AMIDES

3.1. Relative Energetics of N- versus O-protonation

In the gas-phase, unstrained simple amides protonate on oxygen rather than on nitrogen. The proton affinity (PA) of formamide (the negative value for the enthalpy change in Eq. (3.1) is experimentally found to be 196.5 kcal mol^{-1}.[55] The gas phase basicity [*GB*, the negative of the

$$H^+(g) + Amide(g) \rightarrow Amide\ H^+(g) \tag{3.1}$$

free energy of Eq. (3.1)] is less positive (189.1 kcal mol^{-1}) reflecting the negative entropy of Eq. (3.1). These values are in strong agreement with corresponding values of 195.7 kcal mol^{-1} and 188.9 kcal mol^{-1} obtained using ab initio calculations[56] with the 6–31 + G** basis set, electron correlation (MP2) and correction for zero point vibration for E_o^{ref} (MP2) (actually at 0 K) and also including thermal corrections to obtain $\Delta G^{o\ ref}_{298}$ (MP2). Table 3.2 lists a series of experimental values for PA and GA. The calculated values indicate that protonation is at oxygen and that the O-protonated structure (N-anti **16O-Prot**) is 13.9 kcal mol^{-1} [ΔG_{298} (MP2)] more stable than the N-protonated structure (**16N-Prot**).[56] When correction is made for solvation, using a solute-occupied

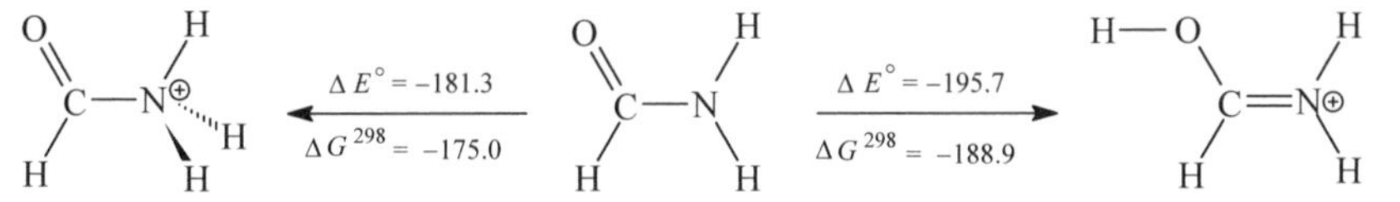

$\Delta E° = -181.3$ / $\Delta G^{298} = -175.0$ (16 → 16N-Prot); $\Delta E° = -195.7$ / $\Delta G^{298} = -188.9$ (16 → 16O-Prot)

16N-Prot **16** **16O-Prot**

(Calculated PA and GB values in kcal mol^{-1})

Scheme 4

cavity and the Onsager model including a dielectric constant [ε = 80 (aqueous model lacking specific H-bonding interactions)], the O-protonated structure (**16O-Prot**) is calculated at about 7 kcal mol^{-1} more stable than the N-protonated in contrast to the ca 14 kcal mol^{-1} gas-phase difference. Thus, this simple model, lacking specific H-bonding interactions, indicates that polar solvents stabilize the N-protonated more than the O-protonated structure. It has also been noted[56–58] that the solution pK_a for N-protonated amides such as **16N-Prot** are estimated at −7 to −8 while the values for the O-protonated amides similar to **16O-Prot** are in the range 0 to −3. If we were to take the mid-ranges of these values (p$K_a \sim -7.5$ for **16N-Prot**; p$K_a \sim -1.5$ for **16O-Prot**) the free energy difference would be about 8.3 kcal mol^{-1} favoring the O protonated form. This is fairly similar to the above calculated difference including the Onsager treatment. Olah et al.[60] nicely summarize the initial ambiguities in solution studies of the protonation sites of amides and note the observation of unambiguously O-protonated amides in strong acid media. Monoprotonation of urea also occurs at oxygen with 6–31G** calculations indicating 5.8 kcal mol^{-1} advantage over N-protonation.[61] Interestingly, reactions of acylium ions with amines initially yield N-protonated amides which are observable and rearrange to the more stable O-protonated isomer.[60] Table 3.2 summarizes the gas-phase

TABLE 3.2. Gas Phase Proton Affinities (PA) and Gas Phase Basicities (GB) (kcal mol^{-1}) Reported for Selected Amides and Lactams[55]

Amides and Lactams	PA(g) (kcal mol^{-1})	GB(g) (kcal mol^{-1})
Formamide (CH_3NO) (**16**)	196.5	189.1
Acetamide (C_2H_5NO)	206.4	199.0
N-Methylformamide (C_2H_5NO)	203.5	196.1
Propionamide (C_3H_7NO)	209.4	202.0
N,*N*-Dimethylformamide (C_3H_7NO)	212.1	204.7
N-Methylacetamide (C_3H_7NO)	212.4	205.0
2-Azetidinone (C_3H_5NO)	203.8	196.4
N,*N*-Dimethylacetamide (C_4H_9NO) (**1**)	217.0	209.6
2-Methylpropionamide (C_4H_9NO)	210.0	202.4
N-Methylpropionamide (C_4H_9NO)	220.0	212.6
N-Ethylacetamide (C_4H_9NO)	214.6	207.2
N-Methyl-2-azetidinone (C_4H_7NO)	210.9	203.5
N-Methylpyrrolidinone (C_5H_9NO) (**2**)	220.7	213.1
N,*N*-Diethylacetamide ($C_6H_{13}NO$)	221.2	213.8
N,*N*-Dimethylbutyramide ($C_6H_{13}NO$)	220.3	212.9
N-Methylpiperidone ($C_6H_{11}NO$)	220.9	213.3
N-Acetylpyrrolidine ($C_6H_{11}NO$)	221.2	213.8

PA and GB data for a variety of relatively simple amides and lactams.[55] All of these are expected to be O-protonated by considerable margins although the difference between O-protonation and N-protonation is smaller for 2-azetidinones (β-lactams).[62] In contrast, while there are no experimental data for gas phase proton affinities for aziridinones (α-lactams), although some of these are quite stable, they are predicted to protonate at nitrogen in keeping with the preference for pyramidalization of nitrogen in the 3-membered ring.[63–67] Specifically, protonation has been calculated to be favored at N by about 4.4 kcal mol^{-1} for aziridinone and by about 3 kcal mol^{-1} in the 1,3,3-trimethyl derivative and its significance is described by Hoffman[66] in understanding the regiochemistry and stereochemistry of acid-catalyzed nucleophilic attack at C-3 in α-lactams*.[66] Consistent with this trend is the prediction of N-protonation in

$\Delta E^{\circ} = -202.4$, $\Delta G^{298} = -196.4$ (**26** → **26O-Prot**)

$\Delta E^{\circ} = -204.2$, $\Delta G^{298} = -198.2$ (**26** → **26N-Prot**)

26O-Prot **26** **26N-Prot**

(Calculated PA and GB values in kcal mol^{-1})

Scheme 5

*Tantillo, D. J.; Houk, K. N.; Hoffmann, R. V.; Tao, J. Unpublished cited in R. V. Hoffmann as Ref. 34.

N-formylaziridine.[56] While the gas phase free-energy difference favoring **26N-Prot** is only about 1.8 kcal mol^{-1}, the calculated difference in water favoring **26N-Prot** is 5 kcal mol^{-1} reflecting predictions of enhanced stabilization of the N-protonated isomer.[56]

Finally, we note that Cassady[67] has described gas-phase protonation studies of simple amides as well as peptides and proteins. In peptides and proteins, the most basic sites are the *N*-terminal amino groups as well as the basic side chains such as lysines. Protonations also occur at peptide linkages. In larger peptides and proteins, coulombic repulsions in multiply protonated species and other inter-residue interactions such as hydrogen bonding will often play dominant roles.[67]

In addition to pyramidalization at nitrogen, torsion about the OC–N bond is another means for favoring N-protonation over O-protonation. The N-protonation of 6,6,7,7-tetramethyl-1-azabicyclo[2.2.2]octan-2-one (**4H**$^+$) is the extreme wherein N-pyramidalization is augmented by a virtually orthogonal amide linkage. We have noted that for unstrained amides O-protonated amides have about twice the resonance energy as their unprotonated conjugate bases.[65] This more than compensates for the higher intrinsic basicity of N versus O. In twisted bridgehead bicyclic lactams, significant resonance energies are lost in the O-protonated forms even as they contort to regain as much resonance energy as possible.[28,29] Table 3.3 lists PA values calculated for the bridgehead bicyclic lactams **17–23** at both N and O using the 6–31G* basis set fully optimized for these large molecules along with zero-point-energy (ZPE) and thermal

TABLE 3.3. Calculated Values (6–31G* + ZPE + thermal corrections) for N and O Proton Affinities of Seven Bridgehead Bicyclic Lactams and Two Model Compounds[28,29]

Lactams	PA at N (kcal mol^{-1})	PA at O (kcal mol^{-1})	Difference (kcal mol^{-1})
trans-Cyclohexene Analogue			
2.2.2 (**17**)	228.9	206.2	22.8
trans-Cycloheptene Analogues			
3.**2**.2 (**18**)	224.7	213.6	11.1
3.2.2 (**19**)	223.6	214.4	9.1
3.3.**2** (**20**)	224.7	215.1	9.6
trans-Cyclooctene Analogues			
3.3.1 (**21**)	219.0	217.6	1.4
3.3.2 (**22**)	214.1	221.2	−7.1
3.3.3 (**23**)	218.1	221.6	−3.5
Model Compounds			
N,N-Dimethylacetamide (**1**)	206.5	218.3 (217.0[a])	−11.8
N-Methylpyrrolidinone (**2**)	203.7	218.5 (220.7[a])	−14.8

[a] See Table 1.2.

corrections. *N,N*-Dimethylacetamide and 1-methylpyrrolidinone are included for the sake of comparison. Table 3.3 indicates that, even though electron correlation was not employed, the calculated PAs are in good agreement with experiment.

Table 3.3 indicates that 1-azabicyclo[3.3.1]nonan-2-one (**21**) may protonate competitively at both N and O sites. If the earlier conclusions of Cho et al.[56] are relevant here, then additional favoring of the N-protonated form in aqueous solutions might well make the O-protonated form undetectable. However, the trends in solvation in the bridgehead bicyclic lactams might not parallel those in the acyclic amides. Obviously, experimentation needs to be done. If aqueous solvation does favor N-protonation, then the presently unknown 3.3.3 system **23** might be the better candidate for the coexistence of N- and O-protonation.

It is interesting to note the changeover in protonation from the **3**.2.2 system (**19**) upon "loosening the lasso," by increasing one of the other "2-bridges" to a "3-bridge," to form the **3**.3.2 system **22**. We suspect that solvation in an aqueous solution will not change this dichotomy (Scheme 6). Of great interest is the experimental finding that **27** methylates at nitrogen, while **29** methylates at oxygen, apparently mimicking protonation (Scheme 7).[68] The authors of this study note that the methylation reaction is extremely slow and it is not clear whether the reaction is under thermodynamic or kinetic control.[68]

In discussing N versus O alkylation of amides and lactams, it is worthwhile making some brief comparisons of the thermodynamic tendencies for alkylation to form N–C and O–C bonds versus protonation to form N–H and O–H bonds. Thus, we can make a crude comparison of these bonds by referring to Eq. (3.2) where the $\Delta H_f^o(g)$ data for the four compounds[55] have been employed. The data indicate a preference for O–H and N–C over N–H and O–C by 5.5 kcal mol^{-1}. Use of the proton affinities of amines and ketones (Eq. 3.3) gives a similar conclusion. If we simply transfer a proton from a protonated ketone to an amine (Eq. 3.4), the enthalpy change is, of course, strongly exothermic

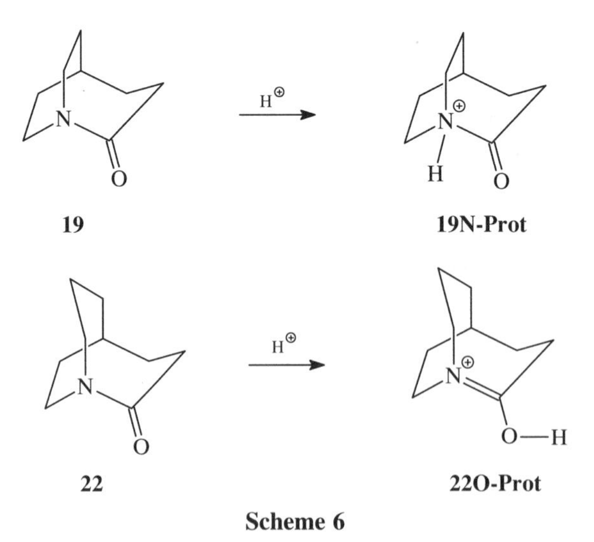

Scheme 6

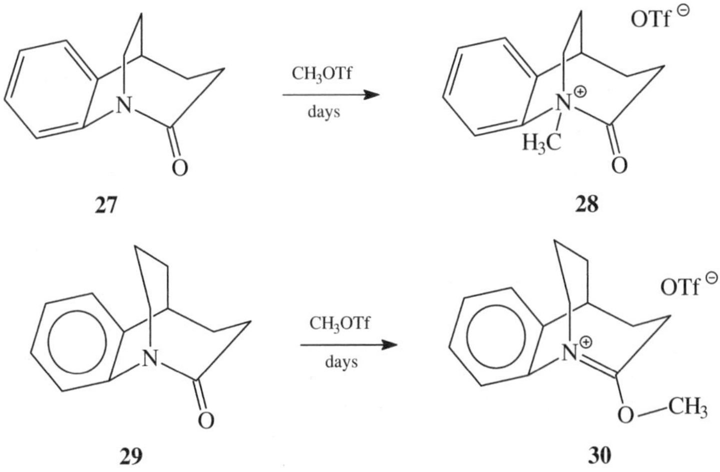

CH_3OTf, days

27 → **28** ($OTf^{\ominus}$)

CH_3OTf, days

29 → **30** ($OTf^{\ominus}$)

Scheme 7

because of the enhanced basicity of the amine. The transfer of a CH_3^+ group from the ketone to the amine is even more exothermic (Eq. 3.5) and the 2.6 kcal mol^{-1} difference is, of course, the value given in Eq. (3.3). Thus, it

$$CH_3\text{-}NH\text{-}CH_3 + CH_3CH_2\text{-}O\text{-}CH_3 \rightarrow CH_3\text{-}N(CH_3)\text{-}CH_3 + CH_3CH_2OH \qquad \Delta H_r = -5.5\ \text{kcal mol}^{-1} \tag{3.2}$$

$$CH_3\text{-}N^+H_2\text{-}CH_3 + CH_3\text{-}C(={}^+OCH_3)\text{-}CH_3 \rightarrow CH_3\text{-}N^+H(CH_3)\text{-}CH_3 + CH_3\text{-}C(={}^+OH)\text{-}CH_3 \qquad \Delta H_r = -2.6\ \text{kcal mol}^{-1} \tag{3.3}$$

$$CH_3\text{-}NH\text{-}CH_3 + CH_3\text{-}C(={}^+OH)\text{-}CH_3 \rightarrow CH_3\text{-}N^+H_2\text{-}CH_3 + CH_3\text{-}C(=O)\text{-}CH_3 \qquad \Delta H_r = -28.0\ \text{kcal mol}^{-1} \tag{3.4}$$

$$CH_3\text{-}NH\text{-}CH_3 + CH_3\text{-}\overset{\overset{+}{O}CH_3}{\overset{\|}{C}}\text{-}CH_3 \rightarrow CH_3\text{-}\underset{\underset{H}{|}}{\overset{\overset{CH_3}{|}}{N^+}}\text{-}CH_3 + CH_3\text{-}\overset{\overset{O}{\|}}{C}\text{-}CH_3$$

$$\Delta H_r = -30.6\,\text{kcal mol}^{-1} \tag{3.5}$$

appears that there is a 2.5–5 kcal mol^{-1} bias favoring N-alkylation versus O-alkylation when compared to the difference in proton affinities.

Another bit of insight may be obtained by comparing the O- and N-protonation and methylation energies of amides. Unfortunately, experimental energies are not available for three out of these four data points, but they can be estimated. The PA data for *N,N*-dimethylacetamide (DMA)[55] may be added to the ΔH_f^o(g) data[55] for CH_3CH_2OH, $CH_3CH_2OCH_3$ and **31** to allow the estimate for **32** shown in Scheme 8. Estimations for the N-protonated and N-methylated species are more problematic. However, let us try two very simple-minded assumptions: a) there is 18 kcal mol^{-1} of resonance energy in DMA that is totally lost upon N-protonation. "Losing" the resonance energy from the amide creates a tertiary "keto-amine" which we will crudely model by a secondary amine since H will be less stabilizing than an alkyl group which would surely be more stabilizing than a keto group. Thus, we remove 18.0 kcal mol^{-1} from ΔH_f^o of DMA[69] to obtain $\Delta H_f^o(g) = -37.6$ kcal mol^{-1}. Use of the proton affinity of dimethylamine (222.2 kcal mol^{-1})[55] for the proton affinity of the "keto-amine" version of DMA leads to the ΔH_f^o(g) for the N-protonated structure **33** shown in Scheme 8. Use of the methyl cation affinity of dimethylamine (+137.9 kcal mol^{-1}—compare ΔH_f^o(g) of dimethylamine and trimethylammonium ion) along with the zero-resonance "keto-amine" yields the ΔH_f^o(g) value for **34**, shown in Scheme 8. The results are interesting. The O-protonated structure is favored over the N-protonated structure by 12.8 kcal mol^{-1} in surprisingly good agreement with the earlier-cited computational data. The methyl cation affinities favor O-alkylation also but only by 2.7 kcal mol^{-1}. Solvation will also effect these results. Reaction of amides by oxonium salts does involve O-alkylation (Eq. 3.6) although one must always be mindful of the need to differentiate kinetic versus thermodynamic effects.[70]

$$R'\text{-}\underset{\underset{O}{\|}}{C}\text{-}NHR'' + R_3O^+BF_4^- \rightarrow R'-\underset{\underset{OR}{|}}{C}=NHR''BF_4^- \tag{3.6}$$

We conclude here that there is an intrinsic 2.5–5 kcal mol^{-1} favoring N-methylation versus O-methylation relative to N versus O protonation in the gas-phase. These points should be noted in viewing the prediction stated in Table 3.3. Thus, **22** may favor O-protonation by only ca 2–4.5 kcal mol^{-1} rather than 7 kcal mol^{-1} before solvent effects are considered.

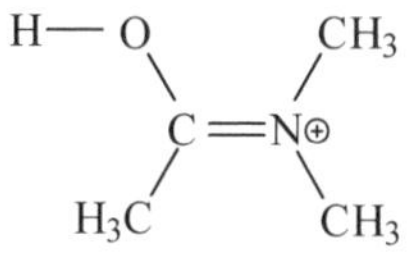

31 ΔH_f^o (g) = +93.1 kcal mol^{-1}

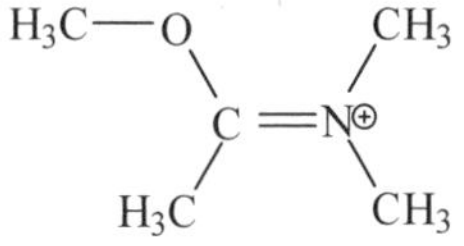

32 ΔH_f^o (g) = +97.6 kcal mol^{-1}

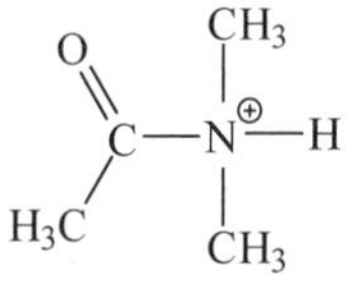

33 ΔH_f^o (g) = +105.9 kcal mol^{-1}

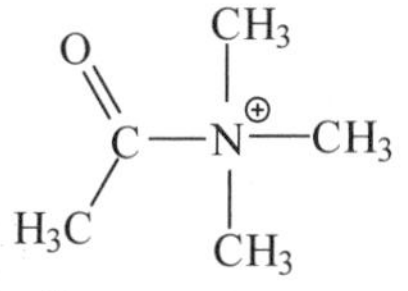

34 ΔH_f^o (g) = +100.3 kcal mol^{-1}

Scheme 8

N-Alkylation of amides is typically performed in basic media, and isotopic proton exchange in basic media is unambiguously due to the acidity of NH (Eq. 3.7).[58]

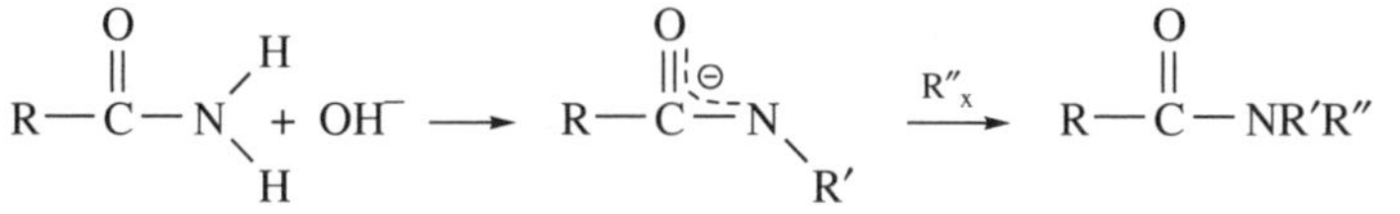

$$R{-}C({=}O){-}NH_2 + OH^- \longrightarrow R{-}C({=}O){=}N^{\ominus}{-}R' \xrightarrow{R''_x} R{-}C({=}O){-}NR'R'' \quad (3.7)$$

3.2. Acid-Catalyzed Proton Exchange and Hydrolysis

Although O-protonation of simple amides (and presumably peptide linkages) is overwhelmingly favored over N-protonation (ca 14 kcal mol^{-1} in the gas phase, ca 8 kcal mol^{-1} in aqueous solution), the mechanisms of proton exchange in mildly acidic solutions and acid-catalyzed hydrolysis under more acidic and higher-temperature conditions are more complex and richer than anticipated.[58,59,71] Scheme 2 of Chapter 4 in this volume describes Brown's generalized mechanism for acid-catalyzed hydrolysis.[59] It is again worth noting that for normal, unstrained amides or lactams low pH and elevated temperatures are required. Brown describes in detail the hydrolysis of strained bridgehead lactams such as **27**.[59,71] The generalized mechanism is written in Brown's Scheme 2[59] for a primary amide although it is also appropriate for secondary and tertiary species. Normally, the rate-determining step is the formation of the neutral tetrahedral intermediate $\mathbf{T_o}$ by attack of water on the O-protonated, activated amide. Relatively, strong basic amino groups favor rapid protonation of $\mathbf{T_o}$ on nitrogen to form $\mathbf{T_{N^+}}$ which rapidly suffers C–N cleavage to products. Remarkably, Kirby et al.[33] have isolated and obtained the crystal structure of the N-protonated hydrate of **5** (**5** · H_3O^+) which would correspond to Brown's $\mathbf{T_{N^+}}$ (Scheme 2, Chapter 4). Probing of most amides through hydrolysis of the ^{18}O amide shows negligible incorporation of ^{16}O from the solvent into unreacted starting material. Thus, the ratio of the exchange of ^{18}O to hydrolysis rates (k_{ex}/k_{hyd}) is extremely low (e.g., 0.01). In contrast, for **35** a $k_{ex}/k_{hyd} \approx 50$ has been

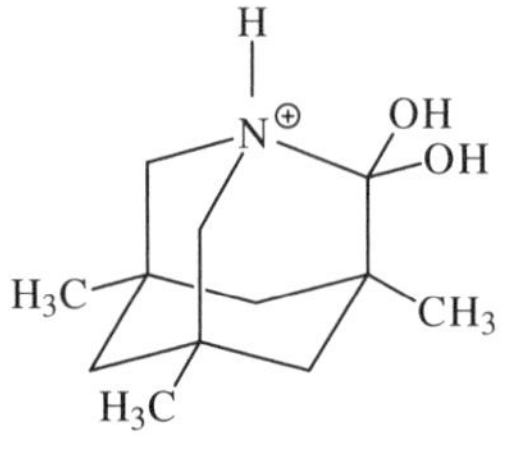

5·H_3O^+ 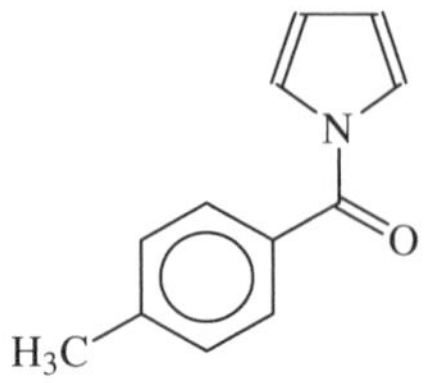 **35**

reported.[59] The weakness of pyrrole as a base makes protonation of $\mathbf{T}_o$ to $\mathbf{T}_{N^+}$ (Brown's Scheme 2) relatively uncompetitive with reversal to starting material. Thus, even though protonation at O, not N, is required en route to acid-catalyzed hydrolysis of unstrained amides, many subtleties of this system are dependent upon the basicity of the nitrogen in $\mathbf{T}_o$.

Under milder conditions of pH and temperature, it is possible to study the protonation of the amide alone, via proton exchange, and avoid the rate-determining attack by water. Here, Perrin[58] discovers a series of "surprises from simple systems." He starts his stimulating discussion by noting that proton exchange, a useful tool in trying to assess the location of amide residues in proteins, was typically described (presumably by physical chemists and biochemists) by equations such as (3.8). In contrast, Perrin notes ironically, a knowledgeable organic chemist knows that the amide oxygen is, as noted earlier,

$$RCONHR_1 + H^+ \rightleftarrows RCONH_2{}^+R_1 \tag{3.8}$$

roughly 6 orders of magnitude more basic than the nitrogen. Thus, a more likely scenario for proton exchange would be via the imidic pathway described in Eq. (3.9). Here, the expected O-protonation would enhance the acidity of N–H to form the imidic acids. In effect, one could add two cul-de-sacs to the initial

$$RCONHR_1 + H^+ \rightarrow RC(OH){=}NHR_1{}^+ \rightarrow RC(OH) = NR_1 + H^+ \tag{3.9}$$

O-protonated intermediate in Brown's mechanism (see Scheme 9). Perrin's clever work anticipated a differing rate of exchange between the amides H_Z and H_E protons since the minute concentration of O-protonated structure (Scheme 10) will keep interconversion of H_Z and H_E slow. Furthermore, more rapid exchange with the solvent (monitored by NMR relaxation times) would be expected for H_Z since it is more acidic in the **O-Prot** structure which would gain and lose protons. Perrin's initial expectation was that extremely rapid C–N rotation in the N-protonated structure would totally equilibrate H_Z and H_E, so no differences in solvent exchange rates would occur. To his evident delight, H_Z and H_E were differentiated, but to his dismay, the wrong proton (H_E) was more

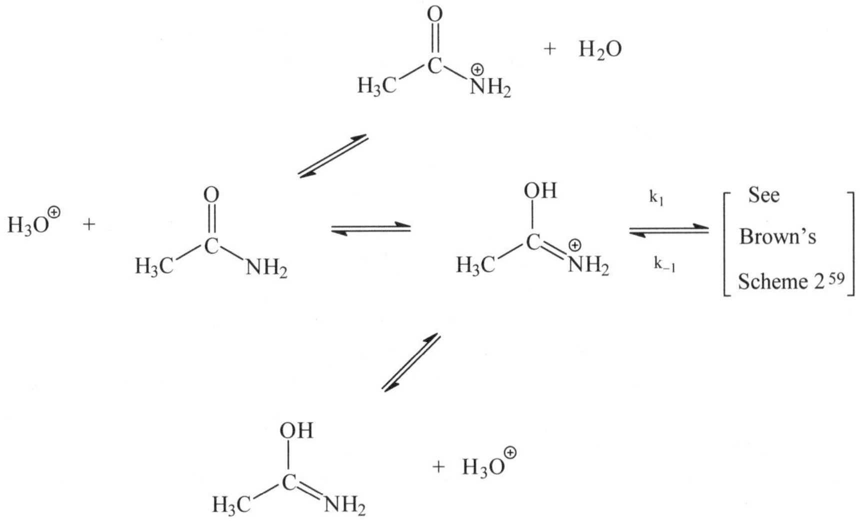

See Brown's Scheme 2[59]

Scheme 9

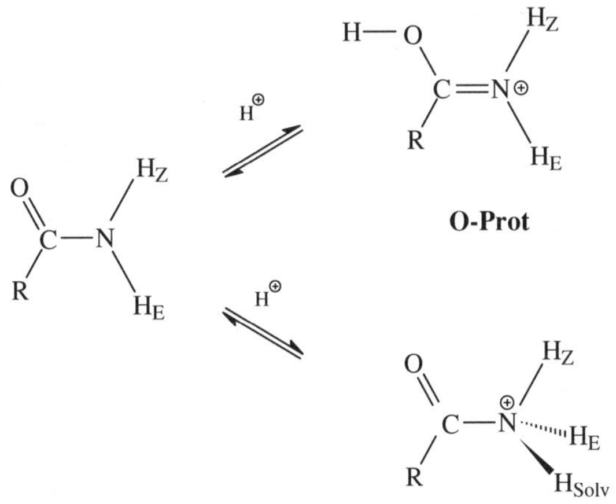

O-Prot

N-Prot

Scheme 10

rapidly exchanged with solvent. Further studies convinced Perrin that C–N bond rotation in the **N-Prot** form was not necessarily faster than proton-exchange and that differentiation of H_Z and H_E and their relative exchange rates were all consistent with the **N-Prot** mechanism (Eq. 3.8) rather than the expected **O-Prot**/imidic acid exchange mechanism (Eq. 3.9) for *primary* amides.

It is important to reemphasize that the point here is *not* that N-protonation is favored thermodynamically over O-protonation. The latter is, of course, favored.

However, the cul-de-sac that leads to NH exchange with solvent is the one in which N is protonated. Indeed, it is at first rather surprising that at acidic pH the rate of rotation about the C–N bond increases. For example, the rate of rotation for *N,N*-dimethylacetamide increases by a factor of 130 as pH is lowered in aqueous solution from 7.0 to 1.8.[72,73] Naively, one might expect protonation to produce a small amount of **31** whose barrier should be higher than that of *N,N*-dimethylacetamide.[29] However, a trace of **33** provides a tiny fraction of a very rapid rotational pathway. Stein estimates the ratio (k_+/k_o) of the rate of C–N rotation in **33** (k_+) to that in neutral *N,N*-dimethylacetamide (k_o) to be over 10^6.[73] Thus, "leakage" through a trace level of **33** is the source of the reduced barrier in acidic media.[73] Perrin's point is that proton exchange with solvent is, nevertheless, competitive with the k_+ process.[58]

However, for secondary amides ($XCH_2CONHCH_3$) and particularly those with electron-withdrawing substituents (e.g. peptides where X = RCONH and CH_3 is replaced by CH_2CONHR_1), the O-protonation/imidic acid mechanism (Eq. 3.9) prevails as Perrin (and presumably other organic chemists would have) had anticipated.[57]

Perrin[58] further notes that protonation at O and deprotonation at N suggests that both oxygen and nitrogen of the relevant peptide linkage need to be exposed to the medium for competitive proton exchange. In any case, O-protonation will be favored by planar amide linkages. Furthermore N–H proton loss to form imidic acids will also be favored by planar linkages. Mock's[8] earlier points about the interplay of binding and distortion of the amide linkage need to be remembered here. A significantly distorted linkage may change the site of initial protonation to N and thus alter the nature of proton exchange as well as proteolysis.

Before leaving this section, it is interesting to note the recent work of Wolfenden et al.[72] on the spontaneous (uncatalyzed) hydrolysis of glycosides. The group notes that gylcosides are the least reactive of the important biological polymers toward spontaneous hydrolysis. Phosphodiester bonds (in DNA) are more reactive by two orders of magnitude, proteins are two orders of magnitude more reactive to spontaneous hydrolysis than DNA and RNA is somewhat less reactive than proteins. Thus, glycosidases appear to be intrinsically even better catalysts than peptidases.[74]

4. INTERACTIONS WITH METAL IONS

While activations of amides and lactams towards hydrolysis normally require extremes of pH, metal ions are capable of acting as Lewis acids and activating the amide ring toward hydrolysis at neutral pH.[75] Thus, the hydrolysis of *N,N*-dimethylformamide (DMF) in the complex $Co(NH_3)_5(DMF)^{3+}$ occurs about 10^4 times more rapidly than that of the free amide in neutral aqueous solution.[76]

Metalloenzymes, as typified by zinc proteases such as carboxypeptidases and thermolysin as well as the zinc-containing carbonic anhydrases, are widely thought to activate peptide linkages toward hydrolysis by complexing with the scissile carbonyl. They are also considered to activate the nucleophilicity of water through complexation. Thus, it is known that zinc coordination of water can lower its pK_a to 7 thus allowing it to behave as hydroxide at physiological pH.[77,78] Hence, the dual role of Zn^{2+} in carboxypeptidases has been illustrated as shown in Scheme 11.[79] Here one sees both the Lewis acid Zn^{2+} complex with the scissile carbonyl as well as the water-activating role of Zn^{2+}. The view that Zn^{2+} complexes with the carbonyl in order to activate it is widely held.[80]

(A) (B)

Scheme 11

However, Christianson and Lipscomb,[78] and Christianson and Fierke[81] posit a mechanism that differs subtly. They note a uniformity in zinc proteases involving a "catalytic tetrad" of Zn^{+2}, a base (ionized Glu or Asp), activated water and an electrophile (protonated Arg, His or Lys). The view is that Zn^{2+} functions primarily to activate water in cooperation with , e.g. Glu, so that it functions as hydroxide. It is also considered to help to electrostatically stabilize the 4-coordinate acyl substitution intermediate. The role of activation of the scissile carbonyl linkage is credited to the electrophile (e.g. Arg) vis hydrogen-bonding. This picture is schematized in Fig. 3.3A for carboxypeptidase A.[78] A rather similar role, activation of the nucleophilicity of water and stabilization of the developing negative charge of the intermediate is postulated for carbonic anhydrase II (Fig. 3.3B).[81]

Despite the varying views of the role of zinc coordination on the peptide linkage, it is still valuable to consider the complexing properties of metal ions and amides or lactams. Indeed, the use of amides such as *N,N*-dimethylacetamide or *N,N*-dimethylformamide as solvents for salts is governed by these interactions. Useful summaries of the structural chemistry of metal ion-amide complexes have been in the literature for many years.[82,83] The evidence overwhelmingly favors complexation at oxygen rather than nitrogen in the majority of cases.

The structural effects of complexation (really considered to be an electrostatic interaction) are explicable in terms of favoring the **1B** type resonance contributor as we note in structure **36** below. For example, in $[Ni(acetamido)_4(OH_2)_2]Cl_2$, crystallographic data indicates very short C–N (1.31 Å) and very long C=O

(A)

(B)

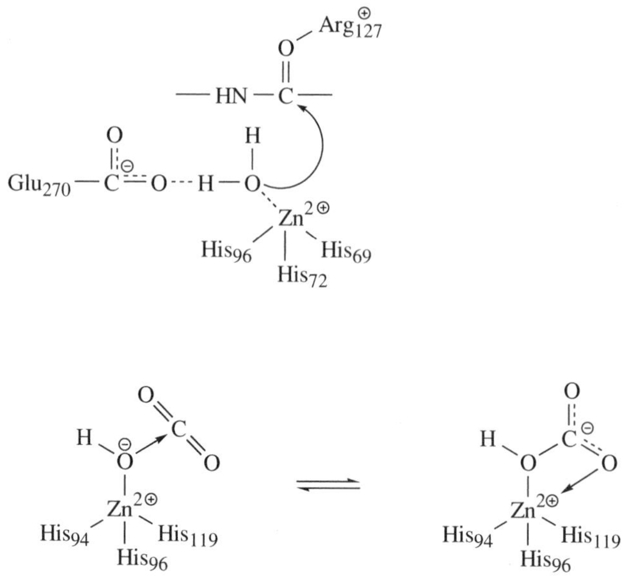

FIGURE 3.3. Role of zinc ion in zinc-containing enzymes schematized for A) Carboxypeptidase A (see Christianson and Lipscomb, ref. 78), and B) Carbonic anhydrase (see Christianson and Ferke, ref. 81).

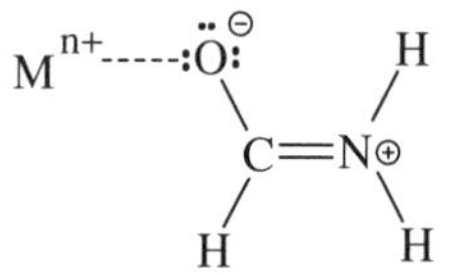

36

(1.25 Å).[82,83] Less pronounced results are seen for $[FeCl_3(formamide)_2]_n$ where C–N is reported at 1.33 Å and C–O at 1.23 Å with greater uncertainty. Since amides are quite weak π-donors, counter ions such as chloride compete with them for the metal site. This can be minimized by the use of classically-weak donor anions such as ClO_4^- or BF_4^-. The corresponding bond lengths in $[Fe(DMF)_6][ClO_4]_3$ are: C=O, 1.26 Å; C–N, 1.30 Å.

Early infrared studies of urea complexes of various transition metals pointed up a dichotomy. Thus, in $Pd(urea)_2Cl_2$ and $Pt(urea)_2Cl_2$ the band primarily assigned to the carbonyl vibration was found at 1725 cm^{-1} [84] (**37**, $M = Pd^{2+}$, Pt^{2+}). In contrast, the bands noted for $Fe(urea)_6Cl_3$ (1655 shoulder, 1625 strong), $Zn(urea)_2Cl_2$ (1660 strong, 1640 strong), $Cu(urea)_2Cl_2$ (1655 shoulder, 1640 strong) and $Cr(urea)_6Cl_3$ (1640 strong) are considered to derive from structure **38** ($M = Fe^{3+}$, Zn^{2+}, Cu^{2+}, Cr^{3+}). For the sake of comparison, the corresponding bands in urea itself are 1683 (primarily C=O stretch, some NH_2 bend) and 1603 (primarily NH_2 bend, some C=O stretch).[84] Although Pt^{2+} and Pd^{2+} were N-bonded in urea, their IR spectra indicated that they were O-bonded

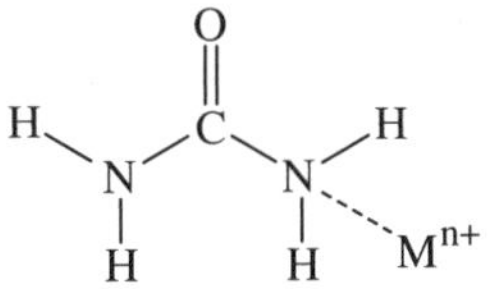

37

($M^{n+} = Pd^{2+}, Pt^{2+}$)

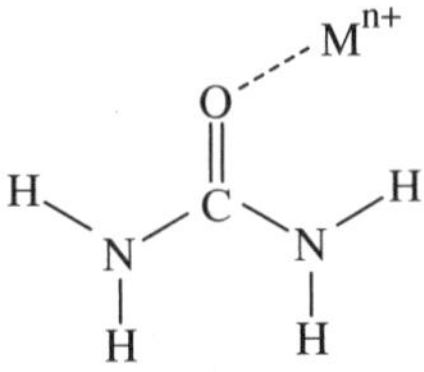

38

($M^{n+} = Fe^{3+}, Zn^{2+}, Cu^{2+}, Cr^{3+}$)

in complexes with dimethylformamide and dimethylacetamide.[85] This is reminiscent of the smaller energy difference calculated and noted earlier for O- versus N-protonation in urea relative to amides.[61]

Studies of the effects of metal cations on the rotational barrier of *N,N*-dimethylacetamide gave increased values of the activation energy (ΔE_a) for all metal ions studied (Li^+, Na^+, Pb^{2+}, Zn^{2+}, Cd^{2+}, Mg^{2+}) except for silver ion (Ag^+).[86] The increases were typically in the range of 1–3 kcal mol^{-1} although one must be careful about the uncertainties in ΔE_a (or $\Delta H^\ddagger$) and log A (or $\Delta S^\ddagger$) in variable temperature lineshape analysis where $\Delta G^\ddagger$ is the most accurately determined parameter.[86] These observations are clearly consistent with enhanced contribution from structure **36**. A linear correlation between the intensity of the electrical field at the surface of the cation and the rotational barrier was observed for six of the eight systems studied (including the uncomplexed amide). The Zn^{2+} and Ag^+ ions had lower rotational barriers than expected on this basis. The Ag^+ ion complex was the only one in which the barrier was actually lower (ca 2 kcal mol^{-1}) than the uncomplexed *N,N*-dimethylacetamide. Subsequent infrared studies[87] of the complexes of the perchlorates of a number of cations with *N,N*-dimethylformamide (Li^+, Ag^+, Mg^{2+}, Ca^{2+}, Sr^{2+}, Ba^{2+}, Co^{2+}, Ni^{2+}, Cu^{2+}, Zn^{2+}, Cd^{2+}, Pb^{2+}, Al^{3+}) and *N,N*-dimethylacetamide (Li^+, Ag^+, Mg^{2+}, Cu^{2+}, Zn^{2+}, Cd^{2+}, Pb^{2+}) indicated that all of these were associated with oxygen rather than nitrogen. The anomaly of the lower barrier for Ag^+ complexation and the IR observation of metallation site was explained by assuming that metallation is predominantly at oxygen with a very small (undetectable by IR) concentration of N–Ag^+ complex which would have a barrier even lower than that of the uncomplexed amide (Scheme 12).[87] This is similar to the explanation in section 3.2 for the reduced C–N barrier for amides at low pH (e.g., pH 1.8). Hoffman[66] explains the regiochemistry of some AgO induced substitution reactions of α-lactams in

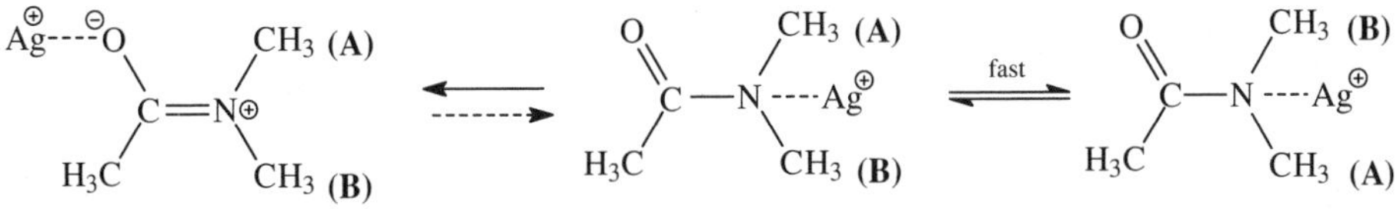

fast

Scheme 12

terms of complexation of Ag^+ with the sp^3-hybridized N in these lactams. Another interesting caveat is that the rotational barrier for the Li^+ complex is concentration dependent, increasing and reaching a maximum at a ratio of $LiClO_4$: DMA (3.85 : 1).[86] Thus, lower concentrations have both complexed and free dimethylacetamide.

We close this section by noting that, to our knowledge, complexation studies with severely distorted amides or lactams have not been done systematically. Ab initio calculations (MP2/6–31G*) predict that β-lactam (azetidinone) has a weaker affinity for protons and for metal monocations than does the acyclic *N*-methylacetamide.[88] It is likely that the intrinsic hardness of Li^+ will favor O-metallation over N-metallation even in systems such as **4** or **5**. However, softer ions such as Ag^+ look like candidates to switch neighbors upon fairly mild distortion. Perhaps species such as **22** could have Ag^+ associated with N even though they are protonated at O. Another recent study compares computational and experimental data for complexation of formamide by Cu^+.[89]

5. AMIDE N-OXIDES

To our knowledge, there are no published examples of N-oxides derived from the nitrogen atom of a tertiary amide (e.g., **39**). Such compounds would be similar in some senses to the previously discussed N-protonated and N-metallated amides and might be interesting oxygen-transfer agents for study of biochemical systems. This is not at all surprising since oxygenation of the nitrogen results in the loss of ca 18 kcal mol^{-1} of resonance energy. Alternatively, one can say that the lone-pair on nitrogen is not available for sharing. However, could the N-oxide derived from a 2-quinuclidone (e.g., **40**) be a stable species? Here there is no loss in resonance energy upon oxygenation. In effect a tertiary keto-amine would be oxidized. Presumably, this species would have some thermodynamic destabilization due to the proximity of the positively polarized carbonyl carbon and the formally positively charged nitrogen. However, this may prove to be advantageous if **40** were to be used as an O-transfer agent. Thus, pyridine N-oxide (**41**) finds use in synthesis. The sp^2-hybrid lone pair is less basic than the sp^3 lone pair in a tertiary amine and the N–O association should be weaker in **41** than in **42**. The situation in **40** would include an sp^3-hybridized lone pair (stabilizing N–O) combined with the electrostatic interaction with the attached

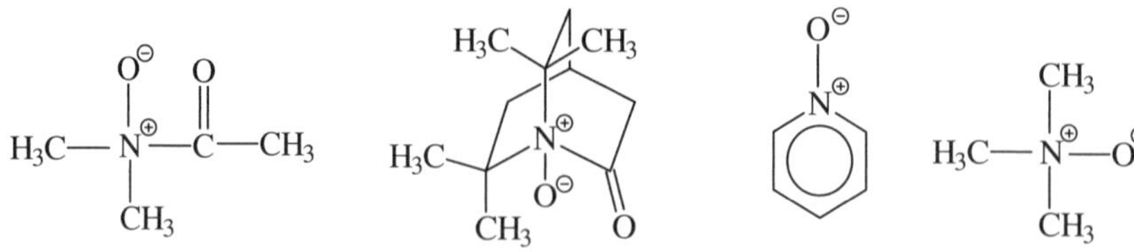

39 **40** **41** **42**

carbonyl (destabilizing N–O). Perhaps the result will be a similarity in N–O bond strengths between **40** and **41**.

However, there are other issues that need to be considered. It is known that tertiary amine oxides may rearrange or suffer elimination if β-hydrogens are present.[90] We have illustrated both of these possibilities for **40** in Scheme 13. The ring-opening elimination will clearly reduce strain energy and also be accompanied by a big gain in resonance stabilization.

The orientation of the β-hydrogens in **40** are not the ideal eclipsed (0°)[90] but closer to staggered (60°) so perhaps there may be some kinetic stability. The formation of **44** via rearrangement would probably decrease strain via expansion of the 2.2.2 to the 3.2.2 system and also pick up the resonance energy of the ester linkage. Before moving forward we comment that the N-oxide of an aziridinone (e.g. **45**) might have some appeal because, as noted earlier, the nitrogen is pyramidal and the sp^3 lone pair might be more available for sharing. The molecule lacks β-hydrogens so the Cope elimination pathway is shut down. Furthermore, ring expansion to **46** would really not significantly decrease the strain energy (cyclobutane and cyclopropane have similar strain energies). However, it is hard to imagine that the system would avoid ring opening to an intermediate such as **47** and subsequent chemistry (Scheme 14) or decarbonylation for that matter.

However, a really interesting possibility would be the formation of the 1-aza-2-adamantanone N-oxide (**48**) derived from oxidation of **5**. The appeal of this compound would lie in the facts that (a) no resonance energy is lost in forming **48** from **5**, (b) there are no β-hydrogens in **48** and thus Cope elimination is shut

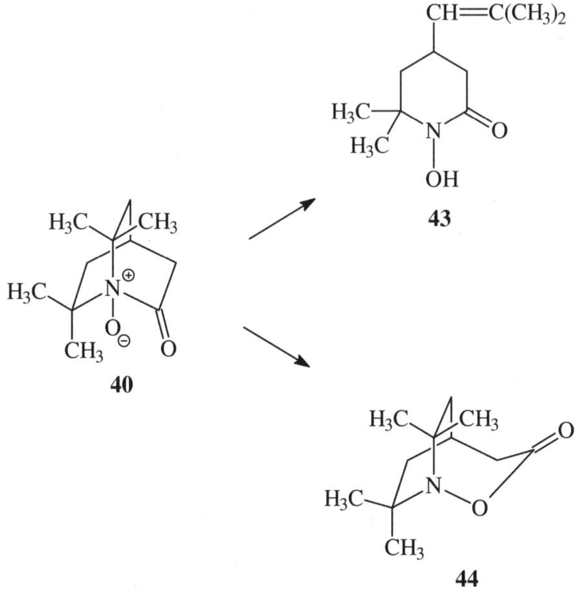

Scheme 13

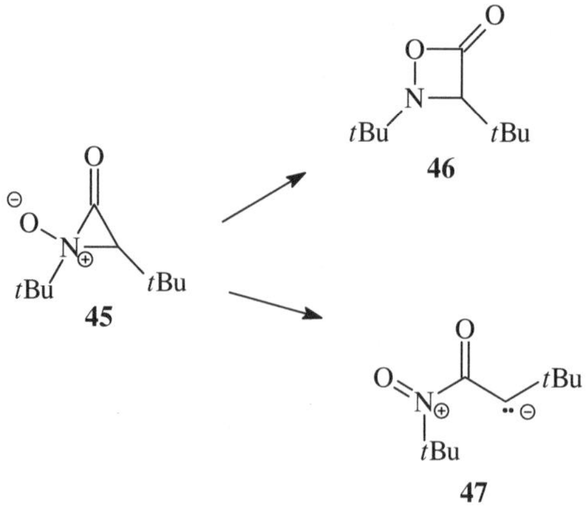

45 → 46, 47

Scheme 14

down, (c) rearrangement to form the ring expanded hydroxylamine ester analogous to **44** would probably significantly increase strain although this would be countered by the resonance energy of the ester linkage. N-oxide **48** might find utility in various reactions observed for the tertiary N-oxide class including glycolization of alkenes in the presence of OsO_4,[91] conversion of primary alkyl halides to aldehydes[90] and oxidation of boranes to alcohols.[91]

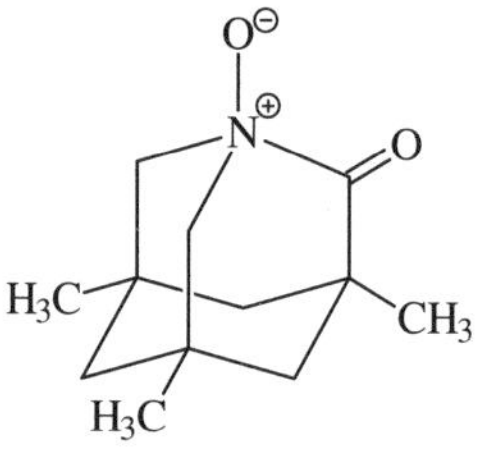

48

Ab initio calculations[92] at the 6–31G* level indicate that the N-oxide of *N,N*-dimethylacetamide (**49**) is 30 kcal mol^{-1} less stable than its hydroxylamine ester **50** and 74 kcal mol^{-1} less stable than the carbonate isomer **51**. The transition states for these rearrangements have not yet been explored. Use of thermochemical data[93] for the $N^{+}–O^{-}$ bond dissociation energy yields a fairly similar set of energy values.

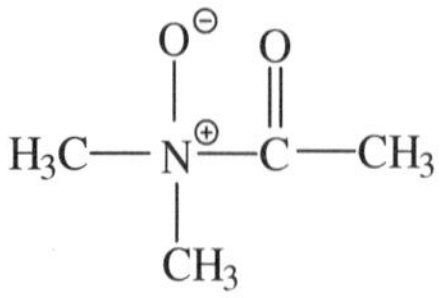

49 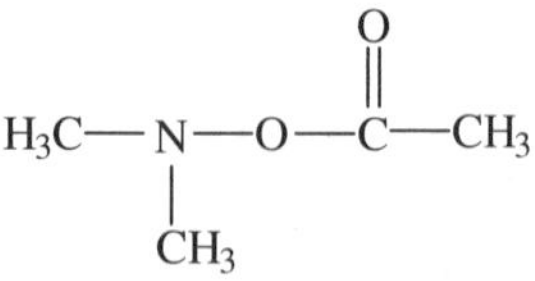 **50** 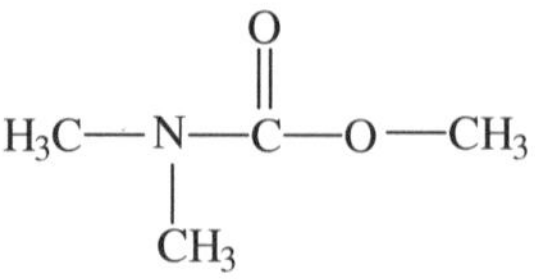 **51**

6. PEPTIDYL-PROLYL *CIS–TRANS* ISOMERASES

Unassisted twisting of the peptide linkage to isomerize *cis*- and *trans*-peptidyl–proline residues (–X_{aa}–Pro–) appears to often be the slow step in the folding of proteins into their native conformations.[73,94–96] However, a recent paper reports investigations of *cis–trans*-isomerizations of secondary nonprolyl peptide linkages and notes that some of these barriers are comparably high and many frequently correspond to the rate-determining step.[97] Thus, is has long been known that unfolded proteins tend to refold through the coexistence of both fast-folding (U_F) and slow-folding (U_s^i) unfolded species that exist in a slowly interconverting equilibrium.[95] If a native protein (N) has been unfolded to an incorrect (*cis/trans*) slow-folding structure (U_s^i), this latter will slowly fold via peptidyl-proline isomerization to the U_F structure which rapidly folds to the native structure N.[95] Unlike other amino acid residues where *trans*-peptidyl–X_{aa} linkages dominate, peptidyl–prolyl (–X_{aa}–Pro–) linkages in folded proteins are typically almost 1 : 1 *cis/trans*. That is why detectable levels of both the U_F and U_s^i proteins may be observed as they approach the native structure.[95] The enzymes which catalyze this isomerization were discovered by Fischer et al. in 1984[94] and are called peptidyl prolyl *cis–trans*-isomerases (PPI, EC 5.2.1.8) and are often commonly called "rotamases" for short. Fischer notes that folding about the peptidyl–prolyl linkage is tantamount to having an intramolecular

$$U_s^i \underset{\text{fast}}{\overset{\text{slow}}{\rightleftarrows}} U_F \underset{\text{slow}}{\overset{\text{fast}}{\rightleftarrows}} N \tag{3.10}$$

switch (the roughly equal amount of *cis* and *trans*-isomers parallels the equal probabilities for "on" and "off").[96] Using this analogy, Fischer equates the function of the PPIs to reducing the resistance of the switch. The *cis* and *trans* geometries at the X_{aa}–Pro linkage lead to very different long-range structures and the switch analogy further connotes a local on/off action and a remote response.[96] Although a protein may have a number of different X_{aa}–Pro peptide linkages, it is thought that perhaps only one such linkage plays the switch role in a given protein.[95,96] It is further noted that there are other mechanisms, besides X_{aa}–Pro rotational isomerism, that may be responsible for the slow folding of a U_s^i structure.[95,96]

The study of PPIs became even much more interesting when it was realized in 1989 that the cytosolic receptor protein for cyclosporin A (CsA) (**52**) an important immunosuppressive drug isolated from microorganisms, is in fact a PPI.[98,99] CsA forms a complex with the receptor, cyclophilin, and thus CsA acts as an inhibitor to its PPI activity. A second immunosuppressive drug (FK506, **53**) as well as third, Rapamycin (**54**), previously isolated from microorganisms were also investigated for binding proteins. Their receptors, called FKBPs (FK 506-Binding PPIases) also showed PPI activity. Interestingly, FKBPs specifically bind macrolides such as **53**, while cyclophilins specifically bind CsA and its derivatives. Binding inhibits PPI activity. Although one might have expected the

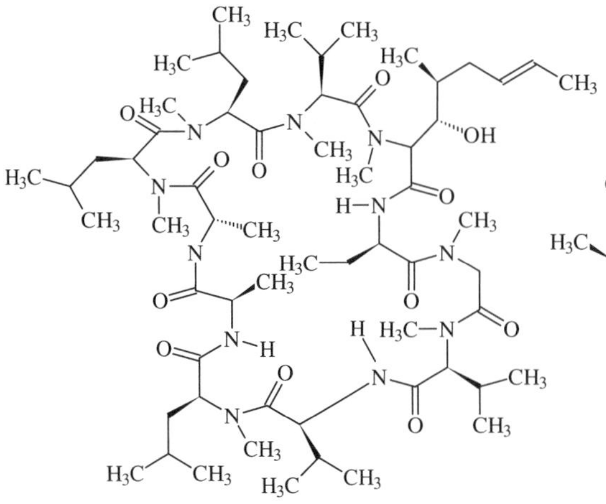

52

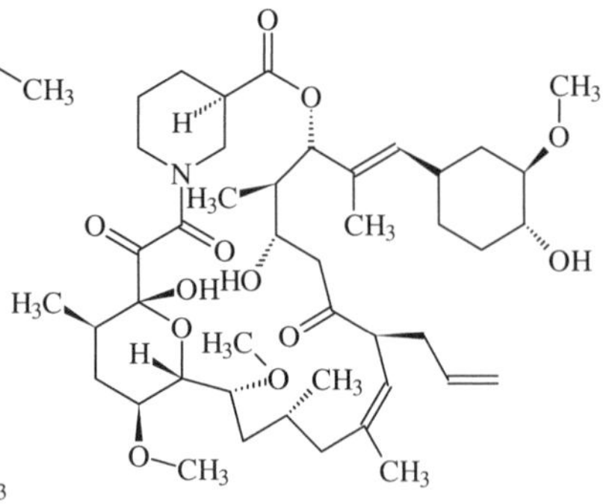

53

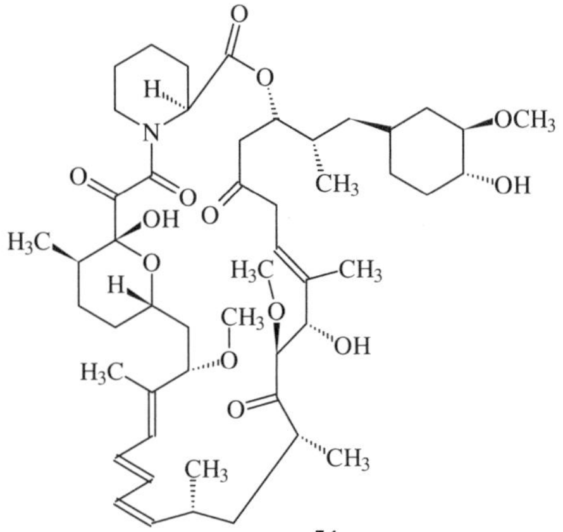

54

immunosuppressive activities of **52–54** to be related to inhibition of PPI activity, e.g., slowing the folding of proteins into their native states, the origin appears to involve suppression of early T cell activation genes in a manner not obviously related to PPI inhibition.

The initial work by Fischer suggested that direct nucleophilic involvement of amino acid residues of PPI were responsible for formation of a four coordinate intermediate with rapid C–N rotation.[94] However, the prevalent view now is that the transition state involves simple distortion of the X_{aa}–Pro peptide linkage upon binding to the enzyme (Scheme 15). Among the evidences favoring simple distortion rather than nucleophilic assistance are the following:[73,96]

1. Normal ($k_H/k_D > 1.0$) secondary deuterium isotope effects in isomerization of -Gly(d_2)–Pro-peptides. The β-hydrogen is involved in hyperconjugative interaction with the adjacent carbonyl group and this interaction disappears in the transition state. The larger effect seen in the PPI (1.12)

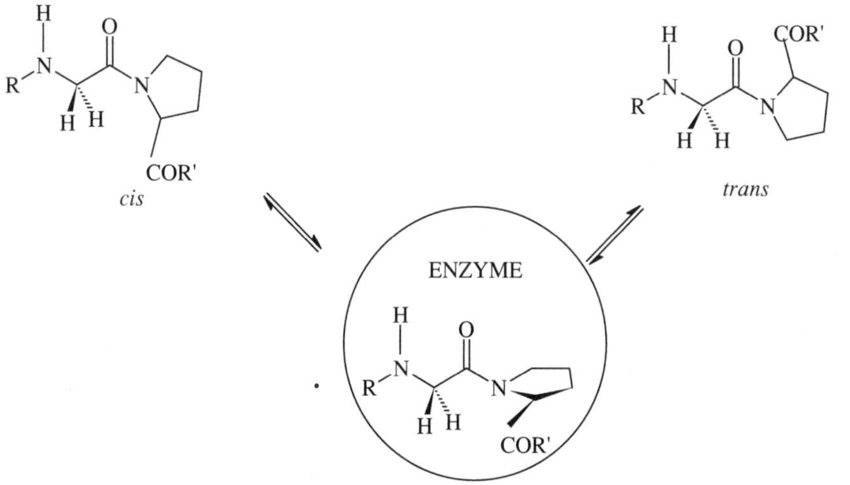

Scheme 15

relative to the uncatalyzed case (1.05) is attributed to the lack of solvation in the enzyme's hydrophobic cavity which would be expected to mitigate hyperconjugation.

2. Substrate specificity—while a series of homologous peptides (in which Pro was replaced by azetidine, piperidine and acyclic analogues) showed the lowest reactivity for *uncatalyzed* reaction of the peptide containing Pro, the *catalyzed* rate (k_c/K_m) was greatest for the Pro-containing peptide by three orders of magnitude. Thus, it is not the intrinsic reactivity of the linkage (e.g. toward nucleophiles) but general shape and binding (including other amino acid residues) that counts.
3. For the investigated PPI isomerizations, k_c/K_m was pH independent between pH 5–10. Presumably, nucleophilic groups in the active site would change their degree of ionization and nucleophilicity in this pH range. (Were k_c/K_m to change that *could* imply mere general conformational charge of the PPI rather than changes at the nucleophilic sites).
4. Solvent isotope effects (k_{H_2O}/k_{D_2O}) are essentially absent implying the absence of proton transfer in the transition state (i.e. absence of general acid or general base catalysis—see Chapter 4 by R. S. Brown).
5. Structural studies of inhibition of FKBP PPI activity by FK506 in the 1 : 1 complex provides an interesting point.[73,96,100–106] It is known that simple α-diketo compounds have the two carbonyl compounds oriented with a roughly 90° torsional angle. FK506 (**53**) for example, shows this effect such that it is the non-amide carbonyl (C_9) that is perpendicular to the approximate place of the amide linkage. Structural evidence indicates that the FK506 : FKBP complex maintains this twist and that this group is a surrogate for a twisted amide (X_{aa}–Pro) linkage.[101–103,105,106] The twisted carbonyl points into a hydrophobic pocket.[73]

twisted peptidyl-prolyl

FK506 substructure

Scheme 16

The barrier ($\Delta G^{\ddagger}$) to peptidylprolyl *cis–trans*-isomerization is typical 19.3–19.7 kcal mol^{-1}.[73] Stein estimates a 10^6 acceleration by PPI corresponding to 8 kcal mol^{-1} of stabilization.[73] This appreciable stabilization is considered to arise largely from (a) binding in the hydrophobic cavity (as previously noted in this chapter, C–N rotation in *N,N*-dimethylacetamide is about 10^2 faster in cyclohexane than in water), (b) binding of other residues in the PPI binding site so as to compensate for the induced strain of twisting, (c) possible stabilization of the twisted structure through H-bonding interaction between, for example, the Tyr-82 residue in FKBP, and the nitrogen of the proline ring. This would of course be a direct consequence of twisting the amide residue and shifting the point of basicity from oxygen to nitrogen.

Syntheses of immunosuppressants and their models have, understandably, focused on α-diketo species reminiscent of FK506. Indeed, compound **55** was synthesized, covalently attached to a carrier protein and a series of 28 monoclonal antibodies were derived from immunization studies.[107] Two of these were found to show PPI activity and significant substrate specificity.[107] Such enzyme-like antibodies offer probes into the nature of the active sites of PPIs and the design of substrates and inhibitors.

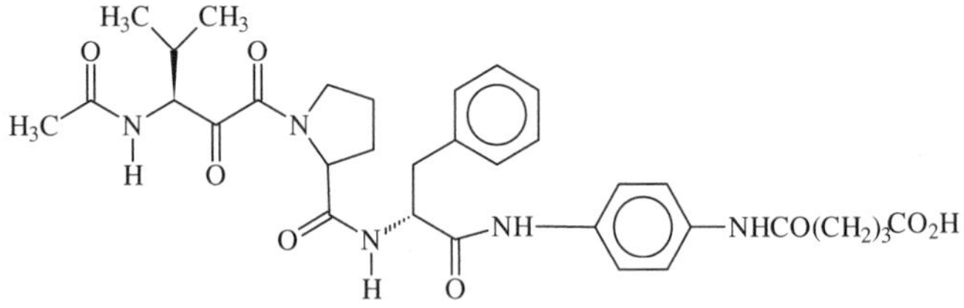

55

It is interesting that none of the PPI inhibitors described here including the natural products **52–54** actually have 90°-twisted amide linkages. Naively, one might think that these might be the best PPI inhibitors. However, acyclic twisted amide linkages must be sterically hindered and are not likely candidates for inhibitors. Derivatives of 2-quinuclidone (e.g. **4**) are hydrolytically unstable and obviously will have to be considerably functionalized to be effective inhibitors. Although the 1-aza-2-adamantanone **5** is an exciting molecule with all sorts of interesting possibilities for scaffolding, it is rapidly hydrolyzed in aqueous solution.

7. SUMMARY

In this chapter we have described the changes in selected ligand properties of amides as a function of distortion. Distortion of the linkage (twisting about C–N, measured by τ, pyramidalization at nitrogen, measured by χ_N-pyramidalization at carbon is usually quite small) significantly increases C–N bond length, very slightly shortens the C=O bond length and reduces resonance energy. Theories which rationalize these properties as well as spectroscopic data are described. Protonation of unstrained amides is very strongly favored at oxygen. However, proton exchange in primary amides as well as reduced C–N rotational barriers of amides in dilute acids are explicable in terms of trace concentrations of N-protonated structures. Twisting of the amide linkage and pyramidalization of nitrogen will ultimately favor N-protonation (and alkylation). Potential "cross-over points" are discussed. The site of protonation (and hydrogen bonding) may well have relevance to proteolysis mechanisms. Similarly, metallation tends to occur overwhelmingly at the oxygen of unstrained amides. This has the effect of increasing the C–N rotational barrier. Interestingly, the soft cation Ag^+ lowers the C–N barrier in amides slightly. Structural investigations indicate attachment to the carbonyl oxygen, but a trace of N-association would produce a tiny concentration of metallated amide with a very low rotational barrier whose impact would be the overall lowering of the barrier. There is evidence suggesting N-complexation by Ag^+ in α-lactams (work of R. V. Hoffman) and this would be consistent with the rotational barrier work. One can expect the possibility of a change in metallation site from O to N as a function of distortion for certain metals. Promising in this regard, based upon the above observations, is Ag^+. In contrast, the hard cation Li^+ is likely to associate with oxygen under all modes of distortion. One other example of ligation would be the attachment of an oxygen atom. Not surprisingly, there are no known amide N-oxides. However, distortion and pyramidalization of a tertiary amide could be expected to allow for synthesis of an α-keto tertiary amide oxide, presumably thermodynamically stabilized relative to a simple tertiary amine oxide, but potentially stable and perhaps usable as a reagent. Perhaps the N-oxide of the newly reported 1-aza-2-adamantanone **5** might be a suitable candidate.

The most obvious biological interest in severely distorted amide linkages derives from the discovery of peptidyl–propyl *cis–trans* isomerases (PPIs) over fifteen years ago. They appear to catalyze what is frequently the rate-determining step in the folding of proteins. The prevalent view is that the transition-state is probably the 90° C–N twisted structure at the X_{aa}–Pro linkage of proteins in which the reduced polarity of the transition state stabilizes binding in the hydrophobic pocket of the PPI. Additional stabilization may derive from donation of an H-bond to the pyramidalized amide nitrogen. Natural cyclic peptides such as FK506 are known immunosuppressants that bind with specific PPIs such as FKBP. They employ α-dicarbonyl linkages attached to piperidino nitrogen to form a unit in which the non-amide carbonyl is twisted by 90°. This structure is considered to mimic the 90°-twisted amide linkage in X_{aa}–Pro.

Studies of inhibitors, PPI binding sites and antibodies that mimic PPIs may provide insights into mechanisms of protein folding and immunosuppression. It is a bit ironic that twisted amides themselves have not been explored as immunosuppression drugs although some potential reasons are described.

REFERENCES

1. Zabicky, J. Ed. *The Chemistry of Amides*, Wiley-Interscience, London, 1970.
2. Zubay, G. *Biochemistry*, 4th edn., Wm. C. Brown Pub., Dubuque, 1996, p. 82.
3. Lorimer, G. (Ed.), *Adv. Protein Chem.*, Vol. 44, Academic Press: San Diego, 1993.
4. Stewart, W. F.; Siddall, T. H. *Chem. Rev.* 1970, **70**, 517.
5. Wiberg, K. B.; Rablen, P. R.; Rush, D. J.; Keith, T. A. *J. Am. Chem. Soc.* 1995, **117**, 4261.
6. Pauling, L. *The Nature of the Chemical Bond*, 3rd edn., Cornell Univ. Press: Ithaca, 1960, pp. 281–282.
7. Pauling, L.; Corey, R. B.; Branson, H. R. *Proc. Natl. Acad. Sci. USA* 1951, **27**, 205.
8. Mock, W. L. *Bioorganic Chem.* 1976, **5**, 403.
9. Benedetti, E.; Di Blasio, B.; Baine, P. *J. Chem. Soc. Perkin Trans.* 1980, **2**, 500.
10. Pracejus, H. *Chem. Ber.* 1959, **92**, 988.
11. Pracejus, H. *Chem. Ber.* 1965, **98**, 2897.
12. Pracejus, H.; Kehlen, M.; Kehlen, H.; Matschiner, H. *Tetrahedron* 1965, **21**, 2257.
13. Levkoeva, E. I.; Nikitskaya, E. S.; Yakhontov, L. N. *Khim. Geterot. Soed.* 1971 (3), 378.
14. Wiberg, K. B.; Laidig, K. E. *J. Am. Chem. Soc.* 1987, **109**, 5935.
15. Bader, R. F. W. *Atoms In Molecules: A Quantum Theory*, Oxford University Press: Oxford, 1990.
16. Breneman, C. M.; Wiberg, K. B. *J. Comput. Chem.* 1990, **11**, 361.
17. Wiberg, K. B.; Breneman, C. M. *J. Am. Chem. Soc.* 1992, **114**, 831.
18. Laidig, K. E.; Bader, R. F. W. *J. Am. Chem. Soc.* 1991, **113**, 6312.
19. Perrin, C. L. *J. Am. Chem. Soc.* 1991, **113**, 2865.
20. Laidig, K. E.; Cameron, L. M. *J. Am. Chem. Soc.* 1996, **118**, 1737.
21. Wiberg, K. B. In *The Amide Linkage: Selected Structural Aspects in Chemistry, Biochemistry and Materials Science*, Greenberg, A.; Breneman, C. M.; Liebman, J. F., Eds.; John Wiley & Sons, Inc., New York 1999, Chapter 2.
22. Hall, H. K., Jr.; El-Shekeil, A. *Chem. Rev.* 1983, **83**, 549.
23. Greenberg, A., In *Structure and Reactivity*; Liebman, J. F.; Greenberg, A. Eds.; Vol. 7; Molecular Structure and Energetics; VCH Pub.: New York, 1988, pp. 139–178.
24. Lease, T. G.; Shea, K. J. In *Advances in Theoretically Interesting Molecules*; Thummel, R. P. Ed.; JAI Press: Greenwich, 1992, Vol. 2; pp. 79–112.
25. Bennet, A. J.; Wang, Q. P.; Slebocka-Tilk, H.; Somayaji, V.; Brown, R. S. *J. Am. Chem. Soc.* 1990, **112**, 6383.
26. Wang, Q. P.; Bennet, A. J.; Brown, R. S.; Santarsiero, B. D. *J. Am. Chem. Soc.* 1991, **113**, 5757.

27. Bennet, A. J.; Somayaji, V.; Brown, R. S.; Santarsiero, B. D. *J. Am. Chem. Soc.* 1991, **113**, 7563.
28. Greenberg, A.; Venanzi, C. A. *J. Am. Chem. Soc.* 1993, **115**, 6951.
29. Greenberg, A.; Moore, D. T.; DuBois, T. D. *J. Am. Chem. Soc.* 1996, **118**, 8658.
30. Yamada, S. *Angew. Chem. Int. Ed. Engl.* 1993, **32**, 1083.
31. Yamada, S. In *The Amide Linkage: Selected Structural Aspects in Chemistry, Biochemistry and Materials Science*, Greenberg, A.; Breneman, C. M.; Liebman, J. F. Eds.; John Wiley & Sons, Inc., New York, 1999, Chapter 8.
32. Kirby, A. J.; Komarov, I. V.; Wothers, P. D.; Feeder, N. *Angew. Chem. Int. Ed. Engl.* 1998, **37**, 785.
33. Kirby, A. J.; Komarov, I. V.; Feeder, N. *J. Am. Chem. Soc.* 1998, **120**, 7101.
34. Lauvergnat, D.; Hiberty, P. C. *J. Am. Chem. Soc.* 1997, **119**, 9478.
35. Glendening, E. D.; Hrabel, J. A., II *J. Am. Chem. Soc.* 1997, **119**, 12940.
36. Greenberg, A.; Winkler, R.; Smith, B.; Liebman, J .F. *J. Chem. Edu.* 1982, **59**, 367.
37. Wiberg, K. B.; Schleyer, P. v. R.; Streitwieser, A. *Can. J. Chem.* 1996, **74**, 892.
38. Greenberg, A.; Plant, C.; Venanzi, C. A. *Theochem* 1991, **80**, 291.
39. Milner-White, E. J., *Protein Sci.* 1997, **6**, 2477.
40. Shirley, D. A. *Electron Spectroscopy*, North-Holland: Amsterdam, 1977.
41. Tsuchiya, S.; Seno, M. *J. Org. Chem.* 1979, **44**, 2850.
42. Greenberg, A.; Thomas, T. D.; Bevilacqua, C. R.; Coville, M.; Ji, D.; Tsai, J.-C.; Wu, G. *J. Org. Chem.* 1992, **57**, 7093.
43. Greenberg, A.; Moore, D. T. *J. Mol. Struct.* 1997, **413–414**, 477.
44. Wiberg, K. B.; Rablen, P. R.; Rush, D. J.; Keith, T. A. *J. Am. Chem. Soc.* 1995, **117**, 4261.
45. Maier, W. F.; Schleyer, P. v. R. *J. Am. Chem. Soc.* 1981, **103**, 1891.
46. McEwen, A. B.; Schleyer, P. v. R. *J. Am. Chem. Soc.* 1986, **108**, 3951.
47. Taha, A. N.; Neugebauer Crawford, S. M.; True, N. S. *J. Am. Chem. Soc.* 1998, **120**, 1934.
48. Drakenberg, T.; Dahlquist, H.-I.; Forsén, S. *J. Phys. Chem.* 1972, **76**, 2178.
49. Dunitz, J. D.; Winkler, F. K. *Acta Crystallogr., Sect. B* 1975, **31**, 251.
50. Bürgi, H. B.; Shefter, E. *Tetrahedron* 1975, **31**, 2976.
51. Gilli, G.; Bertolasi, V.; Bellucci, F.; Ferretti, V. *J. Am. Chem. Soc.* 1986, **108**, 2420.
52. Cieplak, A. *Struct. Chem.* 1994, **5**, 85.
53. Greenberg, A.; Moore, D. T. In *Pauling's Legacy—Modern Modelling of the Chemical Bond*, Maksic, Z. B.; Orville-Thomas, W. J. Eds.; Elsevier Science: Amsterdam, 1999, pp. 321–346.
54. Greenberg, A.; Liebman, J. F. In *Energetics of Organic Free Radicals*, Martinho Simões, J. A.; Greenberg, A.; Liebman, J. F. Eds.; Chapman and Hall: London, 1996, pp. 196–223.
55. Hunter, E. P.; Lias, S. G. *Proton Affinity Evaluation*. In *NIST Chemistry WebBook, NIST Standard Reference Database Number 69*, Mallard, W. G.; Linstrom, P. J. Eds.; National Institute of Standards and Technology: Gaithersburg, MD, March 1998.

56. Cho, S. J.; Cui, C.; Lee, J. Y.; Park, J. K.; Suh, S. B.; Park, J.; Kim, B. H.; Kim. K. S. *J. Org. Chem.* 1997, **62**, 4068.
57. Fersht, A. R. *J. Am. Chem. Soc.* 1971, **93**, 3504.
58. Perrin, C. L. *Accounts Chem. Res.* 1989, **22**, 268.
59. Brown, R. S. In *The Amide Linkage: Selected Structural Aspects in Chemistry, Biochemistry and Materials Science*, Greenberg, A.; Breneman, C. M.; Liebman, J. F. Eds.; John Wiley & Sons, Inc., New York, 1999, Chapter 4.
60. Olah, G. A.; White, A. M.; O'Brien, D. H. *Chem. Rev.* 1970, **70**, 561.
61. Rasul, G.; Prakash, G. K. S.; Olah, G. A. *J. Org. Chem.* 1994, **59**, 2552.
62. Abboud, J.-L. M.; Canada, T.; Homan, H.; Notario, R.; Cativiela, C.; Diaz de Villegas, M. D.; Bordejé, M. C.; Mó, O.; Yánez, M. *J. Am. Chem. Soc.* 1992, **114**, 4728.
63. Lien, M. H.; Hopkinson, A. C. *J. Org. Chem.* 1988, **53**, 2150.
64. Lien, M. H.; Hopkinson, A. C. *J. Am. Chem. Soc.* 1988, **110**, 3788.
65. Greenberg, A.; Hsing, H.-J.; Liebman, J. F. *Theochem* 1995, **338**, 83.
66. Hoffman, R. V. In *The Amide Linkage: Selected Structural Aspects in Chemistry, Biochemistry and Materials Science*, Greenberg, A.; Breneman, C. M.; Liebman, J. F. Eds.; John Wiley & Sons, Inc., New York, 1999, Chapter 6.
67. Cassady, C. J. In *The Amide Linkage: Selected Structural Aspects in Chemistry, Biochemistry and Materials Science*, Greenberg, A.; Breneman, C. M.; Liebman, J. F. Eds.; John Wiley & Sons, Inc., New York, 1999, Chapter 14.
68. Werstiuk, N. H.; Brown, R. S.; Wang, Q. P. *Can. J. Chem.* 1996, **74**, 524.
69. Lias, S. G.; Bartmess, J. E.; Liebman, J. F.; Holmes, J. L.; Levin, R. D.; Mallard, W. G. *J. Phys. Chem. Ref. Data*, **17** (No. 1), *Gas-Phase Ion and Neutral Thermochemistry*, U. S. Dept. of Commerce, Washington, D. C. 1988
70. March, J. *Advanced Organic Chemistry*, 3rd. edn., Wiley: New York, 1985, p. 359.
71. Brown, R. S.; Bennet, A. J.; Slebocka-Tilk *Accounts Chem. Res.* 1992, **25**, 481.
72. Gerig, J. T. *Biopolymers* 1971, **10**, 2443.
73. Stein, R. L. In *Advances in Protein Chemistry*, "Accessory Folding Proteins", Lorimer, G. Ed.; Academic Press: San Diego, 1993, Vol. 44, pp. 1–25.
74. Wolfenden, R.; Lu, X.; Young, G. *J. Am. Chem. Soc.* 1998, **120**, 6814.
75. Polgár, L. *Mechanisms of Protease Action*, CRC Press: Boca Raton, 1989, pp. 25–28, 183–218.
76. Constable, E. C. *Metals and Ligand Reactivity*, VCH Pub: Weinheim, 1996, p. 48.
77. Groves, J. T.; Olson, J. R. *Inorg. Chem.* 1985, **24**, 2715.
78. Christianson, D. W.; Lipscomb, W. N. In *Mechanistic Principles of Enzyme Activity*, Liebman, J. F.; Greenberg, A. Eds.; VCH Pub.: New York, 1988, pp. 1–25.
79. Abeles, R. H.; Frey, P. A.; Jencks, W. P. *Biochemistry*, Jones and Bartlett, Pub.: Boston, 1992, pp. 79–81.
80. Zubay, G. *Biochemistry*, 4th edn., Wm. C. Brown Pub.; Dubuque, 1998, pp. 190–194.
81. Christianson, D. W.; Fierke, C. A. *Accounts Chem. Res.* 1996, **29**, 331.
82. Chakrabarti, P.; Venkatesan, K.; Rao, C. N. R. *Proc. Royal Soc. London Ser. A* 1981, **375**, 127.

83. Wilkinson, G. Ed. *Comprehensive Coordination Chemistry*, Pergamon Press: Oxford, 1987, Vol. 2; pp. 490–494.
84. Penland, R. B.; Mizushima, S.; Curran, C.; Quagliano, J. V. *J. Am. Chem. Soc.* 1957, **79**, 1575.
85. Gioria, J. M.; Susz, B. P. *Helv. Chim. Acta* 1971, **54**, 2251.
86. Waghorne, W. E.; Ward, A. J. I.; Clune, T. G.; Cox, B. G. *J. Chem. Soc., Faraday Trans. I* 1980, **76**, 1131.
87. Waghorne, W. E.; Rubalcava, H. *J. Chem. Soc., Faraday Trans. 1* 1982, **78**, 1199.
88. Bordejé, M. C.; Mó, O.; Yáñez, M. *Struct. Chem.* 1996, **7**, 309.
89. Luna, A.; Amerkraz, B.; Tortajada, J.; Morizur, J. P.; Alcamí, M.; Mó, O.; Yáñez, M. *J. Am. Chem. Soc.* 1998, **120**, 5411.
90. March, J. *Advanced Organic Chemistry*, 4th edn., John Wiley & Sons: New York, 1992, pp. 1018–1019 (cleavage); p. 1102 (rearrangement); pp. 1194–1195 (oxidation of halides to aldehydes).
91. Carey, F. A.; Sundberg, R.F. *Advanced Organic Chemistry*, 3rd edn., *Part B. Reactions and Synthesis*, Plenum: New York, 1990, pp. 343–344 (thermal eliminations), pp. 204–205 (oxidation of boranes); p. 626 (glycolation of alkenes).
92. DuBois, T. D.; Greenberg, A. (unpublished results).
93. Acree, W. E., Jr.; Bott, S. G.; Tucker, S. A.; Ribeiro da Silva, M. D. M. C.; Matos, M. A. R.; Pitcher, G. *J. Chem. Thermodyn.* 1996, **28**, 673.
94. Fischer, G.; Bang, H.; Mech, C. *Biomed. Biochim. Acta* 1984, **43**, 1101.
95. Schmid, F. X.; Mayr, L. M.; Mücke, M.; Schönbrunner, E. R. In *Advances in Protein Chemistry*, "Accessory Folding Proteins", Lorimer, G. Ed.; Academic Press: 1993, Vol. 44; pp. 25–66.
96. Fischer, G. *Angew. Chem. Int. Ed. Engl.* 1994, **33**, 1415.
97. Scherer, G.; Kramer, M. L.; Schutkowski, M.; Reimer, U.; Fischer, G. *J. Am. Chem. Soc.* 1998, **120**, 5568.
98. Takahashi, N.; Hayano, T.; Suzuki, M. *Nature* 1989, **337**, 473.
99. Fischer, G.; Wittmann-Liebold, B.; Lang, K.; Kiefhaber, T.; Schmid, F. X. *Nature* 1989, **337**, 476.
100. Harrison , R. K.; Stein, R. L. *Biochemistry* 1990, **29**, 1684.
101. Rosen, M. K.; Standaert, R. F.; Galat, A.; Nakatsuka, M.; Schreiber, S. L. *Science,* 1990, **248**, 863.
102. Liu, J.; Albers, M. W.; Chen, C.-M.; Schreiber, S. L.; Walsh, C. T. *Proc. Natl. Acad. Sci., USA* 1990, **87**, 2304.
103. Albers, M. W.; Walsh, C. T.; Schreiber, S. L. *J. Org. Chem.* 1990, **55**, 4984.
104. Harrison, R. K.; Stein, R. L. *J. Am. Chem. Soc.* 1992, **114**, 3464.
105. Petros, A. M.; Luly, J. R.; Liang, H.; Fesik, S. W. *J. Am. Chem. Soc.* 1993, **115**, 9920.
106. Holt, D. A.; Luengo, J. I.; Yamashita, D. S.; Oh, H. J.; Konialian, A. L.; Yen, H.-K.; Rozamus, W.; Brandt, M.; Bossard, M. J.; Levy, M. A.; Eggleston, D. S.; Liang, J.; Schultz, L. W.; Stout, T. J.; Clardy, J. *J. Am. Chem. Soc.* 1993, **115**, 9925.
107. Yli-Kauhaluoma, J. T.; Ashley, J. A.; Lo, C.-H. L.; Coakley, J.; Wirsching, P.; Janda, K. D. *J. Am. Chem. Soc.* 1996, **118**, 5496.

CHAPTER 4

STUDIES IN AMIDE HYDROLYSIS: THE ACID, BASE, AND WATER REACTIONS

R. S. BROWN
Department of Chemistry, Queen's University

1. INTRODUCTION

The amide bond is ubiquitous in nature and forms the key chemical linkage in proteins, peptides, enzymes, and antibodies. The huge bulk of the properties of the amide bond, including the short N–C(O) bond length, barrier to N–C(O) rotation, C=O infra-red stretching frequencies, ^{13}C and ^{15}N NMR chemical shifts, and resistance toward nucleophilic attack/hydrolysis is nicely explained through the use of resonance theory,[1] first introduced by Pauling about 60 years ago.[2] (Alternative explanations based on sophisticated calculations are given in Chapters 1 and 2 of this book.)

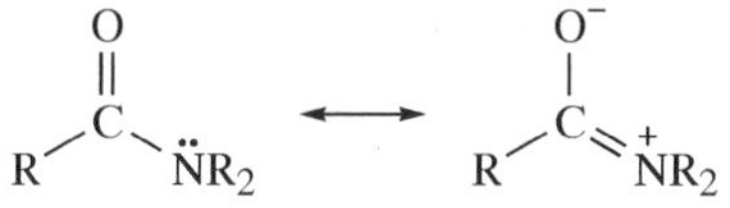

$$R{-}C({=}O){-}\ddot{N}R_2 \longleftrightarrow R{-}C(O^-){=}\overset{+}{N}R_2$$

1

The general physical and chemical aspects of the hydrolyses of carboxylic acid derivatives (RC(O)X) have been extensively documented.[3] There are several treatises that deal with the mechanistic aspects and tools necessary for the study of various aspects of the hydrolysis reactions of carboxylic acid derivatives including amides.[6–13]

In this chapter, we will be dealing with the various mechanisms that have been elucidated for transfer of the RC(O) moiety from amides to water as in the general process outlined in Eq. (4.1). Given the inherent stability of the amide linkage (**1**) toward nucleophilic addition, severe conditions of elevated

The Amide Linkage: Selected Structural Aspects in Chemistry, Biochemistry, and Materials Science, Edited by Arthur Greenberg, Curt M. Breneman, and Joel F. Liebman
ISBN 0-471-35893-2

temperature and pH extremes are necessary for hydrolysis. Regardless of whether the hydrolyses are conducted under acidic, basic or, in rare cases,

$$\mathrm{R{-}C({=}O){-}NR_1R_2} + \mathrm{H_2O} \rightleftharpoons \underset{\mathbf{T}}{\mathrm{R{-}C(O(H))_2{-}NR_1R_2}} \rightleftharpoons \mathrm{R{-}C({=}O){-}O(H)} + \mathrm{HNR_1R_2} \quad (4.1)$$

neutral conditions, the process occurs by the so-called ‘associative mechanism’,[10] whereby the attacking nucleophilic oxygen is bound to the acyl carbon to form one or more unstable tetrahedral addition intermediates, **T**. Depending on whether the reaction is carried out under acidic or basic conditions, the tetrahedral intermediates could be protonated ($\mathbf{T}_{\mathbf{O}^+}$, $\mathbf{T}_{\mathbf{N}^+}$), neutral ($\mathbf{T}_{\mathbf{O}}$), or anionic ($\mathbf{T}_{\mathbf{O}^-}$). This minimally two-step process immediately suggests an interesting mechanistic dilemma since both the formation and the decomposition of the tetrahedral intermediate(s) may require catalysis in order to proceed at an appreciable rate, and a priori it is not clear which step is rate limiting in a given case.

As outlined by Brown et al.[14] for the hydrolyses of amides in acid and in base, there are three key sets of experiments that have been used to distinguish between the possible mechanisms. First, the kinetics of hydrolysis as a function of $[H_3O^+]$ or $[OH^-]$ provide information about the stoichiometry of the rate determining transition states (TS) for the hydrolysis process. Second, the kinetics of $^{18}O{=}C$ exchange in unreacted $RC(O)NR_1R_2$ with solvent $^{16}OH_2$ are determined. This isotopic technique, first introduced by Bender and his coworkers[15–22] is, perhaps, the single most important tool for determining the existence of unstable, but reversibly-formed tetrahedral intermediates during the acid- and base-promoted hydrolyses of carboxylic acid derivatives in water. In a typical experiment, ^{18}O-enriched $RC(O)NR_1R_2$ derivatives of known isotopic composition are subjected to the hydrolysis conditions, recovered at various times after incomplete hydrolysis, and then analyzed by mass spectrometry to determine the decrease in ^{18}O-content as a function of time. If the ^{18}O-content of the unreacted starting material diminishes during the course of the reaction, it may be reasonably concluded that an intermediate or intermediates, having simultaneous bonding of the attacking and departing oxygens (made identical by symmetry or by rapid proton exchange between the oxygens), was reversibly formed during the reaction. This is shown for the associative mechanism in Scheme 1. The analysis requires the reasonable assumptions that: (1) the ^{18}O and ^{16}O-oxygens are in rapid protonic equilibrium relative to the C–O bond cleavages so that only half of the reversal from the tetrahedral intermediate gives ^{18}O loss; and (2) that the intermediate(s) that lead(s) to exchange are on the hydrolytic pathway.

The third general type of experiment to determine the mechanism involves substituting D_2O for H_2O as the hydrolytic medium and determining the effect

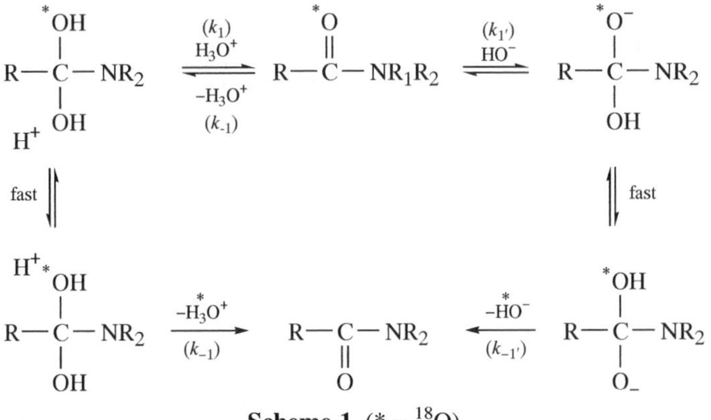

Scheme 1 (* = ^{18}O)

on the reaction rates for hydrolysis and ^{18}O exchange. These deuterium kinetic isotope effect (DKIE) experiments provide information about whether protons are being transferred in the rate-limiting steps. Interpretation of the data can be complicated but is often greatly aided by application of the principles of isotopic fractionation factor analysis.[23–27] Although, a detailed description of this technique is outside the scope of this article and not required for the presentation, there are three highly simplified considerations that must be kept in mind. First, the fractionation factors (ϕ) for hydrogens refer to the tightness of their bonding and are significantly less than unity for H's being transferred or "in flight" (that is, forming a partial bond) between O and N, or O and O as part of the rate-limiting step, as in Structure **2**. In these cases normal primary deuterium kinetic

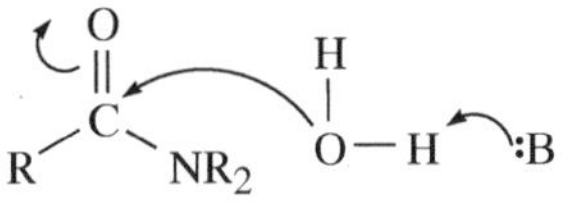

2

isotope effects (DKIEs) of $k_H/k_D > 1$ are expected. Also, in hydrogen bonding situations where the overall bonding is loose, the fractionation factors are less than unity, giving rise to normal DKIEs of $k_H/k_D > 1$. The latter give rise to secondary effects of solvation and can significantly alter the overall DKIE.[28] Second, D_3O^+ is a stronger acid in D_2O than is H_3O^+ in H_2O. This stems primarily from the fact that for the process:

$$L_3O^+ + B: \rightleftharpoons L\text{–}B^{(+)} + L_2O,$$

where L = H or D, the three L–O$^+$ bonds are significantly weaker than the two L_2O bonds, and since the heavier isotope prefers the stonger bond, the equilibrium is shifted to the right in D_2O. (For further reading see references 6, pp. 241–244, and 8, pp. 250–253.)

For specific-acid catalyzed processes where the protonated amide is formed in a rapid preequilibrium and subsequently attacked by solvent, as in Eq. (4.2), L = H, D, the substrate is expected to be more fully protonated, and therefore

$$\mathrm{RC(=O)NR_2} + \mathrm{L_3O^+} \underset{}{\overset{\text{fast}}{\rightleftharpoons}} \mathrm{RC(=O^+L)NR_2} + \mathrm{L_2O} \underset{\text{slow}}{\overset{\mathrm{L_2O}}{\rightleftharpoons}} \mathrm{R{-}C(OL)_2{-}NR_2} + \mathrm{L_3O^+} \tag{4.2}$$

react faster in D_2O than in H_2O unless there are other factors such as protons "in flight" as part of the rate limiting step. Finally, OD^- is about a 2–3-fold stronger base, and correspondingly better nucleophile in D_2O than is OH^- in H_2O. This stems from the fact that hydroxide, despite the fact that the $L{-}O^-$ bond is stronger than the L_2O bond, exists in L_2O as $LO^-(L{-}OL)_3$, and so is more poorly solvated in D_2O than it is in H_2O.[6,8,29,30] The net effect is that a direct nucleophilic attack on the carbonyl by LO^- would be expected to proceed faster in D_2O unless there are other compensating factors such as protons "in flight", or being transferred between solvent and base or substrate, as part of the rate limiting step.

In what follows we will deal explicitly with the simplest hydrolytic processes involving amides and only the components of the aqueous media, namely H_2O, OH^-, and H_3O^+.

2. SPECIFIC-ACID CATALYSIS

The term specific acid catalysis refers to reactions that are catalyzed by the conjugate acid of H_2O, namely H_3O^+. For these reactions the first step involves a preequilibrium protonation of the substrate as in Eq. (4.2). The currently accepted mechanism for H_3O^+-catalyzed amide hydrolysis is termed $A_{Ac}2$, or bimolecular, acid-catalyzed, proceeding with acyl oxygen cleavage. This is depicted in Scheme 2 and involves water assisted attack of H_2O on the O-protonated amide to yield $\mathbf{T_O}$ and H_3O^+.[6–14,31–33]

Scheme 2 is a more detailed version of the acid reaction given in Scheme 1 which includes explicit description of the proton transfers required for the actual addition and breakdown steps. There are several common features of acid catalyzed hydrolysis of a wide variety of amides that are accommodated by this mechanism. Plots of k_{hyd} versus $[H_3O^+]$ are generally of unit slope indicating that the rate limiting transition state for hydrolysis contains the amide and one proton. Generally these plots level off at high $[H_3O^+]$ and then curve downward, consistent with the onset of substantial equilibrium O-protonation of the amide ($A{-}H^+$; pK_a of the O-protonated amide 0 to -3,[34–39] pK_a of the N-protonated amide -7 to -8,[42]), and the reduction in activity of H_2O in concentrated acid

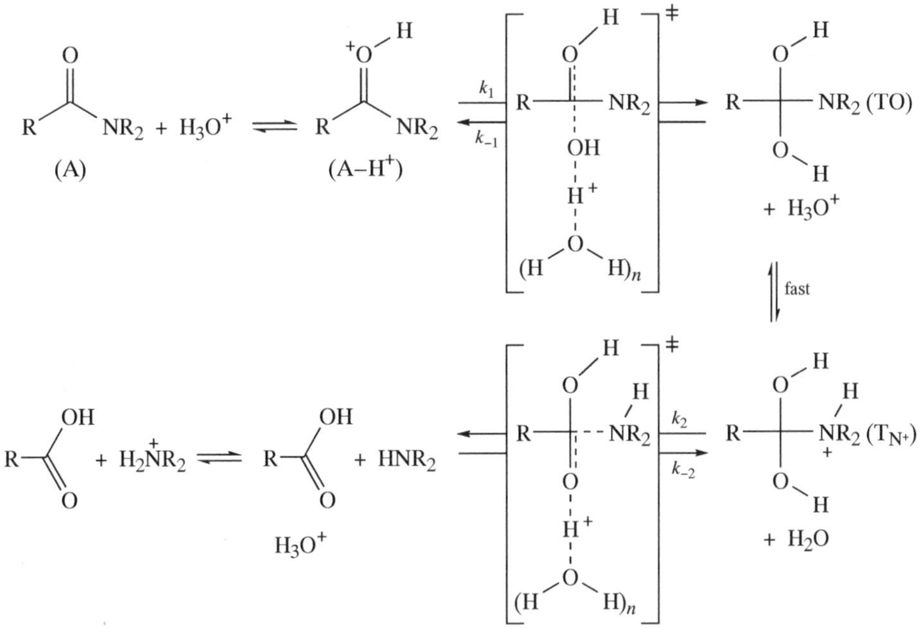

O
R NR$_2$ + H_3O^+ ⇌ R NR$_2$ (with $^+$O–H)
(A) (A–H$^+$)
k_1 / k_{-1}
[R–C(–O–H)(···OH···H$^+$···O(H H)$_n$)–NR$_2$]‡
→ R–C(OH)(OH)–NR$_2$ (TO) + H_3O^+
fast
R–C(OH)(OH)–$\overset{+}{N}$HR$_2$ (T_{N^+}) + H_2O
k_2 / k_{-2}
[R–C(–O–H)(···O···H$^+$···O(H H)$_n$)- - -NHR$_2$]‡
R–C(=O)OH + $H_2\overset{+}{N}R_2$ ⇌ R–C(=O)OH + HNR_2
H_3O^+

Scheme 2

media. Studies of the hydrolyses of simple alkyl amides and some heterocyclic amides and benzamides in concentrated acids, (10–90% H_2SO_4), by Yates and Riordan,[37] Yates and Stevens,[38] Moodie et al.,[42] and Cox and Yates[43] indicated that at moderate [acid] up to about 45–50%, three water molecules were required to convert the protonated amide ($A{-}H^+$) to its transition structure, either by a process where the attack of one of them is assisted by two others (**3**) or by a cyclic concerted process involving the three waters attacking the protonated amide as in **4**. This appears to contrast the situation for the acid catalyzed hydrolyses of esters where only two waters are required to convert the

R–C(–O–H)=$\overset{+}{N}R_2$, HO–H- - $(OH_2)_2$ ⇌ R–C(=O)–NR$_2$ + $H_3O^+(H_2O)_n$ ⇌ R–C(–O–H)=$\overset{+}{N}R_2$, H–OH, HO–H, H–O–H

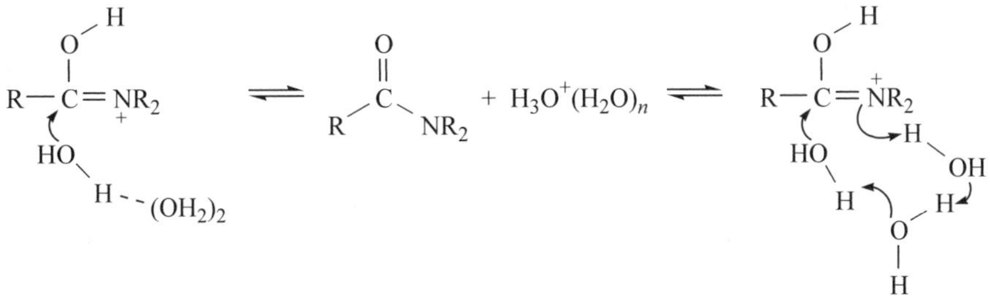

3 **4**

protonated form to the transition structure.[44–46] When the [acid] increases above the 45–50% level, Cox and Yates[43] provide evidence that the mechanism for hydrolysis changes to one involving a single water. Whether the three water mechanism is firmly established, is now open to some debate since Cox has recently reported that the excess acidity data for the hydrolysis of 5, 6, 7 and 8-membered lactams can just as easily be analyzed by a mechanism involving one water molecule attacking via an unknown mechanism.[47] These latest findings are consistent with the mechanism given in Scheme 2.

Previous $^{18}O{=}C$ exchange studies by Bender et al.,[20] Bunton et al.[48] and Smith and Yates[49] failed to detect any loss of ^{18}O from labeled benzamides recovered from highly acidic media after incomplete hydrolysis: accordingly, it was suggested that in Scheme 2, k_{-1} was negligible relative to the rate constant for product formation, k_2. In that case, the proton transfer from the encounter-complex H_3O^+ to $\mathbf{T_O}$ to generate $\mathbf{T_{N^+}}$ in Scheme 2 would proceed faster than diffusional separation. It is reasonable to expect that once the N becomes protonated C–N cleavage is very fast, so that reversal of $\mathbf{T_O}$ to starting material is prohibited. However, in 1975 McClelland subsequently reported[50] that 90% $^{18}O{=}C$ labeled benzamide underwent 0.2% loss of the ^{18}O per half time of hydrolysis when recovered from a 5.9% H_2SO_4 solution at 85 °C. This was an important finding in that it provided the first real evidence that amides hydrolyzed in acid media by an addition mechanism (leading to tetrahedral intermediates) that is similar to that of other carboxylic acid derivatives. Nevertheless, the observation of only small ^{18}O-exchange confirmed the general expectations that in amide hydrolyses, proton transfer from the H_3O^+ to the N of $\mathbf{T_O}$ was fast, and, once $\mathbf{T_{N^+}}$ was formed, the amino functionality immediately departs so that in effect, $k_2 \gg k_{-1}$.

Subsequently, Bennet et al.[53] performed careful studies on $^{18}O{=}C$ exchange and hydrolysis for a variety of amides in H_3O^+ and D_3O^+ in order to assess the factors responsible for controlling the partitioning of the tetrahedral intermediates. For the simplified mechanism given in Scheme 2 where k_1 is redefined to incorporate the amide protonation step and k_2 is redefined to include the $T_O \Leftrightarrow T_{N^+}$ equilibrium and breakdown of T_{N^+} to products, the k_{ex} and k_{hyd} rate constants are given in Eqs. (4.3) and (4.4) with the k_{ex}/k_{hyd} ratio being given in Eq. (4.5). The factor of 2 in the denominator of Eq. (4.3) stems from the fact that if the two oxygens are made equivalent by rapid proton exchange, then only half of the reversal leads to ^{18}O-exchanged starting material.

$$k_{ex} = \frac{k_1 k_{-1}}{2(k_{-1} + k_2)} \tag{4.3}$$

$$k_{hyd} = \frac{k_1 k_2}{(k_{-1} + k_2)} \tag{4.4}$$

$$\frac{k_{ex}}{k_{hyd}} = \frac{k_{-1}}{2k_2} \tag{4.5}$$

The ratio of exchange to hydrolysis indicates the partitioning of the tetrahedral intermediates between reformation of starting materials and product formation. In the case of acetanilide **5a** and N-2,4-trimethylacetanilide (**5b**),

Bennet et al.[51] have shown that the exchange and hydrolyses at 100°C, $\mu = 1.0$ (KCl) are both first order in $[H_3O^+]$ between ~ 0.05 and 1.0 M, the k_{ex}/k_{hyd} ratios being 0.005–0.01 and 0.18–0.23, respectively. In the case of **5a**, the small amount of exchange makes detailed mechanistic conclusions difficult, but in the case of **5b**, where exchange is significant, the solvent DKIEs are

$$(k_{ex}) = 0.96 \pm 0.04$$

and

$$(k_{hyd}) = 0.98 \pm 0.04.$$

The latter value can be compared with the solvent DKIEs on k_{hyd} in 1 M HCl for benzamides **6a–c** of 0.87 (100°C), 0.75 (25°C), and 0.87 (100°C, Bunton et al.[48]). Since $^{18}O{=}C$ exchange in the benzamides was not detectable,[48] the DKIE on that process could not be ascertained, but in this case the lack of exchange indicates that $k_2 \gg k_1$, so $k_{hyd} = k_1$ and the DKIE refers to the addition step.

5a R = H
b R = CH_3

6a $R_1 = R_2 = H$
b $R_1 = H$, $R_2 = CH_3$
c $R_1 = R_2 = CH_3$

7

However, since **5b** exhibits considerable exchange accompanying its hydrolysis, certain mechanistic deductions consistent with the process shown in Scheme 2 can be made. First, there must be one or more intermediates that partition between hydrolysis and reformation of starting material: these could be $\mathbf{T_O}$ or $\mathbf{T_{N^+}}$. Second, since exchange is smaller than hydrolysis, the predominantly rate-limiting step must be k_1, the formation of $\mathbf{T_O}$. Finally, since each of the hydrolysis and exchange processes is first order in acid, each associated transition state must contain one proton: this rules out a neutral or zwitterionic transition state as being responsible for any of the pathways leading away from $\mathbf{T_O}$.

Bennet et al.[51,52] have made a detailed analysis of the DKIEs on the acid catalyzed exchange and hydrolysis of an amide that exhibits considerably more $^{18}O{=}C$ exchange than does **5b**, namely *N*-toluoylpyrrole (**7**, $k_{ex}/k_{hyd} \sim 50$, $T = 72$°C). In both of these cases, the appropriate kinetic term to consider is $(k_{ex}/k_{hyd})_{H/D}$ since this compares directly the DKIE on the two transition states for the partitioning of the intermediate, e.g., $(k_{-1}/2k_2)_{H/D}$. For **5b**, $(k_{ex}/k_{hyd})_{H/D}$

is 0.98 ± 0.04, while for **7** the value is 0.87 ± 0.20. Since the observed ratios for these very different amides is indistinguishable from unity, constraints are placed upon the k_2 and k_{-1} (and by microscopic reversibility, the k_1) transition states with respect to the state of protonation and the number of protons in flight. The data support the partitioning processes given in Scheme 2 in which water acts as a general-base to assist in the delivery of another H_2O to the protonated amide (k_1 step), or, in the reverse process having H_3O^+ act as a general-acid in assisting the decomposition of $\mathbf{T_O}$ to regenerate protonated amide through the same transition state. In the hydrolysis process, water also acts as a general-base to deprotonate the O–H of $\mathbf{T_{N^+}}$ concurrent with C–N bond cleavage. The fact that the solvent DKIE on k_{hyd} for **5a** is unity (Bennet et al.[51,52]) or slightly inverse for **6a,b,c** (Bunton et al.[48]) stems from the superpositioning of an equilibrium deuterium isotope effect on $A \Leftrightarrow A\text{–}H^+$, ($k_H/k_D < 1$), on a compensating primary deuterium kinetic isotope effect for the water promoted delivery of H_2O ($k_H/k_D > 1$). Interestingly, the same conclusion results from the consideration of the data for **7**, where $k_{ex} \gg k_{hyd}$: in this case, Eq. (4.3) can be modified to $k_{ex} = k_1/2$, and the DKIE on this process is, therefore, relegated to a single step (k_1). For **7** $(k_{ex})_{H/D} = 0.81 \pm 0.08$ (Bennet et al.[51,52]), this value being typical for a specific-acid protonation followed by a rate limiting water assisted delivery of H_2O. The fact that the observed DKIE on the k_1 step is close to unity may suggest that only a single water is involved in assisting the nucleophilic attack of a second on the protonated amide since it might be anticipated that the cyclic mechanism of Cox and Yates[43] involving 3 H_2Os having three protons in flight would show evidence of a larger primary DKIE. How many waters are involved in these processes cannot be ascertained with certainty from these experiments, but the available data could be consistent with as few as two,[47,51,52] or with three as proposed by Yates and co-workers,[37,38] Moodie et al.,[42] and Cox and Yates.[43]

2.1. Factors Influencing the k_{ex}/k_{hyd} Ratio

Whether exchange is observed during acid promoted amide hydrolysis depends upon the C–O/C–N cleavage ratio which, in the multi-step process given in Scheme 2, depends in part upon the ability to place the required proton on N prior to the C–N cleavage from $\mathbf{T_{N^+}}$. If the nitrogen cannot be protonated either prior to, or concerted with C–N cleavage then the hydrolysis cannot proceed because amide anion is a very poor leaving group, and the fate of $\mathbf{T_O}$ will simply be reversal leading to large amounts of ^{18}O-exchange. It is therefore reasonable that the 20–40-fold increase in ^{18}O-exchange for **5b** relative to **5a** stems from a protonation difficulty for the N in $\mathbf{T_O}$ because in $\mathbf{T_{N^+}}$, the sp^3 N encounters steric compression from the ortho CH_3 group in **5b** which is absent in **5a**. This sort of steric inhibition of protonation can also rationalize why the pK_a of 2,6-dimethylanilinium is 0.73 units lower than that of anilinium ion,[53] although one must be mindful of solvation effects that could destabilize the protonated form of the methylated-aniline relative to aniline-H^+. Raising the energy of $\mathbf{T_{N^+}}$ should also raise the barriers of the paths leading to, and away from it thereby raising the overall C–N cleavage barrier. Should the above explanation be correct, it seems reasonable that further inhibiting the ability of the N in $\mathbf{T_O}$ to accept a proton should increase the k_{ex}/k_{hyd} ratio. The basicity of the pyrrole in the $\mathbf{T_O}$ of **7** is very low (p$K_a < -3.8$[54*]) and correspondingly the k_{ex}/k_{hyd} ratio is ~50.[52]

3. SPECIFIC-BASE CATALYSIS

Specific base catalysis refers to the reaction promoted by the conjugate base of water, namely HO^-. In base, the hydrolysis of amides generally follows the pathway shown in Eq. (4.6).[14,18,20,32,55] The anionic tetrahedral intermediate ($\mathbf{T_{O^-}}$) can either revert back to starting material, or break down to hydrolysis products via pathways that involve various species in the solution including the acidic and basic forms of any buffers present. The latter species can act as general-catalysts in removing a proton from

$$\mathrm{RC(=O)NR_2} + \mathrm{HO^-} \underset{k_{-1}}{\overset{k_1}{\rightleftharpoons}} \underset{(\mathbf{T_{O^-}})}{\mathrm{RC(O^-)(OH)NR_2}} \left.\begin{array}{l} \xrightarrow{k_2} \\ \xrightarrow{k_3[\mathrm{HO^-}]} \\ \xrightarrow{k_B[\mathrm{B:}]} \\ \xrightarrow{k_{BH}[\mathrm{BH}]} \end{array}\right\} \longrightarrow \mathrm{RCO_2^-} + \mathrm{HNR_2} \quad (4.6)$$

the OH group of $\mathbf{T_{O^-}}$ concurrent with cleavage of the C–N bond (B:), or to protonate the amino group concurrently with C–N cleavage ($B\text{–}H^+$). For unbuffered conditions where the treatment is simplified because there is no B: or $B\text{–}H^+$, steady state treatment of Eq. (4.6) gives:

$$k_{hyd} = \frac{k_1[\mathrm{OH^-}](k_2 + k_3[\mathrm{OH^-}])}{(k_{-1} + k_2 + k_3[\mathrm{OH^-}])} \quad (4.7)$$

$$k_{ex} = \frac{k_1 k_{-1}[\mathrm{OH^-}]}{2(k_{-1} + k_2 + k_3[\mathrm{OH^-}])} \quad (4.8)$$

$$\frac{k_{ex}}{k_{hyd}} = \frac{k_{-1}}{2(k_2 + k_3[\mathrm{OH^-}])} \quad (4.9)$$

For the ^{18}O=C exchange process, the assumptions concerning protonic equilibration of the anionic tetrahedral intermediate(s) are necessary and the factor of 2 in the denominator of Eqs. (4.8) and (4.9) arises from only half of the reversal ejecting ^{18}O. Expulsion of the amide leaving group from $\mathbf{T_{O^-}}$ as its anionic form (R_2N^-) is difficult and so assistance by a second OH^- (in deprotonating the O–H group of $\mathbf{T_{O^-}}$ either prior to, or in concert with C–N cleavage in Eq. (4.10)) is sometimes necessary. Terms second order in $[OH^-]$ have been observed for the hydrolysis of acetanilides,[56] formanilides,[57]

* This pK_a refers to protonation of C_3 of the ring: the pK_a of the protonated nitrogen would be lower.

trifluoroacetanilides,[55,58,59] and acetyl and benzoyl pyrroles.[62,65] At very high $[HO^-]$, as can be seen from Eq. (4.10), the trapping of $\mathbf{T_{O^-}}$ to form $\mathbf{T_{O^{2-}}}$ halts the reversal so that the rate limiting step now becomes k_1, referring exclusively to the formation of the anionic tetrahedral intermediate. The kinetics now become first order in $[HO^-]$ and $^{18}O{=}C$ exchange cannot be observed.

$$RC(=O)NR_2 + HO^- \underset{k_{-1}}{\overset{k_1}{\rightleftharpoons}} (\mathbf{T_{O^-}}) \xrightarrow{k_3[HO^-]} \left[R{-}C(O^-)_2{-}NR_2 \right] (\mathbf{T_{O^{2-}}}) \longrightarrow P \qquad (4.10)$$

Many amides such as benzamides,[15,16,20,21,64,65] toluamides,[68] and simple aliphatic amides,[32,68,69] do not show second order terms in $[HO^-]$, so that the first-formed $\mathbf{T_{O^-}}$ in Eq. (4.6) is sufficiently reactive to break down to products with simple assistance of the solvent, and without the involvement of a second $[HO^-]$. In these cases, Eqs. (4.7)–(4.9) reduce to the simple forms of Eqs. (4.3)–(4.5) respectively. Early ^{18}O-exchange studies of methylated benzamides (**9**) indicated that the k_{ex}/k_{hyd} ratios were independent to $[HO^-]$ with values of ~ 3.5 for **9a**, ~ 0.5 for **9b** and 0 for the tertiary amide **9c**.[15,16,20,21,64,65] In an effort to determine the factors that influenced the partitioning of the tetrahedral intermediates, Slebocka-Tilk et al.[66,67] performed careful exchange and hydrolysis studies in H_2O and D_2O on toluamides **10a–c** and **11a–e**.

9a $R_1 = R_2 = H$
b $R_1 = H$; $R_2 = CH_3$
c $R_1 = R_2 = CH_3$

10a $R = Et$
b $R = iPr$
c $R = tBu$

11a $R_1 = R_2 = CH_3$
b $R_1 = R_2 = -(CH_2)_2O(CH_2)_2-$
c $R_1 = Et$; $R_2 = CH_2CF_3$
d $R_1 = R_2 = -CH_2CF_2CF_2CH_2-$
e $R_1 = R_2 = -CH{=}CH{-}CH{=}CH-$

For each of these the k_{ex} and k_{hyd} values are linearly dependent on $[HO^-]$ and the kinetic data can be analyzed in terms of Eqs. (4.3)–(4.5). Given in Table 4.1 are the various kinetic parameters as well as the pK_a values for the corresponding ammonium ions.[14,64,65] Several trends are apparent from the data. First, for the secondary toluamides, **10a–c**, the k_{ex}/k_{hyd} ratio is fairly constant at ~ 0.4–0.7 but the k_{hyd} value drops by 50-fold in passing through the series. This indicates that the attack of HO^- is subject to a marked steric hindrance, but once the $\mathbf{T_{O^-}}$ is formed, its partitioning is relatively insensitive to the steric bulk of the substituents. It is also to note that the pK_as of the primary ammonium ions of

TABLE 4.1. $^{18}O{=}C$ **Exchange and Hydrolysis Data for Amides 10a–c and 11a–e**[a,b]

Amide	k_{hyd} ($M^{-1}s^{-1}$)	k_{ex}/k_{hyd}	pK_a ($H^+NR_1R_2$)
10a[c]	1.15×10^{-4}	0.49 ± 0.03	10.8
	(1.0×10^{-4})[d]	(0.53 ± 0.04)[d]	
10b[c]	2.46×10^{-5}	0.64 ± 0.07	
10c[c]	2.48×10^{-6}	0.44 ± 0.04	10.8
11a[c,e]	1.15×10^{-3}	0.010 ± 0.002	10.64
	(1.28×10^{-3})[d]	(0.014 ± 0.002)[d]	
11b[f]	8.79×10^{-4}	0.13 ± 0.02	8.33
11c[g]	3.13×10^{-5}	32.2 ± 1.6	6.3
	(2.97×10^{-5})[d]	(35.6 ± 1.4)	
11d[h]	2.47×10^{-3}	9.0	4.05[i]
	(2.91×10^{-3})[d]		
11e[j]	1.29	0.24	< -3.8[k]
	(1.95)[d]		

[a] $T = 100\,°C$ unless otherwise noted; $\mu = 1.0$ (KCl)
[b] pK_a values for CRC handbook of Chemistry and Physics (1967–1968). CRC Press, Cleveland, 48th edn.
[c] Reference 66.
[d] In D_2O.
[e] k_{ex}/k_{hyd} determined at [NaOH] = 1.0 M.
[f] 72 °C, $\mu = 1.0$ (KCl); k_{ex} determined at [NaOH] = 0.19 M and converted to second order rate constant.
[g] k_{ex} and k_{hyd} determined at [NaOH] = 1.08 M; in D_2O determined at [NaOD] = 1.03 M. All values converted to second order rate constant for comparison.
[h] at 73 °C; k_{ex}/k_{hyd} ratio determined at [NaOH] = 0.035 M, where both rate constant are first order in [HO^-], from reference 63.
[i] Roberts, R. D.; Ferrah, H. E., Jr.; Gula, M. J.; Spencer, T. A. *J. Am. Chem. Soc.* 1980 **102**, 7054.
[j] 25 °C; k_{ex}/k_{hyd} ratio determined in the low [HO^-] region where both rate constants are first order in [HO^-]; from reference 63.
[k] Reference 54.

10a–c are relatively constant at 10.8. In the case of tertiary toluamides **11a–e**, which have similar steric demands for the substituents surrounding the N, but widely varying basicity of the amino portion, there is a large variation in the k_{ex}/k_{hyd} ratio from ~0.01–32. These data indicate that the dominant factor that controls the partitioning of the intermediate is the amine basicity. Moreover, there is a maximum in the k_{ex}/k_{hyd} ratio for amide **11c**. Higher and lower pK_as for the ammonium ions lead to less exchange, and the break in the k_{ex}/k_{hyd} versus pK_a plot signifies a change in mechanism that controls k_{ex}, vide infra.

The DKIE data on the exchange and hydrolysis processes for **11c** are particularly informative. Given that the hydrolysis and exchange are both first order in [HO^-], and that $k_{ex} = 32k_{hyd}$, one can conclude that $k_{-1} \gg k_2$, k_3 and that Eq. (4.8) reduces to $k_{ex} = k_1/2$. Therefore, the hydroxide attack step is kinetically isolated and can be probed by determining the DKIE on k_{ex}. The observed DKIE on k_{ex} for **11c** is slightly inverse at 0.90 ± 0.05.[67] Fractionation factor analysis allows one to suggest that the transition state for the HO^- attack resembles **12** in Eq. (4.11), where a desolvated hydroxide (with the loss of one of its 3 solvating H_2Os) attacks the carbonyl with the developing O^- being solvated by 3 waters in the tetrahedral intermediate **13**. The same sort of transition state is suggested for the attack of HO^- on esters and thioesters.[70,71]

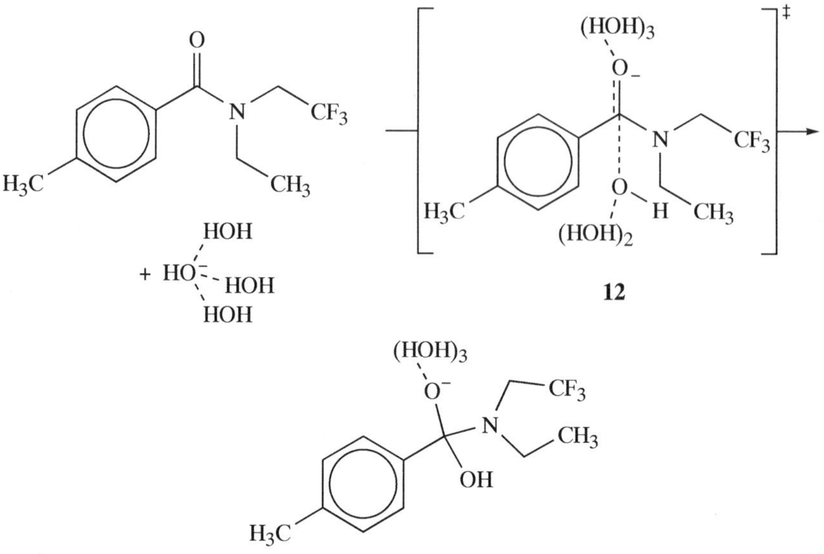

(4.11)

In the case of hydrolysis for **11c**, the rate limiting transition state is clearly the breakdown of $\mathbf{T_{O^-}}$ (**13**) and the observed DKIE on k_{hyd} is 1.05 ± 0.04. This indicates that there cannot be a proton in flight in the breakdown TS, otherwise a strong and normal DKIE > 2 would be expected, contrary to what has been observed. Transition state **14**, as depicted in Eq. (4.12) and having a protonated N and two anionic oxygens, is consistent with that analysis.

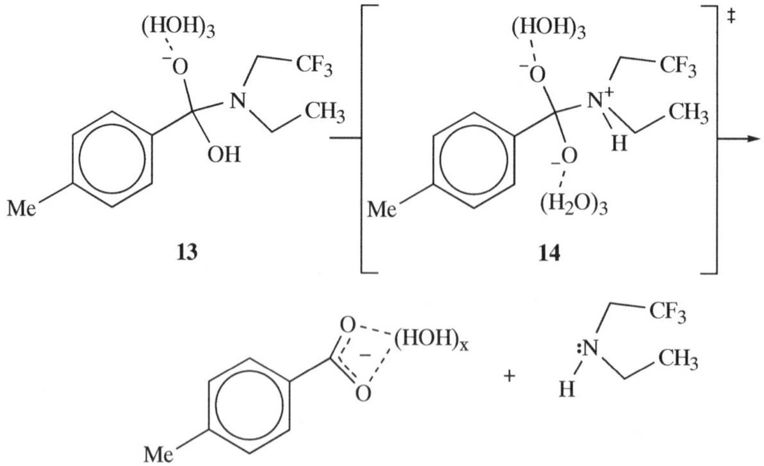

(4.12)

On the basis of solvent DKIEs for the above amides, a picture emerges for both the attack step and breakdown step, the possibilities for the latter being illustrated in Scheme 3. Importantly, for amides **10a–c** and **11a–c**, the data are consistent with breakdown processes where the departing nitrogen must be fully protonated at the time of departure and not with any process where the amino functionality departs with protonation occurring concurrently with C–N bond cleavage.

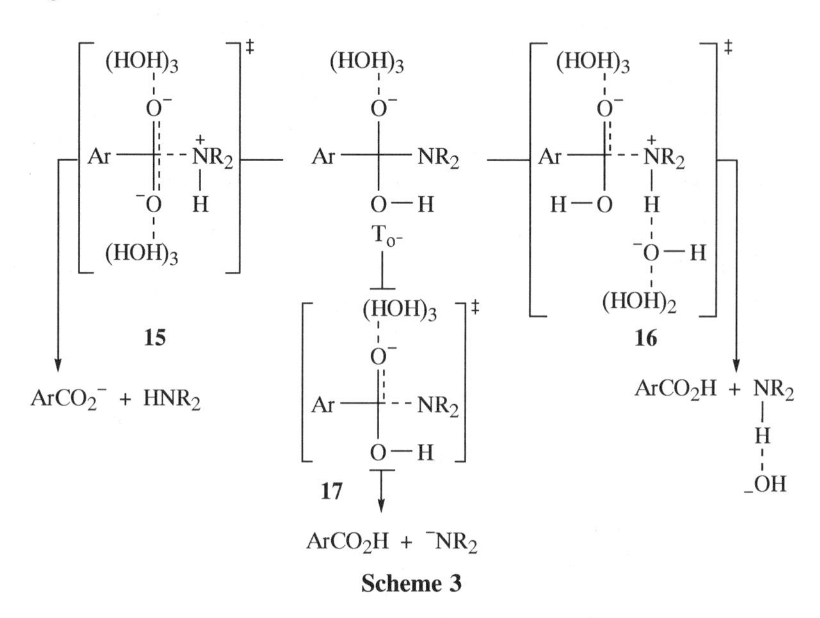

Scheme 3

For these amides which have a relatively basic departing amine, there are two transition states shown in Scheme 3 that are consistent with the data. The first, **15**, involves an unassisted C–N cleavage of an anionic zwitterion where the two anionic oxygens are each solvated by 3 waters to give directly the amine and carboxylate. The second, **16**, which is kinetically indistinguishable from **17**, involves prior protonation of the nitrogen by a solvating H_2O with the C–N cleavage then giving the carboxylic acid and an amine-OH^- encounter complex which must certainly deprotonate the carboxylic acid prior to diffusional separation.[14,51,52,63]

The situation with amides **11d,e** is different and as the basicity of the departing amine is reduced there are important changes that are noticed in the kinetics. It is reasonable that reducing the amine basicity should, at some point, inhibit N-protonation so that the amine cannot be expelled from $\mathbf{T_{O^-}}$ because the N-protonated forms of $\mathbf{T_{O^-}}$, and any transition states for breakdown that involve them, are too high in energy to be viable. In these cases, alternative pathways for breakdown will be followed, and these may involve a second molecule of HO^- to deprotonate the remaining O–H of $\mathbf{T_{O^-}}$ as in Eq. (4.10) or, as shown in Scheme 3, the nitrogen may depart from $\mathbf{T_{O^-}}$ without prior protonation. If the hydrolysis requires two hydroxides, this will be evident from the kinetic dependence on $[HO^-]^2$.

Both **11d** and **11e** exhibit k_{hyd} versus pH profiles that have first and second order in $[HO^-]$ domains.[65] For **11d**, at low $[HO^-]$, where both the hydrolysis and exchange processes are first order in $[HO^-]$, $k_{ex}/k_{hyd} = 9.0$, the latter value dropping as $[HO^-]$ increases. In the low pH region the DKIE for hydrolysis, (where, from the large amount of exchange it can be concluded that breakdown of $\mathbf{T_{O^-}}$ is rate limiting), is slightly inverse at 0.72 but at high pH this value becomes 1.54. The change in DKIE that accompanies the switch in k_{hyd} from first to second order in $[HO^-]$ dependence suggests that the breakdown at low $[HO^-]$ involves expulsion of the amide anion leaving group without general assistance having a proton in flight, while at high $[HO^-]$ the second HO^- deprotonates $\mathbf{T_{O^-}}$ concurrent with departure of the amide (Scheme 4).

$$P \leftarrow \left[Ar-C(^-O)(OH)\cdots N(CH_2CF_2)_2 \right]^{\ddagger} \xleftarrow[{[HO^-]}]{low} Ar-C(O^-)(OH)-N(CH_2CF_2)_2 \ (To^-) \xrightarrow[{[HO^-]}]{high} \left[Ar-C(^-O)(O\cdots H\cdots {}^-OH)\cdots N(CH_2CF_2)_2 \right]^{\ddagger} \rightarrow P$$

Scheme 4

The same sort of behavior is seen with toluoyl pyrrole **11e** in that there is $^{18}O{=}C$ exchange at low $[HO^-]$ where both k_{ex} and k_{hyd} are first order in $[HO^-]$. However, at high pH the hydrolysis exhibits a domain where second order terms in $[HO^-]$ appear and the exchange becomes independent of increasing $[HO^-]$. From inspection of Eq. (4.10), this is a predictable consequence of trapping of the $\mathbf{T_{O^-}}$ by a second HO^- to drive it forward to hydrolysis product, which effectively halts the reversal and ^{18}O-exchange.

In the cases where k_{ex}/k_{hyd} is large the attack and breakdown steps (k_1 and k_2) are kinetically isolated. The exchange provides information about k_1, the attack of hydroxide, and hydrolysis provides information about k_2, the breakdown of the tetrahedral intermediate. Liu and Brown[72] have investigated the activation parameters for the hydrolysis and exchange of **11c,d** under conditions where both processes are first order in $[HO^-]$. Notably, as discussed above, the DKIE evidence indicates that the breakdown pathways for these two amides are different, with $\mathbf{T_{O^-}}$ of **11c** expelling the protonated amine through TS **14**, Eq. (4.12), and $\mathbf{T_{O^-}}$ of **11d** expelling the amide anion as in Scheme 4. Since both amides exchange faster than they hydrolyze, the steps of HO^- attack and breakdown of $\mathbf{T_{O^-}}$ are kinetically isolated so that the activation parameters refer explicitly to these steps. From the activation parameters given in Table 4.2, it can be seen that the $\Delta H^{\ddagger}$ for both exchange and hydrolysis of **11c** are $17.3\,\mathrm{kcal\,mol^{-1}}$ but the $\Delta S^{\ddagger}$ for the two processes are -26.2 and $-33.2\,\mathrm{cal\,K^{-1}\,mol^{-1}}$, respectively. Thus, for this amide, the greater propensity

TABLE 4.2. Activation Parameters for Hydrolysis and $^{18}O{=}C$ Exchange of Amides 11c,d [a,b]

	Amide **11c**[c]		Amide **11d**[d]	
Parameter	Exchange	Hydrolysis	Exchange	Hydrolysis
$\Delta H^{\ddagger}$ (kcal mol^{-1})	17.3 ± 0.3	17.3 ± 0.4	14.5 ± 0.3	16.7 ± 0.2
$\Delta S^{\ddagger}$ (cal K^{-1} mol^{-1})	−26.2 ± 0.8	−33.2 ± 1.1	−25.0 ± 0.8	−23.0 ± 0.6
$\Delta G^{\ddagger}$ (100 °C) (kcal mol^{-1})	27.1 ± 0.6	29.7 ± 0.8	23.8 ± 0.6	25.3 ± 0.4

[a] Data from reference 72.
[b] Determined from Eyring plots of ln (kh/k_BT) vs. $1/T$, where h = Planck's constant, k_B = Boltzmann's constant, and k = second order hydrolysis or exchange rate constant.
[c] 5 temperatures between 72 and 100 °C.
[d] 5 temperatures between 33 and 70 °C.

to exchange is governed entirely by the entropy component. That $\Delta S^{\ddagger}_{hyd} < \Delta S^{\ddagger}_{ex}$ is probably a consequence of greater restriction of solvent in the transition state leading to breakdown of $\mathbf{T_{O^-}}$ in the case of **11c**. For **11d** the situation is different in that the $\Delta H^{\ddagger}$ is 14.5 and 16.7 kcal mol^{-1} for exchange and hydrolysis respectively but the $\Delta S^{\ddagger}$ for the two processes are experimentally the same at −25 and −23 cal K^{-1} mol^{-1}. In the case of **11d**, the greater propensity for exchange is virtually entirely a consequence of the enthalpy differences for the two steps. What is learned from this study is that k_{ex}/k_{hyd} ratio may, as a function of temperature, be constant (as in the case of **11c**), increase, or decrease (as in the case of **11d**), this being controlled by the relative enthalpies of the exchange and hydrolysis processes which, in any given case, is difficult to predict.

4. WATER PROMOTED HYDROLYSES

In the absence of H_3O^+ or HO^- catalysis, the water promoted hydrolysis of nonactivated amides is very slow, and in most cases undetectable. Because esters are more reactive, in general, than are amides, more is known about their water reactions and this provides a useful guideline for the possible reaction mechanisms occurring with amides. Even so, without some sort of structural activation, ester hydrolysis in water is exceedingly slow. For example, at 25 °C the uncatalyzed water reaction for hydrolysis of methyl acetate has been estimated, on the basis of thermochemical analysis[73] to have a rate constant of $3 \times 10^{-10}\,s^{-1}$ corresponding to a half-time of 70 years. The slowness of the water reaction is presumably a consequence of a requirement to avoid the formation of highly charged or zwitterionic species arising from the attack of neutral water on the carbonyl, so that the reactions must involve highly ordered, and possibly cyclic, transition states involving several waters (**18–20**). Whether these processes involve the formation of intermediates (as in the case of **20** to

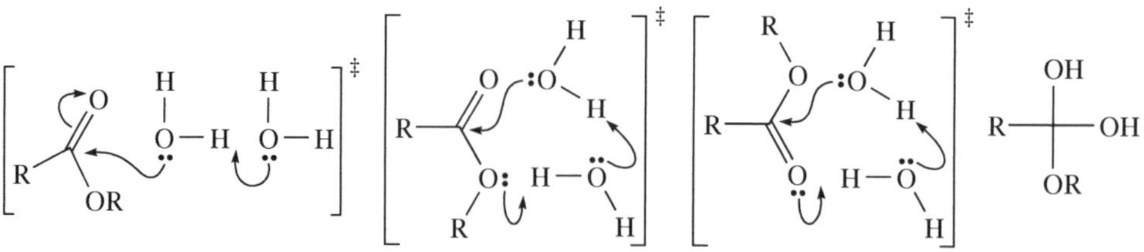

18 **19** **20** **21**

form the hydrate **21**), or are concerted (as in the case of **19**) cannot be established at present. Kirby[76] has summarized the literature in 1972 and noted that there is a paucity of experimental data except for the cases of activated esters such as alkyl and aryl haloacetates, and substituted phenyl acetates. For these, the activation enthalpies are fairly low (7–14 kcal mol^{-1}) but the entropies are large and negative (−40 to −50 cal K^{-1} mol^{-1}) as would be expected for highly ordered transition states. Bunton et al.[75] and Jencks and Carriuolo[76] have determined the solvent DKIE for the hydrolysis of methyl trifluoroacetate (1.8), ethyl dichloroacetate (5), and ethyl difluoroacetate (2.1), respectively. These values, while not at the high end for solvent DKIEs involving several protons in flight, are nevertheless consistent with the transition states **18–20**.

Relatively few amides are known to hydrolyze by a water-promoted reaction; this is probably a consequence of the extremely slow rate of the process for nonactivated amides which makes the detection of hydrolysis difficult unless special techniques are developed. For example, Kahne and Still[77] using a sensitive ^{14}C radioisotope assay, have found that an nonactivated peptide hydrolyzes in aqueous media at pH 7, $T = 25\,°C$ with a k_0 of $3 \times 10^{-9}\ s^{-1}$ corresponding to a half-time of 7 years. Radzicka and Wolfenden[78] used NMR to study the water-promoted hydrolysis of glycylglycine, acetylglycylglycine, and acetylglycylglycine *N*-methylamide at temperatures between 120 and 200 °C in the pH region between 4.2 and 7.8. From the activation parameters, the rate constants for the hydrolyses of these three dipeptides fall in a narrow range from 6.3 to $3.6 \times 10^{-11}\ s^{-1}$ at 25 °C corresponding to half-times of 350–600 years. Using a technique where the free amino acid liberated during hydrolysis is derivatized to form a highly fluorescent tracer, Bryant and Hansen[79] have shown that the hydrolysis rate constant for hippurylphenylalanine at pH 9 in borate buffer is $1.3 \times 10^{-10}\ s^{-1}$ at 25 °C corresponding to a half-time of 168 years. While of the appropriate order of magnitude, the reported value is probably not exclusively attributable to the water reaction since the authors report that they were unable to observe any reaction at pH 7 in a HEPES buffer solution.

Hine et al.[80] have investigated the hydrolysis of the simplest amide – formamide – in H_2O from pH 1–9, at $T = 80\,°C$ and found a possible water term having the value $k_o = 8.4 \times 10^{-8}\ s^{-1}$, ($t_{1/2} = 95$ days). Since this latter term contributes a maximum of 50% to the hydrolysis at the pH/rate minimum, solid information on this process could not be obtained. The same difficulty was encountered by Slebocka-Tilk and Brown[81] in their study of the hydrolyses of *N*-aroylaziridines which apparently showed water terms of $k_o = 0.5-1.7 \times 10^{-6}\,s^{-1}$, ($t_{1/2} = 3-16$ days) at 25°C at their pH/rate minima. Given the anticipated low $\Delta H^{\ddagger}$ and large negative $\Delta S^{\ddagger}$ terms for these water reactions relative to the H_3O^+ or HO^- promoted reactions, one might better separate them from the acid and base hydrolyses and therefore gain more information about them by studying the reactions at lower temperature, but this has the attendant problem of leading to very long reaction times.

Water reactions are most easily observed with activated amides such as **22**,[82–84] **23**,[84] **24**,[85,86] **25**,[87,88] **26**,[89,90] and **27**.[91] For the hydrolysis of this sort of

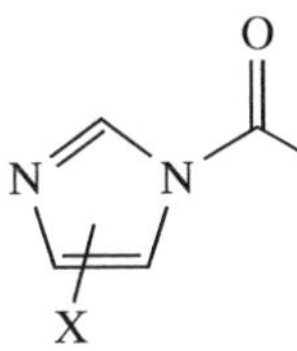

22 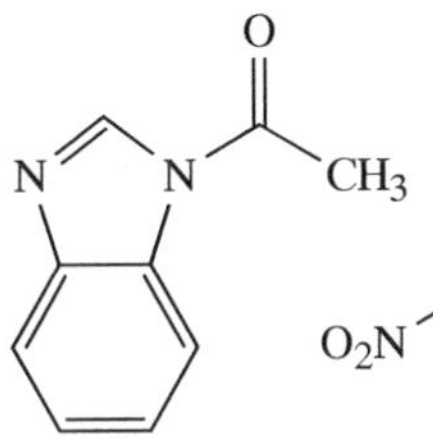 **23** 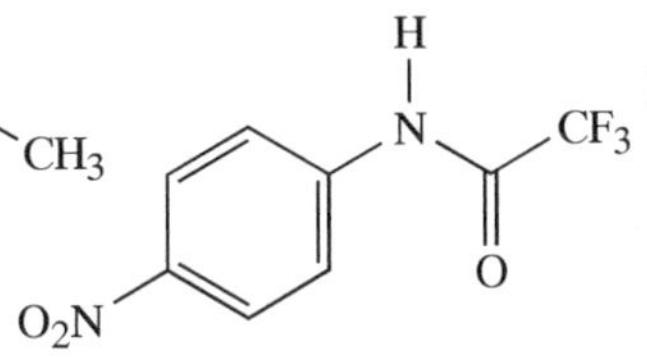 **24** 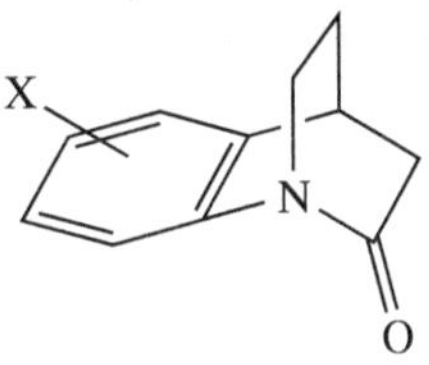 **25**

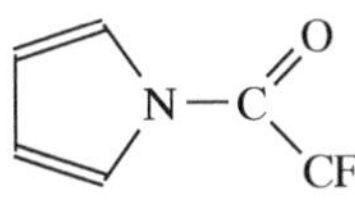

26 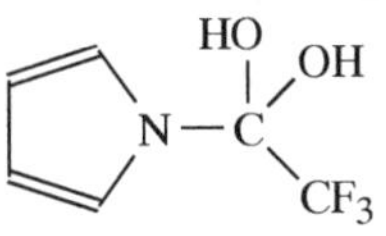 **26-hyd** 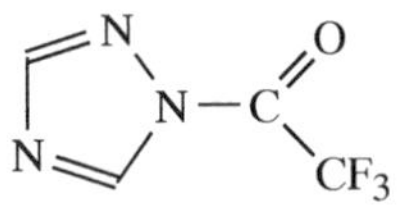 **27**

amide under neutral conditions it is expected that significant DKIEs would be observed due to the large number of O–H bonds that would have to be broken in the TS. For example, Jencks and Carriuolo[83] and Fife et al.[84] have observed that the solvent DKIE on water hydrolysis of *N*-acetyl imidazoles is 2.7–3.5. Fife[82] has suggested that the TS involves the cyclic concerted process shown in **28**. Patterson et al.[91] have determined that the water reaction for 1-acetyl-1,2,4-triazole (**27**) has a solvent DKIE of 3.18, and on the basis of proton inventory experiments, suggested that the transition state for the reaction resembles **28** having four water molecules with four in-flight protons each having a small DKIE of 1.33. However, as is typical with proton inventory experiments,[23–27,92] it would be very difficult to say with any certainty whether the appropriate number of protons in flight is three, four, or more. Komiyama and Bender[85] have found a DKIE of 3.7 for the pH independent H_2O hydrolysis of **24** at 70°C and have subsequently shown using proton inventory analysis[86] that two or more protons are in flight in the rate-limiting step. These authors favor a process where a water molecule exerts general-acid assistance to the breakdown of the tetrahedral intermediate as in **29**. Unfortunately, since $^{18}O{=}C$ exchange studies were not conducted, it was not clear whether the rate limiting step is water attack or breakdown of the tetrahedral intermediate.

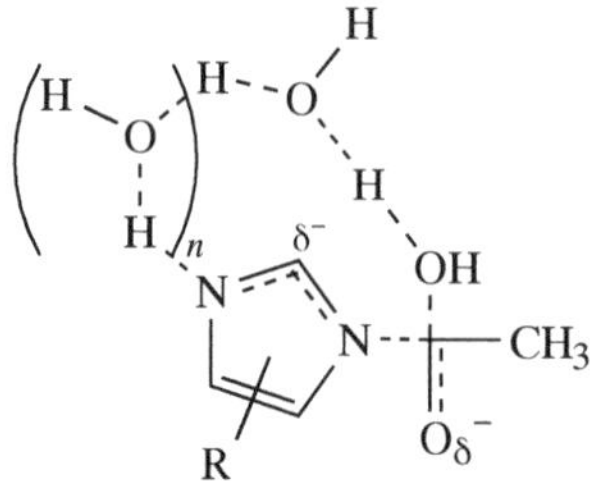

28 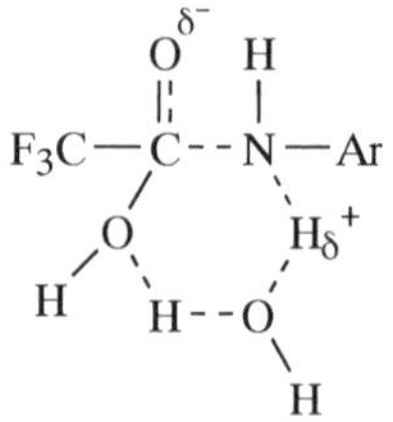 **29** 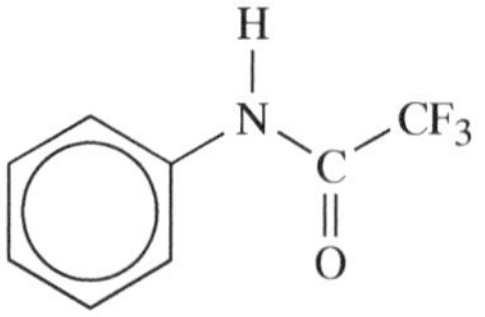 **30**

Of particular importance in any water-promoted hydrolysis would be the detection of intermediates and determining their partitioning between product formation and reversal to starting material. Only in the case of trifluoroacetylpyrrole (**26**) it has been shown that a reversibly formed hydrated intermediate (**26-hyd**) can be formed[89,90] and Cipiciani et al.[93] suggested that a similar hydrate forms during the spontaneous hydrolysis of *N*-trifluoroacetylindole. From simple mechanistic considerations[94] and by analogy with water addition to aldehydes and ketones[95,96] a *gem* diol intermediate is likely to be formed on the reaction pathway for water-promoted amide hydrolysis. Guthrie's thermochemical analyses[94] indicate that the diols of dimethyl formamide and dimethyl acetamide are 19.3 and 19.6 kcal mol^{-1} higher in free energy than the amides: the O^-/N^+–H zwitterions are about 3–4 kcal mol^{-1} higher yet.

In order to cast light on the above question, Brown and co-workers[97] have studied the water promoted hydrolysis and ^{18}O=C exchange of **24** and its parent, **30**. Three essential pieces of information were obtained. First, the activation parameters for the hydrolysis of **24** and **30** were determined to be $\Delta H^{\ddagger} = 14.4 \pm 0.6$ and $11.7 \pm 0.3\ \text{kcal mol}^{-1}$; $\Delta S^{\ddagger} = -36.1 \pm 1.6$ and -52.3 ± 1.3 cal $K^{-1} \text{mol}^{-1}$, respectively. These values are quite close to what is seen for the water reactions of activated esters, vide supra, and the large negative entropies may signify a highly ordered transition state with restriction of several water molecules. Second, solvent DKIEs for the water-promoted hydrolysis of **24** were repeated and found to be 3.3 ± 0.2 at 70.2°C, $[L_3O^+] = 10^{-2}$ M, consistent with the reported value[85,86] of 3.7 at 70°C, pH = 4.0. In addition, proton inventory data,[85,86,97] for the hydrolysis of **24** in the pH independent region indicate that the transition state has two or more protons in flight, consistent with the involvement of two or more water molecules. Finally, no ^{18}O=C exchange was detected in labeled **24** or **30** recovered from the reaction media at times up to 3 $t_{1/2}$ of hydrolysis.

The lack of observable ^{18}O-exchange in recovered staring material comes as a surprise and rules out some mechanisms while placing strict constraints on others. A two- or multistep mechanism given in Eq. (4.13) involving reversible formation of a diol intermediate is easily ruled out. Such a diol may be formed stepwise or in a concerted process involving two or more protons in flight in a cyclic mechanism, but if formed, its reversion to staring material would necessarily be slower than breakdown to product. It may be that the CF_3 group destabilizes the amide thereby promoting the attack of water, and at the same time retards reformation of the amide *via* k_{-1} thus making the more rapid process breakdown, k_2.

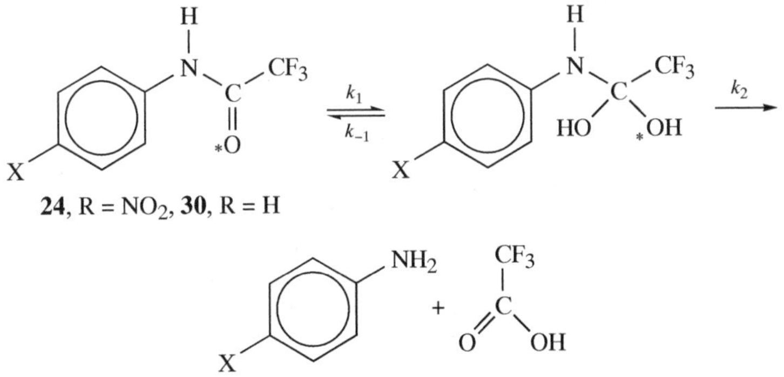

24, R = NO_2, 30, R = H

(4.13)

There is also a possibility that the attack of one water is assisted by a second one as in transition state **31** leading to an anionic tetrahedral intermediate/H_3O^+ encounter complex **32**. In this case, the bold hydrogens are the ones that undergo change in their bonding.

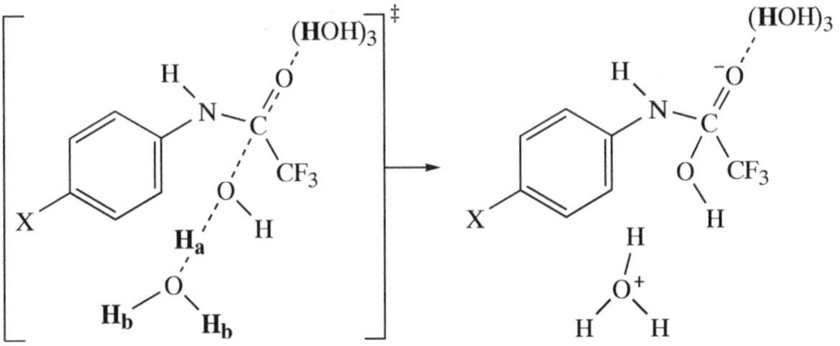

31 **32**

$\mathbf{H_a}$ is the proton in flight that is abstracted from the attacking water; this along with $(\mathbf{H_b})_2O$ ultimately forming the encounter complex hydronium ion in **32**. A virtue of this "immature hydronium ion mechanism"[100] is that the two oxygens attached to carbon are never completely equal. To equilibrate them requires that the O^- becomes protonated by external H_3O^+ or the nascent H_3O^+ in the encounter complex. Proton transfer from external H_3O^+ is thermodynamically favorable and should be diffusion limited[99] which sets the rate constant at pH 1–2 to be 10^9–$10^{10}\,s^{-1}$. Proton transfer from within the encounter complex would require reorientation within the complex with attendant disruption and reformation of H-bonds which could occur competitively with diffusional separation of the complex, i.e. 10^{10}–$10^{11}\,s^{-1}$. Because of the rapidity of both of these processes, the only reason why the two oxygens could not be protonically equilibrated is if the lifetime of the intermediate (**32**) is shorter than 10^{-10}–10^{-11} s. In this case, the attack of water and breakdown of the intermediate to product must be essentially concerted,[100*,102†] but this would require protonation of the leaving amine through an already formed H-bonding network which, in the case of such a weakly basic nitrogen in these delocalized amides seems unlikely. Thus the most favored process for **32** would be to undergo protonation to form the diol, albeit by a more circuitous route than in Eq. (4.13), and the lack of ^{18}O-exchange results from a requirement that $k_2 \gg k_{-1}$.

The bicyclic anilides (benzoquinuclidones, **25**), the parent (**25** X = H) of which was originally prepared and studied by Blackburn et al.,[88] and their higher homologues,[89,109] prove to be interesting molecules for the study of acyl transfer reactions. Because of the orthogonality of the N lone pair of electrons and the carbonyl π-system, **25** is extremely susceptible to hydrolysis promoted by OH^-, H_3O^+, and water, or its kinetic equivalent, hydroxide attack on the protonated amide, in the neutral pH domain. Somayaji and Brown[87] have analyzed the pH/rate profiles obtained at 25 °C for the hydrolysis of **25**, X = H, CH_3, OCH_3, and Cl, and found that the k_o terms for water attack are (2, 2, 2, and 3) $\times 10^{-3}$ s^{-1} respectively, giving rise to half-times for hydrolysis at neutrality of 4–6 min. If one can use the results of Kahne and Still[77] to estimate the rate constant for hydrolysis of a normal amide in the water domain ($k_o = 3 \times 10^{-9}$ s^{-1}) these anilides hydrolyse about six orders of magnitude faster due to the steric inhibition of resonance and other properties associated with the differences in structure. From the temperature study on the hydrolysis of **25** X = OCH_3 at pH 8.0, (the mid-point of the pH independent region), the $\Delta H^\ddagger$ was found to be 10.44 kcal mol^{-1} and the $\Delta S^\ddagger$ −35.9 cal K^{-1} mol^{-1}. Finally, the DKIE for the hydrolysis of **25** X = OCH_3 at 25 °C was reported to be 2.17 at pH 8.0. Although proton inventory data were not collected, the available data support a process (albeit not unambiguously so) where the rate limiting step has considerable restriction in the degrees of freedom of the solvent and has at least one proton in flight.

Kirby and co-workers[108] have found that 1-aza-2-adamantanone **33**, termed the most twisted amide reported to date, reacts rapidly with water. In basic

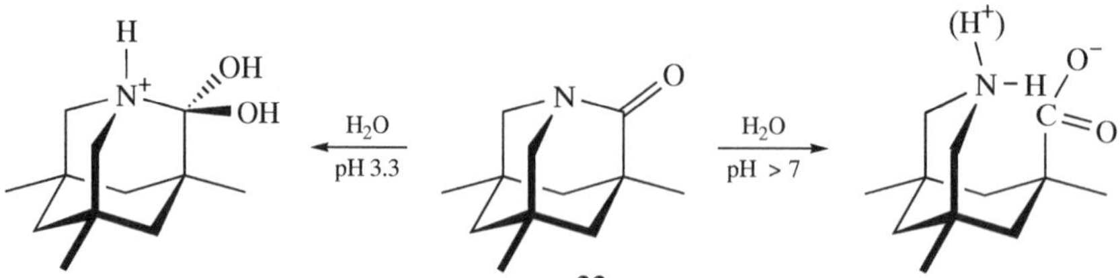

33

media, **33** quickly hydrolyzes to the amino acid, while at low pH, the N-protonated hydrate appears to be stable. Presumably the stability of the protonated

* The question of whether an intermediate has a sufficient lifetime to exist, or last for one vibration before decomposing, is discussed in ref. 100.

† In the case of activated esters, nucleophilic attack by some oxyanions is now considered to proceed without a tetrahedral intermediate being formed, that is, via an S_N2 transition state.

hydrate stems from an enforced proximity of the amino and carboxylic acid groups which does not allow the nitrogen to depart, thereby favouring C–N bond formation.

5. DISTORTED AMIDES AND THEIR HYDROLYSIS

A venerable hypothesis for enzyme mediated amide hydrolysis suggests that a share of the exothermicity of substrate binding can be utilized in a productive way to induce stress or strain in the substrate, the enzyme, or the enzyme: substrate complex, which is relaxed as the transition state for the acyl transfer is approached.[109–114] It is important to realize, however, that efficient binding of the substrate by the enzyme alone cannot lead to enhanced reaction unless the transition state for the acyl transfer is substantially stabilized.[113,115] In effect, the latter statement is equivalent to the widely held notion that enzymes bind transition states better than they do substrates. The question of how much strain can be induced in a substrate by enzyme binding has evoked much debate, and the answer seems to be "not enough to cause a serious structural distortion" of the substrate.[113,116] A comparison of the K_M values for good substrates, and the K_I values for the best transition state inhibitors of an aspartate proteinase shows an expressed difference in binding energy of some 5 kcal mol^{-1} suggesting some strain mechanism to be in operation.[117] Nevertheless, given that the rotational barrier in most amides/peptides[1,3] is 15–20 kcal mol^{-1}, it is difficult to think that simple substrate binding exerts enough energy to rotate a peptide N–C(O) bond by 90°.[113,115] On the other hand, by way of binding, the enzyme could exert stress on the amide bond which would become stronger as the transition state is reached.

An inspection of the amide structure (**1**) with its attendant resonance forms that lead to kinetic stability suggests that the best way to activate an amide toward nucleophilic attack would be to rotate around the N–C(O) bond, or to rehybridize the N from sp^2 toward sp^3. These two distortions, visualized in Scheme 5 as twist (ϕ) and tilt (θ) angles,[87,118*] are conceptually attractive for enzymes to employ since either of them could reduce the amide resonance and lead to enhanced rates of attack on the carbonyl. As a point of reference, if the N is pyramidalized to sp^3, the tilt angle of the lone pair would be 19.5°: twisting the N lone pair completely out of conjugation would generate a twist angle of 90°.

Amides that are distorted by virtue of their molecular geometry have long been of interest from a structural, reactivity, and computational point of view,[122] but there is relatively little accurate kinetic data on their hydrolyses. Hydrolytic studies of most of the known distorted amides is made difficult because they

*The twist and tilt angles are easily conceptualized, but less quantitative than the Dunitz parameters described in ref. 118 that are customarily used for describing amide distortion.

react quickly and seldom contain chromophores which would make studies of the kinetics simpler through the use of UV/vis spectroscopy.

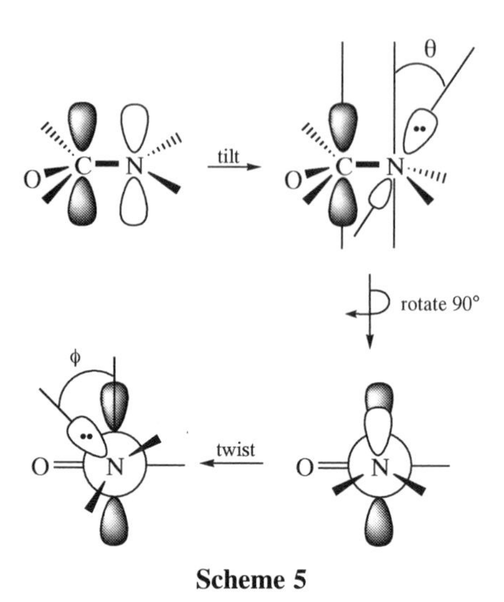

Scheme 5

The first simple examples of this series were the bridgehead lactams, the 1-aza-bicyclo[2.2.2]octan-2-ones, or quinuclidones (**34, 35**) prepared by Pracejus[125,127] which were reported to be very prone to methanolysis. Due to the high reactivity of **34** in water, Pracejus was not able to conduct accurate kinetic studies, but determined a pK_a of 5.3 for the bridgehead protonated N by a half neutralization procedure. Brown and co-workers,[107,125–127,128] have reported the synthesis and detailed kinetic studies of the hydrolyses of the bicyclic anilides **25** X = H, OCH_3, CH_3, Cl, and **36a–c**. Benzoquinuclidone **25,** X = H was prepared and its kinetics of hydrolysis determined between pH 5 and 10 first by Blackburn and co-workers.[88] The bicyclic anilides **25**, **36a–c** make an ideal set for study since increasing the length of the bicyclic straps allows the amide unit to relax from an essentially orthogonal geometry, to that of a nearly planar amide. Anilide **37** is a suitably substituted undistorted amide that was synthesized for comparison purposes.

Given in Table 4.3 are the structural parameters determined for **36a–c** and **37** X = Br by X-ray diffraction.[109,131] The *ortho*-CH_3 substituent in **37** forces the planar amide unit to be nearly perpendicular to the plane of the aromatic ring. The structure of **25** X = H has not been determined but is assumed to be orthogonal with an sp^3-hybridized N. There are several structural features of note passing from the essentially undistorted **37** X = Br to most distorted (**36a,b**) anilide. There is a progressive lengthening of the N–C(O) bond length by 0.06–0.08 Å while the C=O bond length is scarcely affected, the overall decrease in its bond length being 0.02 Å. Exactly this behavior is predicted from ab initio calculations of formamide as a function of rotation about the N–C(O) bond.[129]

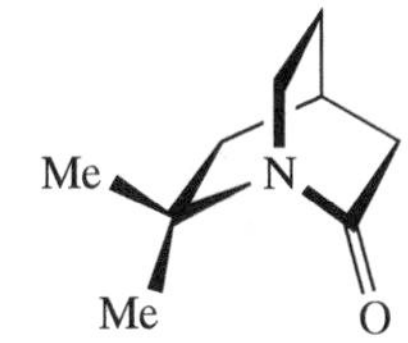

34 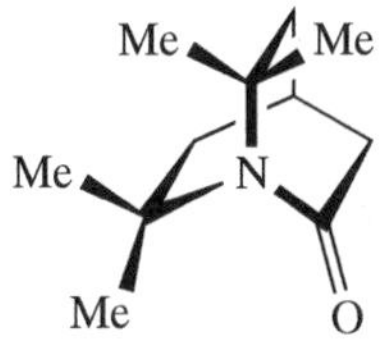 **35**

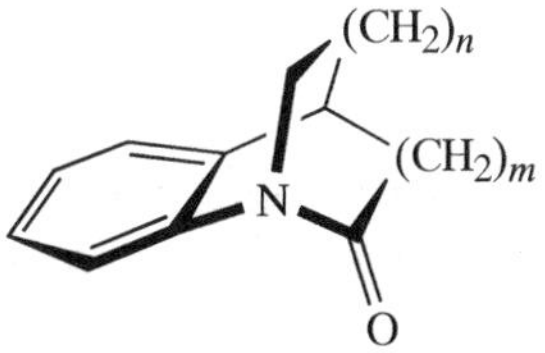

36a $n = 1, m = 2$
b $n = 2, m = 1$
b $n = m = 2$

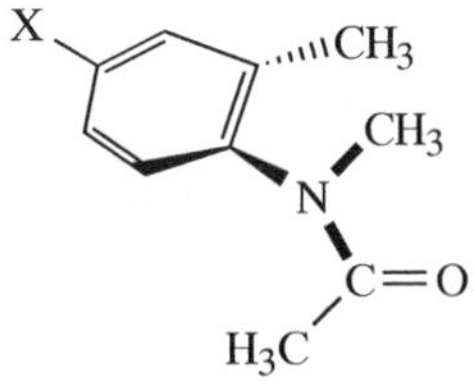

37 X = H, Br

TABLE 4.3. Selected Structural Parameters for Anilides 25 X = H, 36a–c, and 37 X = Br[a]

Amide	*r* (N–C(O) (Å)	*r* (C=O) (Å)	N–C(O)–C (Å)	Twist angle (deg)	Tilt angle (deg)
25 X = H	–	–	–	90[b]	19.5[b]
36a	1.401	1.216	116.5	30.7	15.2
36b[c]	1.413	1.225	116.9	33.2	16.0
	1.419	1.233	117.0	38.9	16.8
36c[c]	1.370	1.233	116.3	15.3	10.5
	1.374	1.241	116.9	17.1	10.6
37 X = Br	1.338	1.235	118.1	1.5	1.0

[a] From reference 107. Twist and tilt angles are defined as in Scheme 5.
[b] Based on expected orthogonality of sp^3 hybridized N lone pair and C=O π-system.
[c] Two crystallographically independent molecules in unit cell.

As expected, there is an increase in the molecular distortion of the amidic unit characterized by *N*-pyramidalization (tilt angle and Dunitz χ_N[107,118]) and rotation about the N–C(O) bond (twist angle and Dunitz τ'[107,118]).

The effect of increasing distortion in passing from **37** to **36c** to **36a,b** to **25** is a marked increase in hydrolytic rate in acid and base. As mentioned above, for the most distorted amides, **25**, the pH/rate profiles show evidence of a water reaction at neutral pH values with $t_{1/2} \sim 5$ min,[87,88] as well as the onset of a pH independent region below pH 3–4 signifying the onset of a pre-equilibrium protonation of the bridgehead N.[87] The general appearance of the pH/rate profiles suggest a mechanism for hydrolysis given in Eq. (4.14).

(4.14)

Given in Table 4.4 are the rate and dissociation constants determined at 25 °C, $\mu = 1.0$ (KCl), for the hydrolyses of these amides along with those for *N*-methyl acetanilide, a comparison anilide which is not distorted. Perhaps a better comparison would have been **37** X = H, but that material was completely inert to hydrolysis. All the amides show a region above pH 8 which is first order in $[HO^-]$, and the data in the table indicate that distortion accelerates the attack of hydroxide by roughly seven orders of magnitude (**25** X = H versus *N*-methylacetanilide). The acceleration in acid hydrolysis is even more striking and if one compares the value for attack on the protonated amide, k_1/K_a, distortion contributes 11 orders of magnitude in passing from *N*-methylacetanilide to **25** X = H. Comparison of the k_3 constants for attack of hydroxide indicates that the very distorted **25** is only 4–12-fold more reactive than the far less distorted **36a,b**: in acid, comparison of the corresponding k_1/K_a values indicates roughly a 1000-fold difference in the reactivities of **25** and **36a,b**. Only in passing to amide **36c** does the reactivity drop markedly.

Also apparent from the data in Table 4.4 is the fact that the most distorted amides, **25** and **36a,b** give evidence of protonation in the accessible pH range. Undistorted amides have pK_a values for their O-protonated forms of 0 to −3,[34–39] and N-protonated forms of −7 to −8.[42] For the most distorted anilides, the pK_a values of ~3.5–3.8 are most easily reconciled as arising from

TABLE 4.4. Rate and Dissociation Constants for the Hydrolysis of Anilides 25a–d, 36a–c and *N*-methyl acetanilide in H_2O, T = 25 °C, $\mu = 1.0$ (KCl)[a]

Anilide	k_1 (s^{-1})	pK_a	k_2 (s^{-1})	k_3, $M^{-1}s^{-1}$	k_1/K_a, $M^{-1}s^{-1}$
25 X = H	4.45	3.71	2.02×10^{-3}	2.62×10^{2}	2.28×10^{4}
25 X = CH_3	4.05	3.81	1.68×10^{-3}	1.75×10^{2}	2.87×10^{4}
25 X = OCH_3	5.31	3.57	1.67×10^{-3}	1.68×10^{2}	1.97×10^{4}
25 X = Cl	3.53	3.38	3.42×10^{-3}	1.65×10^{2}	0.85×10^{4}
36a	1.0×10^{2}	−0.28	–	6.0×10^{1}	5.6×10^{1}
36b	8.3	0.56	–	1.72×10^{1}	2.95×10^{1}
36c	–	–	–	5.12×10^{-4}	1.21×10^{-4}
N-methyl acetanilide	–	–	–	2.2×10^{-5}	2.2×10^{-7}

[a] Data from references 87 and 107.

the N-protonated forms. As the distortion decreases, the pK_as drop towards those of normal amides as expected, but it is not immediately clear for **36a,b** whether the observed kinetic pK_as of -0.28 and 0.56 respectively refer to the O- or N-protonated forms, or a combination of the two. Recent theoretical calculations predict the thermodynamic site of protonation (gas phase) of **25** X = H and **36a,b** to be on N, while **36c** should protonate on O.[127,133]

6. CONCLUSIONS

In the above text we have summarized the various known reactions for the hydrolyses of amides in H_3O^+, HO^-, and water. Outside the scope of this review are reactions of amides that are promoted by other species such as metal ions, or by circuitous mechanisms where the acyl moiety is transiently transferred to another acceptor such as COO^-, ROH, or RSH, processes which may be more relevant to peptide hydrolyses that are believed to occur in the active sites of enzymes.[131] The details of the reaction mechanisms in acid and base are now well understood in terms of the requirements for catalysis at various stages, the formation of tetrahedral intermediates, and the factors that control the partitioning of these. The detailed level of understanding of these reactions reveals that, in most cases, whether the reactions are acid or base catalyzed, breakdown of the intermediates to products requires that the N be protonated prior to its departure. Factors that diminish the ability of the N to become protonated at the stage of the tetrahedral intermediate, such as diminished basicity or steric hindrance, impair C–N cleavage so in these cases the tetrahedral intermediates reverse to starting materials. This can be detected by observing ^{18}O=C exchange of the starting amides recovered from the hydrolysis medium after incomplete hydrolysis. Of note in the HO^- promoted reactions is the fact that if the basicity of the N in the tetrahedral intermediate is so diminished that it cannot readily accept a proton, then the mechanism for C–N cleavage changes to one where the N departs as an amide, with or without assistance of solvent.

The mechanism for amide hydrolysis in pure water in the absence of acid or base catalysis is less well understood at this time. This is because these reactions are very slow and difficult to follow by conventional kinetic techniques. Most of what is presently known about the water reaction come from the studies on amides that are activated in some way either electronically, or through structural distortion, or from the study of non-activated amides and peptides at elevated temperature. It remains to be seen how relevant these studies are to the reactions of normal amides under less drastic conditions.

7. REFERENCES

1. Bennett, A. J.; Somayaji, V.; Brown, R. S.; Santarsiero, B. D. *J. Am. Chem. Soc.* 1991, **113**, 7563–7571.

2. Pauling, L. *The Nature of the Chemical Bond*; Cornell University Press: Ithaca, NY, 1940.
3. Patai, S., Ed. *The Chemistry of Carboxylic Acids and Esters*; Interscience Publishers: London, 1969.
4. Patai, S., Ed. *The Chemistry of Carboxylic Acids and Esters, Parts 1 and 2, Supplement B*; Interscience Publishers: London, 1979.
5. Zabicky, J., Ed. *The Chemistry of Amides*; Interscience Publishers: London, 1970.
6. Lowry, T. H.; Richardson, K. S. *Mechanism and Theory in Organic Chemistry* 3rd edn., Harper and Row: New York, 1987, pp. 710–723.
7. March, J. *Advanced Organic Chemistry*; 4th edn., Wiley-Interscience: New York, 1992, pp. 330–335, 378–386.
8. Jencks, W. P. *Catalysis in Chemistry and Enzymology*; McGraw-Hill: New York, 1969, pp. 7–242, 463–554.
9. Deslongchamps, P. *Stereoelectronic Effects in Organic Chemistry*; Pergamon Press: Oxford, 1983, pp. 54–100.
10. Williams, A. In *Enzyme Mechanisms*; Page, M. I.; Williams, A. Eds.; The Royal Society of Chemistry: London, 1987, pp. 123–139.
11. Fersht, A. *Enzyme Structure and Mechanism*; 2nd ed., W.H. Freeman and Co.: New York, 1985, pp. 47–97.
12. Bamford, C. H.; Tipper, C. F. H. *Comprehensive Chemical Kinetics*; Elsevier: Amsterdam, 1972, Vol. 10.
13. Johnson, S. L. *Adv. Phys. Org. Chem.* 1967, **5**, 237–330.
14. Brown, R. S.; Bennet, A. J.; Slebocka-Tilk, H. *Acc. Chem. Res.* 1992, **25**, 481–488.
15. Bender, M. L. *J. Am. Chem. Soc.* 1951, **73**, 1626–1629.
16. Bender, M. L. *Chem. Rev.* 1960, **60**, 53–113.
17. Bender, M. L.; Ginger, R. D. *J. Am. Chem. Soc.* 1955, **77**, 348–351.
18. Bender, M. L.; Thomas, R. J. *J. Am. Chem. Soc.* 1961, **83**, 4183–4189.
19. Bender, M. L.; Thomas, R. J. *J. Am. Chem. Soc.* 1961, **83**, 4189–4193.
20. Bender, M. L.; Ginger, R. D.; Kemp, K. C. *J. Am. Chem. Soc.* 1954, **76**, 3350–3351.
21. Bender, M. L.; Ginger, R. D.; Unik, J. P. *J. Am. Chem. Soc.* 1958, **80**, 1044–1048.
22. Bender, M. L.; Matsui, H.; Thomas, R. J.; Tobey, S. W. *J. Am. Chem. Soc.* 1961, **83**, 4193–4196.
23. Schowen, K. B. J. In *Transition States of Biochemical Processes*; Gandour, R.; Schowen, R. L. Eds.; Plenum Press: New York, 1978, pp. 225–283.
24. Schowen, R. L. In *Isotope Effects on Enzyme-Catalyzed Reactions*; Cleland, W. W.; O'Leary, M. H.; Northrop, D. B. Eds.; University Park Press: Baltimore, 1977, pp. 64–99.
25. Venkatasubban, K. S.; Schowen, R. L. *CRC Crit. Rev. Biochem.* 1984, **17**, 1–44.
26. Alvarez, F. J.; Schowen, R. L. In *Isotopes in Organic Chemistry*; Buncel, E.; Lee. C. C. Eds.; Elsevier: Amsterdam, 1987, Vol. 7, pp. 1–60.
27. Kresge, A. J.; More O'Ferrall, R. A.; Powell, M. F. In *Isotopes in Organic Chemistry*; Buncel, E.; Lee. C. C. Eds.; Elsevier: Amsterdam, 1987, Vol. 7, pp. 177–273.
28. Kresge, A. J. *J. Am. Chem. Soc.* 1973, **95**, 3065–3067.

29. Gold, V.; Grist, S. *J. Chem. Soc. Perkin Trans.* 1972, **2**, 89.
30. Kwart, H. *Acc. Chem. Res.* 1982, 401.
31. Isaacs, N. *Physical Organic Chemistry*; 2nd edn., Longman Scientific and Technical and J. Wiley & Sons: New York, 1995, pp. 529–531.
32. O'Connor, C. *Q. Rev. Chem. Soc.*, 1970, **24**, 553–564.
33. Talbot, R. J. E. In *Comprehensive Chemical Kinetics*; Bamford, C.H.; Tipper, C. F. H., Eds.; Elsevier: Amsterdam, 1972, Vol. 10, pp. 257–293.
34. Arnett, E. M. *Prog. Phys. Org. Chem.* 1963, **1**, 223–403.
35. Guthrie, J. P. *J. Am. Chem. Soc.* 1974, **96**, 3608–3615.
36. Caplow, M.; Jencks W. P. *Biochemistry*, 1962, **1**, 883–893.
37. Yates, K.; Riordan, J. C. *Can. J. Chem.* 1965, **43**, 2328–2335.
38. Yates, K.; Stevens, J. B. *Can. J. Chem.* 1965, **43**, 529–537.
39. Modro, T. A.; Yates, K.; Beaufays, F. *Can. J. Chem.* 1977, **55**, 3050–3057.
40. Perrin, C. L. *Acc. Chem. Res.* 1989, **22**, 268–275.
41. Williams, A. *J. Am. Chem. Soc.* 1976, **98**, 5645–5651.
42. Moodie, R. B.; Wale, P. D.; Whaite, T. J. *J. Chem. Soc.* 1963, 4273–4274.
43. Cox, R. A.; Yates, K. *Can. J. Chem.* 1981, **59**, 2853–2863.
44. Lane, C. A. *J. Am. Chem. Soc.* 1964, **86**, 2521–2523.
45. Lane, C. A.; Cheung, M. F.; Dorsey, G. F. *J. Am. Chem. Soc.* 1968, **90**, 6492–6494.
46. Yates, K.; McClelland, R. A. *J. Am. Chem. Soc.* 1967, **89**, 2686–2692.
47. Cox, R. A. *Can. J. Chem.* 1998, **76**, 649–656.
48. Bunton, C. A.; Farber, S. J.; Milbank, A. J. G.; O'Connor, C. J.; Turney, T. A. *J. Chem. Soc. Perkin Trans.* 1972, **2**, 1869–1875.
49. Smith, C. R.; Yates, K. *J. Am. Chem. Soc.* 1972, **94**, 8811–8817.
50. McClelland, R. A. *J. Am. Chem. Soc.* 1975, **97**, 5281–5282.
51. Bennet, A. J.; Slebocka-Tilk, H.; Brown, R. S.; Guthrie, J. P.; Jodhan, A. *J. Am. Chem. Soc.* 1990, **112**, 8497–8506.
52. Bennet, A. J.; Slebocka-Tilk, H.; Brown, R. S. *J. Am. Chem. Soc.* 1992, **114**, 3088–3092.
53. Jencks, W. P.; Regenstein, J. In *CRC Handbook of Biochemistry*; Sober, H. A., Ed.; 1st edn., The Chemical Rubber Co.: Cleveland, 1968, pp. J150–J189.
54. Chiang, Y.; Whipple, E. B. *J. Am. Chem. Soc.* 1963, **85**, 2763–2767.
55. Young, J. K.; Pazhanisamy, S.; Schowen, R. L. *J. Org. Chem.* 1984, **49**, 4148–4152.
56. Eriksson, S. O. *Acta Chem. Scand.* 1968, **22**, 892–906.
57. DeWolfe, R. H.; Newcomb, R. C. *J. Org. Chem.* 1971, **36**, 3870–3878.
58. Pollack, R. M.; Dumsha, T. C. *J. Am. Chem. Soc.* 1973, **95**, 4463–4465.
59. Biechler, S. S.; Taft, Jr. R. W. *J. Am. Chem. Soc.* 1957, **79**, 4927–4935.
60. Menger, F. M.; Donohue, J. A. *J. Am. Chem. Soc.* 1973, **95**, 432–437.
61. Cipiciani, A.; Linda, P.; Savelli, G. *J. Heterocycl. Chem.* 1979, **16**, 673–675.
62. Cipiciani, A.; Linda, P.; Savelli, G. *J. Heterocycl. Chem.* 1979, **16**, 677–678.
63. Brown, R. S.; Bennet, A. J.; Slebocka-Tilk, H.; Jodhan, A. J. *J. Am. Chem. Soc.* 1992, **114**, 3092–3098.

64. Bunton, C. A.; Lewis, T. A.; Llewellyn, D. R. *Chem. Ind. (London)* 1954, 1154–1155.
65. Bunton, C. A.; Nayak, B.; O'Connor, C. *J. Org. Chem.* 1968, **33**, 572–575.
66. Slebocka-Tilk, H.; Bennet, A. J.; Keillor, J. W.; Brown, R. S.; Guthrie, J. P.; Jodhan, A. *J. Am. Chem. Soc.* 1990, **112**, 8507–8514.
67. Slebocka-Tilk, H.; Bennet, A. J.; Hogg, H. J.; Brown, R. S. *J. Am. Chem. Soc.* 1991, **113**, 1288–1294 and references therein.
68. de Roo, M.; Bruylants, A. *Bull. Soc. Chim. Belg.* 1954, **63**, 140–157.
69. Calvet, E. *J. Chim. Phys.* 1933, **30**, 140–166.
70. Kellogg, B. A.; Brown, R. S.; McDonald, R. S. *J. Org. Chem.* 1994, **59**, 4652–4658.
71. Kellogg, B. A.; Tse, J. E.; Brown, R. S. *J. Am. Chem. Soc.* 1995, **117**, 1731–1735.
72. Liu, B-Y.; Brown, R. S. *J Org. Chem.* 1993, **58**, 732–734.
73. Guthrie, J. P. *J. Am. Chem. Soc.* 1973, **95**, 6999–7003.
74. Kirby, A. J. In *Comprehensive Chemical Kinetics*; Bamford, C. H.; Tipper, C. F. H. Eds.; Elsevier: Amsterdam, 1972, Vol. 10, pp. 57–207.
75. Bunton, C. A.; Fuller, N. Perry, S. G.; Shiner, Jr. V. *J. Chem. Ind. (London)* 1960, 1130–1131
76. Jencks, W. P.; Carriuolo, J. *J. Am. Chem. Soc.* 1961, **83**, 1743–1750.
77. Kahne, D.; Still, W. C. *J. Am. Chem. Soc.* 1988, **110**, 7529–7534.
78. Radzicka, A.; Wolfenden, R. *J. Am. Chem. Soc.* 1996, **118**, 6105–6109.
79. Bryant, R. A. R.; Hansen, D. E. *J. Am. Chem. Soc.* 1996, **118**, 5498–5499.
80. Hine, J.; King, R. S.-M.; Midden, R.; Sinha, A. *J. Org. Chem.* 1981, **46**, 3186–3189.
81. Slebocka-Tilk, H; Brown, R. S. *J. Org. Chem.* 1987, **52**, 805–808.
82. Fife, T. H. *Acc. Chem. Res.* 1993, **26**, 325–331.
83. Jencks, W. P.; Carriuolo, J. *J. Biol. Chem.* 1959, 1272–1279.
84. Fife, T. H.; Natarajan, R.; Werner, M. H. *J. Org. Chem.* 1987, **52**, 740–746.
85. Komiyama, M.; Bender, M. L. *Bioorg. Chem.* 1978, **7**, 133–139.
86. Komiyama, M.; Bender, M. L. *Bioorg. Chem.* 1979, **8**, 141–145.
87. Somayaji, V.; Brown, R. S. *J. Org. Chem.* 1986, **51**, 2676–2686.
88. Blackburn, G. M.; Skaife, C. J.; Kay, I. T. *J. Chem. Res., Synop.* 1980, 294–295.
89. Cipiciani, A.; Linda, P.; Savelli, G. *J. Heterocycl. Chem.* 1978, **15**, 1541–1542.
90. Cipiciani, A.; Savelli, G.; Bunton, C. A. *J. Heterocycl. Chem.* 1984, **21**, 975–976.
91. Patterson, J. F.; Huskey, W. P.; Hogg, J. L. *J. Org. Chem.* 1980, **45**, 4675–4679.
92. Chiang, Y.; Kresge, A. J.; Powell, M. F.; Wells, J. A. *J. Chem. Soc. Chem. Commun.* 1995, 1587–1588.
93. Cipiciani, A.; Linda, P.; Savelli, G. *J. Heterocycl. Chem.* 1978, **15**, 1541–1542.
94. Guthrie, J. P *J. Am. Chem. Soc.* 1974, **96**, 3608–3615
95. Bell, R. P. *Adv. Phys. Chem.* 1966, **4**, 1.
96. Patai, S. In *The Chemistry of the Carbonyl Group*; Zabicky, J. Ed.; Wiley Interscience: London, Vol. II; 1970, p. 1.
97. Slebocka-Tilk, H.; Rescorla, C. G.; Shirin, S.; Bennet, A. J.; Brown, R. S. *J. Am. Chem. Soc.* 1997, **119**, 10969–10975.

98. Hegazi, M.; Mata-Segreda, J. F.; Schowen, R. L. *J. Org. Chem.* 1980, **45**, 307.
99. Eigen, M. *Angew. Chem. Int. Ed. Engl.* 1964, **3**, 1–72.
100. Jencks, W. P. *Acc. Chem. Res.* 1980, **13**, 161–169.
101. Guthrie, J. P. *J. Am. Chem. Soc.* 1991, **113**, 3941–3949.
102. Stefanidis, D.; Jencks, W. P. *J. Am. Chem. Soc.* 1993, **115**, 6045–6050.
103. Stefanidis, D.; Cho, S.; Dhe-Paganon, S.; Jencks, W. P. *J. Am. Chem. Soc.* 1993, **115**, 1650–1656.
104. Ba-Saif, S.; Luthra, A. K.; Williams, A. *J. Am. Chem. Soc.* 1987, **109**, 6362–6368.
105. Ba-Saif, S.; Luthra, A. K.; Williams, A. *J. Am. Chem. Soc.* 1989, **111**, 2647–2652.
106. Hengge, A. C.; Hess, R. A. *J. Am. Chem. Soc.* 1994, **116**, 11256–11263.
107. Wang, Q.-P.; Bennet, A. J.; Brown, R. S.; Santarsiero, B. D. *J. Am. Chem. Soc.* 1991, **113**, 5757–5765.
108. Kirby, A. J.; Komarov, I. V.; Wothers, P. D.; Feeder, N. *Angew. Chem. Int. Ed. Engl.* 1998, **37**, 785–786.
109. Haldane, J. B. S. *Enzymes*; Longmans Green and Co.: London, 1930.
110. Lumry, R. In *The Enzymes*; Boyer, P. D., Ed.; Academic Press: New York, 1959, Vol. 1, pp. 157–258.
111. Jencks, W. P. *Adv. Enzymol. Relat. Areas Mol. Biol.* 1975, **43**, 219–410.
112. Jencks, W. P. *Adv. Enzymol. Relat. Areas Mol. Biol.* 1980, **51**, 75–106.
113. Fersht, A. In *Enzymatic Structure and Mechanism*; 2nd edn., W. H. Freeman and Co.: San Fransisco, 1985; pp. 311–346.
114. Wolfenden, R. *Acc. Chem. Res.* 1972, **5**, 10–18.
115. Page, M. I. In *Enzyme Mechanisms*; Page, M. I.; Williams, A. Eds., Royal Society of Chemistry: London, 1987, pp. 1–13.
116. Levitt, M. In *peptides, polypeptides and Proteins*; Blout, E. R.; Bovey, F. A.; Goodman, M.; Lotan, N. Eds.; Wiley: New York, 1974, pp. 99–113.
117. Pearl, L. In *Aspartate Proteinases and Their Inhibitors*; Kostka, V. Ed.; Walter de Gruyer: Berlin, 1985, pp. 189–195.
118. Dunitz, J. D.; Winkler, F. K. *Acta Crystallogr.* 1975, **B31**, 251.
119. Chakrabarti, P.; Dunitz, J. D. *Helv. Chim. Acta*, 1982, **65**, 1555–1562.
120. Hall, H. K. Jr.; El-Shekeil, A. *Chem. Rev.* 1983, **83**, 549.
121. Greenberg, A. In *Structure and Reactivity*; "Molecular Structure and Energetics," Liebman, J. F.; Greenberg, A., Eds.; VCH Publishing: New York, 1988, Vol. 7; pp. 139–178.
122. Lease, T. G.; Shea, K. J. In *Advances in Theoretically Interesting Molecules*; Thummel, R. P. Ed.; JAI Press: Greenwich, CT, 1992, Vol. 2, pp. 79–112.
123. Greenberg, A.; Venanzi, C. A. *J. Am. Chem. Soc.* 1993, **115**, 6951–6957.
124. Greenberg, A.; Moore, D. T.; Dubois, T. D. *J. Am. Chem. Soc.* 1996, **118**, 8658–8668 and references therein.
125. Pracejus, H. von. *Chem. Ber.* 1959, *92*, 988–998.
126. Pracejus, H. von; Kehlen, M.; Matschiner, H. *Tetrahedron*, 1965, **21**, 2257–2270.
127. Kostyanovskii, R. G.; Mikhlina, E. E.; Levoeva, E. I.; Yakhontov, L. N. *Org. Mass Spectrom.* 1970, **3**, 1023–1029.

128. Bennet, A. J.; Wang, Q.-P.; Slebocka-Tilk, H.; Somayaji, V.; Brown, R. S.; Santarsiero, B. D. *J. Am. Chem. Soc.* 1990, **112**, 6358–6386.

129. Wiberg, K. B.; Breneman, C. M. *J. Am. Chem. Soc.* 1992, **114**, 831–840.

130. Werstiuk, N. H.; Brown, R. S.; Wang, Q.-P. *Can. J. Chem.* 1996, **75**, 524–532.

131. Bennet, A. J.; Brown, R. S. In *Comprehensive Biological Catalysis*; "Physical Organic Chemistry of Acyl Transfer Reactions," Sinnott, M. L. Ed.; Academic Press Inc.: London, 1998, Vol. 2, pp. 293–326.

CHAPTER 5

THE THERMOCHEMISTRY OF AMIDES

JOEL F. LIEBMAN and HUSSEIN Y. AFEEFY
Department of Chemistry and Biochemistry, University of Maryland, Baltimore County

SUZANNE W. SLAYDEN
Department of Chemistry, George Mason University

1. SOME PATTERNS AND PREDICTIVE REGULARITIES OF THE THERMOCHEMISTRY OF AMIDES

We commence this chapter with definitions of key terms that we will use in the hope that it will help the reader to better understand the scope and goals of our study.

1.1. What Do We Mean by Amides?

The common, quite universal definition of an amide states that these compounds are derivatives of functionalized oxyacids, i.e. species of the type $R_xEO_y(OH)_z$, where R is hydrogen or an arbitrary organic group, E is a nonmetallic element, and $x \geqslant 0$, $y \geqslant 0$, $z \geqslant 1$. Amides have one or more of the –OH groups replaced by $-NH_2$ and/or its organic derivatives $-NRR'$. Although this definition includes species such as sulfonamides and phosphoramidic acids, we choose to restrict our scope to carboxamides, i.e., where E = C. Therefore the simplest definition of an amide is that it is an organic compound with a –C(O)–N$<$ substructure, in which a carbonyl group, –C(O)–, is directly bonded to an amino group, –N$<$, by a formal C–N σ bond.

Accordingly, the species with the generic structure $RCONH_2$, $RCONHR'$ and $RCONR'R''$ are all identified as amides and will be collectively referred to as (simple) amides and individually as primary, secondary and tertiary amides, respectively. Amides in which the R and R′ groups are joined together to form a ring are commonly called lactams. Imides, compounds with the substructure

The Amide Linkage: Selected Structural Aspects in Chemistry, Biochemistry, and Materials Science,
Edited by Arthur Greenberg, Curt M. Breneman, and Joel F. Liebman
ISBN 0-471-35893-2

–C(O)–N(–)–C(O)–, and species that we may call "linked amides" containing –C(O)–N(–)–C(O)–N< groups will also be discussed in this chapter. However, space does not allow us to consider many classes of compounds which are all formally amides: –C(O)–N(–)N< (hydrazides), >N–C(O)–N<, (ureas), and –O–C(O)–N< (carbamates). Thus, somewhat artificially and arbitrarily we have limited the scope to those compounds we believe will be of greatest relevance to the majority of readers of the current and accompanying volume.

1.2. What Do We mean by Thermochemistry?

As has been the approach in most of the authors′ other book chapters and reviews[1–4] on organic thermochemistry, the current chapter will be limited almost exclusively to the enthalpy of formation. We accede here to proper thermochemical practice and use the notation $\Delta_f H_m{}^o$, instead of the commonly used formalism ΔH_f and the erstwhile more correct notation $\Delta H_f{}^o$. Other thermochemical properties such as Gibbs energy, entropy, heat capacity, excess enthalpy and heat of solution will not be discussed. Additionally (following thermochemical convention), the temperature and pressure are tacitly assumed to be 25 °C (298 K) and 1 atmosphere (taken as either 101, 325, or 100,000 Pa), respectively. Admittedly, the majority of experiments are not done exactly under these conditions. However, we will ignore the effects of small temperature and pressure differences from them. Finally, the energy units have been chosen to be kJ mol^{-1} where 4.184 kJ ≡ 1 kcal.

We would prefer that the measured enthalpy-of-formation data to refer to the gas phase where there are no complicating effects of molecular association or aggregation as in the condensed phases. While most organic compounds that we have studied are naturally liquids or solids under the idealized thermochemical conditions, gas phase data may be obtained by using enthalpies of vaporization and of sublimation to "correct" the condensed phase data (Eqs. 5.1 and 5.2).

$$\Delta_v H^o \equiv \Delta_f H_m{}^o(g) - \Delta_f H_m{}^o(lq) \qquad (5.1)$$

$$\Delta_s H^o \equiv \Delta_f H_m{}^o(g) - \Delta_f H_m{}^o(s) \qquad (5.2)$$

This tactic is generally thwarted in the case of amides because disconcertingly few of these necessary phase change enthalpies are to be found in the archival literature of experimental measurements. Parameterized estimation techniques are likewise thwarted by the absence of relevant data. In addition, most amides of interest are solids and the estimation techniques have always been less reliable for deriving thermochemical data for this condensed phase than for the structurally, and otherwise chemically, similar liquid phase.

1.3. Patterns and Predictive Regularities

Patterns and predictive regularities are rules—or should we say procedures—by which one can predict the enthalpy of formation of a compound of interest from the knowledge of the enthalpy of formation of other compounds. The enunciated lack of data, as above, legitimizes our pessimism about finding simultaneously reliable and elementary procedures. Nonetheless, for our past studies, it was very often the absence of experimental measurements that drove us to look for patterns and predictive regularities. It will be found that these amide patterns are coarser and predictions are less reliable than what we have used in our studies before.*

It is not obvious to what extent the resulting discrepancies between "theory" and experiment are caused by the poor quality of thermochemists' samples. For example, we expect amides to be "wetter" than simple aldehydes and ketones, i.e. carbonyl compounds in which the >N– group is replaced by the isoelectronic and isosteric >CH– group. While water, as a minor impurity, generally has little qualitative or even quantitative effect on measurements of enthalpies of reaction and the derived enthalpy of formation, its presence invalidates the corresponding studies of enthalpies of combustion and their derived enthalpies of formation.

Likewise, our concepts are undoubtedly poorer as well because intermolecular interactions within solids may be structurally anisotropic or unique as compared to interactions in liquids. And, almost by definition, these interactions are absent in gases. Unlike some other groups and species of interest, for the amides with their polar functional group we are not able to view condensed phase intermolecular forces *merely* as "complications" and "nuisances". It is the authors' hope that this chapter encourages numerous measurements (and re-measurements) of enthalpies of combustion and of reaction. We also welcome new experimental studies of phase change enthalpies that will allow us to finally and reliably explore the gas phase for which we are most comfortable in understanding, explaining, and predicting. We can then return to the condensed phase that is generally of greater interest and importance to the majority of the chemical community.

1.4. Sources of Data

All enthalpies of formation discussed here are taken from either Ref. 5 or Ref. 6, and phase change enthalpies are taken from Ref. 7. In most cases, there is only one reported measured value for a given compound—indeed, for most compounds of interest, there are no values. If the enthalpy of formation is included in the evaluated archive, Ref. 5, we generally use it here. Many of the more-recently determined values are obtained from Ref. 6. Very often the error bars for the two or more measurements overlap. We accept the error bars as suggested in the archival sources and usually present them when citing a value of

*For one of our very few thermochemical reviews in which solid phase data and associated analysis was dominant (as opposed to that of the gas with a "little" for the liquid), see Afeefy, H.Y.; Liebman, J.F. In *Computational Thermochemistry* Frurip, D.J.; Irikura, K.K.; eds; ACS Symposium Series: Washington, DC, 1998, Vol. 677; p. 94.

interest. However, we do not give error bars when we have summed the values for the enthalpy of formation and of phase change. Although the latter values lack these uncertainties and we are quite confident that they are small, nonetheless, we do not wish to sum thermodynamic quantities measured at different temperatures, whether they be "298 K", or the melting point, or those at which the various condensed phase transitions occur.

2. ENTHALPY OF FORMATION DIFFERENCES BETWEEN AMIDES AND THE CORRESPONDING CARBOXYLIC ACIDS

In an earlier study[1], we showed a nearly constant $110\,\mathrm{kJ\,mol^{-1}}$ difference between the solid phase enthalpies of formation of a carboxylic hydrazide, $RCONHNH_2$ and of the corresponding amide, $RCONH_2$. This value is very much the same for the likewise related thiohydrazides and thioamides, $RCSNHNH_2$ and $RCSNH_2$. The utility of the relationship is severely limited due to the paucity of such compounds with measured enthalpies of formation. Since the chemistry of amides is structurally, synthetically and therefore conceptually related to that of the carboxylic acids—a large class of organic compounds of considerable interest and importance—we will explore their thermochemical connection by making particular use of the formal chemical reaction for amide hydrolysis (Eq. 5.3).

$$\mathrm{RCONH_2(g) + H_2O(lq) \rightarrow RCOOH(g) + NH_3(g)} \tag{5.3}$$

It has been firmly established that enthalpies of formation for a homologous series of compounds show a linear dependence on the number of carbon atoms in the hydrocarbyl group[1–4,6]. Equation 5.4 expresses the general relationship, when all members of the series are in the same physical state, as a function of the total number of carbon atoms in the compound, n_c.

$$\Delta_f H_m{}^{o} = (\alpha \cdot n_c) + \beta \tag{5.4}$$

In general, the function is unreliable[9] when $n_c = 1$ and ideally, the constants α and β are derived from several members of the homologous series for which $n_c > 4$. Only occasionally, however, have enough experimental data been available to fulfill the latter condition. Nonetheless, even restricted data sets which include the lower homologs may yield useful information. Equations 5.5 and 5.6 were produced by applying the method of weighted least squares to the measured enthalpy of formation data[5] for *n*-carboxylic acids (liquid and gas phases) and n-amides (solid and gas phases), respectively.

$$\Delta_f H_m{}^o(\mathrm{RCO_2H, lq}) \pm 1.2 = (-25.4 \pm 0.2 \cdot n_c) + (-432.5 \pm 1.2) \quad (n_c = \mathrm{C_5}\text{–}\mathrm{C_9}\ r^2 > .9999) \tag{5.5a}$$

$$\Delta_f H_m{}^o(\mathrm{RCO_2H, g}) \pm 0.8 = (-21.2 \pm 0.6 \cdot n_c) + (-385.6 \pm 4.3) \quad (n_c = \mathrm{C_5}\text{–}\mathrm{C_9}\ r^2 > .9999) \tag{5.5b}$$

$$\Delta_f H_m{}^o(\mathrm{RCONH_2, s}) \pm 4.9 = (-27.0 \pm 0.7 \cdot n_c) + (-258.8 \pm 3.6) \quad (n_c = \mathrm{C_2}\text{–}\mathrm{C_4}, \mathrm{C_6}, \mathrm{C_8}\ r^2 > .9999) \tag{5.6a}$$

$$\Delta_f H_m{}^o(\mathrm{RCONH_2, g}) \pm 0.9 = (-21.1 \pm 0.3 \cdot n_c) + (-195.9 \pm 1.0) \quad (n_c = \mathrm{C_2}\text{–}\mathrm{C_4}, \mathrm{C_6}, \mathrm{C_8}\ r^2 > .9999) \tag{5.6b}$$

Although the enthalpy of combustion of pentanamide has been measured,[10] its enthalpy of formation is not included here; the value is most certainly inaccurate because the corresponding point on a plot of either Eq. 5.6a or 5.6b is an obvious outlier from the otherwise straight line. The calculated value of $-301\ \mathrm{kJ\,mol^{-1}}$ is $11\ \mathrm{kJ\,mol^{-1}}$ more negative than the experimental value.

Because amides and carboxylic acids are chemically related by Eq. 5.3, we ask what relationship exists between their enthalpies of formation. Since the enthalpies of formation of all homologous series are linearly related to the number of carbon atoms in their structures, the enthalpies of formation of any two series ($\Delta_f H_m{}^o$ and $\Delta_f H_m{}^{o\prime}$) are linearly related to each other with a slope equal to α/α' and y-intercept equal to $(\alpha'\beta - \alpha\beta')/\alpha'$. The relationship can be simplified to Eq. 5.7 which defines the simple, positive difference quantity $\delta_{(5.7)}$ (R,$*$), where $*$ denotes the phase of interest. This quantity will be constant and equal to the difference between the y-intercepts when the slopes of Eq. 5.4 are identical for two series.

$$\delta_{(5.7)}(\mathrm{R}, *) = \Delta_f H_m{}^o(\mathrm{RCONH_2}, *) - \Delta_f H_m{}^o(\mathrm{RCOOH}, *) \tag{5.7}$$

For example, the slopes of Eqs. 5.5b and 5.6b are the same within the uncertainties and so $\delta_{(5.7)}$ (R, g) should be constant for gaseous n-amides and n-carboxylic acids. Equivalently, because the enthalpies of formation of NH_3(g) and H_2O(l) are constant for all hydrolyses, the enthalpies of the formal gaseous hydrolysis reactions (Eq. 5.3) should be constant* for the n-carboxylic acid and n-amide series.

*If the slopes for two homologous series are significantly different, the enthalpies of the formal reaction relating the two series will not be constant but instead will exhibit a regular increase or decrease. This was demonstrated for the enthalpy of hydrogenation of alkenes to alkanes in Slayden, S.W.; Liebman, J.F. In *The chemistry of the double-bonded functional groups*; Supplement A3; Patai, S. Ed.; Wiley: Chichester, 1997.

The gas phase difference quantities are listed in Table 5.1 for the *n*-carboxylic acid and *n*-amide series. The pentanamide outlier produces a difference quantity which is significantly different from the others. The difference quantity for formic acid and formamide (R = H, i.e. $n_c = 1$) is consistent with those of the higher members of the series because, although each deviates from the linear relationship established by their respective series, their individual deviations from linearity are similar. The average of the $n_c = 2$–4, 6, and 8 difference quantities, 193.0 ± 3.5 kJ mol^{-1}, is indeed equal within the uncertainty intervals to the difference between the y-intercepts of Eqs. 5.5b and 5.6b.

Many of the carboxylic acids of interest are liquids while most amides are solids. However, the logical demand of a common phase for amide and acid is relatively easily accommodated by addition of the fusion enthalpy to the liquid phase enthalpy of formation of the latter.* Table 5.1 presents $\delta_{(5.7)}(R, s)$ for the previously-discussed *R* groups and others, which unfortunately are not members of homologous series for which there are thermochemical data. The special symbol, §, conveys that the above solid/liquid correction was explicitly made.

Again, with the exception of pentanamide/pentanoic acid, the difference quantities $\delta_{(5.7)}(n - R, s)$ are relatively constant (181.6 ± 2.3 kJ mol^{-1}). That this is not unexpected is shown by performing a regression analysis on the derived solid enthalpies of formation for the *n*-carboxylic acids. The slope, -27.0, is identical to the slope from Eq. 5.6a, and the intercept equals -440. Both adamantane-1-carboxylic acid/amide and *t*-butyl carboxylic acid/amide have discrepant values (in opposite directions) for $\delta_{(5.7)}(R, s)$ relative to that of other hydrocarbon acids and related amides. While there is no apparent reason for the adamantane discrepancy other than that it is tertiary and the other compounds are not, the explanation for the *t*-butyl case is clear: for all other functionalized *n*-butyl/*t*-butyl isomer pairs of compounds, the branched *t*-butyl isomer is much the more stable. Were the enthalpy of formation of solid *t*-butyl carboxylic acid (2,2-dimethylpropanoic acid) about 20 kJ mol^{-1} more negative than pentanoic acid (a typical difference), $\delta_{(5.7)}(t\text{-Bu}, s)$ would be substantially larger.**

For other amide/acid pairs in Table 5.1, $\delta_{(5.7)}(R, s)$ is generally the same or smaller than that of the (*n-R*, *s*) series. With respect to the electronegativity of attached groups, there seem to be two opposing trends. For the series R=CH_3CO, CH_2ClCO, CCl_3CO, the difference quantity decreases with increasing substituent electronegativity, while for R=H_2NCOCO, HO_2CCO and for

*Actually, we chose to sum all of the condensed phase enthalpy changes to correct for multiple crystal forms and "plastic" crystals, and now we additionally acknowledge that the fusion enthalpy, i.e. the phase transition at the melting point, is usually the major part of this additional "correction". Indeed, we often refer to this sum as the "fusion enthalpy".

**The same anomaly is seen in the gas phase enthalpies of formation of pentanoic acid (-491.9 ± 3.0 kJ mol^{-1}) and 2,2-dimethylpropanoic acid (-491.3 ± 6.6 kJ mol^{-1}). The liquid phase enthalpies are more consistent with those of other functionalized *n*-butyl/*t*-butyl pairs (-559.4 ± 0.7 and -564.5 ± 5.9 kJ mol^{-1}) but the error bars are uncomfortably large in most of the cases.

TABLE 5.1. Enthalpies of Formation[a] and Difference Quantities for Carboxylic Acids and Amides ($kJ\,mol^{-1}$)

RCO	$-\Delta_f H_m^o$ (RCOOH)	$-\Delta_f H_m^o$ ($RCONH_2$)	$\delta_{(5.7)}$(R)
HCO	378.7 ± 0.6 (g)	186 (g)[b]	192.7 (g)
CH_3CO	432.8 ± 1.5 (g)	238.3 ± 0.8 (g)	194.5 ± 2.1 (g)
C_2H_5CO	453.5 ± 0.5 (g)	258.9 ± 0.6 (g)	194.6 ± 0.7 (g)
n-C_3H_7CO	475.8 ± 4.1 (g)	279.1 ± 1.4[b] (g)	196.7 ± 4.3 (g)
n-C_4H_9CO	491.9 ± 3.0 (g)	290.2 ± 1.2 (g)	201.7 ± 4.2 (g)
n-$C_5H_{11}CO$	511.9 ± 2.3 (g)	324.2 ± 1.8 (g)	187.7 ± 2.9 (g)
n-$C_7H_{15}CO$	554.3 ± 1.5 (g)	362.7 ± 3.1 (g)	191.6 ± 3.4 (g)
CH_3CO	496.2§ (s)	317.0 ± 0.8 (s)	179.2 (s)
C_2H_5CO	521.5§ (s)	338.2 ± 0.5 (s)	183.3 (s)
n-C_3H_7CO	545.0§ (s)	364.4 ± 0.7 (s)	180.6 (s)
n-C_4H_9CO	574.2§ (s)	379.5 ± 1.1 (s)	194.7 (s)
n-$C_7H_{15}CO$	657.3§ (s)	473.2 ± 0.9 (s)	184.1 (s)
t-C_4H_9CO	564.6 ± 5.9 (s)	399.7 ± 1.3 (s)	167.9 ± 6.0 (s)
1-$C_{10}H_{15}CO$[c]	643.1 ± 3.6 (s)	427.0 ± 2.4 (s)	216.1 (s)
CCl_3CO	503.3 ± 8.4 (s)	358.0 ± 8.4 (s)	145.3 ± 11.9 (s)
CH_2ClCO	510.8 ± 8.2 (s)	338.4 ± 8.4 (s)	172.4 ± 11.7 (s)
HOOCCO	828.9 ± 0.4 (s)	661.2 ± 0.9 (s)	167.2 ± 1.0 (s)
NH_2COCO	661.2 ± 0.9 (s)	504.4 ± 0.5 (s)	156.8 ± 1.0 (s)
$-C(=O)-C(=O)-$	828.9 ± 0.4 (s)	504.4 ± 0.5 (s)	162.3 ± 0.6 (s)[d]
$-C(=O)-CH_2-C(=O)-$	891.0 ± 0.3 (s)	546.1 ± 1.5 (s)	172.5 ± 1.5 (s)[d]
$-C(=O)-CH_2-CH_2-C(=O)-$	940.5 ± 0.4 (s)	581.2 ± 2.1 (s)	179.7 ± 2.1 (s)[d]
O_2N–(furan-2,5-diyl)–$C(=O)-$	517.1 ± 0.8 (s)	322.9 ± 0.4 (s)	194.2 ± 0.9 (s)
C_6H_5CO	384.8 ± 0.5 (s)	202.1 ± 0.6 (s)	182.7 ± 0.8 (s)
$-C(=O)-C_6H_4-C(=O)-$ (para)	816.1 ± 0.7 (s)	433.1 ± 1.2 (s)	191.5 ± 1.4 (s)[d]
$-C(=O)-C_6H_4-C(=O)-$ (meta)	803.0 ± 0.8 (s)	436.9 ± 1.0 (s)	183.1 ± 1.3 (s)[d]
L-$HOOCCH(NH_2)CH_2CO$[e]	973.3 ± 0.8 (s)	789.4 ± 0.8 (s)	183.9 ± 1.1 (s)
L-$HOOCCH(NH_2)(CH_2)_2CO$[e]	1009.6 ± 0.8 (s)	826.4 ± 0.7 (s)	183.2 ± 1.1 (s)

[a]Data from Ref. 3.
[b]Data from Ref. 4.
[c]These compounds are adamantane-1-carboxylic acid and the corresponding amide.
[d]The difference quantity was halved to account for two functional groups in the compound.
[e]There are seemingly no enthalpy of formation data for the isomeric α-amide, nor for the amino acid diamide.

R=CO$(CH_2)_2$CO, $COCH_2CO$, COCO, the difference quantities increase. No simple explanation of this difference in trend is apparent.

3. AMIDE ISOMERS AND THE EFFECT OF N-SUBSTITUTION

There is very little data available to assess the relative stabilities of amide isomers. We do, however, have enthalpies of formation of three C_4H_9-substituted primary amides: pentanamide (the experimental values, -379.5 ± 1.0 (s) and -290.2 ± 1.2 (g) kJ mol^{-1}, are inaccurate and the values -393.8 ± 4.9 (s) and -301.0 ± 0.9 (g) kJ mol^{-1} are derived from Eqs. 5.6(a,b), 3-methylbutanamide (-390.0 ± 1.7 (s) kJ mol^{-1}) and 2,2-dimethylpropanamide (-399.7 ± 1.3 (s) and $-313.1 + 0.9$ (g) kJ mol^{-1}). The stability order of the *n*-butyl and isobutyl amides is equivocal, but the *t*-butyl amide is clearly the most stable in either phase. The enthalpies of formation of butanamide ($-364.4 + 0.7$ (s) and $-279.1 + 1.4$ (g) kJ mol^{-1}) and 2-methylpropanamide ($-368.6 + 0.9$ (s) and $-282.6 + 0.9$ (g) kJ mol^{-1}) show what is probably a typical tertiary > secondary substituent stability order.

We would like to explore the consequences of structural isomerism in amides by assessing the relative stabilities of primary and secondary amides of the types $CH_3(CH_2)_xCONH_2/CH_3(CH_2)_{x-1}CONHCH_3$, RCONHR′/R′CONHR, and others. We are completely stymied by the lack of data for a given pair, especially for the primary amides. A comparison can be drawn between primary and tertiary amide isomers – 2-methylpropanamide ($-282.6 + 0.9$ (g) kJ mol^{-1}) and *N,N*-dimethylacetamide (-254.5 (g) kJ mol^{-1}). The 30 kJ mol^{-1} difference is quite large, which may indicate steric hindrance to resonance due to branching at trigonal nitrogen.

The three amides $RCONH_2$, RCONHR′, and $RCONR'_2$ form a regular series and so, in principle, there is optimism for finding some pattern in their energetics. In practice, data on the enthalpy of formation are available only for the three species with some given R and R′. The first example is that of $R = R' = CH_3$ for which the gas phase enthalpies of formation are -238.3 ± 0.8, -248.0 ± 5.5 and -254.5 kJ mol^{-1}. We chose this phase because there are seemingly no enthalpy of formation data for dimethylacetamide in the solid phase. The regularity of the enthalpy difference from primary to secondary to tertiary amide is reassuring. With respect to the simple acyclic species, the formal transmethylation amide reaction

$$2CH_3CONHCH_3 \rightarrow CH_3CON(CH_3)_2 + CH_3CONH_2 \tag{5.8}$$

is endothermic by ca. 3 kJ mol^{-1} while the related formal transmethylation amine reaction

$$2CH_3NHCH_3 \rightarrow CH_3N(CH_3)_2 + CH_3NH_2 \tag{5.9}$$

is exothermic by ca. 10 kJ mol^{-1}.

Although there are no complete sets of data in a single phase for any other R, R′ combination, we observe that for primary and tertiary pairs where $R' = CH_3$, the primary amide has the more negative enthalpy of formation. In the gas phase, the differences are: for $R = CH_3$, 5.7 kJ mol^{-1}; for $R = CH_2CH_3$ and $CH_2CH_2CH_3$, ca. 8.5 kJ mol^{-1}; and for $R = (CH_3)_3C$, 27 kJ mol^{-1}. The condensed phase differences are larger: for R = H, 14.6 kJ mol^{-1} (l); for $R = CH_2CH_2CH_3$, 20.8 kJ mol^{-1} (l); and for $R = (CH_3)_3C$, 58 kJ mol^{-1} (s). The gas phase data may again reflect steric hindrance to nitrogen resonance with the carbonyl group, while the condensed phase data reflects hydrogen-bond interactions possible only with the primary amides.

As a further probe into this conclusion, we introduce some parallel formal reactions. Consider first the non-nitrogen analogues of Eqs. 5.8 and 5.9 wherein NH_x (x = 0, 1, 2) is replaced by the isoelectronic, and plausibly isosteric, CH_{x+1}. The first set of reactions replaces the N by CH, NH by CH_2, and NH_2 by CH_3, so we have Eqs. 5.10 and 5.11 respectively.

$$2CH_3COCH_2CH_3 \rightarrow CH_3COCH(CH_3)_2 + CH_3COCH_3 \tag{5.10}$$

$$2CH_3CH_2CH_3 \rightarrow CH_3CH(CH_3)_2 + C_2H_6 \tag{5.11}$$

Equation 5.10 is almost thermoneutral (+2.4 kJ mol^{-1}) while Eq. 5.11 is exothermic by ca. 10 kJ mol^{-1}. The effect of nitrogen on transmethylation in reactions 5.8 and 5.9 as compared with the effect of carbon in reactions 5.10 and 5.11 is nearly identical. Although the tetrahedral α-carbon in ketone is thus compared with the trigonal nitrogen in amides, it does not seem to influence the outcome of the comparison. As a further probe, the enthalpy of Eq. 5.12 is also very nearly thermoneutral (1.9 kJ mol^{-1}).

$$2(CH_3)_2C{=}CHCH_3 \rightarrow (CH_3)_2C{=}C(CH_3)_2 + (CH_3)_2C{=}CH_2 \tag{5.12}$$

There does not seem to be a great difference due to trigonal versus tetrahedral carbon.

An interesting comparison with the above is found with $R = CH_3$, and $R' = C_6H_5$. Paralleling Eq. 5.8 is the new Eq. 5.13.

$$2CH_3CONHC_6H_5 \rightarrow CH_3CON(C_6H_5)_2 + CH_3CONH_2 \tag{5.13}$$

Using solid phase data, we find that this reaction is endothermic by ca. 60 kJ mol^{-1}, some 57 kJ mol^{-1} more endothermic than the aliphatic amide reaction in the gas phase. It is not obvious what additional comparisons can be made using this result. Equation 5.13 might have been compared with the amine as done in Eqs. 5.8 and 5.9 except that the enthalpy of formation of *N*-methylaniline needed for Eq. 5.14 is quite contentious.*

$$2CH_3NHC_6H_5 \rightarrow CH_3N(C_6H_5)_2 + CH_3NH_2 \tag{5.14}$$

*See the discussion in Ref. 3.

We would have liked to use Eq. 5.15 paralleling Eq. 5.12

$$2(CH_3)_2C{=}CHC_6H_5 \rightarrow (CH_3)_2C{=}C(C_6H_5)_2 + (CH_3)_2C{=}CH_2 \qquad (5.15)$$

except that we lack enthalpy of formation data for the 1,1-dimethyl-2,2-diphenylethylene. However, the demethylated versions of Eqs. 5.12 and 5.15, the new Eqs. 5.16 and 5.17, are both found to be essentially thermoneutral.

$$2H_2C{=}CHCH_3 \rightarrow H_2C{=}C(CH_3)_2 + H_2C{=}CH_2 \qquad (5.16)$$

$$2H_2C{=}CHC_6H_5 \rightarrow H_2C{=}C(C_6H_5)_2 + H_2C{=}CH_2 \qquad (5.17)$$

A series of acyl-substituted amides is $(RCO)_3N$, $(RCO)_2NR'$, $RCONR'_2$, and where for completeness, we end the series with NR'_3. $R = R' = CH_3$ is one case where we have essentially all of the data. Although we have only the enthalpy of formation of $(CH_3CO)_2N(CH_2)_3CH_3$, we are confident that the enthalpy of formation for $(CH_3CO)_2NCH_3$ can be readily estimated by assuming that the following reaction is essentially thermoneutral.

$$(CH_3CO)_2N(CH_2)_3CH_3 + CH_3NH_2 \rightarrow (CH_3CO)_2NCH_3 + CH_3(CH_2)_3NH_2 \qquad (5.18)$$

The derived liquid phase value is ca. $-460\,\mathrm{kJ\,mol^{-1}}$. For liquid phase species, the enthalpies of formation for the above series are -610.5, -460, -278.3, and $-45.7\,\mathrm{kJ\,mol^{-1}}$. The sequential replacement enthalpy of acetyl by methyl is increasingly positive: ca. 150, 180, and $230\,\mathrm{kJ\,mol^{-1}}$. Equivalently, the stabilization derived on replacing a methyl group by an acetyl group decreases with increasing number of acetyl groups. This can be understood in terms of both the competing resonance structures (how much the nitrogen lone pair can be delocalized) and steric repulsion between the various acetyl groups.

The other acyl-substituted amide series has $R = CH_3$, and $R' = C_6H_5$ for which we have the enthalpies of formation of -610.5(lq), -362.6, -42.9, and $235.0\,\mathrm{kJ\,mol^{-1}}$. The enthalpy of fusion (solid/liquid correction) is always positive and so the enthalpy of formation of solid triacetylamide must be more negative than the $-610.5\,\mathrm{kJ\,mol^{-1}}$ for the liquid species. The sequential replacement enthalpy of acetyl by phenyl is positive: ca. >250, 320, and $280\,\mathrm{kJ\,mol^{-1}}$. There is no apparent pattern in terms of stabilization of phenyl and acetyl groups. The first replacement of phenyl by acetyl is only slightly more favorable than the first replacement of methyl by acetyl. However, the second acetyl replacement of phenyl is unaccountably much more favorable than methyl replacement. For both phenyl and methyl, the third replacement is less favorable than the second. The results are surprising and we have no simple explanation.

4. LACTAMS AND OTHER CYCLIC COMPOUNDS

In this section we discuss the enthalpies of formation of cyclic lactams and of their *N*-methyl derivatives. As the ring size increases from 5 to 8, the liquid phase enthalpy of formation values for the *N*-methyl lactams (for which we have the most data) become more negative, -265.7 ± 0.5, -293.0 ± 0.5, -306.7 ± 0.5 and $-325.4 \pm 1.3\,\mathrm{kJ\,mol^{-1}}$, respectively. However, their sequential differences are not constant: -26, -13 and $-19\,\mathrm{kJ\,mol^{-1}}$. This nonuniform behavior is not surprising. As with most cyclic compounds, the lactams are surely strained and the strain energy depends non-linearly on the ring size. There are liquid enthalpies of formation for only the 5- and 6-membered ring unsubstituted lactams. Their enthalpy-of-formation difference is $-30\,\mathrm{kJ\,mol^{-1}}$, comparable to their *N*-methyl counterparts.

As there are two trigonal atoms in the amide functional group, we wish to explore their effect on ring strain. A classic measure of cycloalkane ring strain is the enthalpy of formation per methylene group. The calculation for liquid phase cyclopentane through cyclooctane (-21.0, -26.1, -22.4, $-21.0\,\mathrm{kJ\,mol^{-1}}$) shows a maximum exothermicity for cyclohexane and thus it is the least strained cycloalkane. The same calculation* for the *N*-methyl lactams shows a different result: -53.1, -48.9, -43.8, $-40.7\,\mathrm{kJ\,mol^{-1}}$, where the 5-membered ring, *N*-methyl-2-pyrrolidinone, is least strained. The same trends are observed for gas phase cycloalkanes (C_6 least strained) and gas phase *N*-methyl- and unsubstituted lactams (C_5 least strained). Since a ring containing the amide functional group cannot attain the approximately tetrahedral bond angles (and no torsional strain) which define the stability of cyclohexane, it may not be surprising that the 6-membered ring amides are not the most stable, but it is surprising perhaps that the smaller 5-membered ring is less strained than the 7-membered ring. The result is compatible, however, with the calculation of resonance energies in the 7-membered ring lactam of $75\,\mathrm{kJ\,mol^{-1}}$ and in the 5-membered ring lactam of $121\,\mathrm{kJ\,mol^{-1}}$.[11] Choosing cyclic alkenes, ketones, ethers, amines and esters (lactones), we again derive the enthalpy of formation per ring atom. The results fall into three categories, regardless of phase: the 5-membered ring is the least strained in lactams, lactones and ketones; the 6-membered ring is the least strained in cycloalkanes and amines; and cyclooctene is the least strained cycloalkene. (The result for cyclic ethers, for which there are no data for rings larger than six atoms, is inconclusive.)

We recognize the fact that meaningful enthalpy-of-formation comparisons can be made only among compounds in the same category. We further recognize that the difference quantity will be constant only when the sequential differences between enthalpies of formation within each of the two families are the same.

*To divide the enthalpy of formation by ring size is an arbitrary choice for convenience in comparing same-sized cyclic species. The same conclusions for lactams are reached if the enthalpies of formation are divided, for example, by the number of methylene groups in the ring.

The difference quantity is defined as Eq. 5.19 where r denotes the ring size and $*$ is the phase of interest.

$$\delta_{5.19}(\text{cycle A, cycle B}, r^{*}) = {}_{\mathrm{f}}\mathrm{H_m}^{\mathrm{o}}\,(\text{cycle A}, r,^{*}) - \Delta_{\mathrm{f}}\mathrm{H_m}^{\mathrm{o}}(\text{cycle B}, r,^{*}) \tag{5.19}$$

For example, $\delta_{5.19}$ (cyclic amine, cycloalkane, 5–7, g) = 72.9, 76.3, and 72.9 kJ mol^{-1} for the 5-, 6-, and 7-membered rings, respectively. The sequential differences between the enthalpies of formation of the three cyclic amines are -43.8 and 2.0 kJ mol^{-1} and those of the three cycloalkanes are -47.0 and 5.3 kJ mol^{-1}. These are close enough to produce the approximately constant $\delta_{5.19} = 74.0 \pm 2.0$ kJ mol^{-1}.* Unfortunately, of all the cyclic families of compounds explored above, these are the only two which fulfill the requirement for constant differences. All other combinations, which we might have hoped to fruitfully employ, yield sets of either monotonic or varying difference quantities**.

We have often resorted to the assumption of the thermoneutrality of formal reactions of the kind shown in Eq. 5.20 in the hope that variations in ring strain energies would cancel.

(5.20)

For the 6-membered rings shown, the gaseous enthalpy of reaction is -29 kJ mol^{-1}. For the 5-membered rings, it is -12 kJ mol^{-1}. An even better approach to thermoneutrality is to use the enthalpies of formation of the most stable 5-membered ring lactone and lactam and the most (or almost) stable 6-membered ring amine and ether. In such a case the enthalpy of reaction is -7.5 kJ mol^{-1}. Even at their crudest, these reactions are useful estimates.

5. COMPARISON OF SMALL PEPTIDES AND THEIR PARENT AMINO ACIDS

Proteins may safely be considered as the consummate example of polyamides. Given the intricacies of their secondary and tertiary structures, we will not even attempt to derive or discuss the enthalpy of formation of any protein. When discussing the amino acids from which peptides are derived, we do so without

*From this difference quantity, the gaseous enthalpy of formation of cyclooctane (-124.4 ± 1.0 kJ mol^{-1}), and the assumption that sequential differences within the two families will remain the same, we derive a gaseous enthalpy of formation of azacyclooctane of ca. -50 kJ mol^{-1}.

** For example, we find these difference quantities: $\delta_{5.19}$ (*N*-methyl lactam, ketone, 5–8, lq) = -28.3, -20.5, -9.1, $+0.6$ kJ mol^{-1}; $\delta_{5.19}$ (lactam, lactone, 5–7, g) = 169.1, 147.5, 156.4 kJ mol^{-1}.

regard to any zwitterionic character even though these species are more correctly written as $NH_3{}^+CHRCONHCHR'COO^-$ in the solid phase. The enthalpies of formation of some of the parent amino acids are suspect, e.g. two of the archival literature values[5] for the enantiomeric D- and L-alanine differ by ca. 43 kJ mol^{-1} when we know these two numbers must be, in fact, identical.[12] We also note that some of the dipeptides whose enthalpies of formation we quote are reported to contain the "D, L-" amino acid residues while others contain the "L-" residue. The citations on serylserine and valylphenylalanine are silent on the subject of stereochemistry. Thus, the stereochemical composition of some of the dipeptides is uncertain and ambiguous, while the amino acid residue stereochemistry may not be identical to that of the individual amino acid for which measured enthalpies of formation are available. Also for simplicity we use the three-letter abbreviations for the amino acid residues in a peptide, where, by convention, the N-terminal residue is written on the left and the C-terminal residue on the right of the formula.

5.1. Polypeptides

The longest peptide for which thermochemical data are available is the tetrapeptide consisting only of glycine residues: gly · gly · gly · gly. Complete hydrolysis of this (or any) peptide produces the constituent amino acids.

$$\text{gly} \cdot \text{gly} \cdot \text{gly} \cdot \text{gly}\,(s) + 3H_2O\,(lq) \rightarrow 4NH_2CH_2COOH\,(s) \qquad (5.21)$$

The enthalpy of reaction is -68.6 ± 3.4 kJ mol^{-1} or -22.9 kJ mol^{-1} per hydrolysis. The enthalpies of reaction for the tripeptide gly · gly · gly and for the dipeptide gly · gly are -53.7 ± 3.3 kJ mol^{-1} (-26.9 kJ mol^{-1} per hydrolysis) and -21.5 ± 0.9 kJ mol^{-1}, respectively. The mean enthalpy of hydrolysis to a glycine amino acid is thus -23.8 ± 2.8 kJ mol^{-1}. Comparable enthalpies per hydrolytic reaction are found for gly · L-ala · L-phe (-28.2 ± 2.1 kJ mol^{-1}) and L-leu · gly · gly (-22.1 kJ mol^{-1}).

5.2. Dipeptides

There being no other tetra- or tri-peptides with measured enthalpies of formation, we next consider a collection of dipeptides and the enthalpies of their hydrolysis reactions. If these enthalpies are nearly constant, we would conclude that the amide bond is quite independent of its local environment. This seems reasonable indeed, as an approximately constant endothermicity of 26 ± 2 kJ mol^{-1} has been suggested[13] for the reverse of hydrolysis, i.e. the dimerization of two amino acids to form dipeptides (Eq. 5.22). This value is consistent with that for the hydrolysis of the peptides.

$$\begin{aligned} &NH_2CHRCOOH(s) + NH_2CHR'COOH(s) \\ &\rightarrow NH_2CHRCONHR'COOH(s) + H_2O(lq) \end{aligned} \qquad (5.22)$$

TABLE 5.2. Enthalpy of Formation of Some Dipeptides (kJ mol^{-1})

Dipeptide	R	R′	$\Delta_f H_m^o$ (s)
glycylglycine	H	H	-747.7 ± 1.3
D,L-alanylglycine	CH_3	H	-777.4 ± 0.9
D,L-leucylglycine	i-C_4H_9	H	-859.8 ± 1.3
glycyl-L-phenylalanine	H	$C_6H_5CH_2$	-684.3 ± 1.8
glycyl-D,L-valine	H	i-C_3H_7	-834.9 ± 0.4
D,L-alanyl-D,L-alanine	CH_3	CH_3	-807.3 ± 0.8
L-alanyl-L-phenylalanine	CH_3	$C_6H_5CH_2$	-710.1 ± 0.4
valylphenylalanine	i-C_3H_7	$C_6H_5CH_2$	-765.7 ± 1.4
serylserine	CH_2OH	CH_2OH	-1177.7 ± 0.4

The dipeptides for which thermochemical data are available are listed in Table 5.2. The first five dipeptides listed all contain at least one glycine residue and the mean enthalpy of hydrolysis of the first four (to the L-amino acid) is -24.8 ± 3.0 kJ mol^{-1}. The enthalpy of hydrolysis of gly · D,L-val is the disparate -35.7 ± 2.0 kJ mol^{-1}. Although this disparity may be due to the formal hydrolysis of a dipeptide with unknown stereochemical composition, (D,L-), to L-valine, the other D,L-dipeptide hydrolysis enthalpies show no such deviation. In fact, use of the enthalpy of formation of L-valine from Ref. 5 gives an enthalpy of hydrolysis of -25.7 ± 0.9 kJ mol^{-1} for this dipeptide which is consistent with that of others. The suspected inaccuracy of the more recent[6] L-valine enthalpy of formation is also apparent in the calculation of the formal hydrolysis reaction of val · phe for which the enthalpy of reaction is -44.8 ± 2.5 kJ mol^{-1}. If the enthalpy of formation of L-valine from Ref. 5 is used, the enthalpy of hydrolysis is -33.3 ± 1.7 kJ mol^{-1}. This enthalpy of hydrolysis and that from L-ala · L-phe (-31.5 ± 1.9 kJ mol^{-1}) are significantly more negative than that from hydrolysis of gly · L-phe (-24.8 ± 2.0 kJ mol^{-1}), the only other phenylalanine-containing dipeptide. Whether this is the result of inaccurate measurements for the former two dipeptides, or of steric effects, is not known. The enthalpy of hydrolysis of ser · ser is -1.9 ± 0.6 kJ mol^{-1}. That the reaction enthalpies are different for side-chains containing hydrocarbyl versus hydroxyl groups is not surprising, but again we do not know whether the discrepancy is caused by some substituent effect or inaccurate measurement of the dipeptide and/or the amino acid.

5.3. Cyclic Dipeptides

Let us now consider the cyclic dipeptides, also known as amino acid anhydrides or 3,6-disubstituted-2,5-piperazinediones. The equation for formation of the cyclic dipeptide from the open-chain dipeptide is

$$H_2N{-}CHR{-}\overset{O}{\overset{\|}{C}}NH{-}CHR'{-}CO_2H\ (s) \longrightarrow \text{cyclo-}[NH{-}CHR{-}C(O){-}NH{-}CHR'{-}C(O)]\ (s) + H_2O\ (l) \qquad (5.23)$$

Since each of the cyclization reactions forms a 6-membered ring, we might expect the enthalpies of cyclization to have the same dependence on R and R′ as the hydrolysis reactions 5.22. The enthalpies of reaction 5.23 as listed in Table 5.3 fall into three distinct categories: glycylvaline; glycylphenylalanine and alanylphenylalanine; and glycylglycine and serylserine. Although we might have expected the hydroxylic serylserine to differ from the hydrocarbyl-containing substituents, we are very surprised to find its enthalpy of reaction to be the same as glycylglycine.

5.4. Which Values are to be Believed?

Sometimes we find constancy and consistency for some plausible regularity in enthalpy of formation while other times we do not. We may, then, ask which (if any) of the values are correct? Said differently, which enthalpies of formation of which 2,5-piperazinediones and small peptides are reliable? Were we dealing with gas phase species, the following formal reaction would be expected to be essentially thermoneutral (not withstanding the discussion in Section 4).

$$\text{3-R-2-piperidinone} + \text{3-R'-2-piperidinone} \longrightarrow \text{3-R-6-R'-2,5-piperazinedione} + \text{cyclohexane} \qquad (5.24)$$

TABLE 5.3. Enthalpy of Formation of Some Cyclic Dipeptides ($kJ\,mol^{-1}$)

Dipeptide	$\Delta_f H_m{}^o$ (cyclo-,s)	R	R′	ΔH (Eq. 5.23 s)
glycylglycine	-446.5 ± 1.3	H	H	15.4 ± 1.8
glycylvaline	-502.9 ± 0.8	H	i-C_3H_7	46.2 ± 0.9
alanylvaline	-514.3 ± 0.9	CH_3	i-C_3H_7	NA
valylleucine	-628	i-C_3H_7	i-C_4H_9	NA
leucylleucine	-659	i-C_4H_9	i-C_4H_9	NA
glycylphenylalanine	-345.4 ± 1.7	H	$C_6H_5CH_2$	53.1 ± 2.5
alanylphenylalanine	-372.0 ± 2.6	CH_3	$C_6H_5CH_2$	52.3 ± 2.6
serylserine	-875.6 ± 0.9	CH_2OH	CH_2OH	16.3 ± 1.0
phenylalanylphenylalanine	-287.6 ± 1.1	$CH_2C_6H_5$	$CH_2C_6H_5$	NA
glycyltyrosine	-512.3 ± 0.6	H	p-$HOC_6H_4CH_2$	NA

NA: The enthalpy of formation of the open-chain dipeptide has not been measured.

We should like to think that it is nearly thermoneutral for solids as well. In that there are seemingly no formation enthalpies for substituted piperidones, let us derive the enthalpy of formation of the parent compound, i.e., $R = R' = H$. This reaction is very nearly thermoneutral when all species found in this reaction are taken as solids and so we trust the experimental measurement of this value as well as our estimation approaches.

What about the acyclic glycylglycine? The assumption of thermoneutrality for the formal solid phase reaction

$$\begin{aligned} &NH_2(CH_2)_4COOH + BuCONHBu \\ &\quad \rightarrow NH_2CH_2CONHCH_2COOH + BuCH_2CH_2Bu \end{aligned} \qquad (5.25)$$

is discrepant only by ca. $8\,kJ\,mol^{-1}$. We are inclined to trust the enthalpies of formation of both glycylglycine and its cyclic counterpart, and by inference, more than those of serylserine and its cyclic counterpart.

6. HYDANTOINS (IMIDAZOLIDINE-2,4-DIONES)

Hydantoins (**1**), an interesting and important class of polyfunctionalized heterocycles, are also recognizable as polyamides.

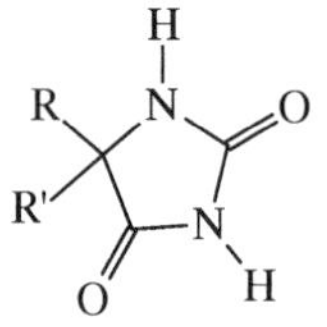

H
N
R O
R'
N
O H

1

The four hydantoins for which enthalpies of formation have been chronicled are the solid parent species and its 5-methyl, 5,5-dimethyl and 5-ethyl-5-methyl derivatives. We find enthalpies of formation[11] of -448 and $-581.9 \pm 1.1\,kJ\,mol^{-1}$ for the parent species from a very old and very new source respectively, and -486.6 ± 1.1, -533.3 ± 1.3 and $-566.1 \pm 1.6\,kJ\,mol^{-1}$ for the three C-alkylated derivatives from a study that is "merely" 40 years old.[15] It is clear that the two values for the parent are incompatible with each other – indeed, we cannot rationalize the nearly $135\,kJ\,mol^{-1}$ discrepancy by any simple assumptions of sample purity or thermochemical error. Similarly, we can offer no explanation for why C-alkylation of the parent hydantoin would result in such a large increase in the enthalpy of formation, yet this is an immediate conclusion if one is to accept the more recent value for the parent species and its three alkyl derivatives.

Can analogies be of use here? A series of similarly substituted carbocycles consists of the corresponding "liquid" phase cyclopentanes with enthalpies of formation of -105.6 ± 0.8, -137.7 ± 0.8, -172.0 ± 1.1 and $-193.8 \pm 1.0\,\text{kJ mol}^{-1}$. From these and fusion enthalpy data, we may derive the corresponding solid phase enthalpies of formation of -111, -145 and -180 for the first three species respectively. That all three enthalpies of fusion are ca. $6\,\text{kJ mol}^{-1}$ suggests that this phase change correction may be applied to the last cyclopentane derivative. This results in a derived enthalpy of formation of solid 1-ethyl-1-methylcyclopentane of ca. $-200\,\text{kJ mol}^{-1}$ – a value nearly identical to that calculated by summing the new group increments for solids[16]. We find the differences between the enthalpies of formation of the solid hydantoin and the corresponding solid cyclopentane to be either -337 or -471, -342, -353 and $-366\,\text{kJ mol}^{-1}$. While the differences are hardly constant, the difference of $-337\,\text{kJ mol}^{-1}$ using the former value for the parent hydantoin is plausible while the difference of -471 is not. This suggests that the earlier value for the enthalpy of formation of the parent hydantoin is roughly consistent with those of its alkylated derivatives.

However, is this earlier value consistent with the enthalpies of formation of other cyclic amides or lactams? Consider now the formal solid state reaction.

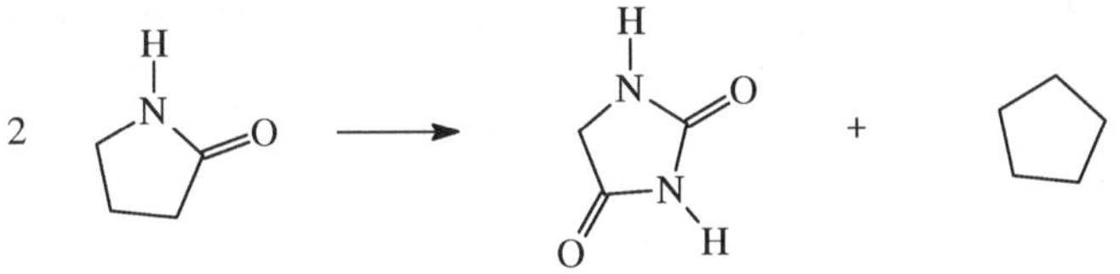

$$2 \quad \longrightarrow \quad + \qquad (5.26)$$

Use of our conventional enthalpy of fusion corrections and of the assumption of thermoneutrality yields a predicted enthalpy of formation of the parent hydantoin of ca. $-500\,\text{kJ mol}^{-1}$. Since the right hand side contains one more amide linkage than the left, we would expect this reaction to result in an enthalpy of formation which is too high. This means that we have underestimated its stability.

Accordingly, the newer value for the enthalpy of formation of the parent hydantoin is more likely to be correct than the earlier. The conclusions are contradictory. We, therefore, strongly suggest re-measurement of all four of the hydantoin enthalpies of formation discussed in this section.

7. AMIDINES AND AMIDES

We have found some regularities in predicting, explaining and understanding the thermochemistry of amides. If disappointingly little is known about the thermochemistry of amides, our corresponding knowledge about amidines is even less. For the simple, acyclic species urea, $(H_2N)_2C{=}O$, and guanidine, $(H_2N)_2C{=}NH$, we find a $226\,\text{kJ mol}^{-1}$ difference between their enthalpies of formation. Indeed, for these simple 1-carbon species, the urea/guanidine transformation may be expressed as either that of –CONH– to –C(=NH)NH–

or –CONH– to –C(NH_2)=N–. There are many more compounds with the –C(NH_2)=N– functionality or substructure than with the isomeric –C(=NH)NH–. We therefore choose to consider the latter transformation.

A series of cyclic, and putative aromatic, species are cyclo-$(CONH)_n$ $(C(NH_2)=N)_{3-n}$: cyanuric acid (**2**), 6-amino-1,3,5-triazine-2,4-dione (**3**), 4,6-diamino-1,3,5-triazine-2-one (**4**), and melamine (**5**) with $n = 3$, 2, 1 and 0 respectively. These have enthalpies of formation in the solid phase of -703.5 ± 1.4, -492.9 ± 4.2, -299.6 ± 2.1, and $-71.7 \pm 0.6\,\mathrm{kJ\,mol^{-1}}$ with sequential differences of 211, 193, and 228 $\mathrm{kJ\,mol^{-1}}$ respectively.

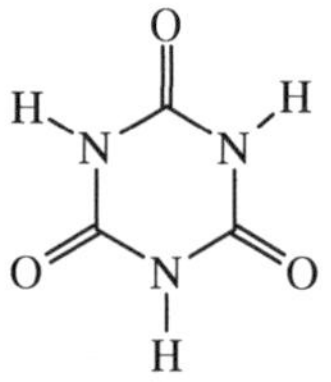

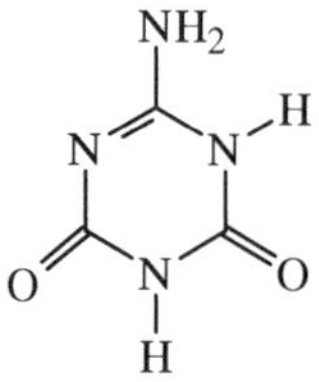

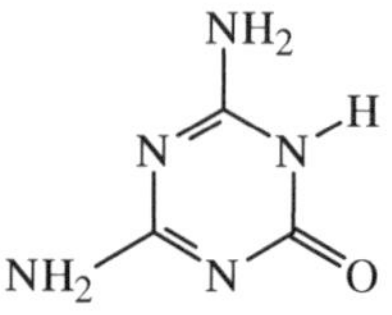

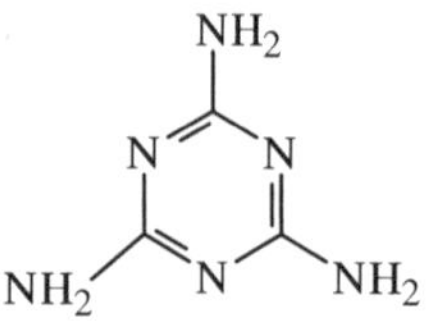

These three numbers are roughly comparable. As these compounds are sequentially related by transforming C(O)NH to C(NH_2)=N, we may consider the related transformation upon going from 2-pyridone (**6**) to 2-pyridinamine (**7**), xanthine (**8**) to guanine (**9**), hypoxanthine (**10**) to adenine (**11**), uracil (**12**) to cytosine (**13**).

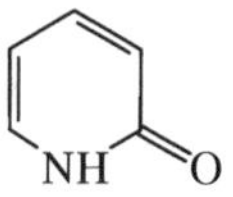

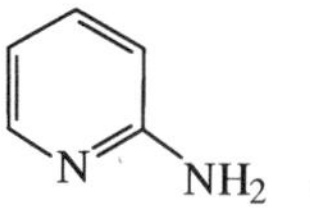

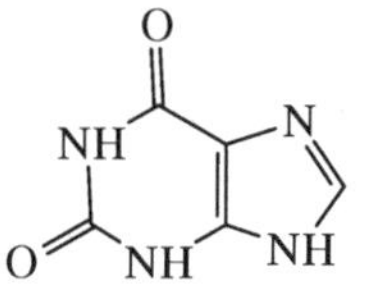

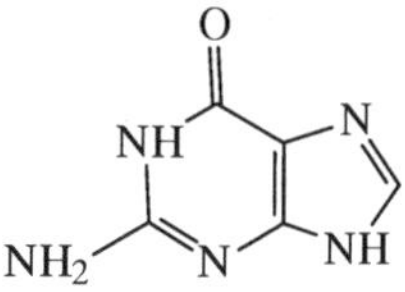

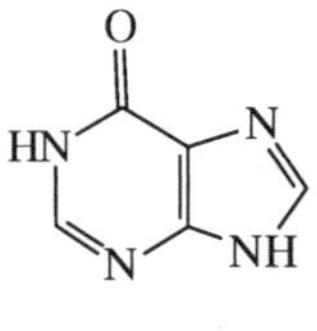

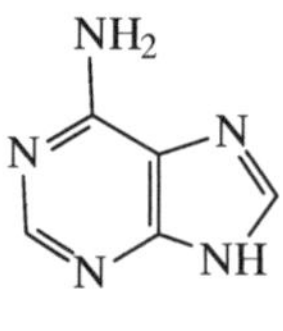

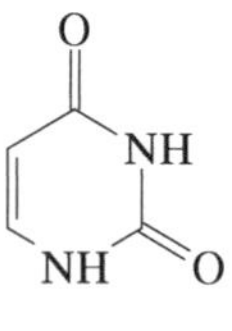

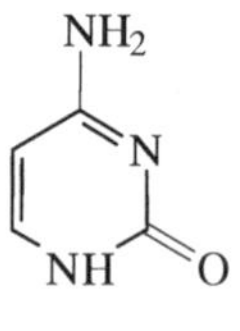

The corresponding differences are numerically rather comparable: 197, 207, 196 and 203 $\mathrm{kJ\,mol^{-1}}$. It would appear that amidines have a more positive enthalpy of formation than the related amides by ca. $220 \pm 20\,\mathrm{kJ\,mol^{-1}}$, i.e.

$$\delta_{(5.27)}(\text{amide, amidine}) = \Delta_f H_m{}^{\circ}(\text{–C(NH}_2\text{)=N–, s}) - \Delta_f H_m{}^{\circ}(\text{–C(O)NH–, s}) \tag{5.27}$$

All of the above amide/amidine comparisons involve species with 6-membered rings. We turn now to comparisons involving species with 5-membered rings. We may likewise compare the solid enthalpies of formation of 1,4-dihydro-5H-tetrazol-5-one (**14**, $6.28 \pm 1.92\,\text{kJ mol}^{-1}$) and tetrazole-5-amine (**15**, $207.8 \pm 2.3\,\text{kJ mol}^{-1}$). Accordingly, we find a 201.5 kJ mol^{-1} difference between the tetrazolone and tetrazolamine in satisfactory agreement with our suggested amide/amidine difference.

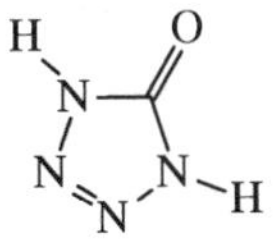

14 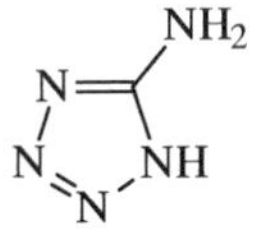 **15** 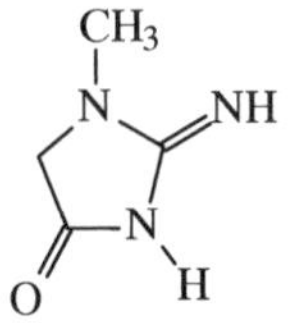 **16**

Now consider the alicyclic creatinine (**16**) with its enthalpy of formation of $-238.5 \pm 0.5\,\text{kJ mol}^{-1}$. We recognize this species as the amidine analog of the N-methylated, but otherwise unsubstituted, parent hydantoin (**1**). Assuming demethylation of N-methylhydantoin would result in but a small change in enthalpy of formation, using the above amide/amidine quasi constant difference, $\delta_{(5.27)}$, we predict that the enthalpy of formation of the parent hydantoin is ca. $-460 \pm 20\,\text{kJ mol}^{-1}$. This is in good agreement with the ancient value and not at all with that from the more modern measurement[14].

A related though much more subtle example involves the substituted urea, ureidoacetic acid ($NH_2CONHCH_2COOH$), and substituted guanidine, creatine (the zwitterionic) $NH_2C(NH)N(CH_3)CH_2COOH$), with solid phase enthalpies of formation of -746.8 and -537.2, respectively. Again assuming that demethylation results in but a small change in enthalpy of formation, and using a literature "dezwitterionizing" energy* of $55\,\text{kJ mol}^{-1}$ for creatine and its demethylated derivative, we deduce the enthalpy of formation of the explicitly non-zwitterionic amidine, $NH_2C(NH)NHCH_2COOH$, is ca. $-485\,\text{kJ mol}^{-1}$. The amide/amidine (or more properly urea/guanidine) solid phase enthalpy of formation difference is ca. $264\,\text{kJ mol}^{-1}$ which is not so very different from the difference quantities derived above, considering the assumptions employed**. It is clear that as little as we seemingly know about the thermochemistry of amides themselves, amide/amidine conversions are somewhat known and predictable.

*In references 16, Domalski and Hearing present group increments for predicting the enthalpy of formation of solids, among which is one for converting a species from the zwitterion to the formally non-polar species.

**We recognize that we are not dezwitterionizing an amino acid but amidino acid. Amidines are more basic than related amines and so the $55\,\text{kJ mol}^{-1}$ would be but a lower bound to the dezwitterionizing energy.

8. ANTIAROMATICITY, AROMATICITY AND AMIDES

In the current section we discuss some of the consequences of extended electronic delocalization in simple amides. In particular, we discuss whether the amide linkage simulates the olefinic one* in antiaromatic and aromatic ring systems. Hydrogenation enthalpies have proven useful in understanding the energetics of carbocyclic rings. We start with the recent discussion[17] on the antiaromaticity of maleimide derivatives, which are formally amide analogs of the antiaromatic cyclopentadienone. The enthalpy of hydrogenation of an arbitrary compound may be equated to the difference between the enthalpy of formation of the compound and its hydrogenated product. From the enthalpies of formation of solid *N*-methylmaleimide (**17**) and solid *N*-methylsuccinimide, the hydrogenation enthalpy of *N*-methylmaleimide is derived to be $-140.5\,\mathrm{kJ\,mol^{-1}}$ while for the nonaromatic cyclopentene, a value of $-112\,\mathrm{kJ\,mol^{-1}}$ is found. The difference is $28\,\mathrm{kJ\,mol^{-1}}$, a value almost equal to what has been recently presented[17] for the corresponding gaseous phase species, and one which shows (**17**) to be relatively unstable as compared to the unfunctionalized cycloalkene.

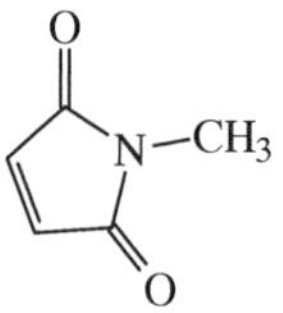

17 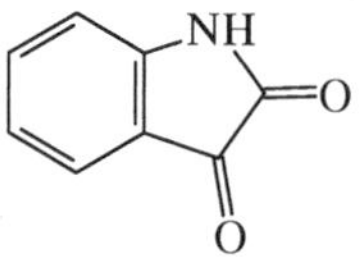 **18** 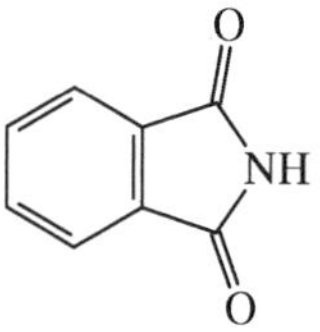 **19**

An interesting comparison would be with *N,N′*-dimethylimidazolidone because this species is formally aromatic as it is an amide analog of *N*-methylpyrrole. However, the desired thermochemical data for the cyclic urea is absent.

Another interesting comparison is that between the isomeric isatin (**18**) and phthalimide (**19**), both of which are nominally amide analogs of the antiaromatic indenone. The enthalpy of formation of isatin is $-268.2\,\mathrm{kJ\,mol^{-1}}$, while that of phthalimide from a very old measurement[18] is $-308.4\,\mathrm{kJ\,mol^{-1}}$. Thus phthalimide is more stable than isatin by ca. $40\,\mathrm{kJ\,mol^{-1}}$. However, it is not obvious to what extent is this due to the presence of another amide linkage in the phthalimide as opposed to the presence of an α-diketo linkage in the isatin.

Benzene, 2-pyridone (**6**), uracil (**12**), and cyanuric acid (**2**) are putatively aromatic with 6π electrons in each 6-membered ring. These species constitute the formally, fully homologous series $(CH{=}CH)_{3-n}(CONH)_n$ with $n = 0$, 1, 2 and 3, respectively. Their respective enthalpies of formation as found in the solid phase are 39.1, -166.3 ± 0.5, -424.4 ± 0.9 and $-703.5 \pm 1.4\,\mathrm{kJ\,mol^{-1}}$. The sequential differences between enthalpies of formation in this series is ca. -205,

*For a discussion on the comparison of amides and olefins, see Greenberg, A. In the current volume Chapt. 3.

-258, and $-280\,\text{kJ mol}^{-1}$. These are changing uniformly, suggesting some sense of homology. In that regard this series is different from the series (solid) benzene, toluene, *m*-xylene and mesitylene (recognizable as isoelectronic and isosteric to the above) with enthalpies of formation 39.1, 5.8, -37.0 and $-73.0\,\text{kJ mol}^{-1}$ with the rather constant enthalpy of formation differences of ca. 33, 43, and $36\,\text{kJ mol}^{-1}$.

But what about the relative aromaticities of benzene, 2-pyridone, uracil and cyanuric acid? Corresponding to the hydrogenation of the non-aromatic 1,3-cyclohexadiene to form cyclohexane are the hydrogenations of the unequivocally aromatic benzene to cyclohexene and plausibly aromatic 2-pyridone to 2-piperidone. Again, the differences between enthalpies of formation of parent and hydrogenated species show that these three reactions are exothermic by 235, 65, and $140\,\text{kJ mol}^{-1}$, respectively. This suggests that 2-pyridone is less aromatic than benzene which is an entirely plausible result.

We would have liked to discuss uracil and its hydrogenation enthalpy, but here we are seemingly thwarted—we know of no enthalpy of formation of dihydrouracil. However, we can consider this analysis in the reverse mode. We expect uracil to be less aromatic than benzene. Again using enthalpies of formation, we find that the enthalpies of mono-hydrogenation reactions of benzene and cyclohexene are $+30$ and $-81\,\text{kJ mol}^{-1}$, respectively. Accordingly, averaging these hydrogenation enthalpies and assigning the full difference as error bars, the hydrogenation of uracil would be exothermic by $25 \pm 55\,\text{kJ mol}^{-1}$. Equivalently, the enthalpy of formation of dihydrouracil would be $-454 \pm 55\,\text{kJ mol}^{-1}$.

We recognize the above species as related to the general generic class of compounds $(\text{CH{=}CH})_m(\text{CONH})_n(\text{CO or NH})_{p=0 \text{ or } 1}$, and their *N*-methyl derivatives. There are many interesting, thermochemically missing, and often experimentally unknown compounds. These include the parent and methylated derivatives of oximide ($m=0, n=1$, CO with $p=1$), azetone ($m=n=1, p=0$) and muconimide ($m=2$, CO with $n=1$, $p=1$) which would be aromatic, antiaromatic, and aromatic, respectively as they are formally related to cyclopropenone, cyclobutadiene, and tropone, respectively.

REFERENCES

1. Liebman, J. F.; Afeefy, H. Y.; Slayden, S. W. In *The Chemistry of Hydrazo, Azo and Azoxy Groups*; Patai, S. Ed.; Wiley: Chichester, UK, 1997, Vol. 2.
2. Slayden, S. W.; Liebman, J. F. *The Chemistry of Doubly-Bonded Functional Groups; Supplement A3*; Patai, S. Ed.; Wiley: Chichester, UK, 1997.
3. Liebman J. F.; Campbell, M. S.; Slayden, S. W. In *The Chemistry of Amino, Nitroso, Nitro and Related Groups, Supplement F2*; Patai, S. Ed.; Wiley: Chichester, UK, 1996.
4. Slayden, S. W.; Liebman, J. F. In *Chemistry of Hydroxyl, Ether and Peroxide Groups; Supplement E2*; Patai, S. Ed.; Wiley: Chichester, UK, 1993.

5. Pedley, J. B.; Naylor, R. D.; Kirby, S. P. *Thermochemical Data of Organic Compounds*; 2nd edn.; Chapman & Hall: New York, 1986.
6. Afeefy, H. Y.; Liebman, J. F.; Stein, S. E. "Neutral Thermochemical Data", In *NIST Chemistry WebBook, NIST Reference Data Base 69*, Mallard, W. G.; Linstrom, P. J. Eds.; National Institute of Standards and Technology: Gaithersburg, MD, (http://webbook.nist.gov) March 1998.
7. Chickos, J. S.; Acree, W. E., Jr.; Liebman, J. F. *J. Phys. Chem. Ref. Data*, (In press).
8. Chickos, J. S.; Acree, W. E., Jr.; Liebman, J. F. In *Computational Thermochemistry*; Frurip, D. J.; Irikura, K. K. Eds.; ACS Symposium Series: Washington, DC, 1998 Vol. 677; p. 63.
9. Cox, J. D.; Pilcher, G. *Thermochemistry of Organic and Organometallic Compounds*, Academic Press: New York, 1970.
10. Young, J. A.; Keith, J. E.; Stehle, P.; Dzombak, W. C.; Hunt, H. *Ind. Eng. Chem.* 1956, **48**, 1375.
11. Greenberg, A. In *Structure and Reactivity*, Liebman, J.F.; Greenberg, A.; Eds.; VCH Publishers, Inc.: New York, 1988.
12. Domalski, E. S. *J. Chem. Phys. Ref. Data* 1972, **1**, 221.
13. Ponamerev, V. V.; Alekseeva, T. A.; Akimova, L. N. *Russ. J. Phys. Chem.* 1962, **36**, 457.
14. El-Sayed, N. I. *Asian J. Chem.* 1993, **5**, 199.
15. Tavernier, P.; Lamouroux, M. *Mem. Poudres* 1955, **37**, 197.
16. Domalski, E. S.; Hearing, E. D. *J. Phys. Chem. Eng. Data* 1993, **22**, 805.
17. Roux, M. V.; Jiménez, P.; Martin-Luengo, M. Á.; Dávalos, J. Z.; Sun, Z.; Hosmane, R. S.; Liebman, J. F. *J. Org. Chem.* 1997, **62**, 2632.
18. Kharasch, M. *J. Res. Nat. Bur. Stand.* 1929, **2**, 359.

CHAPTER 6

STEREOSPECIFICITY IN THE α-LACTAM (AZIRIDINONE) SYNTHON

ROBERT V. HOFFMAN
Department of Chemistry and Biochemistry, New Mexico State University

α-Lactams (aziridinones) were first postulated as reaction intermediates in 1908 by Leuchs[1] (later shown to be erroneous) and in 1949 by Sheehan.[2] They attracted the greatest amount of attention in the 1960s and early 1970s because of their interesting and potentially useful reactivity. This reactivity results from the approximately 41 kcal mol^{-1} of strain energy present in the aziridinone ring.[3] The use of α-lactams as synthetic intermediates has not been widely pursued because their chemistry is for the most part unpredictable and not well understood.

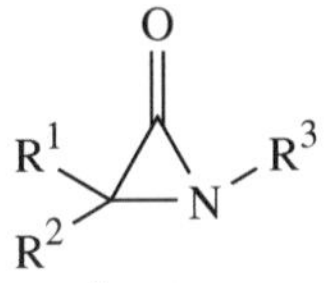

O
R1 R3
N
R2

α-Lactam
Aziridinone

A

Baumgarten was the first to isolate an α-lactam, namely, *N-tert*-butyl-3-phenylaziridone **1**, which was prepared by the reaction of *N-tert*-butyl α-chloro-α-phenylacetamide **2** with base.[4] The structure of **1** was distinguished from several alternate possibilities which include an imino oxirane **3**, and oxaziridine **4**, and an open dipolar structure **5**.[5]

Pertinent to the title of this review is the fact that both **4** and **5** were eliminated as structural possibilities by the preparation of an optically active product from optically active *N-tert*-butyl-α-chloro-α-phenylacetamide. Although the absolute configuration of the α-lactam was not determined, both **4** and **5** have sp^2 hybridized benzylic carbons, and thus are incapable of maintaining chirality at that position.

The Amide Linkage: Selected Structural Aspects in Chemistry, Biochemistry, and Materials Science,
Edited by Arthur Greenberg, Curt M. Breneman, and Joel F. Liebman
ISBN 0-471-35893-2

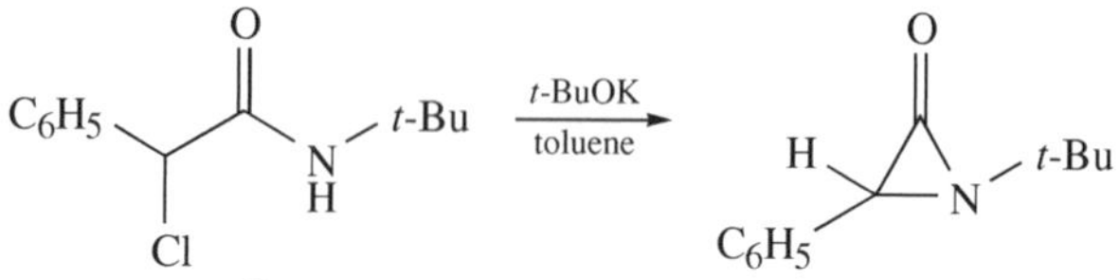

2 $\xrightarrow[\text{toluene}]{t\text{-BuOK}}$ 1

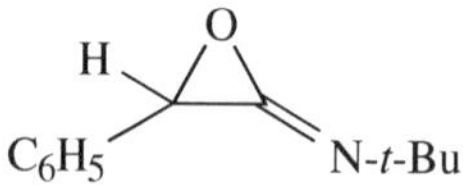

3 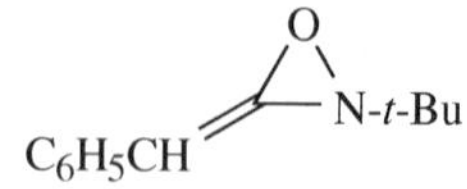 4 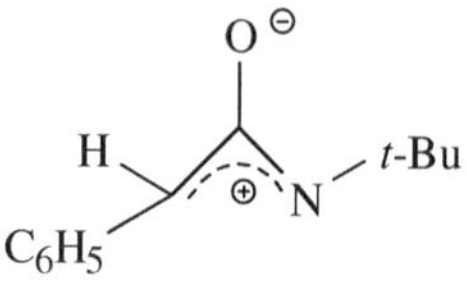 5

Exchanging the carbonyl oxygen of the starting amide for ^{18}O gave a product whose IR spectrum had the normal absorption at $1848\,cm^{-1}$ shifted to $1827\,cm^{-1}$ by the isotopic substitution. This isotopic substitution would not result in a IR band shift if imino oxirane **3** were the correct structure and the $1848\,cm^{-1}$ band arose from the C=N group. These and other data strongly support the α-lactam structure **1**. Subsequently an X-ray structure of a stable α-lactam, 1,3-*bis*(1-adamantyl)aziridinone **6**, confirmed the structural assignment.[6]

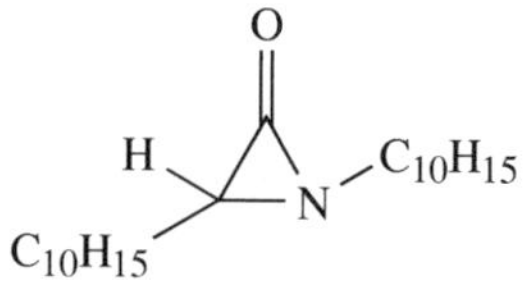

6

$C_{10}H_{15}$ = 1-adamantyl

Since Baumgarten's work, a number of α-lactams have been isolated, but they remain relatively rare. One structural requirement for the isolation of α-lactams is that a bulky group be attached to the nitrogen—the most common group being a *tert*-butyl group, although α-lactams with 1-adamantyl or trityl groups attached to the nitrogen atom have also been isolated. A bulky nitrogen substituent is thought to sterically shield the carbonyl group, retard the rate of nucleophilic addition, and thus kinetically stabilize the α-lactam. α-Lactams with nitrogen substituents smaller than *tert*-butyl are unstable and cannot be isolated.

The chemistry of α-lactams was studied with some intensity in the 1960s and early 1970s and was summarized in a definitive review by Sheehan[7] in 1968, and in a shorter summary by L'abbe[8] in 1980. Most of the data on the reactions and chemistry of α-lactams have been obtained from the studies of isolated α-lactams which must have *tert*-butyl or other bulky groups attached to nitrogen. It is not clear how, or if, the chemistry of less stable α-lactams differs from the kinetically stabilized examples.

The thermal chemistry of α-lactams is interesting but not synthetically relevant. The main thermal pathways include β-elimination with ring opening that occurs when β-hydrogens are present.[9]

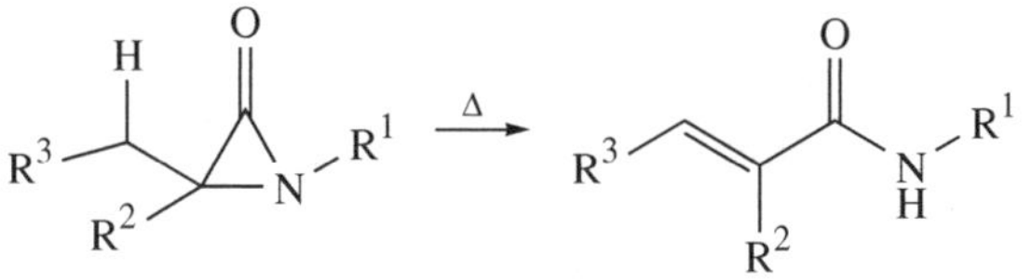

A second major thermolytic pathway is the formation of a carbonyl compound and an isocyanide on heating.[10–13] This transformation is postulated to take place by thermal reorganization of the α-lactam to an imino oxirane which fragments to the observed products. Although the imino oxirane is postulated as an intermediate, it has not actually been observed during the thermolysis.

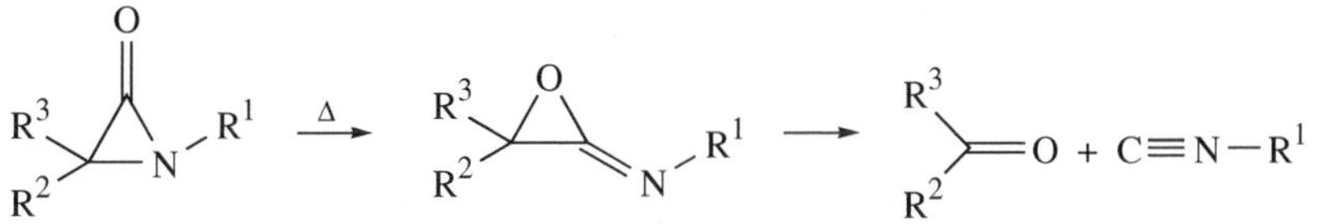

Of greater synthetic interest are the reactions of α-lactams with nucleophiles. In general, nucleophiles open the α-lactam ring by attack either at the acyl (C-2) carbon or at the saturated C-3 position. Nucleophilic addition to the acyl carbon gives C-2 to N-1 C–N bond cleavage and rearranged amino acid derivatives. Nucleophilic attack at C-3 leads to C-3 to N-1 C–N bond cleavage and α-substituted amide products. Depending on the nucleophile, subsequent transformations are possible.

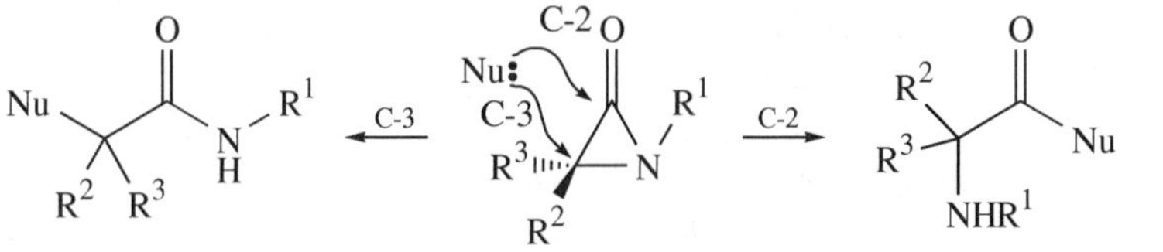

α-substituted amide α-amino acid derivative

A great deal of work was undertaken to understand the structural and/or environmental factors which influence the regiochemistry of nucleophilic attack on α-lactams. In fact, a good portion of the work in this field has been devoted to this issue.[7] In spite of the research efforts that have been undertaken, the mechanisms by which these regioisomers are formed and the factors which influence the regioselectivity remain unsettled.

Based on a suggestion by Sheehan,[7] it is generally assumed that both regioisomeric products originate from competitive nucleophilic addition to a single intermediate, namely, the α-lactam. Moreover, it is generally agreed that strong, aprotic nucleophiles give C-2 attack while protic nucleophiles lead to C-3 attack. This view was reiterated by L'abbe,[8] and has become part of the mechanistic dogma surrounding the chemistry of α-lactams.

The formation of α-lactams by 1,3-elimination of α-halo amides is directly analogous to the formation of cyclopropanones in the Favorski rearrangement.[14] Since a delocalized, dipolar intermediate was postulated in some cases of the

Favorski rearrangement,[15] the analogous dipolar ion **5** was long considered to be a possible intermediate in the chemistry of α-lactams. In fact Sheehan was careful to include the possibility that aziridinone ring-opening to **5** followed by nucleophilic addition to C-3 was the actual source of C-3 products.[7]

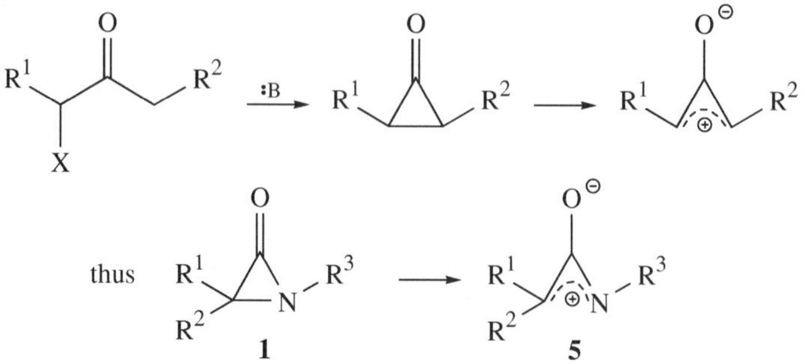

The lack of convincing experimental evidence to support or rule out the involvement of **5** in the reactions of α-lactams with nucleophiles meant that it was difficult to predict the regiochemistry of nucleophilic addition in specific cases. Consequently, the use of α-lactams as synthetic intermediates was never pursued extensively, and relatively little work was reported from the early 1970s until about 1990.

Stereochemical probes provide a powerful tool for investigating the involvement of **5** in the chemistry of α-lactams. As mentioned previously, Baumgarten was the first to prepare a scalemic (chiral, non-racemic) α-lactam.[5] Although the configuration and optical purity of the α-lactam were not known with certainty, the presence of optical activity was used to discount **5** as a possible structure for the product ultimately identified as the α-lactam **1**. Another early study reported that scalemic α-bromopropionanilide **7** reacts with sodium methoxide to give 2-methoxypropionanilide **8** with "considerable racemization" but involvement of an α-lactam or the dipolar intermediate was not discussed.[16]

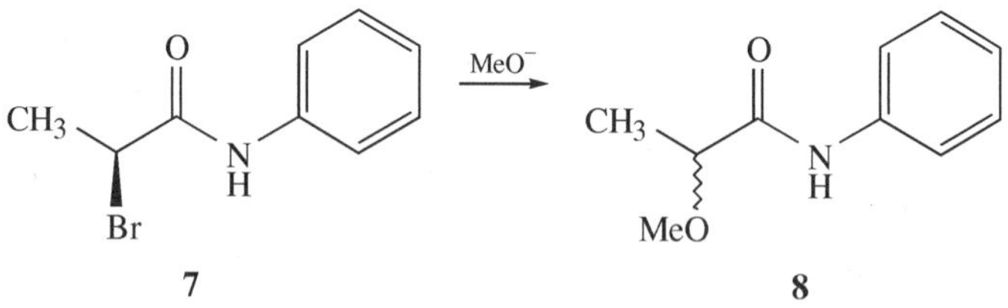

The first detailed stereochemical study of α-lactams was described by Sarel[17] who isolated several scalemic α-lactams in optically pure form and assigned the absolute configuration at C-3 by the Cotton effect in the c.d. spectrum. Since the steroid R^1 is chiral, the recrystallized C-3 epimers **10** were presumably optically pure. The α-lactam structure was confirmed by the 1840 cm^{-1} band in their IR spectra. It was concluded that α-lactam formation occurs with inversion of configuration at C-3.

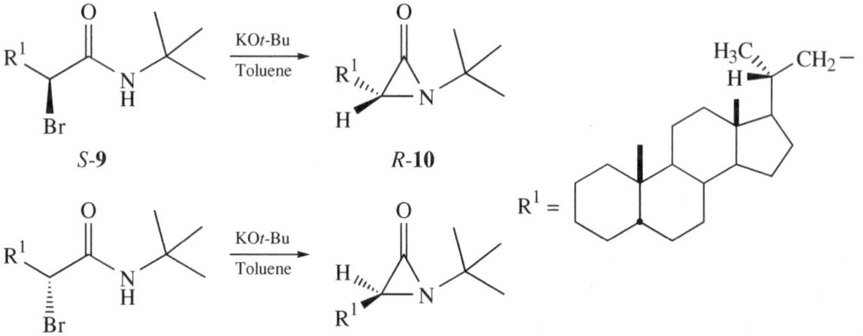

KOt-Bu / Toluene: S-9 → R-10

KOt-Bu / Toluene: R-9 → S-10

R^1 =

Reaction of *R*-**10** with methanol produced α-methoxy amide **11** which was assigned the *R*-configuration at the α-carbon on the basis of the c.d. spectrum. This requires that the methoxy substitution at C-3 takes place with retention of configuration. In the same study a different scalemic α-lactam, the *N*-(1-adamantyl) derivative *R*-**12** was treated with *tert*-butoxide, and an α-aminoester was produced which was assigned by c.d. as having the *R*-configuration at the α-carbon. This product resulted from α-lactam ring opening with retention of configuration at the C-3 carbon.

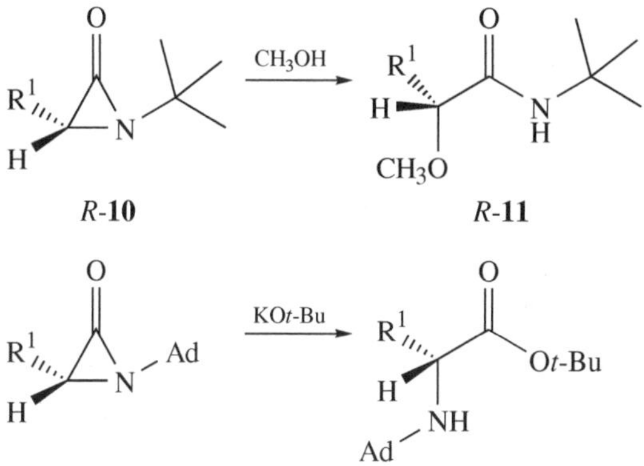

CH_3OH: R-10 → R-11

KOt-Bu: R-12 → R-13

Ad = 1-adamantyl

Although the configurational assignment of **11** was later shown to be in error,[18] this study demonstrated clearly that the configurational integrity of the α-bromo amide is maintained during both the formation of the α-lactam and the attachment of nucleophiles at either the C-2 or C-3 positions. These results rule out the possibility that an acyclic dipolar intermediate such as **5** is involved in either of the transformations.

A reinvestigation of the stereochemistry of the ring opening of aziridinones by nucleophiles was reported in 1991.[18] In that work scalemic chloroamide *S*-**14** (95.2% ee) was converted to [3*R*]-1,3-(di-*tert*-butyl)aziridinone *R*-**15** (92% ee as determined by chiral phase gas chromatography).

S-**14** → *R*-**15** (92% ee)

A substrate whose absolute configuration and optical purity is known with certainty allows the stereochemistry of the ring opening processes to be determined with confidence. Reaction of *R*-**15** with magnesium halides, methanol/tosic acid, and water/tosic acid — all gave incorporation of the nucleophile at C-3 with inversion of configuration. Reaction with sodium methoxide resulted in attack at the acyl carbon and gave retention of configuration at C-3. The absolute configurations of all the products were determined from the chiral chromatographic behavior of authentic samples of the products prepared independently.

The high and relatively uniform ee's indicate that virtually no (<2%) racemization at C-3 takes place. Moreover the data clearly illustrate the general reactivity pattern of α-lactams with nucleophiles. Weak nucleophiles in the presence of acids (Bronsted or Lewis) give nucleophilic addition to C-3 while stronger, anionic nucleophiles add preferentially to the acyl carbon.

R-**15** —MgX_2→ **16a**, X = Cl (89% ee); **16b**, X = Br (89% ee); **16c**, X = I (88% ee)

R-**15** —MeOH / TsOH→ **17** (88% ee)

R-**15** —H_2O / TsOH→ **18** (88% ee)

R-**15** —CH_3O^- / CH_3OH→ **19** (87% ee)

Scalemic *R*-**15** was also used to study the thermal chemistry of α-lactams.[13] At 130°C in benzene solution *R*-**15** undergoes thermolysis to *tert*-butyl isocyanide and pivaldehyde in nearly quantitative yield. Racemization of *R*-**15** under the same conditions is 5-fold slower than fragmentation. It appears that rearrangement of **15** to an imino oxirane and then fragmentation is a process that is distinct from the pathway followed for racemization.

O t-Bu H N t-Bu —Δ→ O t-Bu H Nt-Bu → t-BuCHO + t-BuNC

R-**15**

In an attempt to trap an acyclic, dipolar intermediate that may be involved in either the fragmentation or racemization pathways, the thermolysis of *R*-**15** was carried out in DMF. Decomposition of the α-lactam occurs readily at 80–100°C and oxazolinone **20** was formed in high yield as a mixture of diastereomers. The configuration at C-5 of **19** (which corresponds to C-3 of the α-lactam **15**) was determined to be *S* (85–90% ee) and results from inversion of configuration at C-3 by nucleophilic addition of DMF to the α-lactam. The high stereocontrol indicates that this product cannot result from trapping of an acyclic intermediate. DMF is a relatively poor nucleophile and the conditions are non-acidic, yet the fact that DMF still adds as a nucleophile to C-3 is noteworthy, in spite of the fact that the mechanism is unknown.

O t-Bu H N t-Bu —Δ, DMF→ O H t-Bu 5S Nt-Bu O NMe$_2$

R-**15** **20**

These data obtained with isolated α-lactams demonstrate unequivocally that C-3 addition occurs with inversion of configuration, and C-2 addition gives ring opening with retention of configuration at C-3. Because of the very bulky groups on nitrogen and C-3, which are needed to kinetically stabilize the α-lactam, it is fair to ask if less stable α-lactams would show the same type of stereochemical behavior.

A study of the electrolysis of α-bromo amides **21** provided a tool to address this question.[19] Reduction of the α-bromine generates an enolate **22** which rapidly produces an α-bromoamide anion **23** by proton transfer. The rate of decay of this anion can then be determined. It is possible to evaluate the effects of structure on the decay rate and thus conclude that an α-lactam is being produced by ring closure of **23**. It was found that as the nitrogen substituent R^3 is changed from primary to secondary to tertiary, the rate of decay increases, consistent with an increase in nitrogen nucleophilicity in the series. Moreover as the C-2 carbon is changed from primary to secondary to tertiary, the rate of decay decreases, which parallels the steric congestion at the α-position. These trends suggest that ring closure of the amide anion **23** to an α-lactam is the major decay pathway.

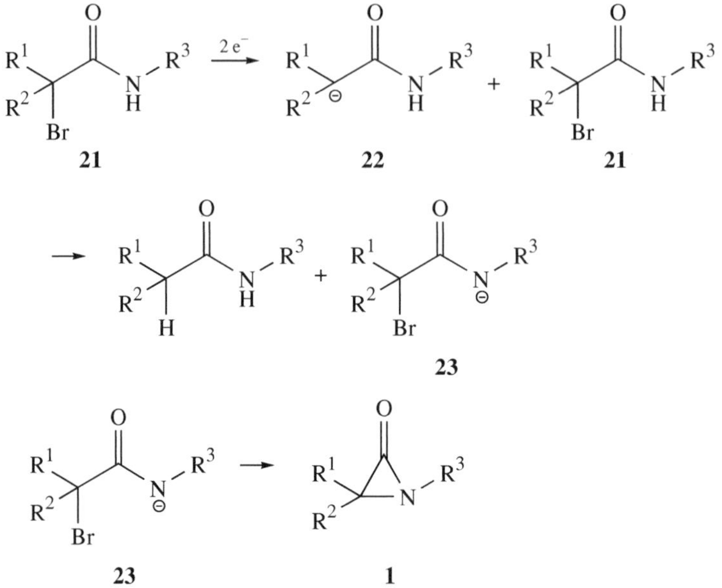

The α-lactams produced by this method are unstable and cannot be isolated. Nevertheless their reactions with nucleophiles can be studied conveniently. When [2*S*]-2-bromo-*N*-phenylpropanamide **24** is treated with *tert*-butylamine, only [2*R*]-2-(*tert*-butylamino)-*N*-phenylpropanamide **25** is produced in 90% yield. Direct Sn2 substitution for bromine occurs with inversion of configuration at C-2. When the same substrate is treated with *tert*-butylamine in DMF under electrolysis conditions, the reaction is much faster. Three products are formed in modest yields. Aminoamide **25** is produced which has the 2*S* configuration (retention). A DMF adduct **26** similar to that observed by Quast[13] which has the 2*S* configuration, is the major product. Finally, a cyclic dimer **27** is formed which is optically active but of uncertain configuration.

The retained C-2 configurations of products **25** and **26** result from two sequential inversions: formation of the α-lactam occurs with inversion at the α-carbon of the bromoamide starting material, and attack by the nucleophile (either *tert*-butyl amine or DMF) on C-3 of the α-lactam occurs with inversion. The stereochemical result is substitution products with net retention of configuration relative to the starting bromide. These results suggest that the chemistry of less stable α-lactams is similar to that of the isolable examples.

An extensive study of the reactions of scalemic α-bromoamides reached similar conclusions. Reaction of scalemic *N*-benzyl-2-bromopropanamide **28**-*R* with benzylamine gave the amino amide **29**-*S* with inversion of configuration at

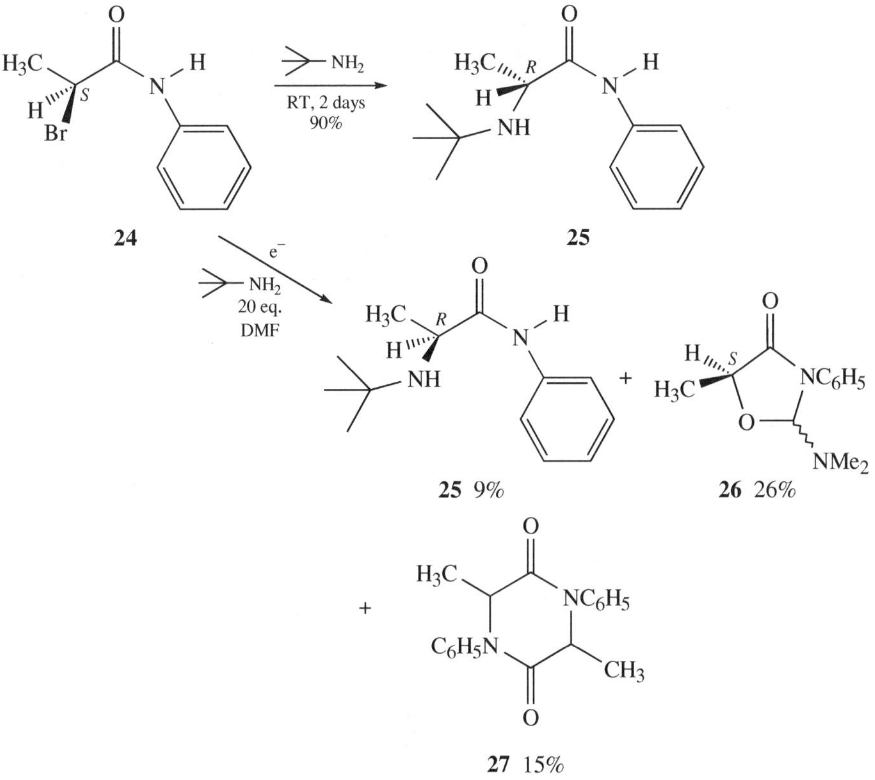

24

25

25 9%

26 26%

27 15%

C-2. In contrast, reaction of **28**-*S* with benzylamine in the presence of suspended silver oxide gave **29**-*S* with retention of configuration at C-2.[20] Because of the identity of the amide substituent and the nucleophile, it was not possible to sort out the regiochemistry of the substitution, but the results require that different stereospecific processes occur under the two sets of conditions.

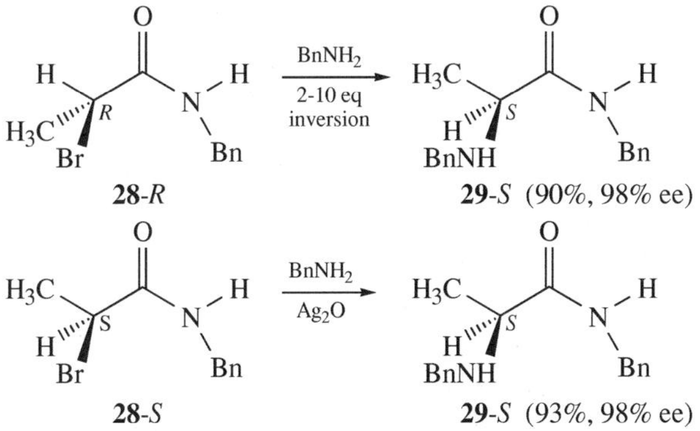

28-*R*

29-*S* (90%, 98% ee)

28-*S*

29-*S* (93%, 98% ee)

A more extensive study using different scalemic amides and different amine nucleophiles appeared soon after.[21] In that study, the absolute configurations of the products were not determined but configurations relative to a standard were used to suggest the stereochemical course of the reaction. Reaction of [*S*]-2-bromo-*N*-substituted propanamides **30a-c** with amine nucleophiles under three sets of conditions gave 2-amino *N*-phenyl amides whose optical rotations were measured. Reaction of **30** with amines in the absence of other reagents was shown to take place with inversion of configuration at the α-carbon, typical of a direct Sn2 substitution. Then the reaction was carried out in the presence of suspended silver oxide or with the addition of silver triflate (soluble Ag^+) to the reaction mixture. Interestingly, different results were obtained with the two different forms of silver[I] used.

30a, R = Ph
b, R = Bn
c, R = *tert*-Bu

The data in Table 6.1 clearly show that the two forms of silver reagent exert very different effects on the reaction. The presence of soluble silver[I] in the reaction mixture gives substitution products with the same configuration as the product of direct reaction of the amine and the α-bromoamide. Moreover, the presence of Ag[I] can decrease the reaction time by a factor of 5–10. Bulky

TABLE 6.1. Reaction of 2-Bromopropanamides 29a–c with Amine Nucleophiles

			Optical Rotation (%yield)		
Entry	Substrate	Amine	A[a]	B[b]	C[c]
1	**30a**	*tert*-$BuNH_2$	+49°(92)	–	+50°(86)
2		$BnNH_2$	+9.3°(94)	+9.2°(75)	−9°(91)
3		Et_2NH	−62°(86)	−65°(94)	+59°(90)
4		Pyrrolidine	−0.8°(100)	–	+0.8°(95)
5		*i*-Pr_2NH	N.R.	N.R.	+48.4°(72)
6	**30b**	*tert*-$BuNH_2$	+8°(43)	–	−9.4°(87)
7		$BnNH_2$	+4.2°(90)	+3.8°(90)	−4.2°(94)
8		Et_2NH	−38.4°(96)	–	+37.5°(95)
9	**30c**	*tert*-$BuNH_2$	N.R.	+20.5°(53)[d]	−26.5°(91)
10		$BnNH_2$	+3.6°(87)	+3.8°(65)	−3.8°(90)
11		Et_2NH	−62.8°(96)	–	+62.7°(95)

[a] Reaction run using 5 equivalents of amine.
[b] Reaction run using 2 equivalents of amine and 1 equivalent of silver triflate (solublesilver[I]).
[c] Reaction run using 2 equivalents of amine in the presence of 1 equivalent of suspended silver oxide.
[d] Reaction time was 700 h.

nucleophiles react much more slowly (Entries 5, 9) with or without soluble Ag[I] being present. It is reasonable to conclude that soluble Ag[I] promotes the direct displacement of bromide by the amine, probably by complexation with bromide to produce a superior leaving group.

In contrast, silver oxide shows a profoundly different effect. The products have retained configuration at the α-position, reaction rates are much faster than for the soluble Ag[I] promoted substitution, and steric bulk appears not to play a significant role in the yield of the reaction (Entries 5, 6, 9). The most simple explanation is that silver oxide functions as a base and produces an α-lactam which then reacts rapidly with the amine nucleophile to give the amine substitution product resulting from C-3 attack. This rationale explains not only the stereochemical results but also the lack of steric influence on C-3 attack. This dichotomy of behavior between soluble Ag[I] and Ag_2O permits a single scalemic α-bromoamide to be converted to an α-aminoamide with either retained or inverted configuration as needed. The only obvious limitation is that products with inverted configurations are subject to steric constraints imposed by the displacement mechanism.

Further support for this scenario is found in a subsequent study.[22] It was found that **30a** does not react directly with methanol but in the presence of soluble Ag[I] it reacts with methanol to give a 2-methoxy amide with inverted configuration. In contrast, **30a** reacts rapidly with methanol in the presence of silver oxide to give the α-methoxy amide with retained configuration. Significantly, **30a** also reacts with soluble Ag[I] and triethylamine to give the α-methoxy amide with retained configuration. These results suggest that either silver oxide or triethylamine provides the basic reaction environment needed to promote formation of the α-lactam. The silver ion may also play a role in influencing the regiochemistry of nucleophilic addition to the α-lactam to favor C-3 attack, but that remains to be proven (vide infra).

CH_3OH, Ag_2O or Ag^+, Et_3N ← *S*-**30a** → CH_3OH, Ag^+

S-**30a**

A very interesting and perhaps related study was carried out on *N*-tosyloxy β-lactams. It was reported that *N*-tosyloxy-β-lactams **31** react with azide ion in the presence of base to give 3-azido-β-lactams **32**.[23] This reaction requires promotion by base and it was proposed that the process takes place by initial enolization of **31** to produce **33** followed by Sn2′ displacement of the tosyloxy group by azide to give **32**.

Subsequent studies showed that a variety of nucleophiles could be incorporated at C-3 as in **35**.[24] Moreover when a substituent was present at C-4 of the β-lactam, the *anti*-diastereomer was favored. For example, azide adds to **34** with a diastereoselectivity of 24:1 *anti; syn*. The diastereoselectivity was attributed to the substituent at C-4 which blocks one side of the enol and forces nucleophilic addition to occur from the opposite side of the ring. This stereoselectivity was developed into a stereo-controlled synthesis of β-lactams and carbapenems.[25]

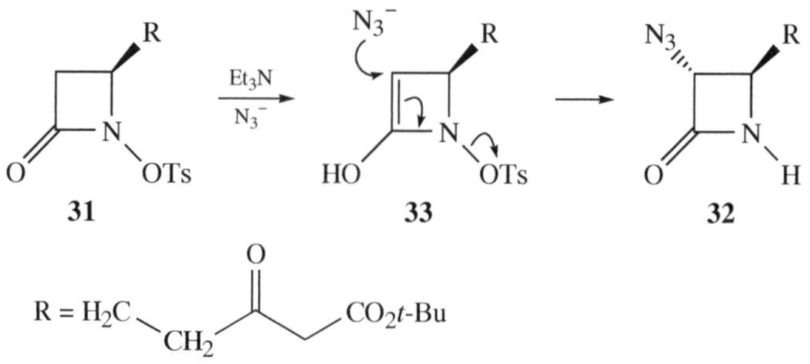

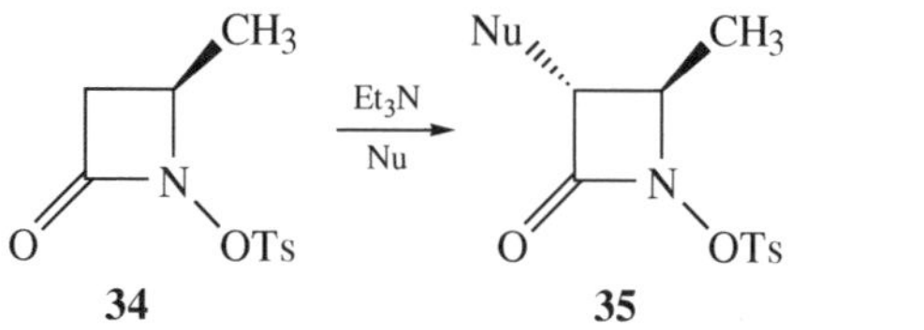

The Sn2′ mechanism proposed for the formation of α-substitution products **35** was favored because it was felt that strain in the azabicyclobutanone intermediate **36** formed by 1,3-elimination (α-lactam mechanism) would preclude its formation. While this argument is very reasonable, an analogous azabicyclobutane intermediate has been proposed as an intermediate in ring expansions of aziridine carboxylates.[26]

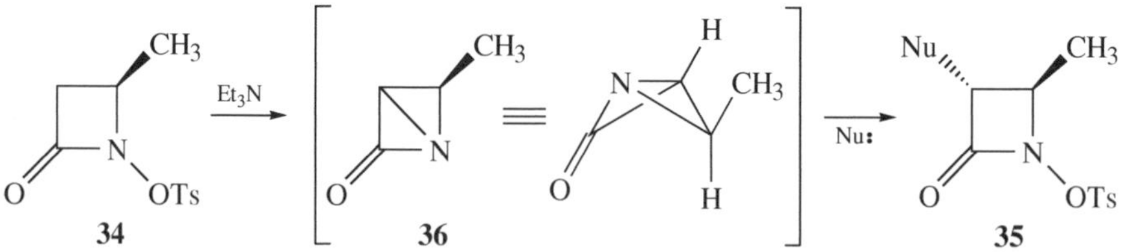

The intermediacy of azabicyclobutane **36** in this transformation would also account for the stereoselectivity of nucleophile incorporation. The geometry of the bicyclic system would favor a pseudoequatorial orientation for the group at C-4, and addition to the central bond of **35** would thus require that the stereochemistry of the product be *anti* as is observed.

In contrast, stereocontrol in the Sn2′ addition to enol intermediate **33** must result from a steric effect. The argument has been made that *syn* addition to **33** leads to an unfavorable eclipsing interaction between the C-4 group and the entering nucleophile and is thus disfavored.[23,24]

R_4 disfavored TsO N Nu HO
33
syn addition

R_4 TsO N HO Nu
33
anti addition

CH_3 CH_3
syn

CH_3 CH_3
anti

In order to achieve the stereoselectivities which have been reported,[24] the energy difference between the transition states for Sn2′ *syn* and *anti* addition to the enol **33** must be at least 3 kcal mole^{-1}. If a planar cyclobutene ring is assumed for **33**, then the transition state energy differences can be approximated by the *syn*/*anti* energy difference between 3,4-disubstituted cyclobutenes. AM1 calculations on *syn* and *anti* 3,4-dimethylcyclobutene show the energy difference to be 1.0 kcal mol^{-1} between these isomers. Because the activated complex for addition to **33** has incomplete bonding to the nucleophile, and the azide nucleophile itself is smaller than a methyl group, the actual difference in transition state energies would be less than 1 kcal mole^{-1}. Thus steric effects operating in an Sn2′ reaction are incapable of accounting for the selectivity exhibited in additions to **33**. A more reasonable origin of the stereoselectivity would be the stereoelectronic effect engendered by the intermediacy of an azabicyclobutanone (bicyclic α-lactam) in the reaction. Assuming that the methyl group adopts a pseudoequatorial position in **36**, then addition of the nucleophile to the σ* orbital on C-3 can give only *anti* product, thus accounting for the high stereoselectivity that is observed.

From these stereochemical studies on α-lactams several facts emerge. First, the chirality of the α-position of the starting materials is maintained throughout the process. Acyclic intermediates cannot be involved either in the formation of α-lactams or in their subsequent reactions with nucleophiles. That is to say, there must be some type of bonding interaction between the amide nitrogen and the α-position throughout the process. This appears to be true for both sterically stabilized examples as well as for examples which are too reactive to be isolated. Second, α-lactam intermediates are significantly more reactive toward nucleophiles than their precursor α-halogenated amides, largely due to the strain in the three-membered ring system. Third, nucleophilic addition to the acyl carbon leads to α-substituted products whose configuration at the α-carbon is the same as that in the precursor α-lactam (retention). Incorporation of the nucleophile at C-3 of the α-lactam gives products whose configuration at the α-carbon is inverted relative to the α-lactam.

While it is possible to reliably predict what will be the stereochemical outcome of nucleophilic addition to either the acyl carbon or to C-3; these stereochemical results do not contribute to an understanding of which process will be followed for a given nucleophile or a set of conditions. The stereochemical results have, if anything, solidified the notion that competitive nucleophilic attack at the two positions of the α-lactam is the correct mechanism,

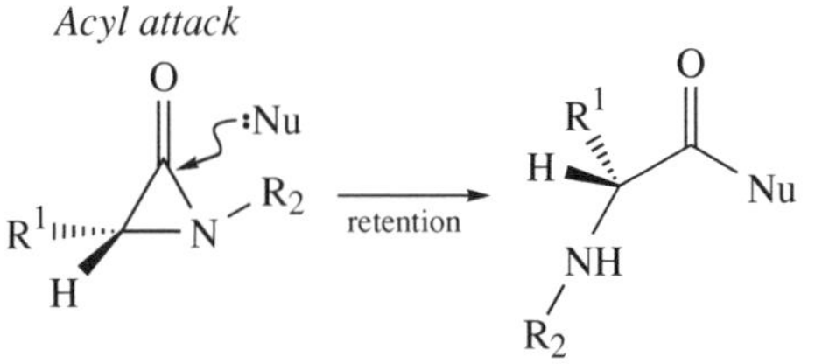

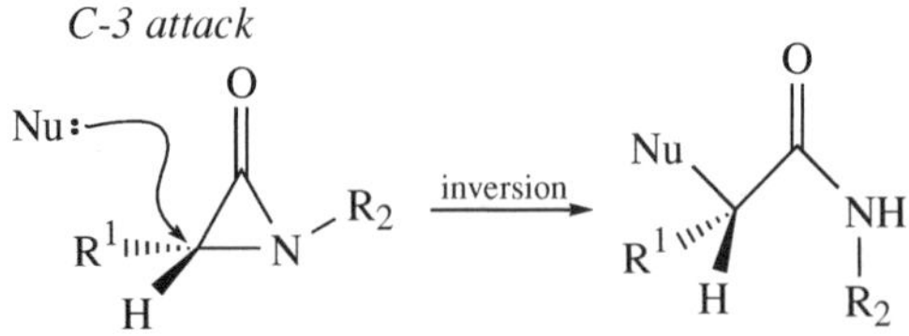

Acyl attack
:Nu
R2
R1
H
retention
R1
H
O
Nu
NH
R2

C-3 attack
Nu:
R2
R1
H
inversion
Nu
O
NH
R1
H
R2

but it is still not possible to predict with any certainty which type of product will be formed.

Since the review by Sheehan in 1968, very few studies of the regiochemistry of nucleophilic addition to α-lactams have been reported. The reaction of stable α-lactams with cyanamide and amines was reported to give only acyl carbon addition.[27] These results were considered "unusual" because these amines were considered to be protic nucleophiles and thus should add to C-3. Another study also using a group of stable, isolated α-lactams reported that both benzylamine and methoxide add exclusively to the acyl carbon.[28]

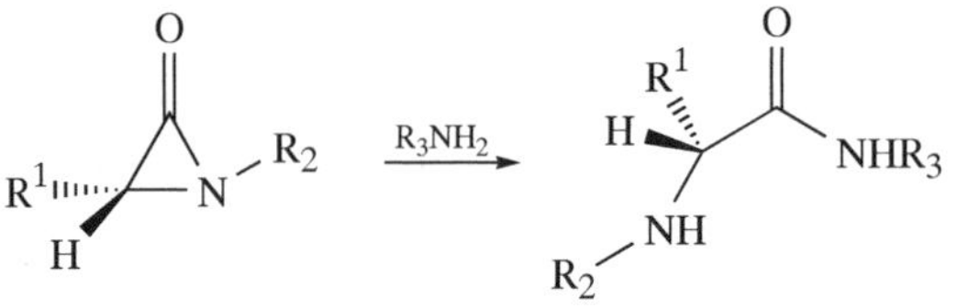

R2
R1
H
R_3NH_2
R1
H
O
NHR_3
NH
R2

R_1, R_2 = *t*-Bu, 1-adamantyl, R_3 = CN, Bn, *i*-Pr, *t*-Bu

The addition of nucleophiles to less stable, non-isolable α-lactams has also been studied recently. The α-lactams were produced by 1,3-elimination in *N*-mesyloxy amides promoted by amines.[29] It was found that the regiochemistry of nucleophilic addition to the α-lactam is dependent not on the protic nature of the nucleophile but on its nucleophilicity. If good nucleophiles such as primary or unhindered secondary amines are present, they are incorporated at the acyl carbon to give **37**. When only poor nucleophiles such as halide, azide, water, or methanol are present in solution (which also contains the ammonium mesylate salt formed during the formation of the α-lactam), they are incorporated at C-3 as in **38**. When bulky secondary amines are present, they add only to C-3 and give **38**.[30–32]

A kinetic model for the regioselectivity was developed from these results.[33] In this model, the two different types of products (**37** and **38**) are derived from two different reaction pathways of the α-lactam **1**. If good nucleophiles are present

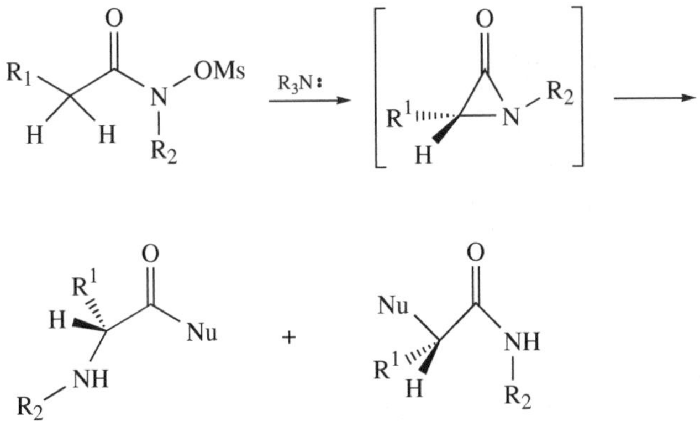

37 38

Nu = RNH_2, R_2NH Nu = X^-, N_3^-, H_2O, ^-OAc, bulky R_2NH

they trap the α-lactam **1** by a reaction at the acyl carbon by a second order process (path a). If only poor nucleophiles are present, an intermediate ***I*** is produced in solution from the α-lactam which reacts rapidly with nucleophiles at C-3.

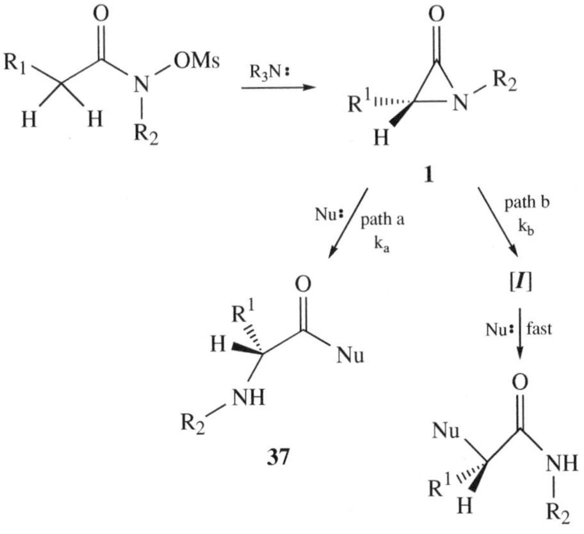

The product ratio of **37** to **38** is given by the ratio of rates by which they are produced. Thus, **37/38** = k_a [**1**] [Nu:]/k_b [**1**] or **37/38** = k_a [Nu:]/k_b. Thus the ratio of **37** to **38** should be dependent on the concentration of nucleophile in solution. If competitive attack on C-2 and C-3 by the nucleophile is operative, the product ratio should be independent of the nucleophile concentration because **37/38** = k_{C-2}[**1**][Nu:]/k_{C-3}[**1**][Nu:] or **37/38** = k_{C-2}/k_{C-3}. Experiments with *tert*-butyl amine, which gives a mixture of products, showed that the ratio of **37** to **38** is linearly dependent on the nucleophile concentration.[33] (In fact, the dependence was essentially first order in the concentration of the nucleophile as predicted by the model.) Thus the products from acyl attack and C-3 attack arise from competitive reaction pathways of different orders.

The nature of the intermediate ***I*** which leads to **38** must be defined and the stereospecificity (inversion) which results from C-3 addition must be accommodated by the structure of ***I***. Moreover, the structure of ***I*** must differ from that of the α-lactam **1** in such a way as to favor C-3 addition over acyl addition. Since C-3 addition is favored in protic environments, a prime candidate for ***I*** is an N-protonated α-lactam and thus path b involves protonation of **1** as the rate determining step. Once protonated, ***I*** is quite activated as an electrophile and reacts rapidly with even weak nucleophiles at C-3.

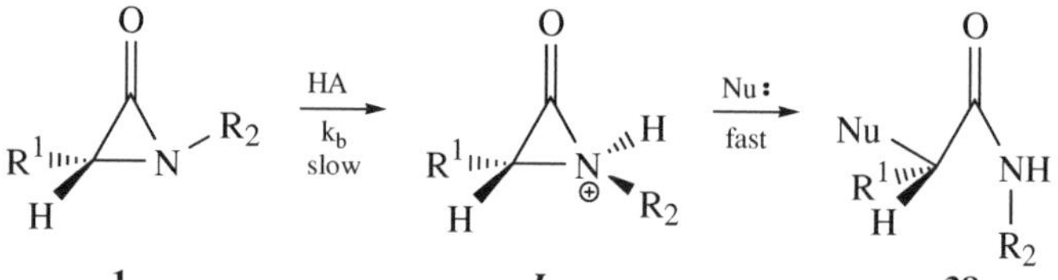

Besides the concentration dependence of the product ratio (vide supra), there are several additional pieces of evidence which support this scenario. Ab initio calculations show that N-protonation of α-lactams is favored over O-protonation by about 4.4 kcal mol^{-1} for the parent unsubstituted case (R_1, $R_2 = H$)[3] and about 3 kcal mol^{-1} for 1,3,3-trimethylaziridinone **39**.[34] The latter results are particularly pertinent because the methyl substituents of aziridinone **39** represent a more realistic model compound, and the calculations were done at a high level of theory (B3LYP/6-31G(d)). The N-protonated form **40** is an energy minimum with the bond between N-1 and C-3 remaining intact. Thus any chirality at C-3 in the α-lactam **38** would be maintained in the N-protonated α-lactam. Reaction of **40** with nucleophiles was also examined by calculation. It was found that addition of chloride to C-3 occurs with inversion of the C-3 carbon via an Sn2-like process.

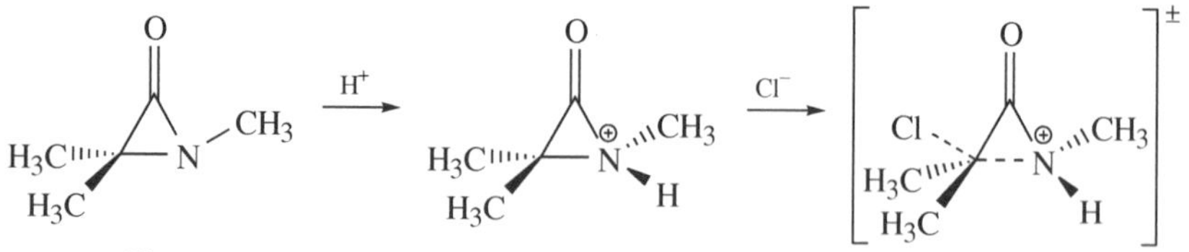

A supportive piece of experimental evidence bolsters this calculational interpretation. Quast found[18] that the reaction of **15** with methanol under various conditions gave variable regiochemistry. Reaction of **15** with methoxide in methanol, an environment in which N-protonation is unlikely and there is an excellent nucleophile present, proceeds exclusively by methoxide addition to the acyl carbon. Reaction of **15** with methanol, a much weaker nucleophile but a more acidic reaction medium, proceeds by nucleophilic addition of methanol at both the C-2 and C-3 carbons with the C-3 product being favored. Addition of as little as 0.1% of tosic acid causes methanol to add exclusively to C-3. A reasonable conclusion is that acid influences the regiochemistry of addition of weak nucleophiles significantly, as would be expected if N-protonation is a key step in the process.

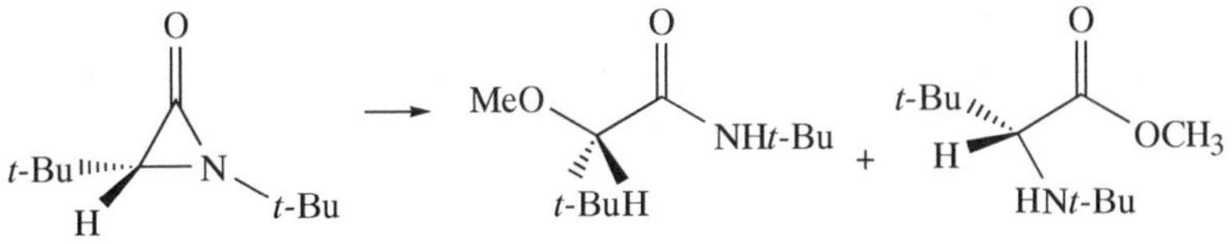

R-**15** → **17** + **19**

	17	**19**
CH_3O^-		100%
CH_3OH	88%	12%
CH_3OH/H^+	100%	

It is also reasonable to expect that metal ions (Lewis acids) present in solution would also complex with the α-lactam nitrogen and thus activate the α-lactam toward C-3 attack, in much the same way as a proton. Support for this hypothesis is available. It has been shown that benzylamine is sufficiently nucleophilic to intercept either stable, isolable α-lactams[28] or unstable α-lactams generated in situ[31] by nucleophilic addition to the acyl carbon.

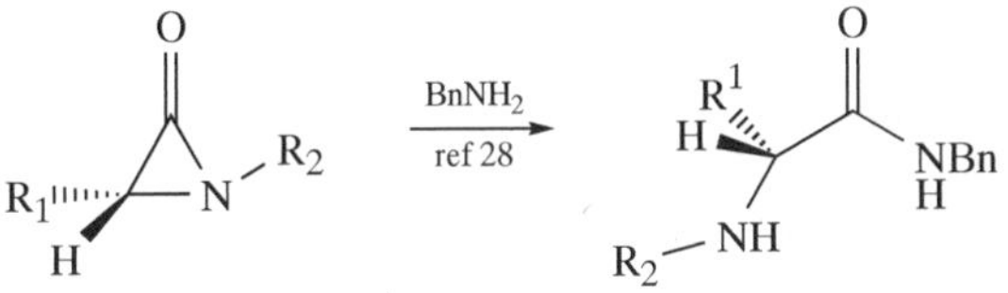

$BnNH_2$, ref 28

R_1, R_2 = bulky groups

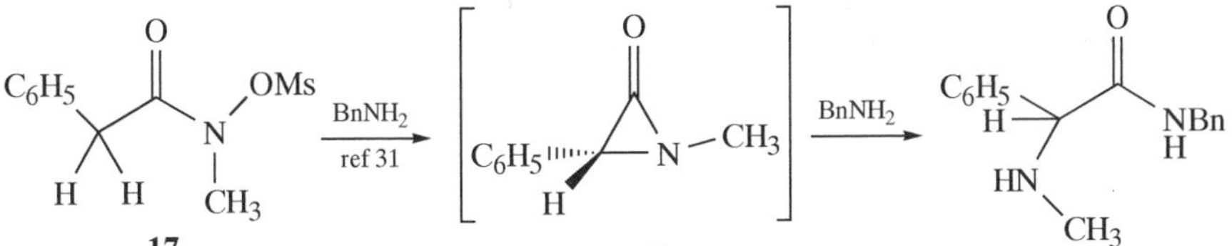

17 → ($BnNH_2$, ref 31) → [unstable] → ($BnNH_2$)

On the other hand, the generation of α-lactam **41** from **30c** using silver oxide in the presence of benzylamine yields C-3 product **42c** with retained configuration.[21] The presence of silver ions in the reaction mixture diverts the normal acyl addition to α-lactams by benzylamine to attack on the C-3 position instead. Silver complexation of the α-lactam nitrogen which activates **41** towards C-3 attack can explain this change in regiochemistry.

30c-*S* $\xrightarrow{Ag_2O}$ [**41**] $\xrightarrow[Ag^+]{BnNH_2}$ **42c**-*S*

The realization that fundamentally different, but competitive, reaction pathways lead from α-lactams to either C-2 or C-3 products allows a great deal of formerly inexplicable data to be correlated and organized. This reaction scenario now constitutes a new starting point for study since additional work is needed to verify it and demonstrate its limits. Hopefully, these studies will result in simple ways to control the regiochemistry of nucleophilic attack on α-lactams. This is quite important because if C-2 versus C-3 addition of nucleophiles to α-lactams can be predicted and controlled, then the stereoselectivity which is inherent in each of these processes can be used to make a variety of scalemic α-substituted acids and amides.

REFERENCES

1. Leuchs, H.; Geiger, W. *Ber. Dtsch. Chem. Ges.* 1908, **41**, 1721.
2. Sheehan, J. C.; Izzo, P. T. *J. Am. Chem. Soc.* 1949, **71**, 4059.
3. Greenberg, A.; Hsing, H.-J.; Liebman, J. F. *J. Mol. Struct. (Theochem)*, 1995, **338**, 83.
4. Baumgarten, H. E. *J. Am. Chem. Soc.* 1962, **84**, 4975.
5. Baumgarten, H. E.; Fuerholtzer, J. F.; Clark, R. D.; Thompson, R. D. *J. Am. Chem. Soc.* 1963, **85**, 3303.
6. Wang, A. H. -J.; Paul, I. C.; Talaty, E. R.; Dupuy, A. E., Jr. *Chem. Commun.* 1972, 43.
7. Sheehan, J. C.; Lengyel, I. *Angew. Chem. Int. Ed. Eng.* 1968, **7**, 25.
8. L'abbe, G. *Angew. Chem. Int. Ed. Eng.* 1980, **19**, 276.
9. Sheehan, J. C.; Beeson, J. H. *J. Am. Chem. Soc.* 1967, **89**, 362.
10. Sheehan, J. C.; Lengyel, I. *J. Am. Chem. Soc.* 1964, **86**, 746.
11. Sheehan, J. C.; Beeson, J. H. *J. Am. Chem. Soc.* 1967, **89**, 366.
12. Baumgarten, H. E.; Parker, R. G.; von Minden, D. L. *Orgn. Mass Spec.* 1969, **2**, 1221.
13. Quast, H.; Leybach, H.; Würthwein, E.-U. *Chem. Ber.* 1992, **125**, 1249.
14. March, J. A. *Advanced Organic Chemistry: Reactions, Mechanisms and Structure*; Wiley: New York, 1992, pp. 1080–1082.
15. House, H. O.; Gilmore, W. F. *J. Am. Chem. Soc.* 1961, **83**, 3972, 3980.
16. El-Abadelah, M. M. *Tetrahedron*, 1973, **29**, 589.
17. Sarel, S.; Weissman, B. A.; Stein, Y. *Tetrahedron Lett.* 1971, 373.
18. Quast, H.; Leybach, H. *Chem. Ber.* 1991, **124**, 2105.
19. Maran, F. *J. Am. Chem. Soc.* 1993, **115**, 6557.
20. D'Angeli, F.; Marchetti, P.; Cavicchioni, G.; Catelani, G.; Nejad, F. M. K. *Tetrahedron Asymm.* 1990, **1**, 155.

21. D'Angeli, F.; Marchetti, P.; Cavicchioni, G.; Bertolasi, V.; Maran, F. *Tetrahedron Assym.* 1991, **2**, 1111.
22. D'Angeli, F.; Marchetti, P.; Bertolasi,V. *J. Org. Chem.* 1995, **60**, 4013.
23. Gasparski, C. M.; Teng, M.; Miller, M. J. *J. Am. Chem. Soc.* 1992, **114**, 2741.
24. Teng, M.; Miller, M. J. *J. Am. Chem. Soc.* 1993, **115**, 548.
25. Guzzo, P. R.; Miller, M. J. *J. Org. Chem.* 1994, **59**, 4862
26. Deyrup, J. A.; Clough, S. C. *J. Org. Chem.* 1974, **39**, 902.
27. Talaty, E. R.; Yusoff, M. M.; Ismail, A.; Gomez, J. A.; Keller, C. E.; Younger, J. M. *Synlett* 1997, 683.
28. Shimazu, M.; Endo, Y.; Shudo, K. *Heterocycles* 1997, **45**, 735.
29. Hoffman, R. V.; Nayyar, N. K.; Chen, W. *J. Am. Chem. Soc.* 1993, **115**, 5031.
30. Hoffman, R. V.; Nayyar, N. K.; Chen, W. *J. Org. Chem.* 1992, **57**, 5700.
31. Hoffman, R. V.; Nayyar, N. K.; Chen, W. *J. Org. Chem.* 1993, **58**, 2355.
32. Hoffman, R. V.; Nayyar, N. K. *J. Org. Chem.* 1995, **60**, 7043.
33. Hoffman, R. V.; Nayyar, N. K.; Chen, W. *J. Org. Chem.* 1995, **60**, 4121.
34. Tantillo, D. J.; Houk, K. N.; Hoffman, R. V.; Tao, J. *J. Org. Chem.* 1999, **64**, 3830.

CHAPTER 7

β-LACTAMS: CYCLIC AMIDES OF DISTINCTION

AJAY K. BOSE, MAGHAR S. MANHAS, BIMAL K. BANIK, and VAIDYANATHAN SRIRAJAN
George Barasch Bioorganic Research Laboratory, Department of Chemistry and Chemical Biology, Stevens Institute of Technology

1. AN OVERVIEW

Nine decades of β-lactam studies have created a vibrant sub-discipline of chemistry that continues to attract large numbers of medicinal and synthetic chemists, biochemists, and biologists.[1–5] Penicillins, cephalosporins, and other β-lactam antibiotics have saved millions of lives since 1945. The worldwide annual sale of these antibiotics runs into many billions of dollars. Publications in technical journals, patents, books, and writings in the popular press are adding to a vast body of information that is of great value to the expert as well as the novice in many areas of science and technology. The following sections highlight special features of the chemistry of this heterocycle rather than present a comprehensive review. The role of the amide group in the synthesis and transformation of various types of β-lactams is accorded special attention.

1.1. Names and Structures

β-Lactams are a class of cyclic amides that are derived from β-amino-propionic acid and are represented by the structural skeleton **1**. They belong to the family of heterocycles named "2-azetidinone" and contain one nitrogen as the heteroatom which is joined to a carbonyl group. The two quaternary carbon atoms in the ring can be chiral; the nitrogen atom carries a hydrogen atom, an aryl, alkyl, or some other group or the nitrogen atom can be part of another ring—as in a fused β-lactam such as penicillin (**2**).

The Amide Linkage: Selected Structural Aspects in Chemistry, Biochemistry, and Materials Science,
Edited by Arthur Greenberg, Curt M. Breneman, and Joel F. Liebman
ISBN 0-471-35893-2

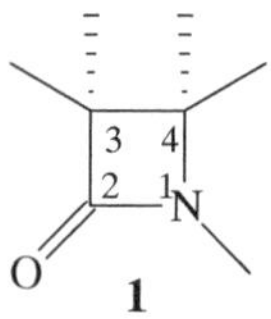

Another type of fused β-lactam that has served as an intermediate during the transformation of a penicillin to a cephalosporin derivative or the total synthesis of cephalosporin by Woodward et al. is shown by the structural skeleton **3** (for several specific examples, see later sections).

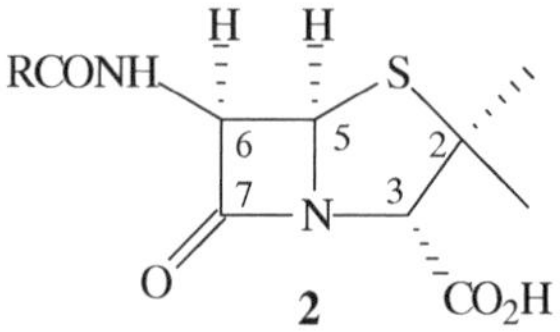

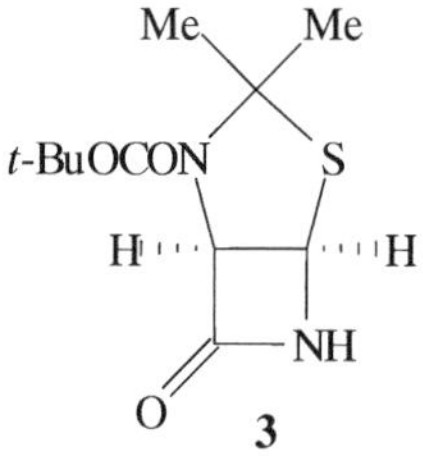

For the sake of convenience, various trivial names such as penams, cephams, carbapenams, oxacephams, etc., are in wide use with an informal numbering system that allows easy comparison with penicillin—the first β-lactam of clinical importance (for other examples, see later sections).

The complicated formal nomenclature and numbering system for bicyclic or tricyclic β-lactams appear to be used mostly by Chemical Abstracts and patent lawyers.

1.2. The Role of Serendipity

The field of β-lactams is a rich area to study the influence of accidental findings and serendipity, on some momentous developments in chemistry and medicine. The first authentic member of the β-lactam family was prepared by accident rather than by design. It was obtained unexpectedly in 1907 in the course of a study on ketenes by Staudinger and his students.[6] Since then the cause of β-lactam chemistry has been advanced at intervals through serendipity. After the initial preparation of several aryl and alkyl substituted β-lactams by variations of the key process of reacting a ketene with an imine (now often referred to as the "Staudinger Reaction"), there was little new activity in the β-lactam field for nearly two decades.

In 1928, Alexander Fleming,[7–8] a bacteriologist in St. Mary's Hospital in London, was culturing a pathogen (various strains of *Staphylococci*) by a new technique that he had just learnt about from a seminar. He allowed the growth of microorganisms to proceed at room temperature instead of at a higher temperature in an incubator. After several days of vacation, when Fleming returned to his work, he found that one of his culture dishes had been spoiled by contamination with a mold—a common occurrence in a microbiology laboratory. At first he discarded this culture dish, but sometime later he made a closer examination when another microbiologist visited him. He discovered that colonies of the bacteria appeared to have undergone lysis around the mold growth. Fleming later noted that, "The appearance of the culture plate was such that I thought it should not be neglected". Photographic evidence of this phenomenon, which was kept on record proved many years later to be of much value (see below).

Fleming cultured this mold which was eventually identified as a strain of *Penicillium notatum*—a common organism that grows on bread and other items of food. Extracts of the mold were shown by systematic experiments to be successful in preventing the growth of several types of pathogenic bacteria. Fleming was unable to purify the extract, but he named the unknown agent responsible for the antibiotic activity as penicillin. He even treated conjunctivitis successfully in the eyes of one of his friends with his penicillin solution.

Harold Raistrick, an expert on isolating natural products and determining their structure, became interested in Fleming's discovery. But penicillin proved to be quite unstable and eluded all efforts by Raistrick to isolate a pure compound from the "mold juice." A few years later, Ernst Boris Chain at Oxford University succeeded in extracting penicillin as an impure brown powder.

More than a quarter century later, Ronald Hare who had worked in another section of St. Mary's Hospital as a contemporary of Fleming tried to reproduce Fleming's historic experiment without initial success. Only after he had studied the well-known sudden temperature variations in England did he come up with a plausible theory about the possible sequence of events.

It would seem that Fleming had left some of the inoculated Petri dishes at room temperature, instead of incubating them at the usual higher temperature. The bacteria would grow rather slowly at the normal room temperature of the laboratory in midsummer in London while he was away, golfing in Scotland for a few days. But, records show that a cold spell occurred between August 28 and September 6, 1928, followed by a warm spell. The cold spell further slowed down the growth of *Staphylococci* colonies while the mold would continued to grow and produce penicillin, which led to lysis or dissolution of the colonies near the growing mold. Using this temperature profile Hare succeeded in duplicating the original experiment and the pattern on the Petri dish in Fleming's photograph.

It is now known that penicillin interacts with growing colonies of bacteria and inhibits their growth by interfering with the formation of new cells. Thus, a clear zone of inhibition is the first normal effect seen in a Petri dish experiment. As a matter of fact, Fleming was never able to reproduce his original experiment when he had observed lysis of colonies. A "replica" of his original Petri dish (the nature of the replica is not explained anywhere) in the British museum shows the normal inhibition zones, but no lysis. Obviously, this was a replica of a later experiment.

Apparently, in the original experiment, through a concatenation of unusual circumstances, penicillin in enough concentration had reached nearly mature colonies of bacteria and had caused lysis—the end stage of penicillin's action.

Since the start of his work in 1921, Fleming was very familiar with lysis caused by the enzyme lysozyme that he had discovered in nasal secretion, tears, other body fluids, and egg white. With his mindset, Fleming was able to appreciate the potential significance of the lysed bacterial cells that he observed. He removed the mold, cultured it, and showed that the "mold juice" inhibited the growth of many pathogens.

Folklore has it that, some of the ancient societies used molds as a means of treating infections and wounds. On a more scientific basis, Pasteur and Jubert studied in 1877 the possibility of using bacterial antagonism for prevention of anthrax infection. In 1877, Garre made observations on the power of molds to destroy bacteria but did not pursue the matter further.

Fleming himself had failed to arouse the enthusiasm of the medical community for using his mold juice/penicillin for curing infections. What kept many doctors interested in his mold, however, was the point made by Fleming in his publication that penicillin destroyed most of the pathogens but it did not inhibit the growth of *B. influenzae* (this latter organism—also known as Pfeiffer's bacillus—is not the cause of influenza). Using the mold juice of Fleming it was thus possible to isolate the elusive *B. influenzae* from bacterial cultures (for example, from throat swabs). Unlike others before him who had found antibacterial activity in certain molds, Fleming had succeeded in keeping his particular strain of penicillin-producing mold in circulation. It is amusing to note that when Chain in Oxford was studying Fleming's work on lysozymes and penicillin, he almost collided one day with a technician in the corridors of his laboratory who was carrying cultures of Fleming's mold for diagnostic tests on *B. influenzae*. Chain did not have to go very far to start his laboratory experiments on *P. notatum* and penicillin!

It is now obvious that the strain of *Penicillium notatum* that had landed in Fleming's Petri dish was special in that it was a comparatively high producer of penicillin. None of the various strains of *P. notatum* in the collections of Charles Thom in the US, who had correctly identified Fleming's mold, produced any noticeable amount of penicillin. Thus, it was a unique combination of unusual circumstances and Fleming's keen power of observation, intuition, and his earlier research that led to the discovery of an antibiotic that changed the history of medicine. There were several instances of serendipity that led to the commercial manufacture of penicillin G as a life-saving drug. During World War II, it became obvious that the large scale preparation of crude penicillin using the extraction and purification methods of Chain at Oxford University was not possible with German bombs falling all over England. Samples of the original *Penicillium* mold of Fleming were brought into the U.S. by Howard Florey and

Norman Heatley, of Oxford University. It was arranged that the most modern facilities of the Northern Regional Research Laboratory in Peoria, Illinois, would be used for developmental work for the large-scale production of penicillin. This laboratory among other projects was doing extensive work on the utilization of corn products.

It was a fortunate accident that corn strip liquor was one of the growth media tested for culturing the mold. The rate of production of penicillin increased greatly. Eventually, a pure crystalline penicillin salt was obtained and its molecular formula established by scientists at Squibb in New Jersey. It became obvious, that this penicillin (named penicillin G) was different from the penicillins studied in England! With pure forms of American and British penicillins available, it was realized that they had the same penam nucleus but differed in their side chain. Widely used penicillin G (**4**) is benzyl penicillin, while British penicillin F is 2-pentenylpenicillin (**5**). Corn strip liquor contains phenylacetic acid (**6**), which induces the biosynthesis of benzylpenicillin.

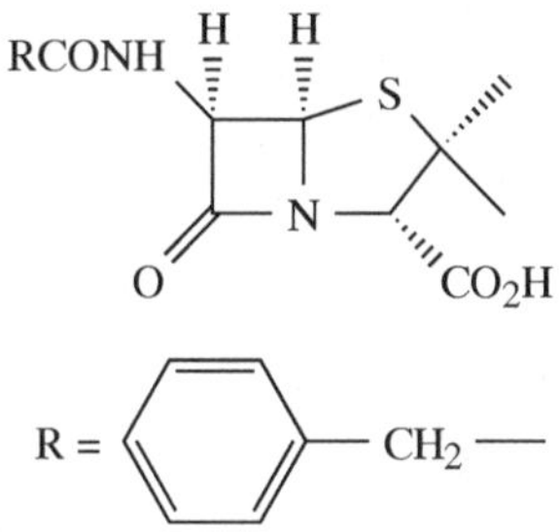

RCONH H H S N O CO_2H

R = $C_6H_5CH_2$—

4

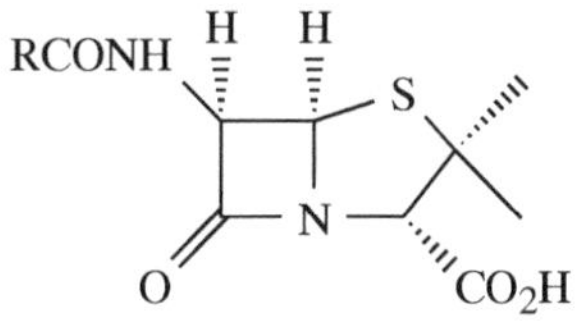

RCONH H H S N O CO_2H

R = $CH_3CH_2CH{=}CHCH_2$—

5

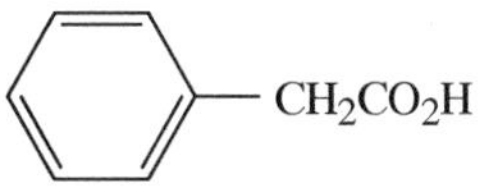

$C_6H_5CH_2CO_2H$

6

When Florey and Heatley had arrived in America with their sample of Fleming's *Penicillium* mold, the U.S. Government could have sent them to the newly built Regional Research Laboratory of the Agriculture Department, which was in the South. This laboratory was developing uses for peanuts and had no supply of corn strip liquor. Had Heatley worked there, the history of penicillin development and β-lactam chemistry would surely have been very different.

After corn strip liquor was found to be ideal for growing the penicillin mold, a search had been started for different strains of *Penicillium notatum* to further increase the yield of penicillin. The best strain was found on a rotten cantaloupe in a local market! This strain was improved by altering it genetically. For the next ten years, most commercial fermentation was carried out with this mold.

Serendipity was very much in evidence in the discovery of cephalosporins.[8] Prof. Guiseppe Brotzu in Sardinia studied in 1945 the microbial flora of seawater near a sewage outfall off the coast of the island. He was able to isolate an organism of the *Cephalosporium* species which produced several antibiotics. Unable to make much progress in his research with the limited resources available to him, he published his findings in an obscure local journal and hoped that further research would be undertaken elsewhere. He failed to create interest in his work in Italy. Taking advantage of a chance acquaintance with a British Public Health Officer, he was able to get the Oxford group interested, and sent a culture of his organism to Florey in 1948. Some twenty years later, cephalosporin antibiotics (**7**) came into wide use clinically. Cephalosporin C (**7**), the compound isolated from Brotzu's culture had to be modified extensively to obtain useful antibiotics (e.g., **8, 9**).

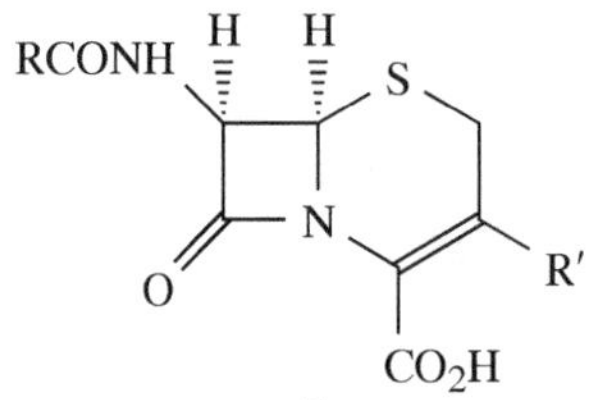

RCONH, H, H, S, N, O, R′, CO_2H

7

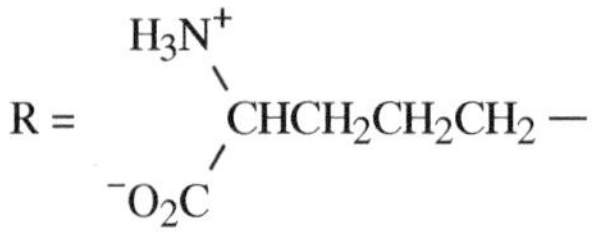

R = H_3N^+(${}^-O_2C$)$CHCH_2CH_2CH_2-$

R′ = $-CH_2OAc$

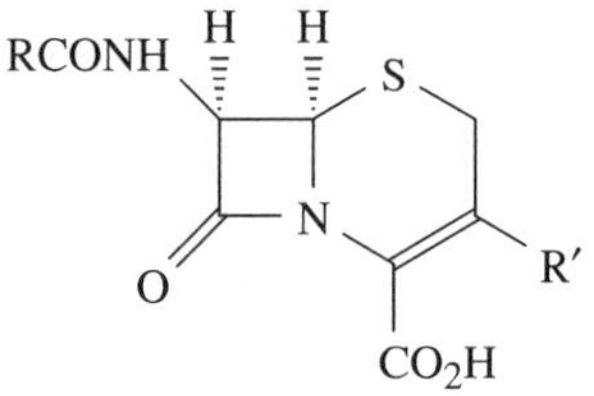

RCONH, H, H, S, N, O, R′, CO_2H

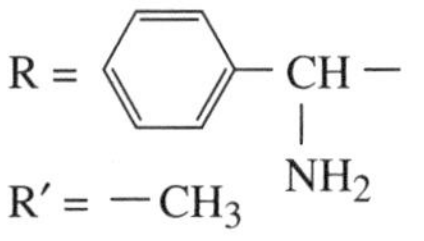

R = C_6H_5–CH(NH_2)–

R′ = $-CH_3$

8

RCONH, H, H, S, N, O, R', CO_2H

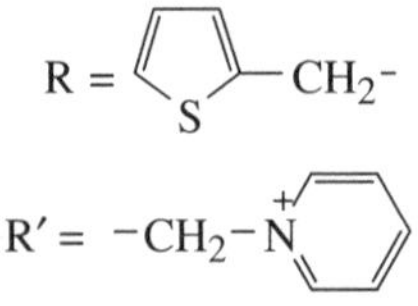

R = 2-thienyl–CH_2–

R′ = $-CH_2-\overset{+}{N}$(pyridinium)

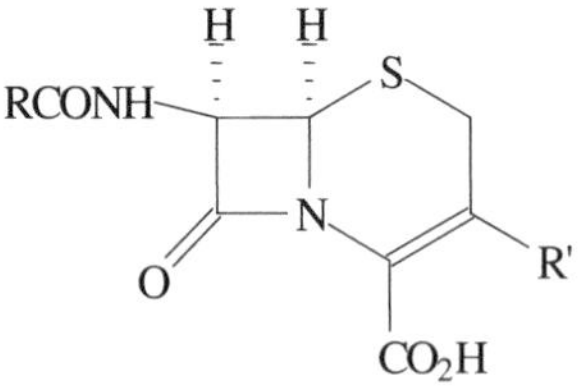

9

A recent instance of serendipity connected with β-lactams may be cited. A small synthetic project at Schering-Plough Company on β-lactam antibiotics did not lead to very promising results. Just before this project was to be terminated, one of the minor β-lactam by-products (**10**) of the main synthesis was found to show dramatic lowering of cholesterol level in the serum.[9] This accidental lead was then pursued vigorously. There appears to be a strong possibility that one or more β-lactams might become a new type of drug for cholesterol lowering.

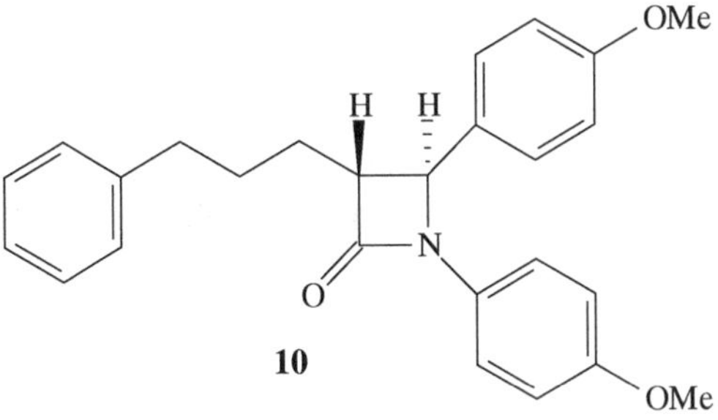

10

1.3. β-Lactams in Nature

Penicillins were the first examples of naturally occurring β-lactams. Cephalosporin C came next with Cephamycins (**11**) following in a few more years. All of these compounds are bicyclic heterocycles (mono aza, mono thio with an α-amido substituted, fused *cis*-β-lactam).

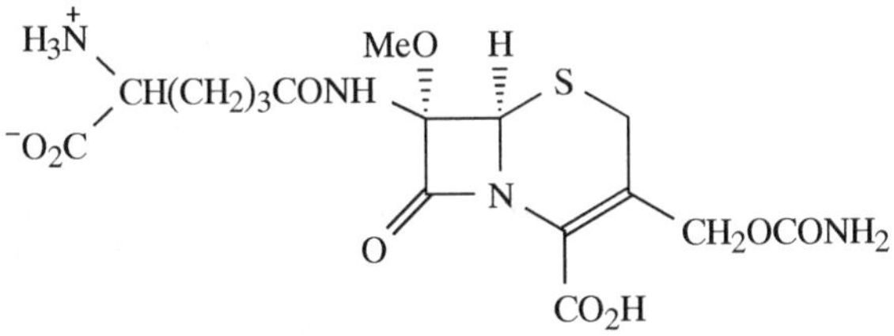

11

As is sometimes the case with scientists, generalization on the basis of a few examples became a dogma. Thus, for many years experts on β-lactams maintained that:

(a) β-lactam antibiotics must have an α-amide side chain,
(b) the β-lactam must have *cis*-stereochemistry,
(c) the antibiotic must be a fused bicyclic β-lactam with a sulfur containing five- or six-membered ring,
(d) a free carboxylic acid must be present, joined by a carbon to the β-lactam nitrogen.

The discovery of thienamycin (**12**) came as a rude shock because this does not have an amide side chain, the β-lactam is of *trans*-stereochemistry and the sulfur is outside the five-membered ring that is fused to the β-lactam ring.

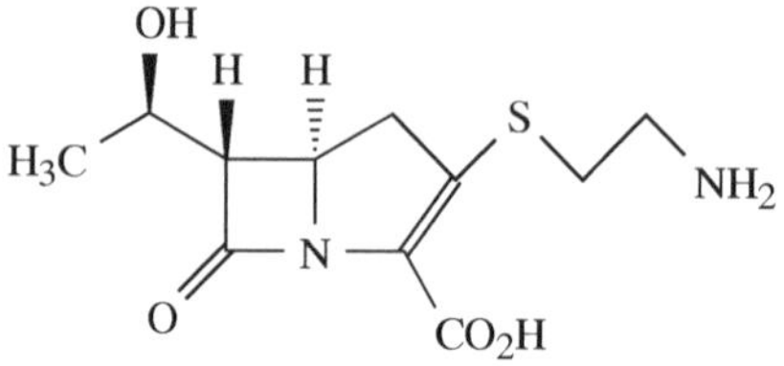

12

Clavulanic acid (**13**) is a fused bicyclic β-lactam with the right type of free carboxylic acid. But, it has no amide chain and sulfur is replaced by oxygen.

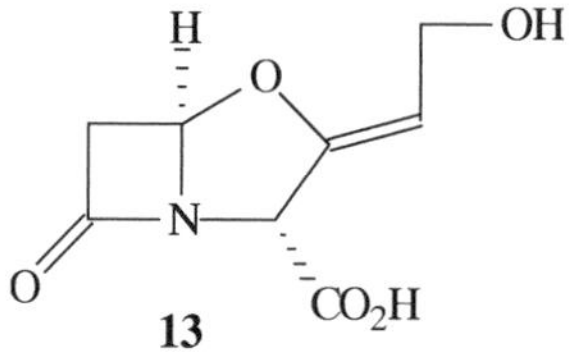

13

Nocardicin (**14**) proved to be a monocyclic α-amido-β-lactam with a free carboxylic acid in the proper place, but there is no substituent at the β-position of the β-lactam, nor is there a sulfur in the molecule.

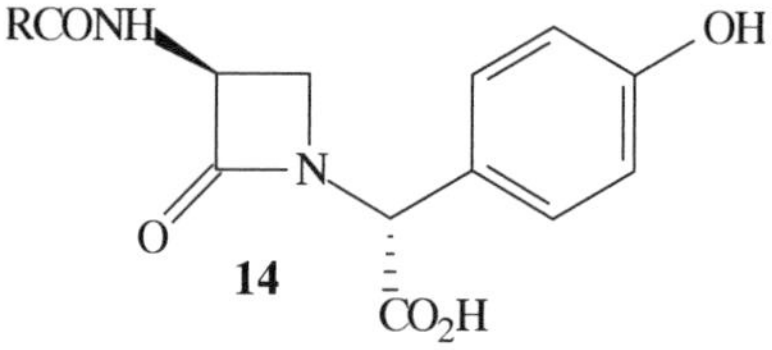

14

Even a simpler antibiotic was found in nature—monobactams (**15**) were discovered among the decaying, acidic pine needles in the Pine Barrens of New Jersey. This family has an α-amido (or α-dipeptide) side chain but no β-substituent on the β-lactam ring; also the cyclic amide nitrogen is substituted by an inorganic acid ($-SO_3H$) group. The same family of natural β-lactams was also discovered simultaneously in Japan.

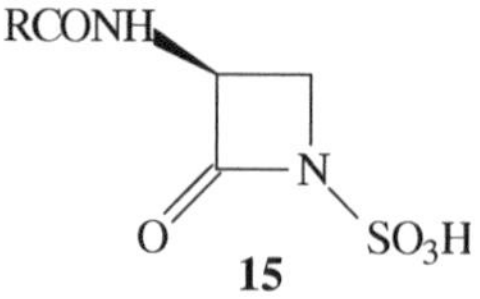

15

Over the decades a few unusual, β-lactam containing compounds have been found in nature which are not antibiotic. For example, a peptide derivative containing a β-lactam ring and named Wildfire Toxin (**16**) is produced by a bacterium *Pseudomonas tabaci* that causes the "wildfire disease" of tobacco leaves. A few other oligopeptide derivatives have been reported that contain a β-lactam ring. Two steroid β-lactams (**17**) were first discovered in 1967 that were produced by the ground cover plant *Pachysandra*.

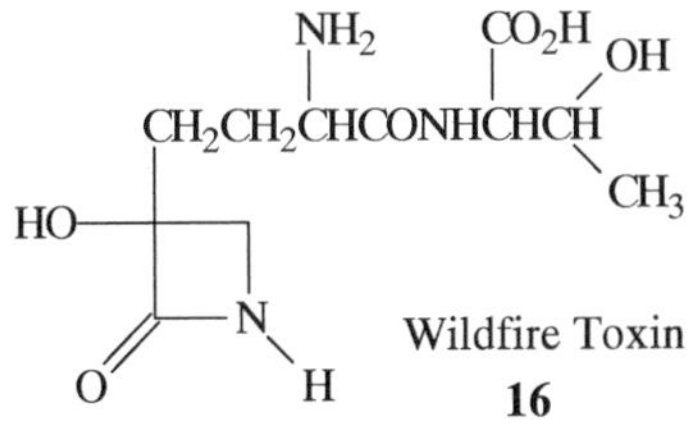

Wildfire Toxin

16

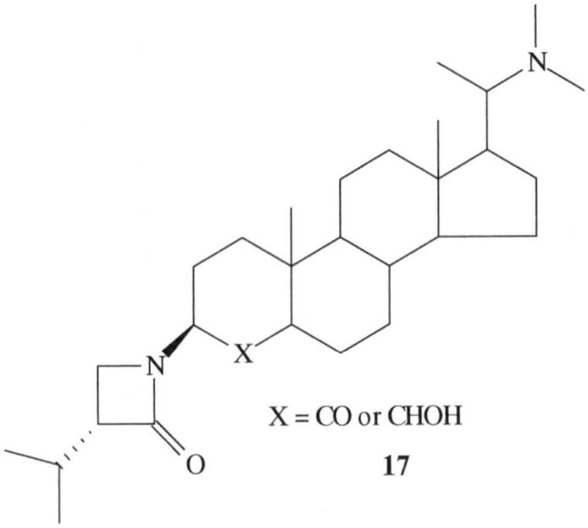

X = CO or CHOH

17

1.4. β-Lactams as Synthons

The few synthetic β-lactams known before World War II had been found to be quite stable chemically. However, penicillins proved to be very unstable to many chemicals. They also, showed a remarkable propensity for molecular rearrangement—much to the discomfiture of chemists who were trying to deduce the structure of these compounds. Interestingly enough, facile molecular rearrangements and the ease of conducting certain types of reactions with penicillins, became the key to the potential use of natural and synthetic β-lactams as synthons for many interesting products.

Manhas et al.[10,11] wrote early reviews that examined molecular rearrangements of variously substituted β-lactams in the context of their use as intermediates for diverse structures, such as amino acids, amino sugars, alkaloids, natural and non-natural peptides, and heterocycles.

Ojima[12,13] has developed a versatile "β-lactam synthon method", which depends on the facile cleavage of the C_4–N bond of β-aryl β-lactams by hydrogenolysis. Using homochiral β-lactams with appropriate substituents, this reductive cleavage method leads to the formation of optically pure aromatic α-amino acids, α-hydroxy acids, dipeptides, and azetidines which in turn can be converted to polyamines, polyamino alcohols, and polyamino ethers.

Presently many β-lactams—racemic or homochiral—are available for acting as versatile synthons. The need for the large-scale production of optically active phenylisoserine for the semisynthesis of Taxol™ and Taxotere™—two important antitumor drugs—has focused attention on α-hydroxy-β-lactams as advanced intermediates for the manufacture of life-saving pharmaceuticals.[14]

Microwave assisted, stereocontrolled synthesis of α-hydroxy-β-lactams has shown promise for providing easy access to a variety of β-lactams.[15] The potential of chemoenzymatic reactions for producing both enantiomers of optically active hydroxy-β-lactams has been demonstrated.[16] There is little doubt that β-lactam chemistry will continue to grow and attract increasing attention from synthetic and medicinal chemists. Several illustrative examples of β-lactam synthons are provided in later sections.

2. MOLECULAR SHAPE OF β-LACTAMS

The amide group has a key role in biological chemistry. For proteins and peptides, the reactivity, structure, and hydrogen bonding ability of this functional group provide a wide scope for many intermolecular and intramolecular interactions. Polarizability of the carbonyl group and the unshared electron pair of the nitrogen bonded to it allow for graded variation of electron sharing between the sp^2 carbon atom and the two hetero atoms linked to it, as expressed by the resonance structures **18** and **19**.

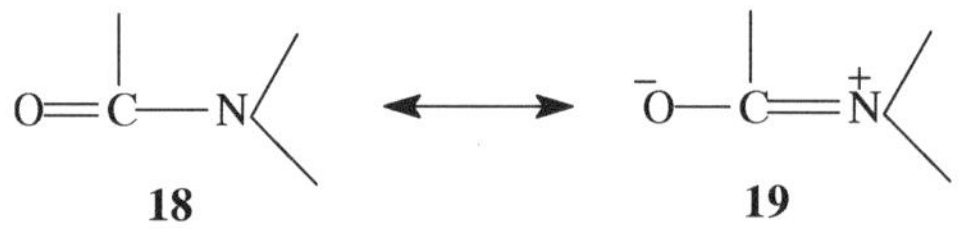

$$O{=}C{-}N\langle \quad \longleftrightarrow \quad {}^{-}O{-}C{=}\overset{+}{N}\langle$$

18 **19**

Cyclic amides have structural features that may modify the relative contributions of the polar and non-polar resonance structures along with other parameters such as bond lengths and bond angles. Such modification may lead to alteration in chemical reactivity. The shape of the cyclic molecule and its lack of flexibility—as compared to an acyclic amide—may also alter the possible contributions of anchimeric assistance from substituents.

Amides that are part of 6-membered rings (δ-lactams) (**20a**) or larger rings suffer little constraint due to ring formation. Therefore, they are very similar to acyclic amides, and the infrared absorption bands of the amide carbonyl in acyclic and δ-lactams are in the same frequency range. In contrast, five-membered γ-lactams (**20b**) have much more limited option for the ring shape; the amide carbonyl of β-lactams absorb at a different frequency range than γ-lactams and acyclic amides.

For β-lactams, the distortion of normal bond angles is severe enough to strongly affect the properties of the amide group. When a β-lactam is fused to a second ring, there are additional distortions that modify the amide group in its geometry and chemical reactivity; the amide carbonyl absorption bands of β-lactams are at much higher frequency than those of acyclic amides (see Section 2.2.2).

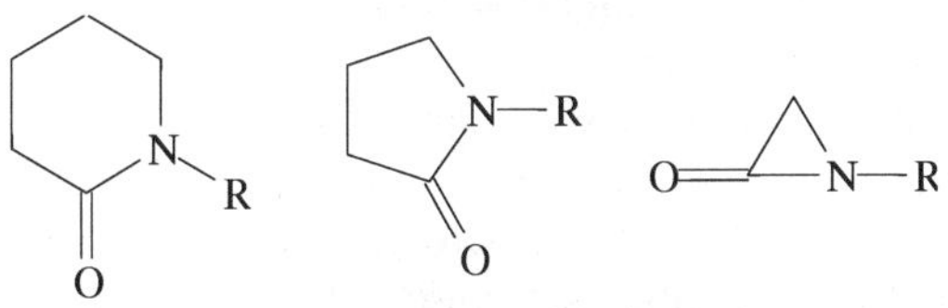

20a **20b** **20c**

The α-Lactams (**20c**) are subject to even more distorting forces. A small number of heavily substituted examples of this family of 3-membered cyclic amides have been synthesized, but, no naturally occurring α-lactams seem to be known.[17]

To state things in an informal way,

- α-lactams are too rigid and unstable to have special "character;"
- γ-lactams and larger lactams are too flexible and stable to be "distinguished;"
- β-lactams have just enough flexibility of structure and chemical reactivity to be cyclic amides of great "distinction."

Or, to paraphrase Goldilocks:

- α-lactams are too hard;
- γ-lactams are too soft;
- β-lactams are just right!

This chapter devotes itself to β-lactams and their special characteristics.

2.1. Structural Features

Until the beginning of World War II, β-lactams had failed to attract special attention. No synthesis of β-lactams was achieved except through accident. Thus, in 1923 Breckpot[18] was conducting further studies on a synthesis of amides using a Grignard reagent that had been reported in 1904; one of the products from the reaction of a β-amino ester (**21**) with a Grignard reagent proved to be a β-lactam (**22**).

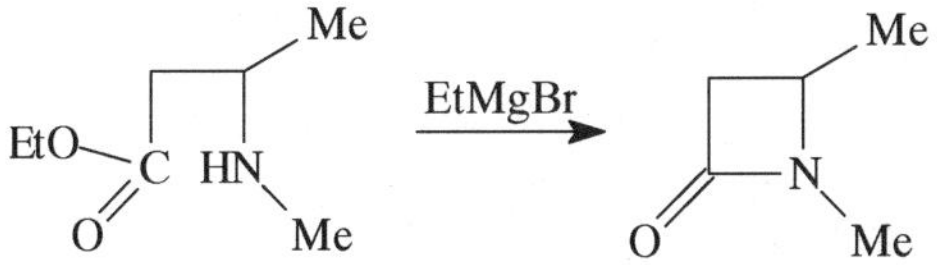

EtMgBr

21 **22**

The β-lactams known until the discovery of penicillin were all heterocycles with only alkyl or aryl substituents. They proved to be fairly stable and thus did not lead to any extensive study of their chemical reactions. As noted earlier, penicillins are very active chemically. The β-lactam ring behaves like an acid chloride rather than an amide: it is cleaved very readily by alcohols, amines, acids, and bases. Also, the penicillins undergo various rearrangements that involve the four-membered ring.

When the fused thiazolidine β-lactam structure (**23**) was proposed as a possibility for penicillins, there was much skepticism.[19] Robinson, the leading figure in penicillin research in Great Britain, was a staunch supporter of the alternative thiazolidine-oxazolone structure (**24**).

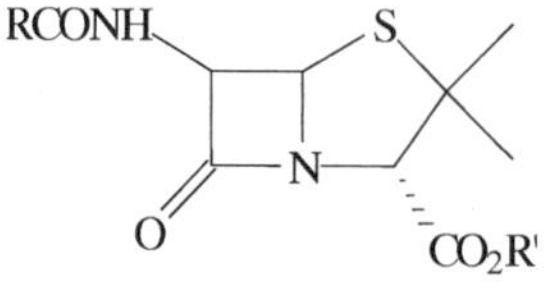

23 $R' = Me$, $R = —CH_2Ph$

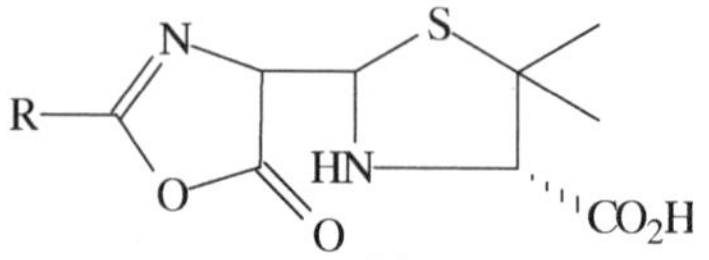

24

Woodward[19] pointed out that in penicillins, the fusion of the β-lactam ring with the five-membered thiazolidine ring in **23** could generate enough strain to make the amide group of the β-lactam very reactive. On the basis of thermochemical data on a monocyclic β-lactam (**25**), and the open chain β-amino ester (**26**), the methyl ester of penicillin G (**23**), and the corresponding β-amino ester (**27**), he estimated that penicillin is strained to the extent of about 6 kcal $mole^{-1}$ compared to the monocyclic β-lactam.

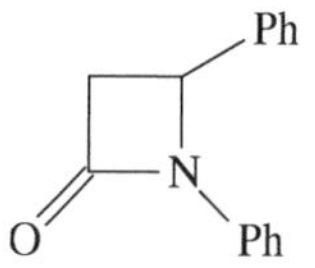

25 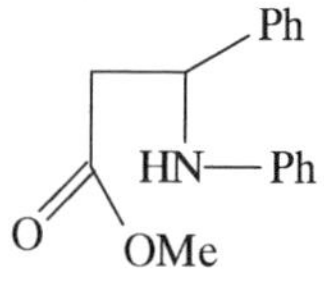 **26** 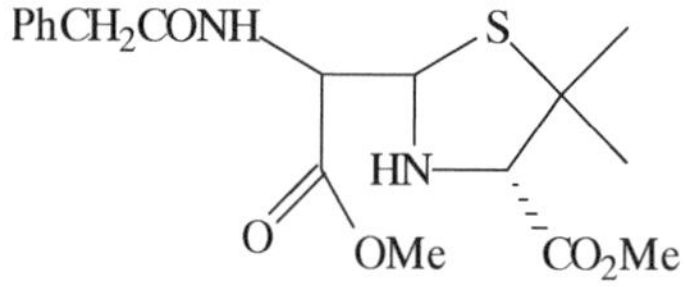 **27**

The validity of the fused β-lactam structure for penicillins was established by single crystal X-ray diffraction studies on penicillin G by Crowfoot and Rogers Low.[20] The fused bicyclic β-lactam structure for cephalosporins (**28**) was also confirmed by single crystal X-ray diffraction studies.[21]

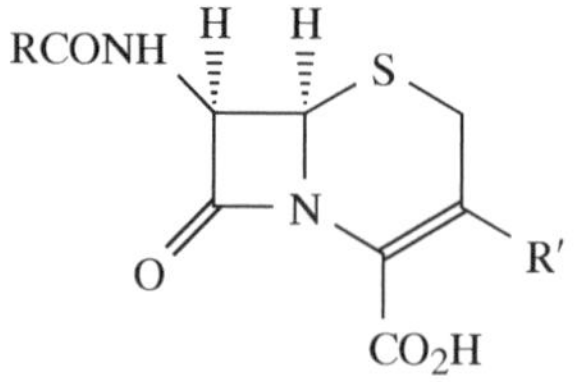

$R = (H_2N)(HO_2C)CHCH_2CH_2CH_2—$

$R' = AcOCH_2—$

28

In the 1950s, it became increasingly apparent that there is a role for conformation—as separate from configuration—for influencing chemical and biological activity of molecules. Conformational evidence for flexible molecules obtained from single crystal X-ray diffraction studies are suspected because the crystal lattice may require a different conformation than solutions of the same compound (see Section 2.2.4). In the case of several β-lactams, however, X-ray diffraction findings and NMR studies have indicated the same conformation.

2.2. Spectral Properties

During the early stages of penicillin research the major instrumental method for structure determination was single crystal X-ray diffraction. In the absence of computers, this method was very time consuming but highly reliable. Infrared spectroscopy was a new tool that became available to organic chemists during World War II. Proton and ^{13}C NMR studies became possible much later. Mass spectrometry, too, is a later addition to the list of tools for structure determination. These and other physical measurements are widely used techniques now for studying both configuration and conformation of β-lactams.

2.2.1. X-ray Diffraction of Single Crystals. Monocyclic β-lactams have been shown by X-ray diffraction studies[22] to be planar. In case of *N*-aryl β-lactams the valence bonds of the nitrogen atom have the shape of a very flat pyramid and the *N*-aryl ring is essentially in the plane of the four-membered ring.

In case of bicyclic β-lactams, the nitrogen is somewhat above the plane of the three carbons in the β-lactam. Thus, the four-membered ring is non-planar. The thiazolidine ring in penicillin is constrained by the β-lactam ring into two stable conformations. One of these conformations (**29**) places the carboxy group at C-3 in nearly axial conformation; the 2β-methyl group is also then in a pseudo axial conformation while the 2α-methyl assumes the pseudo-equatorial conformation. In the other conformation (**30**) of the thiazolidine ring, the carboxy group and the 2β-methyl group are nearly equatorial and the 2α-methyl group is pseudo-axial.

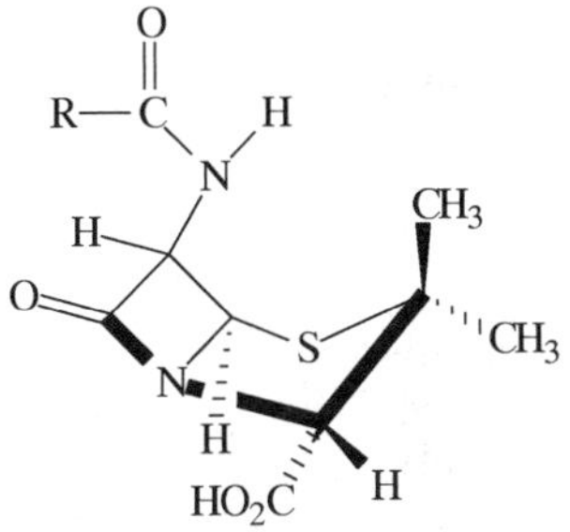

29 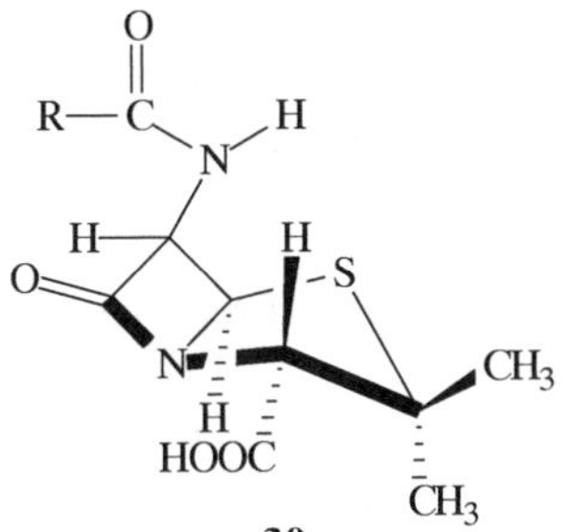 **30**

The carboxy group at C-3 and the 2β-methyl group are switched from pseudo-axial to pseudo-equatorial conformation (as in **30**) when the side chain of penicillin G (**23**, R = $PhCH_2$, R^1 = H) is changed to convert it into ampicillin (**31**).

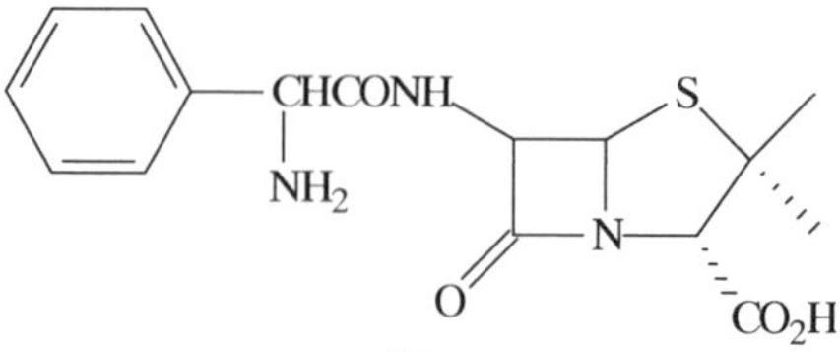

31

Interestingly enough, the same conformational changes occur when penicillin is oxidized to sulfoxide (**32**). Unlike ampicillin, this penicillin sulfoxide is biologically inactive. Attempts to correlate lack of planarity of the β-lactam ring as shown by X-ray diffraction in fused β-lactams with antibiotic activity has not been very successful, either for penicillins or cephalosporins or their analogs.

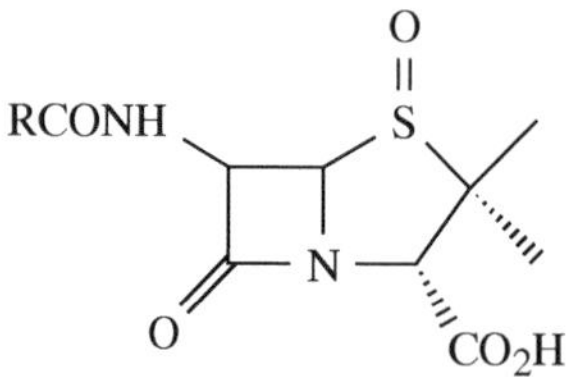

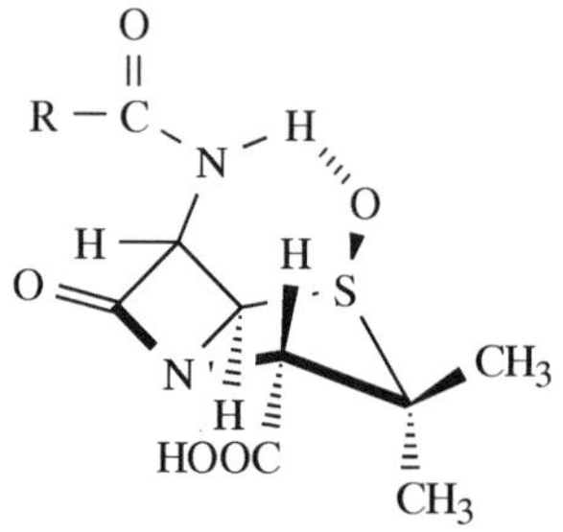

32 **32**

Single crystal X-ray diffraction studies by various laboratories have shown that the amide side chain at the 6β position of penicillins G can be folded over the bicyclic nucleus (as in **33**) or be extended away from it (as in **34**, see Fig. 7.1).

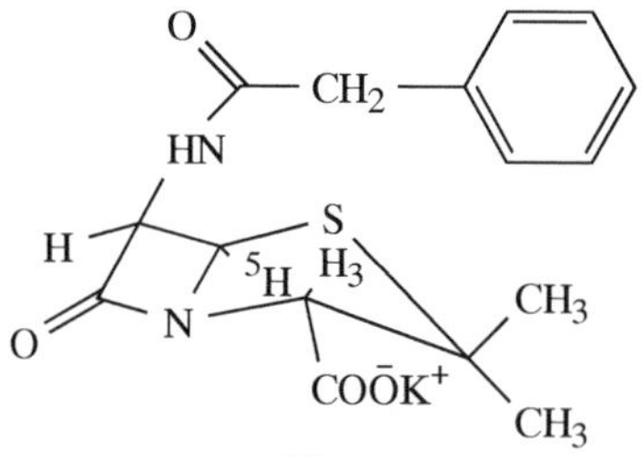

33a 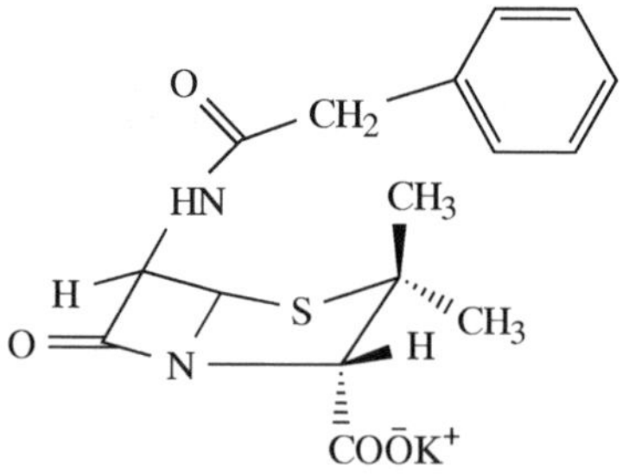 **33b**

In both conformations of the thiazolidine ring, 3β-H is close enough to the 2β-methyl group to experience substantial NOE.[23] Molecular models indicate that in the carboxy-equatorial conformation (as in **30**), the 2α-methyl group would exert NOE on the 5α-H; but in the carboxy-axial conformation (as in **29**) the 2α-methyl group is too far away for interaction with the 5α-H proton.[23]

The conformation of the side chain and the envelope conformation of the thiazolidine ring are variable; but the four-membered β-lactam ring is of fixed,

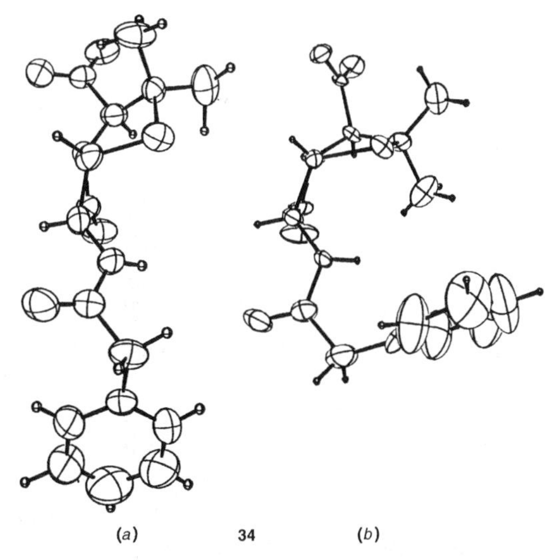

FIGURE 7.1. Overall molecular size and shape comparison of Penicillin G in (*a*) procaine penicillin G and (*b*) potassium penicillin G. The molecules are viewed parallel to the β-lactam ring. The side chain is extended in (*a*) and is coiled in (*b*). In (*b*), the amide proton of the side chain of the potassium salt of penicillin G is shielded by the phenyl group and is not involved in any hydrogen bonding. In contrast, the procaine salt of penicillin G forms head-to-tail 'helical chains' through hydrogen bonding.

nearly planar geometry. Dexter and van der Veen[24] have established that the crystalline potassium salt of penicillin G exists in the carboxy-axial/"folded side chain" conformation. In contrast, the crystalline procaine penicillin G salt assumes the carboxy-equatorial/"extended side chain" conformation[24]. Ampicillin trihydrate has the extended carboxy-equatorial/extended side chain conformation.

Proton NMR studies on penicillin esters (also see Section 2.2.4) in $CDCl_3$ solution have shown that substantial NOE interaction can be observed and the conformation of penicillins can be derived with confidence.[23] In most cases the X-ray data are in accord with the molecular geometry predicted on the basis of NOE studies in $CDCl_3$ or benzene solution.

2.2.2. Infrared Spectroscopy. The carbonyl absorption of amides in their infrared spectra is strongly influenced by the size of the lactam ring. Acyclic amides and δ-lactams (**20a**) show absorption in the same range—at about 1660 cm^{-1}. For γ-lactams (**20b**) the absorption is at about 1720 cm^{-1}. In most cases a β-lactam can be recognized from its infrared band for the amide carbonyl. Substituted monocyclic β-lactams display strong carbonyl absorption in the 1730–1760 cm^{-1} range. Fused β-lactams show a strong band in the 1770–1780 cm^{-1} range. In case of the fused β-lactam, such as anhydropenicillin (**35**), the infrared carbonyl band shifts to 1820 cm^{-1}; this compound is without antibiotic activity.

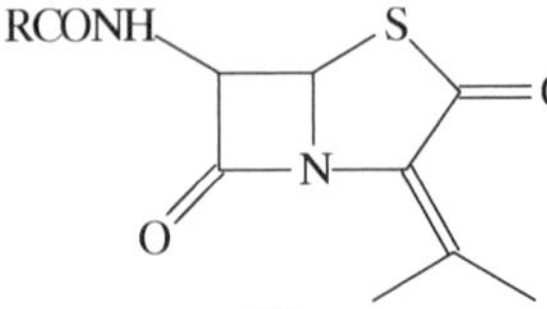

35

The higher amide carbonyl frequency observed for bicyclic β-lactams as compared to monocyclic β-lactams has been ascribed to ring strain resulting in the dominance of the resonance structure (**18**). In 1969, Morin et al.[25] noted that for several bicyclic β-lactams there was a rough parallel between higher β-lactam IR frequency and higher antibiotic activity against *Staphylococcus aureus*. The infrared spectrum by itself, however, is not a reliable indication of useful antibiotic activity. In the early stages of penicillin chemistry during World War II, infrared spectroscopy was in its infancy. Nonetheless, the absorption spectra of synthetic monocyclic β-lactams were very helpful in considering the validity of various structures proposed for penicillin.

Desthiopenicillin (**36**) is a compound in which the sulfur linkage between the two ring systems has been removed. The infrared absorption of the β-lactam carbonyl in **36** was found to be at a much lower frequency than the same carbonyl in penicillins. The carbonyl frequency for **36**, which was within the range of that for monocyclic β-lactams was an indication of the correctness of the fused thiazolidine-β-lactam structure for penicillin.

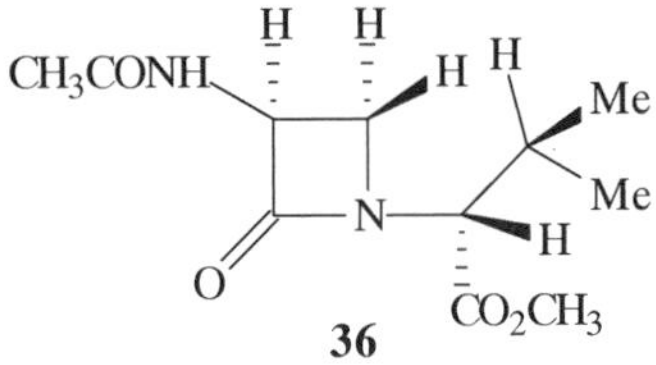

36

The β-lactam carbonyl frequency of penicillins and derivatives provides useful information regarding the presence of the β-lactam ring as well as the state of oxidation of the sulfur in the thiazolidine ring.[26] Oxidation of penams (**2**) to sulfoxide (**32**) shifts the frequency of the β-lactam carbonyl to 1805 cm^{-1}; for the sulfone the frequency is shifted even higher to 1805–1815 cm^{-1}. Similar shifts have been observed for cephalosporins and its derivatives.

Replacing the sulfur atom of a cephem by an oxygen as in 1-oxacephems (**37**) results in higher frequency shift ($1776.8 \rightarrow 1771.8\,cm^{-1}$) for the β-lactam carbonyl; transtition from cephem to carbacephem (**38**) has the opposite effect ($1776.8 \rightarrow 1757.0\,cm^{-1}$). The alkaline hydrolysis rates of these antibiotics varied considerably, the 1-oxa-cephem (**37**) was the fastest and the cephem was the slowest; on the other hand, in tests against certain microorganisms, 1-oxa-cephem (**37**) was a better antibiotic than the cephem while the carbacephem (**38**) was a poor third.

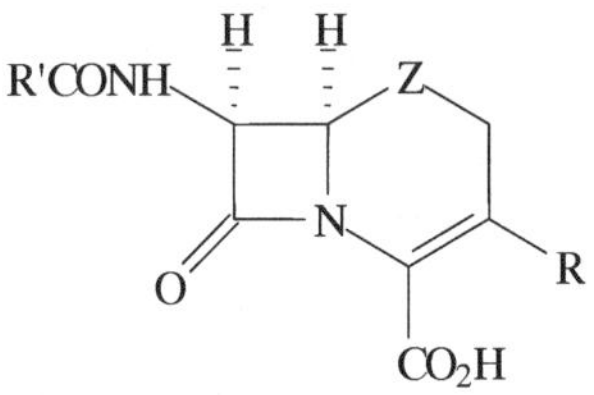

37, Z = O
38, Z = CH_2

2.2.3. Ultraviolet Spectroscopy. The nucleus of penicillins does not display any characteristic UV absorption spectra. The 3-cephem nucleus (**39**) which has several possible resonance structures shows absorption maxima at about 260 nm which is at longer wavelength than that expected for an *N*-acyl-α,β-unsaturated amino acid (**40**).

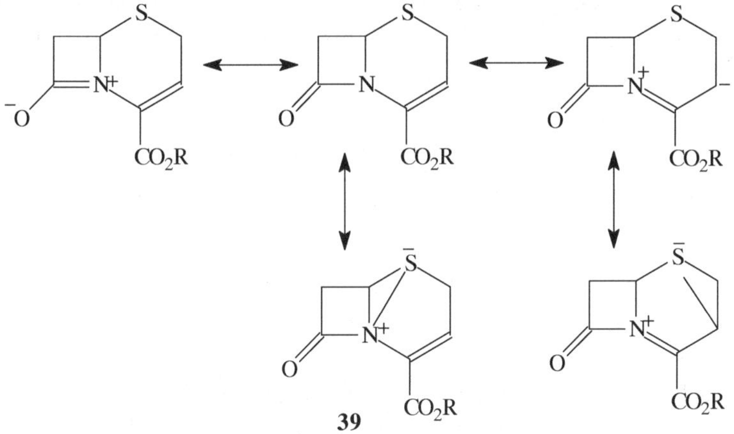

39

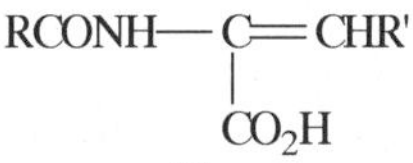

RCONH—C=CHR'
CO_2H
40

Manhas and co-workers[27] have reported that the *N*-aryl-β-lactams (**41**) displayed strong UV maxima at about 250 nm. They suggested that the chromophore for this intense UV absorption must be the aryl-N–CO– group, which allows the overlap of the π-orbitals of the aromatic ring, the p-orbital of the trivalent nitrogen with a lone pair of electrons, and the π-orbital of the amide carbonyl group. This overlap is most effective when the nitrogen valencies are planar and the aromatic ring is in the plane of the β-lactam ring. In support of this conjecture they cited their observation that 1-(*o*-bromophenyl)-2-azetidinone (**42**) shows a much weaker UV absorption than its *p*-bromo isomer (**43**). The *o*-bromo group must move the aryl ring somewhat out of the plane of the β-lactam ring because of steric hindrance.

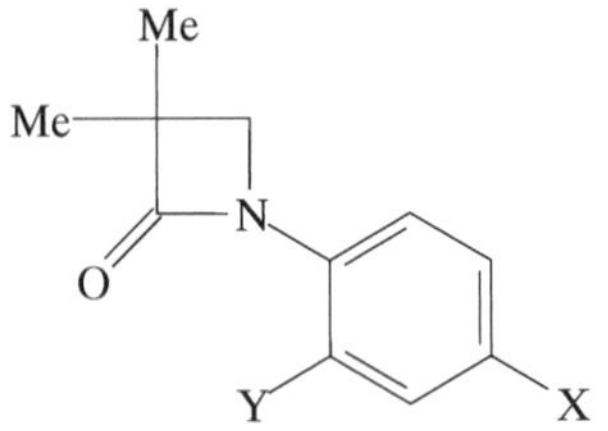

41, X = Y = H
42, X = H, Y = Br
43, X = Br, Y = H

N-aryl γ-lactams (**44**) are also characterized by strong UV maxima near 250 nm whereas the *N*-phenyl-δ-lactam (**46**) shows no such maximum. Substitutions on the γ-lactam ring apparently lead to a conformation which does not allow planarity of the aryl–N–CO– group anymore. Thus, 1-phenylpyrrolidin-2-one shows a strong absorption maximum at 246 nm but 1-phenyl-5,5-dicarbomethoxypyrrolidin-2-one (**45**) shows no maxima above 210 nm.

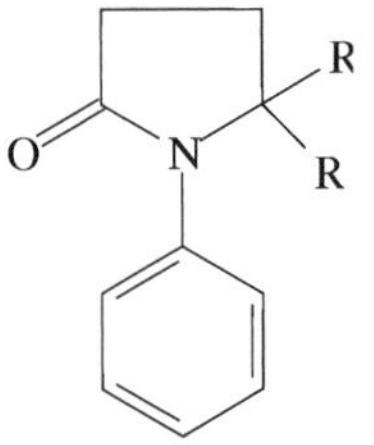

44, R = H
45, R = $-CO_2Me$

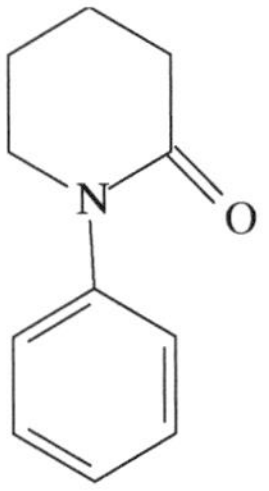

46

The aryl ring in 1-*o*-bromophenyl-2-azetidinones must not be very much out of the plane of the β-lactam ring, however. Additional evidence for the near planarity of the *o*-bromophenyl ring with the β-lactam ring is provided by a study of its proton NMR spectrum.

In Section 2.2.1 it has been noted that X-ray diffraction studies show that at least in the crystalline state the *p*-bromophenyl group in **47** is coplanar with the β-lactam ring.

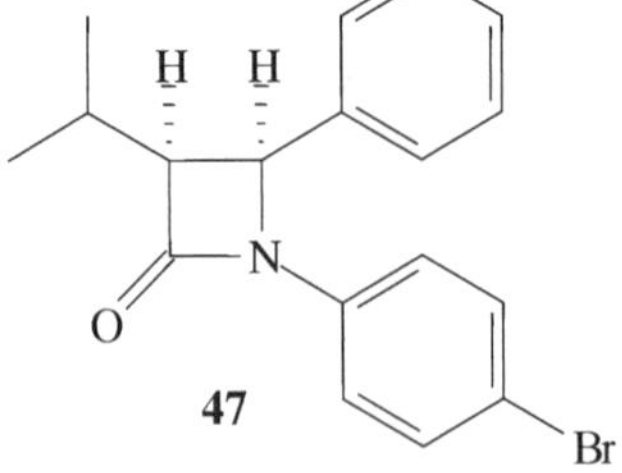

47

2.2.4. NMR Spectroscopy. Proton and ^{13}C NMR spectroscopy are perhaps the most widely used instrumental methods for obtaining structural information about β-lactams. The size of the coupling between the two β-lactam ring protons on adjacent carbons is a highly reliable indication of the stereochemistry of the heterocycle: *cis*-β-lactams show a J value of 4–6 Hz, the *trans*-isomers are characterized by J values of about 1.5–3.0 Hz. The Nuclear Overhauser Effect (NOE) is a specially useful tool for obtaining information about the conformation (as well as configuration) of penicillins and other β-lactams.

Proton NMR spectroscopy is a powerful technique for gathering information about the structure and dynamics of biologically active molecules in solution. But, the direct analysis of small organic molecules in biological fluids by conventional ^{1}H NMR spectroscopy is not possible because their signals will be swamped by proton signals of water. The water peak can be suppressed, however, by special techniques, some of which involve changes in the NMR spectrometer. A water signal suppression method that does not require modification of the NMR instrument, was described by Rabenstein and Fan.[28] These authors showed that by the addition of ammonium chloride, guanidinium hydrochloride or hydroxylamine hydrochloride to an aqueous solution, it is possible to decrease the transverse relaxation time (T_2) of water protons as compared to solute protons. This difference can be exploited by employing the Carr–Purcell–Meiboom–Gill[29,30] pulse sequence during the T_2 relaxation period. Rabenstein and coworkers[31] used chemical relaxation agents and appropriate pulse sequences to perform 2D NMR experiments, such as COSY (Homonuclear Correlation Spectroscopy) and J-Resolved Spectroscopy. Krishnaswamy[32] modified this pulse sequence and was able to record 2D NOESY (Nuclear Overhauser Correlation Spectroscopy) spectra with water peak suppression. It was possible now to obtain information on the conformation of organic molecules in aqueous solution from their proton NMR spectra.

This modified pulse sequence, which was named MODNOESY, was employed by Bose et al.[33] for studying the conformation of penicillin G sodium salt in water solution. The conventional NOESY pulse sequence was also used in D_2O solution for comparison purposes. The MODNOESY spectra were recorded in H_2O using guanidinium chloride as the relaxation agent. In both cases, interaction was observed between 5α-H and the 2α-methyl group, across the lower face (as in **33a**) of the bicyclic nucleus and between 3β-H and 2β-methyl group, across the upper face. Also observed was the interaction between the 2β-methyl group and the benzyl protons of the amide side chain and interaction between 3β-H and the benzyl protons; these interactions indicated that the benzyl side chain was in a "folded conformation" that brought it in the proximity of the β-face or upper face of the bicyclic nucleus.

Interestingly, the MODNOESY/H_2O spectrum revealed an additional interaction between the benzyl protons and the 2α-methyl group, which was absent in the NOESY/D_2O spectrum. Obviously, ion pairing between the guanidinium ion as the relaxation agent and the carboxylate ion at C-3 of penicillin G altered the conformation of the antibiotic molecule.

By reference to the earlier X-ray diffraction studies on penicillin G salts[24] (see Section 2.2.1) it was deduced that penicillin G has as its predominant configuration in aqueous solution the same carboxy-equatorial conformation that is found in the solid state for the procaine salt of penicillin. In presence of the chemical relaxation agent, however, the carboxy-axial conformation (as in **33b**) becomes the preferred one. The NMR spectra also indicated that in solution an equilibrium exists between the two envelope conformations of the thiazolidine ring of penicillin.

3. HIGHLY ENANTIOSELECTIVE SYNTHESIS

A number of books and review articles are available for providing detailed information to synthetic chemists interested in constructing the β-lactam ring with its two potential chiral centers[34]. However, increasing attention is now being paid to synthetic methods with high atom economy (high ratio of desired products to all chemical waste). Also, there is a growing need for obtaining both enantiomeric forms of compounds for studying physiological activity. This section, therefore, will describe almost exclusively the specific or highly enantioselective approaches reported recently (mostly in 1997 and 1998) that proceed from good to excellent yield.

3.1. Synthetic Strategies

In the early days of penicillin chemistry, the high reactivity of this fused β-lactam and its ease of rearrangement had led to the perception that β-lactams, in general, may be unstable. As a matter of fact, even penicillin derivatives in organic solvents can undergo a variety of reactions without disruption of the β-lactam ring. Monocyclic β-lactams, which are essentially planar in structure, are quite stable to several reagents and because of their molecular shape, they undergo a number of predictable, stereoselective reactions.

As noted before, the amide carbonyl of β-lactams is more like a ketone; therefore, it is more efficient in activating the adjacent methylene (or methine) group at C-3 than the amide function in open chain amides including peptides. It is therefore possible to introduce nucleophiles at the C-3 position of azetidin-2-ones with comparative ease to obtain more densely substituted β-lactams. The N–H group in N-unsubstituted β-lactams is much more basic than an acyclic amide N–H; therefore, it is possible to alkylate or acylate easily the N–H group of the β-lactam.

3.1.1. Stereoselective Reactions. Because the flat four-membered β-lactam ring is the scaffold, the substituent at C-4 has strong steric effect on the reactions at C-3. In general, an incoming group at C-3 will be *trans* to a bulky C-4 substituent.

Acyloxy groups at C-4 undergo a special type of reaction: they are easily removed to provide an acyliminium group (**48a**, or ion, **48b**) which behaves like a Michael acceptor that can add an appropriate nucleophile in a stereospecific

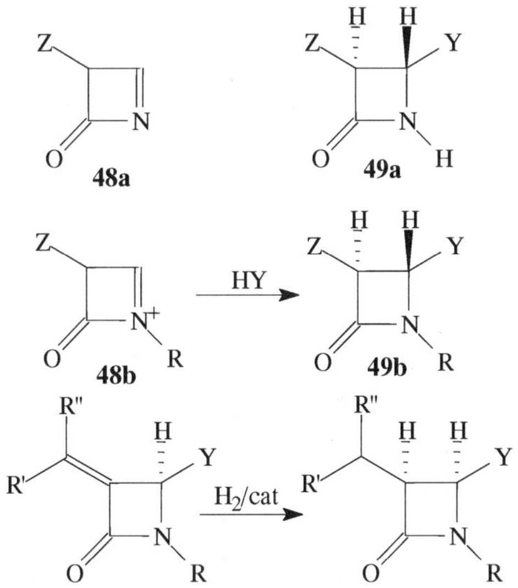

Scheme 1

fashion to give **49**: the incoming group is trans to the substituent at C-3 (Scheme 1).

When a substituent, such as an aryl or alkyl group or a ring segment, is present at C-4, an exo-alkene group at C-3 can be reduced catalytically to provide a *cis*-β-lactam; the catalyst and the hydrogen approach the plane of the ring from the face opposite to that containing the C-4 substituent (Scheme 1).

Advantage can be taken of this steric course of reactions at C-3 for converting a *trans*-α-hydroxy-β-lactam (**50**) to a *cis*-α-hydroxy-β-lactam (**52**) in two steps: oxidation to an α-keto-β-lactam (azetidin-2,3-dione, **51**) followed by sodium borohydride reduction involving *trans* approach to the C-4 substituent by the reducing agent. An alternative approach from 3α-hydroxy to 3β-hydroxy is available based on the observation that good leaving groups at C-3 can undergo S_N2 reactions—thereby providing a convenient method for changing a *cis*-β-lactam to a *trans*-β-lactam and vice-versa. (Scheme 2). Thus, the *cis*-mesyl derivative **53a** from **52** can be allowed to react with sodium acetate (or lithium azide) to provide the *trans*-α-acetoxy-β-lactam **53b** (or **53c**) which can be hydrolyzed to the *trans* compound **50**. Also, the *trans*-mesylate (epimer of **53a**) had been allowed earlier to react with sodium azide to produce a *cis*-azide (epimer of **53c**). The azido group can be reduced easily to an amino group. The α-phthalimido-β-lactam **54a** can be isomerized directly to the thermodynamically more stable *trans*-β-lactam (**54b**) under the influence of a strong organic base such as DBN (1,5 diazabicyclo[4.3.0]non-5-ene). It has been reported that *cis*-β-lactams can be epimerized in good yield at C-3 upon treatment with potassium *t*-butoxide. Since the configuration of C-4 is not altered in the above reactions, an optically pure **50** or **52** will lead to enantiopure products.

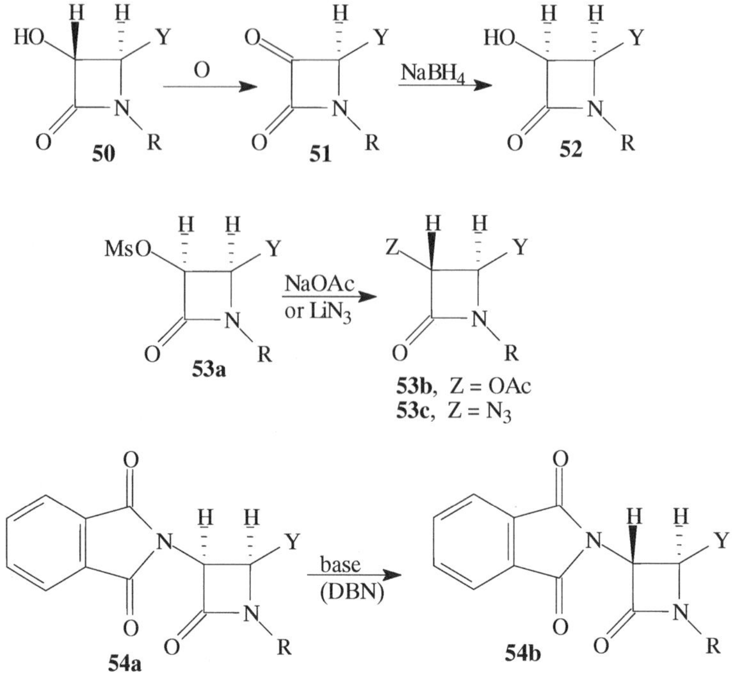

Scheme 2

3.1.2. Homochiral β-Lactams. Two strategies are available for obtaining non-racemic β-lactams and related compounds. It may be more convenient to prepare a racemic product and then conduct optical resolution instead of using a homochiral starting material.

Pasteur had developed techniques for optical resolutions that are still in use for obtaining a synthetic product that would match the absolute configuration of a natural product (or its mirror image). Some pharmaceuticals are manufactured on a large scale by using selective crystallization to separate one enantiomer in a pure form; in many of these cases a way is found to racemize the unwanted antipode and recycle the material for more of the desired enantiomer.

An easier method for some racemic compounds is to prepare salts to provide mixtures of diastereomers, which are usually separated by crystallization. A modern variation of this approach is to prepare a covalent compound with an optically pure reagent and then conduct separation of the diastereomers by newer techniques such as HPLC or other chromatographic methods. Of course, it should be possible to remove easily the covalently bound optically active resolving agent (see later for illustrations in the β-lactam field).

Enzymatic resolution methods have been enhanced by chemoenzymatic reactions *in organic solvents*—first introduced by Klibanov.[35] (Also see Davies, et al.[36])

Many natural products with various functional groups (e.g., amino acids, carbohydrates, chiral carboxylic acids such as tartaric, lactic, malic acids, steroids, and alkaloids) are commercially available in optically pure form to be used as the starting material in a synthetic process that precludes racemization. The use of this ever-increasing chiral pool leads directly to the desired homochiral product and is therefore widely used (see later for illustrations).

Another strategy is to employ a chiral auxiliary that controls the enantioselectivity of a key step. The chiral auxiliary has to be cleaved at the end of the synthesis; it is better for atom economy if the auxiliary can be recycled.

3.1.3. Ring Formation Reactions. Applying a retrosynthetic approach, several key steps can be envisioned for forming a β-lactam ring; of these the following are the most widely used by synthetic chemists:

A. Cyclization (Scheme 3) through formation of the
   (i) N–C-4 bond
   (ii) C-3–C-4 bond
   (iii) N–C-2 bond

B. (2 + 2) Cycloaddition (Scheme 4) involving
   (i) the reaction of an acid chloride (or an equivalent), an imine and a base (the Staudinger Reaction): ketene-imine reaction,
   (ii) the enolate-imine reaction,
   (iii) the reaction of an alkene with *N*-chlorosulfonyl isocyanate (see Section 5).

3.1.4. Cyclization Reactions

3.1.4.1. N–C-4 Bond Formation: The synthesis of variously substituted β-lactams have been described in detail in recent books. Therefore, only a limited number of selected features of contemporary synthetic approaches will be illustrated here.

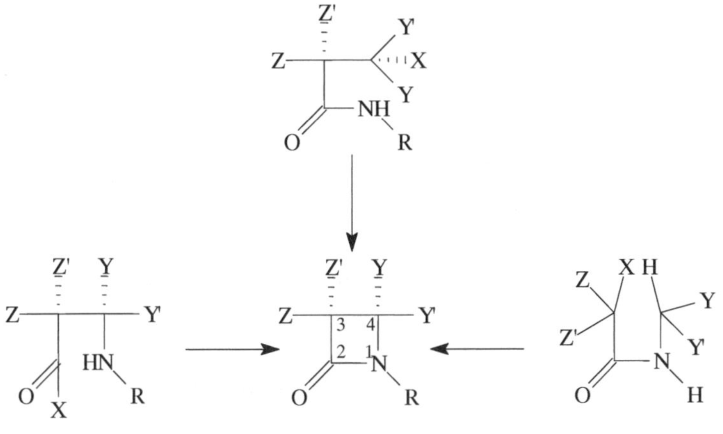

X = good leaving group

Scheme 3

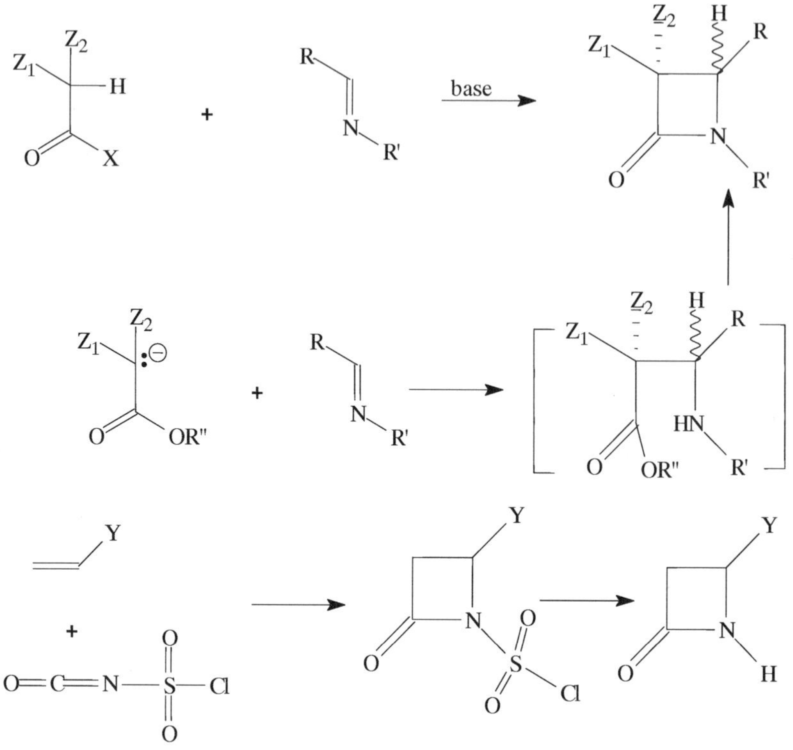

Scheme 4

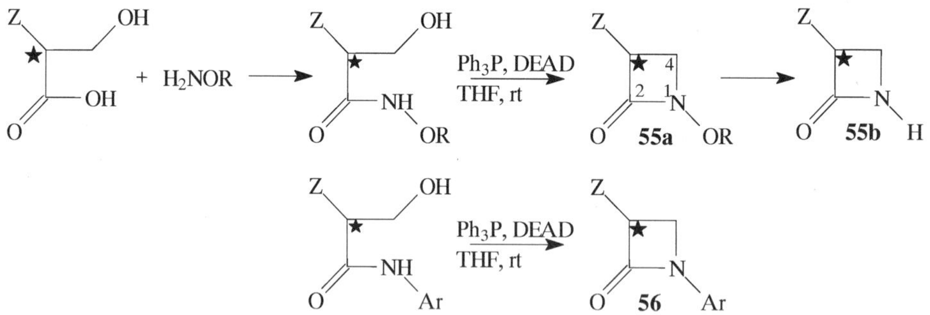

DEAD = **Di**ethyl **azod**icarboxylate = $EtO_2C-N{=}N-CO_2Et$

Scheme 5

Miller[37,38] has developed an effective N–C-4 bond formation approach using the Mitsunobu reaction for cyclization of β-hydroxy amides of a special type. The formation of β-lactams (**55**) in high yield involves the action of the potential N-1 center as a nucleophile and the potential C-4 center as an electrophile (Scheme 5). To increase the nucleophilicity of the amide nitrogen, it was derived

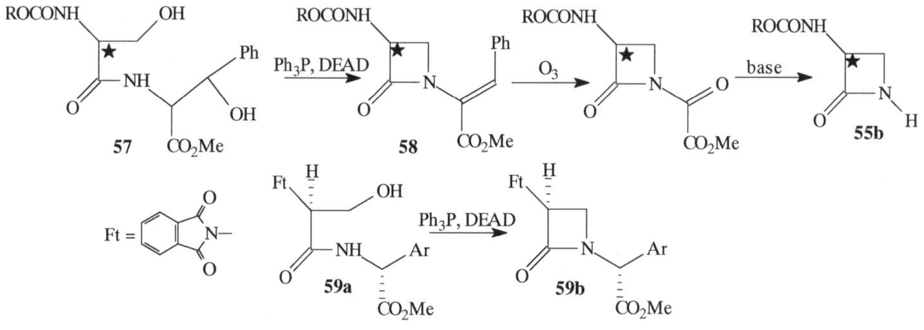

Scheme 6

from a hydroxylamine ether. The β-hydroxy group in presence of Mitsunobu reagents becomes an excellent leaving group that undergoes S_N2 inversion. When the starting material is derived from optically active serine with the α-amino group protected suitably, an optically pure β-lactam (**55**, Z = CbzNH) is obtained in good yield (Scheme 5).

Bose et al.[39,40] showed that a simple aryl amide of a β-hydroxy propionic acid could be cyclized in high yield to a β-lactam (**56**) via intramolecular Mitsunobu coupling (Scheme 5) even though this amide is not acidic. If the aryl group is the p-methoxyphenyl group, the N-unsubstituted β-lactam could be obtained easily by cerium(IV) ammonium nitrate oxidation.

An extension of this approach (Scheme 6) was the cyclization of a protected L-serinylphenylserine (**57**) to the optically active β-lactam (**58**) in good yield. Since the chirality of the phenylserine moiety disappears in the course of β-lactam formation, a dipeptide from an amino-protected L-serine and an ester of inexpensive DL-phenylserine can be used as the starting material. The ozonation of **58** followed by mild base treatment led to an N-unsubstituted β-lactam **55b** which was identical with one of the β-lactams synthesized by Miller (Scheme 5). Obviously, no racemization occurs during the formation of **58** and **55b**. The β-lactams of type **55b** provide access to optically pure nocardicins (**14**) and monobactams (**15**).

Townsend and Nguyen[41,42] have reported a successful use of the Mitsunobu reaction for the formation of *N*-benzyl-β-lactams (**59b**) from a β-hydroxy-*N*-benzylamide (**59a**) the –CO–NH– bond of which is not expected to be particularly acidic.

3.1.4.2. C-3–C-4 Bond Formation: The first example of this type of bond formation was achieved by Sheehan and Bose[43] by the intramolecular cyclization of α-haloacylamino malonate (**60a**) in presence of a mild base to give the β-lactam (**60b**) in high yield (Scheme 7). The question of *cis/trans*-stereochemistry is absent for **60b**.

It is interesting to note that for the dihaloamido compounds (**61a**) and (**62a**), cyclization leads exclusively to a γ-lactam (**61b**) and a β-lactam (**62b**),

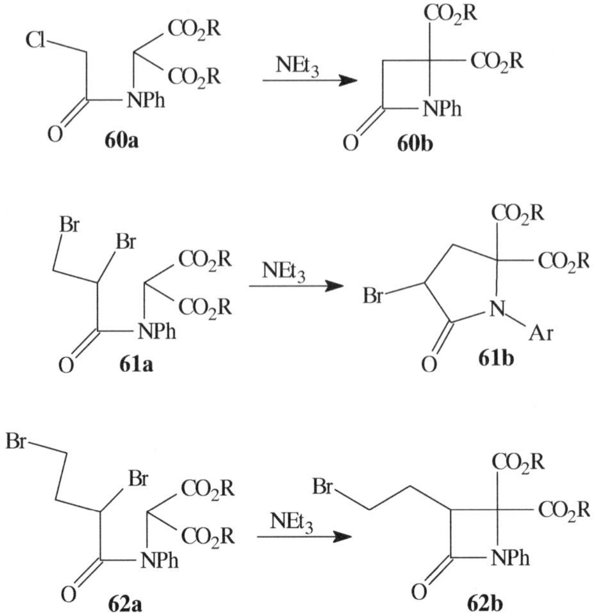

Scheme 7

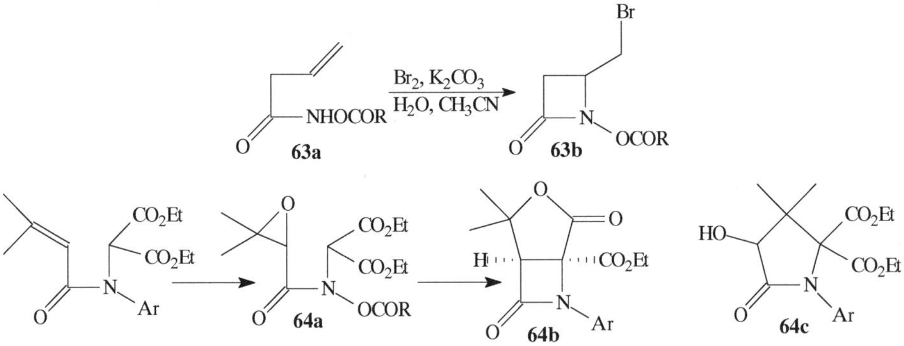

Scheme 8

respectively,[44] even though a δ-lactam from (**62a**) would be much less strained than the β-lactam (**62b**).

Exclusive β-lactam formation as a single stereomer in preference to γ-lactam formation[45] was observed in the cyclization of (**63a**) to (**63b**) (Scheme 8).

In this context, the exclusive cyclization of the epoxide (**64a**) to the β-lactam (**64b**) [instead of a γ-lactam (**64c**)] is worth notice (Scheme 8). Reasons are not obvious for the exclusive formation of strained β-lactams rather than more flexible and less strained γ-lactams.[46]

3.1.4.3. N–C-2 Bond Formation: The β-lactam amide bond formation from a β-amino acid has been aided by various carboxy-activating agents which are well

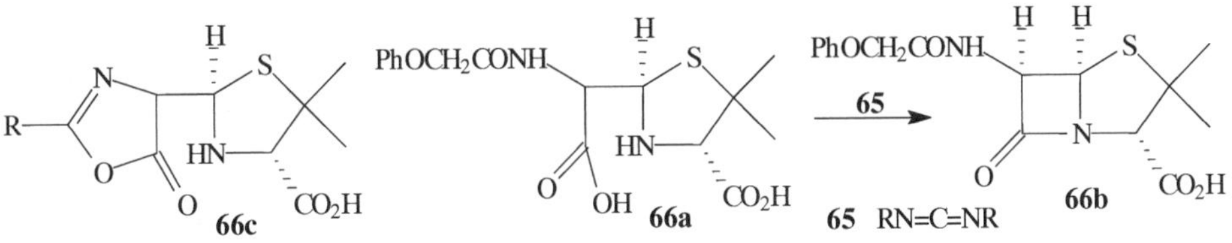

Scheme 9

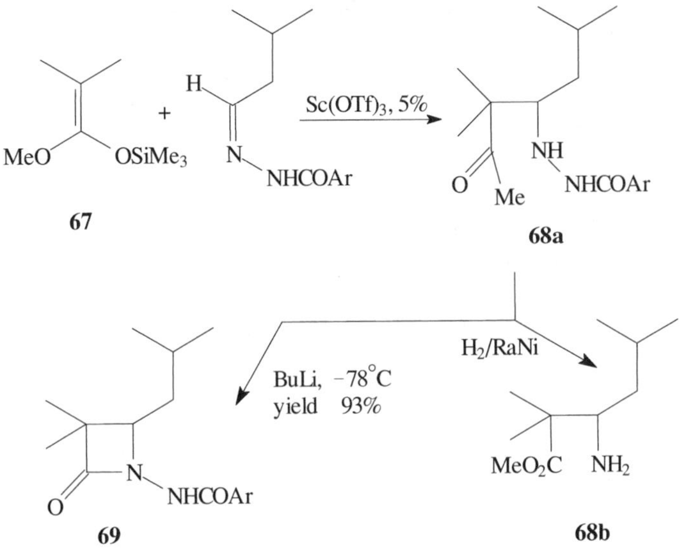

Scheme 10

documented.[47] Cyclization is facilitated if a β-amido acid is the starting material.

The presence of an amino group at C-3 creates special problems for β-lactam synthesis via N–C-2 bond formation. Sheehan et al.[48] were able to solve this difficulty by using carbodiimides (**65**)—new reagents at that time that were devised for the preparation of peptides. A key step in the first rational total synthesis of penicillin achieved by Sheehan and Henry–Logan[49] was carbodiimide induced cyclization of **66a** to the fused bicyclic β-lactam **66b** under very mild conditions. Other cyclizing agents formed an oxazolone **66c** involving the amide side chain (Scheme 9).

An interesting variation of Breckpot's method[18] has been developed that utilizes a β-hydrazino ester rather than a β-amino ester for the preparation of monocyclic β-lactams[50] (Scheme 10). Acylhydrazones, in particular, 4-trifluoromethylbenzoyldrazones were shown to react with silyl enolates in high yield if scandium triflate is used as a catalyst. Thus, the hydrazone prepared from isobutyraldehyde and an acylhydrazine reacted with the silyl enolate **67** in presence of $Sc(OTf)_3$ to give the β-hydrazino ester **68a**. Upon treatment with *n*-

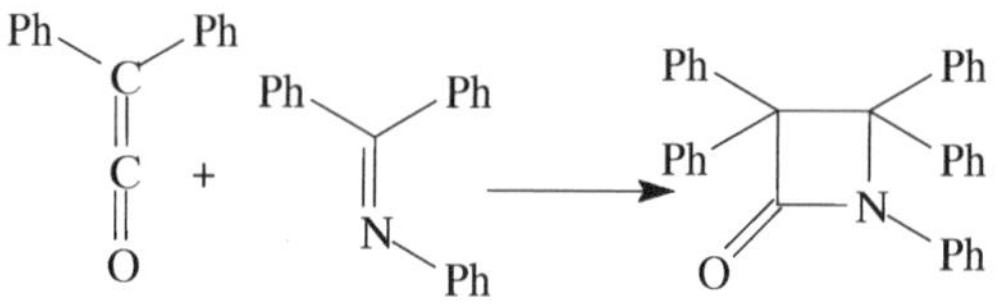

Scheme 11

BuLi at −78°C, **68a** was converted to the β-lactam **69** in excellent yield. Catalytic hydrogenation of **68a** led to N–N bond scission and produced the β-amino ester **68b**. It can be expected that similar hydrogenolysis of **69** would afford the corresponding N-unsubstituted β-lactam in high yield. A recent publication[51] reviews many catalysts for such hydrogenolysis and also describes the use of BF_3.THF for efficient N–N bond cleavage with high enantiomeric purity.

3.1.5. Multiple Bond Formation. Two major synthetic approaches, namely, the Staudinger reaction and the enolate–imine cycloaddition reaction belong to this category. The first authentic synthesis of a β-lactam (**71**) was achieved by Standinger in 1907 by the reaction of a ketene (**70a**) with an imine (**70b**) (Scheme 11).

3.1.5.1. Staudinger Reaction: A variant of the original Staudinger reaction is the use of an acid chloride (or equivalent), and a tertiary amine in place of a ketene. The imine can be acyclic or cyclic so that monocyclic as well as polycyclic β-lactams can be prepared with various functional groups at different sites. The reaction of azido acetyl chloride (or other carboxy-activated forms) (**72**) and triethylamine (or *N*-methylmorpholine and other tertiary bases) for converting imino compounds to α-azido-β-lactams (**73a**), has been termed the Bose reaction[52,53] (Scheme 12). The azido group can be easily reduced to give α-

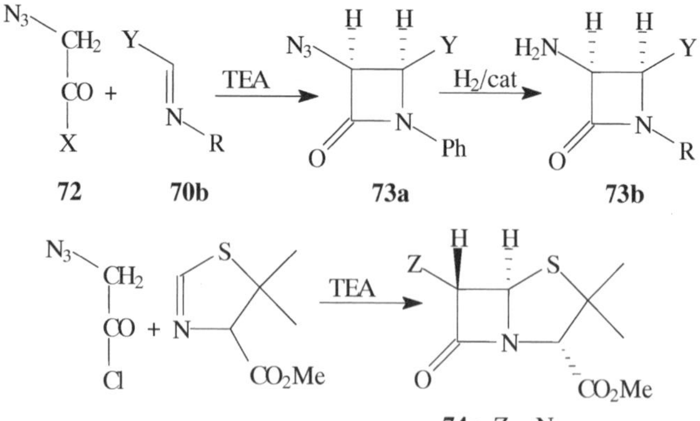

74a $Z = N_3$
74b $Z = NH_2$

Scheme 12

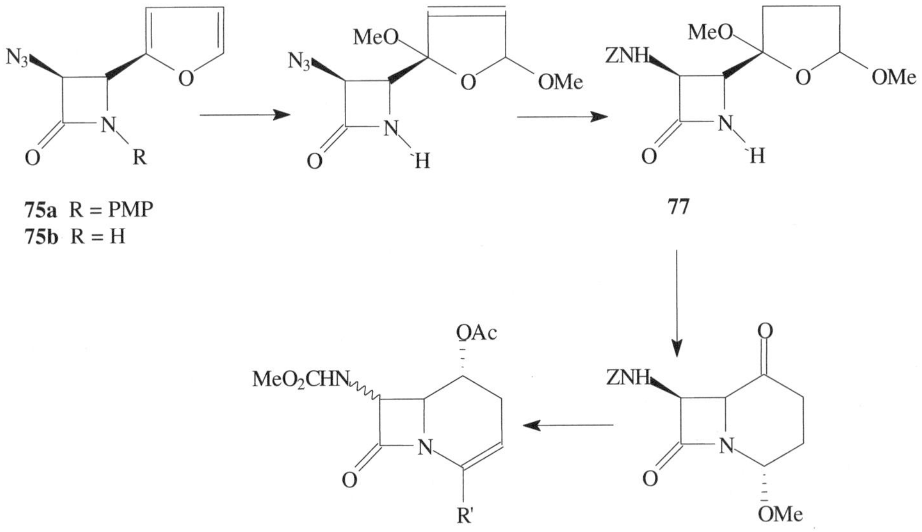

75a R = PMP
75b R = H

77

79a R′ = H
79b R′ = CN

78

Scheme 13

amino β-lactams **73b**. A stereospecific synthesis of 6-epipenicillin ester (**74**) was achieved in 1968 by using the reaction[54] that was simpler and more practical than the classic penicillin synthesis of Sheehan and Henry–Logan.

The Bose reaction has found very extensive application in industrial and academic laboratories for the creation of α-amino-β-lactams (**74**) that are analogs of penicillins, cephalosporins, and other antibiotics[55]. A recent publication[56] has described the synthesis of novel carbacephems by using this reaction (Scheme 13).

A newly developed *aza—Achmatowicz Reaction*[57] has been extended to the synthesis of **79**. In this reaction a furan nucleus is utilized as a latent 1,4-dicarbonyl group. The α-azido-β-furyl-β-lactam **75a**, prepared by the Bose reaction was converted by cerium(IV) ammonium nitrate oxidation to the corresponding N-unsubstituted compound **75b**. Treatment with bromine and methanol oxidized the furan ring to give **76**. Catalytic hydrogenation reduced the azido group as well as the double bond in the furan ring and gave **77** (Z = H) in excellent yield. After the protection of the amino group by forming a carbamate, rearrangement of **77** (Z = $-CO_2Me$) was affected under the influence of trifluoroacetic acid to yield **78** with a carbacephem nucleus. The target compound *cis* **79b** was prepared via the enamide **79a** through multistep reactions. The *trans*-isomer of **79** was also prepared by starting with a *trans*-α-phthalimido-β-furyl-β-lactam instead of *cis*-α-azido-β-lactam **75a**.

Azidoacetyl chloride is a very reactive reagent for the preparation of a variety of monocyclic and fused β-lactams. But because of the risk of explosions if azido compounds are not treated with appropriate caution, an alternative approach to

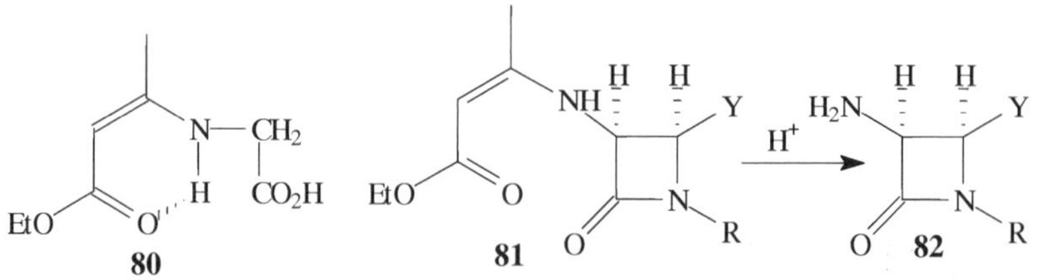

Scheme 14

α-amino-β-lactams (**82**) suitable for multiple kilogram size preparation was developed (Scheme 14). An enamine (**80**) from glycine was used in place of azidoacetic acid. A *cis*-β-lactam (**81**) was obtained in good yield[58]. The protective enamino group was removed under mild acid treatment to obtain (**82**).

Another convenient approach to α-amino-β-lactams has been reported recently[59] that can be conducted on a large scale. Tetrachlorophthalimido acetic acid is used for conducting the preparation of α-tetrachlorophthalimido β-lactams by the Staudinger method. The amino protective group is removed very readily upon reaction with ethylenediamine at room temperature. (See Section 4 for microwave assisted rapid reaction for variants of this approach). The Staudinger reaction can be used with imines derived from ketones and also with various heterocycles containing an imino group. Thus, the reaction of phenoxyacetyl chloride (**83**) with the dihydroquinoline (**84**) in presence of triethylamine led to a single isomer of a fused β-lactam (**85**). X-ray diffraction studies established that the phenyl group at the ring junction and the phenoxy group are *cis* to each other (Scheme 15).

3.1.5.2. Enolate–imine Cycloaddition: In 1980, it was reported[60] that enolates derived from esters (usually under the influence of lithium di-isopropylamide) can undergo cycloaddition reaction at low temperatures with imines to give β-lactams in high yield (Scheme 16). Both *cis* and *trans* geometry are possible for the cyclization product; in some cases only one stereomer is obtained exclusively.

Manhas and co-workers[61] have obtained 3-amido-3-methoxy-2-azetidinones as single isomers (**89**) by this method by starting with α-amido-α-methoxyacetic acid esters (**88**). Several groups including Georg and Kant[62] have employed the dianion from ethyl 3-hydroxybutyrate (**90**) for the preparation of trans β-lactams (**91**) with the thienamycin side chain (Scheme 16).

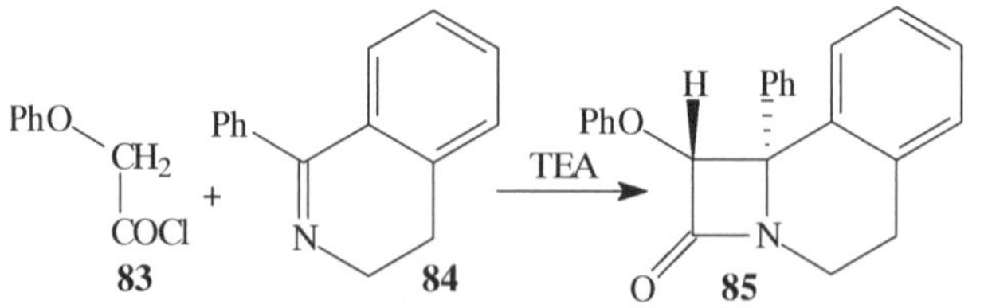

Scheme 15

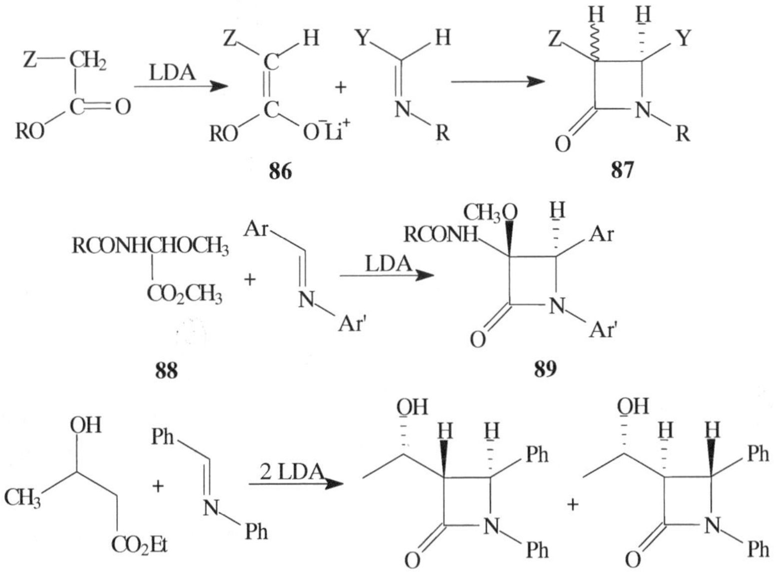

Scheme 16

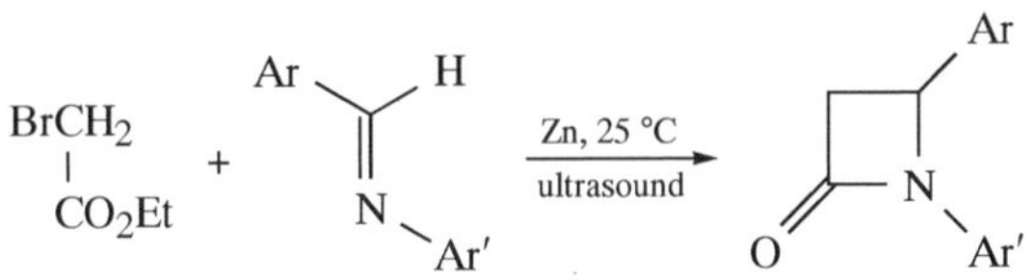

Ar = Ph, C_6H_4Me-*p*, C_6H_4Cl-*p*
Ar′ = PMP, Ph

Scheme 17

The Reformatsky type reaction[63] (Scheme 17) was described in 1943 for obtaining β-lactams from imines, α-bromoesters and zinc. This cyclization may also involve enolates. Although the reaction is conducted at higher temperatures in refluxing toluene, this reaction can be conducted at room temperature under ultrasound irradiation.[64]

The enolate–imine cycloaddition method for assembling the β-lactam molecule is now attracting growing attention. High stereoselectivity can be achieved by (i) using optically active alcohols of special types for the ester function; (ii) using a chiral ligand along with an ester enolate as a chiral auxiliary that could be reusable.

The problem of preparing enantiopure compounds is somewhat simplified by the observation that a solid compound with about 85% ee can be usually obtained optically pure after one recrystallization.

The need for obtaining both enantiomers of a biologically active compound has led to the employment of optical resolution methods—often with the help of chemoenzymatic reactions. A few recent examples will be cited here to illustrate some of the above approaches to enantiopure β-lactams.

3.2. Enantiospecific Synthesis

Of the various methods developed for the synthesis of β-lactams, the ketene–imine cycloaddition reaction is the most widely used because of its simplicity. Also, the stereochemical outcome of this reaction is known in many instances, although the theoretical basis for prediction of the steric course is not adequate yet. Asymmetric induction of high order in this reaction has been achieved in three different ways: (i) using an optically active amine for the preparation of the imine component; (ii) employing a homochiral aldehyde for the preparation of an optically pure Schiff base, and (iii) conducting cycloaddition with an oxazolidinone as a chiral auxiliary for the acid chloride component.

3.2.1. Use of Homochiral Amine and Chiral Aldehyde. Cycloaddition of azidoacetyl chloride in presence of triethylamine to the Schiff base from benzaldehyde and L-alanine methyl ester showed no diastereoselectivity: two optically active α-azido *cis*-β-lactams (**94a**, **94b**) were formed in about 1 : 1 proportion (Scheme 18).

In the course of his studies on "β-Lactam Synthon Method" Ojima[65] converted **94a** to an α-amino β-lactam which was condensed with benzaldehyde to give an optically pure imine (**95**). Cycloaddition of azidoacetyl chloride to **95**

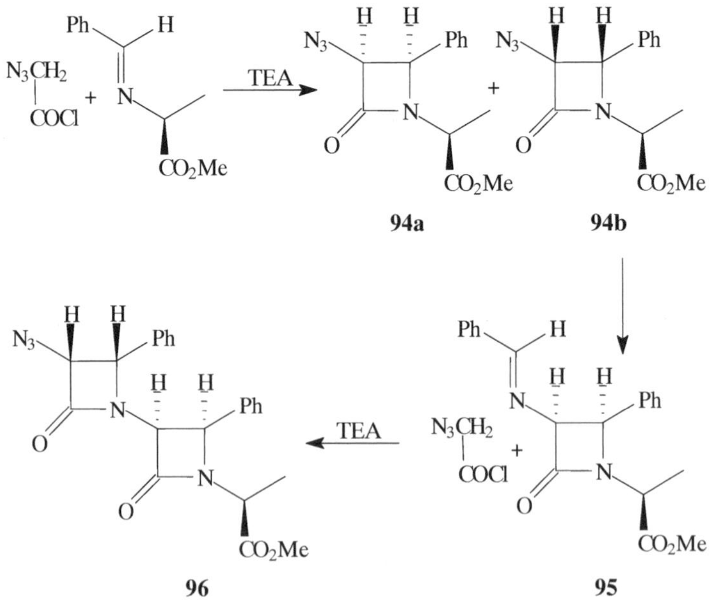

N_3CH_2COCl + Ph–CH=N–CH(CH_3)CO_2Me —TEA→ **94a** + **94b**

94b → **95**; **95** + N_3CH_2COCl —TEA→ **96**

Scheme 18

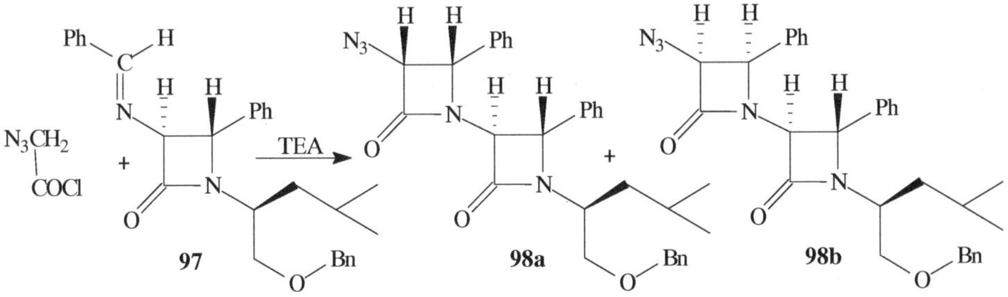

(98a/98b = 81/19)

Scheme 19

produced a single *bis*-β-lactam for which the absolute configuration was shown to be as in the stereostructure **96** (Scheme 18).

When the imine **97** derived from a *trans*-α-amino β-lactam was subjected to cycloaddition with azidoacetyl chloride, poor diastereoselectivity was observed: two isomeric *bis*-β-lactams of stereostructures **98a** and **98b** were obtained (Scheme 19). Ojima has produced evidence to show that the β-lactam carbonyl group was a factor in the loss of high diastereoselectivity when **97** was used as the imine component.

A *bis*-β-lactam[66] (**101**) of a different type than **96** and **98** was prepared as a single isomer from a *cis*-4-formyl-2-azetidinone **99**. The Schiff base **100** from **99** underwent cycloaddition to provide a racemic *bis*-β-lactam (instead of a meso compound) the structure of which was determined by X-ray crystallography (Scheme 20).

Reaction of azidoacetyl chloride with a Schiff base derived from cinnamaldehyde and an *O*-silylated D-threonine ester was reported[67] in 1980 to produce two homochiral *cis*-β-lactams (**103**, **104**) in the proportion of 90 : 10. When the very bulky triphenylsilyl group was employed for the protection of the β-hydroxyl group, the diastereoselectivity was increased to 95 : 5 (Scheme 21).

If the hydroxy group of the threonine ester was left unprotected, the imino compound **102c** produced the two *cis*-β-lactams **103c** and **104c** in about equal amounts. Obviously the bulk of the substituent on the β-carbon of the threonine has an important role in determining the diastereoselectivity of β-lactam formation.

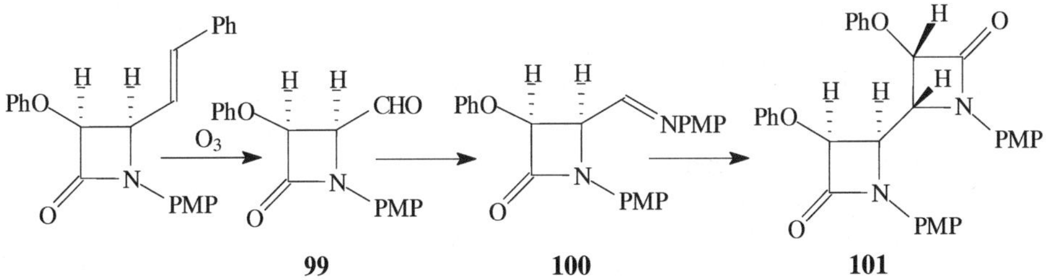

Scheme 20

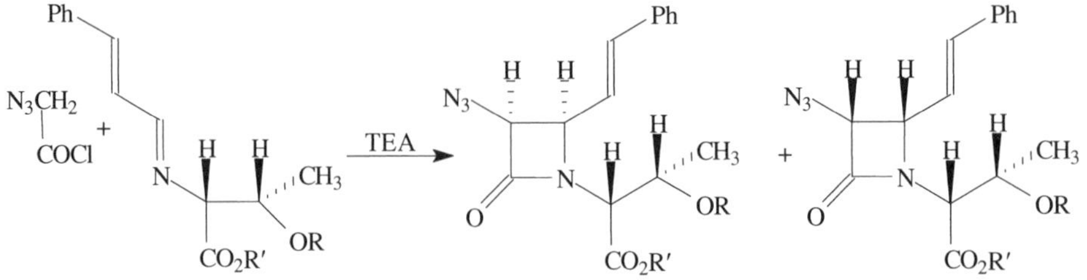

102 → (TEA) → 103 + 104

a, R = TDBMS, **b**, R = –SiPh3, **c**, R = H

Scheme 21

It is interesting to note that if the aim of a study is to establish the correlation between the absolute configuration of the starting material and the resulting β-lactam, then the *racemic form* of the reagents with two or more chiral centers would be adequate as the starting material. The racemic end product should be converted to a crystalline derivative and submitted to single crystal X-ray diffraction studies. Chiral carbons away from centers of chemical reaction would not change their absolute configuration during cycloaddition and later work up. Such fixed chiral centers can be used as points of reference in the X-ray study to predict the absolute configuration of the β-lactam that would have been produced by using an enantiopure starting material.

This approach was tested by submitting a crystalline derivative of racemic **103c** to X-ray diffraction studies. The ORTEP drawings of **103c** corresponding to D-threonine allowed the prediction that the major product from D-threonine should have the 4*S* configuration in **103c**. Chemical correlation between compounds of known absolute configuration and the optically pure β-lactam from D-threonine pointed to the same conclusion.

As has been noted earlier (see Section 3.2.2.1), the reaction of azidoacetyl chloride with a thiazoline ester derived from D-valine produces a single enantiomer corresponding to 6-epi-penicillin (Scheme 12).

3.2.2. Chiral Aldehydes and Achiral Amines. Imines derived from chiral aldehydes have been reported to produce better diastereoselectivity in β-lactam formation than imines derived from chiral amines. If an aldehyde with multiple chiral centers be used (such as an aldehyde derived from a sugar), the absolute configuration of the β-lactam formed can be predicted from the absolute configuration of the chiral center next to the aldehyde employed—the more distant chiral centers do not affect the sense of asymmetric induction. Two laboratories developed this approach independently to homochiral β-lactams and reported the cycloaddition reaction to be enantiospecific.[68,69] It has been shown, however, that if a chiral amine be used with a chiral aldehyde for Schiff base formation, the β-lactam obtained displays poor diastereoselectivity.

A recent use of this approach is described in Scheme 22. A single enantiomer of a β-lactam (**106**) was obtained from the Schiff base (**105**) from a D-galactose

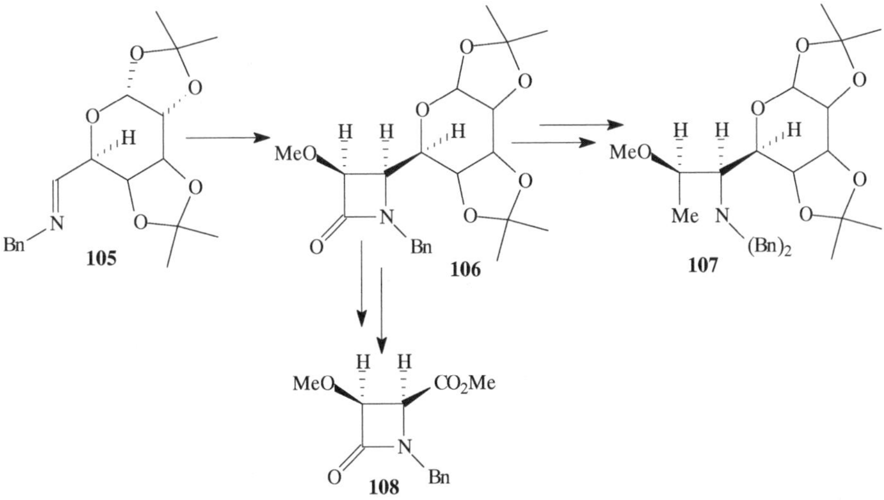

Scheme 22

derived aldehyde.[70] The absolute configuraton of the two new chiral centers of this *cis*-β-lactam was predicted by analogy with the data from Schiff bases from D-glyceraldehyde and other chiral aldehydes. At the end of a series of transformations that did not alter these new chiral centers, an intermediate **107** for a lincosamine derivative was obtained. Single crystal X-ray diffraction studies showed that asymmetric induction had been correctly predicted in the case of **106**. Chemical degradation of **106** to **108** of known absolute configuration had also predicted the same absolute configuration.
Enantiomerically pure *cis*-β-lactams have been prepared in good yield from α,β-epoxyaldehyde derived from *S*-malic acid and (+)-tartaric acid.[71]

Total diastereocontrol of β-lactam formation was also reported for imines derived from N,O-diprotected L-serinal.[72]

3.2.3. Use of a Chiral Acid Derivative. Evans and Sjogren[73] have developed a highly enantioselective method (ee > 90%) for the synthesis of α-amino-β-lactams (**110**). Achiral Schiff bases are allowed to react with the acid chloride (**109**), which contains a chiral auxiliary derived from L-phenyglycine. In most cases the minor product **110b** is produced in a small amount and easily removed from the major product **110a** by chromatography (Scheme 23).

Ojima and Chen[74] have observed that the absolute configuration of the α-amino β-lactam depends only on the chirality of the auxiliary; the imine component may be derived from either a chiral or an achiral amine. It is, therefore, possible to predict the absolute configuration of each asymmetric center of the β-lactams formed to be as shown in the stereostructures **110a** and **110b**. Ojima and coworkers[75] have made extensive use of the chiral acid chloride **109** to devise further extensions of their β-lactam synthon method. The starting point of their work is an optically active amino acid ester, which is condensed

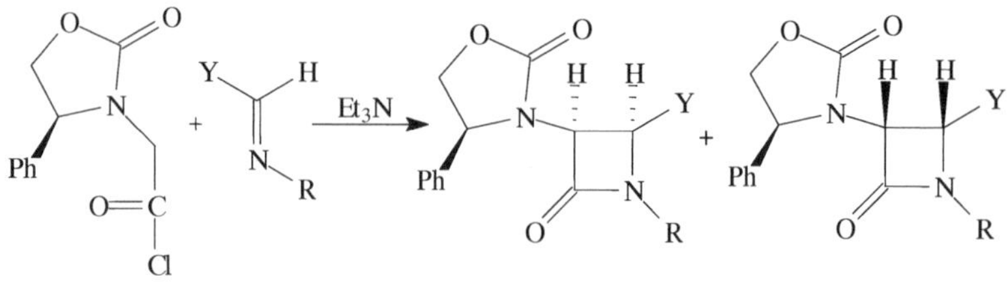

109 110a (major) 110b (minor)

Scheme 23

with an aromatic aldehyde to provide a Schiff base **111**. Reaction with **109** and triethylamine at $-78\,^{\circ}C$ converts **111** to an optically active β-lactam **112** in excellent yield.

Alkylation of β-lactams of **112** is regioselective; alkyl groups are first placed on the side chain on the β-lactam nitrogen in order of priority (Scheme 24) to give a single stereomer **113**. On further alkylation the β-lactam ring is substituted with the alkyl group at the α-position; the entering group is *trans* to the aryl group at C-4 as expected. The densely substituted β-lactam **114** can now be reduced with lithium in liquid ammonia to cause scission of the β-lactam ring as well as the removal of the chiral auxiliary group. An optically active dipeptide **115** is obtained which corresponds to the palnned stereochemistry at each chiral center.

The Evans chiral auxiliary **109** is restricted to the synthesis of α-amino-β-lactams. Recently Ojima[75] has developed a new approach that can produce α-hydroxy β-lactams (**117**) of very high optical purity. The key to this method is the use of a selected optically active ester of an α-hydroxyacetic acid derivative of the type **116** for an enolate–imine cyclocondensation which is enantiospecific (Scheme 25). The absolute configuration of the *cis*-β-lactam formed is predictable.

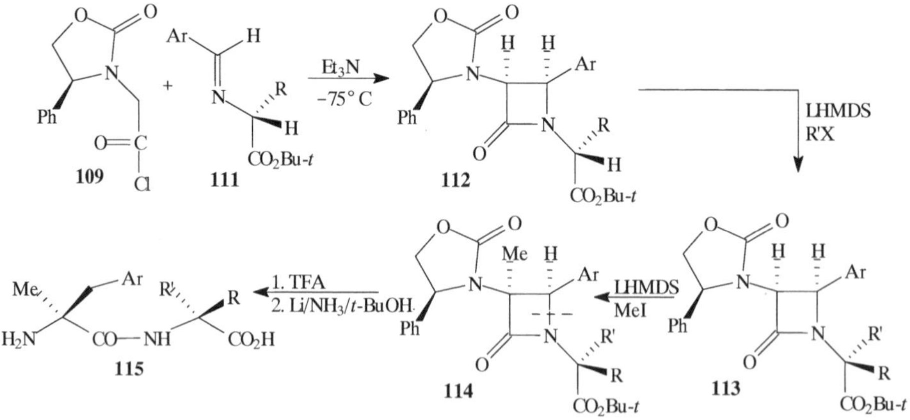

Scheme 24

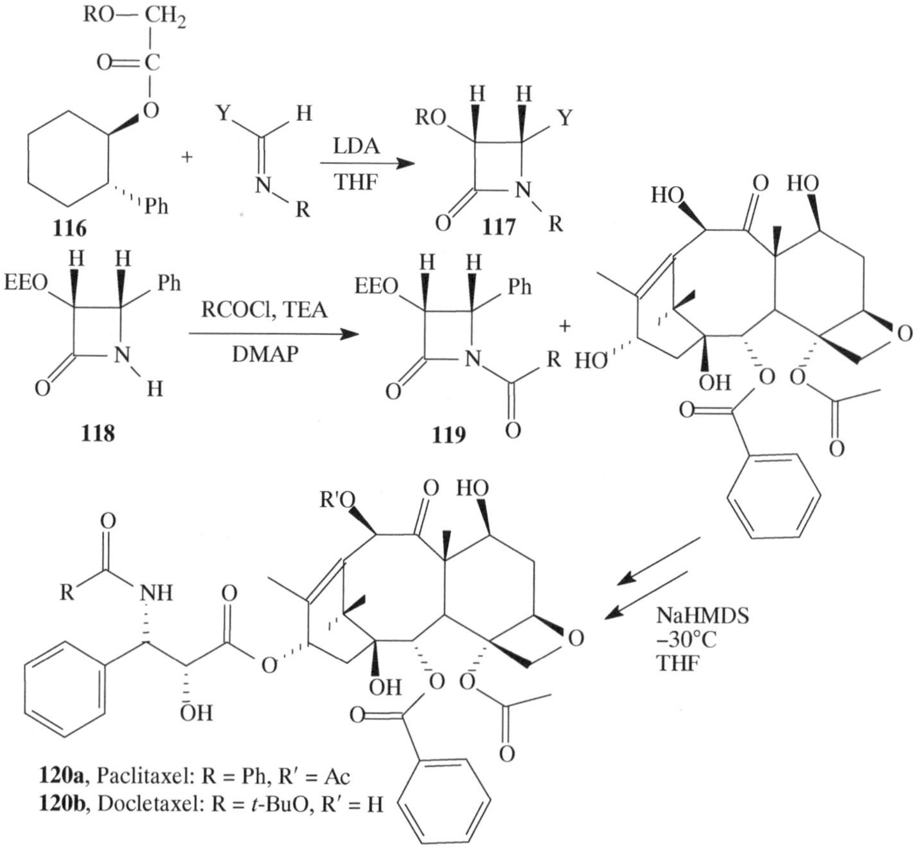

120a, Paclitaxel: R = Ph, R′ = Ac
120b, Docletaxel: R = *t*-BuO, R′ = H

Scheme 25

Diastereoselectivity of the order of about 85–90% has been reported for imines derived from (*S*)-ethyl lactate, optically active mandelate and methyl (*S*)-3 hydroxy-2-methyl proportionate. The protecting group for the hydroxyl moiety played a role in determining the steric course of β-lactam formation.

Ojima et al.[64] have used **116** with great success for the semi-synthesis of Taxol™ and analogs. Thus, the enolate-imine condensation of **116** with a Schiff base derived from silylamine gave access to the optically pure N-unsubstituted β-lactam **118**. Acylation of **118** with an appropriate acid chloride to **119** followed by reaction with baccatin IIIs with proper protective groups led to paclitaxel, docetaxel, and their analogs.

A different approach to optically active β-lactams by the ester enolate–imine condensation method has been reported recently.[76,77] A chiral ether ligand (**121**) is used as a reusable chiral auxiliary. The asymmetric induction step was shown to be the addition reaction between the enolate **122** and an imine, leading to **123** which is then cyclized to the β-lactam **124** of high ee (75–93%). This method is also applicable to the preparation of spiro β-lactams such as **126** from the enolate **125**. Such spiro β-lactams are of interest because they are related to the

Schering–Plough compound **127** which displays potent cholesterol absorption inhibitory activity (Scheme 26).

From the point of view of atom economy, it is desirable to recycle the chiral auxiliary—or better still to use a chiral intermediate, most of which can be a part of the final product. The total synthesis of cephalosporin that Woodward described in his Nobel lecture used the latter approach. The starting point of Woodward's synthesis was L-cysteine (**128**) the entire skeleton of which was incorporated in the target compound.[78] A series of stereocontrolled reactions were conducted for the preparation of the β-amino ester **129**. A key step now was the formation of the fused β-lactam **130** from **129** by variations of the Breckpot's method (see Section 3.2.1). Tri-isobutyl aluminium was utilized instead of a Grignard reagent and cyclization was achieved in good yield (Scheme 27). Modification of the second ring leaving the β-lactam ring intact, led to cephalosporin (**7**). The amino group of cysteine became a part of the amide side chain while the sulfur atom was made into a component of the six-membered ring system.

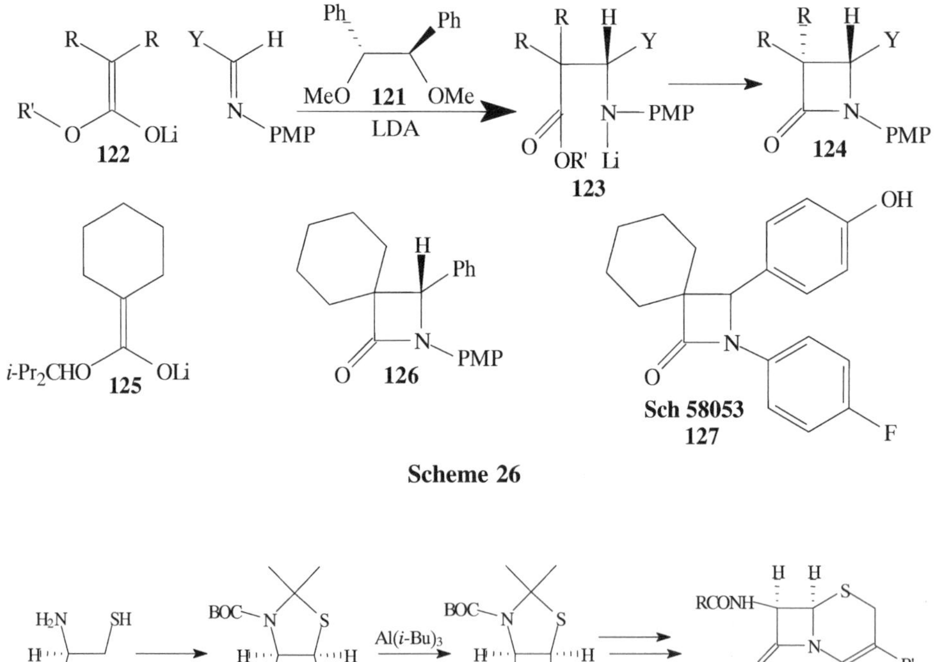

Scheme 26

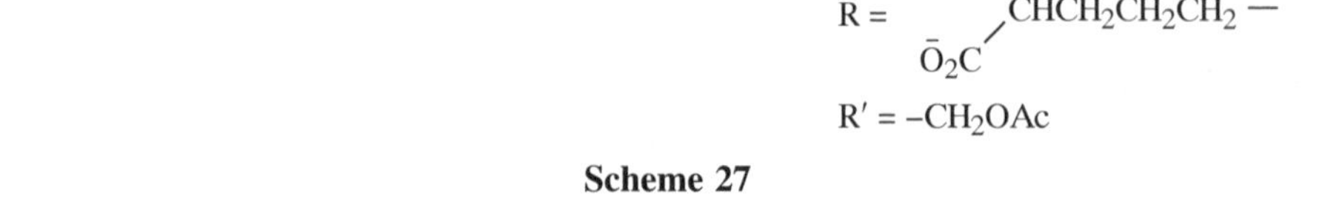

R = $H_2\overset{+}{N}$(⁻O_2C)CHCH$_2$CH$_2$CH$_2$ –

R′ = –CH$_2$OAc

Scheme 27

4. MICROWAVE ASSISTED RAPID SYNTHESIS

Two seminal papers[79,80] appeared in 1986, which reported that a variety of organic reactions could be completed in sealed systems in domestic microwave ovens in a matter of minutes. Since then many laboratories have been studying various organic reactions under microwave irradiation. Microwave assisted synthesis is now an emerging new technology.

Bose and co-workers have shown in a series of publications[81] that microwave assisted reactions can be used very effectively for the preparation of diverse types of substituted β-lactams. Stereocontrol of β-lactam-forming reactions under microwave irradiation has been achieved by them.

At present, there is a divergence of opinion about the possible effect of microwaves on molecular parameters, transition states, energy of activation, etc. Notwithstanding this uncertainty about the theoretical underpinning of microwave assisted reactions, many laboratories are using microwave irradiation for a few minutes to conduct reactions that take hours under conventional conditions. It appears likely that in the coming years microwave technology will be used extensively for conducting eco-friendly and rapid synthetic processes—for research and for manufacture.

4.1. Microwave-induced Organic Reaction Enhancement (MORE) Chemistry

A few explosions due to a rapid rise of temperature and pressure were reported when reactions were conducted in sealed systems in microwave ovens. To avoid such explosions, some laboratories designed "dry" or solventless reactions conducted with the reactants deposited on clay, silica gel, alumina, or other solid supports[82]. At the end of the reaction in a few minutes, the products are extracted with organic solvents. Obviously, such reactions cannot be conducted in a flow system for large-scale synthesis.

Bose et al.[83,84] have taken a different approach and developed non-traditional methods for using microwave irradiation for conducting a wide variety of organic reactions that are fast, safe, inexpensive, and eco-friendly. Their techniques involve reactions in open glass vessels (beakers, conical flasks, baking dishes, etc.) using minimal quantities of high boiling polar solvents (or no solvents if one of the reactants is a liquid). The strategy for conducting MORE chemistry is to provide the reactants with controlled microwave energy input such that an appropriate temperature is reached with minimal vaporization. Reactions are complete in a few minutes even on a molar scale.

The reaction mixture is maintained in an unmodified, domestic microwave oven at a temperature of about 20 °C below its boiling point: reflux condensers are thus unnecessary. For many condensation reactions where water (or some other small molecules) is a reaction product, a reaction temperature of 120–130 °C (easily managed with *N,N*-dimethylformamide, b.p. 154 °C or 1,3-propanediol, b.p. 195 °C) seems to be near optimum. Reactions have been conducted on the scale of several hundred grams. The upper limit of the scale can be readily extended by using larger commercial microwave equipment that is in service in the food industry. Since accessories such as reflux condensers and stirrers are not needed, transfer of this method to other microwave systems is easy.

A minimal of solvents can be used for MORE chemistry because in most cases, a room temperature slurry provides adequate solubility for reactants during the rapid rise in temperature that occurs under microwave irradiation. Less solvent used means less solvent wasted and, therefore, reduced pollution. Reactions can be conducted without the solvent if one of the reactants is a liquid that absorbs microwave energy efficiently. These points are illustrated by a recent application of MORE chemistry techniques to the rapid synthesis of α-amino-β-lactams.[59] Tetrachlorophthaloyl glycine (**132a**) was prepared on a molar scale by allowing tetrachlorophthalic anhydride (**131**) to react with glycine and catalytic amounts of *N*-methylmorpholine in minimal amounts of DMF as the reaction medium. After 8 min of irradiation in an unmodified domestic microwave oven, the reaction mixture was allowed to cool when the reaction product crystallized out in more than 90% yield.

N-Tetrachlorophthaloyl glycine (**132a**) was converted to its acid chloride (**132b**) yield by reaction with thionyl chloride. This acid chloride, a Schiff base and *N*-methylmorpholine were allowed to react in chlorobenzene under microwave irradiation for 3–5 min; near quantitative yield of only the *cis*-β-lactam (**133**) was obtained. The treatment of this β-lactam with 3 equivalents of ethylene diamine (without any solvents) under microwave irradiation for about one minute led to complete deprotection with the formation of the α-amino-β-lactam (**134**) in excellent yield (Scheme 28).

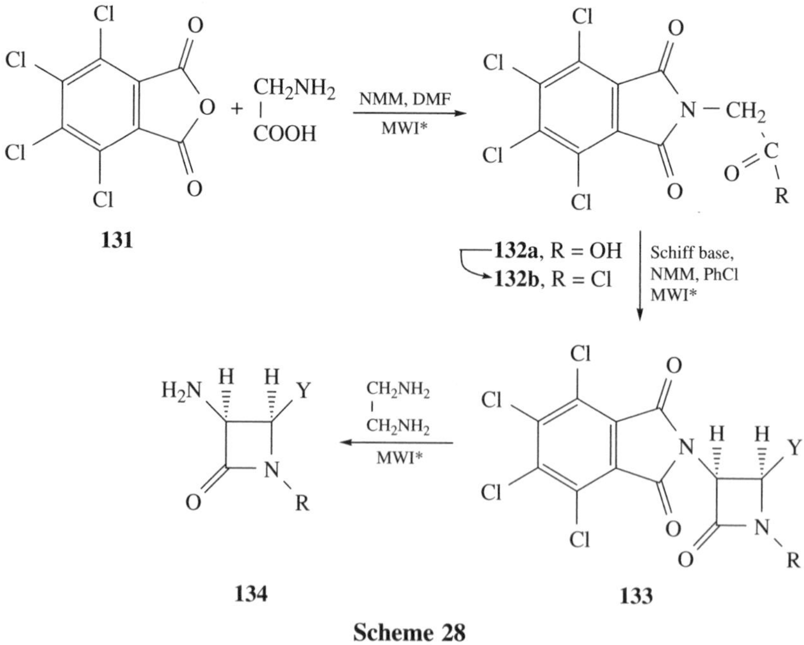

Scheme 28

4.2. Stereoselectivity

The control of *cis*-and *trans*-stereochemistry during β-lactam formation is of considerable importance for the desired configuration of the target synthons. It has been reported that high level of microwave irradiation leads to the preferential formation of *trans* β-lactam derivatives in several cases when the Schiff bases (**136**) are derived from aryl aldehydes.

Thus the reaction between acetoxyacetyl chloride (**135**), triethylamine, and a series of Schiff bases derived from benzaldehyde and aryl amines are known to give mostly the *cis*-β-lactam **137** in low temperature (e.g. 0–5°C) reactions. When the same cycloaddition reaction was conducted at about 110°C under microwave irradiation using chlorobenzene as the solvent, the ratio of the *trans*-β-lactam (**137b**) to the *cis*-β-lactam (**137a**) was nearly 90 : 10. It was demonstrated that the production of the *trans* compound was not due to base-catalyzed isomerization of the *cis* compound.

Interestingly, when the Schiff base was derived from D-glyceraldehyde acetonide and *p*-anisidine, the *cis*-β-lactam (**138**) was formed even at high levels of microwave irradiation. (Scheme 29). It would seem that there are separate pathways to *cis*- and *trans*-β-lactams and the pathway to the *trans* product is rapidly accelerated by microwave irradiation.[85]

The above method was also used for preparing *trans*-α-vinyl-β-lactams under microwave irradiation.[86] It had been demonstrated earlier that reaction of α,β-unsaturated acid chlorides (**139**) with various Schiff bases in the presence of triethylamine in benzene solution gave a mixture of α-vinyl-β-lactams. With aryl–alkyl Schiff base, a mixture of *cis*- and *trans*-isomers (**140a**, **140b**) was formed. The ratio of the isomers depend on the temperature of the reaction mixture and the extent of irradiation. The same reaction but with an aryl–aryl Schiff base produced only the *trans*-α-vinyl-β-lactam (**104c**) under strong microwave irradiation when chlorobenzene was the solvent and *N*-methylmorpholine, the base. (Scheme 30).

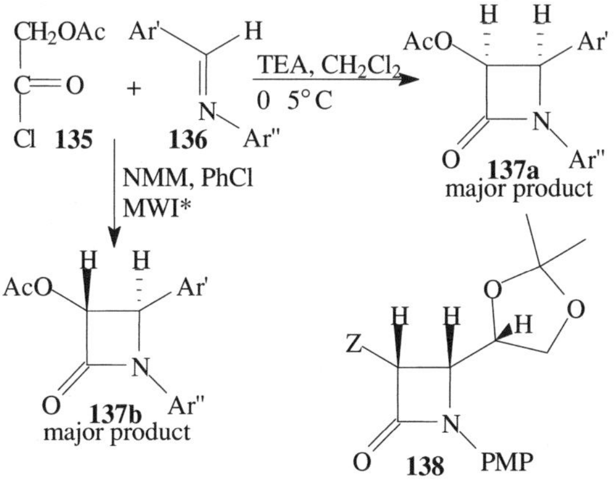

Scheme 29

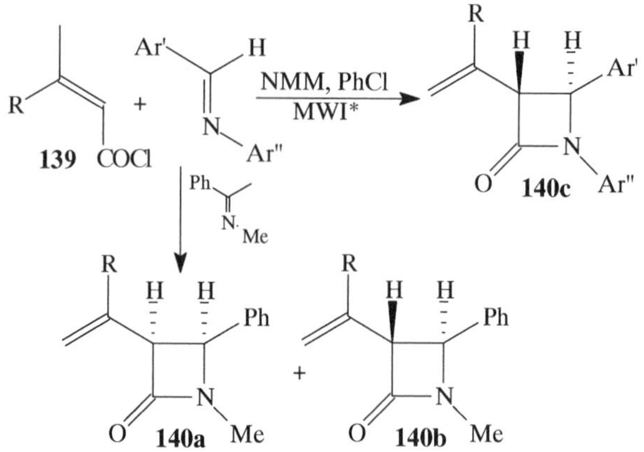

Scheme 30

The observation in the above schemes can be explained by postulating the faster formation of the *trans*-β-lactam compared to the *cis*-β-lactam at higher reaction temperatures. It is possible to get a very high temperature with a polar organic solvent under strong microwave irradiation (110–120°C, 3–5 min depending upon the solvent, power level, presence of a heat sink, etc.).

4.3. Catalytic Transfer Hydrogenation

Catalytic transfer hydrogenation has been studied extensively under microwave irradiation with diversely substituted β-lactams. Ammonium formate and hydrazine hydrate was used as the hydrogen donor; ethylene glycol or 1,3-proanediol of high b.p. (ca. 198°C) served as the energy transfer medium

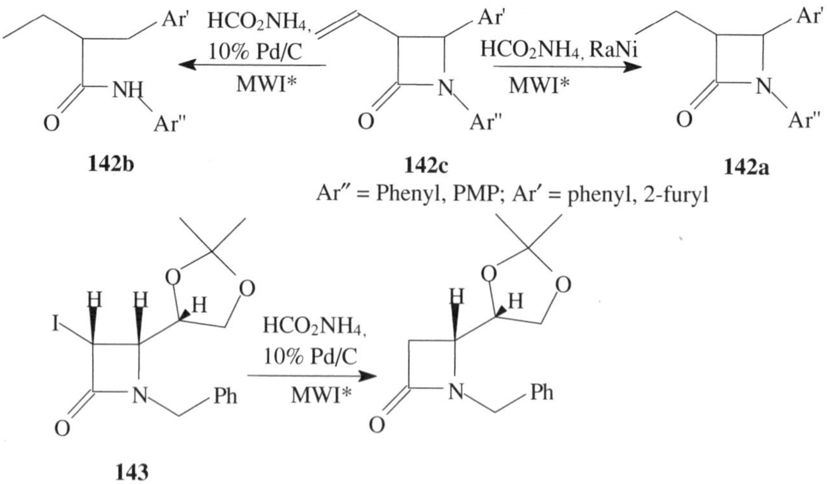

Scheme 31

(solvent), 10% Pd/C or Raney Ni was tested as the catalyst (Scheme 31). Thus, unsaturated β-lactams **142c** were hydrogenated to the saturated β-lactams **142a** with ammonium formate and Raney Ni. The stronger catalyst, 10% Pd/C, reduced double bonds and also caused hydrogenolysis of the N–C-4 bond. An open chain amide (**142b**) was thus produced.[88] The method was extended to the preparation of 3-unsubstituted β-lactam by dehalogenation. Optically active 3-iodo β-lactam (**143**) was used for this study and 3-unsubstituted β-lactam was obtained in good yield (Scheme 31).

4.4. Eco-friendly Approaches: Pollution Control at the Source

As pointed out earlier one of the major advantages of MORE chemistry techniques is the need for only limited amounts of organic solvents. Also, in some reactions, the amount of byproducts is reduced and thus a purer product is obtained. If two or more steps of a multi-step reaction can be combined in a "one-pot" reaction, there is further improvement in atom economy. An example is provided here of this approach to reduction of pollution at the source.

Carbapenem antibiotics have been prepared by using the microwave irradiation method. In a one pot synthesis,[89] the first three reactions (Scheme 32) were carried out without separation of the products. Condensation of diethyl ketomalonate with *p*-anisidine in the presence of DMF was the first step that

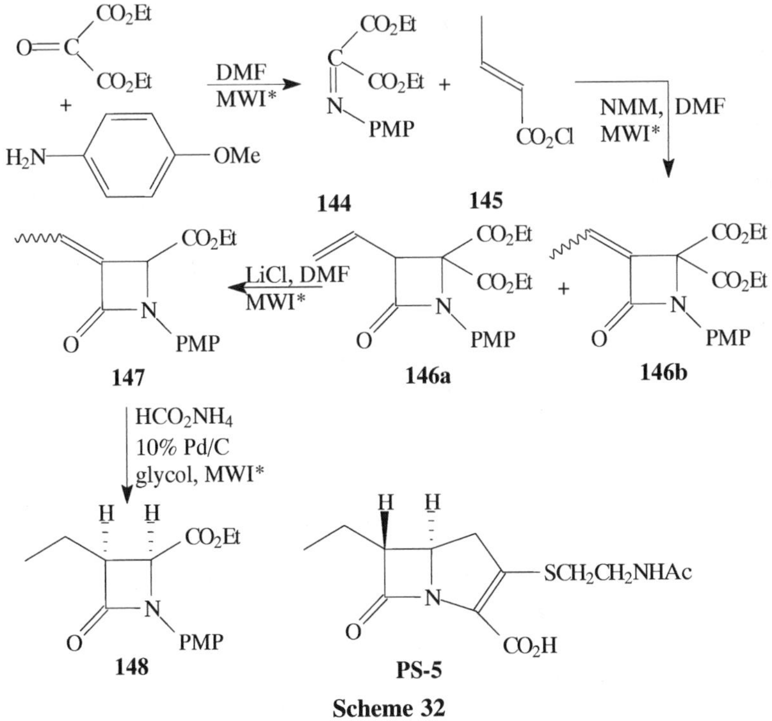
CO2Et
O=C
CO2Et
+
H2N
OMe
DMF
MWI*
C
CO2Et +
N
PMP
CO2Cl
NMM, DMF
MWI*
144
145
CO2Et
LiCl, DMF
MWI*
CO2Et
CO2Et
+
CO2Et
CO2Et
N
O
PMP
147
146a
146b
HCO2NH4
10% Pd/C
glycol, MWI*
H H
CO2Et
148
PS-5
SCH2CH2NHAc
CO2H

Scheme 32

gave the imino compound **144**. The second step was cycloaddition of **144** with an α,β-unsaturated acid chloride (**145**) to give a mixture of two β-lactams (**146a** and **146b**). This mixture was subjected in the third step to decarbethoxylation under microwave irradiation and then filtered through silica gel, and the purified material (**147**) was submitted to catalytic transfer hydrogenation when a single *cis*-β-lactam (**148**) was obtained. This compound is a known intermediate for the antibiotic **PS-5**.

5. NEWER FINDINGS

The selective presentation above, of several special aspects of β-lactam chemistry clearly indicates the expanding boundaries of the field. Some of this expansion is the response by medicinal chemists to the growing threat of resistance to current antibiotics by various types of pathogens. The development of new candidates for β-lactam antibiotics is aided by new synthetic approaches to monocyclic as well as bicyclic and tricyclic fused β-lactams. The restrictions on solvents and reagents imposed by concerns for the environment are placing a strong emphasis on the exploration of novel synthetic techniques and processes that are ecologically more friendly.

The phenomenon of drug resistance is receiving renewed attention from chemists and biochemists. Increasing use is being made of new tools, such as genetic engineering, combinatorial chemistry, molecular modeling, and computational methods. Also of importance is the growing synergy between chemists and biologists in the evolving area of chemical biology. Just a few of these aspects are highlighted in the following sections to provide an intimation of the changes in the making in the field of β-lactams.

5.1. Biosynthesis

A considerable body of biosynthetic data on various types of β-lactams has been reported over the years. Most of the information is based on labeling studies using radioactive and stable isotopes. Recent reports[90,91] illustrate the use of techniques of molecular biology for inducing the production of specific enzymes by a suitable microorganism, to be used for biotransformation of suspected precursors. Thus, a β-lactam synthetase was created which the investigators viewed as a "reverse" β-lactamase. Earlier a β-lactam forming enzyme (isopenicillin N synthase) was described which catalyzes the formation of the first β-lactam during the biosynthesis of penicillins.

In a study on the details of the biosynthesis of clavulanic acid[90] (Scheme 33), it was shown that the recombinant enzyme β-lactam synthetase converts **149a** (derived from arginine and pyruvate) into the C-3 and C-4 unsubstituted β-lactam **150**. It is likely that the mixed anhydride **149b** is an intermediate for the cyclization reaction.

Subsequent steps involve the biotransformation of this monocyclic β-lactam to the bicyclic β-lactam **151** with the 4*S* absolute configuration. By some yet

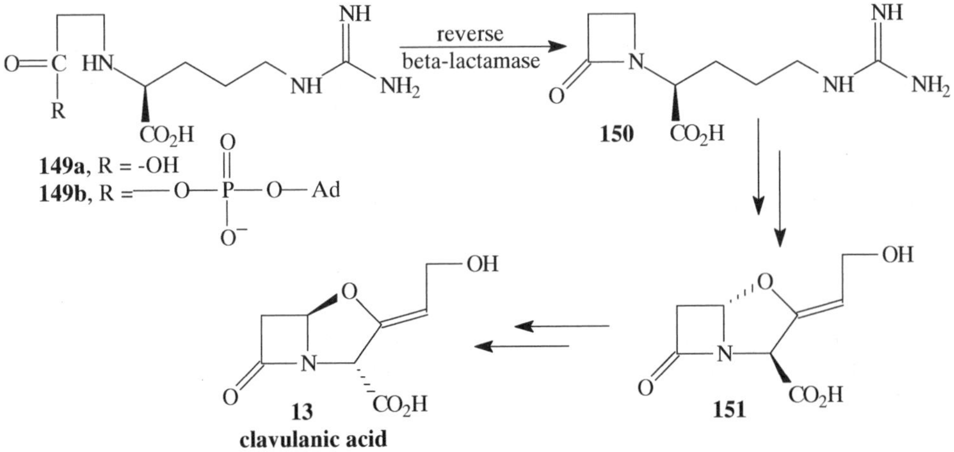

Scheme 33

unknown biosynthetic steps, this chirality is reversed to produce clavulanic acid (**13**). As noted in an earlier section, clavulanic acid is the most important inhibitor of serine β-lactamases.

5.2. Biodegradation

The β-lactamases are widely distributed enzymes of several kinds that catalyze the hydrolysis of the amide bond in the β-lactam ring and produce β-amino acids without antibacterial activity.[92,93] There is growing evidence that β-lactamases are ancient enzymes that existed in nature long before antibiotics were introduced in medicine. These enzymes are produced in increasing amounts by bacteria as they grow more resistant to β-lactam antibiotics. Also, β-lactamases seem to undergo evolution in response to antibiotics.[94]

The X-ray structure of the carbapenem-hydrolyzing β-lactamase complexed to a penam derivative has been studied.[95] It appears that even small structural changes in the β-lactamase through evolutionary techniques might need large modifications in the structure of the inhibitor.

The alarming increase in bacterial resistance to many antibiotics is causing great concern. Now there are very few therapeutic agents (such as, vancomycin) which are effective against methicillin-resistant *Staphylococcus aureus* (MRSA). Strains of MRSA produce a modified penicillin-binding protein (PBP-2′) which shows low affinity for most of the penicillins, cephems and carbapenams in clinical use. Various research groups have discovered a number of new β-lactams which have good anti-MRSA activity and high affinity for the new penicillin binding protein.[96,97] Whether any of these promising β-lactams will become antibiotics of clinical use depends on many factors including the size of the financial investment made and the number of years devoted to the development of a drug candidate.

Some gram-negative strains of bacteria have been observed to produce upto 1 mM concentration of β-lactamase. This level is about ten-fold greater than that used in laboratory kinetic experiments. Such unusually high concentration of enzymes can be very effective in total destruction of β-lactam antibiotics before they attack bacteria.

In the early days of penicillin antibiotics it was assumed that their antibiotic activity was due to the enhanced chemical reactivity of the β-lactam amide bond caused by the strain in the four-membered heterocycles. In contrast, simple amides—as in peptides for example—are far less reactive to nucleophilic attack on the amide carbonyl. Page[93] has discounted the ring strain and the inhibition of amide resonance in penicillins and cephalosporins as the predominant factors responsible for antibiotic activity. It has been pointed out that the rate of alkaline hydrolysis of the simple β-lactam is only three-fold greater than its open chain analog.

The shape and geometry of the essentially planar β-lactam ring do impose certain restrictions to the mode of action in cleaving the amide bond. Thus, in penicillin the attack on the β-lactam carbonyl group involves a tetrahedral intermediate in which the attacking group and the lone pair of electrons on the nitrogen are on the α-face of the β-lactam; the electrons are therefore *syn* to the attacking nucleophile. In case of ordinary peptides, it is assumed that the lone pair on nitrogen is *anti* to the incoming nucleophile.

The pharmaceutical industry has accelerated the search for new antibiotics and analogs of widely used antibiotics—in particular, β-lactam antibiotics. Many laboratories are preparing highly functionalized, novel bicyclic and tricyclic β-lactams.[98] Theoretical calculations, NMR and X-ray studies on the interaction between variously substituted β-lactams and penicillin-binding proteins are being described. Design and synthesis of monocyclic β-lactams as inhibitors of human cytomegalovirus protease have been reported.[99]

5.3. Newer Techniques

5.3.1. Newer NMR Techniques. The current drug discovery process is increasingly involved in the screening of large collections of compounds for their capacity of binding to target proteins. New NMR techniques are being developed for detecting such complexation of a small molecule with a "receptor." A novel technique named "NOE Pumping" has been described by Chen and Shapiro.[100] As an illustration of this "diffusion-assisted Nuclear Overhauser Effect (NOE) pumping" experiment, these authors recorded spectral data on the interactions of the protein receptor of human serum albumin with salicylic acid, L-ascorbic acid, and glucose in D_2O. They were able to correctly identify salicylic acid as the only binding ligand of the three. Undoubtedly, this very convenient NMR technique will prove useful for studying potential β-lactam antibiotics.

5.3.2. Newer Techniques of Natural Product Chemistry. Newer techniques of natural products chemistry—in particular, activity guided fractionation—are being employed increasingly for the detection and isolation of pure compounds present in microscopic amounts in plants and in marine flora and fauna. A recent study[101] directed to cancer chemoprevention used a bioassay-guided fractionation based on inhibition of 3H-tamoxifen binding at the antiestrogenic binding site. Tamoxifen is a very widely used antiestrogenic agent for the treatment of advanced breast cancer.

A native American ground-cover plant, *Pachysandra procumbens*, was subjected recently to a careful separation study and several novel steroid alkaloids including two β-lactam containing steroids (**152a**, **152b**) were isolated and their structure was determined from spectral data. These β-lactams were present in the dried plant material at the level of less than 10 ppm! Also isolated were four amido steroids **153a–d** that are related to **152**. It would appear that the β-lactam carbons in **152** are isoprenoid rather than amino acid derived. All the alkaloids showed bioactivity in the assay used; the most potent was **153d**. Japanese scientists[102] had isolated the first members of this family of β-lactam steroids (**17**) in 1967.

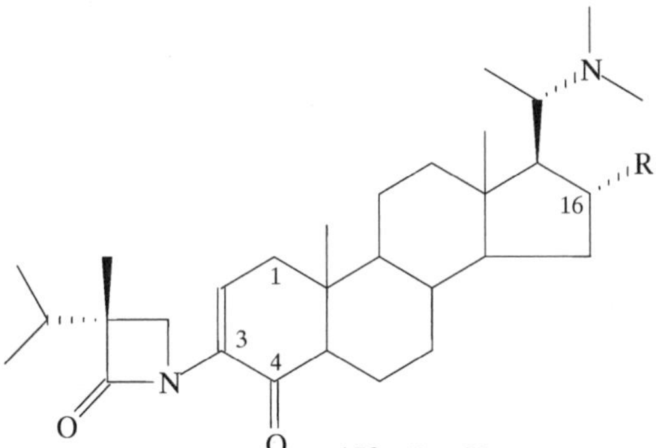

152a R = H
152b R = OH

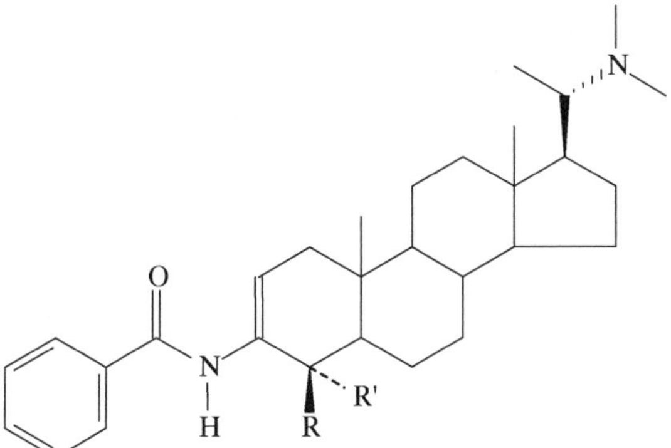

153a R = OAc, R' = H
153b RR' = O

5.4. Synthons for Pharmaceuticals

The U.S. Federal Drug Administration (FDA) now requires information on both enantiomers of drug candidates. Therefore, it is convenient to prepare the racemic form of an advanced intermediate which is submitted to optical resolution.

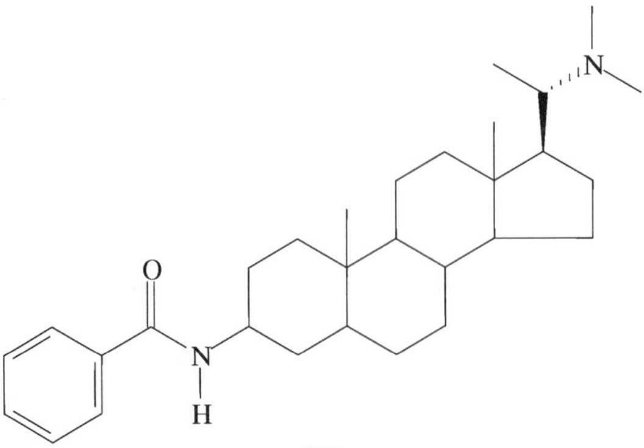

153c

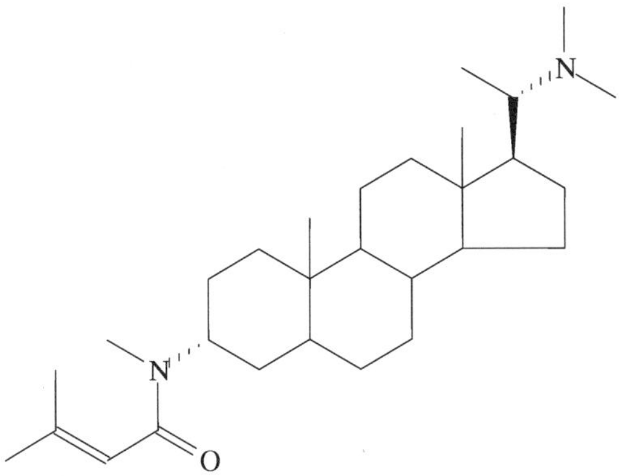

153d

Banik et al.[103] have reported that a room temperature reaction between a homochiral α-hydroxy β-lactam and a peracetylated glycal (for example, **155a**) in presence of iodine as catalyst produces the Ferrier rearrangement product in good yield as a single glycoside. In several cases it was shown that only the α-glycoside is formed.

If a racemic hydroxy β-lactam (**154a**) be used, a mixture of only two diastereomers (**155b**, **155c**) is obtained which can be separated by silica gel chromatography. Mild acid hydrolysis of the two diastereomers provides two optically pure β-lactams (**154b**, **154c**) which are mirror images of each other (Scheme 34).

In a recent publication,[104] optical resolution via iodine catalyzed Ferrier rearrangement has been extended to a racemic β-lactam (**156**) with a thienamycin side chain. Again, only two α-glycosides were formed providing easy access to both enantiomers of a β-lactam synthon for thienamycin analogs.

High atom economy as well as excellent asymmetric induction have been achieved by using dirhodium(II) complexes incorporating a chiral amino acid

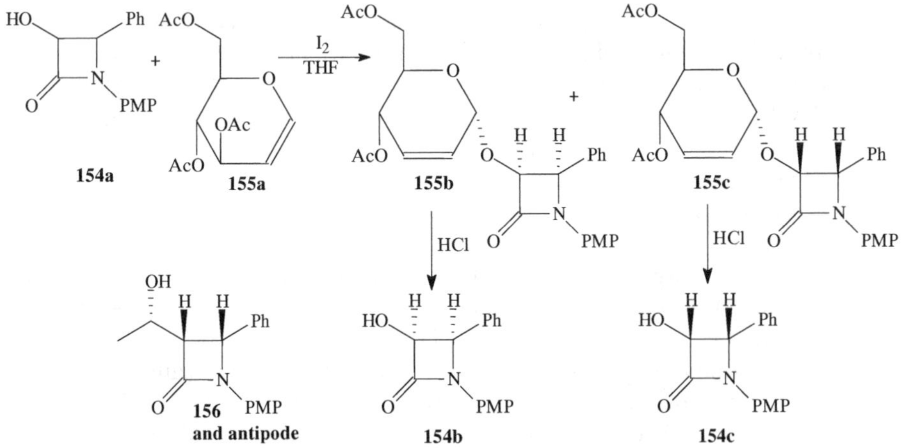

Scheme 34

derivative.[105] The mechanistic details of the reaction reported were not clear but by experimenting with a series of compounds, optimum conditions were found for obtaining an optically pure intermediate for the naturally occurring β-lactam antibiotic **PS-5**.

By extending this methodology it was possible to obtain a key intermediate for 1β-methylcarbapenems (Scheme 35). The starting material was the α-methoxycarbonyl-α-diazoacetonide **157** prepared from achiral compounds.

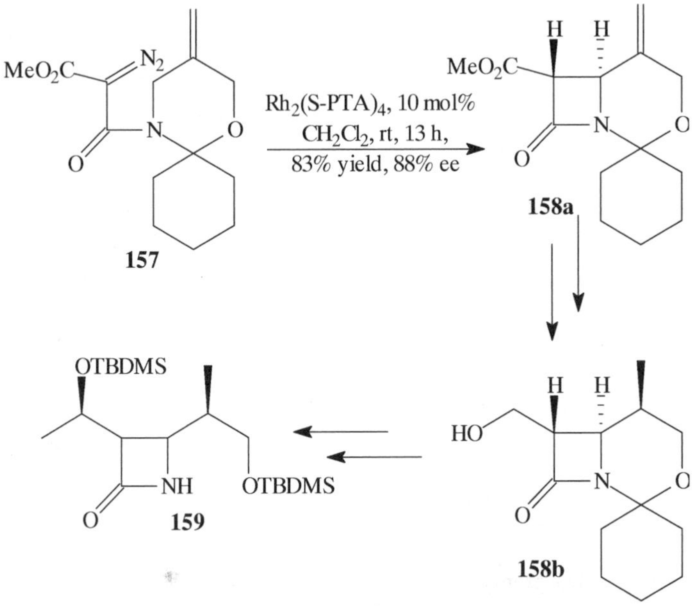

Scheme 35

Cyclization to optically active (88% ee) β-lactam (**158a**) was achieved in high yield by an excellent C–H insertion reaction. By trial and error the best catalyst was found to be a dirhodium(II) complex incorporating *N*-phthaloyl (*S*)-alanine as a chiral bridging ligand. Reduction of the ester group of **158a** with lithium borohydride followed by stereocontrolled hydrogenation produced **158b** as a crystalline compound in the optically pure form. The target compound **159** was obtained from **158b** via a few unexceptional chemical steps.

Pharmaceutical companies are faced with the problem of preparing key intermediates in large quantities while paying attention to all the regulations for the protection of the environment. A recent publication[106] reports an environmentally benign approach to a large-scale process by combining the three conventional steps (deprotection of an amide, diazotization, rhodium acetate catalyzed reaction with methanol) into a one pot reaction (Scheme 36). The starting material was commercially available 7-ADCA (**160a**) which is an inexpensive cephem derivative. The target compounds were **162** and **163**, which are the key intermediates for the preparation of potent inhibitors of mammalian serine protease.

The amino group of **160a** was protected by forming the BOC-derivative (**160b**). Conversion of the carboxy group to the *t*-butyl ketone and periodic oxidation to a sulfone provided **161**—a stable crystalline compound.

Conditions were found that combined three steps in a one-pot reaction by exposing the amido compound **161** to the action of excess boron trifluoride

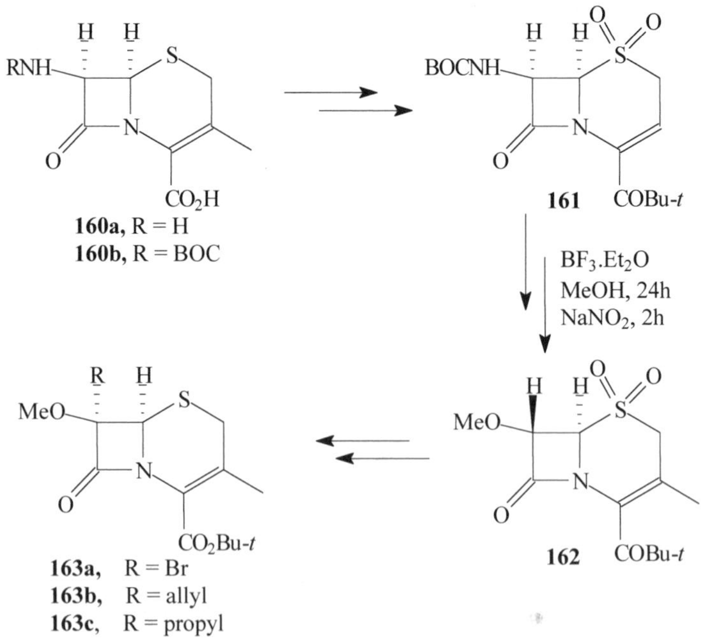

RNH, H, H, S, N, O, CO_2H

160a, R = H
160b, R = BOC

BOCNH, H, H, O, O, S, N, O

161 COBu-*t*

$BF_3.Et_2O$
MeOH, 24h
$NaNO_2$, 2h

MeO, H, H, O, O, S, N, O

162 COBu-*t*

MeO, R, H, S, N, O, CO_2Bu-*t*

163a, R = Br
163b, R = allyl
163c, R = propyl

Scheme 36

etherate in methanol for 24 h at room temperature followed by addition of sodium nitrite. The end product was **162,** one of the target compounds, which could be recrystallized to give an optically pure compound. This was an eco-friendly process because the diazo intermediate did not require direct handling or disposal. The stereochemistry of the major product was as expected: the methoxy group at C-7 was *trans* to the bulky sulfone group at C-6. Since there was no deprotonation at C-6, the absolute configuration of C-6 was unchanged during these transformations.

Analogs (**163b**, **163c**) were obtained via the bromo compound **163a** prepared by reaction of *N*-bromosuccinimide (NBS) with **162**. Inversion of the substituents at C-7 was involved during this step (Scheme 36).

Advances in combinatorial chemistry have led to solid phase synthetic techniques for the preparation of a variety of bioactive compounds. Two reports[107,108] have appeared which describe the development of methods for preparing β-lactams with diverse functional groups by solid phase synthesis.

The earlier report[107] involves the use of an Fmoc protected amino acid tethered through its carboxy group to a resin. After removal of the Fmoc group, the free amine is allowed to react with 10–15-fold molar excess of an aldehyde to form a Schiff base that is bound to a resin.

The second report[108] discloses the use of a resin bound 4-carboxy-benzaldehyde for forming immobilized Schiff bases by reaction with 10 molar excess of a variety of amines.

The Staudinger reaction was used successfully with both types of Schiff bases to form β-lactams which could be detached from the resin without any substantial loss.

A different approach has been reported recently[109]. Soluble polymers such as polyethylene glycol monomethyl ether of molecular weight 5000 (MeOPEG) were employed in place of a solid support. The synthesis of β-lactams on such soluble polymeric matrix was developed. The starting compound was an amide (**166**) immobilized on MeOPEG (Scheme 37). The succinate of MeOPEG (**164**)

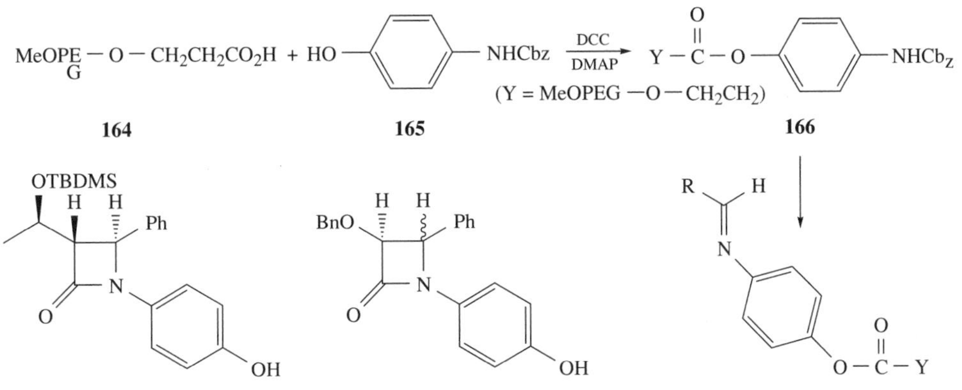

MeOPEG — O — $CH_2CH_2CO_2H$ + HO—C_6H_4—NHCbz $\xrightarrow[\text{DMAP}]{\text{DCC}}$ Y — C(=O) — O—C_6H_4—NHCbz

(Y = MeOPEG —O—CH_2CH_2)

164 **165** **166**

169 **168** **167**

Scheme 37

was converted to **166** by reaction with the phenol **165**. Removal of the *N*-protective group by hydrogenation followed by reaction with an aldehyde led to the immobilized imine **167**. Both imine–acid chloride-*t*-base method and the enolate addition method were used to obtain β-lactam derivatives. Reactions were conducted in methylene chloride solution and the end products were obtained by precipitation with methyl ether. The polymer support could be removed by acid catalyzed methanolysis or by reaction with methanol in presence of catalytic amounts of the base diazabicycloundecane.

The α-hydroxy β-lactam **168** was obtained in this way as a mixture of 85% *trans*- and 15% *cis*- isomer in an overall yield of about 50% by using an appropriate acid chloride and triethylamine.

The optically pure β-lactam **169** was obtained in about 30% yield as a *trans* compound (with 2% of the *cis*-isomer) by employing an appropriate enolate. The major product was shown to be identical with a known intermediate for compounds related to the antibiotic thienamycin (Scheme 37).

A strategy for converting a cephem derivative to its antipode has been described in a recent publication (Scheme 38). The C-6–S bond of a cephem derivative (**170**) was cleaved to give **171**. The bicyclic β-lactam **172** was obtained via reconstitution of the six-membered ring. As pointed out earlier, this reentry of the sulfur atom at C-6 would be *trans* to the bulky phthalimido group at C-7. Therefore, **172** was a *trans*-β-lactam. The C-7 position was epimerized via the sulfoxide of **172** by known chemical reactions. Because of the inversion of both chiral centers, the final compound **173** became the mirror image of **170**.[110]

Newer synthetic methods and variations of old methods are appearing in the literature at frequent intervals. The isocyanate–alkene addition method (see

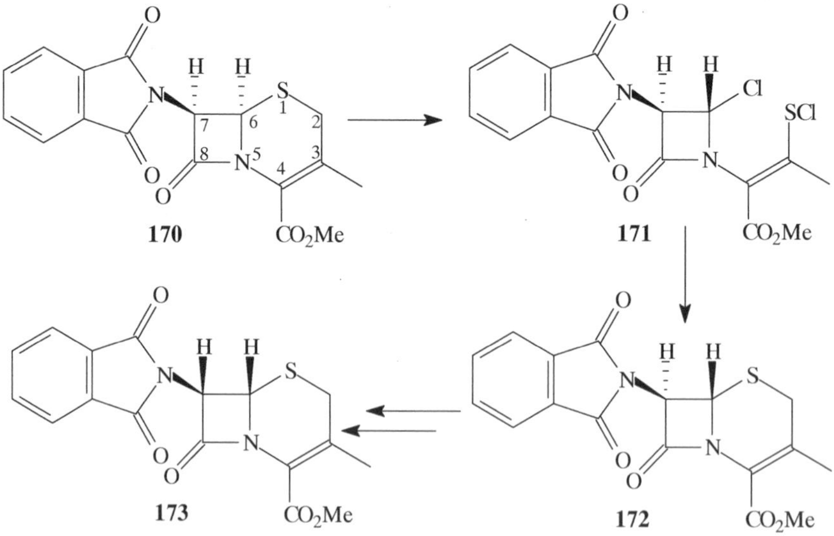

170 CO_2Me 171 CO_2Me 173 CO_2Me 172 CO_2Me

Scheme 38

Section 3.1.3) has been employed for a new pathway to bicyclic β-lactams.[111] A diversity of *bis*-β-lactams has been synthesized both in the racemic and homochiral forms.[112] A stereoselective decarboxylation reaction has been found to provide easy access to 1-β-methylcarbapenem derivatives. The α-alkyloxy-β-lactams with appropriate β-substituents have been used as the starting material for certain dipeptide mimics.[113] Since *trans*-β-lactams were desirable, the Staudinger reaction, which provides mostly *cis*-β-lactams in many cases, was avoided and an alternate method was developed.

The imine—acid chloride—tertiary base approach continues to be a favorite method for the synthesis of β-lactams. Thus, a two-step variant of the Staudinger reaction has been reported.[114] A simplified synthesis of 4-unsubstituted β-lactams has been found.[115] The steric course of the Staudinger reaction has been examined by calculation using density functional theory.[116] Fused tricyclic β-lactams have been prepared[117] from a series of monocyclic β-lactams synthesized by the Staudinger reaction. Thus, the Staudinger reaction, which started the β-lactam field in 1907, is still continuing to be a significant area of research.

5.5. Future Prospects

The drug resistance shown by various pathogens has become a major health issue. Many medicinal chemists believe that newer generations of β-lactam antibiotics will continue to offer protection against many types of infection. It also seems likely that some new β-lactam structures may exhibit a different bioactivity—lowering of cholesterol in plasma by some new mechanism is an example at present. The preparation of antitumor drugs such as Taxol™, protease inhibitors and their analogs will certainly benefit from the availability of various types of β-lactam synthons.

Many unsolved problems still remain about mechanisms of reactions, mode of arrangement and steric course of synthesis, biosynthesis, and biodegradation of β-lactams. Questions are arising about molecular recognition, binding to ligands and the detailed biological roles of these cyclic amides that are so close to amino acids and peptides and are so nimbly transformed into heterocycles of great diversity. Newer research tools including computational methods and molecular modeling will surely answer some questions and pose new ones.

If the current rate of new publications on various aspects of β-lactams is any indication, it is safe to predict that β-lactams will continue to be a lively field for academic and industrial scientists well into the next millenium.

Acknowledgments

We wish to thank M. Jayaraman, A. Bhattacharjee, E. W. Robb, and W. C. Ermler for useful discussions and A. H. Sharma, A. Ghosh-Mazumdar, and U. Shah for excellent technical assistance in the preparation of the manuscript. We also thank Stevens Institute of Technology, Dean P. Flanagan, and the George Barasch Bioorganic Research Fellowship funds for partial support.

Abbreviations

Ac	Acetyl
Ar	Aryl
Bn	Benzyl
BOC, *t*-BOC	*tert*-Butoxycarbonyl
Cbz	Benzyloxycarbonyl
DBN	1,5-Diazabicyclo[4.3.0]non-5-ene
DCC	Dicyclohexylcarbodiimide
DEAD	**Di**ethyl **azod**icarboxylate
DMAP	4-Dimethylaminopyridine
DMF	*N,N*-dimethylformamide
DMSO	Dimethylsulfoxide
ee	Enantiomeric excess
Et	Ethyl
LDA	Lithium diisopropylamide
LHMDS	Lithium hexamethyldisilazide
Me	Methyl
MWI	Microwave irradiation
NBS	*N*-bromosuccinimide
NEt_3, TEA	Triethylamine
NMM	4-Methylmorpholine
NMR	Nuclear Magnetic Resonance
p	*para*
Ph	Phenyl
Ph_3P	Triphenylphosphine
PMP	*p*-methoxyphenyl
R	an organic group
S-PTA	(*S*)-Phthalimidoalanine
TBDMS	*tert*-butyldimethylsilyl
Tf	Trifluoromethane sulfonyl
TFA	Trifluoroacetic acid
THF	Tetrahydrofuran
TMS	Trimethylsilyl

REFERENCES

1. Georg, G.I. Ed.; *The Organic Chemistry of β-Lactams*; VCH Publishers: New York, 1992.
2. Morin, R.B.; Gorman, M. Eds.; *Chemistry and Biology of β-Lactam Antibiotics*, Academic Press: New York, 1982, Vols. 1–3.
3. Flynn, E.H. Ed.; *Cephalosporins and Penicillins, Chemistry and Biology*, Academic Press: New York, 1972.

4. Manhas, M.S.; Bose, A.K. *β-Lactams: Natural and Synthetic, Part* 1; Wiley-Interscience: New York, 1971.
5. Manhas, M.S.; Bose, A.K. *Synthesis of Penicillin, Cephalosporin C and Analogs*; Marcel Dekker: New York, 1969.
6. Staudinger, H. *Liebigs Ann.* 1907, **356**, 51.
7. Macfarlane, G. *Alexander Fleming, The Man and The Myth*; Harvard University Press: Cambridge, MA, 1984.
8. Flynn, E.H. *Cephalosporins and Penicillins*; Academic Press: New York, 1972.
9. Burnett, D.A.; Caplen, M.A.; Davis, H.R.; Burnier, R.E.; Clader, J.W. *J. Med. Chem.* 1994, **37**, 1733.
10. Manhas, M.S.; Amin, S.G.; Bose, A.K. *Heterocycles* 1976, **5**, 669.
11. Manhas, M.S.; Wagle, D.R.; Chiang, J.; Bose, A.K. *Heterocycles* 1988, **27**, 1755.
12. Ojima, I. *Acc. Chem. Res.* 1995, **28**, 383.
13. Ojima, I.; Delaloge, F. *Chem. Soc. Rev.* 1997, **26**, 377.
14. Gennari, C.; Carcano, M.; Donghi, M.; Mongelli, N.; Vanotti, E.; Vulpetti, A. *J. Org. Chem.* 1997, **62**, 4746. and references cited therein.
15. Banik, B.K.; Jayaraman, M.; Srirajan, V.; Manhas, M.S.; Bose, A.K. *J. Ind. Chem. Soc.* 1997, **74**, 951.
16. Brieva, R.; Grich, J.Z.; Sih, C.J. *J. Org. Chem.*, 1993, **58**, 1068.
17. Lengyel, I.; Sheehan, J.C. *Angew. Chem.* 1968, 25.
18. Breckpot, R. *Bull. Soc. Chem. Belg.* 1923, **32**, 412.
19. Woodward, R.B. In *The Chemistry of Penicillin*; Clarke, H.T.; Johnson, J.R.; Robinson, R. Eds.; Princeton University Press: NJ, 1949; p. 440.
20. Crowfoot, D.; Bunn, C.W.; Rogers-Low, D.W.; Turner-Jones, A. In *The Chemistry of Penicillin*; Clarke, H.T.; Johnson, J.R.; Robinson, R. Eds.; Princeton University Press: NJ, 1949; p. 310.
21. Hodgkin, D.C.; Maslen, E.N. *Biochem. J.* 1961, **79**, 393.
22. Luche, J.L.; Kagan, H.B.; Parthasarathy, R.; Tsoucaris, G.; deRango, C.; Zelwer, C. *Tetrahedron* 1968, **24**, 1275.
23. Cooper, R.D.G.; DeMarco, P.V.; Cheng, J.C.; Jones, N.D. *J. Am. Chem. Soc.* 1969, **91**, 1408.
24. Dexter, D.D.; van der Veen, J.M. *J. Chem. Soc. Perkin I* 1978, 185.
25. Morin, R.B.; Jackson, B.G.; Mueller, R.A.; Lavagnino, E.R.; Scanlon, W.B.; Andrews, S.L. *J. Am. Chem. Soc.* 1969, **91**, 1401.
26. Demarco, P.V.; Nagarajan, R. In *Cephalosporins and Penicillins, Chemistry and Biology*; Flynn, E.H. Ed.; Academic Press: New York, 1972; p. 315.
27. Manhas, M.S.; Jeng, S.; Bose, A.K. *Tetrahedron,* 1968, **24**, 1237.
28. Rabenstein, D.L.; Fan, S. *Anal. Chem.* 1986, **58**, 317B.
29. Meiboom, S.; Gill, D. *Rev. Sci. Instrum.* 1958, **29**, 688.
30. Carr, H.Y.; Purcell, E. M. *Phys. Rev.*, 1954, **94**, 530.
31. Rabenstein, D.L.; Srivatsa, G.S.; Lee, R.W.K. *J. Mag. Reson.* 1987, **71**, 175.
32. Krishnaswami, A.; Ph.D. Thesis, Stevens Institute of Technology, 1991.
33. Bose. A.K.; Krishnaswami, A.; Robb, E.W.; Manhas, M.S. *J. Ind. Chem. Soc.* 1999.

34. Lukacs, G. Ed.; *Recent Progress in the Chemical Synthesis of Antibiotics and Related Microbial Products*; Springer-Verlag: New York, 1993, Vol. 2.
35. Klibanov, A.M. *Acc. Chem. Res.* 1990, **23**, 14.
36. Davies, H.G.; Green, R.U.; Kelly, D.R.; Roberts, S.M. Eds.; *Biotransformations in Preparative Organic Chemistry*, Academic Press: New York, NY; 1989.
37. Mattingly, P.G.; Kerwin, J.F.; Miller. M.J. *J. Am. Chem. Soc.* 1979, **101**, 3983.
38. Miller, M.J.; Morrison, M.A. *J. Org. Chem.* 1983, **48**, 4421.
39. Bose, A.K.; Sahu, D.P.; Manhas, M.S. *J. Org. Chem.* 1981, **46**, 1229.
40. Sahu, D.P.; Masava, P.; Manhas, M.S.; Bose, A.K. *J. Org. Chem.* 1983, **48**, 1142.
41. Townsend, C.A.; Nguyen, L.T. *J. Am. Chem. Soc.* 1981, **103**, 4582.
42. Townsend, C.A.; Brown, A.M.; Nguyen, L.T. *J. Am. Chem. Soc.* 1983, **105**, 919.
43. Sheehan, J.C.; Bose, A.K. *J. Am. Chem. Soc.* 1950, **72**, 5158.
44. Bose, A.K.; Manhas, M.S. *J. Org. Chem.* 1962, **27**, 1244.
45. Rajendra, G.; Miller, M. J. *Tetrahedron Lett.* 1985, **26**, 5385.
46. Manhas, M.S.; Bhawal, B.M.; Shankar, B.B.; Bose, A.K. *J. Ind. Chem. Soc.* 1985, **62**, 891.
47. Bose, A.K.; Manhas, M.S.; Mathur, A.; Wagle, D.R. *Recent Progress in the Chemical Synthesis of Antibiotics and Related Products*; Lukacs, G., Ed.; Springer-Verlag: New York, 1993, Vol. 2; p. 551.
48. Sheehan, J.C.; Henery-Logan, J. *J. Am. Chem. Soc.* 1959, **81**, 3089.
49. Sheehan, J.C.; Henery-Logan, J. *J. Am. Chem. Soc.* 1962, **84**, 2983.
50. Kobayasha, S.; Furuta, T.; Sugito, K.; Oyamada, H. *Synlett* 1998, 1019.
51. Eders, D.; Lochtman, R.; Meiers, M.; Muller, S.; Lazny, R. *Synlett* 1998, 1182.
52. Bose, A.K.; Anjaneyulu, B. *Chem. & Ind.* 1966, 903.
53. Bose, A.K.; Anjaneyulu, B.; Bhattacharya, S.K.; Manhas, M.S. *Tetrahedron* 1967, **23**, 4769.
54. Bose, A.K.; Spiegelman, G.; Manhas, M.S. *J. Am. Chem. Soc.* 1968, **90**, 4506.
55. Bose, A.K.; Manhas, M.S.; van der Veen, J.M.; Wagle, D.R.; Hegde, V.R.; Bari, S.S.; Kosarych, Z.; Ghosh, M.; Krishnan, L. *Ind. J. Chem.* 1986, **25B**, 1095.
56. Ciufolini, M.A.; Dong, Q. *J. Chem. Soc., Chem. Commun.* 1996, 881.
57. Ciufolini, M.A.; Hermann, C.Y.W.; Dong, Q.; Shimizu, T.; Swaminathan, S.; Xi, N. *Synlett* 1998, 105.
58. Bose, A.K.; Manhas, M.S.; Amin, S.G.; Kapur, J.C.; Kreder, J.; Mukkavilli, L.; Ram, B.; Vincent, J.E. *Tetrahedron Lett.* 1979, **30**, 2771.
59. Bose, A.K.; Jayaraman, M.; Okawa, A.; Bari, S.S.; Robb, E.W.; Manhas, M.S. *Tetrahedron Lett.* 1996, **37**, 6989.
60. Gluchowski, C.; Cooper, L.; Bergbreiter, D.L.; Newcomb, M. *J. Org. Chem* 1980, **45**, 3413.
61. Bose, A.K.; Khajavi, M.S.; Manhas, M.S. *Synthesis* 1982, 407.
62. Georg, G.I.; Kant, J.; Gill, H.S. *J. Am. Chem. Soc.* 1987, **109**, 1129.
63. Gilman, H.; Speeter, M. *J.* Am. Chem. Soc. 1943, **65**, 2255.
64. Bose, A.K.; Gupta, K.; Manhas, M.S. *Chem. Commun.* 1984, 86.
65. Ojima, I. *Acc. Chem. Res.* 1995, **28**, 383.

66. Bose, A.K.; Womelsdorf, J.F.; Krishnan, L.; Lipkowska, Z.U.; Shelly, D.C.; Manhas, M.S. *Tetrahedron* 1991, **47**, 5379.
67. Bose, A.K.; Manhas, M.S.; van der Veen, J.M.; Bari, S.S.; Wagle, D.R. *Tetrahedron* 1992, **48**, 4831.
68. Hubschwerlen, C.; Schmid, G. *Helv. Chim. Acta.* 1983, **66**, 206.
69. Bose, A.K.; Manhas, M.S.; van der Veen, J.M.; Bari, S.S.; Wagle, D.R.; Hegde, V.R.; Krishnan, L.; *Tetrahedron Lett.* 1985, **26**, 33.
70. Bose, A.K.; Mathur, C.; Wagle, D.R.; Naqvi, R.; Manhas, M.S.; Lipkowska, Z.U. *Heterocycles* 1994, **39**, 491.
71. Evans, D.A.; William, J.M. *Tetrahedron Lett.* 1988, **29**, 5065.
72. Palomo, C.; Cossio, F.P.; Cuevas, C. *Tetrahedron Lett.* 1991, **32**, 3109.
73. Evans, D.A.; Sjogren, E.B.; *Tetradedron Lett.* 1985, **26**, 3783.
74. Ojima, I.; Chen, H.C. *Tetrahedron* 1988, **44**, 5307.
75. Ojima, I. In *The Organic Chemistry of β-Lactams*, Georg, G.I. Ed.; VCH Publishers: New York, 1992; p. 197.
76. Kambara, T.; Hussein, M.A.; Fujieda, H.; Iida, A.; Tomioka, K. *Tetrahedron Lett.* 1998, **39**, 9055.
77. Anklam, S.; Liebscher, J. *Tetrahedron* 1998, **54**, 6369.
78. Woodward, R.B.; Heusler, K.; Gosteli, J.; Naegeli, P.; Oppolzer, W.; Ramage, R.; Ranganathan. S.; Vorbruggen, H. *J. Am. Chem. Soc.* 1966, **88**, 852.
79. Gedye, R.; Smith, F.; Westway, K.; Ali, H.; Baldisera, L.; Laberge, L.; Rousell, J. *Tetrahedron Lett.* 1986, **27**, 279.
80. Giguere, R.J.; Bray, T.L.; Duncan, S.M.; Majetich, G. *Tetrahedron Lett.* 1986, **27**, 4945.
81. Bose, A.K.; Banik, B.K.; Lavlinskaia, N.; Jayaraman, M.; Manhas, M.S. *Chemtech* 1997, **27**, 18, and references cited therein.
82. Caddick, S. *Tetrahedron* 1995, **51**, 10403.
83. Bose, A.K.; Manhas, M.S.; Ghosh, M.; Raju, V.S.; Tabei, K.; Lipkowska, Z.U. *Heterocycles* 1990, **30**, 741.
84. Bose, A.K.; Manhas, M.S.; Ghosh, M.; Shah, M.; Raju, V.S.; Newaz, S, N.; Banik, B.K.; Chaudhary, A.G.; Barakat, K.J. *J. Org. Chem.* 1991, **56**, 6969.
85. Bose, A.K.; Banik, B.K.; Manhas, M.S. *Tetrahedron Lett.* 1995, **36**, 213.
86. Manhas, M.S.; Ghosh, M.; Bose, A.K. *J. Org. Chem.* 1990, **55**, 575.
87. Bose, A.K.; Manhas, M.S.; Banik, B.K.; Robb, E.W. *Res. Chem. Intermed.* 1994, **29**, 1.
88. Bose, A.K.; Banik, B.K.; Barakat, K.J.; Manhas, M.S. *Synlett,* 1993, 575.
89. Banik, B.K.; Manhas, M.S.; Robb, E.W.; Bose, A.K. *Heterocycles* 1997, **44**, 405.
90. Bachmann, B.O.; Li. R.; Townsend, C.A. *Proc. Natl. Acad. Sci. U.S.A.* 1998, **95**, 9082.
91. McNaughton, H.J.; Thirkettle, J.E.; Zhang, Z.; Schofield, C.J.; Jensen, S.E.; Barton, B.; Greaves, P. *Chem. Commun.* 1998, 2325.
92. Page, M.I.; Laws, A.P. *Chem Commun.* 1998, 1609.
93. Page, M.I. *Adv. Phys. Org. Chem.* 1987, **23**, 165.

94. Niccolai, D.; Tarsi, L.; Thomas, R.J. *Chem. Commun.* 1997, 2333.
95. Maveyraud, L.; Mourey, L.; Kotra, L. P.; Pedelacq, J.; Guillet, V.; Mobashery, S.; Samama, J. *J. Am. Chem. Soc. 1998*, **120**, 9748.
96. Ohtake, N.; Imamura, H.; Jona, H.; Sato, H.; Nojamo, M.; Ushijima, R.; Nakagawa, S. *Bioorg. Med. Chem.* 1998, **6**, 1089.
97. Tsushima, M.; Iwamatsu, K.; Tamura, A.; Shibahara, S. *Bioorg. Med. Chem.* 1998, **6**, 1009.
98. Kumagai, T.; Tamai, S.; Abe, T.; Matsunaga, H.; Hayashi, K.; Kishi, I.; Shiro, M.; Nagao, Y. *J. Org. Chem.* 1998, **63**, 8145.
99. Borthwick, A.D.; Weingarten, G.; Haley, T.M.; Tomaszewski, M.; Wang, W.; Hu, Z.; Bedard, J.; Jin, H.; Yuen, L.; Mansour, T. *Bioorg. Med. Chem. Lett.* 1998, **8**, 365.
100. Chen, A.; Shapiro, M.J. *J. Am. Chem. Soc.* 1998, **120**, 10258.
101. Chang, L.C.; Bhat, K.P.L.; Pisha, E.; Kennelly, E.J.; Fong, H.H.S.; Pezzuto, J.M.; Kinghom, A.D. *J. Nat. Prod.* 1998, **61**, 1257.
102. Kikuchi, T.; Uyeo, S. *Chem. Pharm. Bull.* (*Japan*) 1967, **15**, 577.
103. Banik, B.K.; Manhas, M.S.; Bose, A.K. *J. Org. Chem.* 1994, **59**, 4714.
104. Banik, B.K.; Zegrocka, O.; Manhas, M.S.; Bose, A.K. *Heterocycles* 1997, **46**, 173.
105. Anada, M.; Watanabe, N.; Hashimoto, S. *Chem. Commun.* 1998, 1517.
106. Alpegiani, M.; Bissolino, P.; D'Anello, M.; Palladino, M.; Perrone, E. *Synlett* 1998, 322.
107. Ruhland, B.; Bhandari, A.; Gordon, E.M.; Gallop, M.A. *J. Am. Chem. Soc.* 1996, **118**, 253.
108. Singh, R.; Nuss, J.M. *Tetrahedron Lett.* 1999, **40**, 1249.
109. Molteni, V.; Annunziata, R.; Cinquini, M.; Cozzi, F.; Benaglia, M. *Tetrahedron Lett.* 1998, **39**, 1257.
110. De Angelis, F.; Mozzetti, C.; Di Tullio, A.; Nicoletti, R. *Chem. Commun.* 1999, 253.
111. Kaluza, L. *Tetrahedron Lett.* 1998, **39**, 8349.
112. Alcaide, B.; Martin-Cantalejo, Y.; Perez-Castells, J.; Sierra, M.A.; Monge, A. *J. Org. Chem.* 1996, **61**, 9156.
113. Qabar, M.N.; Meara, J.P.; Ferguson, M.D.; Lum, C.; Kim, H-O.; Kahn, M. *Tetrahedron Lett.* 1998, **39**, 5895.
114. Bacchi, S.; Bongini, A.; Panunzio, M.; Villa, M. *Synlett* 1998, 843.
115. Palomo, C.; Aizpurua, J.M.; Legido, M.; Galarza, R. *Chem. Commun.* 1997, 233.
116. Arrieta, A.; Lecea, B.; Cossio, F.P. *J. Org. Chem.* 1998, **63**, 5869.
117. Alcaide, B.; Polanco, C.; Sierra, M.A. *J. Org. Chem.* 1998, **63**, 6786.

CHAPTER 8

STERICALLY HINDERED TWISTED AMIDES

SHINJI YAMADA
Department of Chemistry, Ochanomizu University

1. INTRODUCTION

The unique properties of amides such as planar geometry and low reactivity for nucleophiles as compared to other carboxylic acid derivatives are due to the C(O)–N double bond character arising from amide resonance (Scheme 1).[1,2] The importance of the amide bonds having these properties is obvious from the fact that they are observed in a wide variety of organic compounds. The planarity of the amide bonds restricts the molecular motion and provides the required level of rigidity for the formation of the tertiary structures of peptides, proteins, enzymes, and other polymers. The low reactivity of the amide bond renders it tolerant to hydrolysis.

However, intramolecular steric repulsion around the amide group or intermolecular interaction affects its geometry and, consequently, causes deformation such as C(O)–N bond rotation, C(O)–N bond lengthening, and N-pyramidalization. The steric effects that cause the deformation in amide bonds may be grouped into the three categories described below.

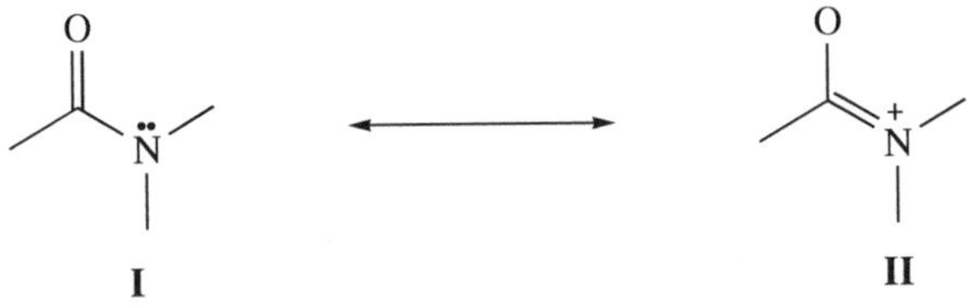

I ⟷ II

Scheme 1

The Amide Linkage: Selected Structural Aspects in Chemistry, Biochemistry, and Materials Science, Edited by Arthur Greenberg, Curt M. Breneman, and Joel F. Liebman
ISBN 0-471-35893-2

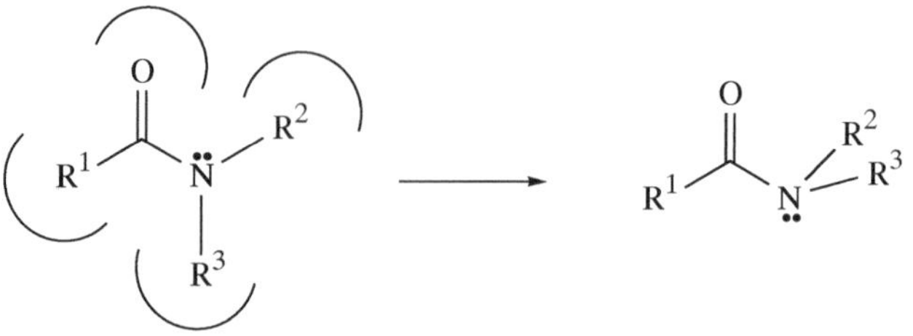

FIGURE 8.1. Distortion of amide bond by steric repulsion.

1.1. Intramolecular Steric Repulsion

An increase in the steric bulkiness of the substituents around the amide group destabilizes its planar geometry and causes deformation in the amide bond (Fig. 8.1). Distorted amides have been found in naturally occurring compounds. For example, nonplanar deformation has been observed in *cis*-X-Pro peptide bonds of proteins and enzymes in relatively higher percentages than in *trans*-X-Pro bonds.[3–6] This may be a consequence of the steric repulsion between the substituents around the *cis*-amide moieties. In this chapter, the structure and reactivity of many sterically hindered twisted amides are reviewed.

1.2. Intramolecular Steric Restriction

Amide bonds in small- and medium-sized lactams and bridgehead bicyclic lactams receive inherent geometrical restriction, and as a result, they have significant distortion (Fig. 8.2). This class of amides has received much attention and has been extensively studied.[7,8]

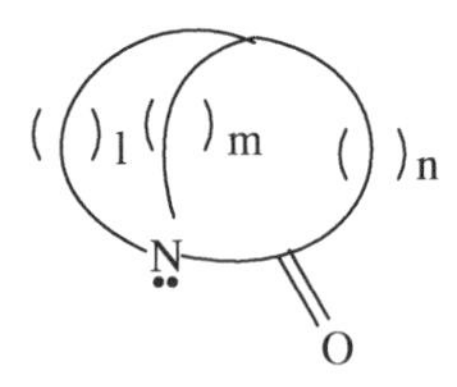

FIGURE 8.2. General structure of bridgehead lactams.

1.3. Intermolecular Steric Interaction

Intermolecular steric interaction also affects the geometry of amides. Deformations in the amide bonds of peptide substrates are postulated to occur during some biochemical processes. The hydrolysis of peptides with proteases is known to proceed through distorted intermediates, the amide bonds of which are sufficiently weakened so as to undergo cleavage.[9,10] Recently, it has been suggested that the orthogonally oriented α-keto amide groups of the immuno-

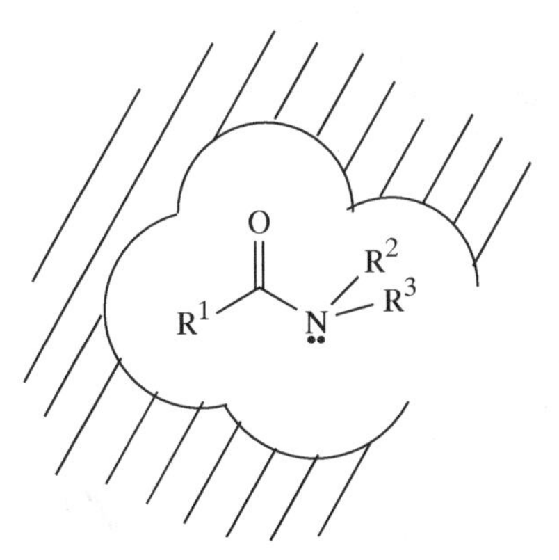

FIGURE 8.3. Schematic complex of amide and enzyme.

suppressive agent FK506 and rapamycin may mimic the twisted amide transition-state of the peptide substrates of FKBP rotamase.[11–14] Such high-energy twisted transition states are stabilized by the formation of enzyme–substrate complexes (Fig. 8.3).

2. CLASSIFICATION OF DISTORTED AMIDES

In order to express the distortion in an amide moiety quantitatively, the Winkler–Dunitz parameters, χ_C, χ_N, and τ, are generally employed.[15] The χ_C and χ_N represent the pyramidalities in C and N atoms, respectively, and τ represents the magnitude of rotation about the C(O)–N bond. Their definitions are described in Fig. 8.4. The τ is also defined to be in the range of $0°$ to $90°$ for easy comparison of the magnitude of the twist angles.[16]

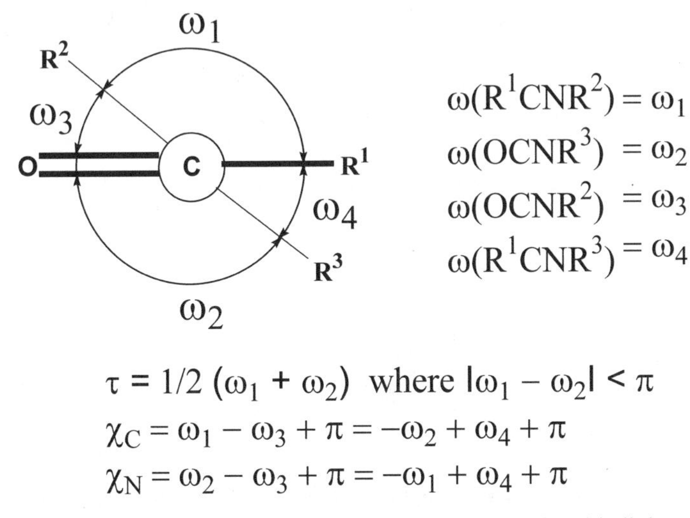

FIGURE 8.4. Dunitz parameters for defining distorted amide linkages.

It is also important to understand the deformation mode qualitatively. The distorted amides can be classified into three representative types, A, B, and C. They are shown in Fig. 8.5 with their projections down the C(O)–N bond. Since the pyramidality in carbonyl carbon is generally very small, it is disregarded here.

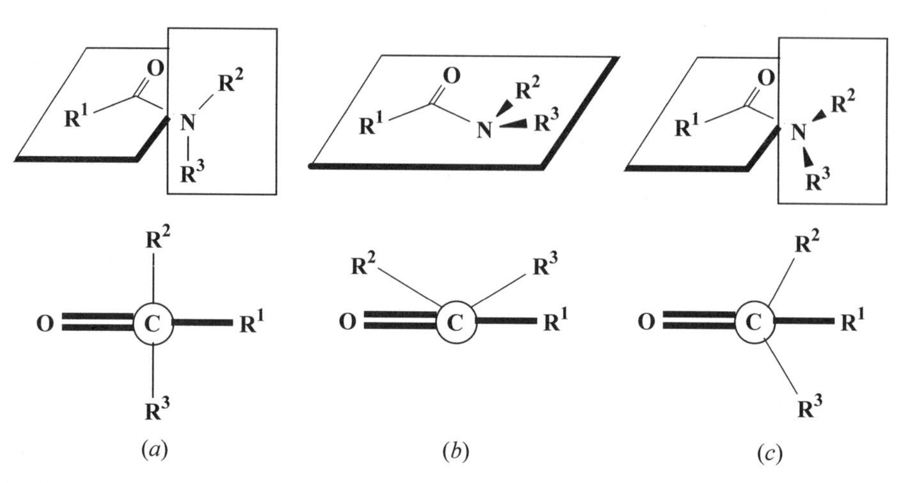

FIGURE 8.5. Extreme structures of three types of distorted amides. (a) twisted amide; (b) nonplanar amide; (c) twisted nonplanar amide.

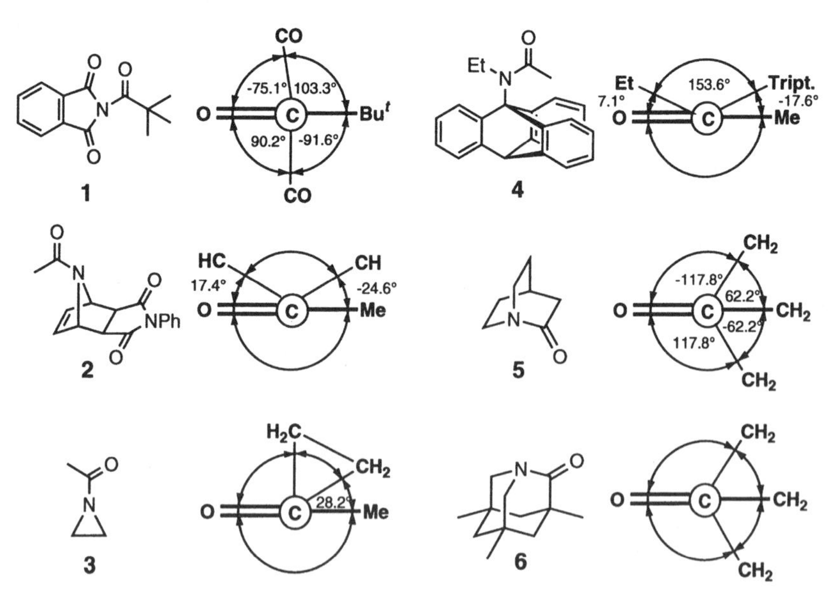

FIGURE 8.6. Representative examples for type A to C and their projections down the C(O)–N bond.

2.1. Twisted Amide (Type A)

Type A has a perpendicularly twisted C(O)–N bond and a virtually non-pyramidalized trigonal N atom. This type of compound may be most suitably called Twisted Amide.[17] *N*-Pivaloylphthalimide (**1**) is a representative example of Type A (Fig. 8.6). The acyl group is orthogonally twisted to the phthalimide plane as shown in the perspective view. The τ and χ_N values are 83.2° and 14.9°, respectively.[18] The extremely large twist angle and relatively small *N*-pyramidalization are due to the *N,N*-diacylamide structure where the C(O)–N rotational barrier is smaller than those of general amides. This class of amides is discussed in more detail in the following sections.

2.2. Nonplanar Amide (Type B)

Type B has a significantly pyramidalized N atom and a planar C(O)–N bond. Three amides **2–4** are shown as examples of Type B (Fig. 8.6). *N*-Acyl-7-azabicyclo[2.2.1]heptane derivatives[19–21] possess extremely pyramidalized N atoms. The deformation parameters of amide[19] **2** are $\tau = 18.9°$, $\chi_C = 2.1°$, and $\chi_N = 39.9°$. The significant *N*-pyramidality would be due to the conformationally rigid bicyclo[2.2.1]heptane system. Thus, replacement of the nitrogen at 7-position with a trigonal atom will cause the destabilization in amide **2**, which is presumed from the fact that 7-norbornanone is some 36 kJ mol^{-1} less stable than its 2-isomer.[22] Despite the large χ_N values, the C(O)–N bond twisting is relatively very small. This type of compounds, in which *N*-pyramidalization is a dominant deformation mode, should be called Nonplanar Amide.

N-Acetylaziridine (**3**) has a strained three-membered aziridine ring, the N atom of which is highly pyramidalized because of its inherent distortion.[23] The C(O)–N bond is a little twisted to avoid steric repulsion between the *N*-acetyl and methylene groups of the aziridine ring.

Recently, an interesting nonplanar amide, *N*-methyl-*N*-9-triptycylacetamide (**4**) has been reported.[24,25] Steric repulsion between the quite bulky *N*-triptycyl and *N*-methyl groups causes *N*-pyramidalization, whereas the twisting of the C(O)–N bond is very small ($\tau = 5.3°$, $\chi_C = 1.7°$, and $\chi_N = 26.4°$).

2.3. Twisted Nonplanar Amide (Type C)

Type C is a combined form of types A and B, which has both a pyramidalized N atom and a twisted C(O)–N bond. This type of amide is often found in bicyclic bridgehead lactams. The N atom located at the bridgehead position is generally highly pyramidalized. 1-Azabicyclo[2.2.2]octane-2-one(2-quinuclidone) (**5**), of which structure has been interested for long time, is the appropriate example of Type C, although the parent remains unknown.[26] Optimized ab initio molecular orbital calculation predicts that the amide bond is orthogonally twisted and the N atom is rehybridized to sp^3 (Fig. 8.6).[27]

Recently, bridgehead lactam 3,5,7-trimethyl-1-aza-adamantan-2-one (**6**) has been synthesized. This is thermally stable and the structure was characterized by

X-ray analysis. The C–N and C=O bond lengths are 1.475(11)Å and 1.196(5)Å, respectively. The twist angle is 90.5°, and the sum of three bond angles at N is 325.7°,[28] which means that the N atom is virtually tetrahedral.

Although the extreme structures of distorted amides are classified into three types as described above, most distorted amides take the intermediate geometries of type A to C.

3. STRUCTURE OF STERICALLY HINDERED TWISTED AMIDES (RCONXY)

Steric repulsion among the substituents R^1, R^2, and R^3 causes deformation of the amide group (Fig. 8.1). The deformation mode and magnitude depend on the steric and electronic effects of the substituents.

In general, C(O)–N bond rotation is accompanied by an increase in energy due to loss of amide resonance (Fig. 8.7a). However, the situation is changed, for sterically congested amides. Thus, when steric repulsion is large enough, the C(O)–N bond will rotate to avoid steric repulsion at the expense of resonance stabilization. This means that the twisted form is more stable than the planar one despite the loss of resonance energy in the C(O)–N bond as shown in Fig. 8.7b.

The electronic effect of the substituents around the amide group also affects the C(O)–N double bond character. Since the C(O)–N rotational barrier is significantly related to the C(O)–N double bond character, it is a useful parameter to study the characteristics of the C(O)–N bond.[29,84] The E_a of *N*-acetylpyrrole (**7**) is only 12.6 kcal mol^{-1},[30] whereas that of *N,N*-dimethylacetamide (**8**) is 19.6 kcal mol^{-1}.[31] Therefore, substituents that delocalize the nitrogen lone-pair electron are expected to facilitate the deformation process by lowering the rotational barrier E_a.

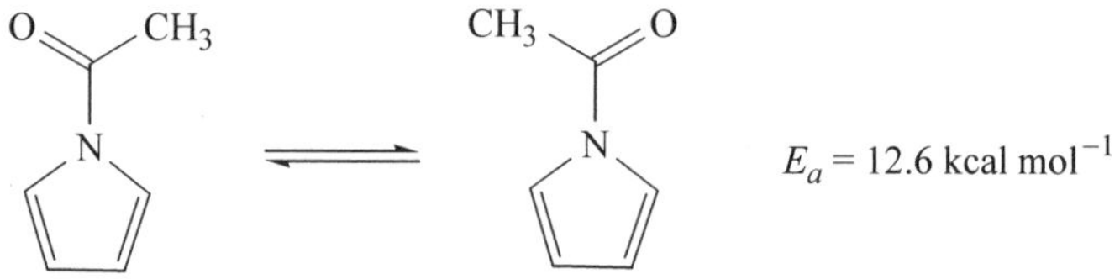

$E_a = 12.6$ kcal mol^{-1}

7

$E_a = 19.6$ kcal mol^{-1}

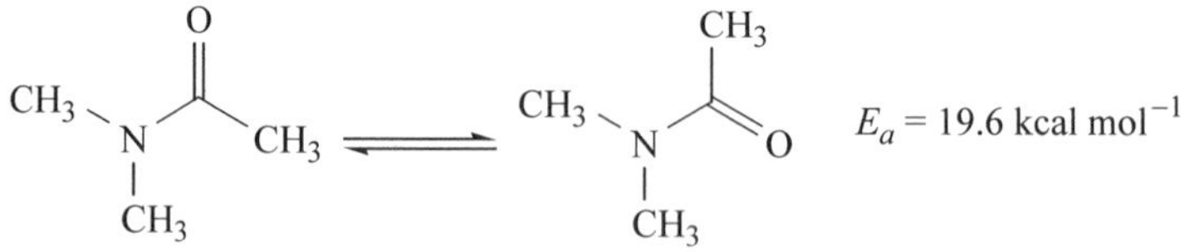

8

3.1. X, Y = Alkyl, Alkenyl

As noted in section 1, *cis*-X-Pro amide bonds in a number of proteins and enzymes have twisted deformations, since the *cis*-X-Pro is more sterically hindered than the *trans* one. When a bulky substituent is present at the 5-position of the proline ring, the amide bond deformation increases because of the steric

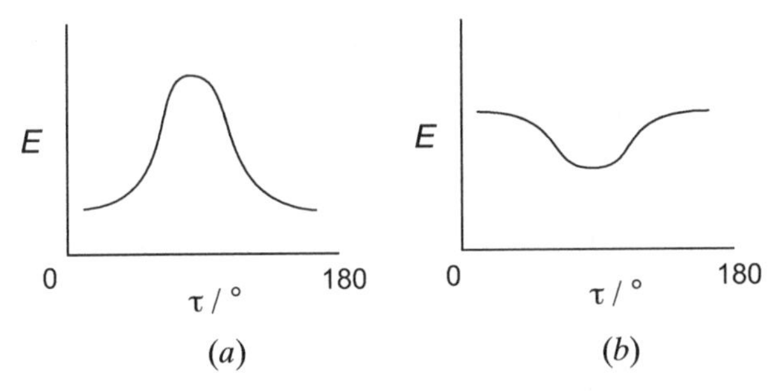

FIGURE 8.7. Schematic representation of relationship between energy vs. twist angle for (a) planar amide; (b) twisted amide.

interactions between the 5-substituent and the *N*-acyl groups. *N*-Acetyl-*trans*-5-*tert*-butylproline-*N'*-methylamide (**9**),[32] which was synthesized as a conformationally restrained peptide mimic, is in 66 : 34 equilibrium of **9a** and **9b**. Molecular mechanics calculations of **9a** and **9b** using the AMBER force field predicted that they have significantly twisted amide bonds, the dihedral angles of which are $-32°$ and $152°$, respectively. Recently, the twisted amide conformation has been proved for the related peptide *N*-acetyl-L-leucyl-5-*tert*-butylproline-*N'*-methylamide[85] by X-ray analysis, the ω-dihedral angle of which is $17.3°$.

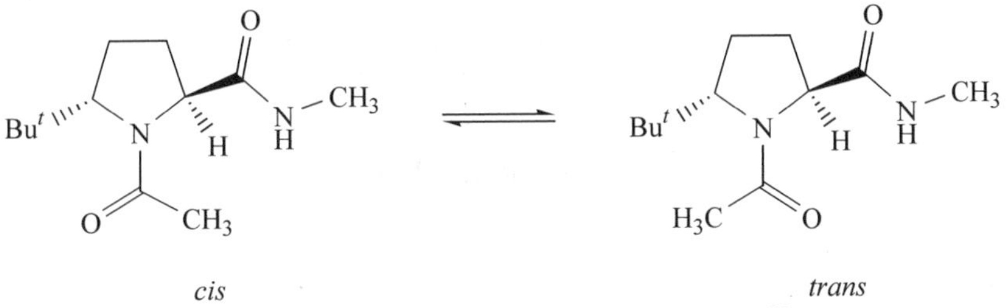

cis

9a

trans

9b

The conformation of *N*-*p*-toluoylpyrrole derivatives significantly depends on the substituents at the 2- and 5-positions. The deformation parameters of amide **10** are $\tau = 7.9°$, $\chi_C = 0°$, and $\chi_N = 10.1°$.[33] On the other hand, the averages of the parameters of amide **11** containing three independent molecules in the crystal, are $\tau = 40°$, $\chi_C = 5.3°$, and $\chi_N = 8.3°$.[33] These results suggest that the methyl groups at the 2- and 5-positions are very effective for twisting the amide bond. In addition to the steric effect, the electronic effect of the pyrrole ring may also be an important twisting factor because it diminishes the C(O)–N rotational barrier as noted above. The C(O)–N bond length of **10** is 1.409 Å, whereas that of **11** is 1.416 Å. The difference in the C(O)–N bond lengths is very small despite the large difference in their twist angles, indicating that the C(O)–N double bond character may be very small even in planar amide **10** because the lone pair electrons of the nitrogen atom contribute to the resonance of the aromatic ring.

The steric effects of the substituents around the amide group on the C(O)–N bond twisting are clearly demonstrated in unsymmetric diamide **12**[34] which possesses both 2,2,6,6-tetramethylpiperidyl and piperidyl groups. The twist angle of the congested amide bond is 23.2°, whereas that of the uncongested amide bond is 5.5°. The C(O)–N bond lengths of the planar and twisted amide moieties are 1.33 Å and 1.37 Å, respectively, which indicate a decreased double bond character in the twisted amide bond.

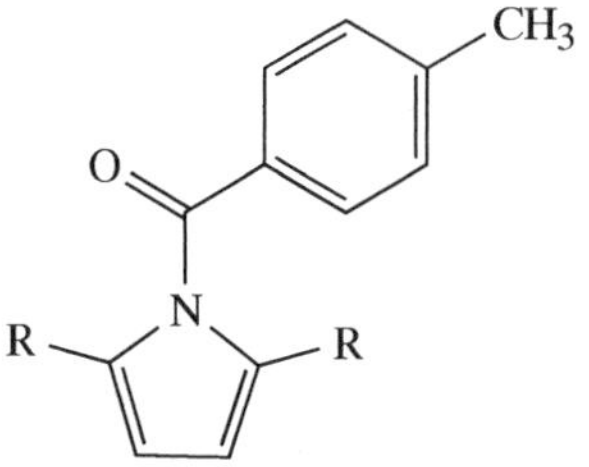

10 R = H
11 R = CH_3

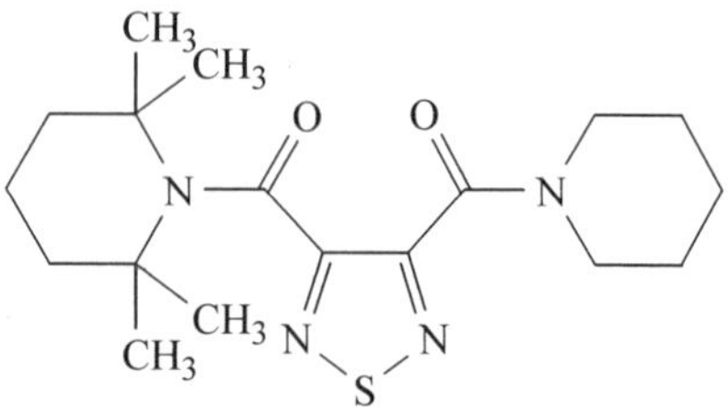

12

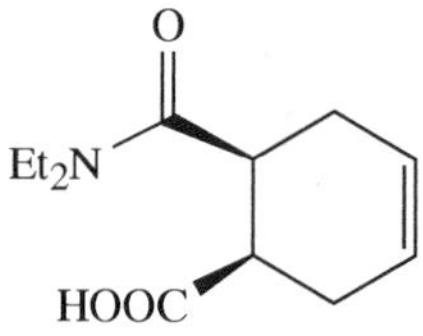

13

Small twisting was observed in the amide bond of 4-diethylcarbamoyl-1-cyclohexene-5-carboxylic acid (**13**) in the crystalline state ($\tau = 4.7°$, $\chi_C = 8.4°$, and $\chi_N = 14.0°$).[35] The steric repulsion between the *cis*-oriented carboxyl and carbamoyl groups may cause the deformation.

3.2. X = Carbonyl, Y = Alkyl

N-Acylamides show considerably different properties from *N,N*-dialkylamides. The rotational barrier of diacetylamine (**14**) in acetone is 10.8 kcal mol^{-1},[36] which is much lower than that of *N,N*-dimethylacetamide (19.6 kcal mol^{-1}). The barrier of amide **15** possessing a methyl group on the nitrogen atom is lower than that of **14**. The C(O)–N bond distance of 1.402 Å in diacetylamide is longer than the distance in general amides of 1.35 Å. These observations can be explained as follows: The nitrogen lone-pair electrons are divided between the two carbonyl groups; as a result, the C(O)–N bond has smaller double-bond character than that in dialkylamides (Scheme 2). It would be expected that when the C(O)–N rotational barrier is very small, intermolecular interactions such as crystal packing and inclusion effect may often be important factors for amide deformations.

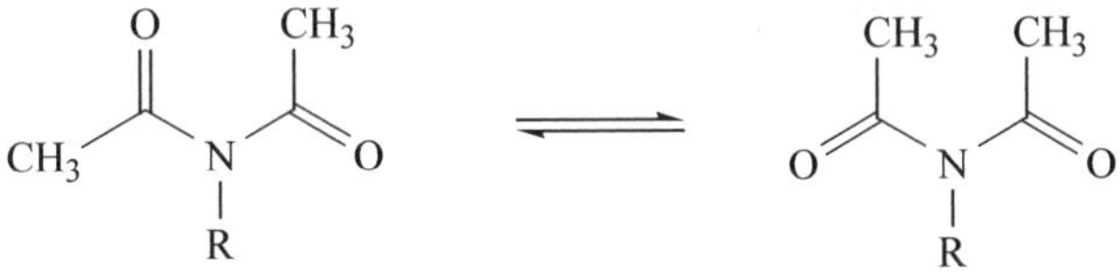

14 R = H $\Delta G^{\ddagger}_{a} = 10.8\ \mathrm{kcal\ mol^{-1}}$

15 R = CH_3 $\Delta G^{\ddagger}_{a} = 8.2\ \mathrm{kcal\ mol^{-1}}$

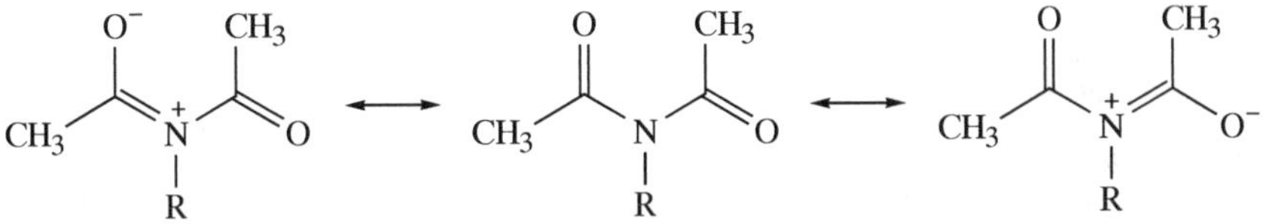

Scheme 2

t-Butyl-*N*-benzoyl-*N*-phenylcarbamate (**16**) possesses a twisted amide in the crystalline state.[37] The *N*-phenyl ring occupies a *cis–cis* position with respect to both carbonyl groups. The deformation parameters of **16** are $\tau = 29.6^\circ$, $\chi_C = 6.2^\circ$, and $\chi_N = -6.0^\circ$. It is interesting to note that the χ_N value is very small, although rotation about the C(O)–N bond is generally accompanied by *N*-pyramidalization. This phenomenon arises from delocalization of the lone-pair electrons with both of the carbonyl groups as shown in Scheme 2. Thus, even if the resonance of one amide is inhibited, the N atom can resonate with the other carbonyl group, and the N atom retains planarity. In contrast to **16**, the amide bond of structurally related molecule **17** is planar, where the *N*-phenyl group occupies a *cis–trans* position with respect to the corresponding carbonyl groups.[37] The factors that cause the structural differences between **16** and **17** are not clear, but the difference in the crystal packings may be one of the reasons.

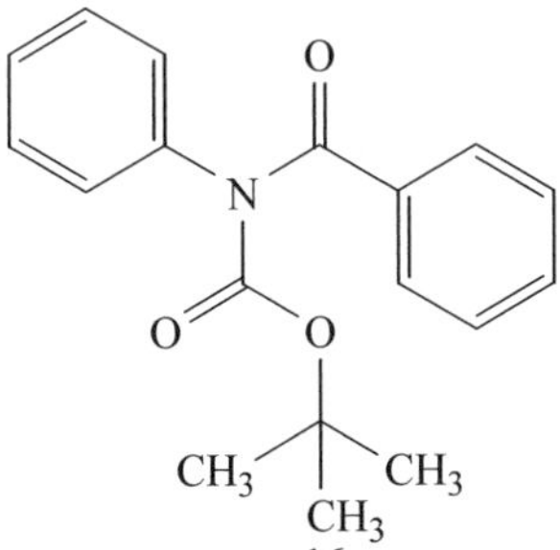

16 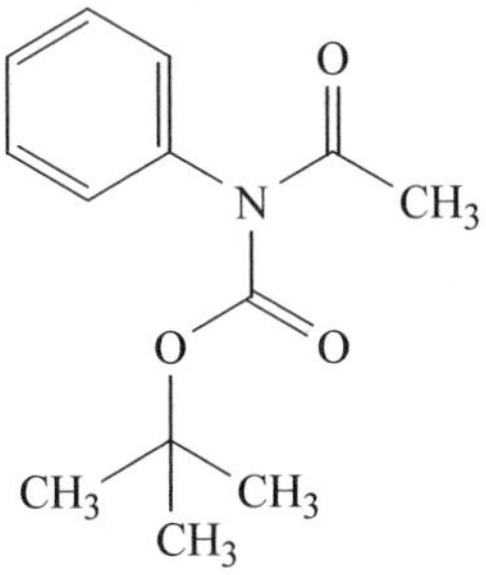 **17**

An eight-membered lactam **18** having a benzyloxycarbonyl group in the N atom has an extremely twisted amide bond in the ring ($\tau = 50.6^\circ$ and $\chi_N = 11.9^\circ$).[38] In contrast, the twist angle of 7-heptanelactam **19** is 4.5°.[39]

The distortion in **18** would be the result of both steric and electronic effects of the *N*-benzyloxycarbonyl group.

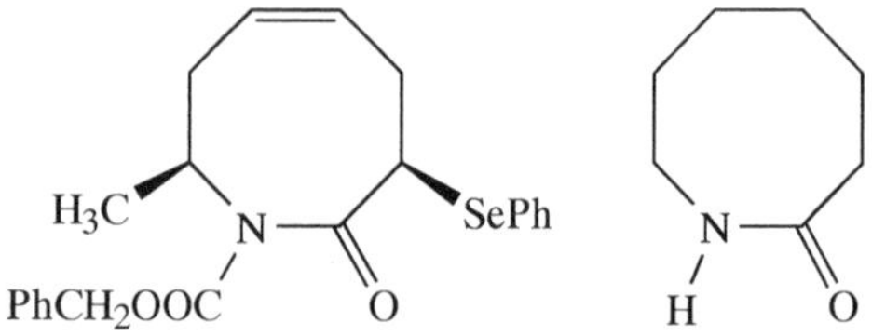

18 **19**

3.3. X = Thiocarbonyl, Y = Alkyl

The amides possessing a thiocarbonyl group on the N atom have similar properties to those of *N*-acylamides. The barrier to rotation of the C(S)–N bond is known to be a little higher than that of the C(O)–N bond.[40] This is due to the greater contribution of the dipolar canonical structure in thioamides (Scheme 3).

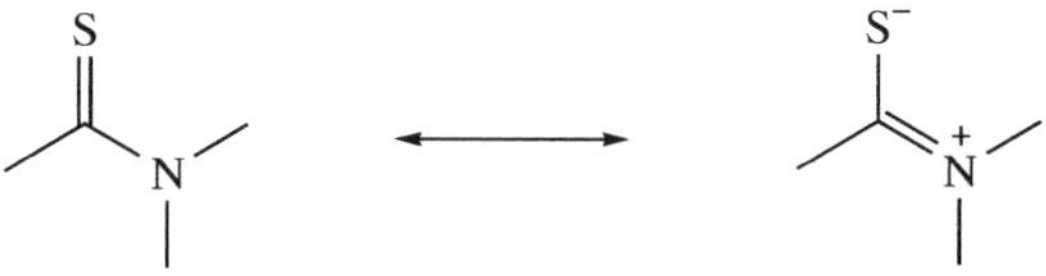

Scheme 3

The geometries of a series of *N*-acylthiazolidine-2-thiones (**20**–**23**) were studied by IR, UV, and ^{13}C NMR spectroscopies[16] (Table 8.1). The absorption band of the carbonyl group of **23** in the IR spectrum appeared at 1726 cm^{-1}, which is much higher wavelength than those of **20**–**22**. In the UV spectra, the λ value of the longest wavelength band in **23** is very different from those in amide **20**–**22**; however, it is equal to the λ_{max} in thiazolidine-2-thione which is the framework structure of the amides. A similar tendency is also seen in the ^{13}C NMR chemical shifts. The $\Delta\delta^{13}$C value is given as the difference in the ^{13}C chemical shifts between **20**–**23** and the corresponding *N*,*N*-dimethylcarboxyamides **30a**–**30d** to compensate for the substituent effects. From comparison of the $\Delta\delta^{13}$C values, it is clear that the $\Delta\delta^{13}$C of **23** is much higher than those of the others. These results strongly suggest the C(O)–N bond of **23** to be extremely twisted.

X-ray analyses of a series of amides **20**, **23**, **24**, **25**, and **28** were performed to study the conformations of their amide moieties. Table 8.2 shows the Dunitz parameters τ, χ_C, and χ_N, and the structural parameters r(C(O)–N), r(C(S)–N), and r(C=O) of these amides. Remarkable differences between **20** and **23** are observed with respect to their C(O)–N bond. The ORTEP structures and their projection of amide groups are shown in Figs. 8.8 and 8.9, respectively. The twist angle τ of **20** is 20.1°, whereas that of **23** is 74.3°. The C(O)–N bond distance of

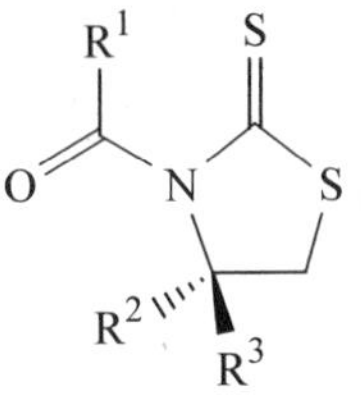

	R^1	R^2	R^3		R^1	R^2	R^3
20 :	Me	H	H	**25** :	Me	Me	Me
21 :	Et	H	H	**26** :	Ph	Me	Me
22 :	*i*-Pr	H	H	**27** :	*t*-Bu	*i*-Pr	H
23 :	*t*-Bu	H	H	**28** :	*t*-Bu	*i*-Bu	H
24 :	Bn	H	H	**29** :	*t*-Bu	*t*-Bu	H

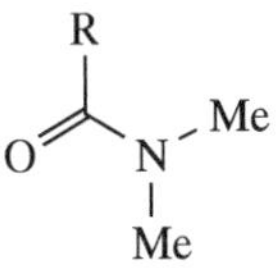

30	R
a :	Me
b :	Et
c :	*i*-Pr
d :	*t*-Bu
e :	Bn

31	R^1	R^2
a :	H	H
b :	Me	Me
c :	*i*-Bu	H

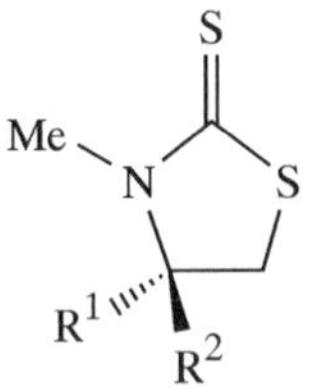

30 **31**

TABLE 8.1. IR, UV and ^{13}C NMR Spectral Data for 20–23

Compound	ν/cm^{-1}	λmax/nm	ε	$\delta_{C=O}$[a]	$\Delta\delta_{C=O}$
20 (30a)	1697	306	11950	171.3 (170.6)	0.7
21 (30b)	1702	306	11730	175.6 (173.8)	1.8
22 (30c)	1701	309	10807	178.7 (177.0)	1.7
23 (30d)	1726	280	8750	187.8 (177.5)	10.3

[a] 100.4 MHz in $CDCl_3$.

TABLE 8.2. Selected Structural Parameters for 20, 23, 24, 25, and 28

Compound	$\chi_C/°$	$\chi_N/°$	$\tau/°$	r(C(O)–N)/Å	r(C(S)–N)/Å	r(C=O)/Å
20	4.3	11.9	20.1	1.413 (9)	1.380 (8)	1.21 (1)
23	8.3	29.5	74.3	1.448 (4)	1.351 (4)	1.196 (4)
24	0.6	13.4	10.2	1.415 (6)	1.372 (6)	1.203 (7)
25	5.8	11.6	36.5	1.432 (3)	1.363 (3)	1.210 (3)
26	8.3	31.4	65.5	1.466 (5)	1.340 (5)	1.195 (5)

23 is longer than that of **20** (1.448 Å and 1.413 Å, respectively). Introduction of a *gem*-dimethyl group at the 4-position of thiazolidine-2-thione moiety successfully twists the amide bond to produce a twist angle of 36.5° (Figure 8.10).[41]

Although **23** and **28** have large twist angles, their χ_N values are much smaller than those of reported distorted amides. The χ_C values are scarcely influenced by τ. The r(C(O)–N) of the amides are longer than the general C(O)–N bond length (1.35 Å); in particular, that of **28** is 1.466(5) Å. An increase in the twist angle causes lengthening of the C(O)–N bond and shortening of the C(S)–N bond.

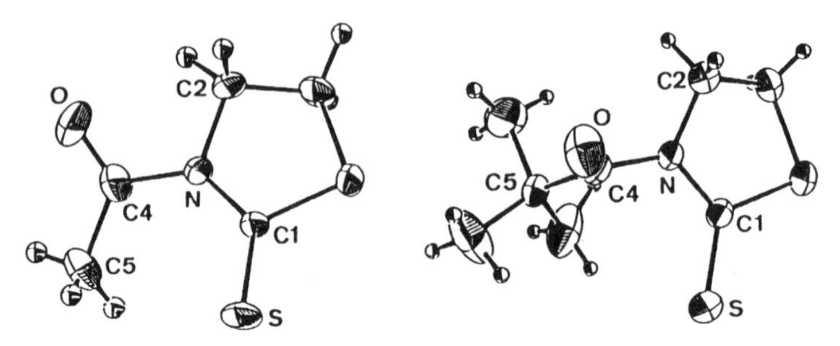

FIGURE 8.8. ORTEP drawings of **20** (left) and **23** (right) (thermal ellipsoids draw at the 50% and 35% probability level, respectively).

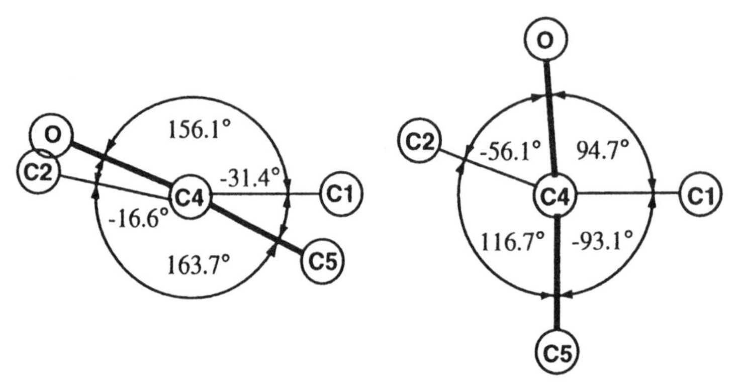

FIGURE 8.9. Projection of the amide groups of **20** (left) and **23** (right) down the C–N bond.

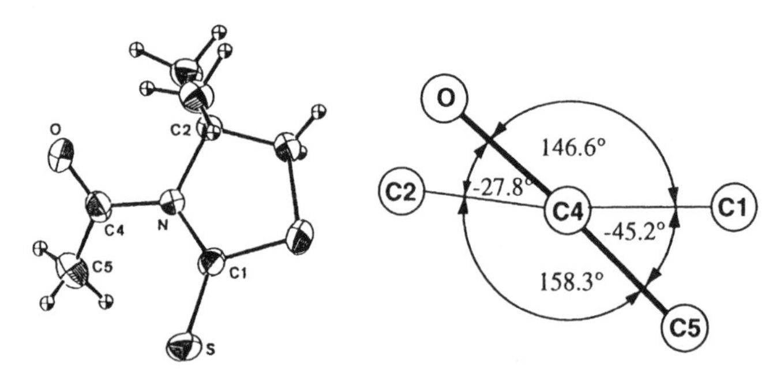

FIGURE 8.10. ORTEP drawing and projection of the amide group of **25** down the C–N bond.

Figure 8.11 shows the plots of r(C(S)–N) vs. r(C(O)–N). As the r(C(O)–N) increases, the r(C(S)–N) decreases linearly. These observations indicate the partial distribution of the nitrogen lone-pair electron to the C=S bond similar to the *N*-acylamides described in Scheme 2.[64] In contrast to the r(C(O)–N), r(C=O) remains virtually unchanged during the C–N bond rotation. Such independence of r(C=O) on the twist angles has also been observed in bicyclic distorted amides.[33]

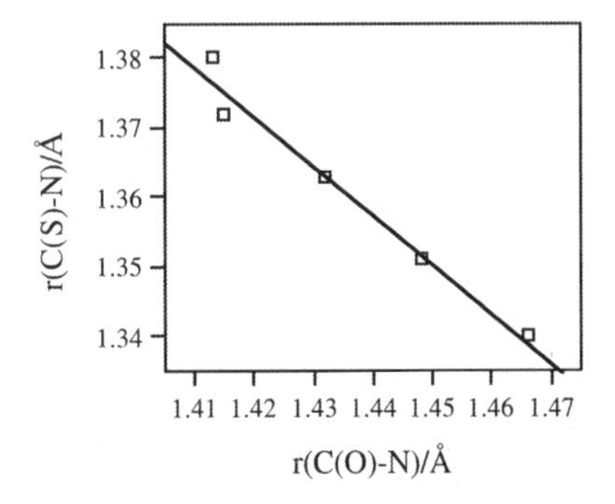

FIGURE 8.11. Plot of *r*(C(S)–N) vs. *r*(C(O)–N).

Recently, an interesting twisted amide possessing both a planar and a twisted amide linkages has been reported. In the X-ray structure of 1,6-diacetyl-3,4,7,8-tetramethyl-2,5-dithioglycoluril (**32**),[43] one acetyl group lies essentially coplanar with the attached thioureido ring, while the other acetyl group is highly twisted relative to the *N*-acetyl-1,3-thiazolidine-2-thione (**20**) described above (Fig. 8.8). Such geometrical difference between the two acetyl groups is due to the steric strain between them.

Me Me Me
S S
Me Me Me
$\tau = 2.6°$ O O $\tau = 55.0°$

32

The other structurally related compounds, 3-acyl-1,3-thiazolidine-2-ones and 3-acyl-1,3-oxazolidine-2-thiones, seem to have similar structural properties to those of 3-acyl-1,3-thiazolidine-2-thiones described above, although their structures have not yet been explored.

3.4. X, Y = Carbonyl

The *N*,*N*-diacylamides that contain three acyl groups attached to the same N atom have much lower E_a values than those in acylamides. The reported $\Delta G_a^{\ddagger}$ value of *N*-acetyldiformylamide (**33**) is 7.5 kcal mol^{-1}.[44] The X-ray-derived structure of tribenzamide (**34**) is reported to have three different twisted amide bonds.[45] The three twist angles are 36.0°, 36.9°, and 42.7°. The C(O)–N bond lengths (1.438 Å to 1.442 Å) are longer than those in diacetylamine (**14**). The average χ_N value is 30.8°, indicating a large pyramidalization in the N atom. The fact that three different τ values are observed in the same molecule suggests the crystal packing to be important in the deformation of **34**.

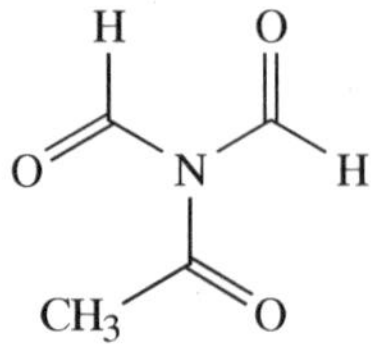

33 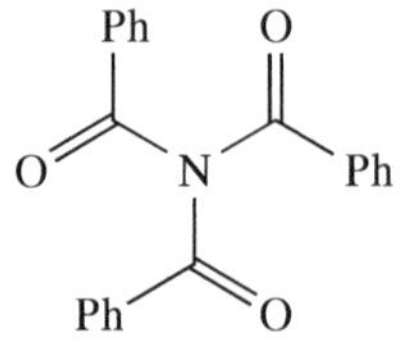 **34**

The structure of 1,3-diacetyl-5-fluorouracil (**35**) has been studied by X-ray analysis in relation to its biological activity.[46] The amide bond in the 3-acetyl moiety is highly twisted; the distortion parameters are $\tau = 89.8°$, $\chi_C = 1.3°$, and $\chi_N = -2.2°$. In contrast, the amide group of the 1-acetyl moiety is coplanar with the 5-fluorouracil ring. The twisted geometry in the 3-acetyl group is explained by steric and electronic effects. Thus, the steric and electronic repulsion between the two carbonyls in the ring and the 3-acetyl groups can be minimized by orienting the acetyl group perpendicular to the 5-fluorouracil plane. Similar twisted geometries have been observed in 1-acetyl-3-*o*-toluyl-5-fluorouracil[47] (**36**) ($\tau = 85.8°$, $\chi_C = 0.05°$, and $\chi_N = 5.7°$) and phenoxycarbonyluridine derivative[48] (**37**) ($\tau = 86.0°$, $\chi_C = 0.8°$, and $\chi_N = 0.9°$).

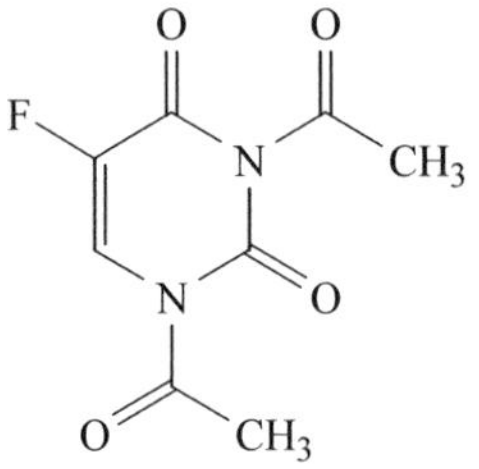

35 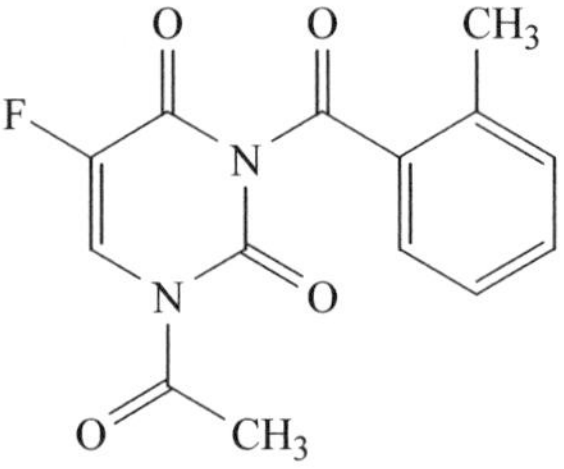 **36**

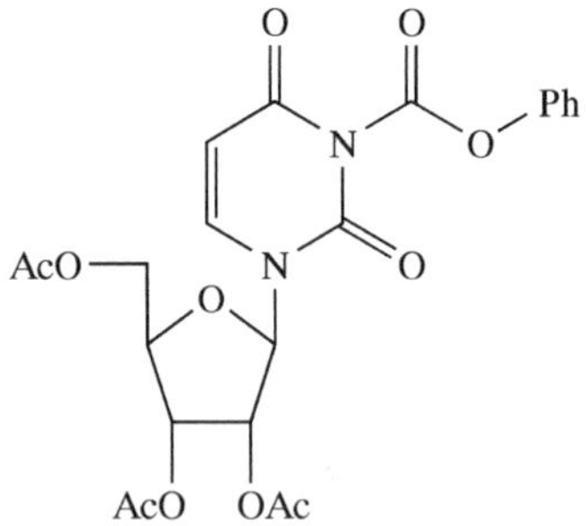

37

As noted earlier, *N*-pivaloylphthalimide (**1**) has a highly twisted amide linkage.[18] The ORTEP structure of **1** is given in Fig. 8.12. The perspective view is shown in Fig. 8.6. Although the *t*-Bu group of *N*-pivaloylphthalimide is highly disordered due to C–C(CH_3)$_3$ bond rotation, it is clear that the acyl group is orthogonally twisted with the phthalimide plane. The twist angle is 83.2°, and the χ_C and χ_N values are 1.4° and 14.9°, respectively (Table 8.3). The C(O)–N bond length of 1.474 Å is much longer than those of C7–N and C8–N. These

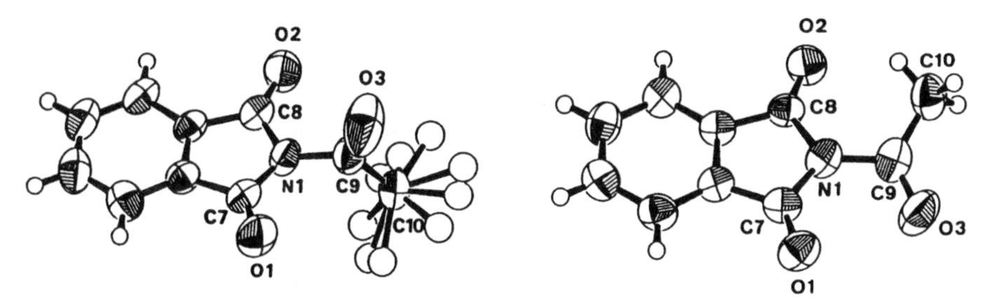

FIGURE 8.12. ORTEP drawing of **1** (thermal ellipsoids draw at the 50% probability level).

TABLE 8.3. Dunitz Parameters for 1 and 38

Compound	Origin	$\tau/^\circ$	$\chi_C/^\circ$	$\chi_N/^\circ$	r(C7-N)/Å	r(C8-N)/Å	r(C9-N)/Å
1	X-ray	83.2	1.6	14.9	1.405 (4)	1.389 (4)	1.474 (4)
1	AM1	70.4	2.6	16.8	1.434	1.429	1.435
38	X-ray	21.5	1.5	0.4	1.432(7)	1.410(7)	1.431(6)
38	AM1	0.1	0.1	0.4	1.442	1.432	1.414

data suggest that the C–N double bond character has been mostly lost. The optimized structure of **1** by AM1 method with MMOK empirical amide correction factor is almost in agreement with the X-ray results except for the bond lengths. On the other hand, the X-ray geometry of *N*-acetylphthalimide[49] (**38**) is virtually planar as shown in Fig. 8.12. The deformation parameters are $\tau = 21.5^\circ$, $\chi_C = 1.5^\circ$, and $\chi_N = 0.4^\circ$. Therefore, the steric repulsion of the *t*-Bu with the imide carbonyl groups is the main factor which causes the amide bond rotation of *N*-pivaloylphthalimide (**1**).

O O N R O

1: R = *t*-Bu
38: R = CH_3

The conformational difference in the amide moieties between **38** and **35** is not clear. However, one probable reason may be the difference in their ring sizes.

3.5. X = Alkyl, Y = Metal

Several twisted amides are known to be in the form of complexes with metal ions. The *N*-acetyl group in bis-(*N*-acetyl-β-ketoamine) chelate of platinum(II) (**39**) makes an angle of 59.2° with the chelating ring.[50] This deformation may be attributable to the steric repulsion between the *N*-acetyl and the adjacent phenyl groups.

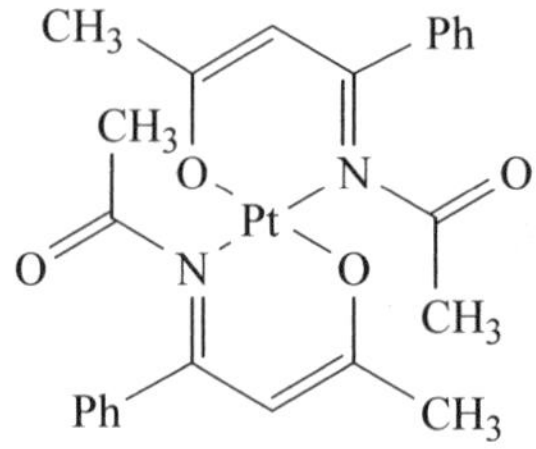

39

The twisted amide moiety is also found in tetradentate complexes. A *trans*-osmium(IV) complex **40** is converted to *cis*-complex **41** having a twisted amide moiety by substitution of triphenylphosphine with carbon monoxide.[51] The isomerization from *trans*- to *cis*-complex is achieved principally by the rotation about the C(O)–N bond. The X-ray analysis of **41** clearly shows the twisted amide structure. The deformation parameters of τ, χ_C, and χ_N of the amide group *trans* to PPh_3 are 73°, 6°, and −33°, respectively. The carbonyl absorption band of the *trans* complex is $1605\,cm^{-1}$, whereas that of the *cis* is 1637 and $1695\,cm^{-1}$. The loss of amide resonance energy may be partially compensated by the resonance of the nitrogen lone-pair electron with the metal ion.

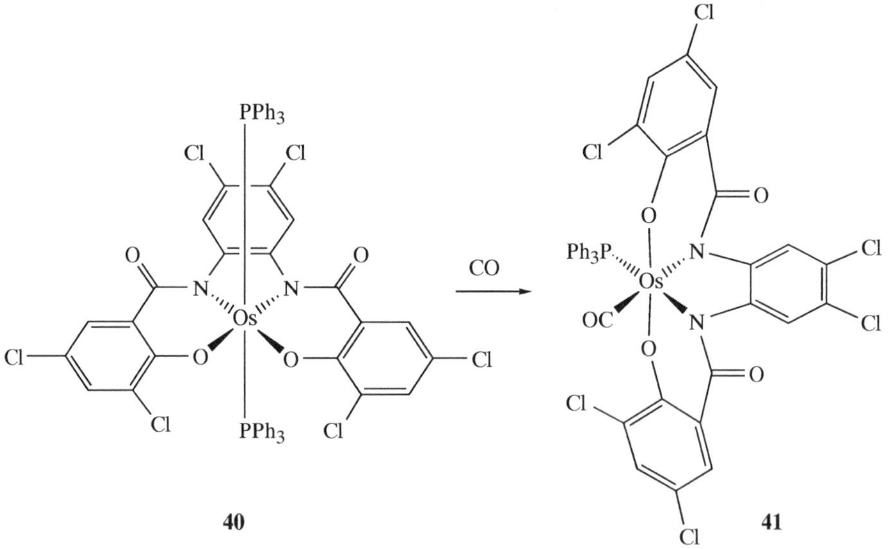

40 **41**

A similar distortion is observed[52] in the *cis*-octahedral cobalt(III) complex **42**. The amide deformation parameters are $\tau = 15°$ and 17°, $\chi_C = 1°$, and $\chi_N = -1°$ and 1°. It is interesting to note that the N atoms remain trigonal planar, indicating the existence of $p \rightarrow d\pi$ bonding between the N atoms and the Co atom. A macrocyclic square-planar cobalt(III) complex **43** has also a twisted amide bond.[53] The twist angle of one of the four amide moieties is −22.3°, and the χ_C and χ_N are −0.6° and −22.3°, respectively.

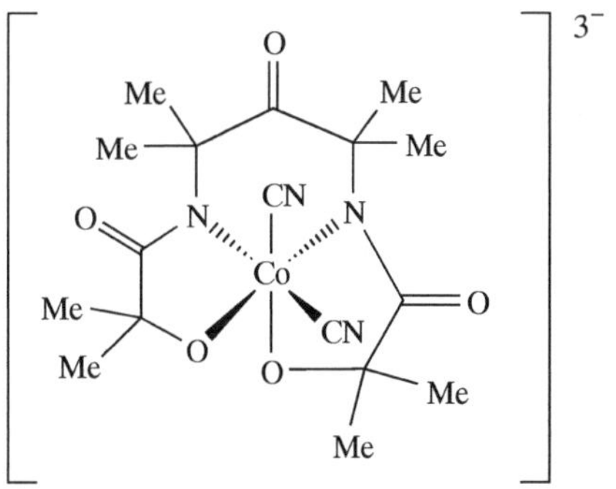

42 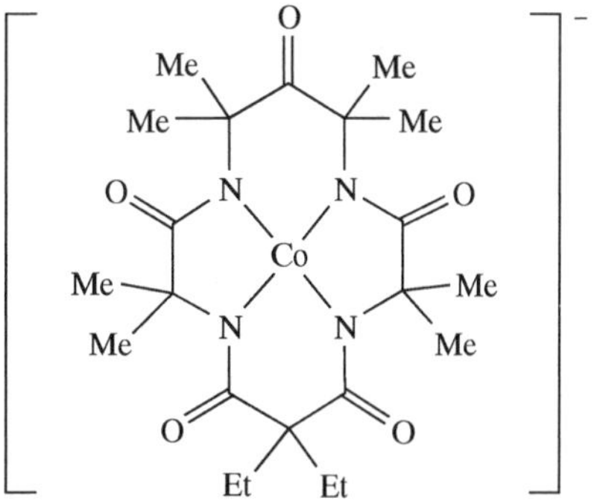 **43**

4. DIRECTIONALITY OF AMIDE BOND ROTATION

When a chiral center is present adjacent to an amide moiety, the conformation of the amide group receives its steric effect from the chiral center. The (2′*S*)-1-(2′-methylbutyryl)aziridine[54] (**44**) exists as a mixture of the two possible diastereoisomeric −*syn*- and +*synclinal* conformers associated with rotation about the C(O)–N bond. Comparing the two rotamers **44a** and **44b**, **44b** is considered to be thermodynamically more preferable than **44a** because of the steric demanding as shown in Scheme 4. This prediction was confirmed by a CD spectrum where a positive Cotton effect for n–π* transition is observed.

For (*S*)-4-alkyl-3-pivaloyl-1,3-thiazolidine-2-thiones (**27–29**), two diastereomeric isomers **a** and **b** are considered to exist (Scheme 5). The X-ray analysis shows that **28** is present in the rotamer **28b** in the crystalline state with a highly

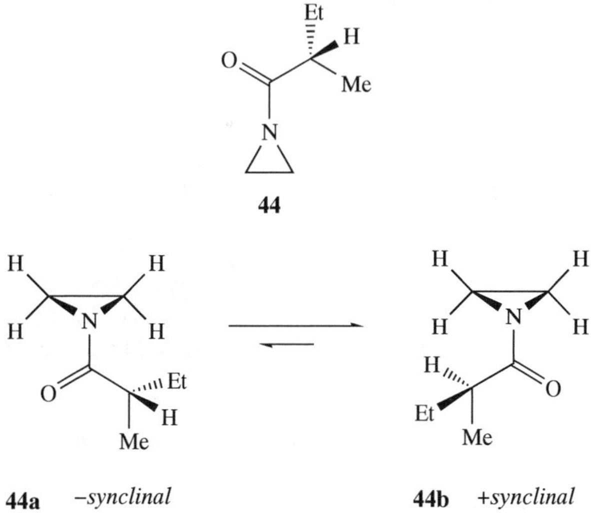

44

44a −*synclinal* **44b** +*synclinal*

Scheme 4

27a–29a ⇌ **27b–29b**

Scheme 5

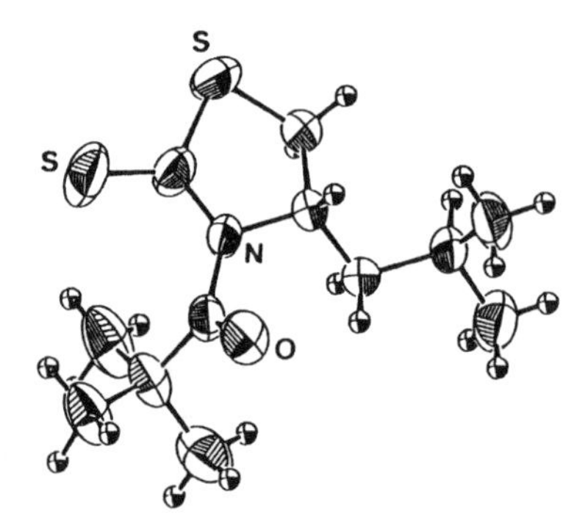

FIGURE 8.13. ORTEP drawing of **28** (thermal ellipsoids draw at the 50% probability level).

TABLE 8.4. Heats of Formation for 27a–29a and 27b–29b Predicted by PM3 Method

a	Heat of Formation/kcal mol^{-1}	b	Heat of Formation/kcal mol^{-1}
27a	−29.23	**27b**	−32.57
28a	−36.49	**28b**	−38.07
29a	−31.74	**29b**	−34.94

twisted amide linkage (Fig. 8.13).[16,64] Heats of formation for **27a–29a** and **27b–29b** obtained by PM3 method are listed in Table 8.4.[56] For each molecule the heat of formation of rotamer **b** is significantly smaller than that of **a**, indicating the thermodynamic preference for rotamer **b**. These predictions are in agreement with the X-ray results described above. These observations indicate that the directionality of the amide bond rotation is controlled by adjacent chiral center.

5. AMIDE BOND ROTATIONAL EFFECT ON SPECTRAL DATA

It is known that rotation of the C(O)–N bond affects various spectroscopic data. However, the quantitative and systematic relationships between those data have not always been explored. The relationships would predict not only the conformation of the amide moiety but also the electronic properties of the C, N, and O atoms, which would provide insight into the chemical and physical properties.

5.1. NMR Chemical Shifts

All three nuclei (C, N, and O) that make up amide groups can be observed by ^{13}C,[57] ^{15}N,[58] and ^{17}O[59,60] NMR spectroscopies, making the NMR methods powerful tools for studying the geometry and properties of amide bonds.

A linear correlation was observed between the δ^{15}N and the activation energy E_a of the C(O)–N rotational process in *N,N*-dimethylcarboxyamide derivatives. The δ^{15}N value shifts to high field with a decreasing E_a value.[61] This relationship can be applied to the prediction of the E_a value of related amides. The relationship between the ^{15}N chemical shifts of bovine pancreatic trypsin inhibitor and the dihedral angles of its amide bonds was studied.[62] However, a clear relationship was not observed between these parameters.

The ^{17}O NMR signal has been recognized to be more sensitive to structural variation than the shifts of the ^{13}C and ^{15}N NMR signals, although the ^{17}O nucleus is disadvantageous for measurement by its quadrupolar properties and low natural abundance.[59,60]

The ^{17}O chemical shift data for *cis*- and *trans*-isomers of *N*-alkylformamides have been reported.[63] The observed abundance of the population of the *cis*-isomer of *t*-butylformamide can be easily rationalized as the result of increasing steric effects between the *t*-Bu and carbonyl groups. The ^{17}O NMR signal of *cis*-*t*-butylformamide appears at a much lower field compared to that of the *trans*-isomer. The difference in chemical shifts between *cis*- and *trans*-isomers can be ascribed to the deformation of the *cis*-amide bond.

Effects of C(O)–N bond rotation for a series of 3-acyl-1,3-thiazolidine-2-thiones on the ^{13}C, ^{15}N, and ^{17}O NMR chemical shifts were studied.[55,64] Table 8.5 shows the ^{13}C, ^{15}N, and ^{17}O NMR chemical shifts for **20**, **23**, **24**, **25**, and **28** and those for standards **30a**, **30d**, and **30e**, and **31a–31c**. The $\Delta\delta^{13}$C and $\Delta\delta^{17}$O values are also given as the differences in the chemical shifts of the twisted amides with the corresponding standards to compensate for the substituent effects around the carbonyl groups.

Figures 8.14a–c show plots of τ vs. the $\Delta\delta^{13}$C, $\Delta\delta^{15}$N, and $\Delta\delta^{17}$O, respectively. Each plot shows slight scatter, but nearly linear relationships are observed. As shown in Figs. 8.14a and 8.14c, the $\Delta\delta^{13}$C and $\Delta\delta^{17}$O increase with increasing τ. In contrast, $\Delta\delta^{15}$N decreases with increasing τ (Fig. 8.14b); the slope in Fig. 8.14b is opposite to that in Figs. 8.14a and 8.14c.

The ^{13}C chemical shift generally reflects the charge density of the C atom. On the other hand, the ^{15}N and ^{17}O chemical shifts exhibit a complex mixture of various effects. In particular, it is known that the ^{15}N chemical shift largely depends on the s-character of ^{15}N nucleus. However, since the series of amides have the same framework structure, the chemical shift seems to depend mainly

TABLE 8.5. ^{13}C, ^{15}N, ^{17}O **NMR Chemical Shifts (δ, ppm) for 20, 23, 24, 25, and 28, and** ^{13}C, ^{17}O **Chemical Shifts (δ, ppm) for 30a, 30d, and 30e, and** ^{15}N **Chemical Shifts (δ, ppm) for 31a–31c**

Compound	$\delta^{13}C^a$	$\delta^{15}N^b$	$\delta^{17}O^c$	Compound	$\delta^{13}C^a$	$\delta^{17}O^c$	Compound	$\delta^{15}N^b$	$\Delta\delta^{13}C$	$\Delta\delta^{15}N$	$\Delta\delta^{17}O$
20	171.3	−184.8	438	**30a**	170.6	338	**31a**	−234.0	0.7	49.2	100
25	174.3	−168.3	476	**30a**	170.6	338	**31b**	−215.4	3.7	47.1	138
23	187.8	−195.2	506	**30d**	177.5	340	**31a**	−234.0	10.3	38.8	166
28	189.0	−182.5	513	**30d**	177.5	340	**31c**	−223.6	11.5	41.0	173
24	172.8	−185.7	436	**30e**	171.0	335	**31a**	−234.0	1.8	48.3	101

[a] Recorded at 100.4 MHz in $CDCl_3$. Chemical shifts are referred to internal TMS.
[b] Recorded at 40.4 MHz in C_6D_6. Chemical shifts are referred to internal CH_3NO_2.
[c] Recorded at 54.1 MHz in CD_3CN. Chemical shifts are referred to external H_2O.

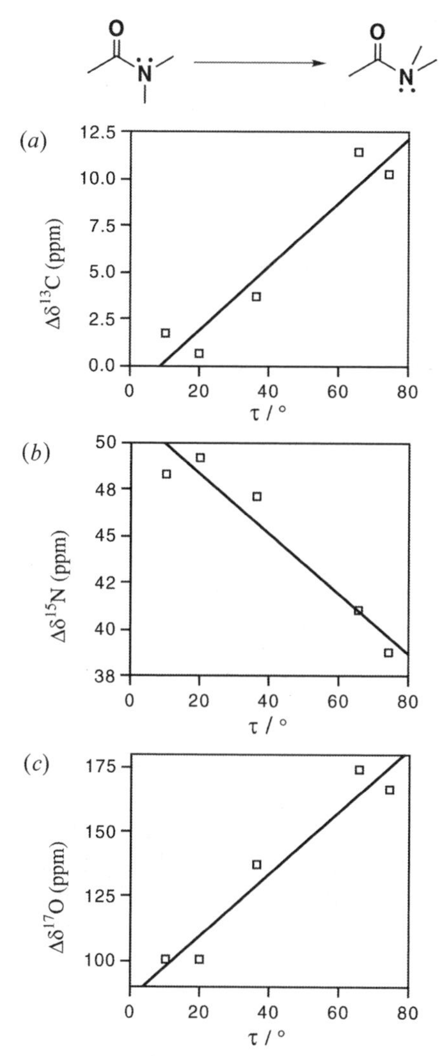

FIGURE 8.14. Plots of (a) $\Delta\delta\,^{13}C$ vs. τ; (b) $\Delta\delta\,^{15}N$ vs. τ; (c) $\Delta\delta\,^{17}O$ vs. τ.

upon the charge densities of the N and O atoms. Therefore, these results indicate that the rotation about the C(O)–N bond increases the charge density of the N atom and decreases those of the C and O atoms. A rough correlation is also observed between $\Delta\delta^{15}N$ and $\Delta\delta^{17}O$ values as shown in Fig. 8.15. This suggests that changes in the charge density of the N atom affect the charge density of the O atom.

These results are in agreement with that expected on the basis of an amide resonance model. Thus, an increase in the twist angle reduces the contribution of the canonical form **II** (Scheme 1); as a result, it leads to a decrease in the charge density of the C, O atoms and an increase in that of the N atom.

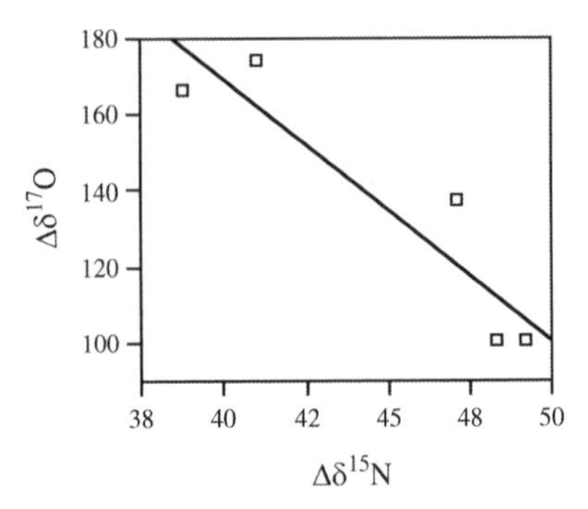

FIGURE 8.15. Plots of $\Delta\delta\,^{17}O$ vs. $\Delta\delta\,^{15}N$.

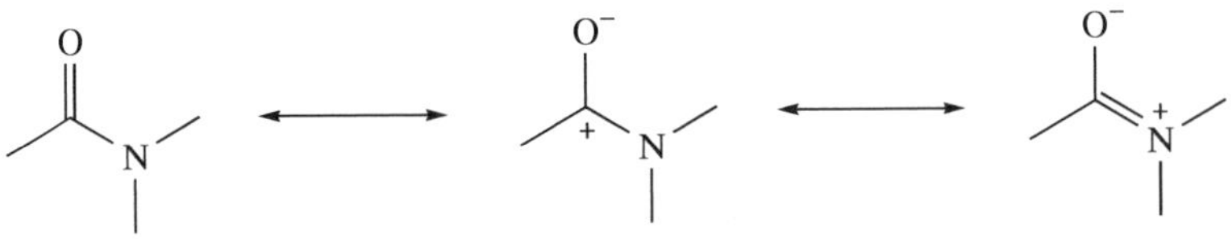

O
N
I

O^-
+
N
III

O^-
+
N
II

Scheme 6

In contrast to the classical amide resonance model, a new model has been proposed as shown in Scheme 6.[65–67] The proposed model, where the contribution of the $C^+{-}O^-$ canonical structure **III** is significant, indicates no correlation in the charge densities between the N and O atoms during the C(O)–N bond rotation and the independence of the charge density of the O atom on the twist angle. However, the observations that the $\Delta\delta^{17}O$ increases with increasing τ and decreases with increasing $\Delta\delta^{15}N$ as described above are not in agreement with the model.

5.2. IR Absorption

IR spectroscopy is the technique most frequently employed to investigate distorted amides. The relationships between loss of resonance stabilization and $\nu_{C=O}$ for monocyclic and bridgehead lactams has been extensively investigated by Greenberg and Liebman.[68] The $\nu_{C=O}$ values increase with increasing resonance energy.

The relationship between the $\Delta\delta^{15}N$ and $\nu_{C=O}$ of toluamides has been studied. Although the plots of $\Delta\delta^{15}N$ vs. $\nu_{C=O}$ show considerable scattering, a rough correlation is observed between them;[33] as the $\Delta\delta^{15}N$ value increases, the $\nu_{C=O}$ value increases.

The amide bond rotational effect of 3-acyl-1,3-thiazolidine-2-thiones on $\nu_{C=O}$ was also studied. Table 8.6 lists the carbonyl stretching frequencies of **20**, **23**, **24**, **25**, and **28**, and *N,N*-dimethylcarboxyamides as standards, and their difference $\Delta\nu_{C=O}$. Figure 8.16a shows the plot of τ vs. $\Delta\nu_{C=O}$. A good correlation was

TABLE 8.6. IR C=O Frequencies (cm^{-1}) for 20, 23, 24, 25, and 28, and 30a, 30d and 30e in $CHCl_3$

Compound	$\nu_{C=O}$	Compound	$\nu_{C=O}$	$\Delta\nu_{C=O}$
20	1704.0	**30a**	1634.6	69.4
25	1712.2	**30a**	1634.6	77.6
23	1738.4	**30d**	1610.6	127.8
28	1727.4	**30d**	1610.6	116.8
24	1711.5	**30e**	1637.6	73.9

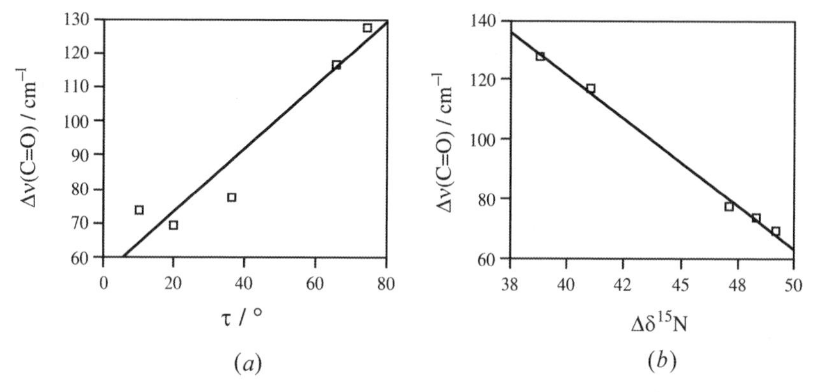

FIGURE 8.16. Plots of Δν(C=O) vs. (a) τ; (b) $\Delta\delta\ ^{15}N$.

observed between them. This relationship also supports the classical resonance model. Thus, an increase in the $\Delta\nu_{C=O}$ with increasing τ indicates that the N atom donates an electron to the carbonyl group. The plot of the $\Delta\nu_{C=O}$ vs. $\Delta\delta^{15}N$ also shows good correlation as shown in Fig. 8.16b, also indicating the role of the N atom as an electron donor.

5.3. NMR Coupling Constant

The C–N coupling constant seems to be another promising parameter for the elucidation of amide bond deformation because it reflects the electronic properties of both C and N atoms.

The $^1J_{C,N}$ values have been measured for a number of amides.[58] It has been reported[69] that the structural variation affects not the $^1J_{C(O),N}$ values, but the $^2J_{C,N}$ values, though the amides employed have planar geometry.[69] An effect of the *cis-trans* geometrical difference on the coupling constants was studied[70] but clear relationship was not observed.[70]

We have studied[71] the relationships between the twist angle and $J_{C,N}$ using a series of twisted amides **45–49** enriched to 99% with ^{15}N.[71] Table 8.7 shows the $^1J_{C,N}$ and $^2J_{C,N}$ values measured by ^{13}C NMR spectroscopy, and their C(O)–N twist angle τ and χ_N values. The plots of $^1J_{C(O),N}$ vs. τ show a slight scattered, but a rough correlation exists between them as shown in Fig. 8.17a; as the twist angle

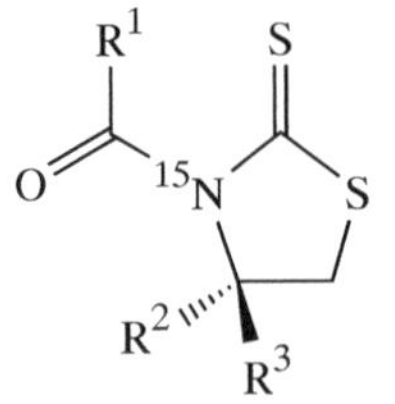

	R^1	R^2	R^3
45:	Me	H	H
46:	Bn	H	H
47:	Me	Me	Me
48:	*t*-Bu	H	H
49:	*t*-Bu	*i*-Bu	H

TABLE 8.7. $J_{C,N}$ Values, χ_N, and Twist Angle τ

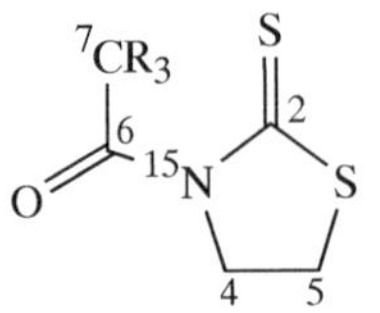

Compound	$^1J_{C(O),N}$[a] /Hz	$^1J_{C(O),N}$[b] /Hz	$^1J_{C(S),N}$[a] /Hz	$^1J_{C4,N}$[a] /Hz	$^2J_{C7,N}$[a] /Hz	χ_N /°	τ /°
45	7.8	7.9	8.9	10.2	5.8	11.9	20.1
46	7.5	7.6	8.7	10.0	5.6	13.4	10.2
47	7.0	7.1	8.7	6.9	6.0	12.2	36.5
48	2.9	3.5	9.1	7.8	7.1	29.5	74.3
49	2.5	3.2	9.1	6.8	6.7	31.4	65.5

[a] In $CDCl_3$.
[b] In DMSO-d_6.

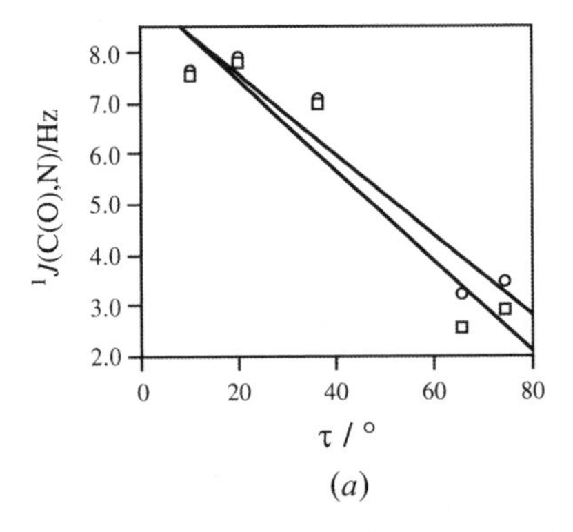

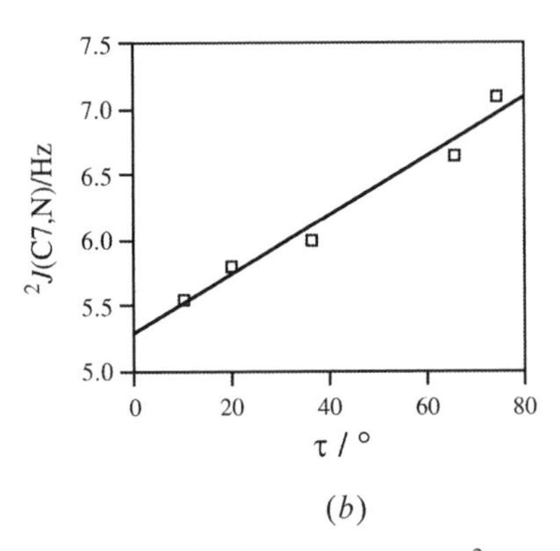

FIGURE 8.17. Plots of (a) $^1J_{C(O)-N}$ (□: in $CDCl_3$; ○: in DMSO–*d*6) vs. τ; (b) $^2J_{C(O)-N}$ vs. τ.

increases, the $^1J_{C(O),N}$ value decreases. It has been reported[72] that $^1J_{C,N}$ is dependent on the s-character of the C and N atoms by assuming the Fermi contact mechanism to be dominant.[72] However, there are no clear-cut relationships between the $^1J_{C(S),N}$ and $^1J_{C4,N}$ values and the χ_N value, which indicate that the contribution of the contact term to $^1J_{C(O),N}$ is not predominant in this case. Therefore, the orbital and dipolar terms that depend on p-electron

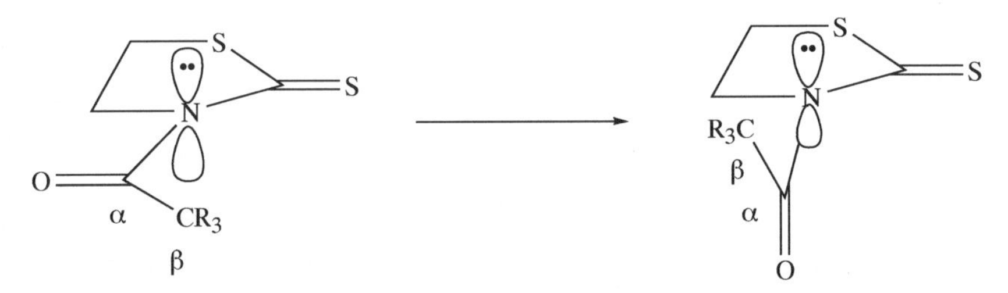

FIGURE 8.18. Schematic planar and twisted amides with lone pair lobe.

distribution may be dominant factors governing $^1J_{C,N}$ similar to the cases of *N,N*-dimethylaniline derivatives.[73] In other words, $^1J_{C(O),N}$ depends on the C–N double bond character. The solvent effect of DMSO is ascribed to the stabilization of canonical form II in Scheme 1.

In contrast to the relationship between $^1J_{C(O),N}$ and τ, the $^2J_{C7,N}$ value increases with increasing twist angle as shown in Fig. 8.17b. It has been assumed that $^2J_{C,N}$ depends on the dihedral angle made by the lone-pair orbital and the Cα–Cβ bond.[74] The present results could also be ascribed to the magnitude of the dihedral angles, which decreases with increasing C(O)–N twist angle (Figure 8.18).

Kainosho and Nagao reported[75] that the $^{13}C-^{15}N$ coupling constant between the Phe-97 carbonyl carbon and the Ser-98 nitrogen in *Streptomyces* subtilisin inhibitor is unusually small, and explained that it is a result of the nonplanarity of the peptide bonds in β-sheets and α-helices. The present result seems to support their hypothesis.

5.4. Other Methods

Although UV,[76] PES,[77] and ESCA[78] spectroscopies have been employed for the investigation of distorted amides and they provide valuable information on the electronic properties of the amide moiety, there is no data on sterically congested amides. These spectroscopies will give more information on twisted amide linkages.

6. REACTIONS OF TWISTED AMIDES

Twisted amides generally have high energy in the ground state as compared to planar ones because of the loss of resonance energy. Therefore, it is predicted that the twisted amides have high reactivity for nucleophiles; moreover, new reactions which are not observed in planar amides are expected. In addition to such electronic properties at the carbonyl carbon, the twisted geometry may cause stereoselectivity in some reactions.

6.1. Reactivity with Alcohols

To compare of the reactivities among a series of 3-acyl-1,3-thiazolidine-2-thiones[79] **20–23**, the rate constants of their solvolysis in ethanol at 75 °C were

23 —(EtOH, 75°C)→ ^{t}BuCOOEt + HN S (S)

Scheme 7

measured at pH ~ 7 (Scheme 7).[16] Although **23** has the largest acyl group, the rate constant is 100–1000 times greater than those of the other members of the series. The high reactivity of **23** can be attributed to its highly twisted structure. Thus, the carbonyl group of **23** is more electrophilic than those of **20–22** because of the loss of amide resonance. The high reactivity of **23** associated with the twisted amide conformation is closely related to the enzymatic acylation reaction, where deformation of the amide group is hypothesized in the transition state.

6.2. Applications for Selective Acylation of Diols

Acylation of a variety of diols with **23** was investigated to explore its selectivity under neutral conditions.[80] From the diols containing primary and secondary hydroxy groups, the primary ones were selectively acylated. For example, acylation of 1,4-octanediol with **23** gives a 44 : 1 mixture of **50** and **51**, whereas the reaction using pivaloyl chloride gives a 4 : 1 mixture of **50** and **51** (Scheme 8). These observations may be explained by the reactivity–selectivity principle; the selectivity of less reactive **23** is higher than that of the more reactive pivaloyl chloride. This method was employed in the synthesis of the spiroketal subunit **53** during the synthesis of (−)-calyculin A[81] (Scheme 9). The monopivalate **53** is not accessible by general methods because the intramolecular acetalization of **52** and the subsequent dehydration proceed.

Interesting results were obtained for the acylation of diols having both phenolic and alcoholic hydroxy groups under neutral conditions.[41,80] Thus, alcoholic hydroxy groups of the diols were acylated to yield alkyl esters in extremely high selectivities. One example of this is shown in Scheme 10. The reaction of 2-hydroxyphenethyl alcohol and **23** proceeds in quite high selectivity to give **54** exclusively. In contrast, the reaction with pivaloyl chloride in the

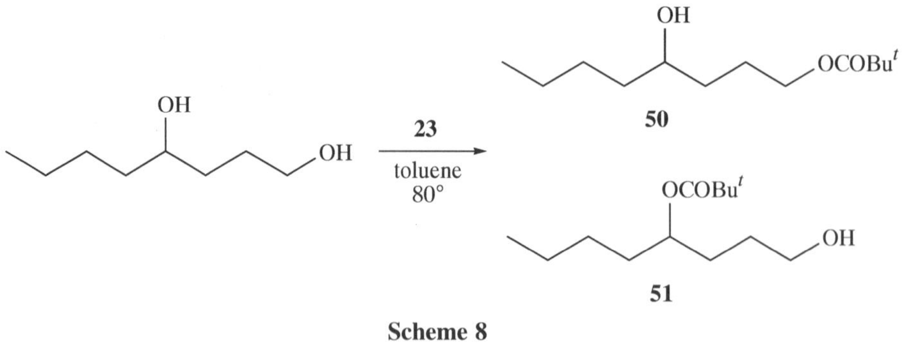

OH ... OH —(23, toluene, 80°)→ **50** (OH, OCOBu^t) + **51** (OCOBu^t, OH)

Scheme 8

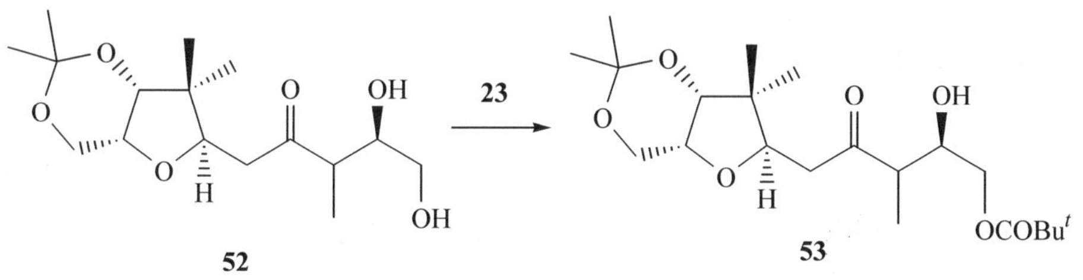

Scheme 9

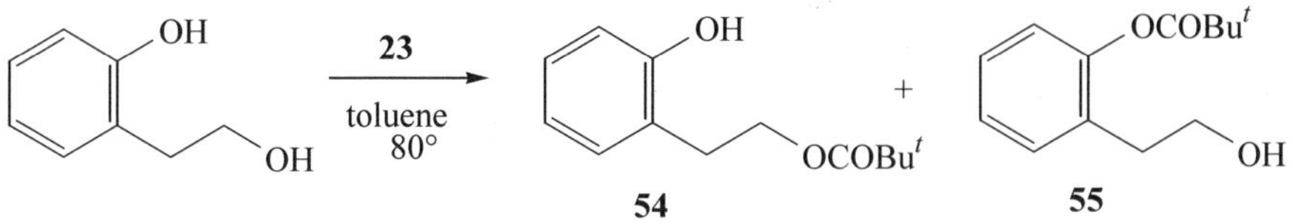

Scheme 10

presence of triethylamine selectively yields aryl pivalate **55**. These differences in the selectivities are explained as follows: The nucleophilicity of the phenolic hydroxy group is less than that of the alcoholic hydroxy group under neutral conditions; in contrast the nucleophilicity is reversed under basic conditions, because the phenolate is more easily produced in basic media. As can be seen from these results, the chemoselectivity of **23** for hydroxy groups is arranged in the order $1° > 2° >$ phenolic OH under neutral conditions.

6.3. Reactions of Axially Chiral Twisted Amides

If the directionality of amide bond rotation is controlled, axial chirality will be generated around the amide bond. (*S*)-4-Alkyl-3-pivaloyl-1,3-thiazolidine-2-thiones (**27–29**) would exist overwhelmingly in the rotamer **B** (Scheme 11) in solution as noted previously, which was predicted by X-ray analysis and PM3 calculations. The preference for the rotamer **B** will arise from the steric repulsion

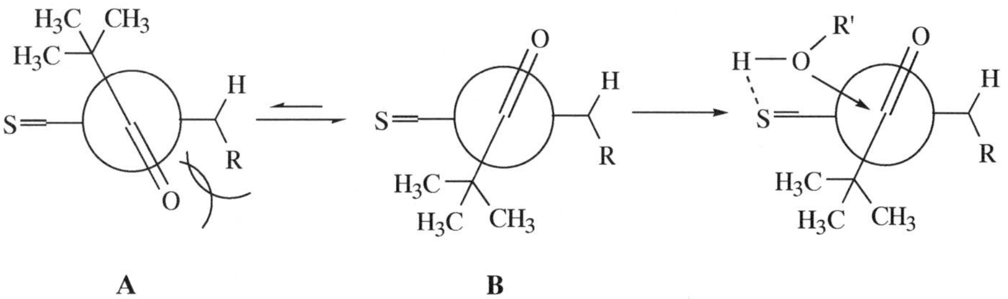

Scheme 11

TABLE 8.8. Enantioselective Acylations of Racemic *sec*-Alcohols[a] with Twisted Amides 27–29

OH → (27-29, toluene, 80°) → OCOBu^t + OH

Entry	Amide	Time/h	Yield[b]/%	*ee*/%	Configuration
1	**27**	14	99	57	*S*
2	**28**	15	96	38	*S*
3	**29**	18	92	80	*S*

[a] Five eq. of alcohol was used.
[b] Isolated yield.

between the *t*-Bu of acyl group and C4 substituent. As a result, these amides have axial chirality about the C(O)–N bond as described in Scheme 11.

Asymmetric acylation of the racemic *sec*-alcohols was performed[82] by taking advantage of the axial chirality. The acylation of 1-tetraloyl with the amide **27–29** under neutral conditions gave (*S*)-1-tetraloyl pivalate as the major product (Table 8.8).[82] The selectivity greatly depends on the bulkiness of the amide substituent at C4; as the steric bulkiness of the C4 substituent increased, the selectivity also increased.

Although the mechanism for the generation of the stereoselectivity is not clear, a plausible working model would be given as follows: a *sec*-alcohol attacks the carbonyl carbon of the rotamer **II** from the less hindered side *via* a 6-membered transition structure as described in Scheme 11. In this stage the racemic alcohols would be discriminated and the (*S*)-isomer preferentially reacts with the amide.

O N CH_3 CH_3 CH_3 **6**

$CH_2{=}PPh_3$

CH_2 N CH_3 CH_3 CH_3 **56**

HO — $(CH_2)_3$ — OH
TsOH

O O N CH_3 CH_3 CH_3 **57**

Scheme 12

6.4. Other Reactions

Because of loss in resonance energy, highly twisted amides behave like amino-ketones and new reactions that are not accessible to planar amides proceed. Indeed, 2-quinuclidones are protonated and are alkylated on the nitrogens.[83] It is noteworthy that 3,5,7-trimethyl-1-adamantan-2-one (**6**) reacts with phosphorous ylids under the standard conditions of the Wittig reaction to give twisted enamine **56**. Furthermore, corresponding acetal **57** was formed when heated with a diol under acidic conditions (Scheme 12).[28]

7. CONCLUSION

As described above, the distorted amides can be classified into three representative types, twisted amide, nonplanar amide, and twisted nonplanar amide, on the basis of the distorted mode. The deformation mode depends on both the steric and electronic effects of the substituents attached to the amide group; the steric bulkiness of the substituents destabilize the planar geometry to cause deformation in the amide linkage, and the substituents having electron withdrawing property reduces the C(O)–N double bond character to facilitate the amide bond twisting. The directionality of amide bond twisting is controlled by the adjacent asymmetric center; as a result, axial chirality is induced about the amide linkage.

The amide bond twisting significantly affects various spectroscopic properties such as IR and UV absorption bands, ^{13}C, ^{15}N, and ^{17}O NMR chemical shifts, $J_{C(O),N}$ values, and ESCA and PES data, which in turn predict the degree of the amide bond deformation. Moreover, the deformation in the amide bond also affects the chemical properties. Several reactions that are not accessible to planar amides such as regio- and stereoselective acylation of alcohols, Wittig reaction, and acetalization are observed for highly twisted amides.

These significantly different properties of twisted amides compared to planar ones suggest that the highly twisted amides behave not like amides but like amino-ketones; therefore, it would be no exaggeration to say that the twisted amide is a new functional group.

REFERENCES

1. Pauling, L. *The Nature of the Chemical Bond*; Cornell University Press, NY, 1960.
2. Robin, M. B.; Bovey, F. A.; Basch, H. In *The Chemistry of Amides*; Zabicky, J. Ed.; Wiley-Interscience: London, 1970, pp. 1–72.
3. Ramachandran, G. N.; Kolaskar, A. S. *Biochim. Biophys. Acta*, 1973, **303**, 385–388.
4. Stewart, D. E.; Sarkar, A.; Wampler, J. E. *J. Mol. Biol.* 1990, **214**, 253–260.
5. Morris, A. L.; MacArthur, M. W.; Hutchinson, E. G.; Thornton, J. M. *Proteins: Structure, Function, and Genetics*; 1992, Vol. 12, pp. 345–364.

6. MacArthur, M. W.; Thornton, J. M. *J. Mol. Biol.* 1996, **264**, 1180–1195.
7. Greenberg, A. Molecular Structure and Energetics. In *Structure and Reactivity*, Liebman, J. F.; Greenberg, A. Eds.; VCH Publishers: New York. 1988, Vol. 7, pp. 139–178.
8. Lease, T. G.; Shea, K. J. *Advances in Theoretically Interesting Molecules*; JAI Press: Greenwich, 1992, Vol. 2, pp. 79–112.
9. Mock, W. *Bioorg. Chem.* 1976, **5**, 403–414.
10. Lipscomb, W. N. *Tetrahedron* 1974, **30**, 1725–1732.
11. Rosen, M. K.; Standaert, B. F.; Galat, A.; Nakatsuka, M.; Schreiber, S. L. *Science* 1990, **248**, 863–866.
12. Albers, M. W.; Christopher, T. W.; Schreiber, S. L. *J. Org. Chem.* 1990, **55**, 4984–4986.
13. Rosen, M. K.; Schreiber, S. L. *Angew. Chem., Int. Ed. Engl.* 1992, **31**, 384–400.
14. Yli-Kauhaluoma, J. T.; Ashley, J. A.; Lo, C. L.; Coakley, J.; Wirsching, P.; Janda, K. D. *J. Am. Chem. Soc.* 1996, **118**, 5496–5497.
15. Winkler, F. K.; Dunitz, J. D. *J. Mol. Biol.* 1971, **59**, 169–182.
16. Yamada, S. *Angew. Chem., Int. Ed. Engl.* 1993, **32**, 1083–1085.
17. Yamada, S. *Rev. Heteroatom Chem.* 1999, **19**, 203–236.
18. Yamada, S.; Nunami, N.; Hori, K. *Chem. Lett.* 1998, 451–452.
19. Drew, M. G.; George, A. V.; Issacs, N. S.; Rzepa, H. S. *J. Chem. Soc., Perkin Trans. 1* 1985, 1227–1284.
20. Fletcher, S. R.; Baker, R.; Chambers, M. S.; Herbert, R. H.; Hobbs, S. C.; Thomas, S. R.; Verrier, H. M.; Watt, A. P.; Ball, R. G. *J. Org. Chem.* 1994, **59**, 1771–1778.
21. Ohwada, T.; Achiwa, T.; Okamoto, I.; Shudo, K. *Tetrahedron Lett.* 1998, **39**, 865–868.
22. Liebman, J. F. (Private communication).
23. Tarasenko, N. A.; Avakyan, V. G.; Belik, A. V. *Zhur. Strukt. Khim.* (*Engl. Transl.*) 1978, **19**, 470–472.
24. Yamamoto, G.; Murakami, H.; Tsubai, N.; Mazaki, Y. *Chem. Lett.* 1997, 605–606.
25. Yamamoto, G.; Tsubai, N.; Murakami, H.; Mazaki, Y. *Chem. Lett.* 1997, 1295–1296.
26. Pracejus, H. *Chem. Ber.* 1959, **92**, 988–994.
27. Greenberg, A.; Moore, D. T.; DuBois, T. D. *J. Am. Chem. Soc.* 1996, **118**, 8658–8668.
28. Kirby, A. J.; Komarov, I. V.; Wothers, P. D.; Feeder, N. *Angew. Chem., Int. Ed. Engl.* 1998, **37**, 785.
29. M. Oki *Applications of Dynamic NMR Spectroscopy to Organic Chemistry*; VCH Publishers Inc.: 1985, pp. 41–106.
30. Dahlgvist, K.-I.; Forsén, S. *J. Phys. Chem.* 1969, **73**, 4124–4129.
31. Neuman, Jr. R. C.; Jonas, V. *J. Am. Chem. Soc.* 1968, **90**, 1970–1974.
32. Beausoleil, E.; Lubell, W. D. *J. Am. Chem. Soc.* 1996, **118**, 12902–12908.
33. Bennet, A. J.; Somayaji, V.; Brown, R. S.; Santarsiero, B. D. *J. Am. Chem. Soc.* 1991, **113**, 7563–7571.
34. Muhlebach, A.; Lorenzi, G. P. *Helv. Chim. Acta.* 1986, **69**, 389–395.

35. Pedone, C.; Benedetti, E.; Immirzi, A.; Allegra, G. *J. Am. Chem. Soc.* 1970, **92**, 3549–3552.
36. Noe, E. A.; Raban, M. *J. Am. Chem. Soc.* 1975, **97**, 5811–5820.
37. Symersky, J.; Malon, P.; Grehn, L.; Ragnarsson, U. *Acta. Cryst.* 1990, **C46**, 683–686.
38. Evans, P. A.; Holmes, A. B.; Collins, I.; Raithby, P. R.; Russell, K. *J. Chem. Soc., Chem. Commun.* 1995, 2325–2326.
39. Winkler, F. K.; Seiler, P. *Acta Crystallogr.* 1979, **B35**, 1920–1922.
40. Walter, W.; Voss, J. In *The Chemistry of Amides*; Zabicky, J. Ed.; Wiley-Interscience: London, 1970, Chapter 8, pp. 383–467.
41. Yamada, S.; Sugaki, T.; Matsuzaki, K. *J. Org. Chem.* 1996, **61**, 5932–5938.
42. Fujita, E.; Nagao, Y. *J. Synth. Org. Chem. Jpn.* 1980, **38**, 1176–1195.
43. Cow, C. N.; Britten, J. F.; Harrison, P. H. *Chem. Commun.* 1998, 1147.
44. Noe, E. A.; Raban, M. *J. Am. Chem. Soc.* 1974, **96**, 1598–1599.
45. Caron, P. A.; Riche, C.; Pascard-Billy, C. *Acta. Cryst.* 1977, **B33**, 3786–3792.
46. Beall, H. D.; Prankerd, R. J.; Todaro, L. J.; Sloan, K. B. *Pharm. Res.* 1993, **10**, 905–912.
47. Jiang, A.; Hu, S.; Wang, Y.; Chen, Q. *Chem. J. Chin. Uni.* 1988, **9**, 307–309.
48. Hirayama, N. *Acta. Cryst.* 1991, **C47**, 215–216.
49. Yamada, S.; Nunami, N., unpublished work.
50. Uchiyama, T.; Takagi, K.; Matsumoto, K.; Ooi, S.; Nakamura, Y.; Kawaguchi, S. *Chem. Lett.* 1979, 1197–1200.
51. Collins, T. J.; Coots, R. J.; Furutani, T. T.; Keech, J. T.; Peake, G. T.; Santarsiero, B. D. *J. Am. Chem. Soc.* 1986, **108**, 5333–5339.
52. Collins, T. J.; Workman, J. M. *Angew. Chem., Int. Ed. Engl.* 1989, **28**, 912–914.
53. Collins, T. J.; Slebodnick, C.; Uffelman, E. S.; *Inorg. Chem.* 1990, **29**, 3433–3436.
54. Shustov, G. V., Kadorkina, G. K., Varlamov, S. V., Kachanov, A. V., Kostyanovsky, R. G.; Rauk, A. *J. Am. Chem. Soc.* 1992, **114**, 1616–1623.
55. Yamada, S. *Angew. Chem., Int. Ed. Engl.* 1995, **34**, 1113–1115.
56. Yamada, S.; Katsumata, H. *Chem. Lett.* 1998, 995–996.
57. Stothers, J. B. *Carbon-13 NMR Spectroscopy*; Academic Press: New York, 1972; pp. 279–310.
58. Levy, G. C.; Lichter, R. L. *Nitrogen-15 Nuclear Magnetic Resonance Spectroscopy*; J. Wiley & Sons: New York, 1979, pp. 58–67.
59. Boykin, D. W.; Baumstark, A. L. *Tetrahedron* 1989, **45**, 3613–3651.
60. Boykin, D. W.; Baumstark, A. L. *^{17}O NMR Spectroscopy in Organic Chemistry*; CRC, 1991, pp. 39–67.
61. Martin, G. J.; Gouesnard, J. P.; Dorie, J.; Rabiller, C.; Martin, M. L. *J. Am. Chem. Soc.* 1977, **99**, 1381–1384.
62. Glushka, J.; Lee, M.; Coffin, S.; Conburm, D. *J. Am. Chem. Soc.* 1989, **111**, 7716–7722.
63. Gerothanassis, I. P.; Troganis, A.; Vakka, C. *Tetrahedron* 1995, **51**, 9493–9500.
64. Yamada, S. *J. Org. Chem.* 1996, **61**, 941–946.
65. Wiberg, K. B.; Laidig, K. E. *J. Am. Chem. Soc.* 1987, **109**, 5935–5943.

66. Wiberg, K. B.; Breneman, C. M. *J. Am. Chem. Soc.* 1992, **114**, 831–840.
67. Wiberg, K. B.; Rablen, P. R. *J. Am. Chem. Soc.* 1993, **115**, 9234–9242.
68. Greenberg, A.; Chiu, Y.-Y.; Johnson, J. L.; Liebman, J. F. *Struct. Chem.* 1991, **2**, 117–126.
69. Lichter, R. L.; Fehder, C. G.; Patton, P. H.; Combes, J. *J. Chem. Soc., Chem., Commun.*, 1974, 114–115.
70. Berger, S. *Tetrahedron* 1978, **34**, 3133–3136.
71. Yamada, S.; Nakamura, M.; Kawauchi, I. *Chem. Commun.* 1997, 885–886.
72. Binsch, G.; Lambert, J. B.; Roberts, B. W.; Roberts, J. D. *J. Am. Chem. Soc.* 1964, **86**, 5564–5570.
73. Axenrod, T.; Watnick, C. M.; Wieder, M. J.; Duangthai, S.; Webb, G. A.; Yeh, H. J. C.; Bulusu, S.; King, M. M. *Org. Magn. Reson.* 1982, **20**, 11–15.
74. Berger, S.; Roberts, J. D. *J. Am. Chem. Soc.* 1974, **96**, 6757–6759.
75. Kainosho, M.; Nagao, H. *Biochemistry* 1987, **26**, 1068–1075.
76. Pracejus, H.; Kehlen, M.; Kehlen, H.; Matschiner, H. *Tetrahedron* 1965, **21**, 2257–2270.
77. Treschanke, L.; Rademacher, P. *J. Mol. Struct.* 1985, **122**, 47–57.
78. Greenberg, A.; Thomas, T. D.; Bevilacqua, C. R.; Coville, M.; Ji, D.; Tsai, J.-C.; Wu, G. *J. Org. Chem.* 1992, **57**, 7093–7099.
79. Yamada, S. *J. Org. Chem.* 1992, **57**, 1591–1592.
80. Yamada, S. *Tetrahedron Lett.* 1992, **33**, 2171–2174.
81. Trost, B. M.; Flygare, J. A. *Tetrahedron Lett.* 1994, **35**, 4059–4062.
82. Yamada, S.; Ohe, T. *Tetrahedron Lett.* 1996, **37**, 6777–6780.
83. Levkoeva, E. I.; Nikitskaya, E. S.; Yakhontov, L. N. *Khim. Geterot. Soed.* 1971, **3**, 378–384.
84. Pedley, J. B. *Thermochemical Data and Structures of Organic Compounds*; Thermodynamic Research Center, College Station, 1994, Vol. 1.
85. Halab, L.; Lubell, W. D. *J. Org. Chem.* 1999, **64**, 3312–3321.

CHAPTER 9

PHOTOELECTRON SPECTROSCOPY OF AMIDES AND LACTAMS

PAUL RADEMACHER
Institut für Organische Chemie, Universität GH Essen

1. INTRODUCTION

Molecular photoelectron spectroscopy (PES) is widely used to study the electronic structure of molecules, and compounds can be characterized by their PE spectra. In this article the results of ultraviolet (UPS) and X-ray PE spectroscopic (XPS) studies of amides and lactams will be summarized. To our knowledge, except for a very brief chapter on UPS spectra of formamide and its *N*-methyl derivatives,[1] until now no extensive review on PES of amides has appeared. UPS spectra of lactams have been discussed by Greenberg.[2]

Many of the investigations on compounds to be included in this chapter, have been performed in the 1970s and 1980s. Since instrumentation has made little progress in UPS in more recent years, there can be no reservation in including these spectra. However, some of the theoretical methods used at that time are no longer adequate today and therefore are generally given lower priority. In a few cases, the results of semiempirical AM1[3] calculations were added. Experience has shown that for organic compounds with heteroatoms of the second period of the periodic table of elements, in particular nitrogen and oxygen, this method can give even better results than high-level ab initio calculations.

Some unpublished PE spectra or spectral data from the author's laboratory are included. These have been measured using a Leybold–Heraeus UPG 200 spectrometer with a He-I radiation source. Orbitals were plotted with the program PERGRA.[4]

In the following sections ionization potentials are given as vertical IP values, if not stated otherwise. The energy unit eV is used; 1 eV = 96.485 kJ mol^{-1}.

The Amide Linkage: Selected Structural Aspects in Chemistry, Biochemistry, and Materials Science,
Edited by Arthur Greenberg, Curt M. Breneman, and Joel F. Liebman
ISBN 0-471-35893-2

2. PHOTOELECTRON SPECTROSCOPY

The basic principles of PES, the experimental methods, the interpretation procedures and the applications have been described in several books.[5–7] There are also some more recent review articles.[8–10]

A schematic diagram of a UV PE spectrometer is depicted in Fig. 9.1. The fundamental principle of PES is the photoelectric effect. Molecules M in the gas-phase are irradiated with monochromatic light, and electrons can be ejected when their binding energy is lower than the photon energy leaving behind radical cations $M^{\cdot+}$ in definite electronic and vibrational states.

$$M \xrightarrow{h\nu} M^{\cdot+} + e \tag{9.1}$$

The kinetic energy of the ejected electrons $E_{\mathrm{kin}}(\mathrm{e})$ is measured and the ionization energy IE or ionization potential IP is obtained from the energy conservation condition.

$$\mathrm{IP} = h\nu - E_{\mathrm{kin}}(\mathrm{e}) \tag{9.2}$$

Measuring a PE spectrum, the number of photoelectrons per time unit (e.g., counts per second, cps) is recorded as a function of the kinetic energy or IP. As is shown schematically in Fig. 9.2, the radical cation may be excited to different vibrational states, and the ionization band exhibits vibrational fine structure. The adiabatic $\mathrm{IP_a}$, i.e. the transition to the vibrational ground state of $M^{\cdot+}$, can be distinguished from the vertical $\mathrm{IP_v}$, that is, the transition with the greatest Franck–Condon factor. The latter property ($\mathrm{IP_v}$) is of higher relevance when studying the electronic structure of a molecule because it is linked with the energy ε of the molecular orbital (MO), from which the electron was ejected, by the Koopmans theorem[11]

$$\mathrm{IP_i} = -\varepsilon^{\mathrm{SCF}}{}_{\mathrm{i}}, \tag{9.3}$$

stating that the vertical ionization energies are equal to the negative values of the SCF orbital energies which are obtained by quantum chemical calculations. This assumption is generally used as the basis for analysis of PE spectra. Actually, this is an approximation which can fail and lead to wrong interpretation of spectral data. Eq. (9.3) assumes that M and $M^{\cdot+}$ can be described with the same MOs, and energy changes upon ionization, associated with electron reorganization and correlation, are neglected. Fortunately, these two effects have opposite signs and largely compensate each other. The correlation between MO theory and PE spectroscopy has both proved convenient and adequate provided possible deviations are borne in mind. These deviations can, for example, result in different ordering of IPs and MO energies.

Quantitative characteristics of a PE spectrum, such as position and intensity of the ionization bands, vibrational structure, Jahn–Teller and spin-orbit effects,

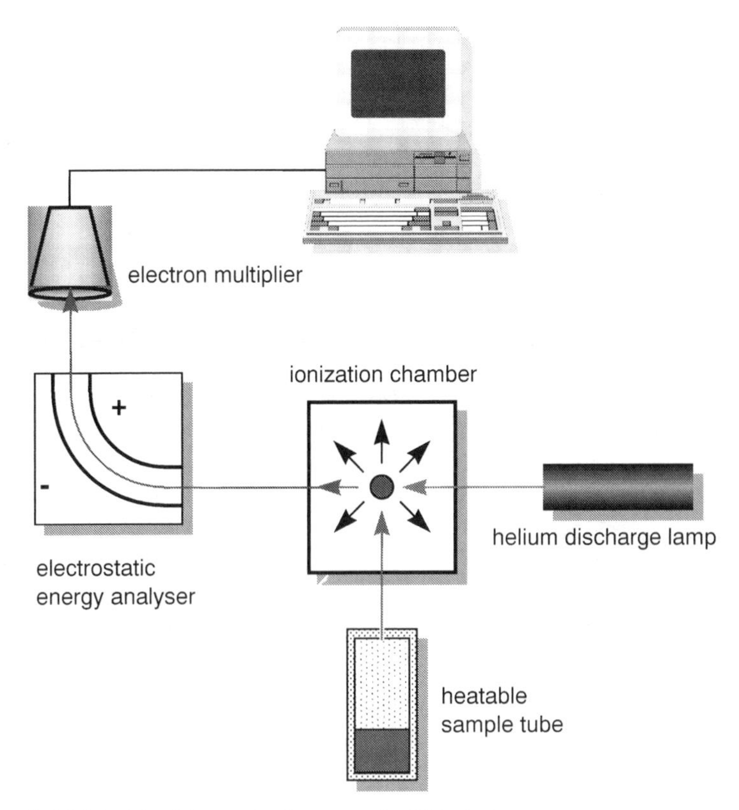

FIGURE 9.1. Schematic diagram of a UV photoelectron spectrometer.

are in general sufficient for reliable assignments of the IPs of simple molecules. The spectra of polyatomic molecules are usually analysed with the aid of quantum chemical calculations making use of the Koopmans approximation.

Compounds with a certain functional group have a certain number of atoms in common and thus their PE spectra should resemble each other. In particular, this holds for the members of a homologous series of compounds. For large molecules, total assignment of all IPs will seldom be intended but for the IPs related to the characteristic MOs of the functional groups. These are, e.g. the π orbitals of double or triple bonds and the n (lone pair) orbitals on heteroatoms like nitrogen, phosphorus, oxygen, sulfur, and the halogens. Of particular value are PE spectroscopic studies of structural effects on functional groups. These can be for example, electronic perturbations by substituents, steric strain, conjugation, and transannular interactions.

In a UV photoelectron spectrometer (Fig. 9.1) a Helium discharge lamp is used as radiation source which emits nearly monochromatic He-I radiation of 21.22 eV ($\lambda = 58.43$ nm) or He-II radiation of 40.81 eV ($\lambda = 30.38$ nm). For XPS an X-ray lamp is used, e.g., Mg Kα, 1254 eV ($\lambda = 0.9890$ nm). A potentially excellent source of radiation is the electron synchrotron. Calibration of the

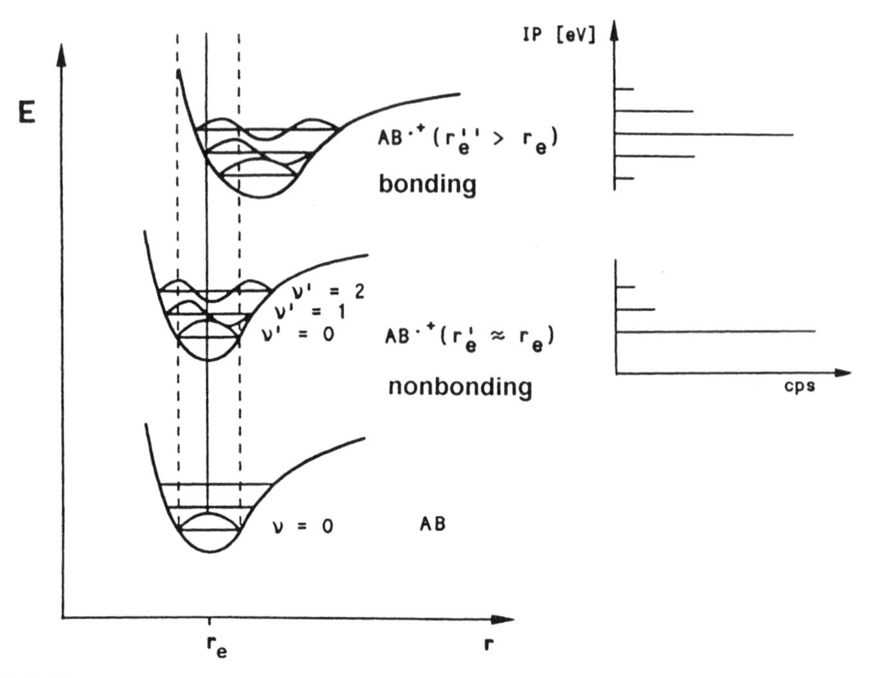

FIGURE 9.2. Potential energy curves for the electronic ground state of a molecule AB and two electronic states of the radical cation $AB^{\cdot +}$. On the right side the corresponding ionization bands are shown.

energy scale is accomplished by adding compounds with known IPs to the sample. For UPS a mixture of the noble gases argon (15.759 and 15.937 eV) and xenon (12.130 and 13.436 eV) is frequently used. The two peaks relate to ionizing transitions to the $^2P_{3/2}$ and $^2P_{1/2}$ states of the ions. XPS spectra are calibrated by, e.g., mixing neon with the samples and using the Ne 1s and 2s lines. Typical uncertainties of ionization energies are about 0.01 eV in UPS and 0.1–0.05 eV in XPS.

The molecular radical cation which is generated upon ionization may be in excited vibrational and rotational states of energies ΔE_{vib} and ΔE_{rot}, respectively. ΔE_{vib} and ΔE_{rot} are much smaller than IP, and usually only vibrational fine structure can be observed because of limited resolution ($\Delta E \approx 0.01$ eV $\cong$ 80 cm^{-1}). Analysis of the fine structure affords vibrational frequencies of the radical cation in its respective electronic state. These frequencies can be related to the corresponding vibrations of the neutral molecule and reveal information about the structure of the molecular ion. The shape of an ionization band reflects the change of the molecular force field of the radical cation as compared to that of the neutral molecule, and some conclusions regarding the bonding properties of the MO from which the electron is removed are possible (Fig. 9.2).

Removal of an electron from a nonbonding MO will have only minor effects on the molecular potential, i.e., the ionization causes only very little geometric changes, and the vertical transition will generate the radical cation in its vibrational ground state, so that this transition becomes identical with the adiabatic ($IP_v = IP_a$). The ionization band consists of a first intense, sharp peak that is followed by peaks of rapidly declining intensities. The ionization of an electron from a bonding MO generates a radical cation with an equilibrium geometry differing strongly from that of the parent molecule. The corresponding ionization band is broad, and the vertical transition relates to a higher excited vibrational state of the derived ion ($IP_v > IP_a$). The ionization band relating to an antibonding MO has a similar shape.

The spacing of the individual peaks of an ionization band correspond to the vibrational frequencies. For small molecules, often several vibrational progressions can be determined. On the other hand, in polyatomic molecules usually the envelope of the individual peaks is recorded. In favorable cases, which are not so rare, the ionization band is dominated by a single vibrational transition, and this can be determined also for large molecules. Lack of vibrational fine structure can also be the result of the presence of individual molecules differing in structure and energy. This may be caused by conformational flexibility.

Electron lone-pairs are usually thought to be localized on single atoms and correspond to nonbonding *n* orbitals. One expects similarity of geometry between molecule and ion and an ionization band typical for such an orbital would have $IP_a = IP_v$. However, nitrogen lone-pairs of amines[12] as well as oxygen and sulfur lone-pairs in certain compounds afford evidence of marked geometrical changes upon ionization.

In UPS, all electrons with binding energies up to the photon energy of the radiation used for ionization are accessible. Typically, photons of 21.21 eV (He-I) are utilized. The valence shell electrons of ordinary organic molecules composed of hydrogen and elements of the second row of the periodic table have energies between about -6 and -50 eV. Of highest interest are the upper occupied levels since they are decisive for the structural and chemical properties of the compound. Therefore, usually He-I spectra are sufficient to study the relevant electronic structure of a molecule. He-II spectra are often used in addition to the He-I spectra in order to assign the ionizations, because the probability that an electron will be "hit" by a photon, is dependent on the shape and size of the orbital in comparison with the wavelength of the photon. Comparison of the He-I and the He-II spectra of the same compound helps to distinguish, ionizations from σ- and π-type MOs.

To ionize core electrons, several hundred eV are necessary, and such energies are provided by X-rays. The experimental technique of XPS became known as electron spectroscopy for chemical analysis (ESCA).[13,14]

Electron transmission spectroscopy (ETS) allows the determination of the energy at which electrons are temporarily captured by an atomic or molecular species in the gaseous phase. By this process a closed shell neutral molecule M is converted to a radical anion $M^{\cdot -}$, and the related energy is the attachment energy E_A. The E_A values are, to a first approximation, the negative of the vertical electron affinities (E_{ea}) of the capturing species. Electron transmission spectroscopy is thus a complementary method to photoelectron spectroscopy (PES). A combination of both methods allows, at least in principle, the experimental determination of the energies of the occupied and unoccupied MOs of a molecule.

3. ELECTRONIC STRUCTURE OF AMIDES

The amide linkage is of high interest not only in structural chemistry, it also forms the backbone of peptides and proteins, and of one of the most important synthetic polymers: nylon. Many natural products and drugs are either amides, or contain an amide group in their structures. Amide bonds are also part of the functional groups in compounds like ureas and urethanes. In proteins and peptides, the amide bond is called the peptide bond. Cyclic amides are termed lactams. Azetidin-2-ones (β-lactams) are the parent compounds of lactam antibiotics like penicillins and cephalosporins.

In amides a carbonyl group is directly connected with an amino group. The conjugation of the nitrogen electron lone-pair with the C=O double bond is usually described with resonance of two Lewis structures **A** and **B**.[15,16]

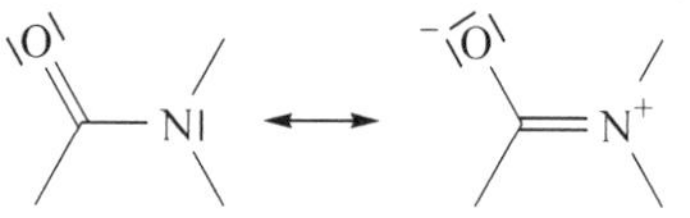

A **B**

Optimal conjugation requires a planar structure of the CO–N unit including their directly bound neighbours. The resonance leads to a weakening of the C=O and to a strengthening of the C–N bond. This is reflected in a low carbonyl stretching vibrational frequency, $\nu_{C=O}$, and hindered rotation about the C–N bond. Typical rotational barriers are $60-90\,kJ\,mol^{-1}$.[1,17] Torsion or another type of distortion of the amide group inhibits the resonance, and relationships between the twist angles and various spectroscopic properties like $\nu_{C=O}$, ^{13}C, ^{15}N, and ^{17}O chemical shifts were found.[16] Amide resonance also explains many other experimental observations like the tendency for O-protonation instead of N-protonation, low carbonyl reactivity and hydrolytic stability, and the acidity of amino hydrogen atoms.

In amides there are three π-type MOs ($\pi_1-\pi_3$) of which two are occupied. In addition, there is an oxygen lone-pair orbital n_O among the high-level MOs. The characteristic MOs of an amide can be composed from those of an aldehyde and an amine, $R-CO-H + H-NR_2 \Rightarrow R-CO-NR_2$. In Fig. 9.3 a correlation diagram is shown for the simplest case, formamide. The orbitals π_1 and π_2 can be considered as being generated by symmetric and antisymmetric combination of the nitrogen 2p orbital (n_N) of ammonia and the $\pi_{C=O}$ orbital of formaldehyde. Since the latter orbital contributes more to π_1 and the former more to π_2, these MOs are often termed π_N and $\pi_{C=O}$ also in amides. Likewise, π_3 can be called $\pi^*_{C=O}$. The interaction of $\pi^*_{C=O}$ with n_N leads to a destabilization of π_3 and a stabilization of π_2 and causes the nitrogen atom to adopt a trigonal planar

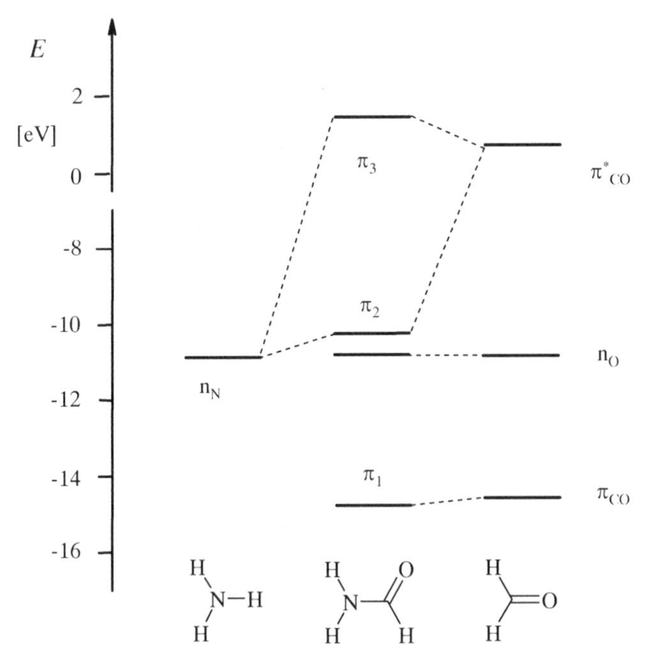

FIGURE 9.3. Orbital correlation diagram for ammonia, formaldehyde, and formamide. (Occupied MOs: PES, unoccupied MOs: AM1.)

geometry and forces all of the substituents on both N and C to be coplanar. The n_O orbital of formamide is a little destabilized relative to its position in formaldehyde and is very close to π_2. These four MOs are depicted in Fig. 9.4. They can be considered as the "characteristic MOs" of amides. Simple substituents like alkyl groups will affect the energies of the MOs of formamide but not disturb their characteristic structure to a major degree.

Formamide is planar as was shown by microwave[18] and electron diffraction[19] structure analyses as well as by quantum chemical calculations.[20] Also other simple aliphatic amides have a planar parent system. This planarity allows for optimal delocalization of the four π electrons. Large substituents can, however, induce deviation from coplanarity of the atoms CO–N and their direct neighbors. In particular, the nitrogen becomes pyramidal and the C–N bond is twisted. This sterically induced distortion is described in the chapter by Yamada in this volume.

It has already been mentioned that—as for other molecules—the electronic structure of amides is closely related to their chemical properties.[21,22] The basicity of the oxygen atom can be expected to be greater than that of the nitrogen atom because n_O is more localized and, therefore, will overlap better with a proton or an electrophile. In addition, O-protonation will lower the energy of the empty MO $\pi^*_{C=O}$ and thus enhance the stabilizing $n_N/\pi^*_{C=O}$ interaction whereas N-protonation would lead to replacement of the n_N by a σ_{NH} orbital which would interact much lesser with $\pi^*_{C=O}$. Because of the high energy of the

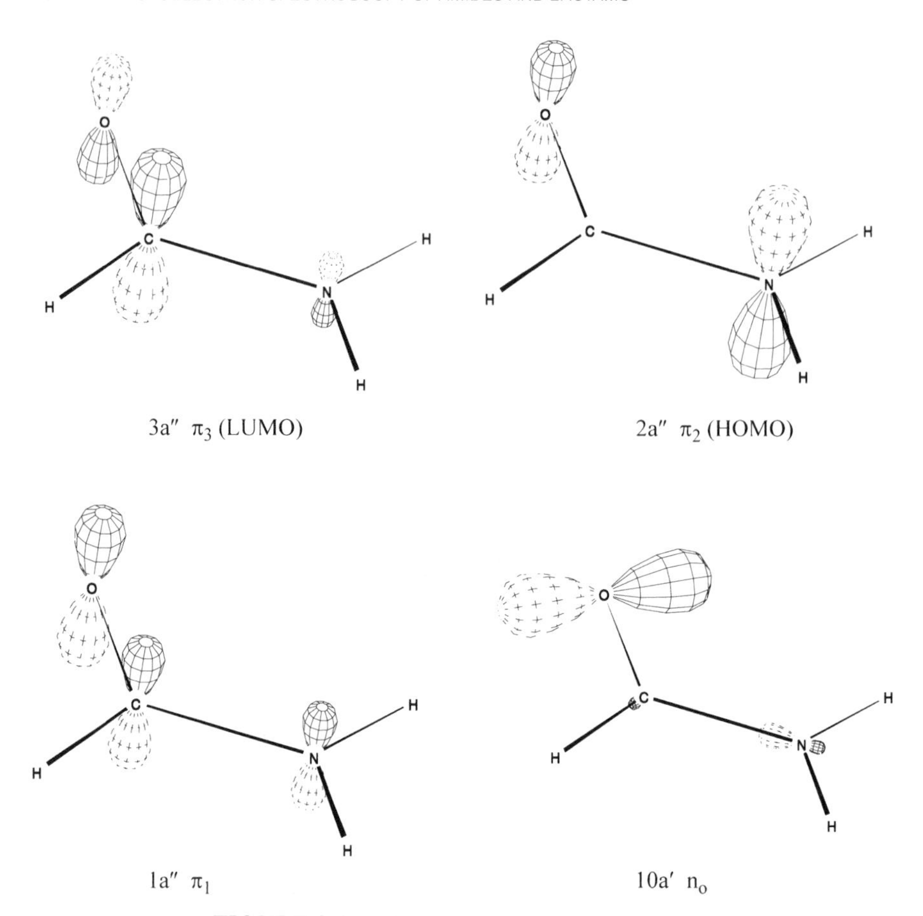

FIGURE 9.4. MOs π_1–π_3 and n_O of formamide.

LUMO (π_3), nucleophilic addition to the carbonyl group is greatly slowed down compared with other carbonyl compounds. If the geometric constraints of the molecular framework force the nitrogen atom to be distorted from planarity, the $n_N/\pi^*_{C=O}$ interaction is reduced, and as a consequence reactivity of the carbonyl group toward nucleophiles is increased as is the nucleophilicity of the nitrogen atom.

4. FORMAMIDE, ACETAMIDE, AND THEIR *N*-METHYL DERIVATIVES

A molecule of formamide, $HCONH_2$, houses 24 electrons which occupy 12 orbitals. Six of these electrons are the 1s core electrons of the atoms C, N, and O which leaves 18 electrons to fill nine valence MOs. In planar formamide, point group C_s, the 12 occupied orbitals transform as $10\,A' + 2\,A''$. The individual orbitals are termed according to their symmetry properties: 1a′–10a′ and 1a″, 2a″. Numbering follows the increase of energy. The MO 3a″ is the LUMO. And the MOs 1a″–3a″ are π-type orbitals. They are antisymmetric to the symmetry plane, while the a′ MOs are symmetric to this symmetry element and can be termed σ-type orbitals.

The PE spectrum of formamide was first investigated by Turner and coworkers.[5,23,24] The spectrum is depicted in Fig. 9.5, IP values, vibrational frequencies, and calculated orbital energies are summarized in Table 9.1. Since formamide is a rather small molecule, one would expect that there are few assignment problems. However, the ordering of the n_O and of the lowest π-type ionization potentials were hotly debated. A second problem was the assignment of the third and the fourth IPs.

The bands corresponding to the first and the second IPs overlap strongly. Because of their different vibrational fine structures and by comparison with the PE spectra of other simple carbonyl compounds and the methyl derivatives of formamide (see below) these bands are assigned to the orbitals n_O (HOMO) and π_2 which are accidentally nearly degenerate. The vibrational interval of $1600\,cm^{-1}$ associated with the first IP is the C=O stretching mode (Amide-I) which has a frequency of $1680\,cm^{-1}$ in the neutral molecule. The mode excited with the second IP could be the O–C–N deformation ($600\,cm^{-1}$ in the molecule) or one of the several modes involving the NH_2 group. The position of the start of this band, i.e., the exact value of IP_a, cannot unambiguously be established.

Semiempirical AM1 calculations (Table 9.1) as well as ab initio SCF MO and CI calculations[25] would lead to a reversal of this assignment since π_2 is calculated as HOMO. On the other hand, the semiempirical HAM/3 method leads to the following ordering of the first three IPs: n_O, π_2, π_1.[26] The same result

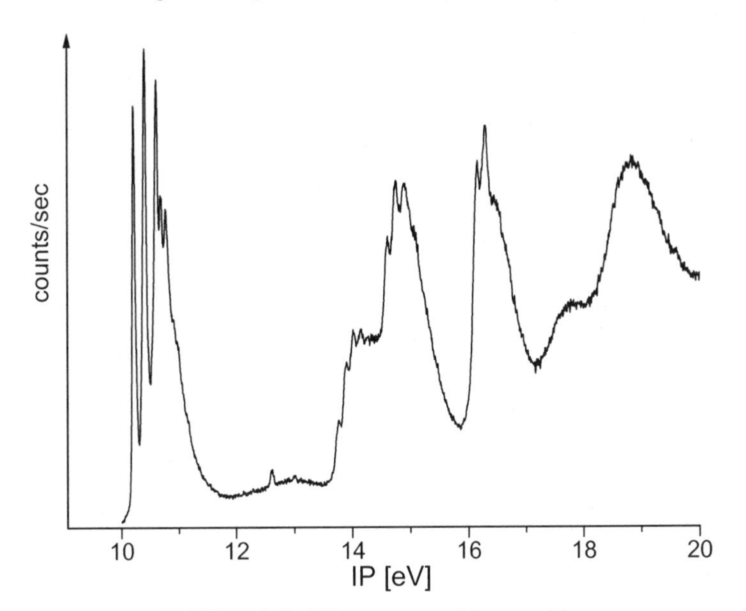

FIGURE 9.5. PE spectrum of formamide.

TABLE 9.1. Ionization Potentials IP [eV], Vibrational Spacings ν [cm^{-1}] and Orbital Energies ε [eV] of Formamide

IP_a [a]	IP_v [a]	ν [a]	$-\varepsilon$ [b]	MO	
10.13	10.32	1600	10.94	10a′	n_O
≤10.52	10.52	640	10.67	2a″	π_2
13.76	14.18	1090	14.81	9a′	
14.62	14.75	1050	14.79	1a″	π_1
16.16	16.30	1100	17.30	8a′	
	≈19		18.81	7a′	

[a] From reference 23.
[b] AM1 results.

was obtained by Oliveros et al.[27] in their CI calculations. An investigation of the ionization potentials using Green's function calculations came to the conclusion that the first two bands cannot unambiguously be assigned.[28] But the third IP was again assigned to π_1.

The series of formamide, acetamide, and their *N*-methyl derivatives permits a systematic study of the influence of methyl groups on the energy of the characteristic amide orbitals. The spectra were recorded by Brundle et al.[23] and by Sweigart and Turner,[24] the relevant data are summarized in Table 9.2. The systematic analysis – together with other carbonyl compounds – was done by McGlynn and co-workers.[29,30]

Comparison of the PE spectra of closely related molecules also permits a safer assignment of the spectra, e.g., the sequence of IP(n_O) and IP(π_2), and the location of IP(π_1), because substituent effects on individual orbitals can be estimated qualitatively. The correlation diagram for the relevant IPs, shown in Fig. 9.6, immediately permits the following conclusions:

- All MOs are destabilized by methyl-substitution.
- The π MOs are more sensitive to methyl-substitution than the n_O orbital. This allows one to distinguish such IPs.

TABLE 9.2. IPs [eV] of the n_O, π_2 and π_1 Orbitals of Formamide, Acetamide, and Their *N*-methyl Derivatives*

	$IP_a(n_O)$	$IP_v(n_O)$	ν [a]	$IP_v(\pi_2)$	$IP_v(\pi_1)$
$HCONH_2$	10.13	10.32	1600	10.52	14.75
HCONHMe	10.05	10.05	1500	9.87	14.30
$HCONMe_2$	≥9.40	9.77	1600	9.25	13.70
CH_3CONH_2	9.80	9.96	1600 [b]	10.32	14.20
$CH_3CONHMe$		≈9.85		9.68	13.7
CH_3CONMe_2		9.43	≈1400	9.09	13.10

* From reference 24.
[a] Vibrational splitting [cm^{-1}].
[b] From reference 31.

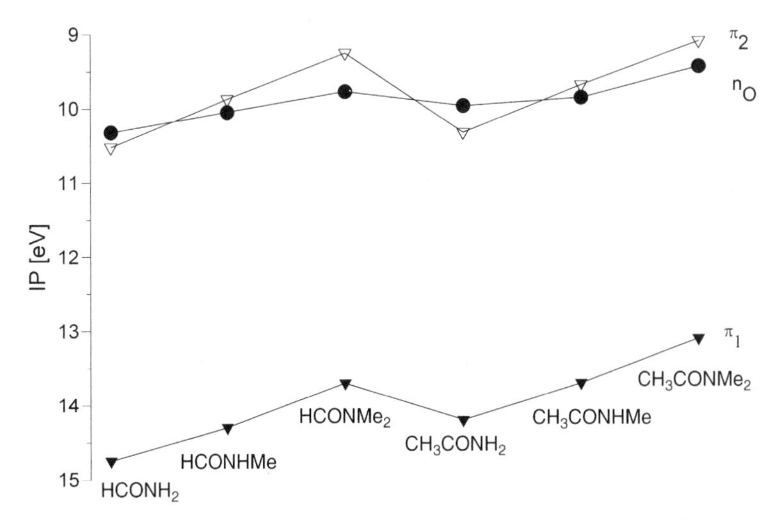

FIGURE 9.6. Correlation diagram for IPs of formamide, acetamide, and their *N*-methyl derivatives.

- In the formamide and the acetamide series the effects of methyl groups are additive; for π_1 the effects are additive in the whole series.

π_1 has coefficients of comparable size on all three atoms of the amide group (Fig. 9.4), therefore C- and N-substitution have nearly the same effect, each methyl group contributes about 0.4–0.6 eV. This requires that the fourth and not the third IP of formamide is assigned to π_1. π_2, has a node or a very small coefficient on the C atom, but a large coefficient on the N atom. Therefore, we find only minor shifts comparing the corresponding IP values of formamides and acetamides substituted with the same number of *N*-methyl groups. On the other hand, within the formamides and the acetamides each *N*-methyl group contributes a lowering of IP(π_2) of about 0.6 eV. The n_O orbital is mainly localized on the O atom, so we can expect that C-substitution will have a somewhat larger effect than N-substitution. But the effects seem to be approximately additive in the whole series: each methyl groups contributes about 0.3 eV. Comparison of this series of closely related compounds facilitates a safer assignment of the IP values than MO SCF calculations.

5. OTHER ALIPHATIC AMIDES

The relevant IPs of some aliphatic amides differing from those covered in Table 9.2 are summarized in Table 9.3. Presumably because there is only little variation in these data with the size of the molecule and its substitution, only few

TABLE 9.3. Ionization Potentials IP [eV] of some Aliphatic Amides

Amide	IP(π_N)		IP(n_O)	Reference
H–CO–NEt_2	8.89			35
Me–CO–NEt_2	8.71		9.20	34
iPr–CO–NEt_2	8.80		9.14	36, 37
FCH_2–CO–NH_2	10.68		10.38	32
F_3C–CO–NH_2		11.23 [a]		32
Cl_3C–CO–NH_2	10.88		10.53	33
$MeSCH_2$–CO–NMe_2	8.83		9.33	34
$MeSO_2CH_2$–CO–NEt_2	9.13		9.60	38
$iPrSO_2CH_2$–CO–NEt_2	9.08		9.50	38

[a] Coincidental merging of the two bands.

compounds were investigated, and indeed there is no unexpected result reported for such compounds which could elicit greater interest. The data of the two fluoroacetamides[32] listed in Table 9.3 show that the orbitals n_O and π_N both are stabilized by the inductive effect of the fluorine atoms, n_O obviously to a somewhat smaller extent. A similar although smaller stabilization is caused by chlorine atoms as the data of trichloroacetamide[33] indicate.

PE spectra of several other, more complicated compounds with an amide function have been recorded and analyzed, but here usually other aspects than the electronic structure of this group were of primary interest.

ETS spectra of a few amides have been measured by Jones et al.[34] For *N,N*-dimethylacetamide and *N,N*-diethylacetamide E_A values, relating to the LUMO $= \pi^*_{C=O}$, of 5.67 and 2.26 eV, respectively, were observed.

6. N-ACYL DERIVATIVES OF CYCLIC AMINES

It has already been mentioned that the interaction of the amino group with the carbonyl group is essential for the structure and the chemical properties of amides which—of course—all depend on the electronic structure. A systematic variation of the amino group can be expected to cause systematic changes of these properties. Such an investigation might reveal the relative donor capacity of different amino groups for which also conformational flexibility in order to escape steric repulsion would be of importance. First of all, one would expect that the $\pi_l \approx \pi_{C=O}$ orbital will be very sensitive to such variation, in a similar way as $\pi_l \approx \pi_{C=C}$ of enamines.[39–41] However, the ionization band associated with π_l of amides is located at about 12–14 eV in a range where it usually is superimposed by other strong bands which prevents its safe identification. On the other hand, as was shown by the PE spectra of the *N*-methyl derivatives of formamide and acetamide, the IP(n_O) is to a certain degree also sensitive to changes on the nitrogen atom.

TABLE 9.4. Vertical Ionization Potentials IP [eV] and Carbonyl Frequencies $\nu_{C=O}$ [cm^{-1}] of *N,N*-Diethylacetamide and *N*-Acetyl Derivatives of Cyclic Amines $R_2N-COCH_3$

R_2N	IP(π_N)	IP(n_O)	IP($\pi_{C=O}$)[a]	$\nu_{C=O}$	References
Et_2N	8.71	9.20	(12.3)	1652	34
N	9.63	10.40	(13.09)	1713	42
N	9.06	9.50	(11.77)	1630	43
N	8.84	9.24	(12.28)	1620	43
N	8.76	9.22	(11.8)	1626	43
O N	8.94	9.62	10.36[b]		
			11.64	1620	43
N	8.76	9.20	(12.0)	1635	43

[a] Data in parentheses indicate tentative assignments.
[b] IP(n_O) of ether oxygen.

In Table 9.4 the IP data of the *N*-acetyl derivatives of cyclic amines from aziridine to hexahydroacepine and of morpholine are summarized. The data of *N,N*-diethylacetamide are included for comparison. With the exception of 1-acetylaziridine and, to a lesser degree, 1-acetylazetidine, the IP(π_2) and the IP(n_O) values cover a narrow range which is close to that of acyclic amides (Table 9.3). In the morpholine derivative it is surprising that relative to 1-acetylpiperidine the IP(n_O) is shifted by 0.40 eV while IP(π_2) is only 0.18 eV larger. The effect of the oxygen atom in the six-membered ring is thus greater at the carbonyl oxygen than at the nitrogen atom, although it is two bonds closer to the latter.

From the data in Table 9.4 it can be concluded that the structure of the amide group is not largely affected by the variation of the amino group. The frequencies $\nu_{C=O}$ and the IP(n_O) values seem to be roughly correlated ($r = 0.89$). A low frequency and a low IP(n_O) both indicate strong $n_N/\pi^*_{C=O}$ interaction or a comparatively large contribution of resonance structure **B** to the electronic structure. According to the $\nu_{C=O}$'s, the pyrrolidine ring acts as a better electron donor than the azetidine and the piperidine ring. This sequence is not reflected in the IPs(n_O); here a continuous decrease with ring size is found. IP(n_O) can thus be regarded as a resonance indicator only with adequate reservation.

7. LACTAMS

7.1. PE Spectra and Electronic Structure

In lactams the amide bond is subjected to greater structural constraints than in open-chain amides or *N*-acyl derivatives of cyclic amines. In particular, the N–C=O bond angle, the configuration of the nitrogen atom and the torsion of the N–CO bond are subject to vary with the ring size of the lactam. Treschanke and Rademacher[44,45] have investigated a series of both secondary (i.e., *N*-unsubstituted) and tertiary (i.e., *N*-substituted) lactams, with a ring size from 4 (β-lactam) to 9 (η-lactam) in the former and 3 (α-lactam) to 13 (λ-lactam) in the latter case, both by PE spectroscopy and semiempirical MNDO calculations. Two bicyclic lactams, **18** and **19**, were included in the studies. The smallest homologue in the lactam series is aziridinone, a molecule which has only been postulated as an intermediate in a multi-step reaction because of its high reactivity.[46] A number of 1,3-disubstituted derivatives have been synthesized with 1,3-di-*t*-butylaziridin-2-one (**17**) as the most stable representative,[47] and this compound was selected to represent an α-lactam. The other tertiary lactams, except for the bicyclic one, are uniformly substituted with an *N*-methyl group.

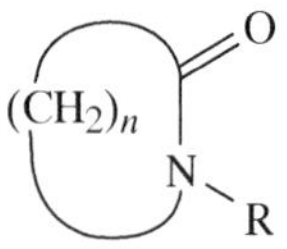

n	2	3	4	5	6	7	8	9	10	11
R = H	**1**	**2**	**3**	**4**	**5**	**6**				
R = CH_3	**7**	**8**	**9**	**10**	**11**	**12**	**13**	**14**	**15**	**16**

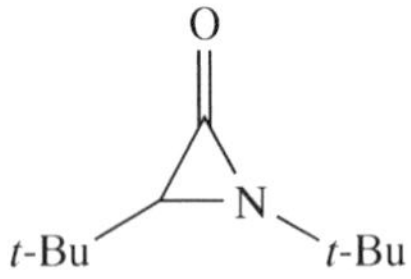

17 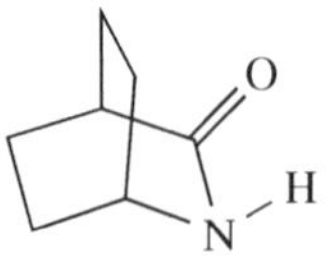 **18** 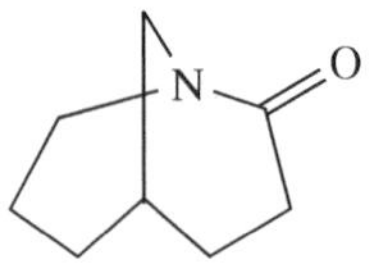 **19**

The MNDO calculations resulted in geometries with planar or nearly planar functional groups in most of the compounds which is in accord with available literature data.[44] As in other aziridine derivatives, the nitrogen atom appears clearly pyramidal in the α-lactam **17**; therefore, the N-substituent must be twisted out of the NCO plane. In the bicyclic compound **19** it is the bridgehead position of the nitrogen atom that is responsible for restricted amide resonance. A chair-boat form is calculated to be the most stable conformer.

Some representative PE spectra are depicted in Fig. 9.7. The PE spectroscopic data are collected in Table 9.5.

In all lactams, the first IP relates to the $\pi_2 = \pi_N$ and the second to the n_O orbital. As in other amides, the first two IPs are very close to each other, $\Delta IP_{1,2}$ values vary between 0.24 and 0.51 eV (except for **17** and **19**, see below) and are a little larger in tertiary than in secondary lactams. A correlation diagram is shown in Fig. 9.8. In general, band shape and vibrational fine structure belonging to IP_2 as well as MNDO calculations served as criteria for the assignments. The stronger influence of ring expansion on $IP(n_O)$ and of nitrogen substitution on $IP(\pi_N)$ can be explained in terms of the calculated electron distribution in the orbitals concerned.

The PE spectra of the unsaturated bicyclic β-lactams 5-azabicyclo[2.2.0]hex-2-en-6-one (**20**) and its 2-methyl derivative (**21**) were investigated by Aitken et al.[48] The ionizations were assigned by recourse to ab initio configuration interaction calculations for the respective states of the radical cations. The data given below the formula should be compared with those of compound **1** in Table 9.5. The $IP(\pi_N)$ values are very close in all three compounds, while the $IP(n_O)$ is 0.6–

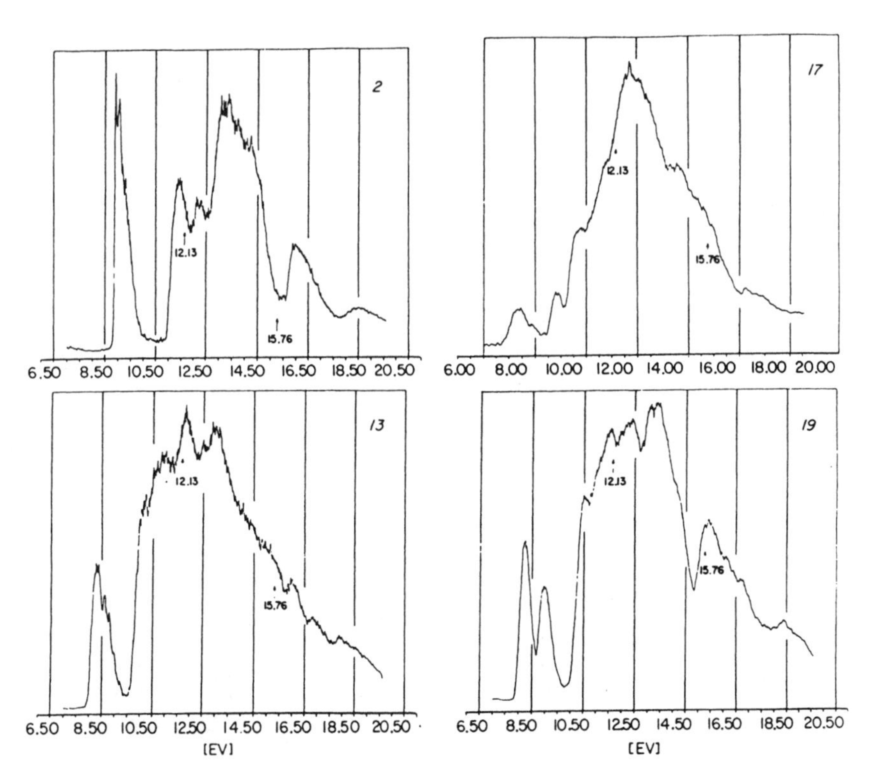

FIGURE 9.7. PE spectra of lactams **2**, **13**, **17**, and **19**. (From reference 45.)

TABLE 9.5. Ionization Potentials IP [eV], Vibrational Fine Structure ν [cm^{-1}], and Carbonyl Frequencies $\nu_{C=O}$ [cm^{-1}] of Secondary and Tertiary Lactams*

		IP(n_O)					
	IP(π_N)	IP_v	IP_a	ν	$\Delta IP_{1,2}$ [a]	IP_3	$\nu_{C=O}$
Secondary Lactams							
1	9.78	10.18	9.78	1600	0.40	12.91	1783
2	9.37	9.73	9.55	1450	0.36	11.92	1717
3	9.30	9.54	9.37	1400	0.24	11.80	1677
4	9.19	9.52	9.33	1550	0.33	11.42	1676
5	9.19	9.48	9.28	1600	0.29	11.14	1672
6	9.12	9.52	9.22	1200	0.40	10.92	1672
18	9.12	9.49	9.17	1300	0.37	11.01	1698
Tertiary Lactams							
7	9.29	9.80	9.40	1600	0.51	12.50	1762
8	9.17	9.68	9.28	1600	0.51	11.65	1698
9	8.92	9.36	9.00	1450	0.44	11.66	1658
10	8.73	9.13	8.77	1450	0.40	11.12	1651
11	8.76	9.16	8.84	1300	0.40	11.05	1649
12	8.78	9.18	8.86	1300	0.40	10.76	1643
13	8.72	9.12	8.80	1300	0.40	10.63	1640
14	8.70	9.10	8.74	1450	0.40	10.71	1641
15	8.74	9.10	8.78	1300	0.36	10.41	1645
16	8.72	9.09	8.77	1300	0.37	10.30	1646
17	8.21	9.76			1.55	10.43	1852
19	8.72	9.46			0.74	11.04	1688

[*] From reference 45.
[a] $\Delta IP_{1,2}$ = Difference between IP(n_O) and IP(π_N).

1.0 eV lower in the bicyclic compounds indicating a destabilization of this orbital by the unsaturated second ring.

	R	n_O	π_N	$\pi_{C=C}$
20	H	9.39	9.9	11.0 eV
21	Me	9.22	9.7	10.8 eV

As is obvious from Fig. 9.8, IP_1 and IP_2 shift uniformly with ring size to lower values up to the seven-membered ring compounds **4** and **10**. Their data can be assumed as "saturation" values. Deviations then amount to only some hundreds of an eV, and only the position of the first alkyl band (IP_3) indicates further growth of ring size. The final value of IP(π_2) in secondary lactams amounts to

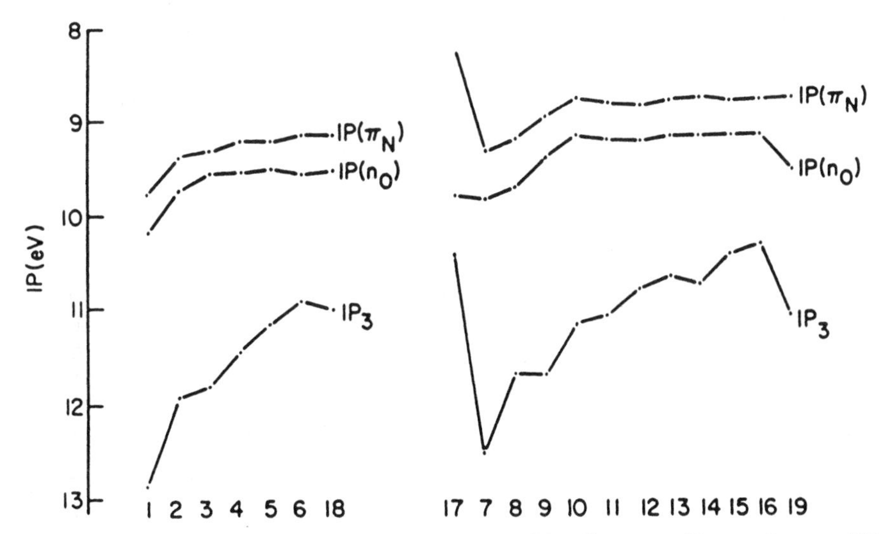

FIGURE 9.8. Correlation diagram for the first three IPs of lactams. (From reference 45.)

about 9.10 eV, in *N*-methyl tertiary lactams to about 8.70 eV. The corresponding data for IP(n_O) are 9.50 and 9.10 eV.

With increasing ring size of the lactam, a change from a *cis*- (synperiplanar, sp) to a *trans*- (antiperiplanar, ap) form of the amide unit can be expected.[2] However, no information about such a change or a coexistence of several conformers can be taken neither from the position of the IPs nor from the band shapes in the spectra of azonan-2-one (η-caprylolactam, **6**), its *N*-methyl derivative **12** and the higher homologues **13–16** for which *cis/trans*-isomerism can be expected. Apparently, different arrangements of the rest of the aliphatic ring and a change from a *cis*- to a *trans*-lactam cause such small modifications in electronic structure that they are not detectable by PES. This is in accord with very small IP differences ($\Delta IP \approx 0.01$ eV) of E- and Z-alkenes.[49]

The spectra of **17** and **19** (Fig. 9.7) are clearly different from those of the other tertiary lactams, and their $\Delta IP_{1,2}$ values are considerably greater. The two *t*-butyl substituents of **17** lower IP_1 to 8.21 eV. IP(n_O), however, remains at 9.76 eV. In **19**, IP(n_O) is quite high with regard to other tertiary lactams. Compared with the first two IPs of **4** and **5**, the IP(π_N) for **6** may be a little too low, and IP(n_O) in **6** and **18** too high.

Since the first two IPs of amides are quite close, to a first approximation their difference $\Delta IP_{1,2}$ can be assumed as an indicator of distortion. Only in some of these compounds can a change of the N–C torsional angle be responsible for a greater $\Delta IP_{1,2}$. It is obvious that in the small rings the bond angle α between the carbonyl carbon atom and its neighboring ring atoms deviates strongly from the normal value of 120°. By X-ray analysis of compounds **6**[50] and **18**[51], α values of 115° or 112° have been found. Hall and Zbinden[52] pointed out some years ago that deviations from the optimum value of α may not necessarily be correlated with an appreciable increase of ring strain but lead to a change of hybridization of the C=O group, and this will influence the corresponding IPs. Change of hybridization also modifies the position of the Amide-I ($\approx \nu_{C=O}$) band, whose dependence on α has been investigated by Cook.[53] A correlation between $\nu_{C=O}$ and IP(n_O) should indicate a relation between α and IP(n_O) and allow one to differentiate between the effects of valence and torsional angles on the PE spectra.

Correlations of IP(n_O) with other spectroscopic parameters like chemical shifts δ-^{13}C of the carbonyl group were observed in other carbonyl compounds[54] and can be expected for amides and lactams also, but to our knowledge, have not been investigated.

7.2. Correlation Between IP(n_O) and the Amide-I Vibration

The data in Table 9.4 indicate that there is a rough linear correlation between IP(n_O) and the frequency of the Amide-I vibration which largely is the carbonyl stretching vibration $\nu_{C=O}$. A more meticulous analysis was done, both for secondary and *N*-methyl tertiary lactams.[45] In these compounds, the amide resonance depends on the ring size, the N-substituent, the configuration of the nitrogen atom, and the torsion of the C–N bond. The primary effect of the ring size concerns the magnitude of the bond angle α. This effect can be eliminated by correcting the $\nu_{C=O}$ values for deviations of α from 120°. By an equation derived by Cook,[53] $\nu^{120}_{C=O}$ values are obtained which correspond to a hypothetical α of 120°.

$$\nu^{120}_{C=O} = \frac{96(\nu_{C=O} - 1439)}{216 - \alpha} + 1439 \; [\mathrm{cm}^{-1}] \qquad (9.4)$$

They are affected strongly in cases of small rings.[53] Figure 9.9 shows a plot of $\nu^{120}_{C=O}$ versus IP(n_O) using the $\nu_{C=O}$ data given in Table 9.5. Except for compounds **17** and **19**, the points are close to the straight line. Without the data for **17** and **19**, the resulting function is:

$$\nu^{120}_{C=O} = 57.81 \, \mathrm{IP} + 1117.8 \; [\mathrm{cm}^{-1}] \quad n = 17, \quad r = 0.978 \qquad (9.5)$$

The coefficient r indicates an excellent linear correlation between IP(n_O) and $\nu^{120}_{C=O}$. Without the corrections according to Eq. (9.4) the correlation coefficient is $r = 0.944$. Thus, actually there is only a minor improvement of the correlation between $\nu_{C=O}$ and IP(n_O) using the angle corrected frequencies according to Cook's method.[53] The IR measurements have been made in very dilute nonpolar solution. Valence angles originate partly from X-ray analyses or from estimates, so the main source of error may be found here. Considering these uncertainties, the deviations appear to be fairly small. An even better result is possible, when the data for secondary and tertiary lactams are correlated separately with $\nu^{120}_{C=O}$.[45]

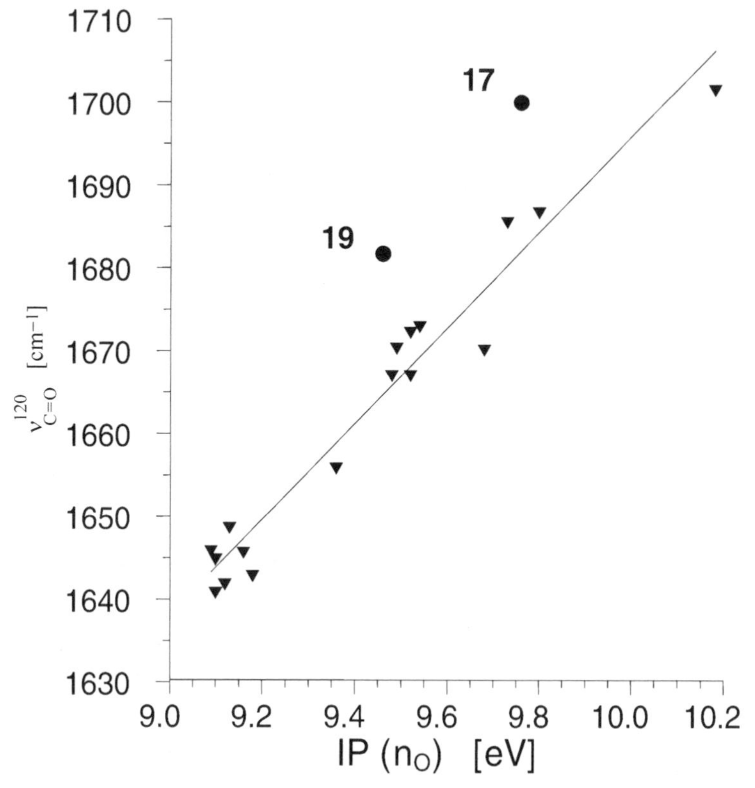

FIGURE 9.9. Correlation of IP(n_O) with the $\nu^{120}_{C=O}$ data from Eq. (9.4) for compounds **1–19**.

Assuming that the influences due to different valence angles α have been eliminated by use of $\nu^{120}_{C=O}$ values, a restricted resonance of the amide group caused by C–N torsion is indicated by points lying above the straight line (Fig. 9.9), and accordingly, the data for the distorted lactams **17** and **19** are found in the expected range.

The reason for the imperfect linear correlation between IP(n_O) and $\nu_{C=O}$ for *all* lactams, i.e., including the distorted lactams **17** and **19**, is that both parameters are dependent in different ways on structural changes of the amide linkage. IP(n_O) can, therefore, only be an indirect indicator of amide resonance with limited reliability. The bonding properties may be described by amide resonance of the two Lewis structures **A** and **B** (see also Section 11.3), or by donor-acceptor interaction of the amino and the carbonyl group, or by $n_N/\pi_{C=O}$ and $n_N/\pi^*_{C=O}$ orbital interactions. One contribution to $\nu_{C=O}$ that is not directly resonance-dependent has been eliminated by corrections for the bond angle α using Eq. (9.4). Also the energy of the n_O orbital is only indirectly affected by the interaction of the two parts of the molecule. Because of their different symmetry properties, n_N and $\pi_{C=O}$ cannot interact directly with n_O (cf. Fig. 9.3). In planar amides they are geometrically orthogonal, and in distorted amides their overlap integral is small.

Probably an even more successful attempt to correlate IP(n_O) and $\nu_{C=O}$ would be to use relative values for both parameters with reference to the corresponding carbonyl compounds *without* amide resonance, i.e. to the cyclic ketones with the same ring size. A linear correlation should be found between ΔIP(n_O) and $\Delta\nu_{C=O}$. For other experimental parameters like chemical shifts δ-^{13}C and δ-^{17}O also such a correlation might be valid. This could be shown for transannular amide resonance in cyclic aminoketones of medium ring size.[54]

8. UNSATURATED AMIDES AND LACTAMS

The previous sections have unveiled that the PE spectra of simple aliphatic amides are rather simple, by exhibiting only two ionizations in the informative low-energy range, and not very exciting, because the most important MO π_1 relates to an ionization in the complex σ range and usually cannot be identified. There are several ways to make the electron system of an amide more complicated and, perhaps, more interesting. One possibility is substitution with heteroatoms and the other extension of the chromophore by addition of multiple bonds.

In the former case we can substitute the hydrogen atoms of formamide, acetamide and their derivatives by halogen atoms. An example which is included in Table 9.3 is trichloroacetamide. The stabilizing effect of the electronegative chlorine atoms becomes evident when the IPs of this compound are compared with those of acetamide. (We will come back to this type of modification of the amide chromophore in Section 10.)

For the second way to modify the electron system of an amide, there are mainly two possibilities: the unsaturated group can be attached to the nitrogen or the carbonyl carbon atom. In this manner, *N*-alkenylamides, anilides or acrylamides, and benzamides are generated. We will discuss PE spectra of some of these compounds in the following sections.

8.1. *N*-Alkenyllactams

N-Alkenyllactams can be considered as a combination of enamines and lactams. The electronic structure of this class of compounds is related to the pentadienyl anion since the π electron system extends over a chain of five atoms and contains six π electrons in its planar configuration. The C=C double bond adds thus a third occupied π MO to the amide orbital system. π_2 and π_3 of the *N*-alkenyllactams are the symmetric and the antisymmetric combination of $\pi_{C=C}$ and π_N. Woydt and Rademacher[55,56] have studied a series of alkenyllactams (**22**) by PE spectroscopy and semiempirical MNDO and AM1 calculations. The PE spectra of three vinyl lactams[55] and of sixteen other alkenyl lactams[56] were recorded. The most stable conformation of nearly all these compounds is an essentially planar arrangement of the conjugated system, but the torsion of the alkenyl group relative to the amide unit seems to be of crucial importance.

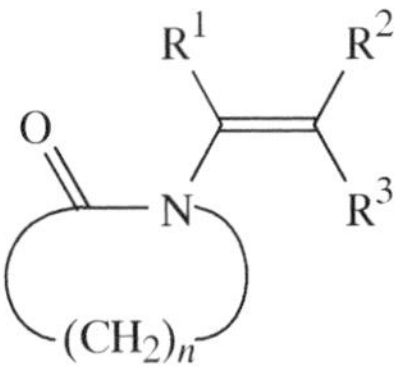

22

$n = 1$–6
R^1 = H, Me; R^2, R^3 = H, Me, Et, iPr, Ph
22a $n = 3$, $R^1 = R^2 = R^3$ = Me

All investigated compounds have three IPs in the region of 8–11 eV which can be assigned to ionizations from the *N*-alkenylamide system. In most compounds, the first ionization band has a rather badly resolved fine structure ($\Delta\nu = 1050$–$1450\,\text{cm}^{-1}$). These frequencies can be assigned to a $\nu_{C=C}$ vibration of the radical cation, and accordingly IP_1 is assigned to π_3. The second ionization band is identified as IP(n_O). This band has a fine structure that can be interpreted as a $\nu_{C=O}$ vibration. The third IP is assigned to the MO π_2.

In **22a** the energy difference of the first three IPs is only 0.73 eV which is an indication for an interrupted π conjugation.

8.2. Benzamides

PE spectra of several benzamides have been included in studies of benzene derivatives.[57–61] Substituent effects on the energies of the benzene π MOs were the primary object in these investigations, which have revealed linear relations between IPs and Hammett σ substituent constants. This holds also for IP(n_O) and IP(π_2) values of benzamides.[57,61]

Some data for benzamides are summarized in Table 9.6. IP(π_N) varies between 10.02 eV for 4-methoxybenzamide and 10.69 eV for 4-nitrobenzamide.[57] The corresponding values for IP(n_O) are 9.57 and 10.33 eV. The IPs of unsubstituted benzamide are roughly in the middle of these ranges. According to these data the "characteristic MOs" of amides seem to be affected approximately to the same extent in substituent benzamides. The introduction of donor groups results in lower IPs (increase of MO energies), and acceptor groups increase the IPs (lower MO energies).

Similar substituent effects have been observed for 4-substituted *N,N*-dimethylbenzamides, however, only IP(n_O) values are reported.[61] The latter are about 0.7–0.9 eV lower than in the corresponding *N*-unsubstituted compounds, which relates well to the findings for formamide and its *N,N*-dimethyl derivative (Table 9.2).

Also some primary benzamides with fluoro substituents in the benzene ring have been investigated by PES, but because of strong overlap of ionization bands in the range of interest it was not possible to determine IP(n_O) and IP(π_2) separately.[59,60]

TABLE 9.6. Ionization Potentials IP [eV] of some Aromatic Amides

Amide	IP(π_N)		IP(n_O)	Reference
Ph–CO–NH_2	10.20		9.78	57,58
C_6F_5–CO–NH_2		10.7 [a]		59,60
2-HO–C_6H_4–CO–NH_2	10.08		10.41	58
3-MeO–C_6H_4–CO–NH_2	10.13		9.77	57
3-Me–C_6H_4–CO–NH_2	10.08		9.72	57
3-Cl–C_6H_4–CO–NH_2	10.31		9.82	57
3-O_2N–C_6H_4–CO–NH_2	10.55		10.05	57
4-MeO–C_6H_4–CO–NH_2	10.02		9.57	57
4-Me–C_6H_4–CO–NH_2	10.08		9.62	57
4-Cl–C_6H_4–CO–NH_2	10.29		9.94	57
4-F–C_6H_4–CO–NH_2	10.28		10.00	57
4-NC–C_6H_4–CO–NH_2	10.60		10.29	57
4-O_2N–C_6H_4–CO–NH_2	10.69		10.33	57
Ph–CO–NMe_2			9.04	61

[a] Coincidental merging of the two bands.

Although the photon probes the electron binding energy of a basic molecule B and the proton probes its ability to share electron density in the formation of a $B–H^+$ bond, attempts were made to use valence shell ionization potentials to predict proton affinities or the Brönsted and Lewis basicities of oxygen-, nitrogen-, or phosphorus-containing molecules. Gal et al.[61] have investigated the Lewis basicities of three series of para-substituted aromatic carbonyl compounds, including *N,N*-dimethylbenzamides, and found linear correlations between the enthalpies ΔH^O of complex formation with boron trifluoride and the ionization potentials of the carbonyl oxygen lone-pair IP(n_O).

X–C_6H_4–C(=O)–NMe_2 + BF_3 ⇌ X–C_6H_4–C(=O^+–BF_3^-)–NMe_2

For the benzamides, three IPs relating to n_O and the highest two benzene π MOs, π_S, and π_A, were identified. In Fig. 9.10 the ΔH^O values are plotted against the IP(n_O) values for the benzamides. A linear correlation between these data is obvious:

$$\Delta H^O = (15.74 \pm 2.31)\mathrm{IP}(n_O) - 243.81\ [\mathrm{kJ\,mol^{-1}}]\quad n = 8,\quad r = 0.941 \qquad (9.6)$$

A plot of the IPs against the Hammett substituent parameters σ_p also revealed a linear relationship. The electron donor or acceptor ability of the substituents is transmitted through the benzene ring to the carbonyl oxygen atom and modifies the electron density at this atom and likewise its nucleophilicity which increases

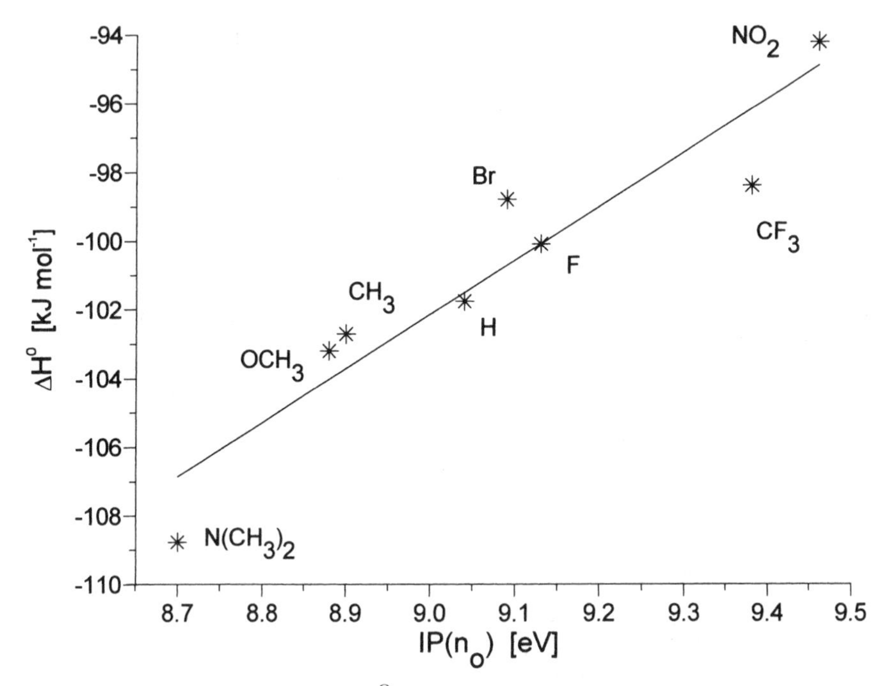

FIGURE 9.10. Relationship of ΔH^{O} and IP(n_O) for BF_3 complexes of *p*-substituted *N,N*-dimethylbenzamides.

with the energy of the n_O orbital or with the lowering of IP(n_O). As a result, the interaction of this orbital with the empty 2p orbital at the boron atom of the electrophile (Lewis acid) BF_3, which is inversely proportional to their energy difference, is increased.

Substituent effects on π MOs have also been studied for aromatic heterocycles, e.g., for furans, thiophenes, selenophenes and tellurophenes,[62] and the dimethylaminocarbonyl group was one of the investigated α-substituents. An ionization between 9.65 and 9.54 eV was attributed to this substituent, which most likely can be assigned as IP(n_O).

8.3. Anilides

UPS is a powerful method for conformational analysis, and many compounds have been studied by this technique.[63–65] The method is based on conformation-dependent interactions of occupied MOs, and the size of this interaction is determined from the IPs related to these MOs. Provided that there is a clear relationship between the IPs and the relevant conformational angle, the latter can be determined from the observed IPs. Usually the difference ΔIP of two IPs is taken as a direct measure of the orbital interaction. The first investigations were made by Maier and Turner[66] on ortho-substituted biphenyls. The torsional angle

ϕ of the two benzene rings is related to the separation of the IPs of two π MOs by the empirical equation:

$$\Delta IP = IP(\pi_6) - IP(\pi_3) = 1.65 \cos \phi + 0.35 \tag{9.7}$$

In a similar way, Maier and Turner[67] have studied "steric inhibition of resonance" in *ortho*-substituted anilines, and the investigation of acetanilides by Nakagaki et al.,[19] can be considered as an extension of this study. The PE spectra of the anilides **23–27** and of oxindole (**28**) were measured and analyzed. The spectra are shown in Fig. 9.11, and the ionization energies are given in Table 9.7.

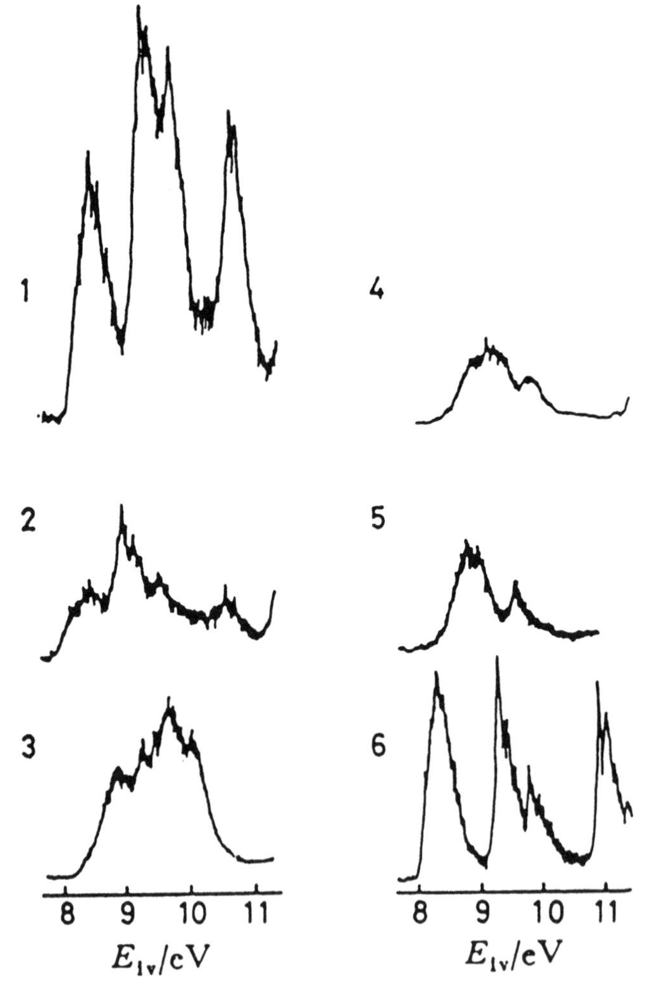

FIGURE 9.11. PE spectra of acetanilide and related compounds. (From reference 68.) 1: acetanilide (**23**); 2: 2′-methylacetanilide (**24**); 3: *N*-methylacetanilide (**25**); 4: *N*,2′-dimethylacetanilide (**26**); 5: *N*,2′,6′-trimethylacetanilide (**27**); 6: oxindole (**28**).

TABLE 9.7. Vertical Ionization Potentials [eV] of Compounds 23–28 and other Anilides

Compound	π^-(S)	π_2^{Ph}	π^+(S)	n_O	Reference
Acetanilide (**23**)	8.46	9.35	10.75	9.70	58,68
2′-Methylacetanilide (**24**)	8.34	8.92	10.59	9.51	68
N-Methylacetanilide (**25**)	8.81	9.28	10.02	9.68	68
N,2′-Dimethylacetanilide (**26**)	8.82	9.08	9.39	9.79	68
N,2′,6′-Trimethylacetanilide (**27**)	8.8		9.0	9.64	68
Oxindole (**28**)	8.36	9.32	10.86	9.79	68
4′-Hydroxyacetanilide (**29**)	8.04	9.35	10.27	9.64	58
4′-Ethoxyacetanilide (**30**)	7.92	9.27	9.95	9.57	58
Trifluoroacetanilide	8.93	9.64	11.25	10.41	69

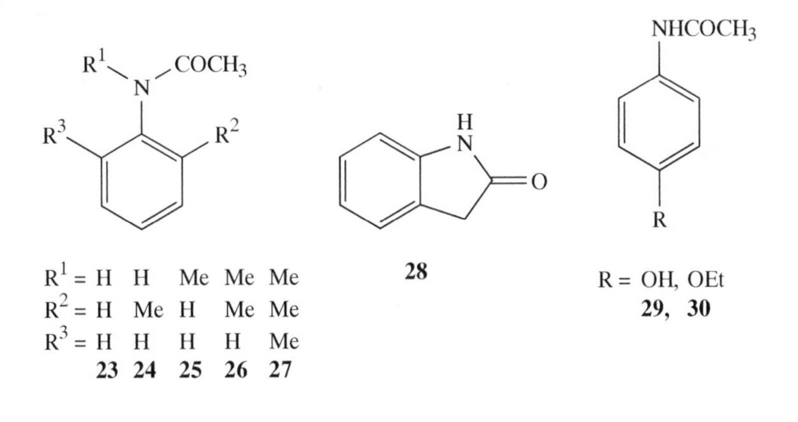

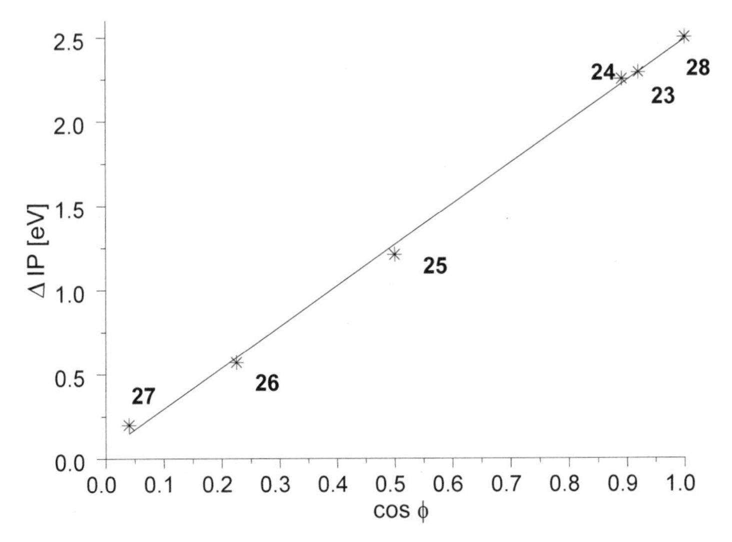

FIGURE 9.12. Plot of ΔIP against cos ϕ for acetanilides **23–28**.

In the region of 8–11 eV, the spectra consist of four bands: two bands correspond to the MOs $\pi^-(S)$ and $\pi^+(S)$, which are interpreted approximately by the antibonding and bonding combinations of the benzene ring MO π_3^{Ph} and the highest occupied π MO of the acetylamino group π_2. The other two bands correspond to π_2^{Ph} of the benzene ring and n_O of the carbonyl oxygen. As expected, IP(n_O) is affected only slightly by the alkylation on the ortho-positions of the ring or on the nitrogen atom and has values similar to those of aliphatic amides, i.e., 9.4–10 eV (Table 9.3).

The ionization energy difference ΔIP between the $\pi^+(S)$ and the $\pi^-(S)$ bands is again found to be a linear function of cos ϕ, as is shown in Fig. 9.12. Oxindole (**28**) was considered to be planar ($\phi = 0°$). The structures of anilides **23** and **25** are known from X-ray analyses; the values of ϕ found in the crystalline state are comparable with those determined by PES.

A similar study including trifluoroacetanilides was performed by Szepes et al.[69]

9. BIOLOGICALLY ACTIVE AMIDES

9.1. Antipyretics

More than a century ago, in 1886, acetanilide (**23**) was introduced into medicine because of its antipyretic action. The less toxic phenacetine (4-ethoxyacetanilide, **30**) was introduced into therapy in 1887. Both compounds are no longer approved in most countries. The use of paracetamol (4-hydroxyacetanilide, 4-acetylaminophenol, **29**) began in 1893, and in 1949 it was shown that paracetamol is the active metabolite of both acetanilide and phenacetine. The biological action of these compounds, as that of other drugs, is certainly, although not in a simple way, related to its electronic structure. It was the motive of Klasinc et al.,[58] to provide some clues to the action of these compounds when they analyzed the PE spectra of these and some related compounds. The results obtained for the anilides **23**, **30**, and **29** are included in Table 9.7.

9.2. Lysergic Acid Diethylamide (LSD)

Lysergic acid diethylamide (LSD, **31**) is the most potent hallucinogen and its PE spectrum was measured by Domelsmith et al.,[36,37] and analyzed together with those of other psychotomimetic drugs in order to establish a correlation between activity and ionization potentials. The PE spectrum of LSD could be partially assigned mainly by comparison with structurally related compounds. For the amide part of LSD, *N,N*-diethylisobutyramide (Table 9.3) was used as reference, and for IP(n_O) a shoulder in the LSD spectrum at 9.08 eV could be assigned. The IP(π_2) band, however, is located in the unresolved 8.5–9.0 eV region, so that no assignment was possible.

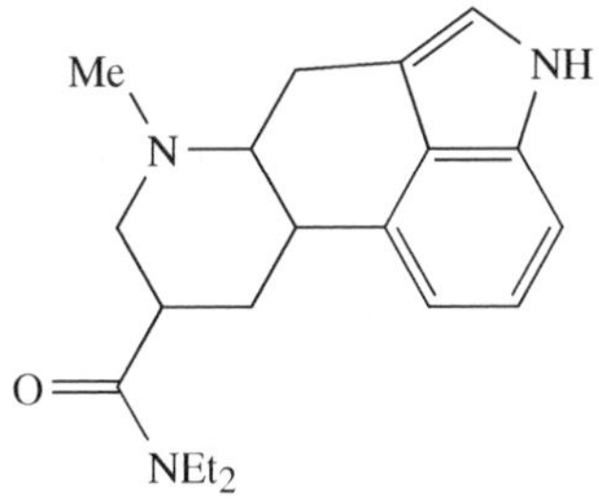

31

9.3. Nicotinamide

The PE spectra of nicotinamide (**32**) and *N,N*-diethylnicotinamide (**33**) were measured by Dougherty et al.[70] Five low-energy ionizations were expected: three IP(π) and two IP(n) events. Two of the π MOs are located on the pyridine ring, while the third is largely on the amidic nitrogen (π_N). One of the n MOs is a σ type lone pair of the pyridinic nitrogen, while the second is n_O of the carbonyl oxygen. Because of strong band overlap, only three IPs of **32** and four of **33** could be identified, and a somewhat safe assignment is only possible for $IP(\pi_N) = 9.4$ eV and $IP(n_O) = 9.98$ eV of **33**.

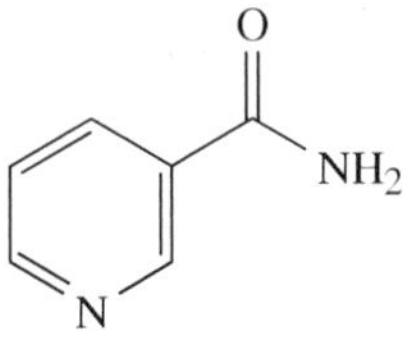

32 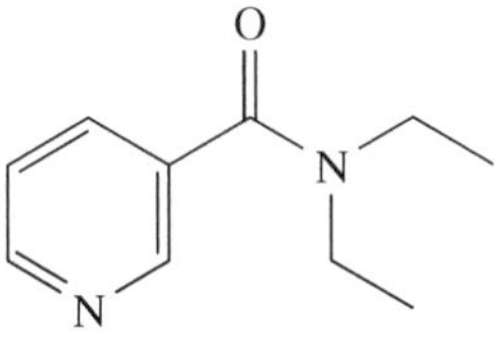 **33**

9.4. Peptides

In many natural biologically active peptides, α,β-unsaturated amino acid derivatives are present.[71,72] They are particularly interesting because of their unusual conformational and electronic features. The PE spectra of some compounds, which can be considered as models for unsaturated peptides, have been studied by Ajò, Granozzi, and co-workers.[73–75] We shall discuss here in some detail the PE spectra and the electronic structure of N-acetyldehydroalanine [2-(acetylamino)prop-2-enoic acid, **34**].[73]

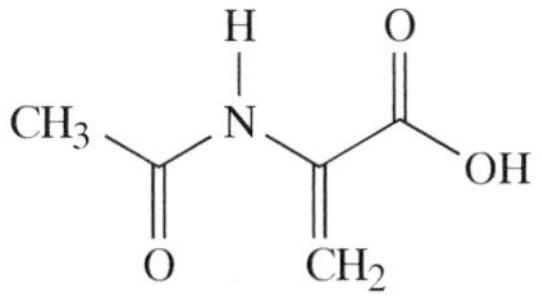

34

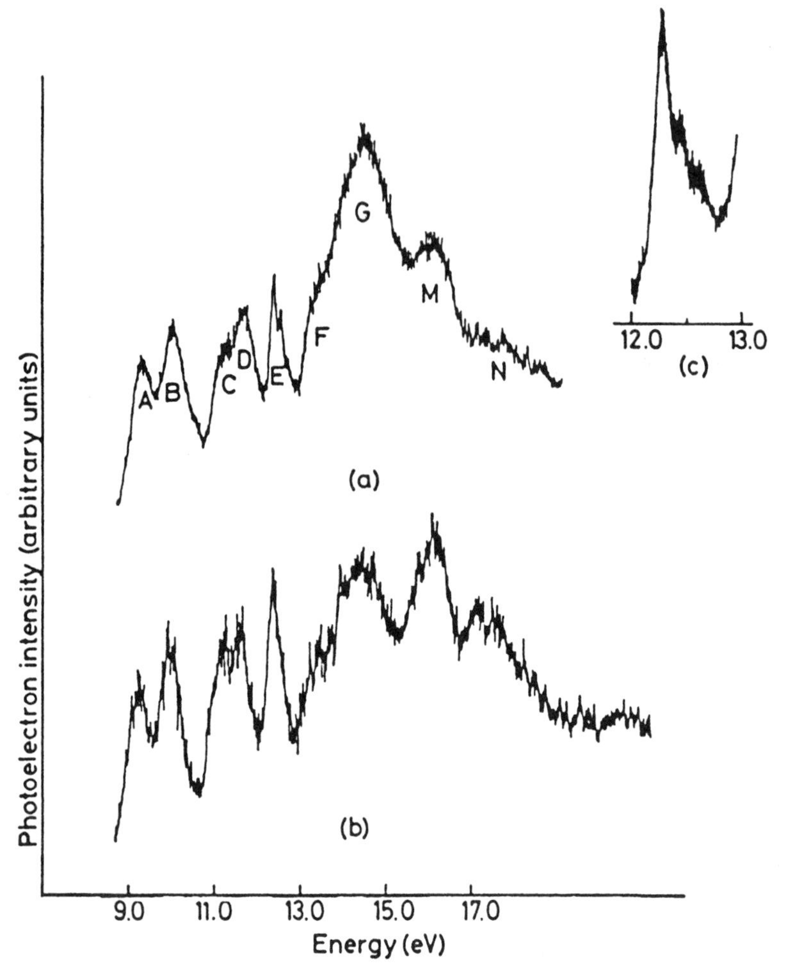

FIGURE 9.13. PE spectra of *N*-acetyldehydroalanine (**34**). (a) He-I spectrum; (b) He-II spectrum; (c) detail of band E (from He-I spectrum). (From reference 73.)

The He-I and He-II PE spectra of **34** are shown in Fig. 9.13. The measured IPs and their assignments are summarized in Table 9.8. The He-I spectrum exhibits a fairly well resolved ionization pattern. Below 13 eV five IPs can be determined. The assignment of the spectrum was based on a correlation with the IPs of related simple compounds, namely acetamide and acrylic acid. Fig. 9.14 shows the correlation diagram.

Bands A and B represent ionizations from MOs which are the counterparts of n_O and π_N of acetamide. The interaction between π_N and $\pi_{C=C}$ of the C=C double bond of the acrylic fragment is responsible for the low-energy shift of IP(π_N) in **34**. IP(n_O) has a value lying in its usual range. Bands C–E match the three lower ionizations present in the PE spectrum of acrylic acid. They correspond to ionizations from the lone pair of the carboxy oxygen (n_O'), from

TABLE 9.8. Ionization Potentials IP [eV] of *N*-Acetyldehydroalanine (34)*

Band	IP	Assignment
A	9.24	π_N
B	9.91	n_O
C	11.11	n_O'
D	11.58	$\pi_{C=C}$
E	12.26	π_{CO_2H}
F	13.40	σ_{CH}
G	14.34	σ_{CC}, $\pi_{C=O}$ (amide)
M	15.83	$\pi_{C=O}$ (carboxy)
N	17.5	

*From reference 73.

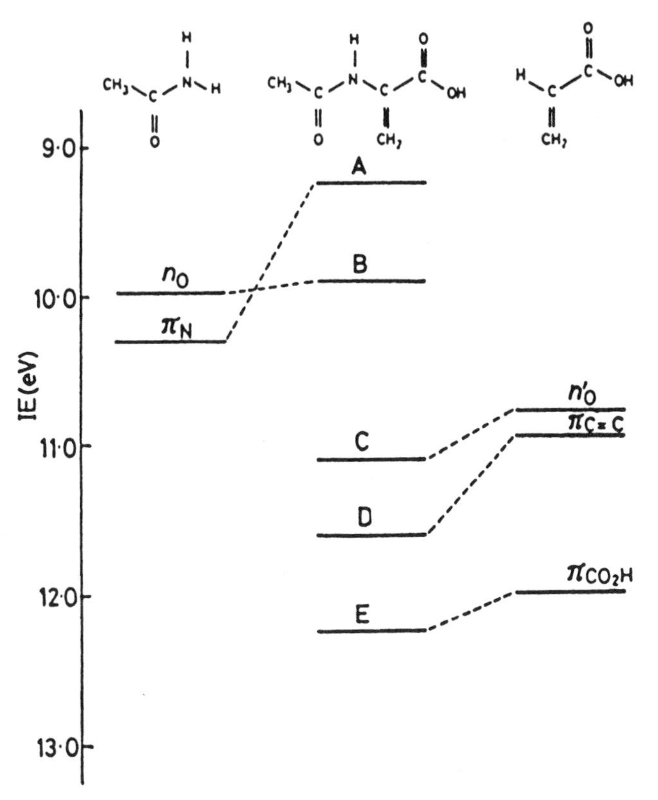

FIGURE 9.14. Correlation diagram of IPs of acetamide, *N*-acetyldehydroalanine (**34**) and acrylic acid. Bands are labelled according to Table 9.8. (From reference 73.)

the $\pi_{C=C}$ and from the antisymmetric π MO localized on the CO_2H group (π_{CO_2H}).

The gas-phase PE spectra of several dipeptides could be recorded by Richer et al.[77] at moderate temperatures (125–155 °C) so that decomposition could be avoided as far as possible. The spectra are generally characterized by very broad ionizations which lead to a more or less continuous band commencing at about 8 eV and revealing few distinct maxima. However, by careful comparison of the spectra with those of the corresponding amino acids, some conclusions regarding the presence of the peptide IP(π_N) and IP(n_O) bands are possible. For example, for glycyl-glycine (gly-gly), the simplest dipeptide, these two IPs are located between 9.5 and 10.0 eV and at about 10.8 eV. It is appropriate to point out here that the dipeptides as well as amino acids exist in the vapor state in the nonpolar (not the zwitterionic) form.

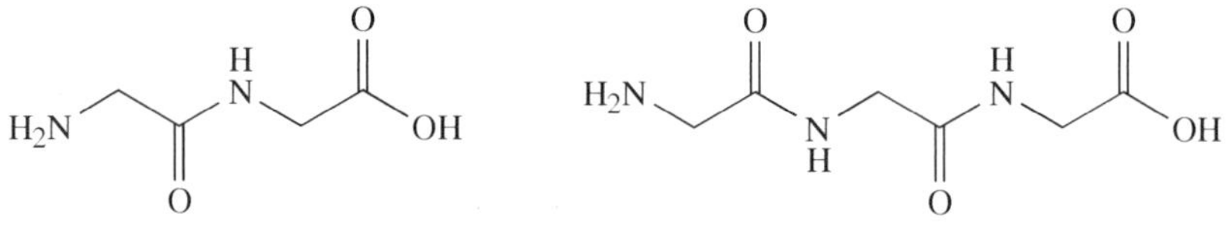

gly-gly

gly-gly-gly

By the same authors[77] an attempt was made to record the spectrum of the tripeptide gly-gly-gly. Only the first bands could be obtained, however, because of incipient decomposition and low volatility. Two greatly overlapping bands were seen, with apparent maxima at about 9.8 and 10.1 eV and a well-defined band at 10.6 eV.

10. UREAS

Urea formally is generated by replacing the formyl hydrogen atom of formamide by a second amino group. Carbamic acid, the parent compound of the urethanes,[78] is obtained by replacing this hydrogen atom by a hydroxy group. Substitution of an amino hydrogen atom by an amino or a hydroxy group leads to hydrazides[79] and to *N*-hydroxyamides.[80] All these compounds have an expanded chromophore, compared to simple amides, because the new heteroatoms introduce additional electron lone pairs. Other interesting compounds formally derived from formamide are oxamide (oxalic acid diamide) and its derivatives[81–83] and the imides.[83]

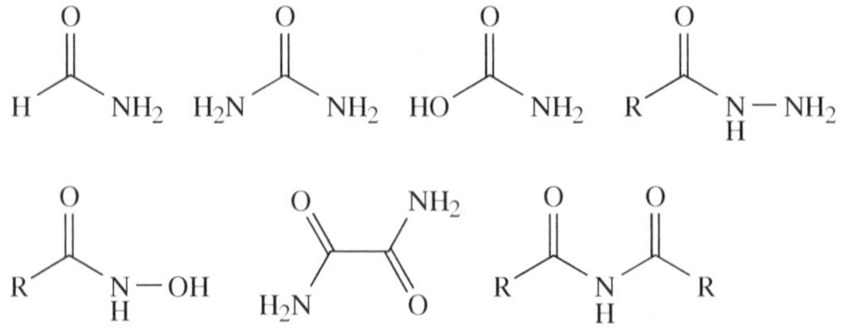

Some of these structures or a combination of them is found in acyl ureas, uracils, and thymine,[85] parabanic acid,[82] and many other heterocyclic compounds[31,75,78,86]. The PE spectra of many of these compounds have been recorded, some literature references are given in the preceding sentences. Since all these compounds are not amides in the precise sense, it would probably lead us too far if we would discuss all these systems in detail, although all of them contain amide bonds. As a representative of such classes of compounds, the ureas are considered a little closer. The essential part of the electronic system of urea consists of three occupied π type MOs and the n_O at the oxygen atom. π_1 can approximately be labeled as $\pi_{C=O}$, and π_2 and π_3 are obtained with respect to the C_2 axis of the planar molecule as the symmetric and the antisymmetric combination, respectively, of the two nitrogen 2p orbitals. The latter are denoted as $\pi_N^- = \pi_2$ and $\pi_N^+ = \pi_3$.

The PES data obtained for urea, some alkylureas, ureas with cyclic amino groups (**35–38**), and cyclic ureas (**39–47**) are summarized in Table 9.9.

TABLE 9.9. Vertical Ionization Potentials [eV] of Ureas

Compound	π_N^+	π_N^-		n_O	Reference
Urea	10.28	10.78		10.28	57
Methylurea	9.66	10.23		9.66	57
1,3-Dimethylurea	9.23		9.73[a]		57
1,1-Dimethylurea	8.96	9.93		9.9	57
Trimethylurea	8.80	9.45		9.82	57
Tetramethylurea	8.64	8.98		9.92	57, 85
1,1′-Carbonylbisaziridine (**35**)	9.56	9.80		10.62	87
1,1′-Carbonylbisazetidine (**36**)	8.74	9.00		9.64	87
1,1′-Carbonylbispyrrolidine (**37**)	8.50	8.84		9.52	87
1,1′-Carbonylbispiperidine (**38**)	8.28	8.68		9.80	87
Imidazolidin-2-one (**39**)	9.65	9.89		10.33	78, 88
Tetrahydropyrimidin-2-one (**40**)	9.29	9.5		9.8	88
1,3-Diazepan-2-one (**41**)	9.17	9.62		10.2	88
1-Methylimidazolidin-2-one (**42**)	8.91	9.64		9.9	88
1-Methyltetrahydropyrimidin-2-one (**43**)	8.79	9.29		9.7	88
1-Methyl-1,3-diazepan-2-one (**44**)	8.83	9.2		9.99	88
1,3-Dimethylimidazolidin-2-one (**45**)	8.66	9.31		9.82	88
1,3-Dimethyltetrahydropyrimidin-2-one (**46**)	8.38	8.96		9.5	88
1,3-Dimethyl-1,3-diazepan-2-one (**47**)	8.59	8.9		9.72	88

[a] Coincidental merging of the two bands.

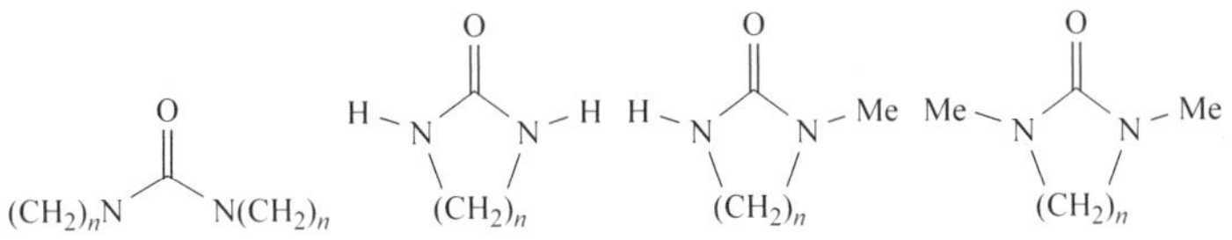

O
$(CH_2)_nN$ — C — $N(CH_2)_n$
n 2 3 4 5
35 36 37 38

H–N(C=O)N–H, $(CH_2)_n$
2 3 4
39 40 41

H–N(C=O)N–Me, $(CH_2)_n$
2 3 4
42 43 44

Me–N(C=O)N–Me, $(CH_2)_n$
2 3 4
45 46 47

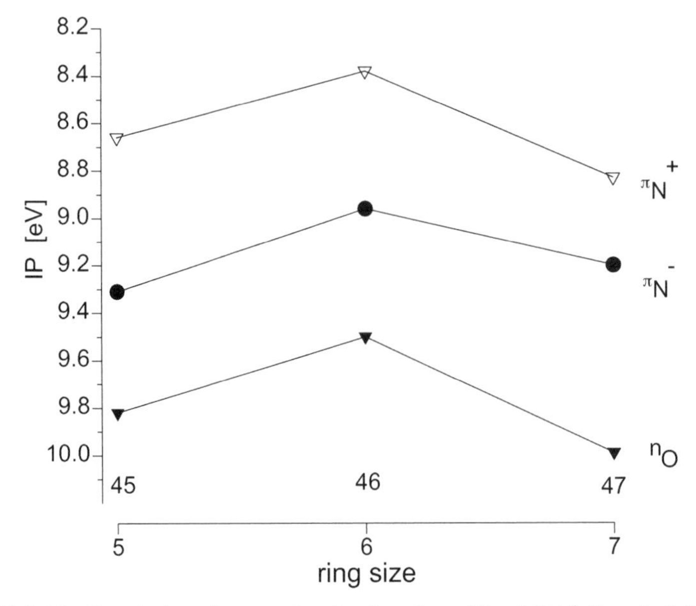

FIGURE 9.15. Correlation diagram for the first three IPs of *N,N'*-dimethyl derivatives of cyclic ureas with different ring size (Compounds **45–47**).

The data allow an analysis of the additivity of substituent effects in the series of urea and its methyl derivatives, and of the ring size in the 1,1′-carbonylbisamines (**35–38**) and the cyclic ureas (**39–47**). With only a few exceptions, the sequence of the four highest occupied MOs was found to be: π_N^+ (HOMO), π_N^-, n_O, $\pi_{C=O}$.

Irregularities in the course of the data with continuous structural variation indicate conformational modification of the chromophore. For example, the changes in spectral appearances, which occur in the pairs methylurea → 1,3-dimethylurea and methylurea → 1,1-dimethylurea, are probably due to a sterically enforced rotation of the dimethylated amino group, thus disrupting its conjugation with the carbonyl group. In the series of cyclic ureas, in most cases the MOs are stabilized in the five- and seven-membered rings relative to the six-membered ring (Fig. 9.15), and this can be explained by different conformations of the urea fragment: The five- and seven-membered rings have a conformation with pyramidal nitrogens, while in the six-membered rings the nitrogens are almost planar.[88]

11. XPS SPECTRA

It is evident that the ease to remove a core electron is dependent on the "environment" of the atom. Accordingly, atoms of the same element in different positions of a molecule give different IP(1s) values, whereas equivalent positions are characterized by the same value. The core electrons, "see" the distribution of the outer electrons in the molecule, and the IP(1s) values, if measured with sufficient accuracy, reveal information on the total electronic structure.

11.1. Core Electron Binding Energy and Atomic Charge

One early result of XPS (or ESCA) studies with special relevance for amides and lactams is the fact that IPs(1s) of atoms like C, N, and O in molecules are correlated with their (partial) charge. XPS has been used to derive relative measures of atomic charge.[13,14] In organic nitrogen compounds, IPs(N 1s) of about 398.6 eV are found for neutral nitrogen. Higher values usually relate to partially positive nitrogen, and protonated amine nitrogens have values above 400 eV.[89]

Binder[90] has measured N 1s energies by ESCA for a series of formyl and acetyl derivatives of ammonia, including formamide and acetamide, and found a linear correlation with a charge parameter obtained from electronegativity considerations. The lowest IP(N 1s) = 399.8 eV was found for $Na^+[N(CHO)_2]^-$, and this corresponds to a charge of $q = -0.90$. On the other hand, for $BrN(CHO)_2$ the IP(N 1s) = 402.4 eV relates to a charge of $q = +0.02$ on the nitrogen. For the two amides the following data were obtained.

$$HCONH_2\text{:}\quad \mathrm{IP(N\,1s)} = 401.5\,\mathrm{eV}, \quad q = -0.29$$

$$CH_3CONH_2\text{:}\quad \mathrm{IP(N\,1s)} = 401.2\,\mathrm{eV}, \quad q = -0.02$$

The N 1s core energy in pyrrole (406.15 eV) is appreciably higher than in pyrrolidine (404.60 eV), and this is readily explained via resonance rather than change from sp^2 to sp^3 hybridization. Also (intramolecular) hydrogen bonding can be observed by XPS: Ethanolamine, $HOCH_2CH_2NH_2$, has a higher N 1s core energy (405.30 eV) than its methyl ether, $CH_3OCH_2CH_2NH_2$, and *n*-butylamine (both 404.88 eV), because the hydrogen bonds lead to a partial positive charge on the N atom which can be represented by a resonance structure having a positive charge on this atom.[91]

11.2. Proton Affinities and Site of Protonation

Cavell and Allison[92] have measured the N 1s ionization energies of a series of amines and of formamide and its *N*-methyl derivatives. They found a linear correlation between the N 1s binding energies and the gas-phase proton affinities PA with large deviations for the amides. This is taken as an indication that the site of protonation is oxygen in these cases rather than nitrogen. This proposal is supported by the good correlation of the proton affinities for the amides with O 1s binding energies and NMR evidence in solution.

Greenberg and Moore[93,94] have calculated gas-phase proton affinities at N and O for several bridgehead bicyclic lactams as well as some tertiary amides and found linear correlations with computed N 1s and O 1s ionization energies. Depending on the degree of distortion, the lactams are either N- or O-protonated.

The relative gas-phase basicities of the three oxygen atoms in *N*-*t*-butoxycarbonylglycine *N′*,*N′*-dimethylamide (**51**), which can be considered as a glycine dipeptide, were estimated from the O 1s ionization energies.[95]

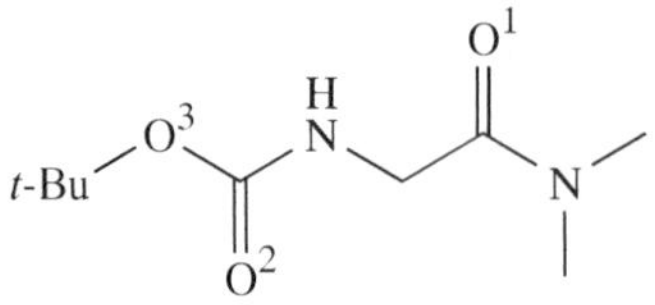

51

From a correlation of IP(O 1s) and proton affinity PA, the equation

$$\mathrm{PA} = 334.79 - 0.607\,\mathrm{IP(O\ 1s)} \tag{9.8}$$

is calculated, and it is found that the relative basicity is $O^3 < O^1 < O^2$. The PA of the urethane carbonyl oxygen O^2 is estimated to be 26–29 kJ mol^{-1} higher than that of the amide carbonyl oxygen O^1 and 85–89 kJ mol^{-1} higher than that of the alkoxy oxygen O^3.

Amides and lactams form double salts with tetrachloroauric acid ($HAuCl_4$). In these salts, the amides and lactams are O-protonated, the proton lying halfway between the oxygen atoms. This was shown by ^{13}C NMR and ESCA spectroscopy.[89] As an example, the $HAuCl_4$-double salt with 1-methylpyrrolidin-2-one is considered. The ^{13}C NMR spectra show that the two pyrrolidinone parts in the salt are magnetically equivalent, suggesting a symmetric structure (**48**). However, this could also be the result of a dynamic process, i.e. a rapid equilibrium between two equivalent structures (**49**, **50**). The

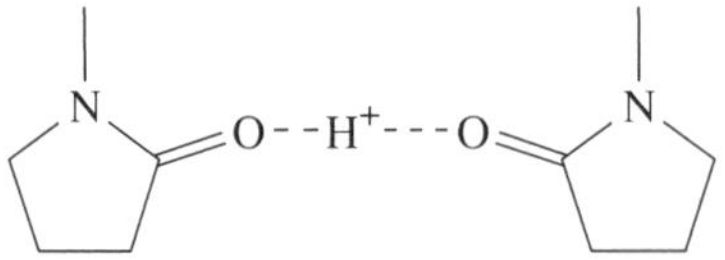

$AuCl_4^-$

48

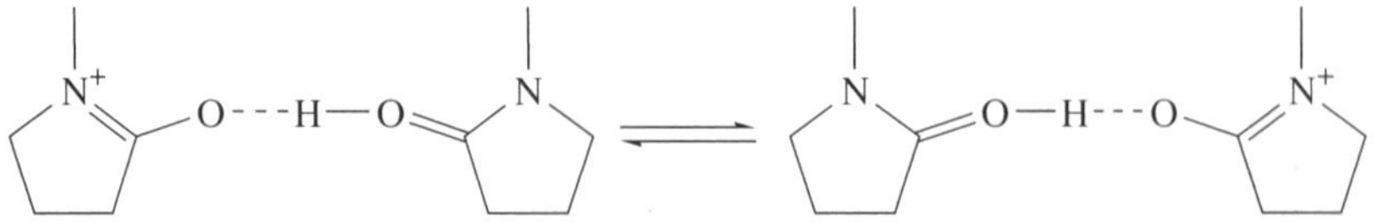

$AuCl_4^-$ $AuCl_4^-$

49 **50**

decision in favor of the symmetrical structure **48** is given by the ESCA results: Only one signal at 399.8 eV is found for N 1s; **49** and **50** should have two signals. Analogous results were obtained for vinylogous lactams (enaminoketones).

11.3. 1s Core Energies in Planar and Distorted Lactams and Amides

Gas-phase XPS was used as a probe for discerning the nature of amide resonance as a function of distortion. In this study by Greenberg et al.,[96] N 1s, O 1s, and some C 1s core ionization energies of a group of planar and distorted amides and lactams were investigated in order to gain a measure of the relative atomic charges. The study focused on distorted lactams and amides as well as suitable model lactams, amines, and ketones. Some experimental data are summarized in Table 9.10.

In general, the high values for N 1s and the low values for O 1s in planar amide (lactam) linkages compared to those in model amines and ketones are consistent with traditional resonance arguments. The N 1s and O 1s data for the distorted lactams 1,3-di-*t*-butylaziridin-2-one (**17**) and 1-azabicyclo[3.3.1]nonan-2-one (**19**) indicate a reduced positive charge on nitrogen and a reduced negative charge on oxygen in accordance with restricted amide resonance. They are also consistent with other spectroscopic data for distorted lactams. The carbonyl C 1s ionization energies are lower in distorted than in planar lactams.

TABLE 9.10. Core Ionization Energies [eV] of Amides and Lactams*

Compound	O 1s	N 1s	C(1) 1s
Formamide	537.71	406.33	294.56
N,N-Dimethylformamide	537.02	405.91	292.17
N,N-Dimethylacetamide	536.60	405.55	291.67
1-Pyrrolidinecarboxaldehyde (**52**)	536.84	405.52	
N,2′-Dimethylacetanilide (**26**)	536.35	405.33	
1,3-Di-*t*-butylaziridin-2-one (**17**)	537.36	405.00	290.69
Azetidin-2-one	537.32	405.76	291.80
Pyrrolidin-2-one	536.98	405.62	291.35
1-Methyl-pyrrolidin-2-one	536.69	405.45	291.48
1-*n*-Butyl-pyrrolidin-2-one (**53**)	536.58	405.20	
1-Azabicyclo[3.3.1]nonan-2-one (**19**)	536.67	405.07	290.81

* From reference 96.

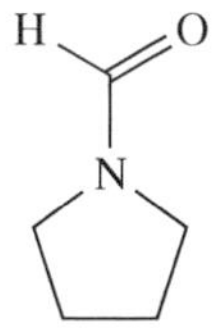

H O N

O N *t*-Bu *t*-Bu

N O

N O

52 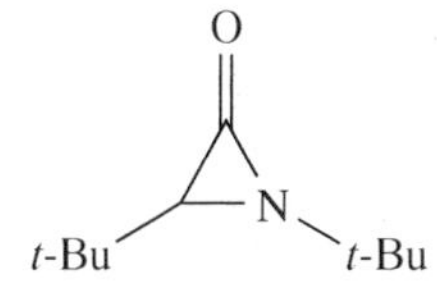 **17** 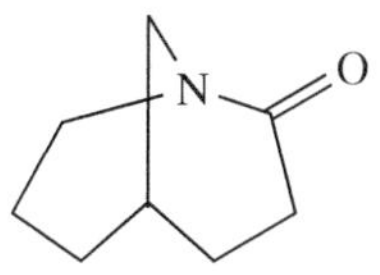 **19** 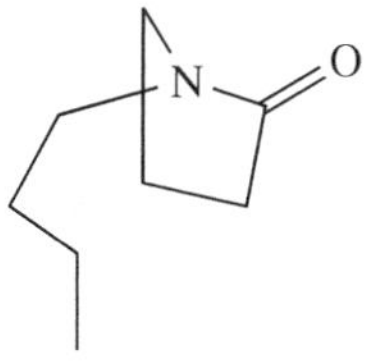 **53**

The XPS data also suggest that the C–O bond in amides is less polar than in ketones.

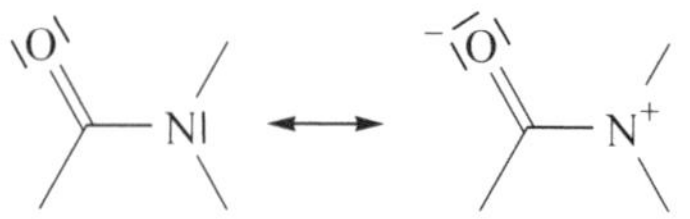

A **B**

In Table 9.11 the N 1s ionization energies of amides and lactams are compared with those of the corresponding amines. The observation that N 1s core energies in amides are considerably higher than those in amines, has been rationalized in terms of the resonance contribution from structure **B**.[97] This is also reflected in the relation between N 1s core energies and IR carbonyl frequencies.[98] Comparison of N 1s in, e.g. *N,N*-dimethylacetamide (405.56 eV) with that estimated for ethyldimethylamine (404.66 eV) indicates that it is much more difficult to remove a core electron from the amide nitrogen. This is consistent with a significant contribution from structure **B**, although part of the effect may be due to the electronegativity of the carbonyl group to which N is attached in the amide.

A comparison between O 1s core ionization energies in amides (or lactams) and those in the corresponding aldehydes and ketones is the most striking. We may, for example, compare the O 1s core energy in *N,N*-dimethylacetamide (536.60 eV) and that for the model ketone 3-methyl-2-butanone (537.66 eV) as well as pyrrolidin-2-one (536.98 eV) and pentan-3-one (537.73 eV). The amide O 1s core ionization energy is ca. 1.0 eV lower than that of the ketone despite the presence of the electronegative nitrogen in place of carbon. This is also consistent with significant contribution from resonance formula **B**.

Azetidin-2-one is absolutely planar in the gas-phase and solid state. The standard enthalpy of formation of this compound has recently been determined

TABLE 9.11. Comparison of N 1s Ionization Energies [eV] Between Amides or Lactams and Amines

Amide or Lactam	N 1s	Amine	N 1s	Δ
$HCONH_2$	406.35	CH_3NH_2	405.15	1.20
$HCONHCH_3$	406.12	$(CH_3)_2NH$	404.92	1.20
$HCON(CH_3)_2$	405.87	$(CH_3)_3N$	404.81	1.06
$CH_3CON(CH_3)_2$	405.56	$C_2H_5N(CH_3)_2$	(404.66)	0.90
$CH_3CON(C_2H_5)_2$	405.26	$(C_2H_5)_3N$	404.35	0.91
Aziridinone	406.13[a]	aziridine	404.96	1.17
Azetidin-2-one	405.76	azetidine	404.70	1.06
Pyrrolidin-2-one	405.62	pyrrolidine	404.60	1.02
1-Methyl-pyrrolidin-2-one	405.45	pyrrolidine	404.60	0.85
1-Methyl-piperidin-2-one	405.40	piperidine	404.48	0.92

* From reference 96.

[a] Calculated value, from reference 94.

by combustion calorimetry and a strain energy of 119.4 kJ mol^{-1} was deduced from this value.[99] Ab initio MO calculations predict only a 8.4 kJ mol^{-1} loss in resonance stabilization for azetidin-2-one in contrast to a 50 kJ mol^{-1} loss in the three membered ring aziridinone.[100] Ab initio MO calculations of unsubstituted aziridinone indicate that this molecule has a pyramidal nitrogen. This suggests that the 1,3-di-*t*-butyl derivative **17** also has a pyramidal nitrogen since it allows a larger dihedral angle between the *t*-butyl groups. The N 1s core energy (405.00 eV) is the lowest of the amides and lactams studied, and the O 1s core ionization energy (537.36 eV) is the highest (except for formamide). For the model amine, 1,2-di-*t*-butylaziridine, the N 1s core energy was estimated to be 404.22 eV. This value is only 0.78 eV lower than that of aziridinone **17** in contrast to the 0.96 eV difference derived for undistorted amides and lactams. This difference is consistent with reduced contribution of resonance structure **B** to aziridinone **17**. It is also consistent with the observation that the carbonyl frequency of this lactam is ca. 13 cm^{-1} *higher* than that of the cyclopropanone, in contrast to the usual situation where the amide (lactam) carbonyl frequency is 50–75 cm^{-1} *lower* than that of the corresponding ketone. Reduced resonance in aziridinones is a consequence of (a) pyramidal geometry at nitrogen and (b) reduced contribution of **B**, since the corresponding zwitterionic aziridinone resonance structure would have cyclopropene character and thus be destabilized to a large enhancement in strain energy relative to the structure **A**.

The N 1s and O 1s core energies of 1-azabicyclo[3.3.1]nonan-2-one (**19**) were compared with those of the corresponding monocyclic lactam 1-butyl-pyrrolidin-2-one (**53**). Compounds **19** and **53** have the same number of C, N, and O atoms and are quite similar in substitution, differing only in the second ring, which leads to considerable distortion of the amide bond in **19**, which coincides with UPS observations.[45] The fact, that the N 1s core energy in the bridgehead lactam **19** is lower by 0.13 eV while the O 1s core energy is 0.09 eV higher, is consistent with the classical resonance theory of bonding in these molecules.

In 1-pyrrolidinecarboxaldehyde (**52**) interplay of various structural factors result in a markedly nonplanar, somewhat twisted amide linkage.[101] It is interesting to note that the >N–CO linkage in proline containing proteins such as collagen is manifestly planar despite the nonplanarity of the corresponding linkage in **52**. The experimental N 1s value (405.52 eV) is 0.08 eV lower than the value calculated for the undistorted linkage, and this is qualitatively consistent with resonance theory. The O 1s value (536.84 eV) is also slightly lower than the calculated value (536.88 eV), but the uncertainties in O 1s data are greater in magnitude. Although aldehyde **52** has a distorted amide linkage, its XPS data are not unambiguously interpretable in terms of reduced resonance.

Resonance structures **A** and **B** both depict carbon in a double-bonded, uncharged state. Thus, to a first approximation, one might anticipate that the carbonyl C 1s ionization energies should vary relatively little as the balance between these two resonance contributions changes as a result of amide distortion. It is noteworthy, however, that the C1s core ionization energies of the carbonyl carbons in amides and lactams are nearly 2 eV *lower* than those in the corresponding ketones, and in the distorted lactams **17** and **19** they are both lower than those of other lactams. In attempting to understand these results, a third resonance structure, **C**, is introduced, and the dominant structures to represent an amide are **B** and **C**. This coincides with the recent results of Wiberg and Breneman[102] that the carbonyl group in amides is best written as $C^+{-}O^-$. The polarity of the CO bond is, however, considerably lower than in ketones.

A ⟷ **B** ⟷ **C**

Greenberg and Moore[94] have recently computed core orbital energies for planar ground-state and rotational transition-state structures of formamide and *N,N*-dimethylacetamide using ab initio molecular orbital calculations at the 6–31G* level. The results are listed in Table 9.12. The values are corrected to reproduce the experimental data for the planar amides. The twisted conformers have pyramidal nitrogens with considerably lower core ionization energies (0.56 and 0.89 eV) than in the planar structures. This is consistent with the absence of resonance contributors similar to **B** in the transition states. The O 1s core ionization energies are lower in the planar conformers by 1.24 and 1.12 eV, respectively. The loss of **B** in the perpendicular structure diminishes the negative charge on oxygen making it harder to ionize. The carbonyl C 1s ionization energy is expected to increase with distortion as the contribution of **B** to the resonance hybrid diminishes. This, too, is observed (0.29 and 0.24 eV, respectively) although the effect is not large.

Distortion of the amide linkage thus decreases the core ionization energy of nitrogen and increases the core ionization energies of oxygen and the carbonyl carbon. The trends observed for bridgehead bicyclic lactams[93,96] are explained accordingly. Bridgehead bicyclic lactams like **19** and its homologs offer a systematic series for probing the effects of distortion of the amide linkage upon structure, energy, and reactivity. Depending on the size of the bicyclic system,

TABLE 9.12. Computed and Corrected Core 1s Ionization Potentials [eV] for Planar and Twisted Conformers of Formamide and *N,N*-Dimethylacetamide*

Molecule	N 1s	C 1s	O 1s
Formamide			
Planar	406.33	294.56	537.71
Twisted 90°	405.77	294.85	538.95
N,N-Dimethylacetamide			
Planar	405.55	291.67	536.60
Twisted 90°	404.66	291.91	537.72

* The twisted structures have pyramidal geometry at nitrogen. See reference 94.

resonance structures **A** and **B** can be either hindered or favored.[2] For 3,5,7-trimethyl-1-azaadamantan-2-one (**54**) which was synthesized recently a twist angle of 90.5° was determined by X-ray analysis.[104] Compound **54** is thus the most twisted amide, and its He-I PE spectrum indicates that it is better described as an aminoketone than a lactam or amide.[105]

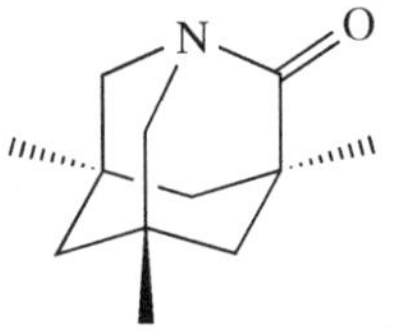

54

REFERENCES

1. Robin, M. B.; Bovey, F. A.; Basch, H. Molecular and electronic structure of the amide group. In *The Chemistry of Amides*; Zabicky, J. Ed.; J. Wiley & Sons: Chichester, 1970, Chapter 1, pp. 1–72.
2. Greenberg, A. Twisted Bridgehead Bicyclic Lactams. In *Structure and Reactivity*; Liebman, J. F., Greenberg, A. Eds.; VCH Publ.: New York, 1988, Chapter 4, pp. 139–178.
3. Dewar, M. J. S.; Zoebisch, E. G.; Healy, H. F.; Stewart, J. J. P. *J. Am. Chem. Soc.* 1985, **107**, 3902.
4. Sustmann, R.; Sicking, W. *Chem. Ber.* 1987, **120**, 1323.
5. Turner, D. W.; Baker, C.; Baker, A. D.; Brundle, C. R. *Molecular Photoelectron Spectroscopy*; Wiley-Interscience: London, 1970.
6. Eland, J. H. D. *Photoelectron Spectroscopy*, 2nd edn., Butterworth: London, 1984.
7. Baker, A. D.; Brundle, C. R. Eds. *Electron Spectroscopy: Theory, Techniques and Applications*; Vol. 1–5, Academic Press: London, 1977–1984.
8. Bock, H. *Angew. Chem.* 1977, **89**, 631.
9. Bock, H. *Angew. Chem. Int. Ed. Engl.* 1977, **16**, 613.
10. Heilbronner, E. Organic chemical photoelectron spectroscopy. In *Molecular Spectroscopy*; West, A. R. Ed.; Heyden: London, 1977, Chapter 20.
11. Koopmans, T. *Physica* 1934, **1**, 104.
12. Rademacher, P. Photoelectron spectra of amines, nitroso and nitro compounds. In *The Chemistry of Amino, Nitroso, Nitro and Related Groups*; Patai, S. Ed.; John Wiley & Sons Ltd.: Chichester, 1996, Vol. suppl. F2, pp. 159–204.
13. Siegbahn, K.; Nordling, C.; Johansson, G.; Hedman, J.; Heden, P. F.; Hamrin, K.; Gelius, U.; Bergmark, T.; Werme, L. O.; Manne, R.; Baer, Y. *ESCA Applied to Free Molecules*; North-Holland Publ. Co.: Amsterdam, 1969.
14. Barr, T.L. *Modern ESCA: The Principles and Praxis of X-Ray Photoelectron Spectroscopy*; CRC Press: Boca Raton, 1994.
15. Pauling, L. *The Nature of the Chemical Bond*; Cornell University Press: Ithaca, NY, 1960.
16. Yamada, S. *J. Org. Chem.* 1996, **61**, 941.

17. Rademacher, P. *Strukturen organischer Moleküle*; VCH Verlagsgesellschaft: Weinheim, 1987.
18. Hirota, E.; Sugisaki, R.; Nielsen, C. J.; Soerensen, G. O. *J. Mol. Spectrosc.* 1974, **49**, 251.
19. Kitano, M.; Kuchitsu, H. *Bull. Chem. Soc. Japan* 1974, **47**, 67.
20. Ostergard, N.; Christiansen, P. L.; Nielsen, O. F. *J. Mol. Struct.* 1991, **235**, 423.
21. Albright, T. A.; Burdett, J. K.; Whangbo, M.-H. *Orbital Interactions in Chemistry*; John Wiley & Sons: New York, 1985.
22. Rauk, A. *Orbital Interaction Theory of Organic Chemistry*; John Wiley & Sons, Inc.: New York, 1994.
23. Brundle, C. R.; Turner, D. W.; Robin, M. B.; Basch, H. *Chem. Phys. Lett.* 1969, **3**, 292.
24. Sweigart, D. A.; Turner, D. W. *J. Am. Chem. Soc.* 1972, **94**, 5592.
25. Kimura, K.; Katsumata, S.; Achiba, Y.; Yamazaki, T.; Iwata, S. *Handbook of HeI Photoelectron Spectra of Fundamental Organic Molecules*; Japan Scientific Societies Press: Tokyo, 1981.
26. Lindholm, E.; Bieri, G.; Fridh, C. *Internat. J. Quantum Chem.* 1978, **14**, 737.
27. Oliveros, E.; Rivière, M.; Teichteil, C.; Malrieu, J.-P. *Chem. Phys. Lett.* 1978, **57**, 220.
28. von Niessen, W. *Chem. Phys.* 1980, **45**, 47.
29. Meeks, J. L.; Maria, H. J.; Brint, P.; McGlynn, S. P. *Chem. Rev.* 1975, **75**, 603.
30. Klasinc, L.; McGlynn, S. P. The photoelectron spectroscopy of double-bonded CC, CN, NN and CO groups. In *The Chemistry of Double-bonded Functional Groups*; Patai, S. Ed.; John Wiley & Sons Ltd.: Chichester, 1989, Chapter 4, pp. 163–238.
31. Mines, G. W.; Thompson, H. W. *Spectrochim. Acta, Part A* 1975, **31A**, 137.
32. Mölder, U. H.; Koppel, I. A.; Pikver, R. J.; Tapfer, J. J. *Org. React. (Tartu)* 1988, **25**, 255.
33. Overman, L. E.; Taylor, G. F.; Houk, K. N.; Domelsmith, L. N. *J. Am. Chem. Soc.* 1978, **100**, 3182.
34. Jones, D.; Modelli, A.; Olivato, P. R.; Colle, M. D.; de Palo, M.; Distefano, G. *J. Chem. Soc., Perkin Trans.* 1994, **2**, 1651.
35. Watanabe, K.; Nakayama, T.; Mottl, J. *J. Quant. Spectrosc. Radiat. Transfer* 1962, **2**, 369.
36. Domelsmith, L. N.; Munchausen, L. L.; Houk, K. N. *J. Am. Chem. Soc.* 1977, **99**, 4311.
37. Domelsmith, L. N.; Munchhausen, L. L.; Houk, K. N. *J. Med. Chem.* 1977, **20**, 1346.
38. Colle, M. D.; Bertolasi, V.; de Palo, M.; Distefano, G.; Jones, D.; Modelli, A.; Olivato, P. R. *J. Phys. Chem.* 1995, **99**, 15011.
39. Müller, K.; Previdoli, F.; Desilvestro, H. *Helv. Chim. Acta* 1981, **64**, 2497.
40. Müller, K.; Previdoli, F. *Helv. Chim. Acta* 1981, **64**, 2508.
41. Rademacher, P. En-amine. In *Methoden der Organischen Chemie: Houben/Weyl*; Schaumann, E. Ed.; Georg Thieme Verlag: Stuttgart, 1993, Vol. E15, pp. 598–717.
42. Rademacher, P.; Würthwein, E.-U. *J. Mol. Struct.* (*Theochem*) 1986, **139**, 315.
43. Irsch, G.; Rademacher, P. (unpublished results).

44. Treschanke, L.; Rademacher, P. *J. Mol. Struct.* (*Theochem*) 1985, **122**, 35.
45. Treschanke, L.; Rademacher, P. *J. Mol. Struct.* (*Theochem*) 1985, **122**, 47.
46. Lengyel, I.; Sheehan, J. C. *Angew. Chem.* 1968, **80**, 27.
47. Sheehan, J. C.; Lengyel, I. *J. Am. Chem. Soc.* 1964, **86**, 1356.
48. Aitken, R. A.; Gosney, I.; Farries, H.; Palmer, M. H.; Simpson, I.; Cadogan, J. I. G.; Tinley, E. *J. Tetrahedron* 1985, **41**, 1329.
49. Masclet, P.; Grosjean, D.; Mouvier, G.; Dubois, J. *J. Electron Spectrosc. Relat. Phenom.* 1973, **2**, 225.
50. Winkler, F. K.; Dunitz, J. D. *Acta Cryst., Sect. B* 1975, **31**, 276.
51. Aubry, A.; Protas, J.; Thong, C. M.; Marraud, M.; Neel, J. *Acta Cryst., Sect. B* 1973, **29**, 2576.
52. Hall, H. K.; Zbinden, R. *J. Am. Chem. Soc.* 1958, **80**, 6428.
53. Cook, D. *Can. J. Chem.* 1961, **39**, 31.
54. Rademacher, P. *Chem. Soc. Rev.* 1995, **24**, 143.
55. Woydt, M.; Rademacher, P.; Kaupp, G.; Sauerland, O. *J. Mol. Struct.* 1989, **192**, 141.
56. Woydt, M.; Rademacher, P. *J. Mol. Struct.* 1992, **265**, 103.
57. McAlduff, E. J.; Lynch, B. M.; Houk, K. N. *Can. J. Chem.* 1978, **56**, 495.
58. Klasinc, L.; Novak, I.; Sabljic, A.; McGlynn, S. P. *Int. J. Quantum Chem., Quantum Biol. Symp.* 1986, **13**, 251.
59. Petrachenko, N. E.; Vovna, V. I.; Furin, G. G. *Zh. Org. Khim.* 1992, **28**, 1218.
60. Petrachenko, N. E.; Vovna, V. I.; Furin, G. G. *J. Org. Chem. USSR* (*Engl. Translation*) 1992, **28**, 953.
61. Gal, J.-F.; Geribaldi, S.; Pfister-Guillouzo, G.; Morris, D. G. *J. Chem. Soc., Perkin Trans.* 1985, **2**, 103.
62. Fringuelli, F.; Marino, G.; Taticchi, A.; Distefano, G.; Colonna, F. P.; Pignataro, S. *J. Chem. Soc., Perkin Trans.* 1976, **2**, 276.
63. Klessinger, M.; Rademacher, P. *Angew. Chem.* 1979, **91**, 885.
64. Klessinger, M.; Rademacher, P. *Angew. Chem. Int. Ed. Engl.* 1979, **18**, 826.
65. Brown, R. S.; Jorgensen, F. S. Conformational Analysis by Photoelectron Spectroscopy. In *Electron Spectroscopy: Theory, Techniques and Applications*; Brundle, C. R., Baker, A. D., Eds.; Academic Press: London, 1984, Vol. 5, Chapter 1, pp. 1–122.
66. Maier, J. P.; Turner, D. W. *Dis. Faraday Soc.* 1972, **54**, 149.
67. Maier, J. P.; Turner, D. W. *J. Chem. Soc., Faraday Trans. 2* 1973, **69**, 521.
68. Nakagaki, R.; Kobayashi, T.; Nagakura, S. *Bull. Chem. Soc. Jpn.* 1980, **53**, 901.
69. Szepes, L.; Distefano, G.; Pignataro, S. *Ann. Chim.* 1974, **64**, 159.
70. Dougherty, D.; Younathan, E. S.; Voll, R.; Abdulnur, S.; McGlynn, S. P. *J. Electron. Spectros. Relat. Phen.* 1978, **13**, 379.
71. Ajò, D.; Granozzi, G.; Tondello, E.; Del Pra, A. *Biopolymers* 1980, **19**, 469.
72. Ajò, D.; Casarin, M.; Busetti, V.; Granozzi, G.; Mayoral, J. A.; Ottenheijm, H. C. J. α,β-Dehydroamino acids as constituents of naturally occuring peptides. In *Chemical Reactivity and Biological Activity*; Stezowski, J. J., Huang, J.-L., Shao, M.-C., Eds.; Oxford University Press: Oxford, 1988, pp. 68–71.

73. Ajò, D.; Granozzi, G.; Ciliberto, E.; Fragala, I. *J. Chem. Soc., Perkin Trans. 2* 1980, 483.
74. Ajò, D.; Casarin, M.; Granozzi, G.; Fragalla, I. *Tetrahedron* 1981, **37**, 3507.
75. Ongania, K. H.; Granozzi, G.; Busetti, V.; Casarin, M.; Ajò, D. *Tetrahedron* 1985, **41**, 2015.
76. Mayoral, J. A.; Cativiela, C.; Lopez, M. P.; Ajò, D.; De Zuane, F. *J. Crystallogr. Spectros. Res.* 1989, **19**, 993.
77. Richer, G.; Sandorfy, C.; Nascimento, M. A. C. *J. Electron. Spec. Relat. Phen.* 1984, **34**, 327.
78. Andreocci, M. V.; Devillanova, F. A.; Furlani, C.; Mattogno, G.; Verani, G.; Zanoni, R. *J. Mol. Struct.* 1980, **69**, 151.
79. Nelsen, S. F.; Blackstock, S. C.; Petillo, P. A.; Agmon, I.; Kaftory, M. *J. Am. Chem. Soc.* 1987, **109**, 5724.
80. Nelsen, S. F.; Thompson-Colon, J. A.; Kirste, B.; Rosenhouse, A.; Kaftory, M. *J. Am. Chem. Soc.* 1987, **109**, 7128.
81. Meeks, J. L.; Arnett, J. F.; Larson, D. B.; McGlynn, S. P. *J. Am. Chem. Soc.* 1975, **97**, 3905.
82. Meeks, J. L.; McGlynn, S. P. *J. Am. Chem. Soc.* 1975, **97**, 5079.
83. Isaksson, R.; Liljefors, T. *J. Chem. Soc., Perkin Trans. 2* 1980, 1815.
84. Ajò, D.; Casarin, M.; Granozzi, G.; Poli, A.; Parasassi, T. *J. Crystallogr. Spectrosc. Res.* 1982, **12**, 227.
85. Dougherty, D.; Wittel, K.; Meeks, J.; McGlynn, S. P. *J. Am. Chem. Soc.* 1976, **98**, 3815.
86. Vondrák, T.; Cauletti, C. *Spectrochim. Acta* 1988, **44A**, 289.
87. Treschanke, L.; Rademacher, P. *J. Mol. Struct.* 1985, **131**, 61.
88. Irsch, G.; Rademacher, P. *J. Mol. Struct.* 1989, **196**, 181.
89. Moehrle, H.; Kamper, C.; Herbke, J.; Nowak, H. J.; Wendisch, D.; Storp, S. *Monatsh. Chem.* 1978, **109**, 1295.
90. Binder, H. *Z. Naturforsch. B* 1977, **32**, 249.
91. Jolly, W. L.; Bomben, K. D.; Eyermann, C. J. *Atom. Data Nucl. Data Tables* 1984, **31**, 433.
92. Cavell, R. G.; Allison, D. A. *J. Am. Chem. Soc.* 1977, **99**, 4203.
93. Greenberg, A.; Moore, D. T.; DuBois, T. D. *J. Am. Chem. Soc.* 1996, **118**, 8658.
94. Greenberg, A.; Moore, D. T. *J. Mol. Struct.* 1997, **413**, 477.
95. Vansweevelt, H.; Vanquickenborne, L.; Van der Vorst, W.; Parmentier, J.; Zeegers-Huyskens, T. *Chem. Phys.* 1994, **182**, 19.
96. Greenberg, A.; Thomas, T. D.; Bevilacqua, C. R.; Coville, M.; Ji, D.; Tsai, J. C.; Wu, G. *J. Org. Chem.* 1992, **57**, 7093.
97. Lindberg, B. J.; Hedman, J. *Chem. Scr.* 1974, **7**, 155.
98. Tsuchiya, S.; Seno, M. *J. Org. Chem.* 1979, **44**, 2850.
99. Roux, M. V.; Jiménez, P.; Dávalos, J. Z.; Castano, O.; Molina, M. T.; Notario, R.; Herreros, M.; Abboud, J.-L. M. *J. Am. Chem. Soc.* 1996, **118**, 12735.
100. Greenberg, A.; Chiu, Y.-Y.; Johnson, J. L.; Liebman, J. F. *Struct. Chem.* 1991, **2**, 117.

101. Lee, S. G.; Hwang, K. W.; Bohn, R. K.; Hillig, K. W.; Kuczkowski, R. L. *Acta Chem. Scand.* 1988, **A42**, 603.

102. Wiberg, K. B.; Breneman, C. M. *J. Am. Chem. Soc.* 1992, **114**, 831.

103. Kirby, A. J.; Komarov, I. V.; Wothers, P. D.; Feeder, N. *Angew. Chem.* 1998, **110**, 830.

104. Kirby, A. J.; Komarov, I. V.; Wothers, P. D.; Feeder, N. *Angew. Chem. Int. Ed. Engl.* 1998, **37**, 785.

105. Kirby, A. J.; Komarov, I. V.; Kowski, K.; Rademacher, P. *J. Chem. Soc., Perkin Trans. 2* 1999, 1313.

CHAPTER 10

THE ROLE OF AMIDES IN THE NONCOVALENT SYNTHESIS OF SUPRAMOLECULAR STRUCTURES IN SOLUTION, AT INTERFACES AND IN SOLIDS

G. TAYHAS R. PALMORE
Department of Chemistry, University of California

JOHN C. MACDONALD
Department of Chemistry, Northern Arizona University

1. INTRODUCTION

Hydrogen bonding between amides is perhaps the most important noncovalent intra- and intermolecular force in determining recognition, geometry, and modes of association between biological molecules.[1] The role of amide groups in promoting the assembly of molecules into larger structures through hydrogen-bonding interactions was recognized early in the twentieth century. Pauling first introduced the importance of hydrogen bonding between amides in the first edition of *The Nature of the Chemical Bond* in 1939.[2] The acceptance of hydrogen bonding as a strong cohesive force between amides played a key role in understanding the structure and the biological function of the α-helix and the β-pleated sheet of proteins in 1951,[3] as well as the Watson–Crick base-pairing in the DNA double-helix in 1953.[4]

Advances in research show that amides are able to organize molecules precisely and selectively into larger, more complex supramolecular structures, having properties that are often quite different from those of the individual molecules. Consequently, a number of research programs have been established in the last two decades that use the amide functional group as an element for designing both natural and non-natural supramolecular structures that self-

The Amide Linkage: Selected Structural Aspects in Chemistry, Biochemistry, and Materials Science,
Edited by Arthur Greenberg, Curt M. Breneman, and Joel F. Liebman
ISBN 0-471-35893-2

assemble through hydrogen-bonding interactions. One of the goals of these studies is to explain how the structure and properties of individual molecules are related to those of supramolecular aggregates composed of many molecules. This difference in behavior between molecules and their aggregates can be illustrated by examining other aggregates containing molecules that are not amides, but the behavior of which is understood. For example, the properties and function of a micelle differ considerably from those of the individual long-chain sulfonates from which it is made. Analogously, does the behavior of an aggregate of amides differ from that of the individual amide? How does the structure of supramolecular aggregates of amides relate to their function? Can the structures of aggregates of amides be tailored at the molecular level to produce specific properties and functions?

The majority of research, investigating the self-assembly of amides in the past two decades has focused primarily on two areas. The first is the design of supramolecular aggregates in solution that show catalytic activity, recognize specific substrates, self-replicate, or exhibit photoinduced electron transfer. A number of reviews have appeared in the literature on these subjects recently.[5–10] The second area of research is the design of crystalline materials, or crystal engineering. Crystal engineering is a relatively young field of research.[11] Research involving the design of organic and organometallic crystals has grown exponentially in recent years, with hundreds of papers having appeared since Schmidt first introduced the term "crystal engineering" in 1971.[12] Much of this work with amides has been summarized in several reviews.[13–18] In addition to the work with amides in solution and in the solid state, a number of groups are exploring the role of amides in supramolecular chemistry at surfaces and in liquid-crystalline media. An underlying theme of this research is to understand the role of the amide functional group in creating and controlling supramolecular structure and order in two dimensions.

The supramolecular chemistry of amides, which encompasses the behavior of amides in all phases and media, now has progressed to the point that all of the structural motifs of amides—as defined by different patterns of hydrogen bonds[15,19–23]—are known, well characterized, and can be generated at will. For example, amides have been used successfully to form a wide variety of supramolecular nanoscale architectures that are described as capsules[24,25] and spheres,[7] rosettes,[26,27] rods,[28] helices,[16,29] tapes,[15,30–35] ribbons,[36,37] tubes,[38–41] channels,[13,14,42,43] and sheets or layers.[19,33,42–54] In addition, amides have been used to create three dimensional motifs that generate longer-range order (e.g., diamondoid lattices[55] in crystals and liquid crystals[56]). Supramolecular chemists now have the tools necessary to advance beyond the creation of specific supramolecular structures and to begin tailoring the *function* of these structures to specific applications. As the field of supramolecular chemistry develops, researchers are beginning to determine what types of properties and functions supramolecular aggregates exhibit, and how these properties and functions differ from those of the individual molecules that comprise them.

A number of studies have appeared recently, featuring aggregates of amides that display a wonderfully broad range of properties and functions. The goal of this chapter is to give a brief survey of some of the different types of supramolecular properties and functions that have been achieved with amides. It is not our intent to provide a comprehensive review of the literature on the supramolecular chemistry of amides. Instead, this chapter highlights a few selected examples of supramolecular materials in solution, at interfaces and on surfaces, in liquid crystals and in solids. These examples demonstrate that a variety of different properties and functions can be deliberately controlled and modified with the appropriate choice of amide as the fundamental building block. The potential of amides for the design of materials with specific functions has been heralded for years. The studies outlined here establish quite clearly that this potential is in fact being realized.

Following this introduction, the chapter is divided into two sections. In Section 2, we describe the amide functional group, chemical structures that we consider to be amides, patterns of hydrogen bonds and supramolecular aggregates formed by amides in the solid state, and selected information from previous studies of amides that are relevant to the discussion. Section 2 highlights studies of crystalline solids that explore different strategies to create specific supramolecular structures that can be characterized by X-ray diffraction, but that generally do not exhibit useful properties or functions. In Section 3, we focus on examples of supramolecular aggregates in different media and phases that display useful properties and functions. We describe the specific properties and functions these systems exhibit, and discuss their significance in the context of the development of the field of supramolecular chemistry.

2. THE AMIDE FUNCTIONAL GROUP

For the purpose of this chapter, a compound is considered to contain an amide group if it has a carbonyl group bonded directly to a nitrogen atom, regardless of what other atoms are connected to the carbonyl group and nitrogen atom. This definition includes a broad range of different structures, many of which contain functional groups that normally are not classified as amides. For example, an amide is a subset of groups such as imides or ureas. Fig. 10.1 gives examples of different functional groups that we have included as amides.

Amides serve as a fundamental building block from which a variety of different supramolecular structures can be assembled. The type of structure formed depends on several different factors, often in combination. These factors include the geometry of the amide group, the number and type of substituents attached to the nitrogen atom, the number of different amide groups that are present in a molecule, and the proximity of neighboring functional groups that may alter the hydrogen-bonding capacity of amides both sterically and electronically. Entropic effects also contribute significantly to the process of self-assembly, especially when more than one amide group is present on a

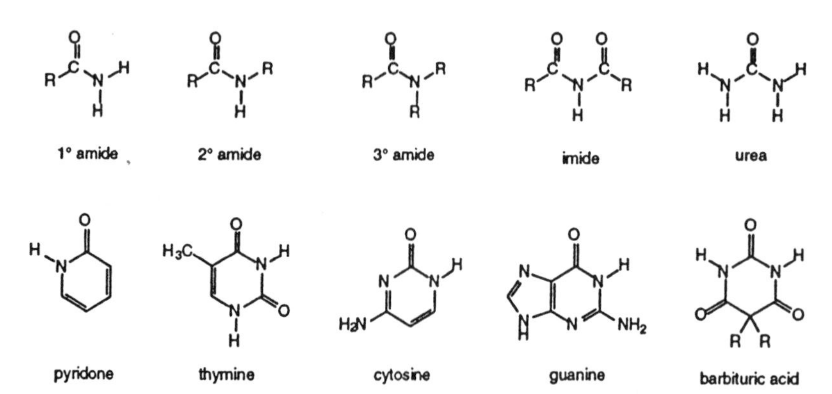

FIGURE 10.1. Examples of different functional groups considered as amides.

flexible molecule. To illustrate this point, consider the role that amides play in controlling structure in large biological molecules such as proteins. Intramolecular hydrogen bonding between many different amide groups is responsible for generating and stabilizing structures such as α-helices, 3_{10}-helices, parallel and antiparallel β-sheets, β-bulges and β-turns that define the tertiary structure of proteins.[1]

While the structural motifs generated by amides via *intramolecular* hydrogen bonding in biological molecules has been recognized for decades, the first serious attempt to define systematically structural motifs generated by amides through *intermolecular* hydrogen bonding was that of Leiserowitz and Schmidt in 1969.[19] This seminal study examined patterns of hydrogen bonds between primary amides in organic crystals, and established that primary amides form predominantly one type of hydrogen-bonded structure, as shown in Fig. 10.2. This structure, which might be described as a ribbon, contains two different types of hydrogen bonds that give a ring motif and a chain motif. Together, these two motifs form an infinite chain-of-rings. Subsequent studies by Leiserowitz on the crystal packing of achiral secondary *N*-methyl amides,[20] chiral secondary amides,[22,23] cocrystals between amides and dicarboxylic acids,[57] and a computational study[21] on the packing patterns of amides revealed that secondary amides simply form chains, as shown in Fig. 10.2. Tertiary amides lack any acidic hydrogen atoms on heteroatoms that can act as donors; they do not form hydrogen bonds with themselves unless additional functional groups capable of forming hydrogen bonds are present on the molecule. Taft has shown that tertiary amides act as very strong hydrogen-bonding acceptors in solution,[58] and that they often form hydrogen-bonded complexes in the presence of guest molecules that contain acidic hydrogen atoms.

The geometry of the two hydrogen atoms on primary amides determines the type of structure that is formed through self-aggregation or by aggregation with a guest molecule. Leiserowitz defines the orientation of these hydrogen atoms as having either *syn* or *anti* geometry (Fig. 10.3), depending on whether the hydrogen atom is located on the same or the opposite side of the C–N bond as

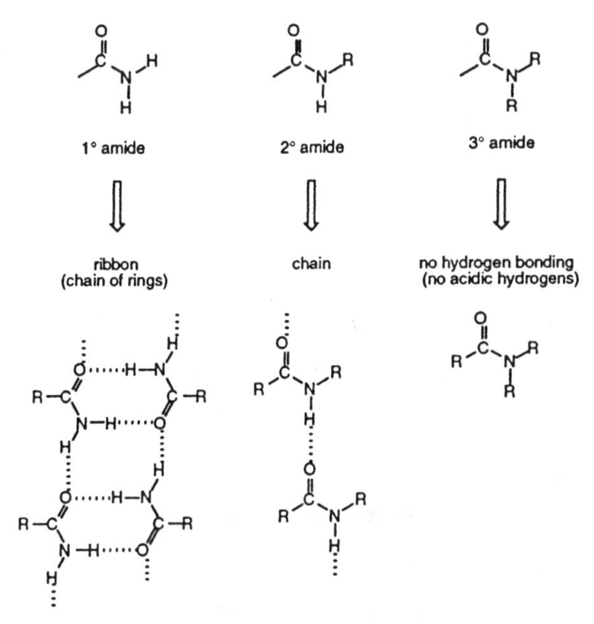

FIGURE 10.2. Patterns of hydrogen bonds formed by primary, secondary, and tertiary amides.

the carbonyl group.[19] When primary amides self-associate, hydrogen bonding between the *syn* hydrogen atom and carbonyl group forms dimers joined by eight-membered rings. This motif is analogous to the ring motif formed between two carboxylic acids. The *anti* hydrogen atoms make single point contacts with carbonyl groups on neighboring amides to form chains. Since acyclic secondary amides prefer to adopt a conformation that places the amido hydrogen atom in the *anti* position, acyclic secondary amides generally form chains. MacDonald and Whitesides have shown that cyclic secondary diamides in which the amido hydrogen atom is restricted to the *syn* position generally form hydrogen-bonded rings, although other patterns of hydrogen bonds were also observed for these compounds.[15]

Etter, MacDonald, Bernstein and Davis have developed a nomenclature called graph sets to describe patterns of hydrogen bonds.[59–61] Graph sets provide a convenient language with which to identify and compare patterns of hydrogen bonds. Surprisingly, even the most complicated networks of hydrogen bonds can be reduced to combinations of four simple patterns that are specified by a designator: chains (**C**), rings (**R**), intramolecular hydrogen-bonded patterns (**S**), and other finite patterns (**D**). Each designator contains a subscript and superscript that indicate the number of donors, **d**, and acceptors, **a**, respectively. In addition, the number of atoms in the pattern, **n**, including hydrogen atoms, is specified in parenthesis. For example, the graph set for the eight-membered ring formed by primary amides is $R_2^2(8)$, as shown in Fig. 10.3. The chains of hydrogen bonds linking the rings have the graph set $C_1^1(4)$, or more simply C(4) (the superscript

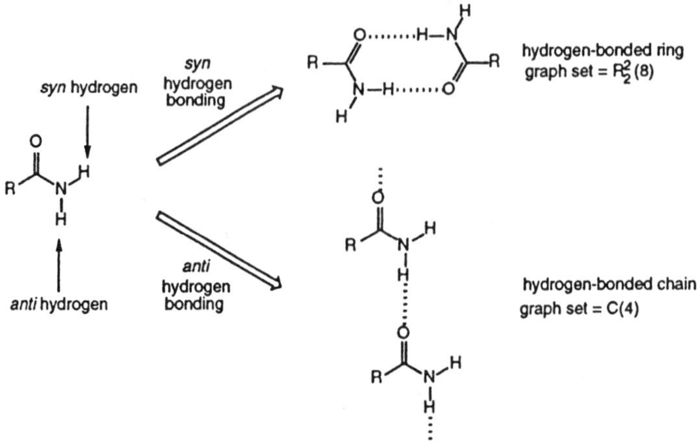

FIGURE 10.3. Ring and chain motifs generated by hydrogen-bonding interactions involving the *syn* and *anti* hydrogen atoms, respectively, of primary and secondary amides.

and subscript are omitted when there is just one donor and one acceptor). Combining the two graph sets gives $R_2^2(8)C(4)$, which describes completely the chain-of-rings pattern formed by primary amides. The chains formed by secondary amides simply have the graph set C(4). We will make use of graph sets occasionally throughout the rest of this chapter when comparing or describing patterns of hydrogen bonds.

Early studies on the patterns of hydrogen bonds between amides and the analysis of their graph sets found that rings and chains are recurring structural motifs—a term referred to by Bernstein and Davis as "hydrogen-bond pattern functionalities"[61]—with different properties that can be generated deliberately with the appropriate choice of amide. For example, rings and chains can be used to control symmetry within a hydrogen-bonded structure. A ring motif allows a center of symmetry between a pair of amide molecules. Indeed, Leiserowitz has shown that dimers of primary amides frequently sit on centers of symmetry within crystals.[19] On the other hand, a chain motif precludes the formation of a center of symmetry, thereby creating a polar arrangement of molecules. This feature is attractive from the standpoint of engineering materials for applications such as second harmonic generation, where the absence of a center of symmetry in the bulk material is essential for expression of second-order nonlinear optical behavior.[62] Several examples of studies are presented later in this chapter where researchers attempt to use the symmetry of individual aggregates to control the symmetry of bulk materials.

2.1. Studies of Hydrogen-bonded Structures in the Solid State

Interest in the development of new materials, the ease with which X-ray crystal structures are determined, and the availability of crystallographic databases such as the Cambridge Structural Database[63] have produced a large body of literature in the last twenty years on the structures of amides and their aggregates in the solid state. More recently, a number of research groups use amides to exercise control on the packing arrangements of molecules in the solid state. The goal of many of these studies is to engineer crystalline materials that exhibit particular properties or functions such as porosity and zeolitic behavior, nonlinear optical behavior, conductivity of electrons or protons, magnetism, or selective enclathration of guest molecules during crystallization. All these studies use the ability of amides to form dimers and chains to lock molecules into aggregates with specific structures (e.g., chains, tapes, ribbons, helices, or layers). Designs for supramolecular aggregates have included a range of different amide–amide interactions, as well as interactions between amides and other functional groups or guest molecules.

The present state of the field of crystal engineering is such that a wide variety of supramolecular aggregates have, in fact, been created, and their crystal packing patterns and supramolecular structure characterized by X-ray diffraction and other solid-state analytical techniques. The goal of achieving functional materials from organic crystals, however, has not yet been realized, especially when compared to materials based on organic polymers that are now available. While crystalline organic materials with many interesting structures have been produced, none are currently used for device applications or sold commercially as materials that serve a specific function (e.g., organic molecular sieves). One area in which crystal engineering has had a limited commercial impact is the pharmaceutical industry, where the study of polymorphism in crystalline materials has become increasingly important in the manufacture and marketing of drugs.[64–69] For example, polymorphs of pharmaceuticals are regarded as different materials in courts of law. Consequently, pharmaceutical companies spend considerable resources isolating and characterizing polymorphs of drugs that form solids. Polymorphs often differ in physical properties that are important pharmaceutically, such as rates of dissolution, solubility, stability, melting point, taste, and color.

Despite the lack of functional crystalline materials based on amides, it is worthwhile to examine the types of supramolecular structures that have been achieved in organic solids for several reasons. First, from the standpoint of developing strategies for the self-assembly of molecules in any medium, it is useful to understand the evolution in strategies for the design of aggregates, which range from those with finite structure, to those with one-dimensional, two-dimensional, and three-dimensional structure. Second, from the standpoint of designing functional materials, it is instructive to understand the different types of structures that have been made, especially those incorporating more than one type of molecule that can be interchanged with another (i.e., cocrystals). Third, and perhaps the most important, many of the supramolecular aggregates first characterized in crystals have been used successfully as models, or platforms, upon which to build analogous structures in solution, on surfaces and at interfaces. Accordingly, selected examples of several different types of supramolecular aggregates studied in the solid state are presented in Sections 2.2.–2.4. All these examples represent supramolecular aggregates whose structures were determined unambiguously by using X-ray methods. These structures illustrate the variation in supramolecular architectures that can be achieved deliberately using amides. Most of the supramolecular structures discussed have been studied because of their potential as building blocks for crystal engineering.

The structures discussed in this section represent only a few of the compounds that have been examined. Our goal is to provide a broad sample of the variety of approaches being used to understand and control supramolecular aggregation in crystals. The compounds discussed are grouped into four classes of supramolecular aggregates: (1) finite, or zero-dimensional, (2) one-dimensional, (3) two-dimensional, and (4) three-dimensional.

2.2. Finite Aggregates

A number of groups have used hydrogen bonds to form discrete, or finite, aggregates of amides. Etter has provided several examples of systems in which two different compounds form complexes that cocrystallize together.[18] Most of Etter's work focused not on generating motifs for engineering crystalline materials, but instead these studies probed the selectivity and preferences of different functional groups in forming hydrogen bonds with other functional groups during crystallization. Fig. 10.4 shows two examples of cocrystalline complexes that were studied. The structure on the left shows the complex that forms when 1,3-*bis*(3-nitrophenyl)urea and *N,N′*-dimethylnitroaniline are cocrystallized. The diarylurea acts as a strong hydrogen-bonding donor that cocrystallizes with a variety of molecules containing hydrogen-bonding acceptor groups (e.g., ketones, ethers, alcohols, phosphine oxides, and compounds with nitro groups).[70] The structure on the right is that of a cocrystal between diacetimide and 4-nitrophenol. In this example, Etter showed that acyclic imides such as diacetimide use one of the two carbonyl groups of the imide to dimerize

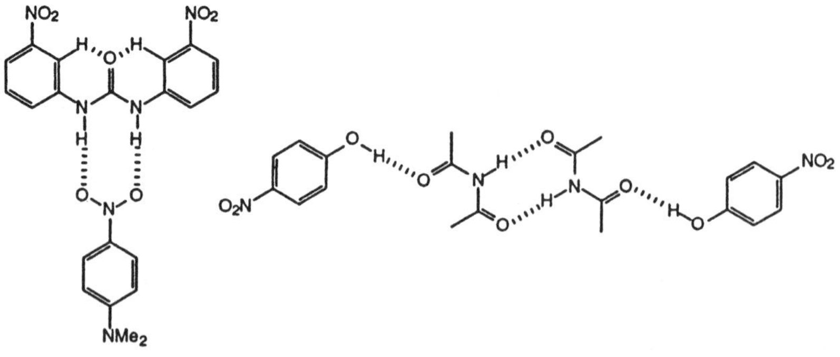

FIGURE 10.4. A 1 : 1 complex between 1,3-*bis*(3-nitrophenyl)urea and *N,N′*-dimethylnitroaniline (left)[70] and a 2 : 2 complex between diacetimide and 4-nitrophenol (right).[71]

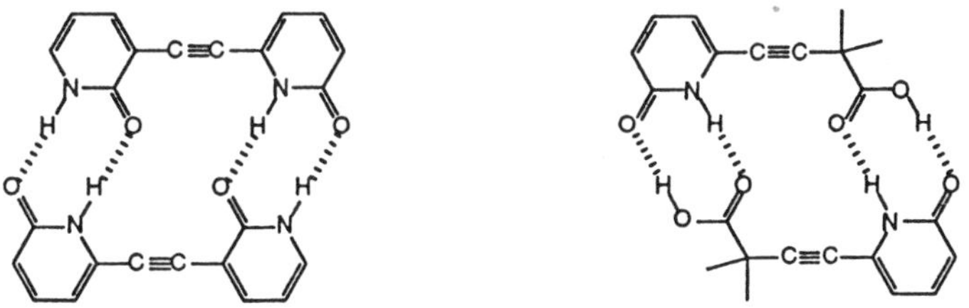

FIGURE 10.5. Dimers of an asymmetric di-2-pyridone (left)[72] and a carboxy-2-pyridone (right).[73]

[$R^2_2(8)$], leaving the second carbonyl group free to bind to phenols and other guest molecules that possess acidic hydrogen atoms.[71]

Wuest has provided an elegant example of controlled self-assembly of di-2-pyridones in which two different structures were generated purposefully simply by changing the relative orientation of the two carbonyl groups. The structure on the left in Fig. 10.5 is a dimer of unsymmetric di-2-pyridones in which both carbonyl groups point towards one end of the molecule. Wuest showed that symmetric di-2-pyridones preclude the formation of dimers and form chains of molecules instead (an example of which is illustrated in Fig. 10.8 in Section 2.3. on one-dimensional aggregates).[72] The 2-pyridones in both the dimer and the chain structures associate through an $R^2_2(8)$ interaction. Lightner has also demonstrated the formation of a dimer of similar type.[73] This structure differs from the di-2-pyridone dimer in that the two molecules are held together by hydrogen bonds between 2-pyridone and carboxylic acid groups. The structure of this complex is shown on the right in Fig. 10.5.

Hamilton has created many different complexes by cocrystallizing *bis*-acylaminopyridine receptors with dicarboxylic acids.[16] These complexes form a broad range of aggregates whose structures can be changed as a function of the separation and relative orientation of the acylaminopyridine units. In one study, Hamilton demonstrated that discrete cyclic complexes containing two receptors and two diacids could be formed and characterized, both in solution and in the solid state (Fig. 10.6).[74] This structure was part of a larger effort in the solid-state community to design aggregates with molecular weights in the intermediate range ($1-20 \times 10^3$ Da) between those of individual small molecules and polymers or proteins.

Strategies for creating aggregates containing many molecules and having relatively high molecular weights and stability have been explored by several research groups. Whitesides produced a wonderful example of a hexameric rosette structure containing three molecules of *bis*(4-*tert*-butylphenyl)melamine and three molecules of 5,5-diethylbarbituric acid (barbital) joined by eighteen hydrogen bonds.[27] The rosette structure is shown at the top of Fig. 10.7. In addition to this remarkable example of a rosette motif, cocrystals with diphenylmelamines and barbital also form two one-dimensional motifs called linear tapes and crinkled tapes that will be discussed later in Section 2.3. Whitesides has shown that the type of motif formed can be controlled sterically

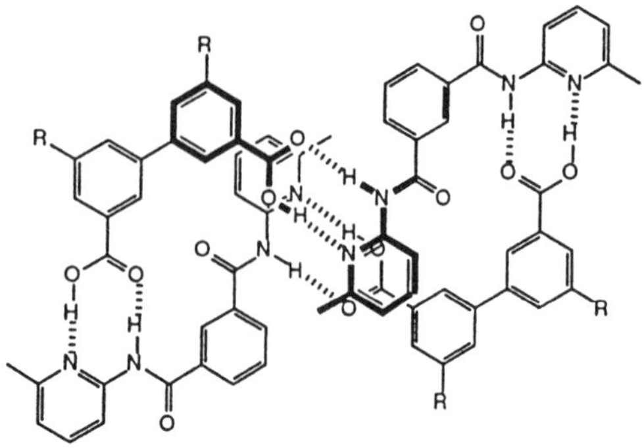

FIGURE 10.6. A cyclic tetramer of molecules based on a 2 : 2 complex between a *bis*-acylaminopyridine receptor and a dicarboxylic acid.[74]

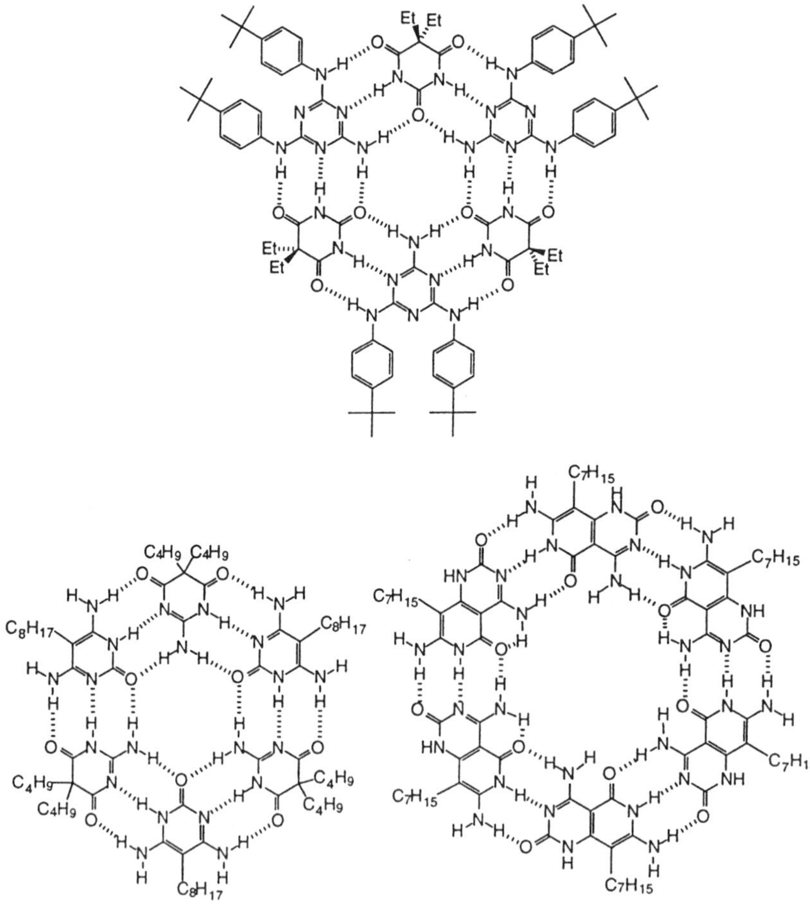

FIGURE 10.7. Rosette formed by the 3 : 3 complex between *N,N′*-bis(4-*tert*-butylphenyl)melamine and barbital (top).[27] Hypothetical structure of the rosette as predicted by Lehn (bottom left).[37] Rosette formed by a DNA-base hybrid (bottom right).[26]

by placing substituents of different sizes at the *para* positions on the phenyl rings of the melamine molecule. Small substituents (e.g., methyl groups) and medium substituents (e.g., carboxymethyl groups) produce linear and crinkled tapes, while large substituents such as *tert*-butyl groups give the rosette motif.[27]

Other groups have worked towards developing systems that eliminate the formation of tapes. For example, Lehn experimented with changing the relative order of donors (D) and acceptors (A), thereby altering the complementarity on either side of the two components from [A-D-A/A-D-A] · [D-A-D/D-A-D] to [D-D-A/A-A-D] · [A-A-D/D-D-A]. The change in complementarity eliminated the possibility of forming tapes if the two components cocrystallized. Unfortunately, attempts at cocrystallization gave a mixture of crystals containing only the individual components as linear tapes. A hypothetical structure of the rosette as predicted by Lehn is shown in the bottom left of Fig. 10.7.

Mascal took an alternative approach to that of Whitesides and Lehn by using a single self-complementary molecule rather than two different complementary molecules. Mascal combined the A-A-D sequence of donors and acceptors of cytosine on one side of a molecule and the D-D-A sequence of guanine at an angle of 120° to each other to form the rosette shown in the bottom right of Fig. 10.7.[26] An interesting feature in the crystalline solid of this compound is that the rosettes coincide and form channels with a diameter of 10.5 Å. These channels extend throughout the crystal.

2.3. One-dimensional Aggregates

A number of different studies have linked amides into infinitely long aggregates with one or more contacts between molecules as a strategy to control molecular aggregation and thereby influence crystal packing. The simplest one-dimensional motif is a chain composed of amides, where each molecule is joined by a single hydrogen bond to neighboring molecules. Leiserowitz showed that simple chains form between secondary amides.[20,22,23] Chains have limited utility as a motif with which to control crystal packing because the molecules within a chain have considerable conformational and rotational freedom. Attempts have been made to reduce the flexibility of chains by introducing amide functional groups such as 2-pyridones that form cyclic dimers with an $R_2^2(8)$ pattern. For example, Wuest has shown that relatively rigid chains can be generated using symmetric di-2-pyridones with an acetylenic linker, as shown in Fig. 10.8.[72]

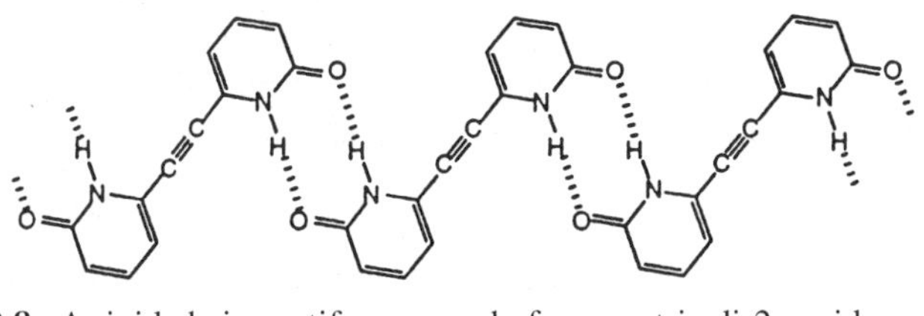

FIGURE 10.8. A rigid chain motif composed of symmetric di-2-pyridone molecules.[72]

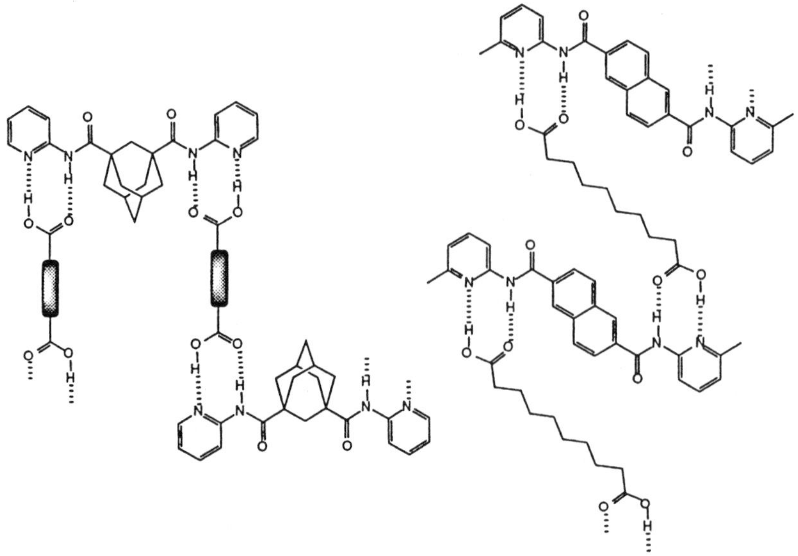

FIGURE 10.9. Chain motifs formed in cocrystals containing *bis*-acylaminopyridines and dicarboxylic acids.[49,75]

Karle[75] and Hamilton[49] used cocrystallization as a method for making chain motifs containing complexes of *bis*-acylaminopyridines and dicarboxylic acids, as shown in Fig. 10.9. Both researchers established that *bis*-acylaminopyridine receptors form 1 : 1 complexes with a range of different dicarboxylic acids. Hamilton showed that the chains crystallize in sheet-like structures, and by matching the lengths of the aminopyridine and diacid, the slip-angle between the components, and thus the overall shape of sheets, could be controlled.

Hamilton[29] also tailored the structure of one *bis*-acylaminopyridine to form helical chains when cocrystallized with heptanedioic acid, as shown in Fig. 10.10. Formation of the helical structure is promoted by using a dicarboxylic acid that is significantly longer than the aminopyridine. It is interesting to note that helices of opposite handedness form and interlock in the crystal.

By tethering three acylaminopyridines to a central cyclohexane ring, Hamilton[28] also formed an unusual rod-like structure, as shown in Fig. 10.11. The rods form when molecules stack on top of one another and are joined to neighboring molecules above and below by three hydrogen bonds between the secondary amides. The rods sit on a 3-fold axis of symmetry within the crystals and pack in a manner that is qualitatively similar to that of urea. When tested for nonlinear optical properties, the crystals showed modest second-harmonic generation (0.06 of that of urea).

A family of secondary arenedicarboxamides was studied by Lewis to determine the effect of arene–arene interactions in conjunction with hydrogen bonding on crystalline structure.[33] Lewis discovered that these compounds form

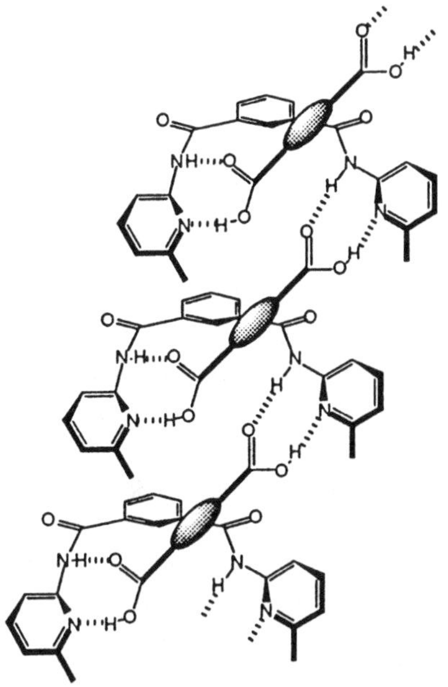

FIGURE 10.10. Schematic representation of the helical structure formed in cocrystals containing a *bis*-acylaminopyridine and heptanedioic acid.[29]

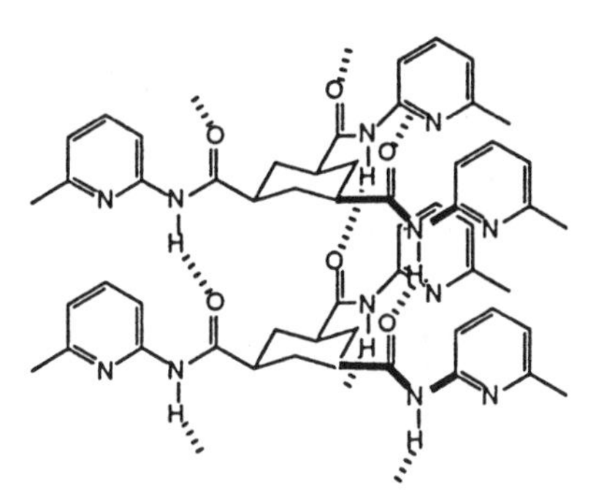

FIGURE 10.11. Rod-like structure formed by stacking *tris*-acylaminopyridine molecules.[28]

two layered (two-dimensional) patterns and one tape motif in which the molecules are separated at 5 Å intervals. The tape motif is shown in Fig. 10.12. Molecular packing and intermolecular interactions between aryl rings in this series of secondary arenedicarboxamides suggest that the choice of arene can influence whether tapes or layers will assemble. For example, tapes are favored when "narrow" arenes (e.g., biphenyl) are chosen. Layers are favored when "wide" arenes (e.g., naphthalene) are chosen. Lewis found that edge-to-face interactions between adjacent aryl rings was a common feature in the structures with tapes, but that this interaction was absent in the structures with layers.

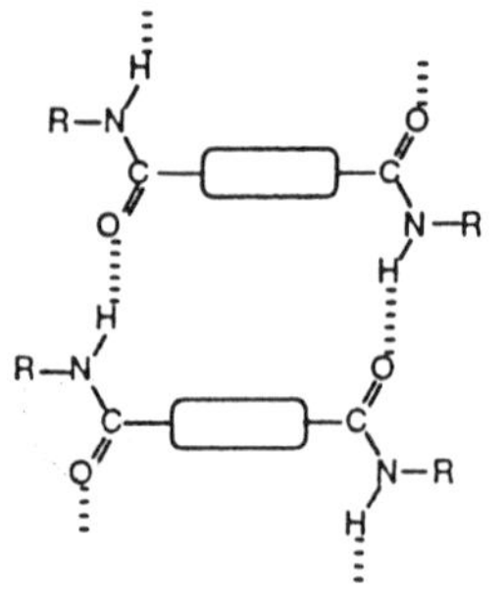

FIGURE 10.12. Schematic representation of the tape motif formed by secondary arenedicarboxamides.[33]

Lehn[76] and Whitesides[77] simultaneously developed strategies for creating one-dimensional motifs called tapes by using molecules with complementary sets of donors and acceptors that self-assemble through triads or diads of hydrogen bonds. Fig. 10.13 gives examples of some of these tape motifs. Whitesides demonstrated that the shape of tapes—as defined by the configuration of molecules within the tape—composed of substituted diphenylmelamines and barbital can be selectively switched from "linear" structures [Fig. 10.13(a)]

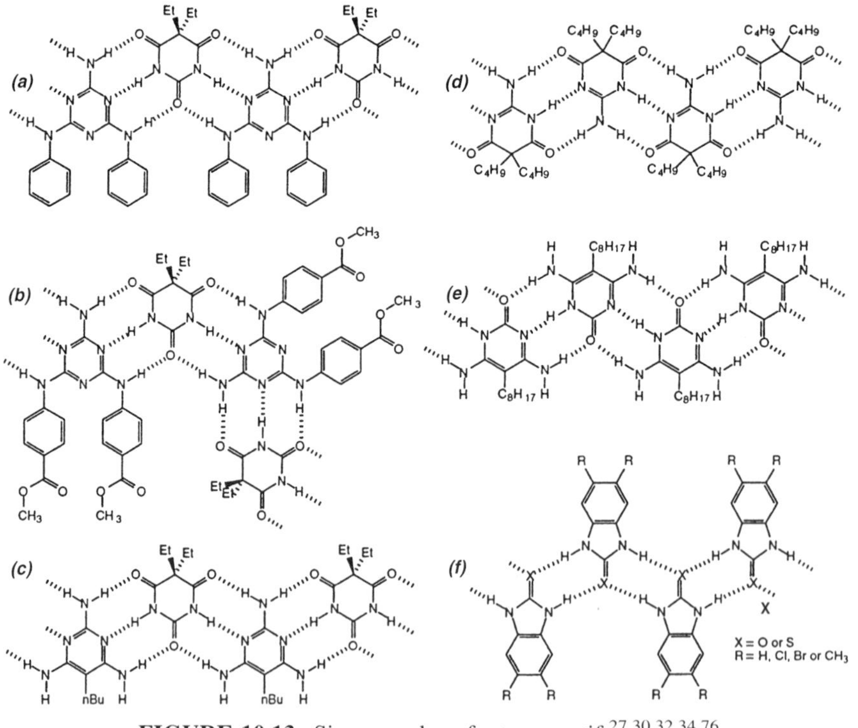

FIGURE 10.13. Six examples of a tape motif.[27,30,32,34,76]

to "crinkled", or wavy, structures [Fig. 10.13(b)] with the appropriate choice of substituents.[27,30,31,77] Fig. 10.13(d) and (e) give examples where the arrangement of donors and acceptors was chosen to preclude the formation of crinkled tapes.[37] Fig. 10.13(f) gives an example of a tape motif containing a single component where each molecule is joined to adjacent molecules through a pair of hydrogen bonds rather than a triad of hydrogen bonds.[32,34]

2.4. Two-dimensional Aggregates

A number of research groups have pursued crystal engineering by designing two-dimensional, or layered, structures. Layers are attractive as a motif for controlling structure because the translational freedom of molecules is eliminated when they are locked into a two-dimensional lattice. The ability to predict the packing of molecules in crystals simplifies from that of predicting the packing of individual molecules to that of the packing of layers. Leiserowitz[19] and Bridson[78] have shown that primary amides are ideal building blocks for the design of layered structures because the *syn* and *anti* hydrogen atoms form tapes that are cross-linked. For example, primary dicarboxamides generally crystallize in layers having the structure shown in Fig. 10.14(a). Leiserowitz and Tuval classified the structures of secondary dicarboxamides as having either of the two layered motifs shown in Figs. 14(b) and (c).[20] These two layered motifs differ only in that one motif contains a center of symmetry and is non-polar [Fig. 10.14(b)], while the second motif lacks a center of symmetry and is polar [Fig. 10.14(c)]. In a recent work with secondary arenedicarboxamides, Lewis demonstrated that both motifs could be formed, but that controlling which motif formed was difficult.[33]

Lauher and Fowler demonstrated several examples of layered structures that were formed by cross-linking chains of ureas with other functional groups. Their initial study of ureylenedicarboxylic acids established that layers could be formed by coupling ureas with dicarboxylic acids, and that the spacing between the ureas could be controlled by varying the chain length of the dicarboxylic acids (Fig. 10.15a).[45,46] A variety of structures subsequently have been engineered in which guest molecules were introduced to control structure within layers [Figs. 15(b) and (c)].[52,53] Hollingsworth has designed a similar layered structure in which chains of ureas are cross-linked by α,ω-dinitriles [Fig. 10.15(d)].[50,51]

Several different layer motifs have been developed recently using cyanuric acid and trithiocyanuric acid as the principal components to control the organization of molecules (Fig. 10.16). Harris designed the interesting and rather complex layered structure shown in Fig. 10.16(a) by cocrystallizing cyanuric acid with the dicarboxamide molecule, biuret.[48] This structure is unique in that molecules of water were incorporated into the structure, and they appear to be an integral component in maintaining the structure of the layer. Rao discovered that when crystallized with 4,4′-bipyridyl in benzene, trithiocyanuric acid forms hydrogen-bonded layers that stack and create large channels that trap molecules of benzene [Fig. 10.16(b)].[42]

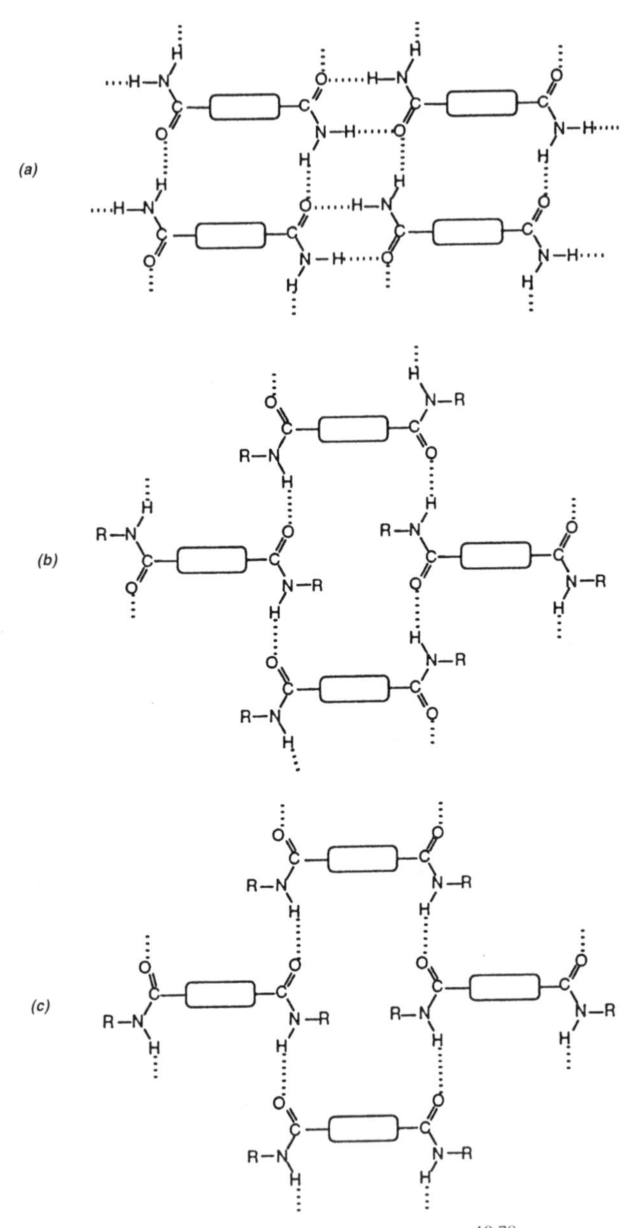

FIGURE 10.14. (a) Layer of primary dicarboxamides;[19,78] (b) Nonpolar layer of secondary dicarboxamides;[19,33] (c) Polar layer of secondary dicarboxamides.[19,33]

Aakeröy has taken a different approach by using ionic hydrogen bonding between amides to generate very robust anionic layers.[54] Two such layers based on networks of hydrogen bonds between oxamate anions are shown in Fig. 10.17. Both layers contain dimeric pairs of oxamate ions that are linked into

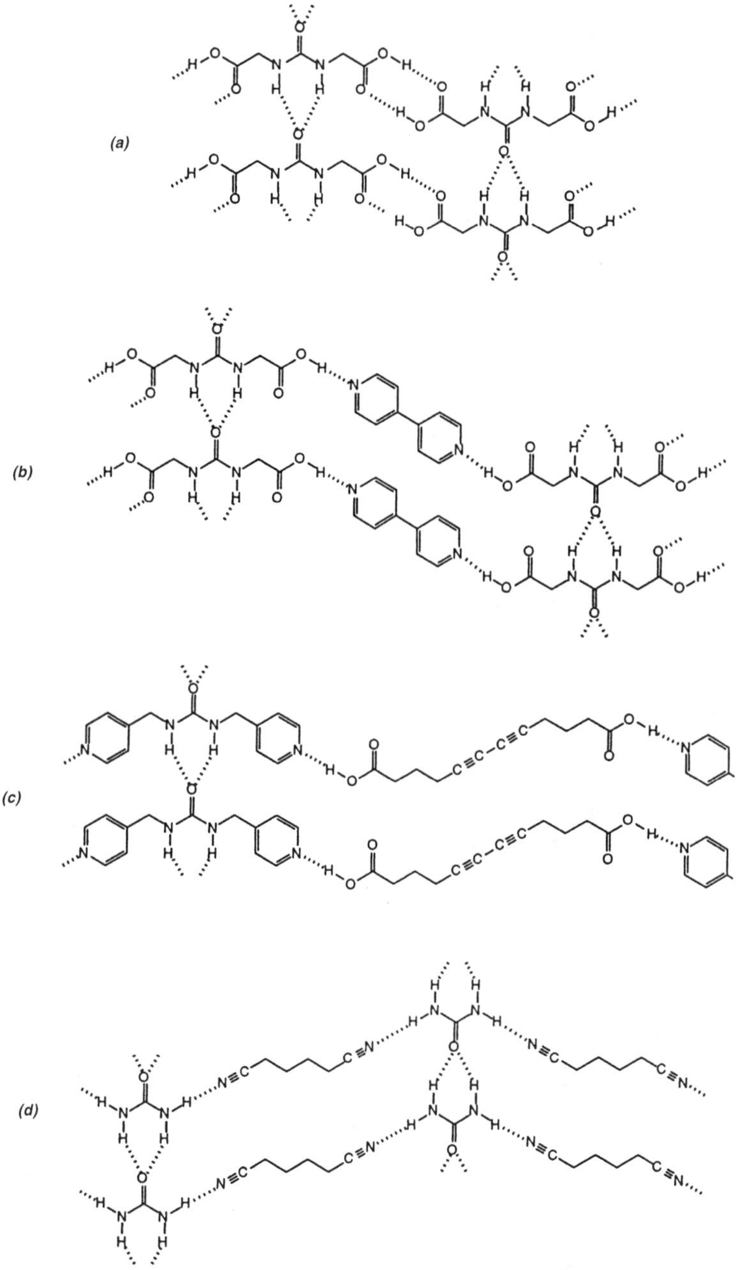

FIGURE 10.15. (a) Layer of a ureylenedicarboxylic acid.[45,46] (b) Layer of a ureylenedicarboxylic acid cocrystallized with 4,4′-bipyridyl.[53] (c) Layer of a pyridyl-derivatized urea cocrystallized with a dicarboxylic acid.[52] (d) Layer of urea cocrystallized with an α,ω-dinitrile.[50,51]

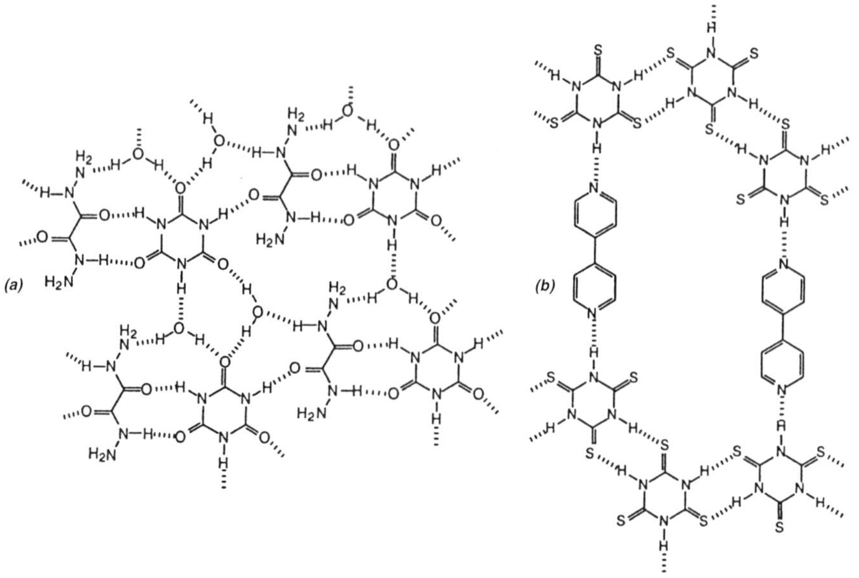

FIGURE 10.16. (a) Layer composed of cyanuric acid, biuret, and water molecules.[48] (b) Layer composed of trithiocyanuric acid and 4,4′-bipyridine.[42]

layers by a second set of hydrogen bonds. Structural differences occur in the type of cyclic dimers that form between oxamate anions; the dimers of anions in Fig. 10.17(a) form an $R_2^2(10)$ pattern, while those in Fig. 10.17(b) form an $R_2^2(8)$ pattern. Larger rings with $R_6^6(22)$ and $R_6^6(24)$ patterns are also present in Figs. 10.17(a) and (b) respectively, due to hydrogen bonds between donors and

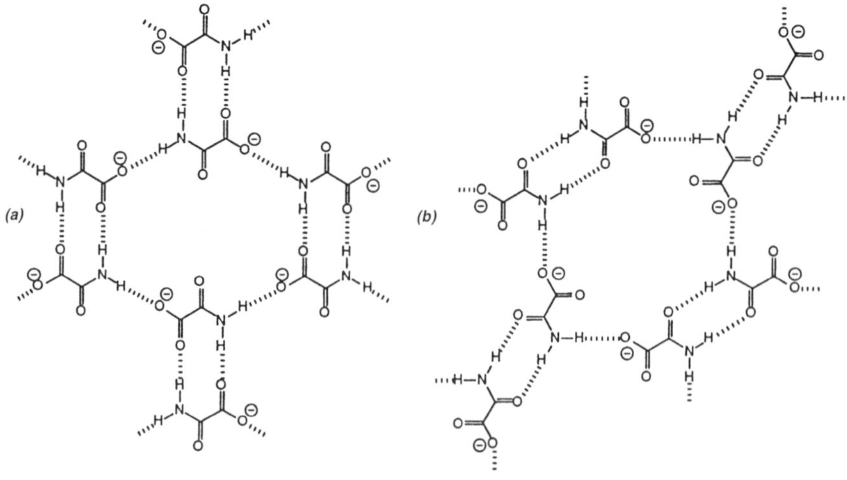

FIGURE 10.17. Layers composed of oxamate anions.[54]

acceptors on six different anions within the same layer. The layer motif that forms depends, in part, on the size and shape of the cations (not shown).

2.5. Three-dimensional Aggregates

Molecules that generate three-dimensional networks of hydrogen bonds have tremendous potential for engineering structure in crystals. Infinite aggregation of this type creates long-range order throughout a crystal and provides a rigid framework with well-defined structure. Presently, there are few examples of three-dimensional supramolecular aggregates composed of amides. Wuest has provided an example of three-dimensional aggregation of this type using the tetrapyridone molecule shown on the right in Fig. 10.18 as the fundamental building block.[55] The goal in designing this molecule was to create a rigid framework with which to build extended three-dimensional networks with large internal cavities. Wuest demonstrated that this tetrapyridone—referred to as a tecton—did, in fact, produce the organic diamondoid network shown schematically on the left in Fig. 10.18. Moreover, molecules of carboxylic acids were selectively enclathrated from mixtures of solvents during crystallization. Although, large chambers separated by 20 Å were formed within a given lattice, the cavity within was partially filled by six interpenetrating lattices.

Lauher and Fowler have taken a unique approach to designing three-dimensional networks by using a combination of hydrogen bonds and metal–ligand interactions to assemble *bis*-pyridylureas and *bis*-pyridyloxalamides.[43] The structure of a complex between a *bis*-pyridylurea and silver cations is shown in Fig. 10.19. In this structure, formation of metal–ligand bonds between the pyridyl rings and silver cations generates layers of ligands. These layers are linked in the third dimension by hydrogen bonding between the urea groups. The layers coincide and form narrow channels throughout the crystal.

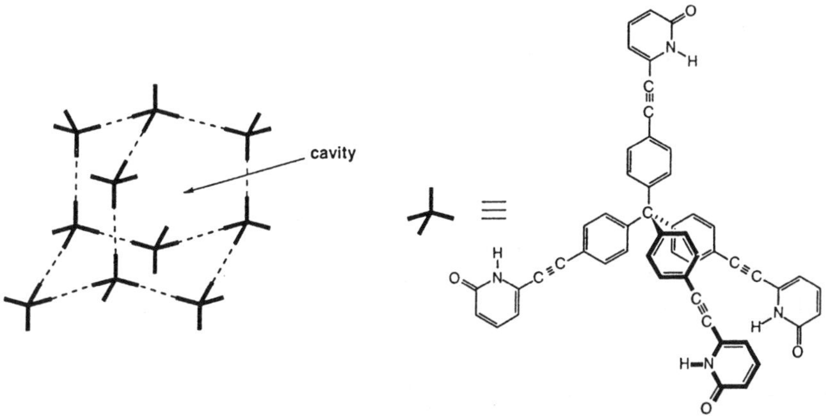

FIGURE 10.18. Three-dimensional diamondoid lattice formed by a tetrapyridone host. The cavity is filled by six interpenetrating lattices and by enclathrated molecules of carboxylic acids.[55]

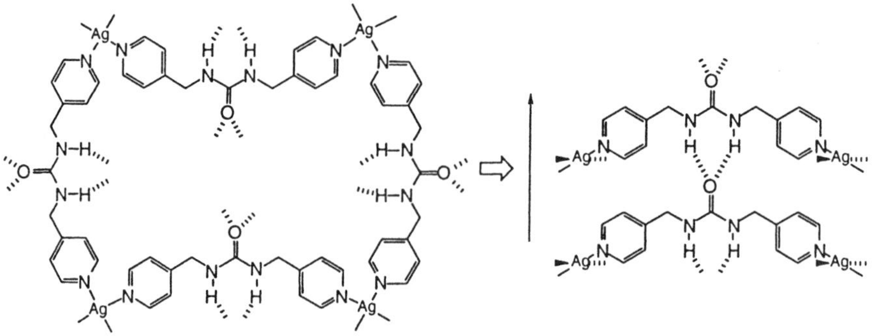

FIGURE 10.19. Puckered layers are formed by metal-ligand interactions between bis-pyridylurea molecules and silver cations (left). Hydrogen bonding between urea groups link adjacent layers in the third dimension (right).[43]

3. EXAMPLES OF AMIDES IN BIOMIMETIC STRUCTURES AND MATERIALS

Until recently, our understanding of structure in supramolecular aggregates has come almost exclusively from the analysis of X-ray crystal structures. Advances in instrumentation and techniques for characterizing supramolecular structure in media other than solids has enabled researchers to investigate supramolecular aggregation in solution, on surfaces and at interfaces as well. An increasing number of studies focus on the design and characterization of supramolecular structures in these environments. While our current understanding of the supramolecular properties and functions that can be achieved with amides is still fairly limited, researchers are producing a growing array of structures based on amides that display novel functions and behavior. In this section, we highlight several recent examples of supramolecular aggregates that exhibit specific properties and functions. Emphasis in this section is placed on those supramolecular systems with unique structures that enable researchers to investigate problems that are intractable using conventional synthetic approaches. Examples are chosen to highlight the broad range of areas of research in which supramolecular aggregates of amides are useful. Moreover, these examples represent many new directions and applications of supramolecular chemical research, and therefore, may provide inspiration to new endeavors. This section is divided into three subsections that include structures in solution (Section 3.1.), structures at interfaces and in semi-solids (section 3.2.), and structures in the solid state (Section 3.3.).

3.1. Structures in Solution

Protein activity and specificity depend on both the sequence of amino acids along the polypeptide chain, and their collective folding into secondary and tertiary structures. Two types of secondary structure with long-range order are observed in proteins—α-helices and β-sheets—which when combined with turns, loops, and coils produce proteins with diverse shape and function. Although, the exact sequence of amino acids along the polypeptide chain of a protein can be determined, it remains a mystery as to how this linear sequence of amino acids codes for the secondary and tertiary structures of proteins. Current experimental data suggest that folding of a protein is initiated with the formation of discrete secondary structures (conceptually similar to the transition states of chemical reactions) that subsequently reorganize to interact with each other, thus forming the final tertiary structure of the protein. If in fact protein folding is initiated by the formation of discrete secondary structures, then it is important to understand what role the primary sequence of amino acids has in the nucleation and stabilization of these structures. Several approaches to solving this problem have emerged: NMR studies of polypeptide conformations; selective substitution along the polypeptide chain to identify amino acid residues responsible for folding; and more recently, the design and synthesis of peptidomimetic model systems.

3.1.1. Beta-sheets and Alpha-helices. Peptidomimetic systems are interesting not only in the context of protein folding but because they may provide a means toward an "artificial protein" end. For example, Gellman has demonstrated recently that short oligomers of β-amino acids can function as β-peptide foldamers. By including an L-proline-glycolic acid segment in the structure of the β-peptide foldamer, a tight turn forms that is stabilized by intramolecular hydrogen bonds [i.e., $S_1^1(11)$ and $S_1^1(19)$]. Consequently, the β-peptide foldamer adopts a stable β-sheet structure (Fig. 10.20).[79]

A distinguishing characteristic of a β-sheet structure comprising β-amino acids is the possibility of forming a sheet with a net dipole. The polarity of a β-sheet derives from the parallel alignment of carbonyl groups, which occurs when the NC_β–$C_\alpha C(=O)$ torsion angle of each residue is in the *anti* conformation. If the NC_β–$C_\alpha C(=O)$ torsion angle of each residue is in the gauche conformation,

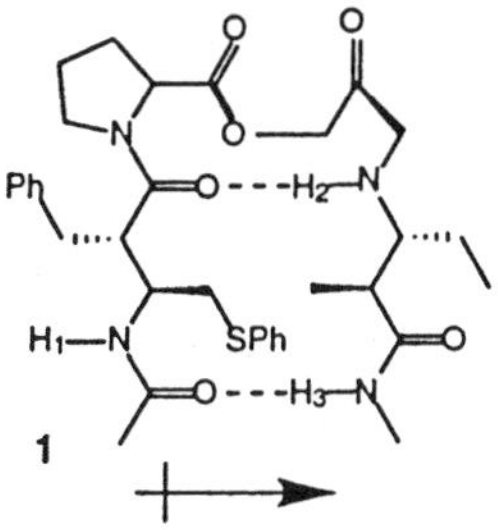

FIGURE 10.20. ^{1}H NMR data of **1** in CD_2Cl_2 indicate extensive intramolecular hydrogen bonding at NH(2) (δ 7.40) and NH(3) (δ 7.27) and not at NH(1) (δ 5.81). Furthermore, $^3J(\alpha,\beta)$ values for each β-amino acid are 10–11 Hz, consistent with torsion angles in *anti* conformation. IR data of **1** (CH_2Cl_2) in the N–H stretching region indicates extensive intramolecular hydrogen bonding (3423, 3367, 3347 cm^{-1}).[79]

the carbonyl groups align antiparallel and the β-sheet exhibits no net dipole. A polar β-sheet does not exist among sheets formed from α-amino acids because only an antiparallel arrangement of carbonyl groups is possible. By placing substituents at both the C_α and C_β of the β-amino acid, torsion angles in the *anti* conformation are promoted due to the two substituents minimizing their steric interactions.

Another type of artificial β-sheet, comprising a "molecular scaffold" and peptide groups, has been pursued by Nowick.[80–82] The molecular scaffold is an oligourea, which can adopt a conformation that brings several peptide strands into close proximity, and consequently, induce β-sheet formation (Fig. 10.21). Using a combinatorial approach, Nowick determined the relative propensities of four amino acids to form β-sheets when tethered to a diurea scaffold. These results (leucine, valine > alanine > glycine) are similar to those from earlier studies by other research groups with the exception of the indistinguishable behavior of leucine and valine.

Seebach discovered the formation of stable helices when β^3- and β^2-oligopeptides containing as few as six β-amino acids were dissolved in methanol (Fig. 10.22, $n = 2$ or 3 in β^n indicates substitution at the α- or β-carbon atoms, respectively, of a β-amino acid).[83,84] This result was surprising at the time of its discovery because it was commonly believed that the additional carbon atom in β-amino acids would lower the tendency towards highly ordered structures by introducing too much conformational freedom. Moreover, most helical structures in proteins consist of 10–15 α-amino acids. In addition to the higher stability of helices of β-oligopeptides compared to α-oligopeptides, β-oligopeptides are resistant to cleavage by peptidases—a property that holds tremendous potential for the treatment of disease.

3.1.2. Hybrid "Genes." The theories of genetics and natural selection provide a mechanism by which life could evolve from simplicity to its present complexity.

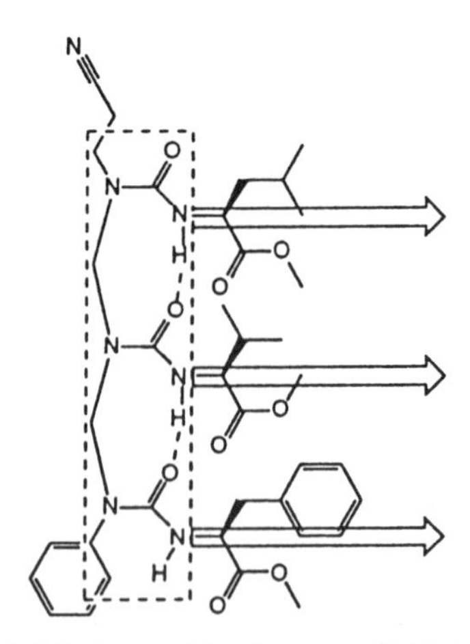

FIGURE 10.21. Artificial β-sheet with triurea scaffold (within dashed box).[80–82]

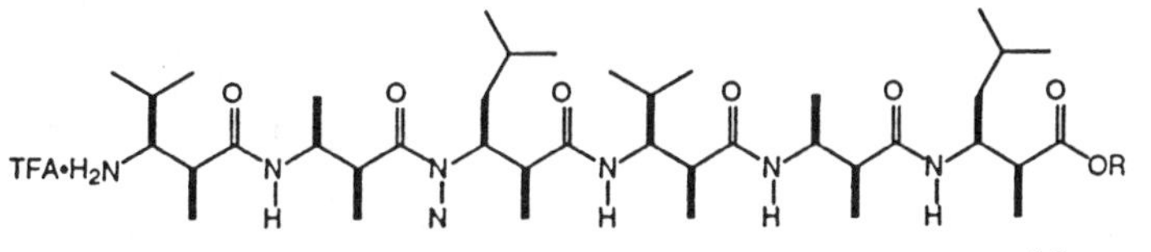

FIGURE 10.22. One of several β-hexapeptides studied (in this case, a $\beta^{2,3}$-hexapeptide substituted with a methyl group in the 2-position and the side chains of Val, Ala and Leu in the 3-position). CD absorptions in methanol occur at 200 nm for the maximum and 215 nm for the minimum. This pattern of absorptions in the CD spectrum is independent of both concentration (0.02–1.0 mM) and the presence or absence of terminal protecting groups.[84]

The Nobel Prize winning discovery of catalytic RNA (i.e., ribozymes) by Cech provides support to the theory that life previously existed in a simpler "RNA-world" where coded proteins and DNA were absent.[85–88] Existence of an "RNA-world" necessitates a chemical means by which RNA can be prepared. Under conditions of primitive Earth, the lack of adequate chemistry for the synthesis of the backbone of RNA has led to a proposal that a simpler, more easily prepared form of genetic material preceded RNA, which when conditions became favorable, relinquished to a genetic system based on RNA.

What did the pre-RNA genetic system look like, and how did biology change from this system to one based on RNA? Orgel, Nielsen and colleagues have shown recently that mimics of DNA and RNA, for which they coined the term "peptide nucleic acids" or PNAs, can form double-helical complexes with complementary RNA and DNA that are stabilized through hydrogen-bonding interactions between base pairs.[89] Instead of the β-ribofuranoside-5′-phosphate and β-D-deoxyribofuranose-phosphate backbones of RNA and DNA, respectively, the backbone of PNA is a pseudopeptide comprising *N*-(2-aminoethyl)-glycine units (Fig. 10.23). The nucleotide bases in PNA are attached to the pseudopeptide backbone via carbonyl methylene linkers to the glycine nitrogen atom.

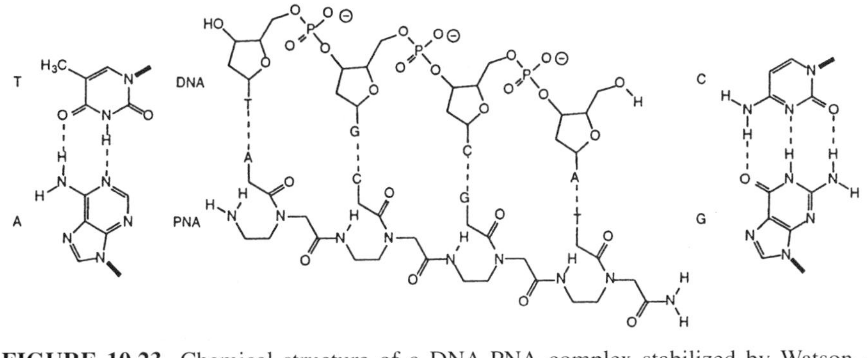

FIGURE 10.23. Chemical structure of a DNA-PNA complex stabilized by Watson–Crick base pairing between nucleotides.[89]

Although Nielsen and Orgel do not claim that PNA is the precursor of RNA, their work has shown that PNA can function as a template to facilitate the synthesis of complementary RNA or DNA strands and vice versa.[90–92] These remarkable results are significant because they demonstrate that information can be transferred between two "genetic" systems that have very different backbone structure.

3.1.3. Nanotubes. Hollow, tube-like filaments of varying diameters have been detected in many plant and animal cells (i.e., mitotic spindles, axons of nerve fibers, and the cilia and flagella of eukaryotic cells). These tube-like filaments provide cells with both structural support and a pathway for the transport of material across their membranes. Ghadiri and his colleagues have prepared nanotubes through the self-assembly of cyclic peptides of specific design.[93] The cyclic peptides are constructed from an even number of alternating D- and L-amino acids, which allows the polypeptide ring to adopt a planar conformation (Fig. 10.24). This conformation positions the amide groups perpendicular to the plane of the ring and the side-chains pointing radially from its center. The stacking of rings into nanotubes is promoted in this conformation, and the nanotubes are stabilized by intermolecular hydrogen bonds between the amide groups of adjacent rings in a manner similar to that of β-sheets. The diameter of the nanotube depends on the number of amino acids in the peptide ring. For example, cyclic octa-, deca- and dodecapeptides have pore diameters of 7, 10 and 13 Å, respectively.

3.1.4. Molecular Containers. Molecules that encapsulate other molecules have the potential to function as "Trojan horses" in the treatment of disease. In

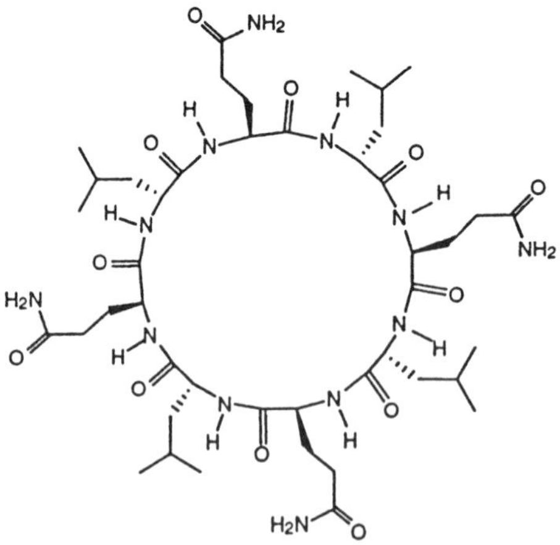

FIGURE 10.24. Cyclic octapeptide comprised of alternating L-Gln and D-Leu. FT-IR spectra in the Amide **I** and **II** regions (1629, 1688, 1541 cm^{-1}) are similar to that of hydrogen-bonded β-sheets; the N–H stretch at 3280 cm^{-1}, and electron diffraction data indicate that the intrapeptide distance is *ca.* 4.72 Å.[93]

addition to their medical promise, molecular containers exhibit interesting physical organic chemistry (i.e., stabilization of reactive intermediates, chiral microenvironments, enhancement of reaction rates). A recent report by Kim describes the properties of cucurbituril, a barrel-shaped molecule, in the presence of alkali metal salts.[94] Although cucurbituril possesses a hollow core ca. 5.5 Å in diameter, it is insoluble in all but very acidic aqueous solutions. This insolubility at biological pH makes cucurbituril unattractive as a synthetic host. Kim and his colleagues discovered, however, that cucurbituril dissolves readily in a neutral solution of sodium sulfate (0.2 M) due to interactions between the metal cation and the carbonyl oxygen atoms of cucurbituril. Moreover, they found that cucurbituril traps one or more guest molecules within its hollow core (i.e., $K_f = 5.1 \times 10^2\,M^{-1}$ for THF guests as estimated from the relative ratio of free and bound THF in ^{1}H NMR spectra). Guest molecules get trapped when two sodium atoms and five water molecules form a hydrogen-bonded "lid" with the carbonyl oxygen atoms of cucurbituril. The two "lids" on the cucurbituril barrel can be reversibly removed and replaced by lowering and raising the pH, respectively (Fig. 10.25).

Rebek has constructed "softballs" in organic solvents that are held together by hydrogen bonds between two concave-shaped monomers.[24] This work examines new derivatives of glycolurils that when dimerized present cavities large enough to contain guests the size of ferrocene and adamantane (Fig. 10.26). When R_1 and R_2 are different, the result is a racemic mixture of capsules with external chiral surfaces. The NMR of the complex with encapsulated camphor exhibits doubled peaks of equal intensity, which confirms the racemic nature of the solution of capsules as well as the lack of steric interaction between the chiral guest and external chiral surface of the host.

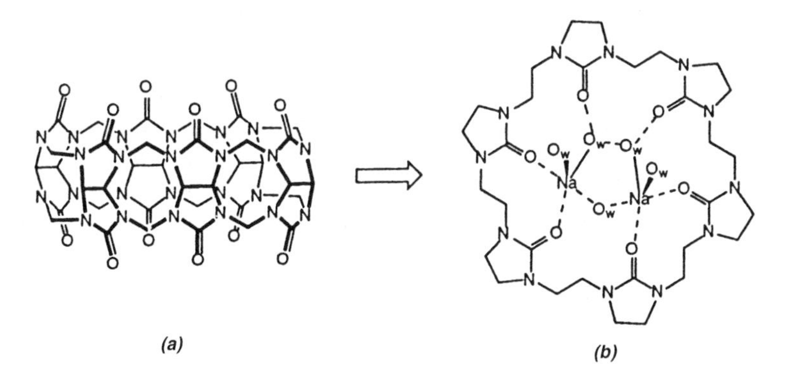

FIGURE 10.25. (a) Cyclization of six diurea monomers gives the barrel-shaped curcubituril; (b) top-half of curcubituril showing the $2Na^+/5H_2O$ "lid". Each sodium ion is coordinated to two adjacent carbonyl oxygen atoms of cucurbituril and three water molecules in a distorted square pyramidal geometry. One molecule of water bridges the two sodium ions, two of the water molecules are hydrogen bonded to each other and the remaining two adjacent carbonyl oxygen atoms of cucurbituril, which effectively "seals" the core of cucurbituril.[94]

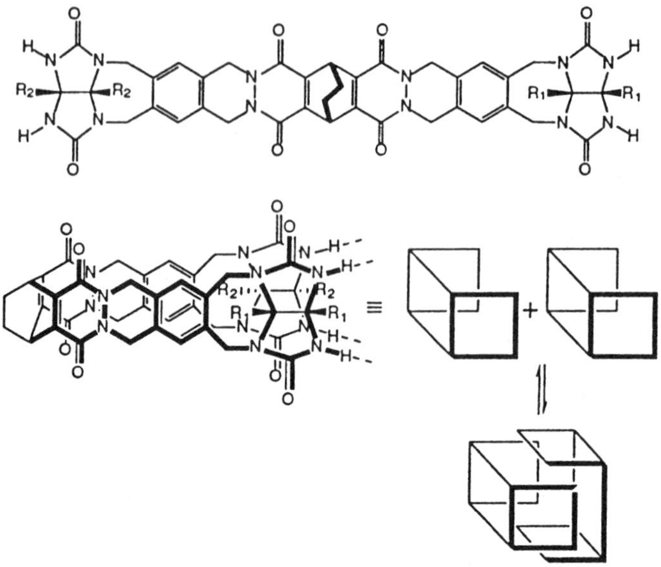

FIGURE 10.26. Derivative of *bis*-glycoluril that self-assembles into a "softball" in $CHCl_3$. $R_1 = R_2 =$ Ph or $CO_2{}^iPen$, or $R_1 = CO_2{}^iPen$ and $R_2 = p\text{-}C_6H_4OC_6H_{13}$.[24]

3.1.5. Receptors. Numerous research groups have designed and synthesized receptors that target biologically relevant compounds such as urea and barbiturates. Both biosignaling and biocatalysis require some mechanism by which to recognize a substrate. This recognition process often involves hydrogen-bonding interactions between the receptor and its substrate. Bell reports a receptor capable of binding urea solely through hydrogen-bonding interactions.[95] The receptor was designed to encircle urea and form a hydrogen bond with each of its four donor and two acceptor sites, making a total of six hydrogen bonds between receptor and substrate (Fig. 10.27). Since the receptor has a high degree of conformational preorganization present in its fused-ring structure, and an exact complement of donor and acceptor sites for the binding of urea, this receptor does not bind alkylated ureas.

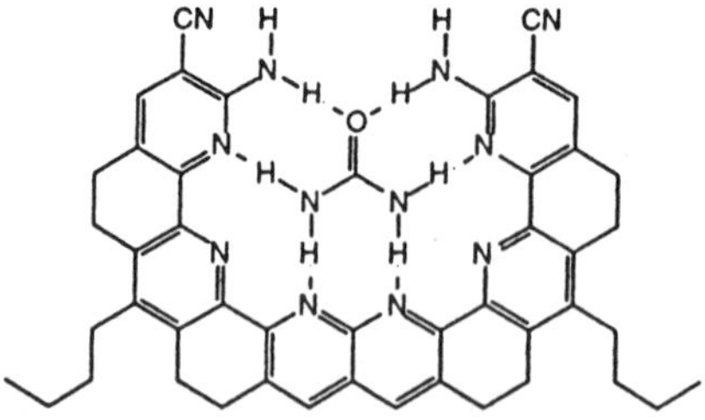

FIGURE 10.27. A receptor for urea that exhibits a stability constant of $8 \times 10^3\ M^{-1}$ (from UV–Vis data) or $1.4 \pm 0.2 \times 10^4\ M^{-1}$ (from NMR data).[95]

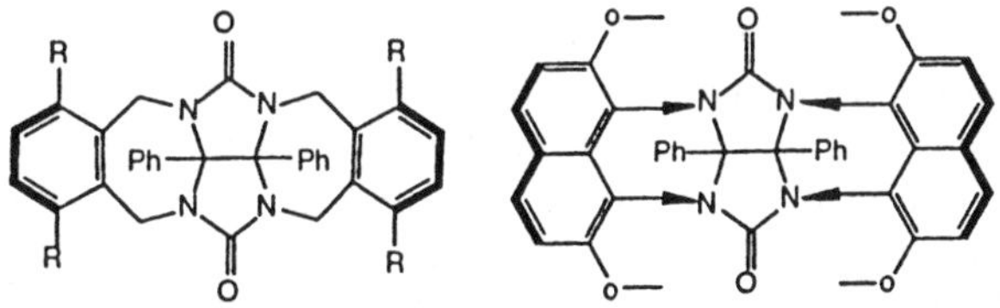

FIGURE 10.28. Two of 20 compounds used to examine systematically the influence of different factors on the binding of derivatives of 1,3-benzendiol substituted at the 5-position. R = –H, $-CH_3$ or $-OCH_3$.[96]

Clip-shaped molecules based on diphenylglycoluril are conformationally preorganized to bind planar aromatic guest molecules through hydrogen-bonding and π–π interactions (Fig. 10.28). Nolte has prepared a series of molecular "clips" to determine the influence that each factor (i.e., strength of hydrogen-bonding donors and acceptors, size and shape of the cavity, substituents) has on the binding affinity between host and guest.[96] The change from oxygen to sulfur acceptors in the host dramatically reduces the binding constant for guest molecules (i.e., $K_a = 2.6 \times 10^3\,M^{-1}$ versus $5.1 \times 10\,M^{-1}$ for 1,3-dihydroxybenzene). When the *para*-substituents of the guest were varied, the binding constants were found to vary over a wide range ($K_a = 0$–$10^5\,M^{-1}$) and to correlate with Hammett σ-values. The cavity effect appears to have only a minor role in the binding of guest especially when hydrogen-bonding interactions or π–π interactions contribute significantly.

Two equivalents of 1,3,5-triaminomethyl-2,4,6-triethylbenzene mixed with three equivalents of 2,6-pyridine dicarbonyl dichloride gives a bicyclic cyclophane product that has a cavity (78.3 Å^3) lined with six hydrogen-bonding donors, three hydrogen-bonding acceptors, and a pair of aromatic rings separated by 7.0 Å (Fig. 10.29). This receptor, designed by Anslyn, tightly binds anions dissolved in CD_2Cl_2 with $K_a > 10^4$–$10^5\,M^{-1}$.[97] In a more polar mixture of

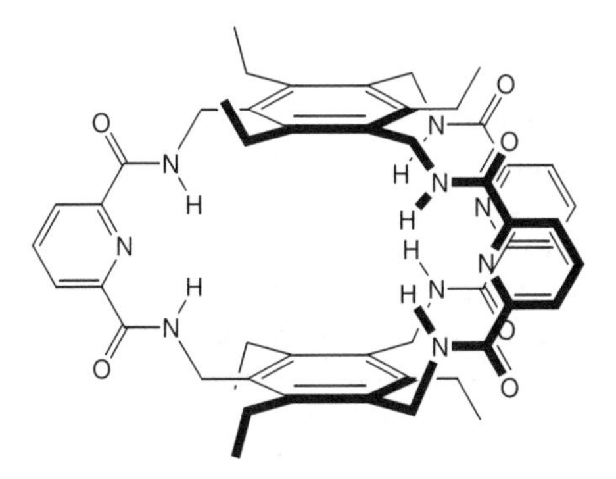

FIGURE 10.29. Bicyclic cyclophane receptor designed to bind anions. In the absence of anions, the cavity of the receptor is occupied by two molecules of water and one molecule of CH_3CN.[97]

solvents (25% CD_2Cl_2 in CH_3CN), $K_a = 770 \pm 120$ and $300 \pm 30\,M^{-1}$ for acetate and nitrate, respectively.

Lehn reported recently a receptor that combines two different types of sites of recognition.[98] One of the recognition sites, melamine, uses hydrogen bonds to interact with barbiturate substrates. The other site of recognition, which contains two benzo-18-crown-6 ethers, uses carbonyl oxygen atoms to coordinate alkali metal ions. The receptor exists in three different rotational conformations (i.e., closed, half-open, and open depending on the orientation of the two crown ethers relative to each other). Interconversion between conformations requires rotation around C–N bonds of the receptor with an activation barrier between conformations of at least 74 kJ mol^{-1}. The presence of large alkali metal ions such as Cs^+ or derivatives of barbiturate in solution increases the percentage of the closed conformation due to complexation or formation of hydrogen bonds between these species and the receptor (Fig. 10.30). This behavior mimics that of many substrate-specific structural changes that occur in biological molecules such as transport proteins, which become permeable to cell membranes upon binding of both substrate (i.e., glucose) and metal ion (i.e., sodium).

3.1.6. "Signaling" Receptors or Sensors. To reduce the number of synthetic steps needed in the construction of a receptor, Hamilton has designed a series of receptors that self-assemble.[99] The receptor shown in Fig. 10.31 self-assembles when its two bifunctional molecular components, which contain binding sites for a metal and a dicarboxylic acid substrate, are exposed to metal ions. The metal ion induces dimerization of the bifunctional molecules, which brings the substrate binding sites into close proximity. A receptor designed to bind glutaric acid signals the binding of glutaric acid ($K_a = 4.3 \times 10^4\,M^{-1}$) with a hypsochromic shift in the absorption maximum of the copper complex.

Fabbrizzi has designed molecular sensors that derive their function through the linking of a receptor with a signaling unit.[10] For example, the binding of metal ions to a diamide–diamine receptor is indicated when the intensity of fluorescence of the linked anthracene decreases (Fig. 10.32). Specificity for metal ions (i.e., discrimination of Cu^{2+} ions in the presence of Ni^{2+} ions) is achieved through the control of pH, which reversibly deprotonates the two amide groups of the receptor. The discrimination between Cu^{2+} and Ni^{2+} ions is due to the greater stability of Cu(II)-polyaza complexes in solution compared to that of Ni(II).

Smith has designed two receptors whose strength of association with substrate can be controlled electrochemically (Fig. 10.33).[100] For example, when the receptor based on *o*-quinone is reduced, the binding constant for urea is 660 M^{-1}. When the receptor is oxidized, however, the binding constant for urea drops to 1 M^{-1}. The difference in binding strengths between the reduced and oxidized versions of the receptor is due to the anionic versus neutral character of their carbonyl oxygen atoms, respectively. Moreover, the reduction potential of the receptor shifts positively upon binding of a substrate with the maximum shift in potential related to the ratio of K_{red}/K_{ox}.

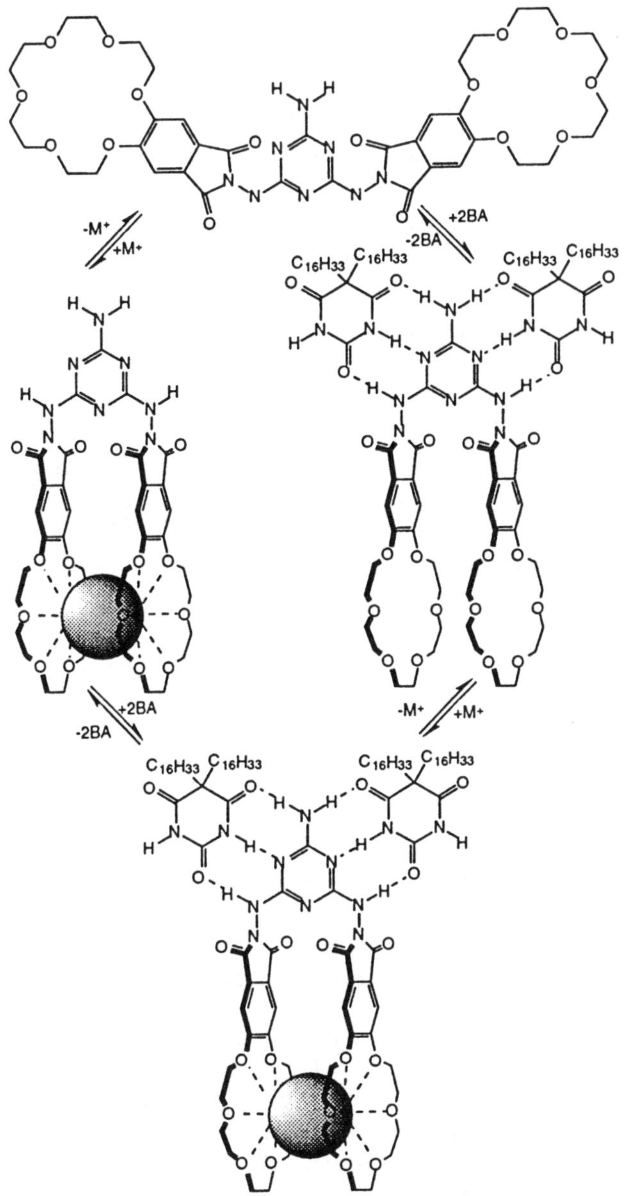

FIGURE 10.30. Double receptor shown in open and closed conformations. Large cations such as Cs^+ form a sandwich-type complex with the two crown ethers, thus favoring a closed conformation of the receptor. Hydrogen bonding of a barbiturate derivative has a similar effect. No synergistic effect of metal ion and complementary hydrogen bonding could be observed because a common solvent could not be found in which the two effects occur.[98]

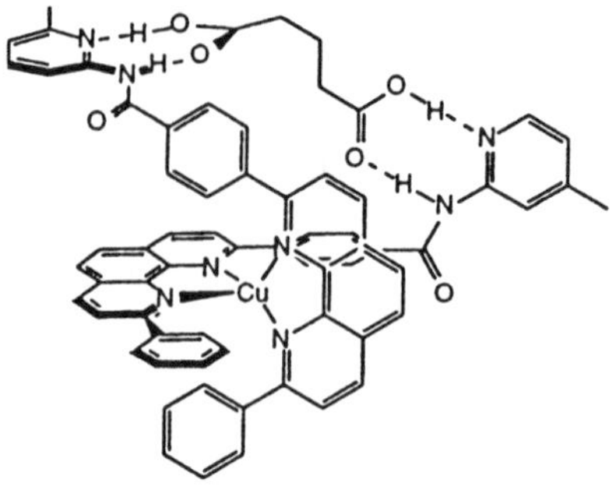

FIGURE 10.31. The presence of Cu^+ ion induces the self-assembly of a receptor for glutaric acid.[99]

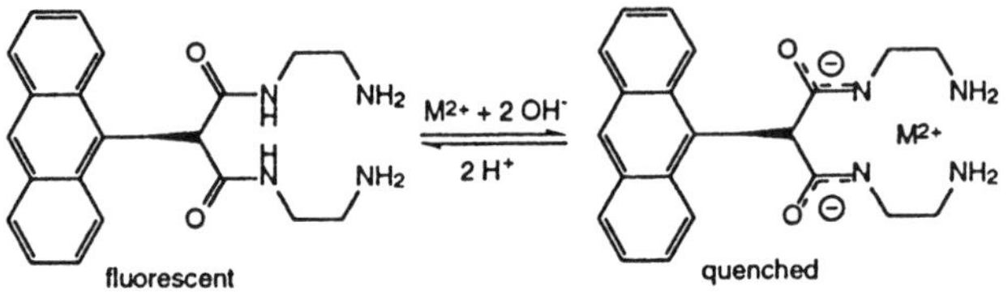

FIGURE 10.32. Reduction of intensity of fluorescence occurs when an electron is transferred from a chelated redox active metal ion to the excited fluorophore. Very low levels of analyte ($\geq 10^{-7}$ M) are detectable.[10]

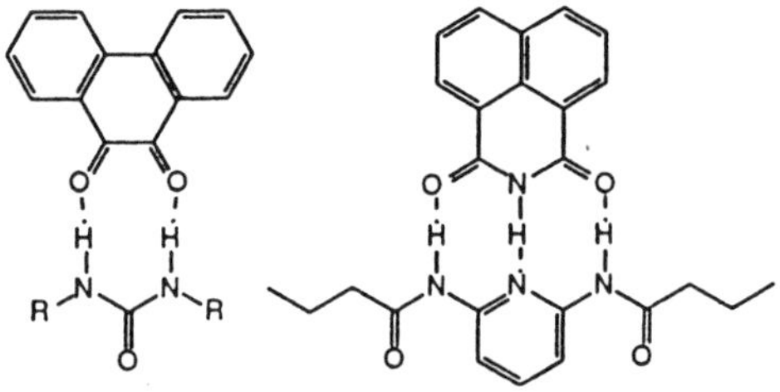

FIGURE 10.33. The strength of association between receptor and substrate (urea and derivatives of diaminopyridine) depends on the state of oxidation of the receptor.[100]

A series of anion receptors capable of optical and electrochemical sensing were recently reported by Beer.[101,102] For example, a tris-bipyridyl ruthenium (II) complex tethered to the lower-rim of a calix[4]arene framework through an amide linkage forms a cavity with two amide and two phenolic hydrogen-bonding donors (Fig. 10.34). This receptor is moderately selective for $H_2PO_4^-$ anion in DMSO, which can be discerned in the presence of a ten-fold excess of HSO_4^- and Cl^-. The binding of anions in this cavity produces a substantial cathodic perturbation of the amide-substituted 2,2′-bipyridyl redox couple as well as the optical properties of the complex. The emission from the metal-to-ligand charge transfer band is blue shifted by 16 nm and has a high quantum yield. The change in quantum yield is postulated to arise from the stiffening of the receptor upon binding of anion, which inhibits vibrational and rotational relaxation modes of non-radiative decay.

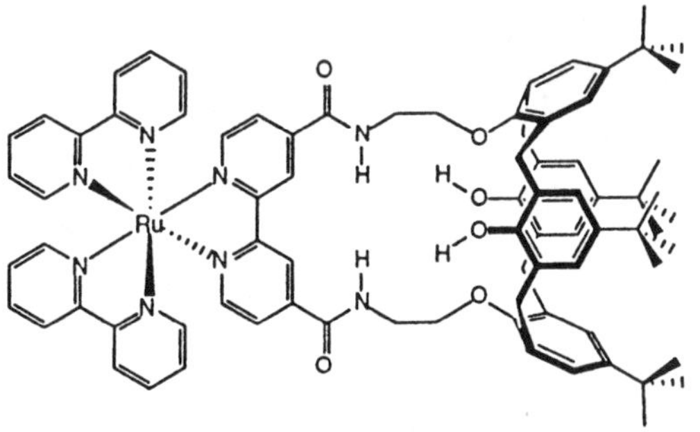

FIGURE 10.34. An optical and electrochemical sensor for anions. The binding constant for $H_2PO_4^-$ in DMSO is $2.8 \times 10^4\,M^{-1}$.[101,102]

3.1.7. Photoactive Aggregates. In an effort to understand better the mechanism of photosynthesis, numerous research groups over the past two decades have studied the photophysics and photochemistry of energy and electron transfer between covalently linked donor, chromophore, and acceptor subunits.[103–105] More recently, the desire to prepare molecular-scale electronic devices has resulted in the need for greater variety in the choice of subunits. Because of the synthetically challenging methods of preparation, other methods for generating photoactive assemblies were sought. The self-assembly of donor, chromophore, and acceptor subunits through complementary hydrogen-bonding interactions is an attractive approach to building these types of molecules because the synthetic complexity is reduced and the modularity of the photoactive system is increased. Ward has recently reported on the energy transfer across three hydrogen bonds between two chromophores, ruthenium(II)- and osmium(II)-polypyridine (Fig. 10.35).[106] The hydrogen-bonding interactions are the result of hydrogen-bonding complementarity between cytosine and guanine subunits, which are tethered to one of the bipyridyl ligands on each of the chromophores.

Fitzmaurice describes another example in which molecular recognition and self-assembly is used to associate a chromophore and an acceptor

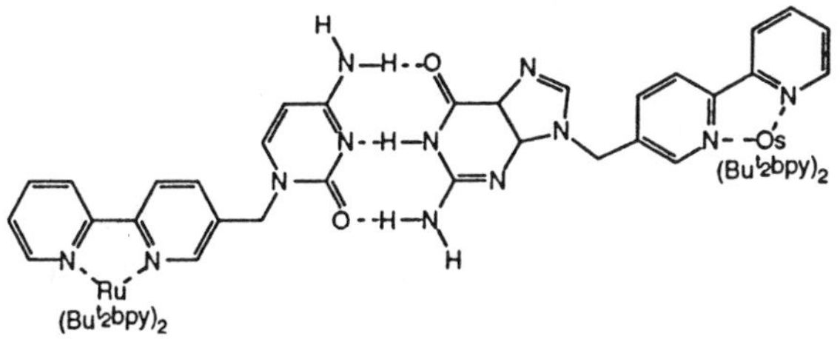

FIGURE 10.35. Hydrogen-bonding complementarity between cytosine on the Ru(II) complex and guanine on the Os(II) complex drives the association between the two chromophores ($K_a > 5 \times 10^3\,M^{-1}$ in CH_2Cl_2). The results of time-resolved luminescence experiments conducted at three different concentrations ([Ru–C] = [Os–G] = 1.0, 2.2, and 4.5×10^{-4} M) gave $k_{en} = 9.5 \times 10^7\,s^{-1}$10%.[106]

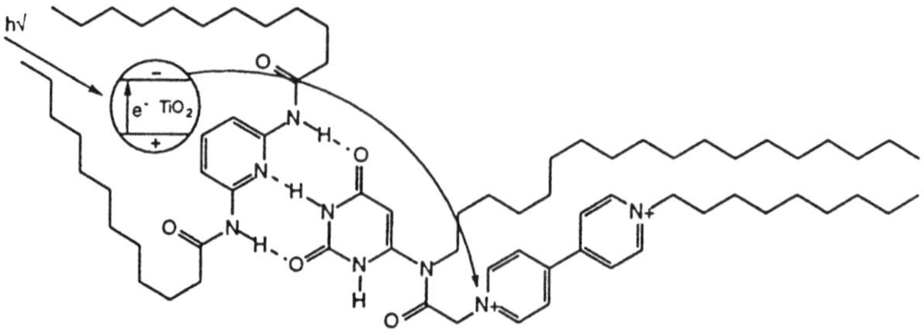

FIGURE 10.36. Electron transfer from TiO_2 chromophore to a viologen acceptor.[107]

(Fig. 10.36).[107] The association between the two subunits is caused by the formation of three hydrogen bonds between a derivative of diamino pyridine that entangles a nanocrystallite of TiO_2 ($\sim$ diameter of 22 Å) and a uracil derivative tethered to viologen. Electron transfer from TiO_2 to viologen occurs rapidly upon irradiation of the solution with 355 nm light. Control experiments indicate that electron-transfer does not occur either through space or as a consequence of collisions between the donors and acceptors. The hydrogen-bonding interaction, therefore, is essential for electron transfer to occur between these two subunits.

3.2. Structures at Interfaces and Semi-solids

Living cells are surrounded by an outer membrane (phospholipid bilayer with a thickness of 60–100 Å), which serves as a barrier between internal components of the cell and the external environment. The fluidity of the phospholipid bilayer allows for various proteins to reside within or at the surface of the membrane. Membrane proteins can function as receptors for signaling changes in the environment, or as channels for the transport of ions and other small molecules across the membrane. To better understand the properties of signal and ion conduction in membrane proteins, several groups have prepared and characterized peptidomimetics of membranes.

3.2.1. Receptor Films. Kunitake and coworkers are exploring the properties of molecular recognition in Langmuir–Blodgett films that function as models for receptor proteins found on the surface of cell membranes. In particular, they have prepared Langmuir–Blodgett films from oligopeptide amphiphiles containing different dipeptides (i.e., $-XYNH_2$ where X = Gly and Ala, and Y = Gly, Ala, Val, Leu, and Phe (Fig. 10.37(a)).[108] These LB-films were examined for their ability to recognize and bind aqueous dipeptides using π-A isotherm techniques, FT-IR spectroscopy and XPS. It was found that in the absence of guest dipeptides, a LB-film comprised of two different amphiphiles is well mixed. In the presence of guest dipeptides, however, the components of the LB-films reorganize to produce a specific binding site for the dipeptide. This phenomenon suggests that combinatorial recognition may occur at membrane surfaces through substrate-induced reorganization of the receptor proteins in

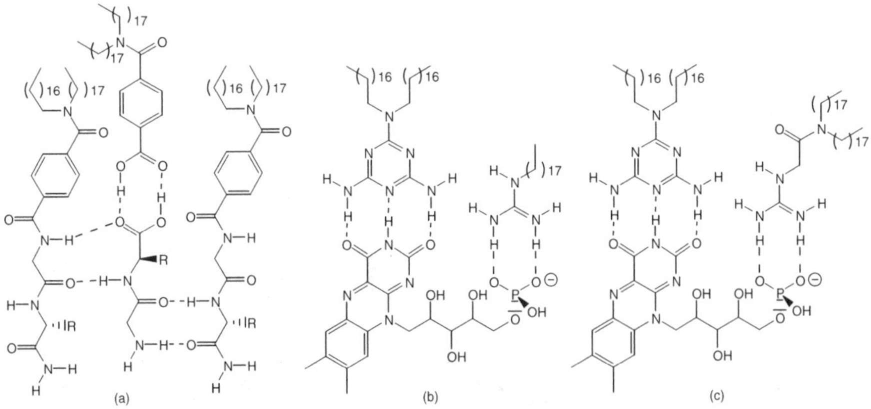

FIGURE 10.37. Amphiphiles of the LB-film that recognize (a) aqueous dipeptides, (b) and (c) aqueous FMN.[108,109]

membranes. Kunitake has also shown that LB-films comprised of two amphiphiles will bind aqueous flavin mononucleotide (FMN). The amphiphiles used in these studies contain head groups that are the hydrogen-bonding complement to the isoalloxazine ring and the phosphate moiety of FMN (Figs. 10.37(b) and (c)).

3.2.2. Ion-selective Films. Ghadiri has elaborated further his design of self-assembling nanotubes by replacing D and L-α-amino acids in the cyclic peptide monomers with β^3-amino acids (α-unsubstituted-β-chiral-β-amino acids).[41] In contrast to the previously reported nanotubes based on cyclic D,L-α-peptides, cyclic β^3-peptides form nanotubes that possess a net dipole, which ultimately may exhibit interesting behavior related to rectification of current.[110,111] Studies of molecular modeling show that the size of channels in these nanotubes is predicted to be 2.6–2.7 Å in diameter (Fig. 10.38). The conductance of the nanotubes embedded in planar bilayers of Type II-S soybean phosphatidylcholine was measured. The rate of channel mediated transport of potassium ions is 1.9×10^7 ions s^{-1} (conductance of potassium ions is 56 pS under the conditions of measurement), which is faster than that for the channel-forming protein, gramicidin A. In addition to the polarity of the nanotube, the resistance to cleavage by peptidases may render these nanotubes useful as cytotoxic agents.

Ghadiri has also examined the properties of size selection in nanotubes built with cyclic D,L-α-peptides.[40] The nanotubes were embedded in alkanethiol monolayers on gold and characterized by grazing-angle FTIR, cyclic voltammetry and impedance spectroscopy (Fig. 10.39). FTIR data indicate that the orientation of the nanotube depends on the sequence of assembly. The nanotubes are parallel to the gold surface if they are coadsorbed with the alkanethiols. If the nanotubes are added subsequent to SAM formation, their channel axes are perpendicular to the surface. The movement of different redox-

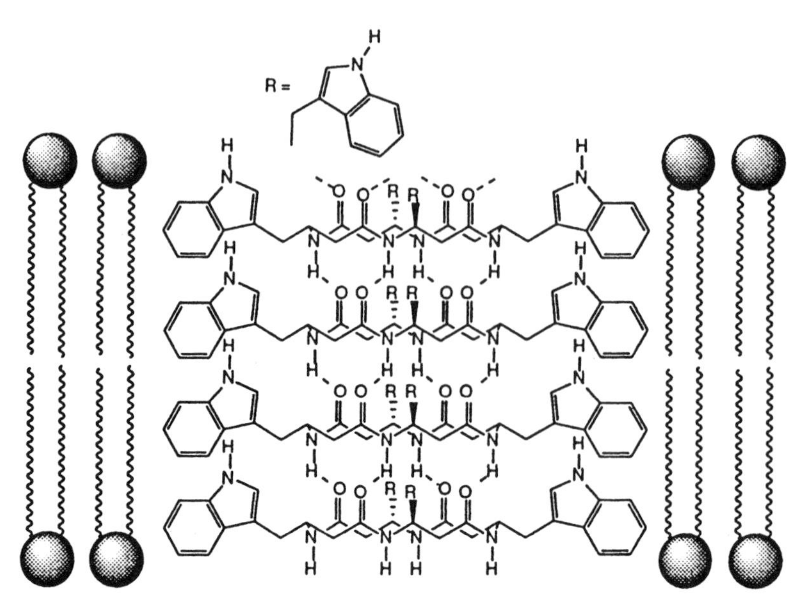

FIGURE 10.38. Ion-conducting channel comprised of cyclic β^3-peptides. FT-IR data indicate extensive hydrogen-bonded networks (N–H, $3289\,cm^{-1}$; C=O, $1607\,cm^{-1}$ and $1549\,cm^{-1}$).[41]

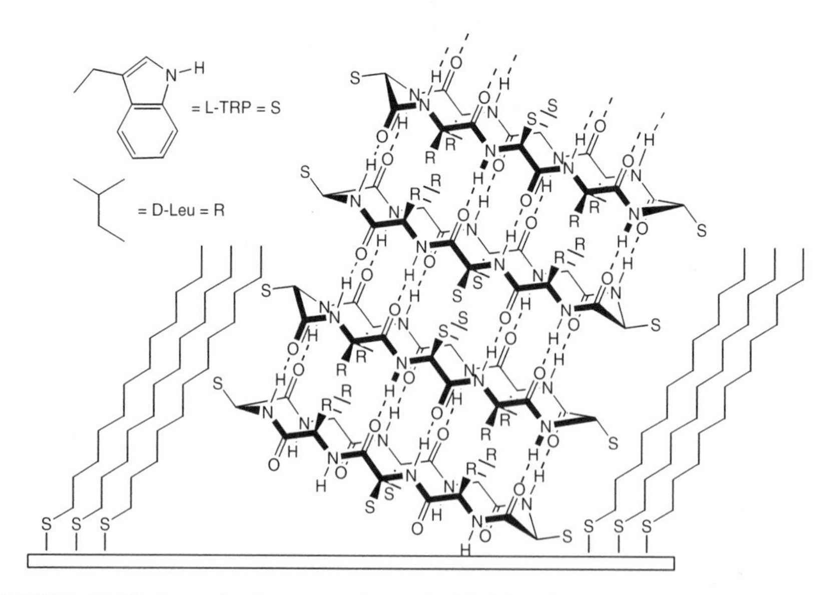

FIGURE 10.39. Ion-selective nanotube embedded in alkanethiol monolayer on gold surface. The N–H stretch at $3280\,cm^{-1}$ corresponds to an intermolecular N–O distance of 2.8–2.9 Å.[40]

active ions through the tubular channels was compared using cyclic voltammetry: $[Fe(CN)_6]^{3-}$ and $[Ru(NH_3)_6]^{3+}$ ions move through the channels whereas the larger $[Mo(CN)_8]^{4-}$ ions do not. Impedance spectroscopy was used to measure the capacitance of the various films. The capacitances of bare gold, SAM coated gold, and nanotubes-SAM coated gold in 10 mM $Ca(NO_3)_2$ are 33.92, 1.06 and 1.22 $\mu F\,cm^{-2}$, respectively. The slight increase in capacitance observed when nanotubes are present in the SAM relative to when they are absent is thought to be the result of ion penetration through the nanotubes.

Biomimetic structures are the primary application of amides in studies in solution. In the absence of solvent, the primary application of amides begins to shift from that of biomimetic structures to that of new materials. For example, amides are found as integral components in "soft" structures such as self-assembled monolayers, polymers, and liquid crystals, and "hard" structures such as molecular solids. The remainder of this section highlights some recent examples of the role of amides in the development of materials.

3.2.3. Self-assembled Monolayers (SAMs). The enormous utility of SAMs derives from their simplicity of preparation (i.e., dipping a gold surface into a solution containing alkanethiols) and their seemingly limitless surface chemistry. Consequently, SAMs have been used to study the properties of wetting on surfaces, rates of electron transfer across defined distances, mechanisms of corrosion, and kinetics of protein adsorption. In addition to these studies, SAMs are being used to prepare biosensors, or to fabricate small structures using "soft lithography", a methodology developed by Whitesides.[112] Alkanethiols with terminal functional groups ($X{-}(CH_2)_n{-}SH$) are the compounds most often used to prepare SAMs on gold surfaces. Van der Waals interactions between adjacent alkyl chains stabilize these monolayers in a manner analogous to LB-films. Hutchinson has further stabilized SAMs by incorporating amides in the internal segments of the alkyl chains (Fig. 10.40), which form hydrogen bonds with amides of three neighboring thiols.[113]

3.2.4. Liquid Crystals. Hydrogen-bonding interactions can be used to promote the aggregation of molecules into liquid crystalline phases. For example, molecules shaped as half-disks with a pyridone core dimerize into disk-like aggregates that form discotic liquid crystalline phases (Fig. 10.41).[56] Despite changes in the length of the alkyl chains ($n = 8$, 10, 12), these compounds all undergo transitions from crystalline to mesophase phase (40–70 $J\,g^{-1}$), and from mesophase to isotropic phase (2–3 $J\,g^{-1}$). Over a range of temperature that spans all three phases (70–170°C), the frequency of the C=O band (1670 cm^{-1}) in the IR spectrum of the *O*-dodecyl derivative remains constant. This evidence, in conjunction with the absence of a sharp N–H band near 3440 cm^{-1} or a high-frequency carbonyl band in the IR spectrum, indicates that the hydrogen-bonded pyridone disks remain intact in all three phases. Furthermore, conic textures appear in the mesophase, which provides further evidence that the pyridone dimers stack into hexagonally packed columns.

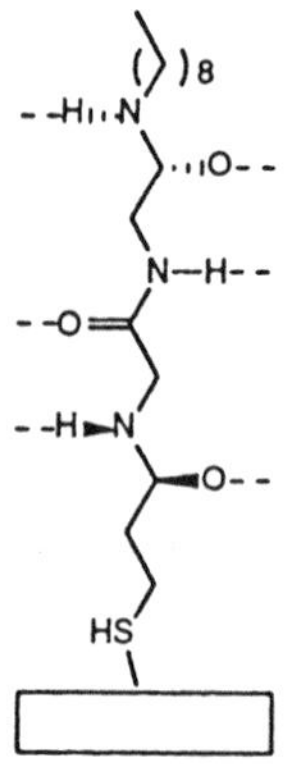

FIGURE 10.40. A thiol monomer used to prepare a SAM stabilized through intermolecular hydrogen bonds, which desorbed from gold in argon at 175°C as compared to 50°C for SAMs made from octadecanethiol.[113]

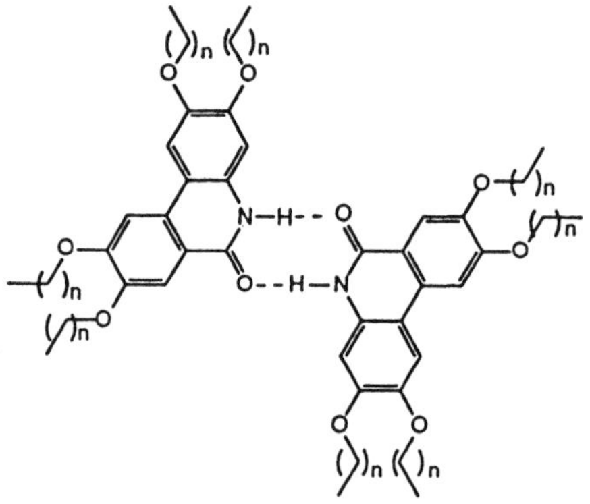

FIGURE 10.41. Formation of hydrogen bonds between two derivatives of pyridone results in the formation of disk-shaped molecules that stack into columns.[56]

3.2.5. Polymers. Although many electrical applications of polymeric materials rely on their insulating properties, recent efforts have successfully produced highly conductive polymers (e.g., the conductivity of polyacetylene is $1 \times 10^5\ \mathrm{S\,cm^{-1}}$ at room temperature). The insolubility of many of the most conductive polymers, however, is an unattractive characteristic from a practical standpoint. Increasing the solubility of conducting polymers with various substituents is often accompanied by a decrease in conductivity. This reduction in conductivity occurs because the polymer adopts a conformation to minimize the steric interactions between substituents, which consequently misaligns the pathway for the conduction of electrons and holes (i.e., parallel π-orbitals). In an effort to increase the solubility of polyarylenes without interrupting coplanar sequences of fused rings, Meijer has exploited strong intramolecular N–H–N hydrogen bonds to force adjacent aromatic rings to become coplanar

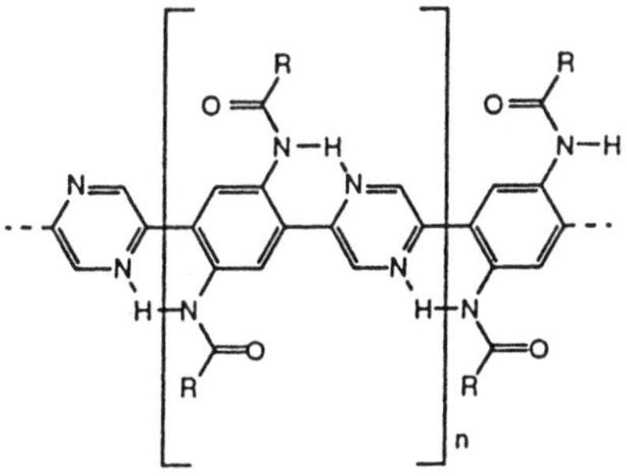

FIGURE 10.42. Ladder polymer (uninterrupted sequence of rings) with adjacent aryl rings forced to be coplanar using strong intramolecular hydrogen bonds. R = –O–tBu or $-C(CH_3)_2(CH_2)_7CH_3$.[114]

(Fig. 10.42).[114] Spectroscopic data for the polymer where R = $-C(CH_3)_2(CH_2)_7CH_3$ supports the effectiveness of intramolecular hydrogen bonds to force adjacent rings coplanar: the amide N–H proton absorption in the ^{1}H-NMR spectrum ($CDCl_3$) occurs at low field (δ 11.7 ppm); the frequency of the N–H stretch in the IR spectrum occurs at 3291 cm^{-1}; the redox potential in THF is −1.32 V versus SCE. The wavelength of the visible absorption band of the ladder polymer can be tuned from 446 to 369 nm simply by adding trifluoroacetic acid, which disrupts the intramolecular hydrogen bonds and decreases conjugation.

3.3. Structures in Solids

As the technology of information storage moves toward increasing densities, new approaches to the fabrication of small devices are required. One approach is to build devices from molecular building blocks, in a manner analogous to building a house from bricks and mortar. Crystalline solids are attractive in this respect because they comprise millions of molecules that self-assemble with high levels of precision into well-defined structures. Understanding the principles necessary to fabricate a device by this approach would have a profound impact on the fields of chemistry, materials science, and engineering by making possible the construction of such devices by deliberate design. Moreover, an understanding of the relationship between molecular structure and crystal structure would provide insight into the design of other types of solids with properties that depend on the arrangement of their constituent molecules. Numerous solids have been investigated, with a few examples that show interesting new properties or function highlighted below.

3.3.1. Fibers and Plates. Shimizu has prepared a series of 1-glucosamide bolaamphiphiles to study the influence that the hydrocarbon link has on the resulting self-assembled supramolecular structure (Fig. 10.43).[115] For example, fibrous assemblies form when the hydrocarbon link contains an even number of

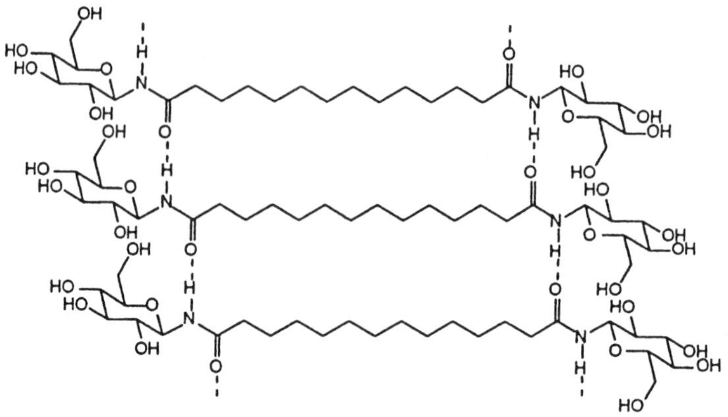

FIGURE 10.43. A 1-glucosamide bolaamphiphile with an even-numbered hydrocarbon link. The amide carbonyl groups within the same molecule are pointed in opposite directions, promoting the formation of a hydrogen-bonded fiber.[115]

methylene units, while planar platelets or amorphous solids form when the number of methylene units is odd. The strength of hydrogen bonds in the fibrous assemblies is weaker than that in the platelets or amorphous solids, as indicated by IR studies. The C=O stretching band (Amide I) and the N–H deformation band (Amide II) in the IR spectra of even-numbered bolaamphiphiles appear at 1657–1659 and 1534–1536 cm^{-1}, respectively. In comparison, the C=O stretching band (Amide I) and the N–H deformation band (Amide II) in the IR spectra of odd-numbered bolaamphiphiles appear at 1638–1640 and 1547–1549 cm^{-1}, respectively.

3.3.2. Ribbons, Tapes, and Layers. Mascal has prepared hydrogen-bonded ribbons in crystalline solids that are assembled from *N*-(3-hydroxypropyl)cyanuric acid and 5-butyl-2,4,6-triaminopyrimidine.[36] The molecules used to assemble these ribbons are not only complementary in terms of their hydrogen-bonding donors and acceptors, but their relative pK_a's promote the transfer of a proton from cyanuric acid to pyrimidine. The result is a robust ribbon that consists of alternating pyrimidinium cations and cyanurate anions that are hydrogen bonded to each other (Fig. 10.44). Consequently, this solid should act as a conductor of protons, although the measurement of this property was not reported.

Another robust motif that has appeared in molecular solids is the hydrogen-bonded tape derived from 2,5-diketopiperazines.[35] These tapes have been shown to withstand the steric pressure of bulky spirocycloalkyl groups that occupy a volume as large as 290 $Å^3$ (Fig. 10.45). Moreover, by tethering carboxylic acids to the hydrogen-bonded tapes, the tapes can then function as scaffolds with which to position guest molecules within the crystalline lattice (Fig. 10.46).[44] Depending on the guest, tapes can be cross-linked into layers or simply used to interdigitate guests on adjacent tapes. This modularity—i.e., the ability to

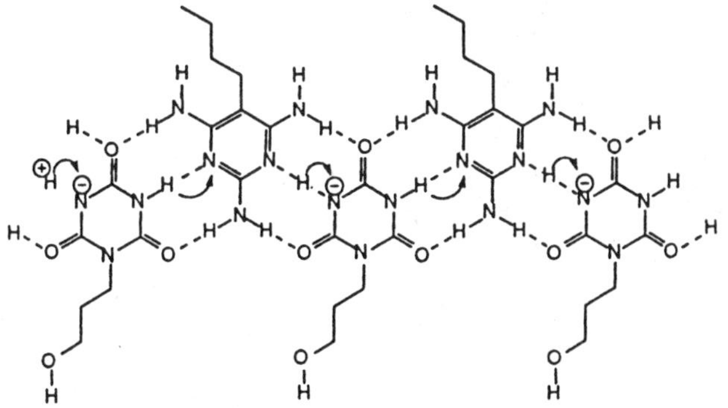

FIGURE 10.44. Derivatives of cyanuric acid and triaminopyrimidine form a proton-conducting ribbon (p$K_a = 7.29 \pm 0.02$ and 7.06 ± 0.10, respectively).[36]

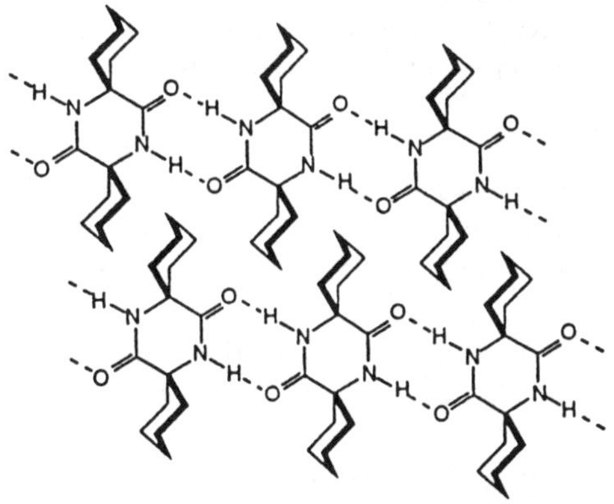

FIGURE 10.45. Two hydrogen-bonded tapes from 2,5-diketopiperazine with their spirocyclohexane rings interdigitated.[35]

interchange one guest molecule for another—provides a powerful and convenient method to manipulate the physical properties of these crystalline materials without significantly altering the framework of tapes that define the supramolecular structure within the crystalline lattice. Understanding what patterns of packing these tapes and layers adopt will provide insight into how tapes composed of diketopiperazines can be used to engineer the structure and function of crystalline solids (i.e., solids with magnetic, conductive, or optical properties).

3.3.3. Diamondoids. Three-dimensional supramolecular structures can be achieved using tetrahedral or octahedral building blocks. Munakata has used the amide functional group in pyridinone to prepare coordination polymers from copper(I) complexes.[14] The ligand, 3-cyano-6-methyl-2(1H)-pyridinone (Hcmp), possesses both a coordination group (CN) and a hydrogen-bonding site

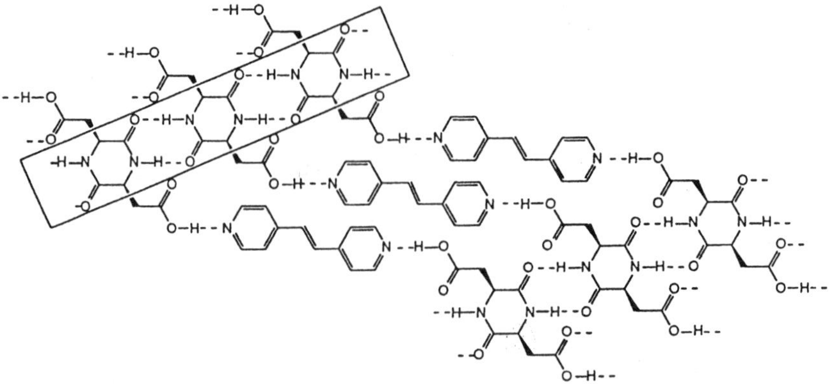

FIGURE 10.46. Two hydrogen-bonded tapes from the aspartic acid derivative of 2,5-diketopiperazine (the box outlines one tape). The tapes function as scaffolds for positioning guests through hydrogen-bonding interactions between the carboxylic acid donor and the pyridyl acceptor. In this example, the guest molecule, (E)-1,2-*bis*(4-pyridyl)ethene, cross-links tapes to form a layered structure.[44]

(O--H–N). The reaction between Hcmp and a copper(I) salt results in a three-dimensional framework of tetrahedral $[Cu(Hcmp)_4]X$ centers, where $X = ClO_4^-$, BF_4^-, PF_6^- and $CF_3SO_3^-$. The tetrahedral units are linked by intermolecular hydrogen bonds between pyridone nitrogen and oxygen atoms (Fig. 10.47). Two different types of frameworks result depending on the pattern of hydrogen bonds between the pyridones. When the pattern is $R_2^2(8)$, a diamondoid framework forms; when the pattern is C(4), square channels form. This polymorphism is interesting because it demonstrates how a simple change in the pattern of hydrogen bonds dramatically alters the overall framework, even when the same building block is used. In this system, it appears that the counter anions serve as templates around which the hydrogen-bonded network assembles.

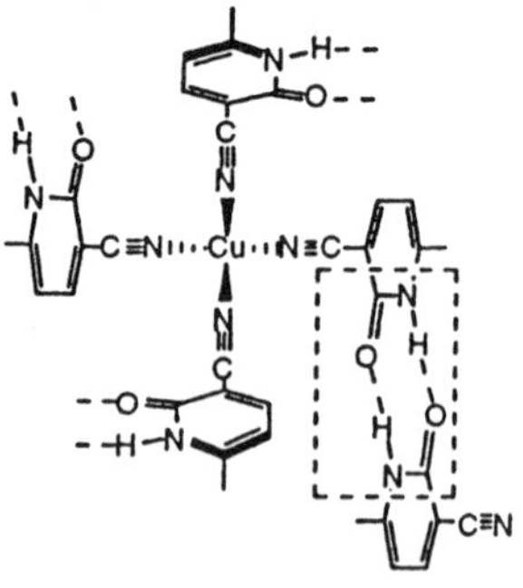

FIGURE 10.47. Tetrahedral units of $[Cu(Hcmp)_4]$ are linked by intermolecular hydrogen bonds between pyridone nitrogen and oxygen atoms. The $R_2^2(8)$ pattern of hydrogen bonds between pyridones (outlined dashed box) generates a diamondoid framework.[14]

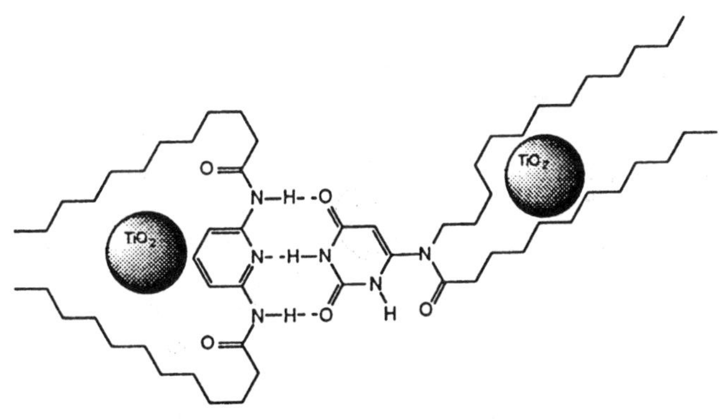

FIGURE 10.48. Alkylated monomers used to assemble a superlattice of TiO_2.[116]

3.3.4. Ordered Arrays of Nanocrystallites. An interesting approach to assembling nanocrystallites into ordered arrays has been achieved by exploiting the complementarity between the hydrogen-bonding groups of a 2,6-diamidopyridine and uracil.[116] Nanocrystallites of TiO_2 entangled within the alkyl chains of derivatives of 2,6-amidopyridine and uracil self-organize in the manner illustrated in Fig. 10.48 into loosely packed mesoaggregates. The result is an aggregate greater than 1000 nm in diameter with 12 ± 2 Å nanocrystallites organized into planes that are separated by 6 ± 2 Å. This level of aggregation and order is possible only with complementary components since the self-aggregation of the individual components never reaches diameters greater than 50 nm.

4. CONCLUDING REMARKS

The supramolecular chemistry of amides is already having a significant impact on the development of new directions of research. Studies highlighted in this chapter show emerging applications for supramolecular aggregates of amides that include peptidomimetic systems that function as hybrid "genes", non-natural α-helices and β-sheets to control the tertiary structure of proteins, artificial membranes that mimic receptor proteins found on the surface of cell membranes, ion-channels with variable pore-sizes, receptors for biosignaling and biocatalysis, molecular "containers" for site-specific recognition and delivery of drugs, as well as aggregates that mimic natural photosynthetic systems. Recognition of biomimetics as powerful new tools for biotechnology and the treatment of disease is evident by their recent commercialization (e.g., PerSeptive Biosystems has commercialized an instrument capable of synthesizing PNA).

The tremendous potential of supramolecular aggregates for the design of materials also is beginning to emerge. Already new methods exist for preparing liquid crystals with properties defined by the shape of aggregates instead of individual molecules, polymers that conduct electrons or protons, ion-selective films, and molecular sieves. In addition, supramolecular aggregates of amides can serve as substrates for the nucleation of crystals or as interfaces with biological systems. Moreover, they exist via self-assembly, as novel nanostructures such as fibers and plates, and as stabilized structures within membranes, at interfaces and on surfaces. Although the commercialization of materials based on supramolecular aggregates is only in its infancy, applications of these materials should continue to grow as our understanding of supramolecular synthesis develops.

Clearly, research has only begun to explore the power of self-assembly and its manifestation as supramolecular structures in solution, at interfaces, and in solids. These examples establish, however, that the supramolecular chemistry of amides, in fact, is being exploited to produce supramolecular structures with remarkable properties and function that are a consequence of the hydrogen-bonding behavior of amides. No doubt, the continued exploration and subsequent mastery of the complex and rich chemistry of supramolecular structures will revolutionize science in the next century.

REFERENCES

1. Jeffrey, G. A.; Saenger, W. *Hydrogen Bonding in Biological Structures*; Springer-Verlag: Berlin, Heidelberg, New York, 1991.
2. Pauling, L. *The Nature of the Chemical Bond*; Cornell University Press: Ithaca, NY, 1939.
3. Pauling, L.; Corey, R. B.; Branson, H. R. *Proc. Nat. Acad. Sci.* 1951, **37**, 205–211.
4. Watson, J. D.; Crick, F. H. C. *Nature* 1953, **171**, 737–738.
5. Demming, T. J. *Adv. Mater.* 1997, **9**, 299–311.
6. Lawrence, D. S.; Jiang, T.; Levett, M. *Chem. Rev.* 1995, **95**, 2229–2260.
7. Rebek, J. *Chem. Soc. Rev.* 1996, 255–264.
8. Paleos, C. M.; Tsiourvas, D. *Adv. Mater.* 1997, **9**, 695–710.
9. Philp, D.; Stoddart, J. F. *Angew. Chem. Int. Ed. Engl.* 1996, **35**, 1155–1196.
10. Fabbrizzi, L.; Poggi, A. *Chem. Soc. Rev.* 1995, 197–202.
11. Desiraju, G. R. *Crystal Engineering: The Design of Organic Solids*; Elsevier: New York, 1989, Vol. 54.
12. Schmidt, G. M. J. *Pure Appl. Chem.* 1971, **27**, 647.
13. Harris, K. D. M. *Chem. Soc. Rev.* 1997, **26**, 279–289.
14. Munakata, M.; Wu, L. P.; Kuroda-Sowa, T. *Bull. Chem. Soc. Jpn.* 1997, **70**, 1727–1743.
15. MacDonald, J. C.; Whitesides, G. M. *Chem. Rev.* 1994, **94**, 2383–2420.
16. Fan, E.; Vicent, C.; Geib, S. J.; Hamilton, A. D. *Chem. Mater.* 1994, **6**, 1113–1117.
17. Aakeröy, C. B.; Seddon, K. R. *Chem. Soc. Rev.* 1993, 397–407.
18. Etter, M. C. *J. Phys. Chem.* 1991, **95**, 4601–4610.
19. Leiserowitz, L.; Schmidt, G. M. J. *J. Chem. Soc.* 1969, **A**, 2372–2382.
20. Leiserowitz, L.; Tuval, M. *Acta Crystallogr.* 1978, **B34**, 1230–1247.

21. Leiserowitz, L.; Hagler, A. T. *Proc. R. Soc. Lond.* 1983, **A338**, 133–175.
22. Weinstein, S.; Leiserowitz, L. *Acta Crystallogr.* 1980, **B36**, 1406–1418.
23. Weinstein, S.; Leiserowitz, L.; Gil-Av, E. *J. Am. Chem. Soc.* 1980, **102**, 2768–2772.
24. Tokunaga, Y.; Rebek, J., Jr. *J. Am. Chem. Soc.* 1998, **120**, 66–69.
25. Bergeron, R. J.; IV, O. P.; Yao, G. W.; Milstein, S.; Weimar, W. R. *J. Am. Chem. Soc.* 1994, **116**, 8479–8484.
26. Mascal, M.; Hext, N. M.; Warmuth, R.; Moore, M. H.; Turkenburg, J. P. *Angew. Chem. Int. Ed. Engl.* 1996, **35**, 2204–2206.
27. Zerkowski, J. A.; Seto, C. T.; Whitesides, G. M. *J. Am. Chem. Soc.* 1992, **114**, 5473–5475.
28. Fan, E.; Yang, J.; Geib, S. J.; Stoner, T. C.; Hopkins, M. D.; Hamilton, A. D. *J. Chem. Soc., Chem. Commun.* 1995, 1251–1252.
29. Geib, S. J.; Vicent, C.; Fan, E.; Hamilton, A. D. *Angew. Chem. Int. Ed. Engl.* 1993, **32**, 119–121.
30. Zerkowski, J. A.; MacDonald, J. C.; Seto, C. T.; Wierda, D. A.; Whitesides, G. M. *J. Am. Chem. Soc.* 1994, **116**, 2382–2391.
31. Zerkowski, J. A.; MacDonald, J. C.; Whitesides, G. M. *Chem. Mater.* 1994, **6**, 1250–1257.
32. Schwiebert, K. E.; Chin, D. N.; MacDonald, J. C.; Whitesides, G. M. *J. Am. Chem. Soc.* 1996, **118**, 4018–4029.
33. Lewis, F. D.; Yang, J.-S.; Stern, C. L. *J. Am. Chem. Soc.* 1996, **118**, 12029–12037.
34. Simanek, E. E.; Tsoi, A.; Wang, C. C. C.; Whitesides, G. M.; McBride, M. T.; Palmore, G. T. R. *Chem. Mat.* 1997, **9**, 1954–1961.
35. Palacin, S.; Chin, D. N.; Simanek, E. E.; MacDonald, J. C.; Whitesides, G. M.; McBride, M. T.; Palmore, G. T. R. *J. Am. Chem. Soc.* 1997, **119**, 11807–11816.
36. Mascal, M.; Fallon, P. S.; Batsanov, A. S.; Heywood, B. R.; Champ, S.; Colclough, M. *J. Chem. Soc., Chem. Commun.* 1995, 805–806.
37. Lehn, J.-M.; Mascal, M.; DeCian, A.; Fischer, J. *J. Chem. Soc., Perkin Trans. 2* 1992, 461–467.
38. Ghadiri, M. R.; Granja, J. R.; Milligan, R. A.; McRee, D. E.; Khazanovich, N. *Nature* 1993, **366**, 324–327.
39. Khazanovich, N.; Granja, J. R.; McRee, D. E.; Milligan, R. A.; Ghadiri, M. R. *J. Am. Chem. Soc.* 1994, **116**, 6011–6012.
40. Motesharei, K.; Ghadiri, M. R. *J. Am. Chem. Soc.* 1997, **1997**, 11306–11312.
41. Clark, T. D.; Buehler, L. K.; Ghadiri, M. R. *J. Am. Chem. Soc.* 1998, **120**, 651–656.
42. Pedireddi, V. R.; Chatterjee, S.; Ranganathan, A.; Rao, C. N. R. *J. Am. Chem. Soc.* 1997, **119**, 10867–10868.
43. Schauer, C. L.; Matwey, E.; Fowler, F. W.; Lauher, J. W. *J. Am. Chem. Soc.* 1997, **119**, 10245–10246.
44. Palmore, G. T. R.; McBride, M. T. *Chem. Commun.* 1998, 145–146.
45. Zhao, X.; Chang, Y.-L.; Fowler, F. W.; Laugher, J. W. *J. Am. Chem. Soc.* 1990, **112**, 6627–6634.
46. Chang, Y.-L.; West, M.-A.; Fowler, F. W.; Lauher, J. W. *J. Am. Chem. Soc.* 1993, **115**, 5991–6000.

47. Toledo, L. M.; Lauher, J. W.; Fowler, F. W. *Chem. Mater.* 1994, **6**, 1222–1226.
48. Harris, K. D. M.; Stainton, N. M.; Callan, A. M.; Howie, R. A. *J. Mater. Chem.* 1993, **3**, 947–952.
49. Garcia-Tellado, F.; Geib, S. J.; Goswami, S.; Hamilton, A. D. *J. Am. Chem. Soc.* 1991, **113**, 9265–9269.
50. Hollingsworth, M. D.; Santarsiero, B. D.; Oumar-Mahamat, H.; Nichols, C. J. *Chem. Mater.* 1991, **3**, 23–25.
51. Hollingsworth, M. D.; Brown, M. E.; Santarsiero, B. D.; Huffman, J. C.; Goss, C. R. *Chem. Mat.* 1994, **6**, 1227–1244.
52. Kane, J. J.; Liao, R.-F.; Lauher, J. W.; Fowler, F. W. *J. Am. Chem. Soc.* 1995, **117**, 12003–12004.
53. Coe, S.; Kane, J. J.; Nguyen, T. L.; Toledo, L. M.; Wininger, E.; Fowler, F. W.; Lauher, J. W. *J. Am. Chem. Soc.* 1997, **119**, 86–93.
54. Aakeröy, C. B.; Hughes, D. P.; Nieuwenhuyzen, M. *J. Am. Chem. Soc.* 1996, **118**, 10134–10140.
55. Simard, M.; Su, D.; Wuest, J. D. *J. Am. Chem. Soc.* 1991, **113**, 4696–4698.
56. Kleppinger, R.; Lillya, C. P.; Yang, C. *J. Am. Chem. Soc.* 1997, **119**, 4097–4102.
57. Leiserowitz, L. *Acta Crystallogr.* 1977, **B33**, 2719–2733.
58. Kamlet, M. J.; Abboud, J.-L. M.; Abraham, M. H.; Taft, R. W. *J. Org. Chem.* 1983, **48**, 2877–2887.
59. Etter, M. C. *Acc. Chem. Res.* 1990, **23**, 120–126.
60. Etter, M. C.; MacDonald, J. C.; Bernstein, J. *Acta Crystallogr.* 1990, **B46**, 256–262.
61. Bernstein, J.; Davis, R. E.; Shimoni, L.; Chang, N.-L. *Angew. Chem. Int. Ed. Engl.* 1995, **34**, 1555–1573.
62. Prasad, P. N.; Williams, D. J.; John Wiley & Sons, Inc.: New York, 1991, pp. 78.
63. Allen, F. H.; Kennard, O. *Chemical Design Automation News* 1993, **8**, 31–37.
64. Bernstein, J. *J. Phys. D: Appl. Phys.* 1993, **26**, B66–B76.
65. Bernstein, J.; Etter, M. C.; MacDonald, J. C. *J. Chem. Soc., Perkin Trans. 2* 1990, 695–698.
66. Bernstein, J. In *Computer-Assisted Modeling of Receptor-Ligand Interactions: Theoretical Aspects and Applications to Drug Design*; Rein, R.; Golombek, A. Ed.; Liss: New York, 1988, pp. 203–215.
67. Bernstein, J. In *Organic Solid State Chemistry*; Desiraju, G. R. Ed.; Elsevier: New York, 1987; Vol. 32, pp. 471–518.
68. Bernstein, J.; Etter, M. C.; MacDonald, J. C. *J. Chem. Soc. Perkin Trans. 2* 1990, 695–698.
69. Zerkowski, J. A.; MacDonald, J. C.; Whitesides, G. M. *Chem. Mater.* 1997, **9**, 1933–1941.
70. Etter, M. C.; Urbanczyk-Lipkowska, Z.; Zia-Ebrahimi, M.; Panunto, T. W. *J. Am. Chem. Soc.* 1990, **112**, 8415–8426.
71. Etter, M. C.; Reutzel, S. M. *J. Am. Chem. Soc.* 1991, **113**, 2586–2598.
72. Ducharme, Y.; Wuest, J. D. *J. Org. Chem.* 1988, **53**, 5787–5789.
73. Wash, P. L.; Maverick, E.; Chiefari, J.; Lightner, D. A. *J. Am. Chem. Soc.* 1997, **119**, 3802–3806.

74. Yang, J.; Fan, E.; Geib, S. J.; Hamilton, A. D. *J. Am. Chem. Soc.* 1993, **115**, 5314–5315.
75. Karle, I. L.; Rangathanan, D.; Haridas, V. *J. Am. Chem. Soc.* 1997, **119**, 2777–2783.
76. Lehn, J.-M.; Mascal, M.; DeCian, A.; Fischer, J. *J. Chem. Soc., Chem. Commun.* 1990, 479–481.
77. Zerkowski, J. A.; Seto, C. T.; Wierda, D. A.; Whitesides, G. M. *J. Am. Chem. Soc.* 1990, **112**, 9025–9026.
78. Bridson, J. N.; Schriver, M. J.; Zhu, S. *J. Chem. Crystallog.* 1995, **25**, 11–14.
79. Kräuthauser, S.; Christianson, L. A.; Powell, D. R.; Gellman, S. H. *J. Am. Chem. Soc.* 1997, **119**, 11719–11720.
80. Nowick, J. S.; Mahrus, S.; Smith, E. M.; Ziller, J. W. *J. Am. Chem. Soc.* 1996, **118**, 1066–1072.
81. Nowick, J. S.; Holmes, D. L.; Mackin, G.; Noronha, G.; Shaka, A. J.; Smith, E. M. *J. Am. Chem. Soc.* 1996, **118**, 2764–2765.
82. Nowick, J. S.; Insaf, S. *J. Am. Chem. Soc.* 1997, **119**, 10903–10908.
83. Seebach, D.; Overhand, M.; Kühnule, F. N. M.; Lengweiler, U. D. *Helv. Chim. Acta* 1996, **79**, 913–941.
84. Seebach, D.; Matthews, J. L. *Chem. Commun.* 1997, 2015–2022.
85. Bass, B. L.; Cech, T. R. *Nature* 1984, **308**, 820–826.
86. Cech, T. R. *Gene* 1993, **135**, 33–36.
87. Zaug, A. J.; Cech, T. R. *Science* 1985, **229**, 1060–1064.
88. Zaug, A. J.; Cech, T. R. *Science* 1986, **231**, 470–475.
89. Nielsen, P. E.; Haaima, G. *Chem. Soc. Rev.* 1997, 73–78.
90. Schmidt, J. G.; Nielsen, P. E.; Orgel, L. E. *Nucleic Acids Res.* 1997, **25**, 4797–4802.
91. Schmidt, J. G.; Christensen, L.; Nielsen, P. E.; Orgel, L. E. *Nucleic Acids Res.* 1997, **25**, 4792–4796.
92. Böhler, C.; Nielsen, P. E.; Orgel, L. E. *Nature* 1995, **376**, 578–581.
93. Hartgerink, J. D.; Granja, J. R.; Milligan, R. A.; Ghadiri, M. R. *J. Am. Chem. Soc.* 1996, **118**, 43–49.
94. Jeon, Y.-M.; Kim, J.; Whang, D.; Kim, K. *J. Am. Chem. Soc.* 1996, **118**, 9790–9791.
95. Bell, T. W.; Hou, Z. *Angew. Chem. Int. Ed. Engl.* 1997, **36**, 1536–1538.
96. Reek, J. N. H.; Priem, A. H.; Englekamp, H.; Rowan, A. E.; Elemans, J. A. A. W.; Nolte, R. J. M. *J. Am. Chem. Soc.* 1997, **119**, 9956–9964.
97. Bisson, A. P.; Lynch, V. M.; Monahan, M.-K. C.; Anslyn, E. V. *Angew. Chem. Int. Ed. Engl.* 1997, **36**, 2340–2342.
98. Otsuki, J.; Russell, K. C.; Lehn, J.-M. *Bull. Chem. Soc. Jpn.* 1997, **70**, 671–679.
99. Linton, B.; Hamilton, A. *Chemtech* 1997, 34–40.
100. Ge, Y.; Lilienthal, R. R.; Smith, D. K. *J. Am. Chem. Soc.* 1996, **118**, 3976–3977.
101. Szemes, F.; Hesek, D.; Chen, Z.; Dent, S. W.; Drew, M. G. B.; Goulden, A. J.; Graydon, A. R.; Grieve, A.; Mortimer, R. J.; Wear, T.; Weightman, J. S.; Beer, P. D. *Inorg. Chem.* 1996, **35**, 5868–5879.
102. Beer, P. D. *Acc. Chem. Res.* 1998, **31**, 71–80.

103. Gust, D.; Moore, T. A.; Moore, A. L.; Lee, S.-J.; Bittersmann, E.; Luttrull, D. K.; Rehms, A. A.; DeGraziano, J. M.; Ma, X. C.; Gao, F.; Belford, R. E.; Trier, T. T. *Science* 1990, **248**, 199–201.
104. Collin, J.-P.; Guillarez, S.; Sauvage, J.-P.; Barigelletti, F.; Cola, L. D.; Flamigni, L.; Balzani, V. *Inorg. Chem.* 1991, **30**, 4230–4238.
105. Christ, C. S.; Yu, J.; Zhao, X.; Palmore, G. T. R.; Wrighton, M. S. *Inorg. Chem.* 1992, **31**, 4439–4440.
106. Armaroli, N.; Barigelletti, F.; Calogero, G.; Flamigni, L.; White, C. M.; Ward, M. D. *Chem. Commun.* 1997, 2181–2182.
107. Cusack, L.; Rao, S. N.; Fitzmaurice, D. *Chem.-A Eur. J.* 1997, **3**, 202–207.
108. Cha, X.; Ariga, K.; Kunitake, T. *J. Am. Chem. Soc.* 1996, **118**, 9545–9551.
109. Ariga, K.; Kamino, A.; Koyano, H.; Kunitake, T. *J. Mater. Chem.* 1997, **7**, 1155–1161.
110. Palmore, G. T. R.; Smith, D. K.; Wrighton, M. S. *J. Phys. Chem. B* 1997, **101**, 2437–2450.
111. Wrighton, M. S.; Palmore, G. T. R.; Hable, C. T.; Crooks, R. M. In *New Aspects of Organic Chemistry*; Yoshida, Z.; Shiba, T.; Ohshiro, Y.; Ed.; VCH Publishers: New York, 1989; Vol. 4, pp. 277–302.
112. Xia, Y.; Whitesides, G. M. *Angew. Chem. Int. Ed.* 1998, **37**, 550–575.
113. Clegg, R. S.; Reed, S. M.; Hutchinson, J. E. *J. Am. Chem. Soc.* 1998, **120**, 2486–2487.
114. Delnoye, D. A. P.; Sijbesma, R. P.; Vekemans, J. A. J. M.; Meijer, E. W. *J. Am. Chem. Soc.* 1996, **118**, 8717–8718.
115. Shimizu, T.; Masuda, M. *J. Am. Chem. Soc.* 1997, **119**, 2812–2818.
116. Cusack, L.; Rizza, R.; Gorelov, A.; Fitzmaurice, D. *Angew. Chem. Int. Ed. Engl.* 1997, **36**, 848–851.

CHAPTER 11

β-LACTAM ANTIBACTERIAL AGENTS: COMPUTATIONAL CHEMISTRY INVESTIGATIONS

DONALD B. BOYD
Department of Chemistry, Indiana University—Purdue University at Indianapolis (IUPUI)

1. INTRODUCTION

An amide bond in a four-membered ring is an interesting moiety in its own right, but additionaly the β-lactam ring is a key structural component of a major class of antibacterial agents. (The common term "antibiotics," when used correctly, refers to antimicrobial compounds isolated from natural sources, whereas the term "antibacterial agents" includes both natural products and the tens of thousands of analogs that are manmade.) The first members of the β-lactam family were penicillins: these date back to the discovery by the Scottish bacteriologist Alexander Fleming in 1928. The β-lactam family expanded following the 1945 discovery of cephalosporins by the Sardinian bacteriologist Guiseppe Brotzu. The discoveries were turned into useful drugs by large teams of scientists in the United Kingdom and the United States.[1] Both penicillins and cephalosporin are based on a bicyclic nucleus (Fig. 11.1), and it was thought in the 1960s and 1970s that the geometrical distortion of the β-lactam moiety caused by the fused five- or six-membered ring was essential for biological activity.

By the early 1980s, scientists had learned that a bicyclic system was not required; certain monocyclic β-lactams also exhibit respectable levels of antibacterial activity. Recall that a pharmacophore is defined as the minimal features of molecular structure essential for evoking an intended biological response. As additional β-lactam antibacterial agents were discovered in the 1970s and early 1980s, the concept of their pharmacophore simplified until it

The Amide Linkage: Selected Structural Aspects in Chemistry, Biochemistry, and Materials Science, Edited by Arthur Greenberg, Curt M. Breneman, and Joel F. Liebman
ISBN 0-471-35893-2

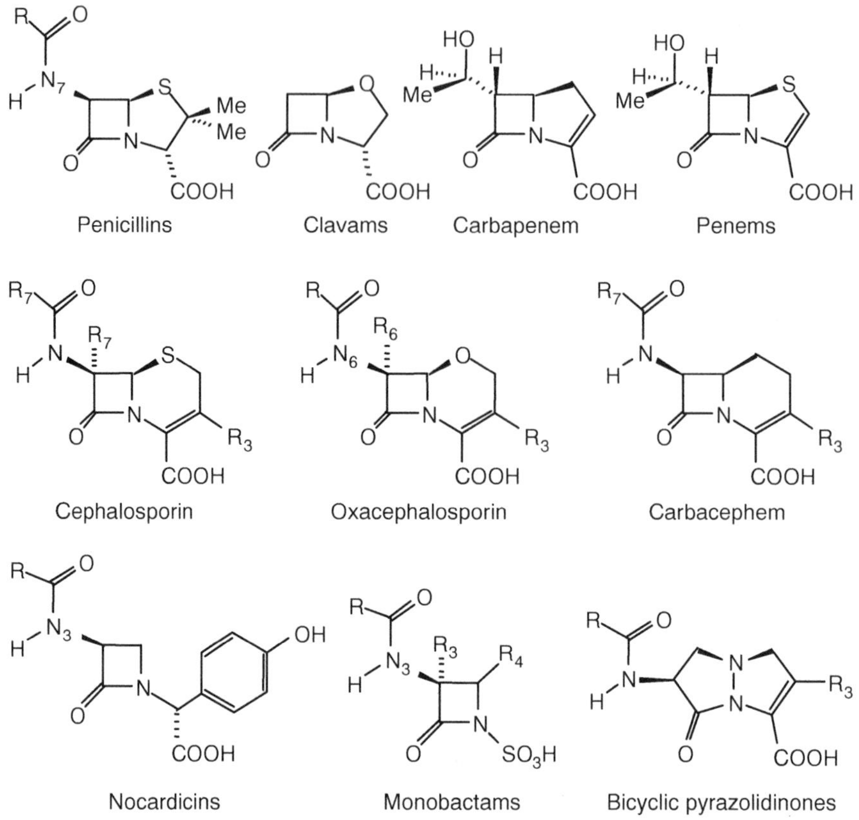

Penicillins Clavams Carbapenem Penems

Cephalosporin Oxacephalosporin Carbacephem

Nocardicins Monobactams Bicyclic pyrazolidinones

FIGURE 11.1. Examples of some types of β-lactam antibacterial agents. In general, of all the possible diastereomers of these compounds, those shown are the ones that exhibit significant levels of antibacterial activity. Stereochemical requirements became clear when three-dimensional structures were determined and when isomers of the naturally occurring antibiotics were made and found to be poorly active. Penicillins differ by acyl groups in the 6β-position side chain of the penam nucleus. The clavams are oxapenams and can be traced to the discovery of clavulanic acid, a β-lactamase inhibitor. Carbapenems, which have a double bond in the five-membered ring, include thienamycin and imipenem. Cephalosporins have a hydrogen in the 7α position, whereas cephamycins is the name applied to cephalosporins with $R_{7\alpha} = $ methoxyl. Alternation of the atom at the 1 position of the cephem nucleus gives the oxacephalosporins (e. g., moxalactam) and carbacephems (e. g., loracarbef). Members of 4-6 ring systems are distinguished by different substituents at the 3 and 7 positions, and there are literally thousands of them. Prior to inroads by generics in the last 5 years, the most widely sold antibacterial agent was cefaclor, a cephalosporin. Nocardicins are monocyclic β-lactam antibiotics. The monobactams are made noteworthy by sulfazecin and aztreonam, which differ by the side chains on the 3 and 4 positions. Although not a β-lactam, the pyrazolidinone analog is a bio-isostere with antibacterial activity that can be comparable to that of third-generation cephalosporins.

was just a chemically reactive lactam ring and an acidic group A in a certain spatial vicinity of the lactam **1**.[2] (Note that in the preceding sentence we purposely used the generalized term "lactam ring" here because one particular γ-lactam analog has been found[3] to function as a bio-isostere of a β-lactam.[4,5] In other words, the five-membered pyrazolidinone ring shown in Fig. 11.1 is able to inhibit the same receptors as do the β-lactam compounds.)

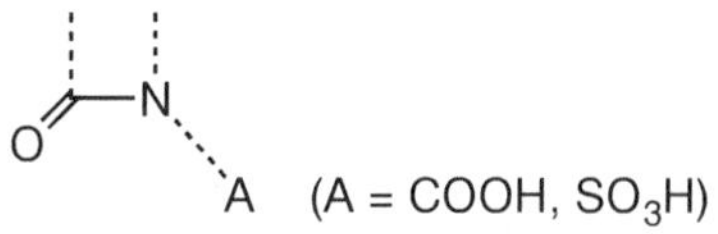

(A = COOH, SO_3H)

1

Designating part of a molecular structure as the pharmacophore does not mean that the rest of the molecule is unimportant. The rest of the molecule can play a role in holding the pharmacophore in the right conformation and can also have a large influence on the level of biological activity and the efficacy of the compound as a medicine. Thus, the appendages associated with a pharmacophore are not without consequence. The β-lactam nucleus, its substituents and side chains influence both potency (the minimum concentration needed to stunt bacterial multiplication) and spectrum (the set of bacterial species susceptible to an inhibitor).

The β-lactam ring not only imparts structural and, therefore, conformational characteristics on the antibacterial agents, but it also plays the lead role in the mode of action. The β-lactam ring acts as an acylating agent that covalently binds to bacterial transpeptidases and carboxypeptidases. These enzymes operate near the cell surface of bacteria and are essential for normal bacterial growth and multiplication. If the enzymes are inhibited, the organisms produce defective cell walls, burst open, and die. The enzyme targets are collectively called penicillin-binding proteins (PBPs) for historical reasons, although these are the same targets that are blocked by cephalosporins, monocyclic β-lactams, etc. We use the term "β-lactam recognizing enzymes" to encompass all bacterial enzymes to which β-lactam antibacterial agents interact, including PBPs, β-lactamases, and related proteins. (Whereas inhibitors with four- or five-membered lactam rings block PBPs, other antibiotics with even larger rings of amide (peptide) bonds, such as vancomycin, have completely different modes of action.[6] These other antibiotics are outside the scope of this review.)

Usually each bacterium has several functioning PBPs. If one PBP is blocked by an antibacterial agent, one of the other enzymes in the cell wall construction machinery can be invoked by the organism to keep it going. Thus, an effective β-lactam antibacterial agent must be able to block several enzymes, each with slightly different receptor site requirements. The target receptor sites also differ somewhat from one bacterial species to another. This multiplicity of targets complicates a totally rational design of one compound that will inhibit all targets and be an effective medicine.

Whereas most other classes of drugs have a mode of action that involves noncovalent binding to a receptor, the β-lactam antibacterial agents are unusual because they attach to their target receptors via a chemical reaction. The mechanism of action of the β-lactam antibacterial agents is basically reaction **2** wherein the hydroxyl group of a serine in the active site of the PBPs becomes acylated. The covalent binding is reversible to some degree, but lasts long enough to interfere with the construction of peptidoglycan, a mesh-like macromolecule that forms the bacterial cell wall. (The reversibility arises from slow hydrolysis of the acyl function, thereby regenerating free enzyme. The remnant of the antibacterial agent undergoes extensive rearrangement both before and after release from the active-site serine.)

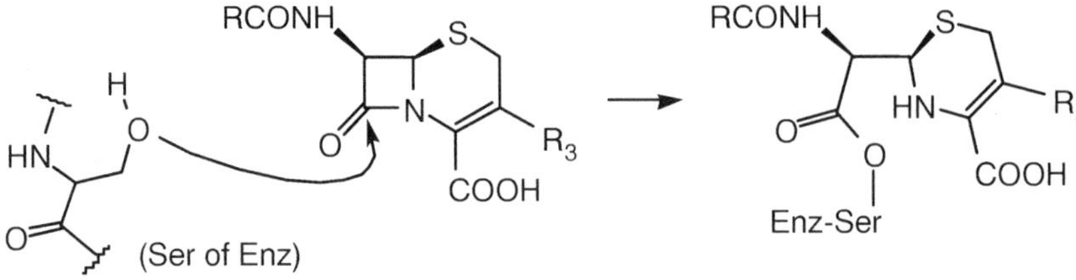

RCONH S H O HN O (Ser of Enz) O N COOH R_3 → RCONH S HN R_3 O O COOH Enz-Ser

2

The pharmacophore of the β-lactam antibacterial agents resembles key features of the natural substrates of the transpeptidases and carboxypeptidases. The substrates, which are oligopeptides terminating in D-alanyl–D-alanine, are precursors to crosslinked peptidoglycan (**3**). The resemblance was first noticed with simple molecular models[7] in 1965 and has been reviewed numerous times.[8] When computerized molecular modeling became possible, it was used to understand the geometrical correspondence more exactly: penicillins and cephalosporins resemble the tetrahedral transition state formed as the peptide bond of D-Ala–D-Ala breaks.[9,10] Additional comparisons of β-lactams and D-Ala–D-Ala have been reported.[11,12] The β-lactams are thus able to deceive the bacterial enzymes into recognizing and binding them. The β-lactam agents can be thought of as peptidomimetics.[13] The agents can also be thought of as transition state analogs[14] because they resemble the tetrahedral intermediate formed when the D-Ala–D-Ala peptide bond is attacked by a nucleophile.[2]

(D-Ala—D-Ala precursor)

X—NH—CH(Me)—CONH—CH(Me)—COOH + Enz—OH ⟶

X—NH—CH(Me)—COOEnz + H_2N—CH(Me)—COOH $\xrightarrow{-\text{D-Ala}}$

X—NH—CH(Me)—COOEnz $\xrightarrow{+\ H_2N-Y\ \text{(peptidoglycan)}}$

(crosslinked peptidoglycan)

X—NH—CH(Me)—CONH—Y + Enz—OH

3

Undoubtedly, experimental medicinal chemistry research on β-lactam antibacterial agents reached its acme in the 1960s and 1970s. (As already mentioned, all that early research is documented in older books and reviews, and is beyond the scope of this chapter.) In the 1960s and 1970s, the traditional empirical approach to molecular design was prevalent, although a few theoretical calculations proved useful in revealing some insights and in supplying bits and pieces of information. By the 1980s, when the theoretical methods and computers had improved considerably from the earlier period, interest of pharmaceutical discovery scientists in β-lactam antibacterial agents was waning.

The decline in interest was not because of difficulty in finding biologically active β-lactam compounds, but rather because the new compounds were only incrementally different in their potency and antibacterial spectrum from the existing pharmaceutical products. From experience with structure–activity relationships, recorded in a very large scientific and patent literature, medicinal chemists learned that many different side chains could be hung on the known β-lactam frameworks (nuclei), and an active compound would most likely result. But the activities and other properties of the new compounds were not significantly different from those of compounds that had already been synthesized. Medicinal chemists had worked on scores of different β-lactam templates and had synthesized tens of thousands of compounds. Few truly novel lactam structures seemed possible. Drug discovery researchers were faced with the prospect of producing nothing but more and more "me-too" drugs. There was declining economic incentive to do so because of the increasing availability of less expensive generic drugs. Moreover, by the mid- to late 1980s, the perception that the medical need for antibacterial agents was well satisfied led most pharmaceutical companies to attenuate or terminate their β-lactam research programs. Currently, few companies are still pursuing new β-lactam antibacterial products. Already pharmacy shelves are crowded with about 100 different β-lactam compounds, mostly cephalosporins and penicillins.

With the emergence of more and more bacteria that are resistant to the pharmaceutical arsenal, interest in anti-infectives is reawakening somewhat in the 1990s but in different directions for reasons explained here. The widespread use — and overuse — of β-lactam antibacterial agents for the last half century have given microorganisms ample time and opportunity for the less susceptible members of their population to proliferate and to spread resistance genes to other bacterial species. There has been a "chemical pressure" on the bacterial populations to evolve toward mutations conferring resistance to current antibacterial agents. Commonly encountered bacterial pathogens that in the past responded to antibacterial therapy no longer always respond. Even less common bacterial strains causing infections in serious illnesses, like lung infections in cystic fibrosis patients or blood poisoning in leukemia patients, are being encountered that are no longer affected by the most powerful drugs of last resort. The latter last-defense agents include the carbapenem imipenem, the penicillin methicillin, and the glycopeptide vancomycin. When pathogens exhibit resistance to these compounds, physicians are virtually powerless to fight the infection. Many scientific publications have discussed the emerging resistance problem,[15–20] and the problem has even been proclaimed loudly by the media.[21] Thus far, public alarm about the threat does not seem to have risen to a perceptible level, probably because relatively few people have to date succumbed to one of the resistant infections.

Bacterial resistance comes in one or more of three forms: (1) bacterial production of β-lactamases which hydrolyze the β-lactam ring and render the agent impotent, (2) permeability changes which hamper the agents from reaching their sites of action in the bacterial cell wall, and (3) target modifications which prevent those inhibitor molecules reaching the active sites from being fully effective in disrupting bacterial growth. To counter the first mechanism, some pharmaceutical researchers have resorted to combining a β-lactamase inhibitor with a β-lactam agent, but even these are less than fully up to the task of treating serious infections. A more general strategy is desirable. Therefore, much of the new effort in antibacterial discovery research is directed at finding compounds with modes of action completely different from those of the current agents, so as to circumvent the bacterial defense systems.[22,23] In light of this reality, it must be said that many of the theoretical (i.e., computational) and structural studies of β-lactam antibacterial agents in the last 15 years are of only academic interest and have limited potential in facilitating meaningful drug discovery.

An abundance of excellent books and reviews on β-lactam antibacterial agents is available, and we cite a sampling of them.[24–28] One of these reviews dealt rather comprehensively with the physical chemistry and computational chemistry of these compounds.[2] The present chapter is devoted mainly to computational research that has been conducted since 1982. The reader will find many of the older references in the earlier review,[2] but occasionally we do include some of the older references here to fill in the required background information.

Although, all the basic understanding of the β-lactams was well established by the time of the 1982 review, there has been no shortage of papers published in the intervening years. In the last 10 years, many computational chemists, especially those in Spain, have turned their attention to the β-lactam ring and β-lactam antibacterial agents as interesting subjects for investigation (Fig. 11.2). We review as many of these papers as space allows.

The present chapter also partially summarizes some of the many previously unpublished calculations on β-lactams undertaken by the author when he was at Lilly Research Laboratories. Quite frequently in pharmaceutical discovery research, industrial scientists do not publish all their work. Contrary to the common misconception, failure to publish is not always because the work is proprietary. Rather, the failure to publish arises because of an intense pressure on industrial scientists to move on to the next project instead of wrapping up details of a previous project. The surge to remain competitive with other companies in the fast-moving pharmaceutical business has the side effect that research on a prior project becomes passé quickly.

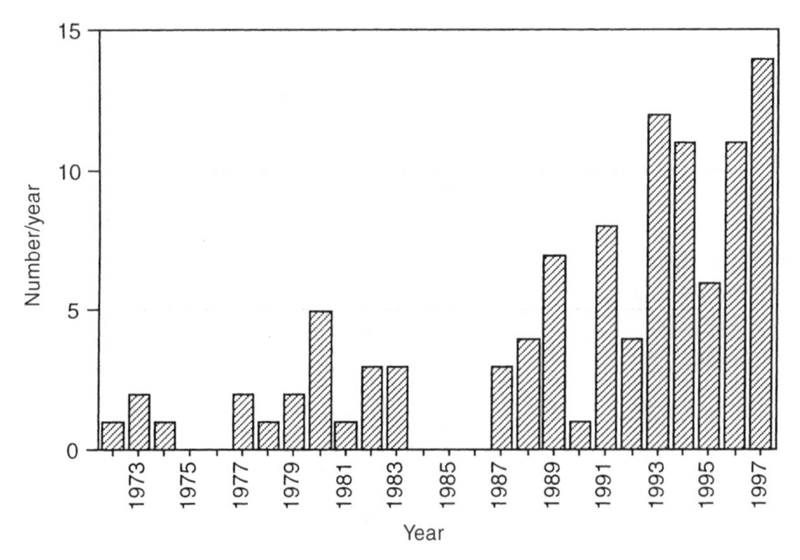

FIGURE 11.2. Number of items pertaining to theoretical calculations on β-lactams abstracted by the Chemical Abstracts Service (CAS/STN). The oldest of such paper in their database is of 1972. The first MO calculations on β-lactams were performed in the late 1960s, but the work was not immediately published. There are a few references reporting theoretical calculations which did not get indexed as such in CAS. Hence the data in the figure slightly underestimate the actual numbers.

2. GRAPPLING WITH THE FOUR-MEMBERED RING

The β-lactam ring is a deceptively simple moiety, being just an amide bond constrained in a four-membered ring. One might think that this azetidin-2-one ring should be rather straightforward to model by computational chemistry methods. However, the story is not so simple. The problem is that in the bicyclic β-lactam antibacterial agents, the four-membered ring is sometimes puckered, and the β-lactam nitrogen is intermediate between being sp^3 hybridized and sp^2 hybridized. Consequently, most empirical force fields used in molecular mechanics and molecular dynamics calculations are not adept at accurately reproducing the geometry of the compounds. Semiempirical molecular orbital methods are also tripped up by this geometrical situation and predict it less than satisfactorily. Even ab initio methods fail unless a sufficiently large basis set is employed. Any of these different theoretical approaches models reality at some level of approximation.[29]

Traditionally, the geometry of the β-lactam ring has been described as dominated by the influence of amide resonance **4**. The resonance form with a C=N double bond favors a planar (sp^2) geometry at the nitrogen. When the β-lactam ring is fused to a five-membered ring — as in the case of penicillins and carbapenems — or to a six-membered ring — as in the case of cephalosporins and carbacephalosporins — the exocyclic bond to the β-lactam nitrogen cannot be coplanar. This nonplanarity is often expressed in terms of the so-called Woodward *h* value[30] which is the distance of the nitrogen from the plane defined by its three substituent atoms (shown schematically in **5**). The nonplanarity decreases the degree of amide resonance, which reduces the double bond character of the amide bond. Despite the influence of amide resonance, most molecular orbital (MO) methods, which are described later, yield high negative net atomic charges on both the oxygen and nitrogen of the β-lactam ring, and some methods make the nitrogen the more negative of the two heteroatoms. However, net atomic charges can be misleading due to the nature of electron population analyses.[31,32]

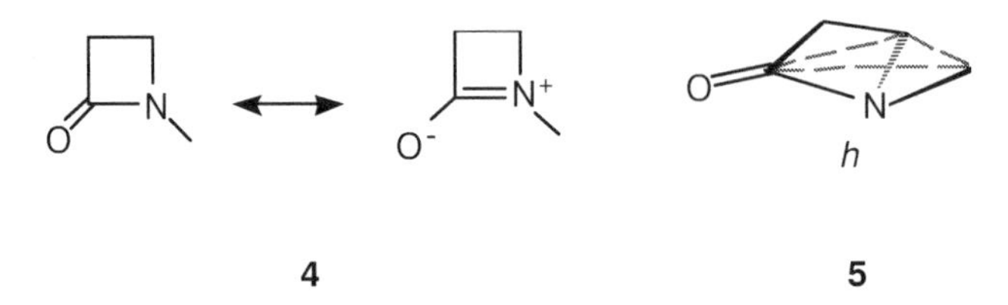

4 **5**

The relationships of the geometry at the β-lactam nitrogen, the biological activity, and the physicochemical properties are manifest in many well established ways.[33,34] For instance, bicyclic nuclei that cause the β-lactam C–N bond to be longer also exhibit a shorter carbonyl C=O bond length. A shorter C=O bond length results in a higher carbonyl stretching frequency in infrared (IR) spectra. The same relation shows up in theoretically predicted optimized C–N and C=O bond lengths. The interconnectedness is also seen in net atomic charges and overlap populations. The relation is seen in theoretically modeled[35,36] and experimentally observed rates of base-catalyzed hydrolysis.[33,37,38] The rate of nucleophilic attack on the β-lactam carbonyl carbon is, in turn, related to the ability of the ring to act as an acylating agent. Biological activity is generally better when the inhibitors are able to acylate and remain attached to their targets. All these interrelationships are qualitatively summarized in Table 11.1.

The relationships in Table 11.1 are generally linear, except at some "extrema," which results in a roughly U-shaped curve. At extrema of the physicochemical properties, other factors can take over. For instance, it would be naive to think that making the β-lactam more and more reactive would result in better and better antibacterial activity. Δ^1-Carbapenem **6** has one of the most pyramidal β-lactam nitrogens known ($h \approx 0.53$ Å), but is essentially inactive.[39] When the β-lactam becomes too reactive, stability is decreased, so that the compound decomposes before being able to reach the active site of the PBPs, and hence its biological activity is low. On the other hand, if a molecular design can partially protect a very pyramidal β-lactam ($h \approx 0.53$ Å) from premature decomposition, modest activity can be observed.[38] It should also be pointed out that the qualitative relationships in Table 11.1 are generalizations. The side chains on the cephalosporins and other β-lactams, as mentioned, can and do

TABLE 11.1. Qualitative Relationships Between Geometrical, Electronic, and Experimental Properties of β-Lactam Antibacterial Agents

Antibacterial Activity	High	Low
β-Lactam C=O bond length, *r*(C=O)	short	long
β-Lactam C–N bond length, *r*(C–N)	long	short
β-Lactam C=O overlap population, *n*(C=O)	high	low
Carbonyl stretching frequency, ν(C=O)	high	low
Net atomic charge on β-lactam O, Q_O	less negative	more negative
Rate of alkaline hydrolysis of ring	high	low
Sum of bond angles at lactam nitrogen	$\ll 360^\circ$	$\leq 360^\circ$
Woodward *h* value	high	low

affect both potency and spectrum of activity because steric fit at each receptors, transport to the active sites, and other factors also come into play in determining the observed biological activity.

O N COOH

6

3. DEVIATIONS FROM PLANARITY

Many papers have been written in which it has been stated or assumed that an unfused β-lactam ring is planar and that the substituent on the lactam nitrogen is in a coplanar arrangement due to amide resonance. Such a viewpoint is only approximately correct. Whereas puckering and nonplanarity should be expected when the β-lactam ring is fused in a bicyclic nucleus, it was unexpected that many monocyclic β-lactams turned out to be nonplanar also.

A significant advance in computational chemistry occurred in the early 1970s with the development of a molecular orbital program that could do automatic geometry optimization. (The first such program, GEOMO,[40] was followed by others which rapidly became widely used. Optimization methods have been reviewed elsewhere.[41,42]) Among the many MO calculations on β-lactams made possible by the new software, some were performed on simple monocyclics to determine their equilibrium structures.[43] Both semiempirical and ab initio methods were evaluated. The theory underlying these methods is well known.[44–47] MOPAC[48] and Gaussian 82[49] were the programs used to obtain results (Table 11.2) on model system **7** with various small substituents on nitrogen. The restricted Hartree-Fock (RHF) calculations were done with one of the standard basis sets of Gaussian functions indicated in the table. Additional calculations by the AM1 and PM3 methods were performed with the Spartan computational

TABLE 11.2. Energy Optimized β-Lactam C–N Bond Length (Å) and Dihedral Angles (deg.) to Show the Degree of Coplanarity of the Ring and the *N*-Substituent of 7 [a,b,c]

Method	R_6	r(C–N)	$\lvert\phi(C_4{-}C_3{-}C_2{-}N_1)\rvert$	$\lvert\phi(O_5{=}C_2{-}N_1{-}R_6)\rvert$
MINDO/3	H ($\Delta H_f = -47.4$)	1.37	0°	0°
MNDO	H ($\Delta H_f = -28.1$)	1.43	6°	44° ($h = 0.35$ Å)
AM1	H ($\Delta H_f = -9.1$)	1.41	4°	34° ($h = 0.24$ Å)
PM3	H ($\Delta H_f = -24.7$)	1.46	5°	49° ($h = 0.37$ Å)
RHF/STO-3G	H	1.45	10°	44° ($h = 0.40$ Å)
RHF/4-31G	H	1.36	0°	0°
RHF/6-31G	H	1.36	0°	0°
RHF/6-31G(d) [d]	H	1.36	0°	0°
RHF/6-31G**	H	1.36	1°	4°
X-ray [e]	H	1.33	0°	0°
Exptl [d]	H	1.34–1.38	0°	0°
RHF/4-31G	F	1.36	0°	0°
RHF/4-31G	Cl	1.37	0°	0°
RHF/4-31G	CH_3	1.36	0°	2°
RHF/4-31G	$CH{=}CH_2$	1.37	0°	0°

[a] The optimized dihedral (torsional) angles tabulated have an uncertainty of ±1° (±4° in the case of STO-3G) because of the small gradient in the vicinity of the minimum and the different optimization methods in available ab initio computer programs.

[b] From reference 43, except the AM1 and PM3 results which are current work.

[c] Also tabulated are the heat of formation (kcal mol^{-1}) predicted by the semiempirical MO methods and the Woodward h value (Å) for the more pyramidal optimized geometries.

[d] From reference 52.

[e] Low temperature (170 K) data from reference 51.

chemistry program.[50] Experimental data[51,52] are included in the table for comparison.

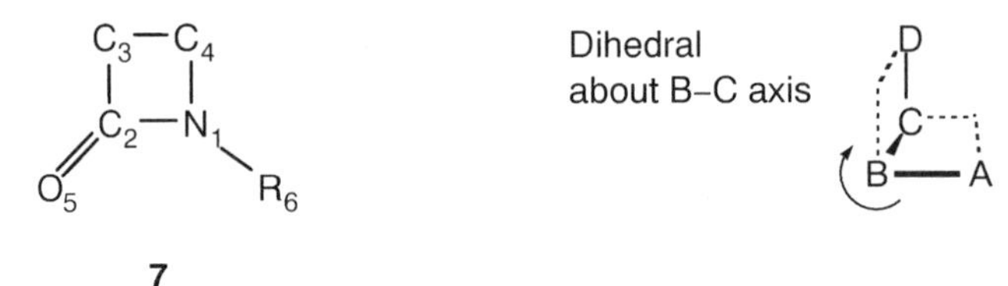

As other investigators subsequently also learned, the oldest of the widely used semiempirical MO methods, MINDO/3, does well in predicting the geometry of the β-lactam ring. However, MINDO/3 overestimates the stability (heat of formation) of four-membered rings in general. This systematic weakness is obvious for the azetidin-2-one in Table 11.2. MNDO was the next generation method that was designed by the Dewar group[53] to improve on MINDO/3 for molecules in general, but it actually is worse for β-lactam geometries. MNDO makes the β-lactam C–N bond much too long and predicts the nitrogen to be highly pyramidal. The still newer parameterizations, AM1 and PM3, also predict the azetidin-2-one to have the same defects as MNDO, i.e., the optimum β-lactam C–N bond is too long and the nitrogen is too pyramidal. Only the two newest semiempirical methods are able to describe the existence of hydrogen bonds adequately, so none of the semiempirical methods is fully satisfactory for trying to understand β-lactam structures and how they might interact with a protein receptor. At best, investigators can hope for a rough cancellation of errors in the approximate methods, so that trends, if not absolute values, might be predictable. However, this assumption needs to be rigorously tested in each specific application.

Turning to the ab initio results in Table 11.2, the smallest basis set, STO-3G, which was used by many investigators for a wide range of research problems in the 1970s and 1980s, does as poorly as MNDO and the other semiempirical methods for β-lactams. This is an example illustrating the fact that an ab initio calculation is not necessarily an improvement over (much faster) semiempirical calculations.

Going from the minimal basis set STO-3G to the split-valence basis sets in Table 11.2 improves the results that ab initio geometry optimizations are able to give for the monocycle. Basis sets 4-31G, 6-31G, and 6-31G** (polarization functions on both first-row atoms and hydrogens) give reasonable agreement with experiment. The data in Table 11.2 also indicate that whereas the smaller split-valence basis sets predict the β-lactam ring to be planar, the largest basis set examined[43] predicts a lower energy when the ring and exocyclic substituent are bent very slightly out of coplanarity. Thus, one cannot automatically assume that amide resonance will dictate a coplanar geometry. Conversely, a still higher level ab initio calculation might favor a flat nitrogen for an azetidin-2-one in vacuum.

From the MO results presented here, it is clear that a split-valence basis set, preferably with polarization functions, is necessary for any hope of reliably modeling the structures and reactions of a β-lactam ring compound at the receptor sites. To further judge the capability of the basis sets, it is worth considering the theoretically predicted barrier to inversion in ammonia. Ab initio calculations[43] were performed using geometry optimization for the pyramidal and planar geometries. With the 6-31G basis set, the barrier to inversion is grossly underestimated at 0.4 kcal mol^{-1}, and with the 6-31G* basis set the barrier is overstated at 6.5 kcal mol^{-1}. However, introducing p functions on the hydrogens and d functions on the first-row atoms, as in 6-31G**, gives an excellent result (5.5 kcal mol^{-1})[43] agreeing well with experiment.[54]

Optimized structures tell us only about the energy minimum on a potential energy surface. Consider next the shape of these surfaces. Potential energy curves for bending the hydrogen on the β-lactam nitrogen out of coplanarity are shown in Fig. 11.3 for various basis sets. In three curves, all geometrical variables of **7** (R = H) were optimized except the $O_5{=}C_2{-}N_1{-}H_6$ dihedral angle under study.[43] In the fourth curve marked "expt," averaged[55] crystallographic bond lengths and bond angles for various β-lactam antibacterial agents were used. The minima are essentially at a dihedral angle of 0° for the 6-31G//6-31G and 6-31G**//6-31G optimizations but are at 4° for 6-31G**//6-31G** and

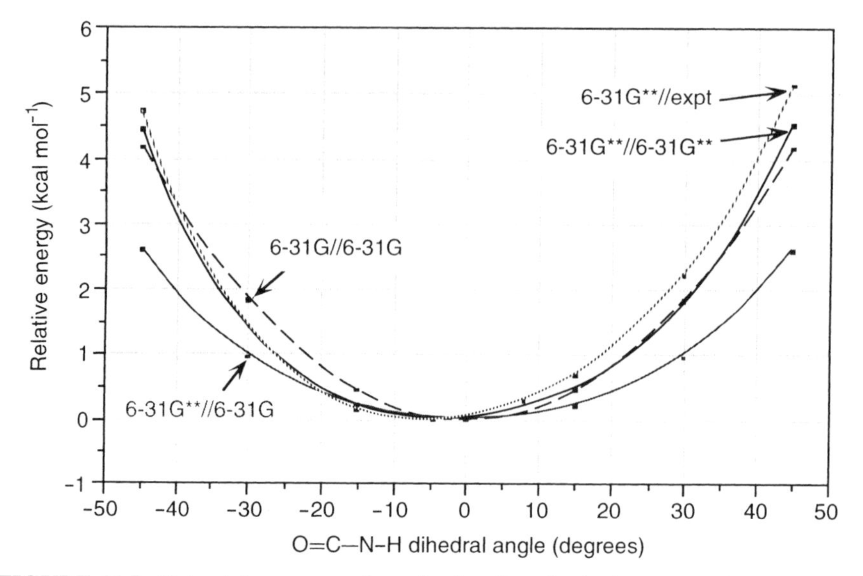

FIGURE 11.3. Potential energy surfaces for bending the hydrogen on nitrogen out of coplanarity by varying the $O_5{=}C_2{-}N_1{-}H_6$ dihedral angle in **7** (R = H). The standard atom numbering system for a monocyclic β-lactam ring is used. All calculations were obtained from Gaussian 82 running on VAX superminicomputers. Energies are plotted relative to the lowest energy obtained by the designated basis set. Following standard notation, the basis set listed before the double slash is used for the energy calculations, whereas the basis set listed after the double slash was used to optimize the molecular geometry. The 6-31G**//expt curve corresponds to single point (i. e., no optimization) energy calculations using bond lengths and angles from an average of X-ray structures, except for the N–H bond length which comes from 6-31G optimization. The 6-31G**// 6-31G** curve corresponds to complete geometry optimization except for the dihedral angle being varied. The 6-31G**//6-31G curve corresponds to single point energies obtained at the 6-31G** level with geometries optimized at the 6-31G level. The 6-31G// 6-31G curve corresponds to 6-31G energies obtained by complete geometry optimization except for the dihedral angle being varied. A quadratic curve fits the data points for 6-31G//6-31G well ($r^2 = 0.9998$), whereas a fourth order polynomial is required to fit the other sets of data well ($r^2 \geq 0.9999$ vs. $r^2 \approx 0.995$ for a quadratic polynomial fit. Please note that these statistics are not meaningful because the number of independent variables is large compared to the number of dependent variables.)

6-31G**//expt. For the latter two curves, the relative total Hartree–Fock molecular energy at 0° is 0.05 kcal mol^{-1}. Thus, it is likely that the exocyclic substituent on the nitrogen of monocyclic β-lactams can easily bend or vibrate out of coplanarity. Fairly weak intermolecular forces, such as may occur in crystal packing or in a receptor site, or intramolecular interactions such as with substituents at the 4 position of the β-lactam ring should easily be able to induce the *N*-substituent to bend in or out of the plane of the amide according to the dictates of free energy. In fact, supposedly planar experimentally "observed" β-lactam structures could be a thermal average of two slightly nonplanar conformations, one with the R_6 group bent "up" from the β face and the other bent "down" from the α face.

Compounds that dramatically make the point that monocyclics need not be coplanar are the thiamazins **8**.[56,57] These have an h value of 0.18 Å, which is almost as large as the value observed in the crystal structures of cephalosporins ($h \approx 0.2$ Å). Amide resonance is thus insufficient to prevent this monocycle from finding its lowest free energy in a nonplanar conformation.

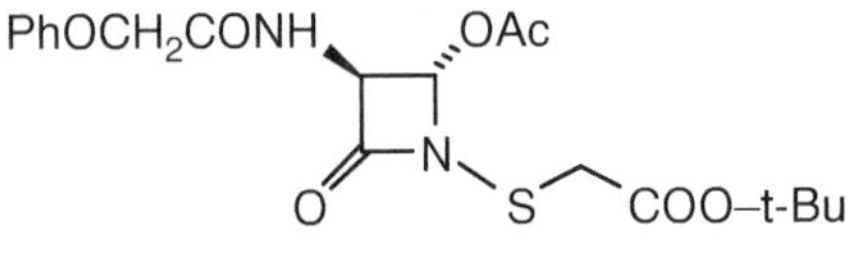

$PhOCH_2CONH$, OAc, N, S, COO–t-Bu, O

8

$PhCH_2CONH$, Me, N, O, COOH, O

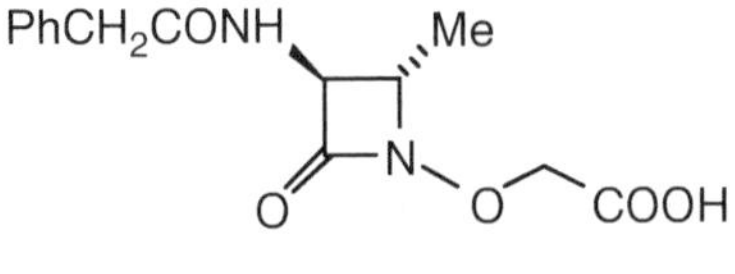

9

It is interesting to note that the thiamazins do not have significant biological activity despite their pyramidal geometry. In contrast, the oxa analogs, oxamazins **9**, do exhibit low levels of antibacterial activity. The difference in activity level is ascribed to the fact that the distance between the components of the pharmacophore is too great in thiamazins.[56] The interatomic distance between the β-lactam carbonyl carbon and the carboxyl carbon is ca. 3.5 Å in oxamazin and 3.9 Å in thiamazin. The latter is outside the range of 3.1–3.6 Å as exhibited by most of the biologically active β-lactams.[56] Thus, thiamazins may be inactive because they do not align optimally with the active site residues in the PBPs.

A wider study of the crystal structures of other monocyclic β-lactams is revealing in regard to the nonplanarity of the amide nitrogen in monocyclics. The well-known Cambridge Structural Database (CSD)[58,59] is a valuable resource of geometrical data on mostly small organic molecules.[60] A search of the database reveals 38 monocyclic β-lactam compounds with atomic co-ordinates available. These structures entered the database because they formed crystals suitable to X-ray diffraction analysis, rather than because of their biological activity. In fact, most of these compounds lack substituents and other properties appropriate for significant antibacterial activity. Of the 38 compounds, 7 have a spiro connection at the 4 position carbon or are complexed to mercury. Eliminating these compounds, so as to avoid any possible special effects on the geometrical parameters, leaves 31 monocycles. The distribution of pertinent geometrical parameters are shown in the histograms of Fig. 11.4. It can be seen that most of the β-lactam amides are not coplanar. Almost two-thirds have h values greater than 0.05 Å, and the mean h value for all 31 monocyclic β-lactams is 0.10 Å. The maximum h value reported for a monocycle is 0.42 Å, yet its crystallographic R factor (5.6%) gives no indication that this structure is of poor resolution. The data in Fig. 11.4 clearly show that monocyclic β-lactams are nonplanar more often than commonly thought.

Finally, we present crystallographic data for the set of 31 monocycles showing the relationship between the lengths of the β-lactam C–N bond and the

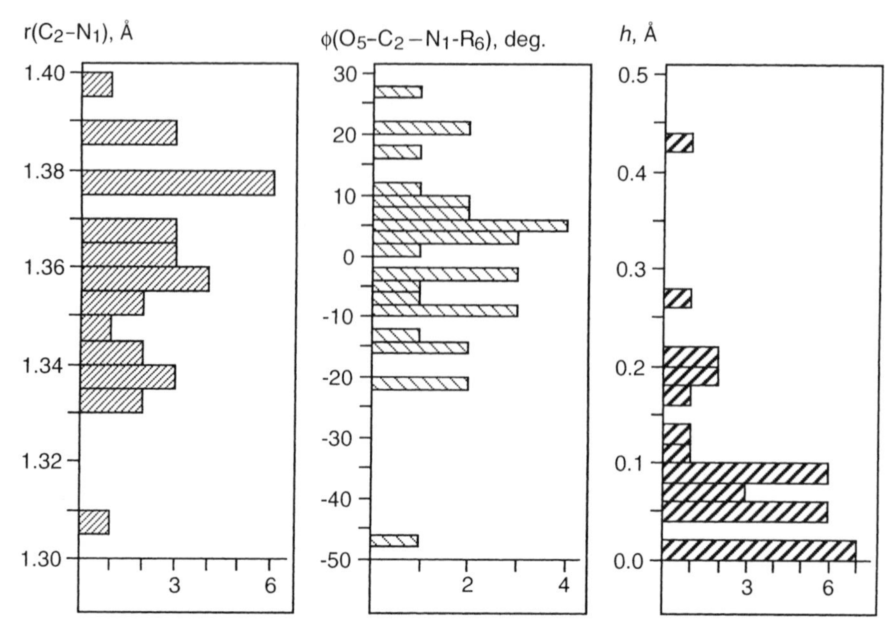

FIGURE 11.4. Histograms of the distribution of selected geometrical variables for 31 monocyclic β-lactams from the Cambridge Structural Database. The abscissa of each histogram indicates the count in the population. The most prevalent β-lactam C–N bond length is just under 1.38 Å, but values between 1.33 and 1.37 Å are common. The dihedral angle to indicate the degree of planarity/nonplanarity of the β-lactam nitrogen, $\phi(O_5{=}C_2{-}N_1{-}R_6)$, is widely scattered but is most prevalent at $+5°$. Both positive and negative dihedral angles are reported because it is measured in a clockwise direction, and the *N*-substituent can be either "above" or "below" the plane of the three amide atoms O=C–N. The Woodward *h* values are a function of the dihedral angle $\phi(O_5{=}C_2{-}N_1{-}R_6)$ and the bond lengths to the three substituents on nitrogen. The fact that the exocyclic N–S bond in the monobactams is about 0.25 Å longer than a corresponding N–C bond in the other monocyclics means that the *h* values of the former would be higher even if the bond angles at the nitrogen were the same.

carbonyl C=O bond (Fig. 11.5). One would expect these bond lengths to vary inversely because of the amide resonance idea (**4**). This is in fact what we see. However, there is a great deal of scatter in the data points, suggesting that other effects in the molecules and crystals can overpower amide resonance.

In the case of the monobactams, the sulfur attached directly on the nitrogen causes the amide C–N bond to be rather long (1.375–1.39 Å) and to vary over a relatively small range (Fig. 11.5). The lengthening of the $C_2{-}N_1$ bond in the monobactams might be explainable by electron density on nitrogen back-bonding into the nominally empty 3d orbitals on sulfur (**10**), thereby decreasing the amount of electron density available for the amide π bond. This effect is reminiscent of, but not the same as, enamine resonance (**11**) often invoked to understand the electronic effects in β-lactams with an α,β-unsaturated *N*-

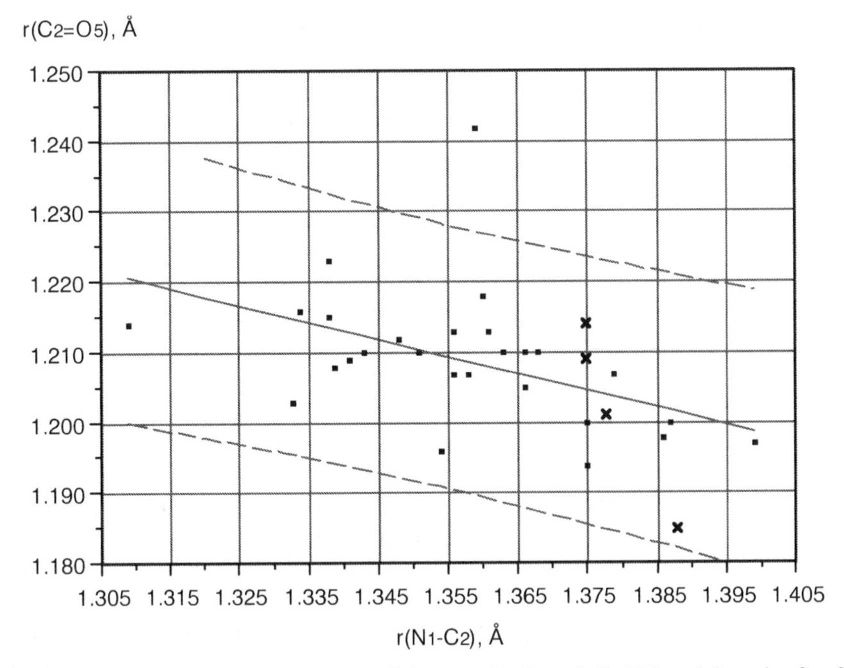

FIGURE 11.5. Relationship between β-lactam C=O and C–N bond lengths for 31 monocyclic β-lactams from the Cambridge Structural Database. A regression line is fit to the data: $r(C_2{=}O_5) = 1.537 - 0.242\ r(N_1{-}C_2)$. The dashed lines outline the 95% confidence interval for prediction of additional data points. The one obvious outlier has the second highest crystallographic R factor (8.4%) in this set of compounds, a relatively long carbonyl bond (1.24 Å), a relatively long $C_3{-}C_4$ bond length (1.61 Å), a relatively large $N_1{-}C_2{-}C_3$ bond angle (94.2°), and a relatively small $C_2{-}C_3{-}C_4$ bond angle (83.5°), therefore, this experimental result (VELVOA) is not too reliable. The four monobactams, that have an exocyclic nitrogen-sulfur bond are indicated by "X".

substituent.[2,39] The sulfur attachment in monobactams (and thiamazins) makes the nitrogen more amine-like. Curiously, the C=O bond length in the four monobactams varies over a relatively wide range, and, in principle, a line with a very high slope could be fit to these four points in Fig. 11.5. The monobactam data, if it can be confirmed by more data points, imply that r(C=O) is varying almost independently of amide resonance. The r(C=O) values are apparently more influenced by crystal packing interactions than by intramolecular electronic effects in the monobactams.

O N+ S- R'

10

O N X R ⟷ O N+ X - R

11

4. FORCE FIELDS FOR β-LACTAMS

Molecular mechanics is a widely used computational chemistry technique for studying conformation and other manipulations of three-dimensional molecular structures. To perform these calculations, the investigator (or the molecular modeling program that the investigator is using) must assign an atom type to each atom in a molecule. These assignments are required so that the program knows what standard bond lengths and angles and force constants to use.[61] The β-lactam ring is problematic because its atom types are rather unique and were not programmed into the early molecular mechanics programs. Even today many of the molecular modeling programs simply employ crude generic (wild card) parameters for unrecognized atom types and combinations of atom types.

Computational chemists who wanted to use molecular mechanics on β-lactam antibacterial agents and have some degree of confidence in the results were faced with the task of first determining specialized force field parameters that would reasonably reproduce experimentally known geometrical data. Unfortunately, full details of many of these efforts, some dating from about 20 years, have never been published, so it is impossible to compare and fully evaluate them.

One of the first efforts was based on empirical potential functions, rather than on a full-fledged molecular mechanics force field. Bond lengths were frozen, but potential energy terms were included for bond angle bending, plus nonbonded, electrostatic, and torsional interactions.[62] Although a few details about the potential functions were published by the Rao group, the exact form of the functions and the parameters were never revealed in the accessible literature. Nevertheless, the empirical potential functions were used in a series of papers[63–66] dealing with various penicillins, cephalosporins, and other β-lactams. These papers have been summarized by Rao and Vasudevan.[67] As pointed out elsewhere,[2] the molecular models used in these calculations were oversimplified and unrealistic. Whereas the Bangalore group correctly modeled rotation about single bonds when modeling side chain conformations, their conformational analyses of the five- and six-membered rings were limited to the bending of one atom out of the plane of its two attached atoms, much like the flap of an envelope moving up and down. Such flapping might be how a molecule vibrates, but this is not how these complex ring systems flip from one conformation to another. Examination of high quality, hand-held Dreiding (wire) or CPK (spacefilling) models of these ring systems shows that interrelated conformational motions involving simultaneous rotations about several ring bonds should be modeled on the computer.

Another early effort at using molecular mechanics involved a molecular modeling program called SCRIPT. It was used in-house at Roussel Uclaf (Romainville, France), but not many details were available.[68,69] Modeling was applied to Δ^3-cephalosporins **12** and Δ^2-cephalosporins **13** and to penicillins in a comparison of the spatial positions of the pharmacophoric groups, i.e., the β-lactam and the carboxyl group at the 4 position (or 3 position in penicillins).

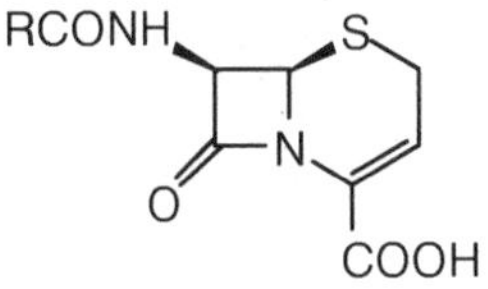

12 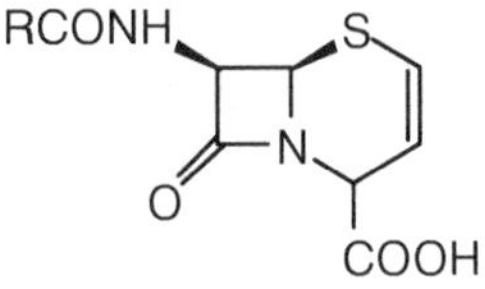 **13**

An effort at extending a well-known, standard parameter set involved Allinger's original MMI molecular mechanics program.[70] MMI was widely used in the 1970s and early 1980s. New (unpublished) parameters were developed at Pfizer (Groton, Connecticut) for modeling of penicillin and cephalosporins[71] and were used in-house. Although the parameters were derived from a fairly small database of experimental (X-ray) structures, the force field was reported to have given a quite accurate prediction of the three-dimensional structure of thienamycin about a year prior to the publication of X-ray structure in 1981.[38]

Another case of extending the capabilities of an existing molecular mechanics program involved the then popular MM2 program.[72] Allinger's standard MM2 force field was known to work well for ordinary organic compounds, and it was better than MMI.[73] The missing parameters for β-lactam compounds were obtained starting with MM2(82) parameters for similar atom types and bonding situations and then varying the parameter values by trial and error.[74] Parameters were found that reproduced the β-lactam ring bond lengths and the foldedness of the penam and cephem ring systems. The resulting force field was called MM2.1 and was used by chemists in-house at Lilly Research Laboratories, but was not published for proprietary reasons. Additional missing parameters for other pharmaceutically interesting molecules were developed and used in the same way.[74]

The strategy of supplementing an existing force field was also used by Wolfe and coworkers[75] in their development of a force field specifically for penicillins. Their new parameter set, termed MMPEN, had the capability of reproducing crystallographic bond lengths to ± 0.005 Å and bond angles to $\pm 0.5°$. The parameters worked in conjunction with Allinger's MMP2(85) force field. The preferred conformations of penicillins were predicted to curl up into balls when a dielectric constant mimicking water was used. MMPEN was not used extensively either by the investigators who developed it or by others. One paper[76] used it to study a model oligopeptide substrate and penicillin inhibitors vis-á-vis a β-lactam recognizing enzyme, the so-called DD-peptidase from *Streptomyces* R61.

We briefly interject here a few words about the R61 DD-peptidase. The R61 enzyme was a model system popular among several groups of antibacterial researchers in the 1970s and 1980s because it performed a transpeptidase reaction on a D-Ala–D-Ala substrate similar to that of PBPs, and it was inhibited by β-lactam antibacterial agents. A great deal was learned about the R61 protein biochemically, kinetically, and structurally.[77–80] In terms of protein folding, the R61 DD-peptidase has a structural motif very similar to that of other β-lactam recognizing enzymes. β-Lactam recognizing enzymes fold into two domains: one consisting of a β-pleated sheet of at least five strands; helices constitute the second domain adjoining one edge of the sheet as well as covering both faces of the sheet. The active site is a cleft at the junction of the domains and involves residues on the edge strand plus a serine which is the residue acylated by ligands. Unfortunately for those interested in designing new β-lactam drugs, the model enzyme has not been useful. A reason for this is that the inhibitory profile of the R61 enzyme is unrelated to the profiles of the targets in bacterial pathogens.[81,82]

Force field parameters and a new atom type to handle β-lactams were developed for the variant of the MM2 force field used by the MODEL molecular modeling program.[83,84] As with any force field, it is up to the developers to refine new parameters to the extent necessary for handling whatever research question they are interested in. In general, a set of force field parameters may seem adequate or satisfactory for some research problems, but not for others. It has been pointed out[85] that the β-lactam parameters in MODEL give errors in bond angles as large as 8°.

More recently, force field parameters were developed[85] for azetidin-2-one and some bicyclic nuclei without sulfur: carbacepham, oxacepham, carbapenam, and clavam. (Refer to Fig. 11.1 for nomenclature; a cepham has a saturated six-membered ring in contrast to a cephem.) These parameters were designed to work in conjunction with Allinger's MM3(92) force field. Structures optimized with the parameters were compared to available X-ray data and empirically corrected 4-21G optimized values. (Although the 4-31G basis set is used more widely, the Santiago, Spain, authors[85,86] invoked 4-21G claiming it to be roughly as good as 6-31G** in reproducing the β-lactam geometry.) Molecular mechanics energy minimization gave bond lengths generally within ± 0.015 Å and bond angles generally within $\pm 1.5°$ of the reference values. At the time of writing this review, neither applications of these force field parameters, nor extensions of the parameter set to penicillins or cephalosporins, have been reported.

In concluding this section on force field parameters, we make some brief general observations. Considering what is known from crystallographic structures of both bicyclic and monocyclic β-lactams, an "ideal" geometry for the β-lactam nitrogen atom would be essentially planar, but with force constants that allow it to bend into pyramidal configurations without great energy cost. The energy of bending should be comparable to that shown in Fig. 11.3.

5. MODELING RING FORMATION BY [2 + 2] CYCLOADDITION

A mechanism for formation of the β-lactam ring has been the subject of several papers. The easiest method for synthesizing the ring commercially is by fermentation of fungi that produce the antibiotics naturally. However, one way to synthesize a β-lactam ring in the laboratory is via the Staudinger reaction. This involves reaction **14** of a ketene and an imine.

14

Some of the earliest calculations[37,87] on this reaction were by the semiempirical MNDO MO method. An intermediate structure with a bond between the ketene's carbonyl carbon and imine's nitrogen was obtained. The potential energy surface for the two-step reaction is complex because not only are two new bonds formed, but also the R_1 and R_2 substituents undergo conrotatory motion resulting in a chiral product.

More recent semiempirical calculations[88] also favor a two-step mechanism for the Staudinger reaction in agreement with experimental evidence. First, the C–N bond forms, and then the ring closes in a second step under stereoselective control. A transition state for the latter step was computed to have biradical character. Configuration interaction calculations were applied to understand the periselectivity of the reaction.[89] Ab initio MO calculations[90] employing a continuum solvent model were able to reproduce the high stereoselectivity of the reaction known from experiments. Whereas both the conrotatory and disrotatory motions in the ring closure were found to be possible, the former is favored. Importantly, the French and Spanish collaborators[90] found that the role of the solvent was crucial because it modifies the reaction mechanism as modeled quantum mechanically: the reaction is said to be concerted in the gas phase and stepwise in solvent. This difference perhaps explains why Beijing workers concluded from Hartree–Fock calculations that the reaction goes through a concerted mechanism and not the two-step mechanism.[91,92] High level ab initio calculations were performed on β-lactam ring formation from an imine and an analog of a ketene.[93] A semiempirical AM1 study on the cycloaddition reaction of a larger ketene-imine system favored a two-step mechanism.[94]

6. MODELING β-LACTAM STRUCTURES

In the days before modern software, workstations, and desktop computers, the structures of the β-lactam antibacterial agents were too large to handle by the more reliable theoretical methods. Hence in many calculations, simplified models of the compounds were adopted to keep computer times reasonable. Various force field, semiempirical MO, and ab initio MO calculations have been reported for simple azetidin-2-ones and assorted bicyclic structures. As mentioned, the focus of this chapter is on papers that were published since the

1982 review.[2] Some of the new papers have reported data comparable to that in Table 11.2, so these papers are cited without much further commentary.

Molecular mechanics calculations on penams, cephems, and monobactams were reported using extensions of the MM2 and AMBER force fields.[95,96] The latter force field is one designed for proteins, so relatively few atom types were in it specifically for organic compounds like penicillins. A pyramidal nitrogen was chosen for the β-lactam, which, of course, would preordain the results. The authors[95] proposed in their 1989 paper that the bioactive conformation of the monobactams may have the nitrogen pyramidal, but the solid state structure of sulfazecin, a monobactam, was already known to be pyramidal from a 1981 X-ray study.[97] More recently, tests of the relatively new Merck molecular force field (MMFF)[98] found that X-ray-determined bond lengths of penams and cephems could be reproduced usually within ± 0.02 Å and bond angles usually within 2.5°.[99]

In the semiempirical arena, MINDO/3 calculations were performed on model β-lactams.[100] Frau, Muñoz, et al. published comparisons of MINDO/3, MNDO, and AM1 for penicillins[101] and cephalosporins[102] as part of a long series of papers. These calculations confirmed what we have already described in regard to the semiempirical methods. The methods perform adequately for some geometrical and conformational questions, but no method is universally accurate. Electrostatic potential fields obtained by AM1 and corrected to include solvent effects show the expected large negative potential lying in the β-lactam plane beyond the end of the C=O bond in the region of the oxygen lone pairs.[52] AM1 calculations were used to investigate the 1,3-diazetin-2-one ring and some bicyclic analogs as potential β-lactamase inhibitors.[103–105] It was speculated that such structures could acylate the active site serine and the resulting carbamoyl linkage would be more resistant to hydrolysis than an ester linkage; thereby the aza-β-lactams might be better inhibitors. However, the designs have not been reported to yield better activities. The AM1 and PM3 methods were tested on carbacephem and penem nuclei and applied to investigate novel tricyclic bridged β-lactams, e.g., **15**.[106] Some of the novel structures were found to have optimized geometries at the β-lactam nitrogen similar to those observed in penams **16** and penems **17**.

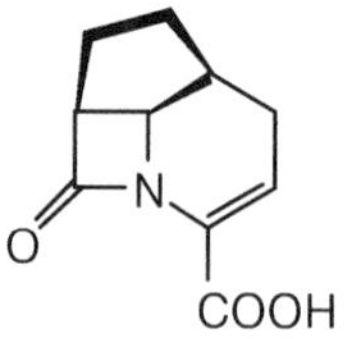

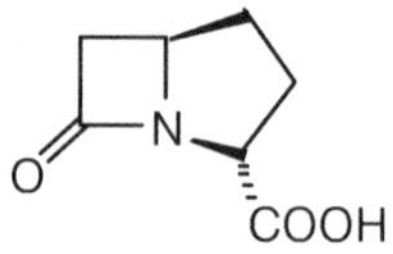

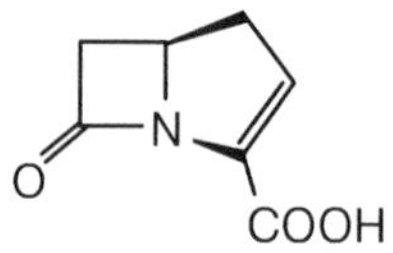

Early ab initio studies on model β-lactam structures employed the STO-3G and 4-31G basis sets.[100,107,108] As mentioned, the former basis set is inadequate for predicting the β-lactam ring correctly, and the latter basis is adequate, but barely so. Ab initio calculations on azetidin-2-one with a split-valence basis set indicate that the molecule is interconverting between a conformer with a pyramidal nitrogen and another one with a planar nitrogen; the conversion barrier is 1.7 kcal mol^{-1} or less.[109] Ab initio calculations at the 6-31G** level on azetidin-2-one were performed in conjunction with microwave and electron diffraction analyses.[110] The results were interpreted to conclude that the molecule (except, obviously, for the methylene hydrogens) is planar. Ring puckering vibrations were also analyzed.

More recently, a group in Madrid published a detailed analysis of the optimized structures and fundamental vibrations of azetidin-2-one and *N*-methylazetidin-2-one.[111] The authors found the *N*-H substituted β-lactam to be planar at the 6-31G** and MP2/6-31G** levels, whereas the *N*-Me substituted β-lactam is slightly nonplanar at the 6-31G** level, which is the only method they used for the larger structure. The sum of the bond angles at the nitrogen in *N*-methylazetidin-2-one is 358.9°, the $N_1-C_2-C_3-C_4$ dihedral angle is 3°, and the $O{=}C_2-N_1-R_6$ dihedral angle is 5°. Meanwhile, a group at Alcalá, Spain, reported MP2 calculations at the 6-31G(d) level for azetidin-2-one; besides reporting a planar ring, the authors also determined the strain energy in the β-lactam ring to be 28.5 kcal mol^{-1}.[112]

In conjunction with Table 11.2, we have already cited the work of the Barcelona group who reported high-level ab initio calculations on azetidin-2-one and some small (CH_3, $COOCH_3$) 4-substituted analogs.[52] The 6-31G(d) basis set was used for geometry optimization. The 4-substituent puckers the ring 4° out of coplanarity. Similarly, with an (*S*)-4-cyclohexyloxycarbonyl side chain on 2-azetidinone, the four-membered ring puckers so the $N_1-C_2-C_3-C_4$ dihedral angle is 1–2° according to Hartree–Fock geometry optimizations at the 3-21G level.[113]

7. MODELING REACTIVITY OF THE β-LACTAM RING TOWARD NUCLEOPHILES

When a β-lactam antibacterial agent reaches the vicinity of its target protein, the drug molecule is pulled into the active site by positively charged residues in or near the receptor cleft. These residues exert an electrostatic attraction on the negatively charged acidic group of the pharmacophore. Once drawn into the active site, the inhibitor is maneuvered by hydrogen bonds and the complementarity of steric and electrostatic interactions with the protein. In the case of penicillins, which can exist in two interconverting conformations,[2] the molecule must either be in the bioactive conformation or able to flip easily into that conformation with the assistance of the intermolecular interactions with the receptor during the docking process. Once the inhibitor is positioned near the active site serine, the serine's hydroxyl group acts as a nucleophile that attacks the β-lactam carbonyl carbon. Computational chemists are able to study this binding process at its various stages. Conformational analysis can compare docking of the inhibitors and the natural substrate, X-D-Ala–D-Ala. As mentioned in a later section, there have been many papers speculating on the role of the residues in the active site. Here we summarize papers on the nucleophilic attack step.

The tools of the computational chemist allow detailed modeling of the reaction path for nucleophilic attack. Quantum chemistry is the obvious tool of choice because bond making and breaking are involved. By way of a brief background, interest in using MO theory to model the reaction between a nucleophile and the β-lactam ring began in the mid-1970s.[114] Those calculations were fairly simple by today's standards. The CNDO/2 MO method, which as the reader may recall was parameterized to reproduce low level ab initio calculations, was used. A simple nucleophile, OH^-, was placed on a trajectory to attack the β-lactam carbonyl carbon. Because no solvent was used, no energy barrier confronts the approach, and a sharp energy minimum was obtained at about 1.5 Å from the α face of the carbonyl carbon of cephalosporins. It was discovered that for cephalosporins differing by the substituent at the 3 position the more stable the OH^- complex, the better the Gram-negative antimicrobial activity of the compounds.[2,35] The CNDO/2 energy of complexation was called the transition state energy (TSE).

The correlation between TSE and Gram-negative activity made possible setting up what we would now call a "virtual screen." New 3-substituted cephalosporin structures **18**, which the medicinal chemists at Lilly conceived and were considering for synthesis, could first be evaluated by computer to determine if their prospect for Gram-negative activity was good. The virtual screen was used for several years at Lilly until the 3-position chemistry had pretty much been played out. The course of cephalosporin research turned to large thiols and quaternary heterocycles at the 3′ position **19** ($R_{3'} = S–R''$, where R'' is a heterocycle, or $R_{3'} = R^+$, where R^+ is a heterocycle with a quaternary nitrogen). These compounds had β-lactams of similar acylating ability, but levels of bioactivity differed because of factors other than those that could be theoretically predicted. The details of the TSE correlation and modeling were disclosed in 1980.[35] This work was an early case of using a quantum mechanically derived descriptor for a rational quantitative structure-activity relationship (QSAR).

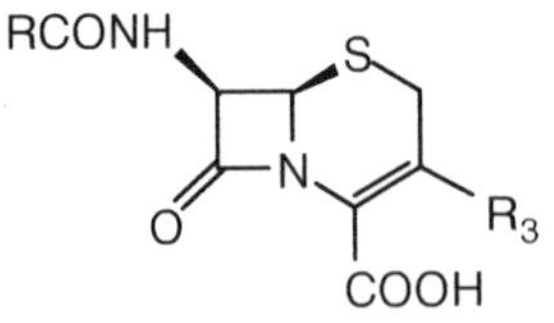

18 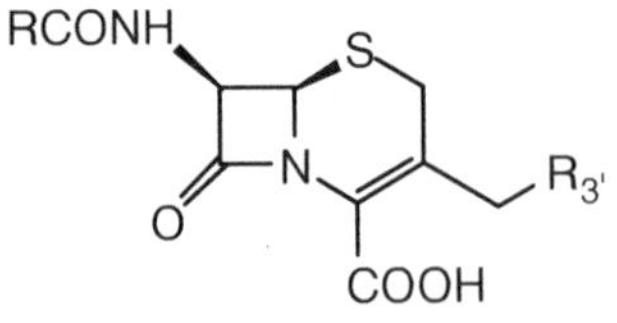 **19**

Having given a synopsis of the early research, we now review the many other computational chemistry studies that also have been designed to model the acylating ability of the β-lactam ring. As in the case of ab initio calculations at the small basis set level, it must be cautioned that semiempirical results may or may not correctly model reality. Computations at various levels of theory can, however, be valuable for testing the methods and extending our understanding of their reliability.

Close on the heels of the CNDO/2 work,[35] an extensive series of ab initio calculations was performed[115–118] to plot the reaction coordinate proceeding from model reactants through a tetrahedral intermediate to ring-opened structures. The computations were for the gas-phase reactions of OH^- with azetidin-2-one and simple 3-cephems (i.e., Δ^3-cephems). These impressive ab initio calculations, which were state-of-the-art at the time, employed the STO-3G and other small Gaussian basis sets. So in hindsight, they probably were not a great deal more accurate for β-lactam structures than semiempirical results. And, like most calculations of that day, they neglected solvent effects.

In the early 1980s, the premise underlying the TSE model was extended to investigate novel bicyclic β-lactam and γ-lactam nuclei.[119] It was at that time that a few medicinal chemists began to realize that if they were going to discover truly novel lactam antibacterial agents, they would have to investigate nuclei other than the penams, cephems, and other templates that were then known. This computer modeling in the early 1980s allowed virtual screening of totally hypothetical structures. In fact, many of the structures examined in this exploratory research were beyond routine synthetic feasibility, so the calculations provided a "peek beyond the horizon" of what experiment could then provide. The MO calculations modeled the interaction of the lactam carbonyl with a simple nucleophile, either water or hydroxyl ion. The calculations initially used the MINDO/3 method, but were extended to MNDO as soon as it became available. The calculations helped forge the trail toward investigating γ-lactams, and some of this work was published.[120–122] Still other unpublished lactam structures were predicted by the calculations to have potential for antibacterial activity,[119] but the organic chemists working on the project saw no way to synthesize them. (It is a common problem in pharmaceutical research that structures that look good "in computero" are not easily attainable in the laboratory.)

More recently, other investigators have used CNDO/2 calculations to develop a concept of the interaction of β-lactams with serine.[123] The AM1 semiempirical MO method was applied to the methanolysis of a cephalosporin[124] and the determination of a geometry for a tetrahedral intermediate.[125] A group in Madrid used the PM3 method to model nucleophilic attack on β-lactams.[126] Ab initio calculations at the STO-3G level were used to model the interaction of a methanol and water with β- and γ-lactams.[127] The TSE concept was reworked at the AM1 semiempirical level and extended to more cephalosporins.[128] This Korean work beautifully corroborated the earlier CNDO/2 correlations.[35]

The most extensive series of computational studies has been pursued by Frau, Muñoz, et al. These authors published MINDO/3, MNDO, and AM1 calculations on OH^- attacking azetidin-2-one,[129] MINDO/3, MNDO, and AM1 calculations on OH^- attacking penicillin,[130] AM1 calculations on OH^- attacking cephalothin (a cephalosporin with 3′ acetoxy),[131] AM1 calculations on

OH^- attacking clavulanic acid[132], PM3 calculations on OH^- attacking β-lactams in the gas phase and with solvent effect included.[133,134] ab initio (6-31+G*) calculations on OH^- attacking azetidin-2-one,[135] and finally ab initio (6-31+G* and 6-31G**) calculations on OH^- and H_3O^+ attacking azetidin-2-one.[136] In the former ab initio calculations, solvation was modeled by the self-consistent reaction field (SCRF) method. Hartree–Fock calculations at the 6-31G** level were used to model acidic hydrolysis of azetidin-2-one, whereby the nitrogen is first protonation and then water adds to the ring.[137]

Ab initio calculations on the neutral hydrolysis and methanolysis of *N*-methyl-azetidin-2-one and simple penam models have been reported by Wolfe and co-workers.[138] The Canadian group used 3-21G and 3-21G* basis sets for geometry optimization plus single point calculations at the MP2/6-31G* level. Calculations compared attack on the α (convex) and β (concave) faces of the β-lactam. The former is energetically preferred as found by earlier authors.[2] This preference stems from the α face being less hindered sterically as can be seen from examination of space-filling molecular models.[2] Coincidentally, X-ray data suggest that when inhibitors are in the receptor sites the attack from serine must come on the convex face. Wolfe et al. predicted that protonation at the β-lactam nitrogen, rather than at the carbonyl oxygen, provides the lower energy pathway leading to ring opening. The Canadian group learned that the MINDO/3 semiempirical method was best at reproducing trends predicted at the split-valence ab initio level.[139] Predicted reactivity of Δ^3-cephalosporins was less than that of penicillins which is consistent with the latter having a nitrogen that is more pyramidal. Whereas many authors pay lip service to how their calculations will be useful for drug design, the Canadian group is to be commended for actually using the insight from their modeling to propose some new designs.[140] Synthesis of their 1,4-thiazine structures, which were non-β-lactams, yielded one compound with low activity, but it was chemically unstable. This effort illustrates the difficulty encountered all too often in computer-aided molecular design. It is easier to do a theoretical calculation than to design a useful, novel compound, and unfortunately too many computational chemists stop after the easy part.

Semiempirical PM3 calculations examined the effects of the 1-position atom and the 3-substituent on the reactivity of cephem, oxacephem, and carbacephem.[141] The group at Burjassot, Spain, reported ab initio calculations on neutral and alkaline hydrolysis of *N*-methylazetidinone.[142] The calculations were carried out at the Hartree–Fock and MP2 levels using 3-21G, 6-31G*, and 6-31+G* basis sets. Solvent effects were included by means of a polarizable continuum model. Further calculations with 6-31G* and 6-31+G* basis sets explored the effect of an ancillary water molecule in hydrolysis.[143] The extra water in the modeling affects the activation energy predicted at the MP2 level. The Spanish group tried to sort out whether β-lactam ring opening and proton transfer occur in a stepwise or concerted manner. The two pathways are of similar quantum mechanical energy; a molecular dynamics simulation suggested that the stepwise process may occur in an aqueous environment.

8. MODELING THE LEAVING GROUP PHENOMENON

A number of papers have discussed the mechanism of what happens to the 3′ side chain of cephalosporins after a nucleophile attacks and opens the β-lactam ring. Many cephalosporins have at the 3′ position a substituent that is a good leaving group (nucleofuge), such as acetoxy or a heterocyclic thiol. There is ample experimental evidence that 3′ substituents can depart[2] in association with opening of the β-lactam ring. A leaving group phenomenon was conceived by physical organic reasoning[144] and was supported by—rather than predicted by—theoretical calculations. CNDO/2 MO calculations indicated a weakening of the $C_{3'}$–$X_{3''}$ bond when a nucleophile approaches the β-lactam carbonyl carbon.[145] As has been pointed out,[146] experiments on nonenzymatic hydrolysis show the rate of appearance of free leaving groups was comparable to the rate of ring opening.[147] These two transformations are associated, but need not be concerted. As an aside, we note that the pK_a's of the conjugate acids of leaving groups are very roughly related to their "leavability" (nucleofugacity).[148] Cephalosporins with 3′ groups that have low pK_a's should be able to undergo the leaving group phenomenon more easily, and, in fact, these compounds tend to exhibit better bioactivity.[144]

In two early publications describing the leaving group phenomenon,[2,145] the word "concerted" was used in a sense akin to that discussed by Lowe,[149] namely, a reaction is concerted if one bond is breaking (lengthening) while a second bond is forming (shortening). The definition of concerted as advanced by Lowe is based on geometrical changes along the reaction pathway. He pointed out that his definition differs from the standard one based on the energy profile of the reaction pathway. Conventionally, a reaction is concerted if it involves simultaneous making and breaking of several bonds in the transition state, with no intermediates being generated. The use of the word "concerted" in the geometrical sense *vis*-a-*vis* the leaving group phenomenon was ill-advised because one or two investigators misinterpreted the comment.

To investigate the mechanism of β-lactam ring opening in cephalosporins, various subsequent test tube experiments — none, however, performed in the environment of the PBPs in bacterial cell walls — have clearly shown that ring opening is definitely not synchronous with departure of the 3′ leaving group (**20**). In addition, AM1 calculations[131] addressed the question of concertedness in the conventional sense and found no support for it either in the gas-phase or when modeled as a supermolecule whereby the reactant was surrounded by five water molecules to mimic the role of solvent.

Arguments on the mechanism of an organic reaction can often consume inordinate attention, pivot on pedantic semantics, and distract attention from the actual utility of a reaction. In the case at hand, cephalosporins with a good leaving group are observed experimentally to release their 3′ substituent in association with β-lactam ring opening, and this phenomenon has proven useful in designing new antibacterial agents and other drugs.

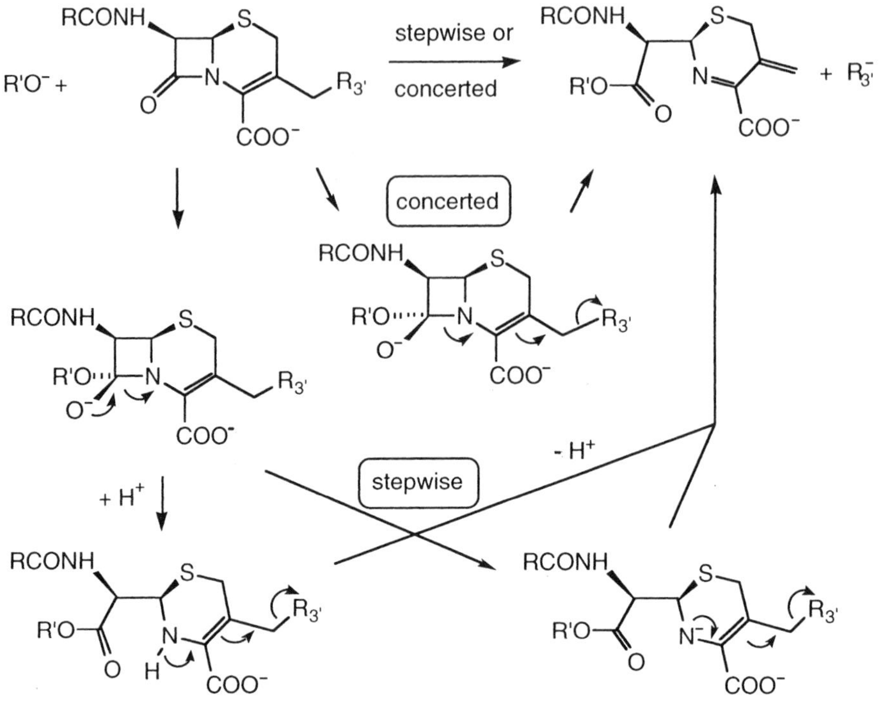

20

One way to capitalize on the leaving group phenomenon involves attaching an antibacterial agent with a different mode of action at the 3′ position of a cephalosporin. The cephalosporin is targeted at the PBPs, where it releases the second agent. The latter can then diffuse from the bacterial cell wall to its target which may be in the cytoplasm of the organism. Various second agents have been tried at the 3′ position,[150–155] but the fluoroquinolone antibacterials have been a favorite.[153–155] Extensive microbiological testing has been reported on more than 15 of these so-called dual-action cephalosporins; none has reached clinical use to the author's knowledge, but one went as far as Phase II clinical trials. (Phase III clinical trials must be passed in order for a compound to become a pharmaceutical product.) Interestingly, the leaving group idea can be extended to antibacterial β-lactams other than traditional 3′-substituted cephalosporins. Sanfetrinem **21**, a trinem (called a tribactam in the earlier literature), was selected as a clinical candidate.[156,157] The isodethiaazacephem **22** and related compounds have been reported recently to be very potent antibacterials.[158] In both types of structure, a leaving group phenomenon may contribute to the observed excellent biological activities.

In another application of the leaving group idea, a cephalosporin can be linked through the 3′ position to known anti-cancer drugs.[159,160] The

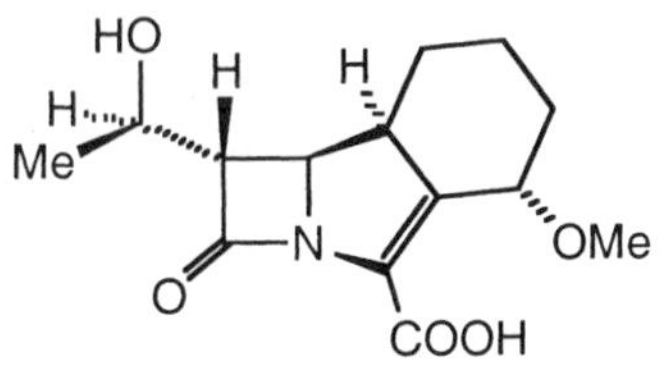

21 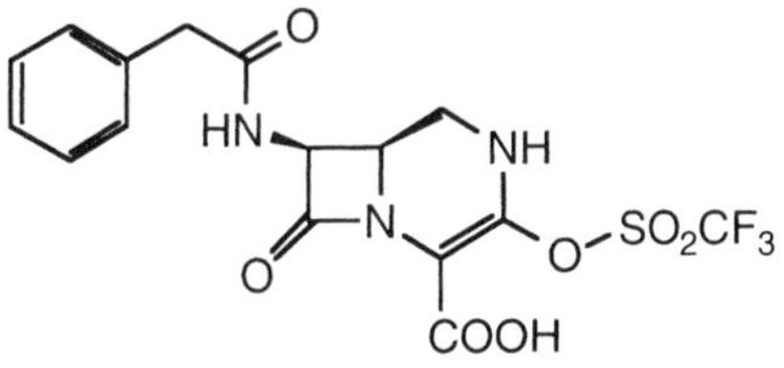 **22**

combination molecule can then be used as a prodrug for delivering the oncolytic to a tumor in the body. The tumor is marked by an antibody to which has been conjugated a β-lactamase. When the β-lactamase acts upon the prodrug circulating in the blood, the oncolytic is released in the immediate vicinity of the tumor. So, in conclusion, the leaving group phenomenon has proven quite useful and general.

9. MODELING RECEPTORS, SUBSTRATES, AND INHIBITORS

Nowadays, almost every protein crystallography study uses simulated annealing in the refinement process.[160,161] As a consequence, protein crystallography involves theoretical calculations. However, a review of all the crystallographic papers on β-lactam recognizing proteins is beyond the scope of the present chapter. This section is confined to papers where modeling was emphasized, usually using crystallographic atomic coordinates as the starting point of the calculations.

One of the earliest attempts[27] to determine the shape of a ligand attached to a β-lactam recognizing enzyme involved the ring-opened form of cephalosporin C **23**. When cephalosporin C (**24**) was cocrystallized with the previously mentioned R61 enzyme, the electron density difference map revealed the inhibitor was in the active site pocket.[163] Interestingly, the 3′ side chain was missing, clearly indicating the enzyme had been acylated by the β-lactam and the leaving group reaction had occurred. The resolution of the diffraction data (2.8 Å resolution; R factor 7.4%) did not permit an accurate estimate of the conformation of the ring-opened form of the inhibitor. No separate structure determination of **24** was available, so this geometry had to be obtained either by

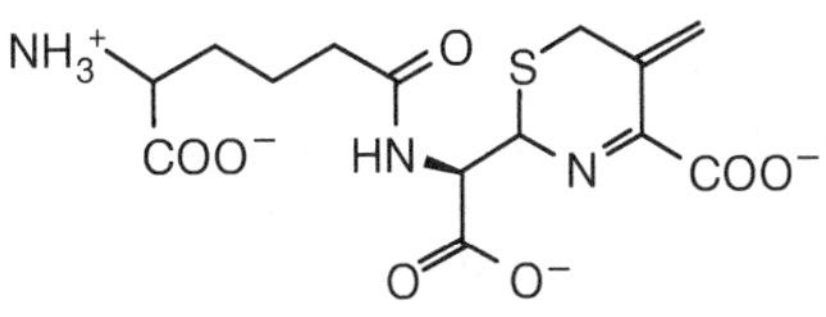

23 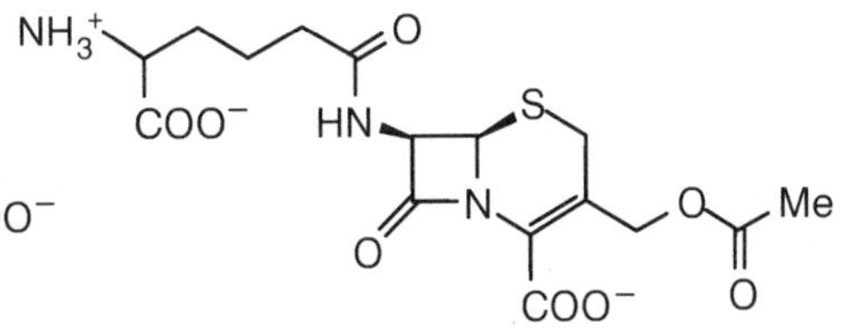 **24**

simple molecular modeling (standard bond lengths and bond angles) or by energy minimization. The latter method should be more reliable if the computational chemistry method is good enough. MNDO calculations[74] were tested to determine the optimized bond lengths and angles and preferred conformations of **24**. The computed structures could be compared to the electron density difference map to determine which conformer best fit the observed (crystallographic) density. In the end, the MNDO geometries were judged to be not distinctly superior to standard bond lengths and angles.

In 1991 molecular dynamics simulations were reported for the R61 DD-peptidase.[164] This paper also pointed out the inadequacies of modeling β-lactam structures with the TRIPOS force field in the SYBYL molecular modeling program.[165] The atomic coordinates of this protein, which at that time were not well established, were refined[166] by simulations using the Quanta/CHARMm program.[167] Ill-aligned parts of the backbone chain and overly close contacts in the X-ray model of R61 were cleaned up.

The group at Liege, Belgium, has reported some of the most extensive modeling of β-lactam recognizing enzymes. Energy minimization was done to support conjectures on the acyl transfer mechanism in *Streptomyces albus* G β-lactamase.[168] Molecular mechanics was also used to investigate mutants of this same enzyme.[169] Three class A β-lactamases were modeled with a cephalosporin, an oxacephalosporin, and a penicillin in the active sites.[170] (Class A β-lactamases rupture the amide bond in penicillins easily, whereas class C β-lactamases work mainly on cephalosporins.) The crystallographic structure of the TEM-1 β-lactamase was solved and refined by simulated annealing.[171] As mentioned, all the β-lactam recognizing enzymes have the same basic folding pattern, and the TEM-1 enzyme is no exception. More recently, the Belgian group has studied mutants of the class A β-lactamases, *Streptomyces albus* G and TEM-1.[172]

Molecular dynamics simulations were used to analyze the folding of a β-lactamase,[173] but the emphasis was on characteristics of the simulation and stability of the secondary structure, rather than on the mode of action of the enzyme. In other work, a monocyclic 4-alkoxy-2-azetidinone was modeled in the active site of the class A β-lactamase from *Bacillus cereus* 569/H.[174] The groups at Wayne State and Connecticut collaborated on modeling the interactions of β-lactam recognizing enzymes with penicillins, cephalosporins, clavulanic acid, and sulbactam (the sulfone of a penam with inhibitory activity against β-lactamases).[175–178] Mobashery and coworkers also used molecular modeling in conjunction with their experimental studies on inhibitors and mechanisms of β-lactamases.[179–182] The hydrogen bonds between β-lactam ligands and class A β-lactamases and the roles of the active site residues have been investigated at Suntory Ltd.,[183] where the authors used the variant of MM2 (MM2*) in the MacroModel modeling program[184] for energy minimization of penicillins, but no details were reported on how satisfactory the predicted geometries were. Calculations examined the influence of electrostatics on the docking of penicillins, clavulanate, and a carbapenem in *Staphylococcus aureus* PC1 β-lactamase.[185]

Protein homology modeling has been applied to four different β-lactam recognizing enzymes.[186] In this study, the reference protein to build models of each of these proteins was the X-ray crystallographic structure of a chromosomally-mediated class A β-lactamase from *Bacillus licheniformis* 749/C. This 29.5 kDa, 264-residue protein was selected as reference because its three-dimensional structure had been solved using data of good resolution (2 Å) and refined to a low crystallographic R factor (16%).[187] Sequence alignment and threading the other four proteins to the backbone of the reference protein were followed by energy minimization and molecular dynamics simulations to relieve close contacts and improve interatomic interactions. These calculations employed Quanta 3.3/CHARMm 22.[167] Of the four proteins built, crystal structures were known for two, R61 DD-peptidase and a class A β-lactamase from *Staphylococcus aureus* PC1, so they provided a check on the reliability of the homology modeling procedure.

For the PC1 β-lactamase, which had high primary sequence homology to the reference protein (72% similar residues; 42% identical residues), modeling produced a three-dimensional structure in excellent agreement with experiment. For R61, which had low sequence homology to the reference protein (32% similar residues; 14% identical residues), the loops and strands were not well predicted, especially when insertions and deletions occurred. However, the active region in the center of the proteins was modeled better than the outer regions. For the R61 DD-peptidase, it was possible to find a sequence alignment that gave a better three-dimensional fit to the known X-ray structure than a published alignment.[77] The latter alignment proposed by the Liege group[77] was based on characteristics of the amino acids in the primary sequence. In contrast, our new sequence alignment was based on the known three-dimensional structure. Our alignment gave lower primary sequence alignment (24% similar residues; 11% identical residues), but a better distance match between corresponding residues (only 15% of residues had C_α's within 3 Å of the observed crystallographic positions in the Liege alignment,[77] whereas 32% were within this distance in our alignment.)

The third and fourth proteins built by homology[174] were β-lactamases associated with resistance to cephalosporins, i.e., class C enzymes. One was the β-lactamase produced by *Enterobacter cloacae* P99, and the other was the ampC gene product. The former has 22% similar residues and 9% identical residues, whereas the ampC gene product has 19% similar residues and 7% identical residues as compared to the *Bacillus licheniformis* reference β-lactamase. Because of the low homology, the models that were built could not be expected to be highly accurate in detail, even if the gross features of the folding might be satisfactory. A PBP associated with resistant organisms, PBP 2a, has greater homology with the reference protein (37% similar residues and 15% identical residues), so the prospect of building an adequate model of 2a is somewhat better. Further work on these β-lactam recognizing proteins was limited due to shifting corporate priorities.

The modeling studies, supported by an increasingly large amount of crystallographic data, are a rich source of information for speculating on mechanistic details. The sequence of steps for all β-lactam recognizing enzymes interacting with an inhibitor is first noncovalent binding, then covalent attachment, and finally detachment of the (modified) ligand from the receptor. A crucial kinetic difference between the β-lactamases and PBPs is that turnover is much greater in the former enzymes. Both types of enzymes have an active site serine that is acylated by β-lactam antibacterial agents. However, β-lactamases are more efficient at deacylation, so that the inhibitor is released in its decomposed form. In the PBPs, deacylation is slower, so that the machinery of bacterial cell wall construction is upset.

Inevitably, computational chemists have recently reached the point of trying to model the acylation step by quantum methods in an environment closer to what a β-lactam would experience in a receptor site. Ab initio calculations were performed with amino acid residues around a β-lactam substrate so as to resemble the inhibitor in the clutches of a β-lactamase.[188] A hybrid quantum mechanics/molecular mechanics approach[189] was recently used to map the reaction mechanism of a β-lactamase acting on a penicillin.[190] With atomic coordinates readily available for so many such β-lactam recognizing enzymes (but no PBPs as yet), further calculations of this type can be expected to be carried out.

10. CONCLUSIONS

The amide bond of the β-lactam ring is fascinating and well-studied. The reactivity of this bond is, to some extent, related to the level of antibacterial activity exhibited by the compounds in which it occurs. However, we have learned that the relationship between structure of the β-lactam ring and biological activity is not simple, and, as expected from the earliest calculations, the relationship is not the only factor to consider in searching for new medicines. Besides the obvious electronic and geometric factors associated with the β-lactam, other physicochemical properties of the inhibitors play important roles in determining whether a compound with be therapeutically useful. The appendages on the β-lactam antibacterial agents affect transport between pharmacokinetic compartments in the body, metabolic stability, excretion through the kidney and intestine, as well as steric fit and intermolecular interactions in the receptors. A good inhibitor should be able to acylate a variety of similar but slightly different bacterial receptor sites; it also should resist deacylation from the active sites of the target enzymes and resist hydrolysis by destructive bacterial enzymes. Many additional factors, most of them not predictable theoretically, ultimately determine whether an inhibitor will be an effective medicine.[191]

In the last two decades many theoretical (computational) studies of β-lactam structures have been performed. The justification for many, if not most, of the studies has been that the computations would somehow be useful in designing new, more potent compounds. Such promises have not been fulfilled. To the author's knowledge, no new β-lactam compounds of sufficient interest to warrant clinical investigation have emanated from computational chemistry studies. Also, none of the many X-ray crystallographic studies on β-lactam recognizing enzymes has led to the design of a novel, clinically useful compound.

Superficially, it might appear that rational design methods[192] have not lived up to their potential in the case of β-lactam antibacterial agents. However, the computational and structure-determination methods are not inherently deficient. On the contrary, there are now many documented cases of inhibitors or ligands being discovered with input from computer modeling, and some of these compounds have even advanced to the pharmaceutical market.[193,194]

The shortfall with rational design attempts was not that they could not discover biologically active β-lactam structures, but rather that very few compounds designed from computer calculations or protein crystallography were synthesized and tested. β-Lactam compounds from traditional approaches of natural product isolation and medicinal chemistry were readily available. When the research field was active, new β-lactam compounds from medicinal chemistry were never in short supply. The few β-lactam (or related) compounds that have been synthesized based on modeling were not sufficiently advantageous compared to the plethora of other available compounds. As mentioned, many factors besides molecular geometry and ligand–receptor interaction determine the suitability of a compound for becoming a marketable medicine.

Recall too that the odds of producing a new pharmaceutical product are small. For every experimental compound reaching advanced pre-clinical testing in the 1970s, probably 50 or 100 had been synthesized and screened. Over the last four decades, 200 compounds in the penicillin family have undergone advanced pre-clinical or clinical testing, and in the cephalosporin family, more than 1200 compounds have undergone advanced pre-clinical or clinical testing. In general, only about one compound in ten that undergoes clinical testing is sufficiently promising to be marketed as a pharmaceutical product.

In terms of new molecular entities (NMEs) approved by the U.S. Food and Drug Administration (FDA), the number of β-lactam antibacterial agents has generally been declining since 1983. The number of these NMEs peaked at nine in 1987 and at five in 1992, and the average number for the period 1983–1996 was 2.5 per year. Because it typically takes 8–15 years for an NME to advance through the rigorous testing before the compound can become a pharmaceutical product, an NME approved in 1992 was probably discovered in 1984 or before. Given the current situation, as pharmaceutical discovery researchers seek totally new antibacterial targets initially free from resistance problems,[195] it would be very difficult to design a novel lactam antibacterial structure valuable enough to reach medical practice.

To a certain extent, many of the computational and structural studies have been independently rediscovering the same ideas published by earlier authors. This is satisfying because the scientific method requires that work be reproducible. The recent work has also been scientifically valuable for interpreting retrospectively existing β-lactam antibacterial compounds, for testing computational methods, and for shedding light on possible mechanisms occurring in the active sites of the target enzymes.

ACKNOWLEDGMENTS

The author expresses gratitude to his many former coworkers and collaborators, many of whom are cited in the references, for their brilliant work. Among those coworkers was Dr. Lowell D. Hatfield (1940–1999) who had the courage to step outside the "box" of the β-lactam ring and look at γ-lactams and other totally novel β-lactam surrogates. G. Pearl and K. B. Lipkowitz collaborated on the searches of the CSD. Also appreciated are the many preprints and reprints that authors sent to me over the years. The editors provided helpful comments and asked good questions, which this chapter tries to answer.

REFERENCES

1. Abraham, E. P. In *Beta-Lactam Antibiotics: Chemistry and Biology*; Morin, R. B.; Gorman, M., Eds.; Academic Press: New York, 1982, Vol. 1, pp. xxi–xxxvii.
2. Boyd, D. B. In *Beta-Lactam Antibiotics: Chemistry and Biology*; Morin, R. B.; Gorman, M., Eds.; Academic Press: New York, 1982, Vol. 1, pp. 437–545.
3. Jungheim, L. N.; Sigmund, S. K.; Fisher, J. W. *Tetrahedron Lett.* 1987, **28**, 285.
4. Jungheim, L. N.; Boyd, D. B.; Indelicato, J. M.; Pasini, C. E.; Preston, D. E.; Alborn, W. E. Jr. *J. Med. Chem.* 1991, **34**, 1732.
5. Boyd, D. B. *J. Med. Chem.* 1993, **36**, 1443.
6. Williams, D. H. *Acc. Chem. Res.* 1984, **17**, 364.
7. Tipper, D. J.; Strominger, J. L. *Proc. Natl. Acad. Sci. U. S. A.* 1965, **54**, 1133.
8. Blumberg, P. M.; Strominger, J. L. *Bacteriol. Rev.* 1974, **38**, 291.
9. Boyd, D. B. *Proc. Natl. Acad. Sci. U. S. A.* 1977, **74**, 5239.
10. Boyd, D. B. *J. Med. Chem.* 1979, **22**, 533.
11. Lamotte-Brasseur, J.; Dive, G.; Ghuysen, J.-M. *Eur. J. Med. Chem.* 1984, **19**, 319.
12. Labischinski, H.; Barnickel, G.; Naumann, D.; Rönspeck, W.; Bradaczek, H. *Biopolymers* 1985, **24**, 2087.
13. Damewood, J. R. Jr. In *Reviews in Computational Chemistry*; Lipkowitz, K. B.; Boyd, D. B., Eds.; VCH Publishers: New York, 1996, Vol. 9, pp. 1–79.
14. Kalman, T. I., Ed. *Drug Action and Design: Mechanism-based Enzyme Inhibitors*; Elsevier North Holland: New York, 1979.
15. Livermore, D. M. *Drugs* 1987, **34** (Suppl. 2), 64.

16. Moellering, R. C., Jr. *J. Antimicrob. Chemother.* 1993, **31** (Suppl. A), 1.
17. Silver, L. L.; Bostian, K. A. *Antimicrob. Agents Chemother.* 1993, **37**, 377.
18. Coleman, K.; Athalye, M.; Clancey, A.; Davison, M.; Payne, D. J.; Perry, C. R.; Chopra, I. *J. Antimicrob. Chemother.* 1994, **33**, 1091.
19. Moosdeen, F. *Clin. Infect. Dis.* 1997, **24**, 487.
20. Medeiros, A. A. *Clin. Infect. Dis.* 1997, **24** (Suppl. 1), S19.
21. *Newsweek*, March 28, 1994, p. 47.
22. Gadebusch, H. H.; Stapley, E. O.; Zimmerman, S. B. *Crit. Rev. Biotechnol.* 1992, **12**, 225.
23. Domagala, J. M.; Sanchez, J. P. *Annu. Rev. Med. Chem.* 1997, **32**, 111.
24. Flynn, E. H., Ed. *Cephalosporins and Penicillins: Chemistry and Biology*; Academic Press: New York, 1972.
25. Morin, R. B.; Gorman, M., Eds. *Beta-lactam Antibiotics: Chemistry and Biology*; Academic Press: New York, 1982, Vol. 1–3.
26. Martin, Y. C.; Fischer, E. W. In *Quantitative Structure-activity Relationships of Drugs*, Topliss, J. G., Ed.; Academic Press: New York, 1983, pp. 77–135.
27. Umezawa, H., Ed. *Frontiers of Antibiotic Research*. Academic Press: Tokyo, 1987.
28. Neuhaus. F. C.; Georgopapadakou, N. In *Emerging Targets in Antibacterial and Antifungal Chemotherapy*, Sutcliffe, J.; Georgopapadakou, N., Eds.; Chapman and Hall: New York, 1991, pp. 206–273.
29. Boyd, D. B. In *Ullmann's Encyclopedia of Industrial Chemistry*, Wiley-VCH: Weinheim, 1998, 6th edn. on CD-ROM.
30. Woodward, R. B. *Phil. Trans. R. Soc. Lond.* 1980, **B289**, 239.
31. Williams, D. E. In *Reviews in Computational Chemistry*; Lipkowitz, K. B.; Boyd, D. B., Eds.; VCH Publishers: New York, 1991, Vol. 2, pp. 219–271.
32. Bachrach, S. M. In *Reviews in Computational Chemistry*, Lipkowitz, K. B.; Boyd, D. B., Eds.; VCH Publishers: New York, 1994, Vol. 5, pp. 171–227.
33. Boyd, D. B. *J. Med. Chem.* 1983, **26**, 1010.
34. Boyd, D. B. *J. Med. Chem.* 1984, **27**, 63.
35. Boyd, D. B.; Herron, D. K.; Lunn, W. H. W.; Spitzer, W. A. *J. Am. Chem. Soc.* 1980, **102**, 1812.
36. Boyd, D. B. *Drug Inf. J.* 1983, **17**, 121.
37. Boyd, D. B. In *Frontiers of Antibiotic Research*; Umezawa, H., Ed.; Academic Press: Tokyo, 1987, pp. 339–356.
38. Pfaendler, H. R.; Gosteli, J.; Woodward, R. B.; Rihs, R. B. *J. Am. Chem. Soc.* 1981, **103**, 4526.
39. Pfaendler, H. R.; Hendel, W.; Nagel, U. *Z. Naturforsch.* 1992, **47b**, 1037.
40. Rivail, J.-L.; Maigret, B. In *Reviews in Computational Chemistry*; Lipkowitz, K. B.; Boyd, D. B., Eds.; VCH Publishers: New York, 1998, Vol. 12, pp. 373–380.
41. Schlick, T. In *Reviews in Computational Chemistry*; Lipkowitz, K. B.; Boyd, D. B., Eds., VCH Publishers: New York, 1992, Vol. 3, pp. 1–71.
42. McKee, M. L.; Page, M. In *Reviews in Computational Chemistry*; Lipkowitz, K. B.; Boyd, D. B., Eds.; VCH Publishers: New York, 1993, Vol. 4, pp. 35–65.
43. Boyd, D. B. Unpublished work, 1985.

44. Feller, D.; Davidson, E. R. In *Reviews in Computational Chemistry*; Lipkowitz, K. B.; Boyd, D. B., Eds., VCH Publishers: New York, 1990, Vol. 1; pp. 1–43.
45. Stewart, J. J. P. In *Reviews in Computational Chemistry*; Lipkowitz, K. B.; Boyd, D. B., Eds.; VCH Publishers: New York, 1990, Vol. 1, pp. 45–81.
46. Boyd, D. B. In *Reviews in Computational Chemistry*; Lipkowitz, K. B.; Boyd, D. B., Eds.; VCH Publishers: New York, 1990, Vol. 1, pp. 321–354.
47. Zerner, M. C. In *Reviews in Computational Chemistry*; Lipkowitz, K. B.; Boyd, D. B., Eds.; VCH Publishers: New York, 1991, Vol. 2, pp. 313–365.
48. Stewart, J. J. P. *Quantum Chemistry Program Exchange Bull.* 1993, **13**, 40–43. MOPAC, QCPE Program 455, Indiana University, Bloomington, IN.
49. Hehre, W. J.; Radom, L.; Schleyer, P. v. R.; Pople, J. A. *Ab Initio Molecular Orbital Theory*; Wiley: New York, 1986.
50. Hehre, W. J.; Burke, L. D.; Shusterman, A. J. *A Spartan Tutorial*; Wavefunction, Inc., Irvine, CA, 1993. Spartan version 4. 1. 1.
51. Yang, Q.-C, Seiler, P.; Dunitz, J. D. *Acta Crystallogr., Sect. C* 1987, **43**, 565.
52. León, S.; Alemán, C.; García-Alvarez, M.; Muñoz-Guerra, S. *Struct. Chem.* 1997, **8**, 39.
53. Dewar, M. J. S.; Thiel, W. *J. Am. Chem. Soc.* **1977**, *99*, 4899.
54. Rauk, A.; Allen, L. C.; Clementi, E. *J. Chem. Phys.* 1970, **52**, 4133.
55. Boyd, D. B. *J. Chem. Educ.* 1976, **53**, 483.
56. Boyd, D. B.; Eigenbrot, C.; Indelicato, J. M.; Miller, M. J.; Pasini, C. E.; Woulfe, S. R. *J. Med. Chem.* 1987, **30**, 528.
57. Boyd, D. B.; Smith, D. W.; Stewart, J. J. P.; Wimmer, E. *J. Comput. Chem.* 1988, **9**, 387.
58. Allen, F. H.; Kennard, O.; Taylor, R. *Acc. Chem. Res.* 1983, **16**, 146.
59. Taylor, R.; Kennard, O. *Acta Crystallogr., Sect. B* 1983, **39**, 517.
60. Nangia, A.; Biradha, K.; Desiraju, G. R. *J. Chem. Soc., Perkin 2* 1996, 943.
61. Boyd, D. B.; Lipkowitz, K. B. *J. Chem. Educ.* 1982, **59**, 269.
62. Vasudevan, T. K.; Rao, V. S. R. *Biopolymers* 1981, **20**, 865.
63. Vasudevan, T. K.; Rao, V. S. R. *Int. J. Biol. Macromol.* 1982, **4**, 219.
64. Vasudevan, T. K.; Rao, V. S. R. *Int. J. Biol. Macromol.* 1982, **4**, 347.
65. Vasudevan, T. K.; Rao, V. S. R. *Curr. Sci.* 1982, **51**, 402.
66. Vasudevan, T. K.; Rao, V. S. R. *J. Biosci.* 1982, **4**, 209.
67. Rao, V. S. R.; Vasudevan, T. K. *CRC Crit. Rev. Biochem.* 1983, **14**, 173.
68. Cohen, N. C. Drug Information Association Symposium on Computer Assisted Chemistry; McNeil Pharmaceutical: Spring House, PA, March 12, 1981.
69. Cohen, N. C. *J. Med. Chem.* 1983, **26**, 259.
70. Allinger, N. L., et al. *Quantum Chemistry Program Exchange* 1976, **11**, 318. MMI/MMPI, QCPE Program 318, Indiana University, Bloomington, IN.
71. Dominy, B. W. *Abstracts for the Molecular Mechanics Symposium*; Lipkowitz, K. B.; Boyd, D. B., Eds.; Indiana University-Purdue University at Indianapolis, June 23–24, 1983.
72. Bowen, J. P.; Allinger, N. L. In *Reviews in Computational Chemistry*; Lipkowitz, K. B.; Boyd, D. B., Eds. VCH Publishers: New York, NY, 1991, Vol. 2, pp. 81–97.

73. Allinger, N. L. *J. Am. Chem. Soc.* 1977, **99**, 8127.
74. Boyd, D. B. 1983–1984 (Unpublished work).
75. Wolfe, S.; Khalil, M.; Weaver, F. *Can. J. Chem.* 1988, **66**, 2715.
76. Wolfe, S.; Yang, K.; Khalil, M. *Can. J. Chem.* 1988, **66**, 2733.
77. Joris, B.; Ghuysen, J.-M.; Dive, G.; Renard, A. Dideberg, O.; Charlier, P.; Frère, J.-M.; Kelly, J. A.; Boyington, J. C.; Moews, P. C.; Knox, J. R. *Biochem J.* 1988, **250**, 313.
78. Kelly, J. A.; Knox, J. R.; Zhao, H.; Frère, J.-M.; Ghuysen, J.-M. *J. Mol. Biol.* 1989, **209**, 281.
79. Ghuysen, J.-M. *Annu. Rev. Microbiol.* 1991, **45**, 37.
80. Kelly, J. A.; Kuzin, A. P. *J. Mol. Biol.* 1995, **254**, 223.
81. Boyd, D. B.; Ott, J. L. *Antimicrob. Agents Chemother.* 1986, **29**, 774.
82. Boyd, D. B.; Ott, J. L. *J. Antibiot.* 1986, **39**, 281.
83. Durkin, K. A.; Sherrod, M. J.; Liotta, D. *J. Org. Chem.* 1989, **54**, 5839.
84. Durkin, K. A.; Sherrod, M. J.; Liotta, D. *Adv. Mol. Model.* 1990, **2**, 93.
85. Fernández, B.; Rios, M. A. *J. Comput. Chem.* 1994, **15**, 455.
86. Fernández, B.; Carballeira, L.; Rios, M. A. *Biopolymers* 1992, **32**, 97.
87. Cooper, R. D. G.; Daugherty, B. W.; Boyd, D. B. *Pure Appl. Chem.* 1987, **59**, 485.
88. Cossio, F. P.; Ugalde, J. M.; Lopez, X.; Lecea, B.; Palomo, C. *J. Am. Chem. Soc.* 1993, **115**, 995.
89. Arrastia, I.; Arrieta, A.; Ugalde, J. M.; Cossio, Fernando, P.; Lecea, B. *Tetrahedron Lett.* 1994, **35**, 7825.
90. Lopez, R.; Ruiz-Lopez, M. F.; Rinaldi, D.; Sordo, J. A.; Sordo, T. L. *J. Phys. Chem.* 1996, **100**, 10600.
91. Fang, D.-C.; Fu, X.-Y. *Chin. J. Chem.* 1996, **14**, 97.
92. Fang, D.-C.; Fu, X.-Y. *Int. J. Quantum Chem.* 1992, **43**, 669.
93. Wang, X.; Lee, C. *Tetrahedron Lett.* 1993, **34**, 6241.
94. Barcza, M. V.; de M. Carneiro, J. W.; Serra, A. A.; Barboza, J. C. S. *THEOCHEM* 1997, **394**, 281.
95. Chung, S. K.; Chodosh, D. F. *Bull. Korean Chem. Soc.* 1989, **10**, 185.
96. Chung, S. K. *Bull. Korean Chem. Soc.* 1989, **10**, 216.
97. Kamiya, K.; Takamoto, M.; Wada, Y.; Asai, M. *Acta Crystallogr., Sect. B* 1981, **37**, 590.
98. Halgren, T. A. *J. Comput. Chem.* 1996, **17**, 490.
99. Won, Y. *Bull. Korean Chem. Soc.* 1995, **16**, 944.
100. Glidewell, C.; Mollison, G. S. M. *J. Mol. Struct.* 1981, **72**, 203.
101. Frau, J.; Coll, M.; Donoso, J.; Muñoz, F.; Garcia Blanco, F. *THEOCHEM* 1991, **77**, 109.
102. Frau, J.; Donoso, J.; Muñoz, F.; Garcia Blanco, F. *THEOCHEM* 1991, **83**, 205.
103. Nangia, A. *THEOCHEM* 1991, **251**, 237.
104. Nangia, A. *Proc. Indian Acad. Sci.* (*Chem. Sci.*) 1993, **105**, 131.
105. Nangia, A.; Chandrakala, P. S.; Balaramakrishna, P. V.; Latta, T. V. *THEOCHEM* 1995, **343**, 157.

106. Bruton, G. *Bioorg. Med. Chem. Lett.* 1993, **3**, 2329.
107. Scanlan, M. J.; Hillier, I. H.; Hodgkin, E. E.; Sidebotham, R. P.; Warwick, C. M.; Davies, R. H. *Int. J. Quantum Chem., Quantum Biol. Symp.* 1983, **10**, 231.
108. Vishveshwara, S.; Rao, V. S. R. *THEOCHEM* 1983, **92**, 19.
109. Sedano, E.; Ugalde, J. M.; Cossio, F. P.; Palomo, C. *THEOCHEM* 1988, **43**, 481.
110. Marstokk, K. M.; Moellendal, H.; Samdal, S.; Uggerud, E. *Acta Chem. Scand.* 1989, **43**, 351.
111. Alcolea Palafox, M.; Núñez, J. L.; Gil, M. *J. Phys. Chem.* 1995, **99**, 1124.
112. Roux, M. V.; Jiménez, P.; Dávalos, J. Z.; Castaño, O.; Molina, M. T.; Notario, R.; Herreros, M.; Abboud, J.-L. M. *J. Am. Chem. Soc.* 1996, **118**, 12735.
113. León, S.; Martínez de Ilarduya, A.; Alemán, C.; García-Alvarez, M.; Muñoz-Guerra, S. *J. Phys. Chem. A* 1997, **101**, 4208.
114. Boyd, D. B.; Hermann, R. B.; Presti, D. E.; Marsh, M. M. *J. Med. Chem.* 1975, **18**, 408.
115. Petrongolo, C.; Ranghino, G.; Scordamaglia, R. *Chem. Phys.* 1980, **45**, 239.
116. Petrongolo, C.; Ranghino, G.; Scordamaglia, R. *Chem. Phys.* 1980, **45**, 279.
117. Petrongolo, C.; Pescatori, E.; Ranghino, G.; Scordamaglia, R. *Chem. Phys.* 1980, **45**, 291.
118. Petrongolo, C.; Ranghino, G.; Scordamaglia, R. In *Medicinal Chemistry Advances* (Proc. 7th Int. Symp. on Med. Chem.); De las Heras, F. G.; Vega, S., Eds.; Pergamon: Oxford, 1981, pp. 103–116.
119. Boyd, D. B. 1981 (Unpublished work).
120. Boyd, D. B.; Elzey, T. K.; Hatfield, L. D.; Kinnick, M. D.; Morin, J. M. Jr. *Tetrahedron Lett.* 1986, **27**, 3453.
121. Boyd, D. B.; Foster, B. J.; Hatfield, L. D.; Hornback, W. J.; Jones, N. D.; Munroe, J. E.; Swartzendruber, J. K. *Tetrahedron Lett.* 1986, **27**, 3457.
122. Allen, N. E.; Boyd, D. B.; Campbell, J. B.; Deeter, J. B.; Elzey, T. K.; Foster, B. J.; Hatfield, L. D.; Hobbs, J. N. Jr.; Hornback, W. J.; Hunden, D. C.; Jones, N. D.; Kinnick, M. D.; Morin, J. M. Jr., Munroe, J. E.; Swartzendruber, J. K.; Vogt, D. G. *Tetrahedron* 1989, **45**, 1905.
123. Lu, L.; Liu, J.; Du, F.; Hua, W. *THEOCHEM* 1990, **65**, 17.
124. Nahm, K. *Bull. Korean Chem. Soc.* 1991, **12**, 674.
125. Nahm, K. *Bioorg. Med Chem. Lett.* 1992, **2**, 485.
126. Smeyers, Y. G.; Hernandez-Laguna, A.; Gonzalez-Jonte, R. *THEOCHEM* 1993, **106**, 261.
127. Marchand-Brynaert, J.; Couplet, B.; Dive, G.; Ghosez, L. *Bioorg. Med. Chem. Lett.* 1993, **3**, 2303.
128. Choi, J.-H.; Kim, H. *Bull. Korean Chem. Soc.* 1993, **14**, 631.
129. Frau, J.; Donoso, J.; Muñoz, F.; Garcia Blanco, F. *J. Comput. Chem.* 1992, **13**, 681.
130. Frau, J.; Donoso, J.; Vilanova, B.; Muñoz, F.; Garcia Blanco, F. *Theor. Chem. Acta* 1993, **86**, 229.
131. Frau, J.; Donoso, J.; Muñoz, F.; Garcia Blanco, F. *J. Comput. Chem.* 1993, **14**, 1545.
132. Frau, J.; Donoso, J.; Muñoz, F.; Garcia Blanco, F. *Helv. Chim. Acta* 1994, **77**, 1557.

133. Frau, J.; Donoso, J.; Muñoz, F.; Garcia Blanco, F. *Helv. Chim. Acta* 1996, **79**, 353.
134. Frau, J.; Donoso, J.; Muñoz, F.; Garcia Blanco, F. *THEOCHEM* 1997, **390**, 247.
135. Frau, J.; Donoso, J.; Muñoz, F.; Vilanova, B.; Garcia Blanco, F. *Helv. Chim. Acta* 1997, **80**, 739.
136. Frau, J.; Coll, M.; Donoso, J.; Muñoz, F.; Vilanova, B.; Garcia Blanco, F. *Electron. J. Theor. Chem.* 1997, **2**, 56.
137. Coll, M.; Frau, J.; Donoso, J.; Muñoz, F. *THEOCHEM* 1998, **426**, 323.
138. Wolfe, S.; Kim, C.-K.; Yang, K. *Can. J. Chem.* 1994, **72**, 1033.
139. Wolfe, S.; Hoz, T. *Can. J. Chem.* 1994, **72**, 1044.
140. Wolfe, S.; Jin, H.; Yang, K.; Kim, C.-K.; McEachern, E. *Can. J. Chem.* 1994, **72**, 1051.
141. Lee, J. C.; Koh, H.-Y.; Chang, M.-H.; Lee, Y. S. *Bull. Korean Chem. Soc.* 1996, **17**, 604.
142. Pitarch, J.; Ruiz-Lopez, M. F.; Pascual-Ahuir, J.-L.; Silla, E.; Tunon, I. *J. Phys. Chem. B* 1997, **101**, 3581.
143. Pitarch, J.; Ruiz-Lopez, M. F.; Silla, E.; Pascual-Ahuir, J.-L.; Tunon, I. *J. Am. Chem. Soc.* 1998, **120**, 2146.
144. Lunn, W. H. W. 1976–1978 (Unpublished work).
145. Boyd, D. B.; Lunn, W. H. W. *J. Med. Chem.* 1979, **22**, 778.
146. Boyd, D. B. *J. Org. Chem.* 1985, **50**, 886.
147. Coene, B.; Schanck, A.; Dereppe, J.-M.; Van Meerssche, M. *J. Med. Chem.* 1984, **27**, 694.
148. Boyd, D. B. *J. Org. Chem.* 1985, **50**, 885.
149. Lowe, J. P. *J. Chem. Educ.* 1974, **51**, 785.
150. Fountain, R. H.; Russell, A. D. *J. Appl. Bacteriol.* 1969, **32**, 312.
151. O'Callaghan, C. H.; Sykes, R. B.; Staniforth, S. E. *Antimicrob. Agents Chemother.* 1976, **10**, 245.
152. Lin, H.-S.; Boyd, D. B. 1988 (Unpublished work).
153. Albrecht, H. A.; Beskid, G.; Chan, K.-K.; Christenson, J. G.; Cleeland, R.; Deitcher, K. H.; Georgopapadakou, N. H.; Keith, D. D.; Pruess, D. L.; Sepinwall, J.; Specian, A. C. Jr.; Then, R. L.; Weigele, M.; West, K. F.; Yang, R. *J. Med. Chem.* 1990, **33**, 77.
154. Matera, G.; Berlinghieri, M. C.; Foti, F.; Barreca, G. S.; Foca, A. *J. Antimicrob. Chemother.* 1996, **38**, 799.
155. Bryskier, A. *Expert Opin. Invest. Drugs*1997, **6**, 1479.
156. Di Modugno, E.; Broggio, R.; Erbetti, I.; Lowther, J. *Antimicrob. Agents Chemother.* 1997, **41**, 2742.
157. Hanessian, S.; Griffin, A. M.; Rozema, M. J. *Bioorg. Med. Chem. Lett.* 1997, **7**, 1857.
158. Hwu, J. R.; Tsay, S.-C.; Hakimelahi, S. *J. Med. Chem.* 1998, **41**, 4681.
159. Vrudhula, V. M.; Svensson, H. P.; Senter, P. D. *J. Med. Chem.* 1985 **38**, 1380.
160. Meyer, D. L.; Law, K. L.; Payne, J. K.; Mikolajczyk, S. D.; Zarrinmayeh, H.; Jungheim, L. N.; Kling, J. K.; Shepherd, T. A.; Starling, J. J. *Bioconjugate Chem.* 1995, **6**, 440.

161. Brunger, A. T.; Kuriyan, J.; Karplus, M. *Science* 1987, **235**, 458.
162. Lybrand, T. P. In *Reviews in Computational Chemistry*; Lipkowitz, K. B.; Boyd, D. B., Eds.; VCH Publishers: New York, 1990, Vol. 1, pp. 295–320.
163. Kelly, J. A.; Knox, J. R.; Moews, P. C.; Hite, G. J.; Bartolone, J. B.; Zhao, H.; Joris, B.; Frère, J.-M.; Ghuysen, J.-M. *J. Biol. Chem.* 1985, **260**: 6449.
164. Boyd, D. B.; Snoddy, J. D.; Lin, H.-S. *J. Comput. Chem.* 1991, **12**, 635.
165. Tripos, Inc.: St. Louis, MO.
166. Boyd, D. B.; Snoddy, J. D. In *Molecular Aspects of Chemotherapy*, Shugar, D.; Rode, W.; Borowski, E., Eds.; Springer-Verlag: New York, 1992; pp. 1–22.
167. Molecular Simulations, Inc.: San Diego, CA.
168. Lamotte-Brasseur, J.; Dive, G.; Dideberg, O.; Charlier, P.; Frére, J.-M.; Ghuysen, J.-M. *Biochem. J.* 1991, **279**, 213.
169. Lamotte-Brasseur, J.; Jacob-Dubuisson, F.; Dive, G.; Frére, J.-M.; Ghuysen, J.-M. *Biochem. J.* 1992, **282**, 189.
170. Matagne, A.; Lamotte-Brasseur, J.; Dive, G.; Knox, J. R.; Frére, J.-M. *Biochem. J.* 1993, **293**, 607.
171. Fonze, E.; Charlier, P.; To'th, Y.; Vermeire, M.; Raquet, X.; Dubus, A.; Frére, J.-M. *Acta Crystallogr., Sect. D: Biol. Crystallogr.* 1995, **51**, 682.
172. Guillaume, G.; Vanhove, M.; Lamotte-Brasseur, J.; Ledent, P.; Jamin, M.; Joris, B.; Frére, J.-M. *J. Biol. Chem.* 1997, **272**, 5438.
173. Vijayakumar, S.; Vishveshwara, S.; Ravishanker, G.; Beveridge, D. L. *Biophys. J.* 1993, **65**, 2304.
174. Ahluwalia, R.; Day, R. A.; Nauss, J. *Biochem. Biophys. Res. Commun.* 1995, **206**, 577.
175. Juteau, J.-M.; Billings, E.; Knox, J. R.; Levesque, R. C. *Protein Eng.* 1992, **5**, 693.
176. Imtiaz, U.; Billings, E.; Knox, J. R.; Manvathu, E. K.; Lerner, S. A.; Mobashery, S. *J. Am. Chem. Soc.* 1993, **115**, 4435.
177. Imtiaz, U.; Billings, E. M.; Knox, J. R.; Mobashery, S. *Biochemistry* 1994, **33**, 5728.
178. Billings, E. M. Ph. D. Thesis, University of Connecticut, 1996.
179. Miyashita, K.; Mobashery, S. *Bioorg. Med. Chem. Lett.* 1995, **5**, 1043.
180. Bulychev, A.; Massova, I.; Lerner, S. A.; Mobashery, S. *J. Am. Chem. Soc.* 1995, **117**, 4797.
181. Bulychev, A.; O'Brien, M. E.; Massova, I.; Teng, M.; Gibson, T. A.; Miller, M. J.; Mobashery, S. *J. Am. Chem. Soc.* 1995, **117**, 5938.
182. Taibi, P.; Mobashery, S. *J. Am. Chem. Soc.* 1995, **117**, 7600.
183. Ishiguro, M.; Imajo, S. *J. Med. Chem.* 1996, **39**, 2207.
184. Mohamadi, F.; Richards, N. G. J.; Guida, W. C.; Liskamp, R.; Lipton, M.; Caufield, C.; Chang, G.; Hendrickson, T.; Still, W. C. *J. Comput. Chem.* 1990, **11**, 440.
185. Frau, J.; Price, S. L. *Theor. Chim. Acta* 1997, **95**, 151.
186. Yang, J.; Klimkowski, V. J.; Boyd, D. B. *Abstracts of the Fourth International Conference on Chemical Synthesis of Antibiotics and Related Microbial Products*, Nashville, IN, September 11–16, 1994.
187. Knox, J. R.; Moews, P. C. *J. Mol. Biol.* 1991, **220**, 435.

188. Wladkowski, B. D.; Chenoweth, S. A.; Sanders, J. N.; Krauss, M.; Stevens, W. J. *J. Am. Chem. Soc.* 1997, **119**, 6423.

189. Gao, J. In *Reviews in Computational Chemistry*; Lipkowitz, K. B.; Boyd, D. B., Eds.; VCH Publishers: New York, 1995, Vol. 7, pp. 119–185.

190. Cunningham, M. A.; Makinen, M. W.; Bash, P. A. *Abstracts of the 215th National Meeting of the American Chemical Society*, Dallas, TX, March 29–April 2, 1998, COMP045.

191. Lipkowitz, K. B.; Boyd, D. B.; *Reviews in Computational Chemistry*, Wiley-VCH: New York, 1997, Vol. 11, pp. v–x.

192. Boyd, D. B. In *Rational Molecular Design in Drug Research*, Liljefors, T.; Jørgensen, F. S.; Krogsgaard-Larsen, P., Eds.; Munksgaard: Copenhagen, 1998; pp. 15–23.

193. Boyd, D. B. *CHEMTECH* 1998, **28**, 19.

194. Boyd, D. B. In *Rational Drug Design: Novel Methodology and Practical Applications*, Parrill, A. L.; Reddy, M. R., Eds.; ACS Symp. Series 719, American Chemical Society: Washington, DC, 1999, pp. 346–356.

195. Katz, L.; Chu, D. T.; Reich, K. *Annu. Rev. Med. Chem.* 1997, **32**, 121.

CHAPTER 12

THREE-DIMENSIONAL DESIGN OF ENZYME INHIBITORS WITH HETEROCYCLIC AMIDE BOND MIMICS

REGINE S. BOHACEK and WILLIAM C. SHAKESPEARE
Ariad Pharmaceuticals, Inc.

1. INTRODUCTION

One of the most challenging aspects of the drug discovery process is the design of molecules which not only bind to their biological targets but have in vivo stability and are orally bioavailable. In fact, in the era of high throughput screening and combinatorial chemistry, the bottleneck in drug development remains the creation of compounds with appropriate absorption, distribution, metabolism and elimination profiles.

The design of enzyme inhibitors often begins with compounds which resemble the natural substrate of the target and, therefore, are frequently peptides or molecules containing significant peptidic portions. Although, such compounds have contributed significantly to our understanding of complex biological functions through their high affinity interactions, they can have serious shortcomings as therapeutic agents. Peptides typically have a short duration of action because they are rapidly degraded by peptidases and, therefore, exhibit poor oral bioavailability.[1] Even cyclic peptides which are small and resistant to cleavage by peptidases may exhibit poor oral bioavailability due to poor transport properties in vivo and/or because of biliary excretion.

Numerous techniques have been developed to modify the peptide linkages that are susceptible to enzymatic degradation. Such methods include the systematic replacement of amide bonds with amide bond surrogates, the use of unnatural D-amino acids or *N*-Me analogs or the introduction of conformational constraints by forming bridges or cyclic peptides. Such pseudopeptides,

The Amide Linkage: Selected Structural Aspects in Chemistry, Biochemistry, and Materials Science, Edited by Arthur Greenberg, Curt M. Breneman, and Joel F. Liebman
ISBN 0-471-35893-2

although employed to increase the half life of compounds, often result in analogs which display poor oral bioavailability.

There has been a continued search for amide bond mimics which improve the pharmacokinetic properties of a compound without a significant loss of in vitro binding. Recently, there have been a number of publications exploring the utility of heterocycles as amide bond mimics. In theory, this strategy might yield compounds that are less susceptible to cleavage by proteolytic enzymes, more orally bioavailable, and in addition, would impose a measure of conformational constraint, which, depending on the target protein, may or may not be beneficial.

In this chapter we will focus on enzyme inhibitors for which the three dimensional structure of the target was used to design amide bond surrogates, which are able to position hydrogen bond donor and acceptor groups to form hydrogen bonds with the binding site atoms in a manner similar to those of the amide bonds which they replace. Wherever possible, we used X-ray crystal structure data for our analysis. If this information is missing, we modeled the compounds of interest in the binding site of the target enzyme using the most relevant crystal structures available. (Wherever possible the Brookhaven Data Base[2] entry code for the crystal structures are given).

The likely binding mode of compounds was determined by first constructing the molecules with the FLO96 software[3] using the crystal structure of the ligand from the complex of interest as a starting point. Each molecule was then minimized in the presence of the binding site using the QXP docking module of FLO96. If the compound to be modeled was significantly different from the starting point, the Monte Carlo / energy minimization procedure of FLO96 was used to obtain the predicted binding conformation.

Using examples from HIV protease, elastase, interleukin-1β-converting enzyme (ICE), papain, matrix metalloproteases, neutral endopeptidase (NEP), endothelin converting enzyme (ECE), thermolysin, and pepsin we wanted to explore the success of this strategy. What is the effect of amide bond replacement on in vitro binding? Do these modifications improve pharmacokinetic properties such as cell penetration?

2. HIV PROTEASE INHIBITORS

Recently Hirschmann et al.[4–8] described the synthesis of several potent inhibitors of HIV-1 protease based on the novel 3,5-linked pyrrolinone scaffold. This scaffold was chosen for its ability to adopt a backbone conformation mimicking a β-strand, an orientation observed crystallographically in a host of diverse inhibitors bound to the target protease.[9–12] The removal of amide bonds, in addition to increasing proteolytic stability, was thought to increase bioavailability by reducing the number of hydrogen bonds with water.

To determine the interactions that the pyrrolinone containing inhibitor **2** may form with HIV-1 protease, **2** was docked into the HIV-1 binding site. The energy minimized conformation of **2** had the pyrrolidinone NH displaced from the backbone relative to the P_2' amide in **1** (L-697,807),[13] thereby precluding it from hydrogen bonding to Gly[27] (Fig. 1). The ketone of the pyrrolinone did, however, hydrogen bond to water in a manner identical to **1**. Ultimately, both of these predictions were borne out crystallographically.[8] Interestingly, although the pyrrolinone NH in **2** did not bind to Gly[27], it did hydrogen bond to Asp[25], a catalytic residue in the active site (Fig. 12.1(a)). Additionally, the indanol hydroxyl, which normally forms a hydrogen bond with Asp[29], was rotated up slightly away from the catalytic sight allowing for the inclusion of an entropically disfavorable water molecule. An IC_{50} of 2 and 0.03 nM for compounds **2** and **1** respectively, indicate that the pyrrolinone is indeed a good inhibitor of HIV-1. Certainly, the loss of the hydrogen bond to Gly[27] (although this appears to be partially offset by an alternate one to Asp[25]) is significant, but, perhaps, more important is the effect that the ring has on the conformation of the indanol hydroxyl and its effect on the hydrogen bond to Asp[29] (the authors speculate that this might cost ca. 2 kcal mol^{-1}).

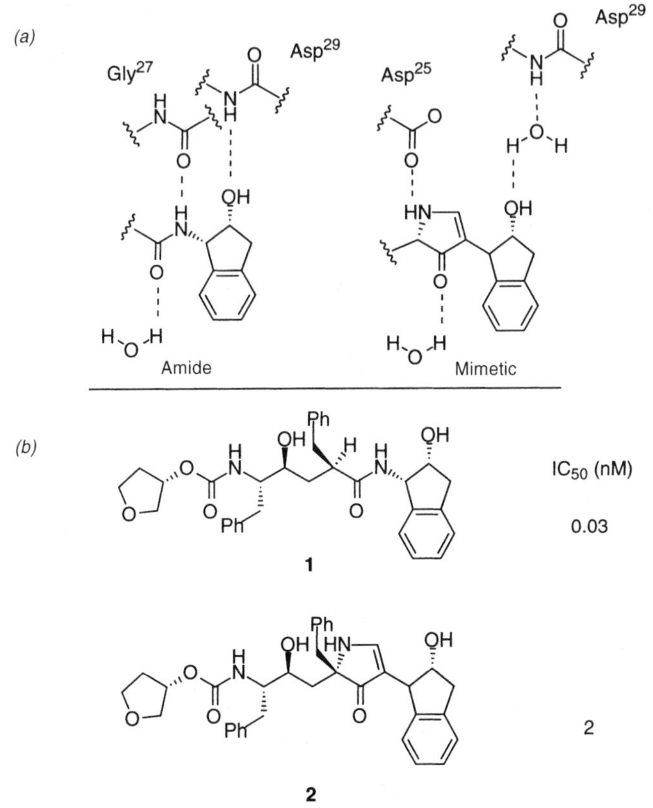

FIGURE 12.1. (*a*) Hydrogen bonding network between the amide (mimetic) and the target protein HIV-1; partial structure; (*b*) Comparison of the amide and mimetic structures and their corresponding inhibition data.

Finally, the CIC_{95} (cellular antiviral activity) of **2** was 100 nM relative to 3 nM for the corresponding peptide **1**. More significant, however, is the ratio of the cellular inhibition to the enzyme inhibition. This ratio, an indirect measure of transport properties, was lower for **2**, indicating that **2** is more readily transported into HIV infected lymphocytes than its peptidyl counterpart.

Thompson et al.[14,15] have recently described a very successful construction of a hydroxyethylene-based HIV-1 protease inhibitor, **4**, containing an imidazole $P_1'-P_2'$ amide bond surrogate. This mimic was chosen for its ability to function as a hydrogen bond acceptor for the bound water molecule and a hydrogen bond donor to the carbonyl oxygen of Gly^{27} (Fig. 12.2(a)). Their modeling studies suggested that the ring could indeed form these interactions and also provide a

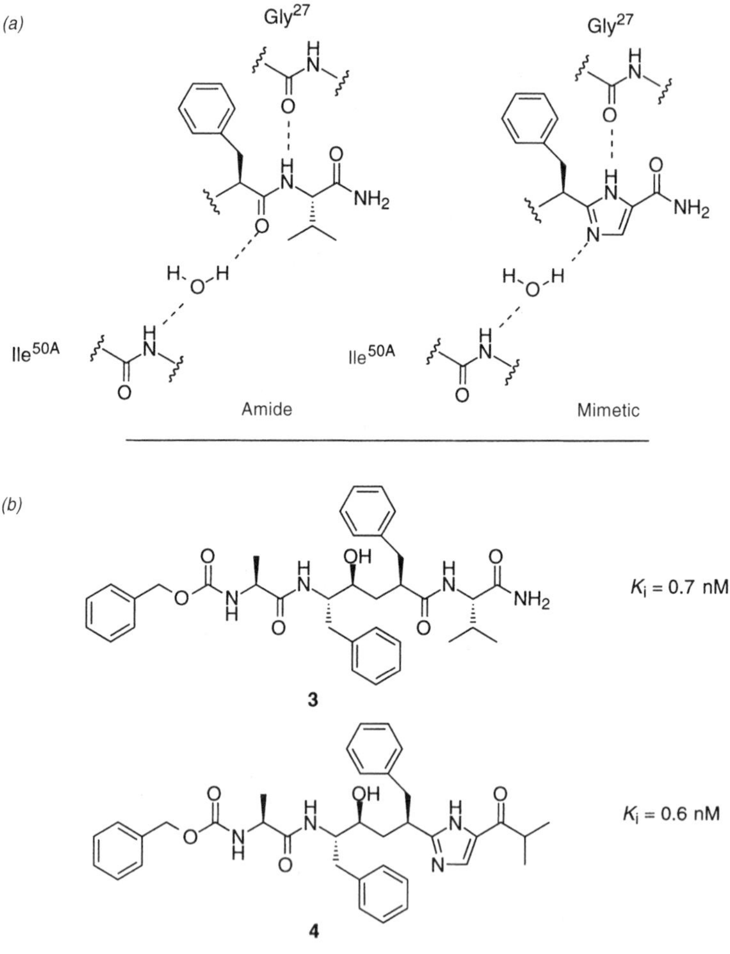

FIGURE 12.2. (*a*) Hydrogen bonding network between the amide (mimetic) and the target protein HIV-1; partial structure; (*b*) Comparison of the amide and mimetic structures and their corresponding inhibition data.

template for backbone extension. Inhibitor **4** (SB 206343) had an IC_{50} of 0.6 nM which is in good agreement with its peptidic counterpart **3** (0.7 nM).

The crystal structure of **4** bound to HIV-1 protease was determined.[14] It showed that the inhibitor binds to the active site of the enzyme in an extended conformation with every heteroatom of the inhibitor forming hydrogen bonds to the protein. More importantly, the imidazole binds precisely in the manner for which it was designed (Fig. 12.2(a)). N-3 of the imidazole ring forms a water mediated hydrogen bond to Ile^{50} and Ile^{150} and N-1 forms a hydrogen bond with Gly^{127}. All these interactions are identical to those observed for the P_1–P_2 amide groups in most HIV-1 protease inhibitors (Fig. 12.2(a)). Unfortunately, no cellular data was available for this compound.

In a similar example, Abdel-Meguid et al.[16] reported the successful substitution of the P_2'–P_3' amide bond with an imidazole ring (Fig. 12.3(b)).

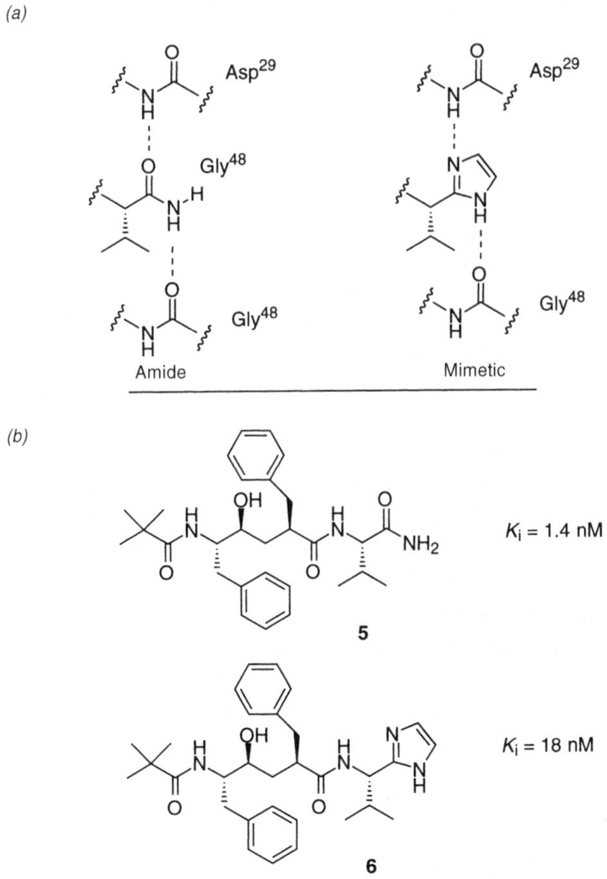

FIGURE 12.3. (*a*) Hydrogen bonding network between the amide (mimetic) and the target protein HIV-1; partial structure; (*b*) Comparison of the amide and mimetic structures and their corresponding inhibition data.

This compound, **6**, has a K_i of 18 nM—about one order-of-magnitude higher than its peptidic counterpart **5** (1.4 nM). The authors propose that the decrease in affinity may be due to the protonation of the imidazole nitrogen, making it unable to function as a hydrogen bond acceptor.

The crystal structure of **6** bound to HIV-1 protease reveals that the two essential hydrogen bonds observed in other peptide analog inhibitors are maintained; N-1 of the imidazole ring is positioned in such a manner that it mimics the amide nitrogen of P_3', while N-3 mimics the carbonyl oxygen of P_2' (Fig. 12.3(a)). Importantly, replacement of the C-terminal amide by imidazole conferred benifits on the pharmacokinetic properties of the inhibitor. The CIC_{50} of **5** and **6** were 570 and 28 nM respectively, roughly proportional to their K_i values. Compound **5**, however, demonstrated a three-fold increase (relative to **6**) in the elimination phase half-life in rat and showed a $t_{1/2\beta} = 84$ min in conscious cynomolgous monkeys. Additionally, the calculated oral bioavailability of **6** in rats was 30%, relative to 1% for **5**. Thus, the substitution of the carboxamide by the C-terminal imidazole, while not as effective for in vitro inhibition, has a profound effect on the pharmacokinetic properties of the molecule.

3. HLE INHIBITORS

Human leukocyte elastase (HLE) and human neutrophil elastase (HNE) are proteolytic enzymes that are able to degrade a variety of endogenous and exogenous proteins. As a result, these enzymes are believed to contribute to the pathological effects associated with pulmonary emphysema,[17] adult respiratory distress syndrome,[18] rheumatoid arthritis,[19] atherosclerosis,[20] cystic fibrosis[21] and other inflammatory disorders.[22] It has been proposed that low molecular weight inhibitors of elstase could be useful in the treatment of these diseases.

Trifluoromethyl ketone **7** (Fig. 12.4) is a potent inhibitor of HLE which has been shown to provide long lasting protection against elastase induced lung damage in animals following intratracheal administration;[23,24] however, this compound is not orally bioavailable. Therefore, researchers at ZENECA sought to design an orally active elastase inhibitor by reducing the peptidic character of **7.** The crystal structure of **8** complexed to a closely related enzyme, porcine pancreatic elastase, had been determined.[25] This structure shows some of the important interactions formed by the Inhibitor **8** with elastase. Hydrogen bonds are formed by the following peptidic inhibitor (**8**) atoms: NH of Val (**8**), and the

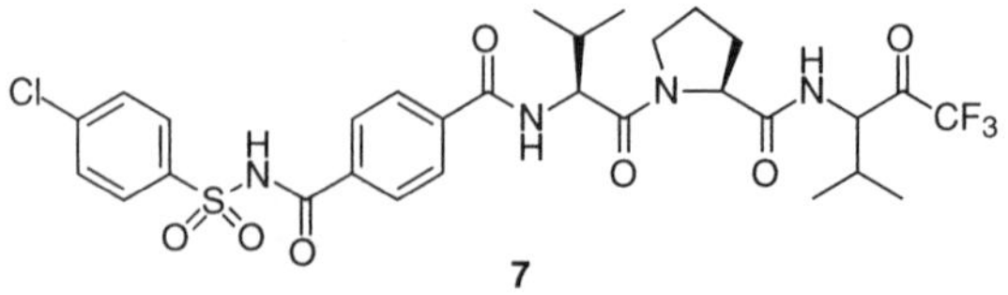

Cl O H N O O O H N O N H O N O H N O CF_3

7

FIGURE 12.4. ICI-200,880. Peptidic trifluoromethyl ketone inhibitor of HLE.

NH and carbonyl of Ala. The inhibitor's proline lies in the S_2 subsite within van der Waals contact of the binding site atoms of His[57], Val[99] and Phe[228]. Using this and related crystal structures, the group at ZENECA[26,27] designed a pyridone dipeptide **9** (Fig. 12.5(b)) which replaces the P_2 proline and P_3 alanine of peptidic inhibitor **8**.

Compound **9** was modeled into the active site of HLE and predicted to maintain similar hydrogen bonding interactions to those of **8**; the carbonyl of the pyridone and the 3-amino group form hydrogen bonds to Val[216] (Fig. 12.5(a)). Compound **9** was synthesized and found to inhibit HLE with an IC_{50} of 2800 nM. This represents a 13-fold decrease in binding affinity when compared to **8** ($IC_{50} = 210$ nM). The binding hypothesis was confirmed by crystallography.

The decrease in affinity of **9** was attributed to the loss of the hydrophobic interactions made by the proline ring with atoms of the S_2 subsite. This was ultimately addressed by incorporation of a phenyl ring in the 6-position, which fills the S_2 subsite, and, according to our modeling studies, stacks against His[57]. Upon replacement of the N-terminus acetyl group with a benzyl carbamate (Cbz N-cap), this pyridone based inhibitor proved to be a potent elastase inhibitor with a $K_i = 4.5$ nM. Unfortunately, this compound was not orally active at a dose of 20 mg kg^{-1} in an acute hemorragic assay. It was determined that the compounds of this series were not well-absorbed from the gastrointestinal tract, which was proposed to be a result of their poor aqueous solubility.

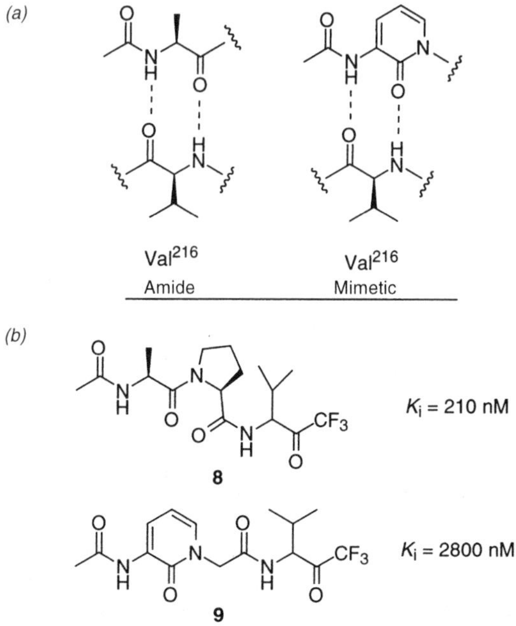

FIGURE 12.5. (*a*) Hydrogen bonding network between the amide (mimetic) and the target protein HLE; partial structure; (*b*) Comparison of the amide and mimetic structures and their corresponding inhibition data.

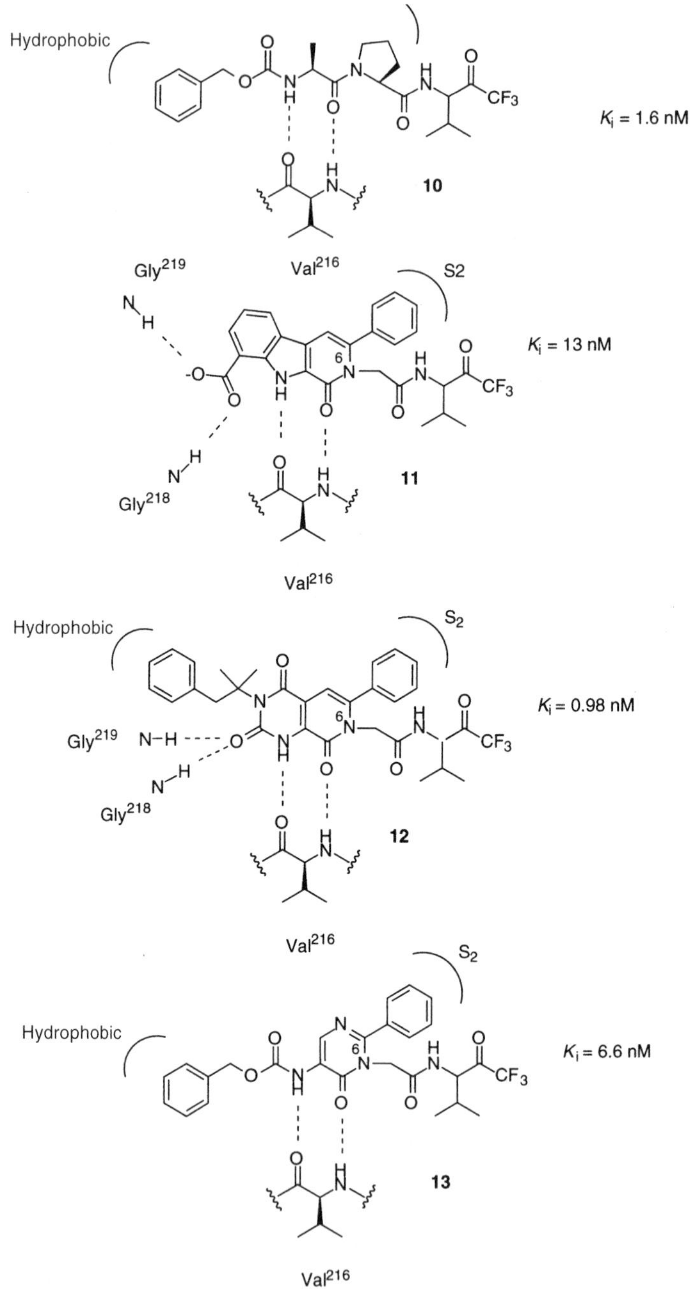

FIGURE 12.6. Proposed hydrogen bonding and hydrophobic interactions formed by a peptidic elastase inhibitor, **10**, and additional elastase inhibitors (**11–13**) with novel heterocycles that incorporate the pyridone design scaffold.

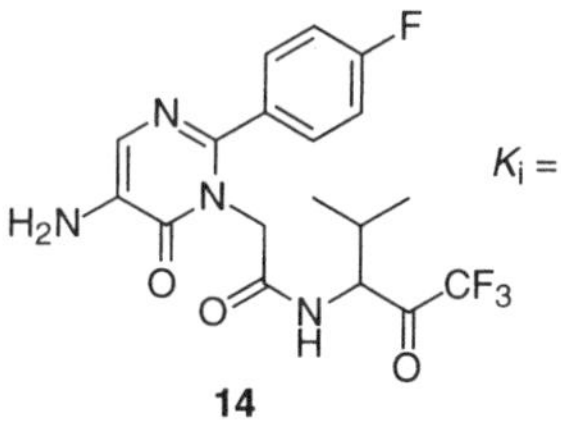

K_i = 101 nM

14

FIGURE 12.7. Orally active inhibitor of HLE.

Subsequently, other heterocycles incorporating the pyridone design features were explored (Fig. 12.6).[28] Some of these compounds are excellent inhibitors of HLE. Compound **12**, with a K_i of 0.98 nM, has essentially the same affinity as the closest peptidic analog **10** which has a K_i of 1.6 nM. Compounds **11** and **12** also incorporate solubilizing groups directly into the bicyclic moieties.

To examine the interactions that these inhibitors form with elastase, we docked these molecules into the active site using the binding site model from the elastase structure 1HNE[29]. In our model, these compounds form a beautiful network of hydrogen bonds. In addition to the hydrogen bonds observed in the 1HNE crystal structure, Compounds **11** and **12** make additional hydrogen bonds with Gly^{218} and Gly^{219}. The C(6) phenyl extends into the S_2 subsite and forms hydrophobic interactions with Leu^{99} and His^{57} and the terminal phenyl forms hydrophobic interactions with the side chain of Arg^{217}. Our modeling studies are consistent with the modeling results described by the ZENECA modelers.[28]

The pyrimidones with a 4-fluorophenyl substituent exhibited especially good pharmacokinetic profiles. Although pyrimidones with a Cbz N-cap displayed the best in vitro inhibition ($K_i = 8.2$ nM), the 3-aminopyrimidone analog **14** exhibited the best pharmacokinetic profile (Fig. 12.7). Compound **14** ($K_i =$ 102 nM) exhibited excellent oral bioavailability and sustained duration of action extending over 4 hours.[26]

4. ICE INHIBITORS

Interleukin-1β-converting enzyme (ICE) is a cysteine protease responsible for the production of interleukin-1β (IL-1β) in monocytes.[30,31] ICE functions by cleaving a precursor of IL-1β into the biologically active form of IL-1β, a cytokine, which is purported to have a role in the pathogenesis of inflammatory diseases such as rheumatoid arthritis.[32–34] Although, a series of ICE inhibitors with high affinity in vitro have been reported, their peptidic nature is likely to make them poor drug candidates.

The manner in which ICE inhibitors bind was revealed through a series of *N*-methyl scans of the Val–Ala–Asp backbone and crystallography (Fig. 12.8).[35] Like many proteases, the enzyme and the inhibitor adopt a β-sheet motif, making critical hydrogen bonds between the backbone of the enzyme and the P_3 and P_1 NHs and the P_3 carbonyl. Based on this information, researchers[36] at Sanofi

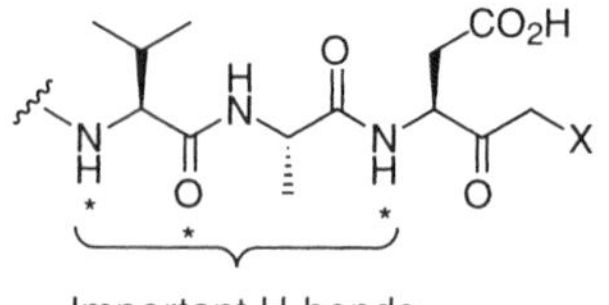

Important H-bonds

FIGURE 12.8. Hydrogen bonds required for high affinity binding to ICE.

Winthrop incorporated a pyrimidone, previously descibed in the design of effective HLE inhibitors[26], as a P_3–P_2 amide bond mimic resulting in Compound **16**. Although, the provision for correct hydrogen bonding exists in **16**, incorporation of the pyrimidone resulted in a two-fold loss in binding affinity compared to the tripeptide **15**. In addition, the potency in this class did not rigorously conform with that of the peptide. For example, the rate of inactivation

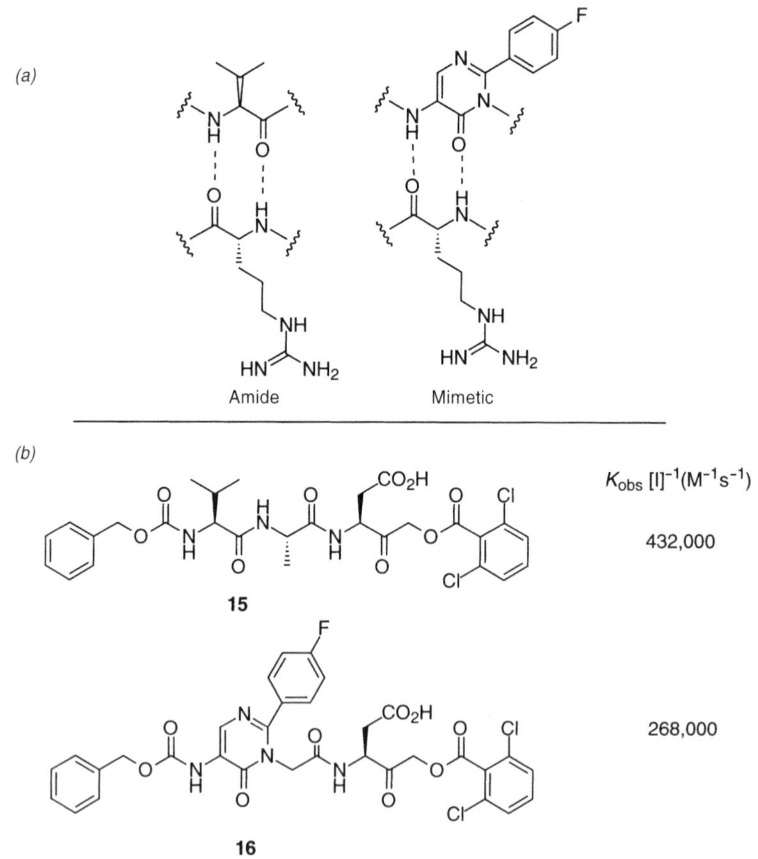

(a)

Amide

Mimetic

(b)

K_{obs} $[I]^{-1}(M^{-1}s^{-1})$

15 432,000

16 268,000

FIGURE 12.9. (*a*) Hydrogen bonding network between the amide (mimetic) and the target protein ICE; partial structure; (*b*) Comparison of the amide and mimetic structures and their corresponding inhibition data.

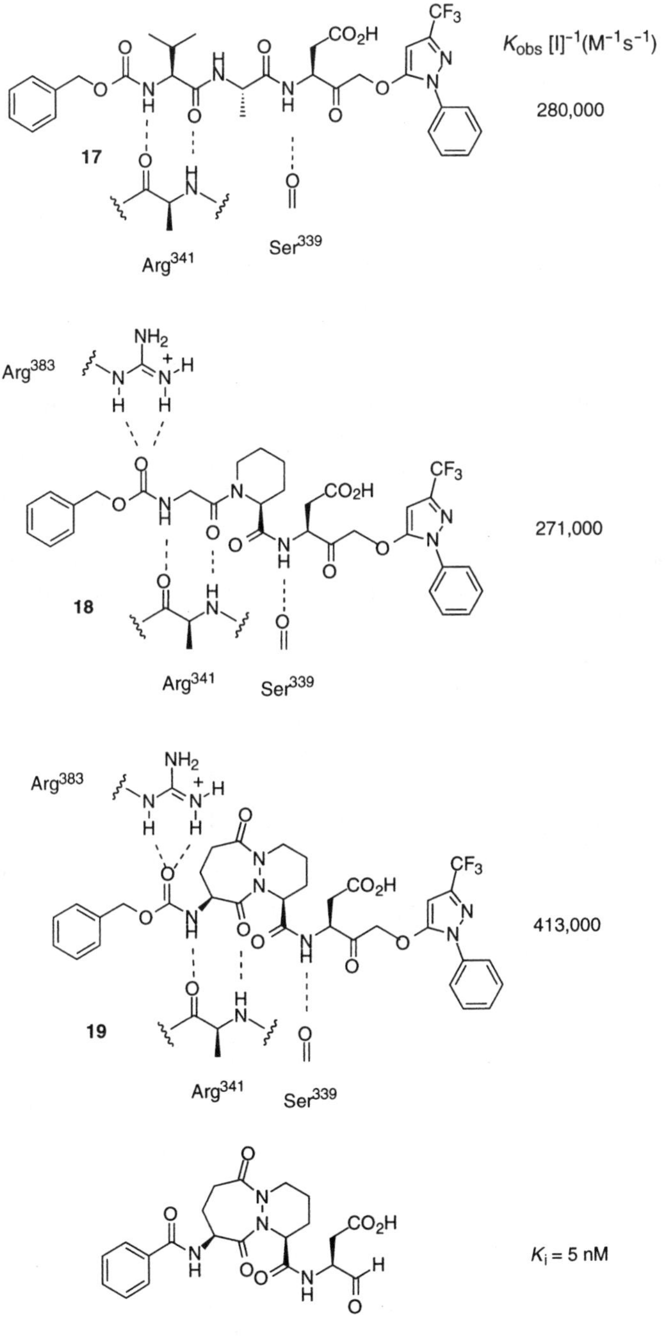

K_{obs} $[I]^{-1}$($M^{-1}s^{-1}$)

17 280,000

18 271,000

19 413,000

20 K_i = 5 nM

FIGURE 12.10. Pyridazinodiazepine based inhibitors of ICE.

in the peptidyl series increases in going from a Cbz N-cap to a 4-(methylthio)benzoyl group, but a similar trend in potency is not observed in the pyrimidone series (Fig. 12.9). The authors attribute this divergence in SAR to the conformational constraint imposed by the sp^2 center in the pyrimidone series and its impact on presentation of the P_3 amide and side chain to the enzyme.

Since no three-dimensional structures of these compounds bound to ICE are available, we docked Compounds **15** and **16** into the active site of ICE, using the crystal structure 1ICE[37]. We find that the hydrogen bonding network of **16** has shifted relative to that of Compound **15**. In addition, as the authors suggested, the terminal aromatic group cannot adopt an ideal conformation for interaction with the S_3 site. Compound **15**, on the other hand, has a sp^2–sp^3 bond between the Cα and the carbonyl carbon. This offers enough flexibility so that terminal aromatic group can fit nicely into the S_3 site interacting with His^{342} and Val^{348}. In a subsequent publication[38], the authors describe a series of second generation dipeptide mimics in which the N-terminus is attached to a sp^3 hybridized center. This modification resulted in compounds with higher in vitro inhibition constants as shown in Fig. 12.10.

Since no three-dimensional structures of these compounds bound to ICE are available, we docked Compounds **18** and **19** into the active site of ICE. In our model, the piperidine ring of Compound **18** lies above Trp^{340} forming hydrophobic interactions with the indole ring. The N-terminal phenyl ring occupies the S_3 subsite forming hydrophobic interactions with His^{342}, Met^{345} and Val^{348}. The binding affinity of compound **18** (271,000 $K_{obs}[I]^{-1}$ (M^{-1} s^{-1})) is almost the same as that of its closest peptide **17** (280,000 $K_{obs}[I]^{-1}$ (M^{-1} s^{-1})).

Ultimately, the search for the ideal mimic resulted in Compound **19** which incorporates a pyridazinodiazapine scaffold. Compound **19** displayed rates of inhibition (413,000 $K_{obs}[I]^{-1}$ (M^{-1} s^{-1})) *twice* that of peptidic compound **17**. The SAR of the P_3 N-terminal capping groups in this series parallels that of the tripeptides, suggesting that the orientation for binding is similar.

In our docking studies, **19** forms a beautiful network of hydrogen bonds. The pyridazinodiazapine ring lies above the indole of Trp^{340} forming a number of

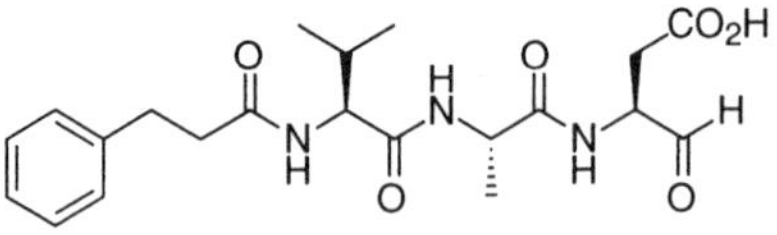

21 K_i = 54 nM

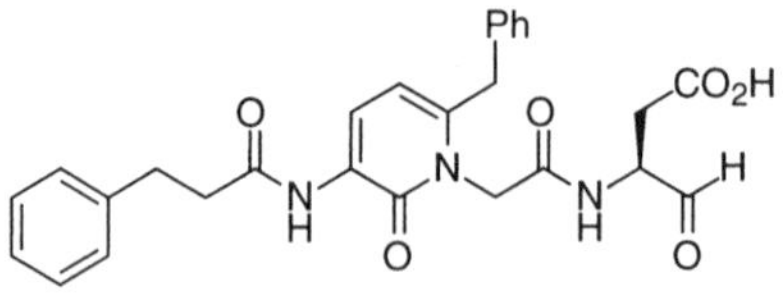

22 K_i = 150 nM

FIGURE 12.11. Pyridone based inhibitors of ICE containing an aldehyde functionality.

contacts. An analog of **19**, **20** was selected for pharmacokinetic evaluation. Inhibitor **20**, which had a K_i of 5 nM, was stable in ex vivo liver and intestinal slice assays as well as being 14% orally biaoavailable in dogs.

In a similar fashion, Golec et al.[39] recently disclosed a series of non-peptidic pyridone containing aldehydes as reversible inhibitors of ICE. Through an iterative cycle of modelling, synthesis and crystallography, the authors ultimately arrived at **22** ($K_i = 150$ nM) which is three-fold less active than its corresponding peptide-based inhibitor **21**. The crystal structure of a closely related analog revealed that the hydrogen bonding motif of the pyridone isostere closely resembled the peptide after which it was designed (Fig. 12.11).

5. PAPAIN

Cysteine proteases have been implicated in promoting diseases such as muscular dystrophy, myocardial infarction, cancer, and arthritis.[40] A group of researchers[41] at the University of Toronto used the crystal structure of papain to design generic cysteine protease inhibitors with amide bond isosteres. The goal of this study was to design conformationally restricted cysteine protease inhibitors containing heterocyclic peptide isosteres. The crystal structure of an irreversible, chloromethyl ketone inhibitor, **23** (Fig. 12.12), bound to papain was the starting point of this study.[42] Cheng et al. [41] examined a number of different five- and six-member heterocycles to find one that closely approximated the dihedral angles of the inhibitor's backbone without disturbing any key hydrogen bonding interactions or the orientation of the side chains. Compound **25** (Fig. 12.13(a)) was selected and synthesized and found to be a competitive inhibitor of papain with a K_i of 790 nM. Thus confirming that the amidine group is a viable conformationally restricted amide bond replacement.

Although, Compound **25** represented a good starting point, the design was not yet ideal; **25** is significantly weaker than its closest peptidyl analog **24**, which has a K_i of 46 nM. The authors suggested that one reason for the decrease in binding affinity may be due to the protonation state of the tetrahydropyrimidine moiety. The kinetic assays were carried out at pH 6.5; very likely at this pH the nitrogen of the tetrahydropyrimidine is protonated.

Since the three dimensional structure of **25** bound to papain is not available, we docked this compound into the active site of papain using the crystal structure 1PAP.[42] Our docking results indicated that the tetrahydropyrimidine nitrogen of **25** is exposed to solvent and not involved in hydrogen bonding with papain.

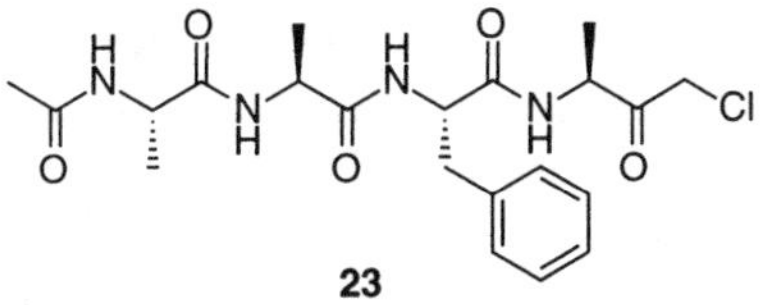

FIGURE 12.12. Irreversible chloromethyl inhibitor of papain.

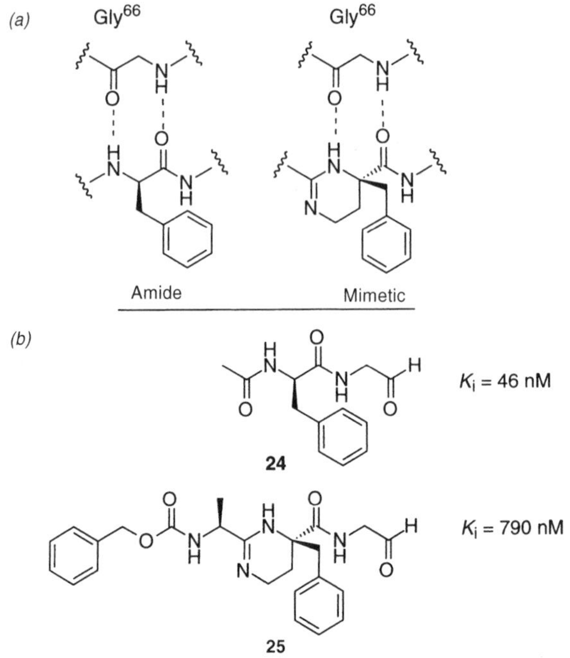

(a) Gly^{66} Gly^{66}

Amide Mimetic

(b)

24 K_i = 46 nM

25 K_i = 790 nM

FIGURE 12.13. (*a*) Hydrogen bonding network between the amide (mimetic) and the target protein papain; partial structure; (*b*) Comparison of the amide and mimetic structures and their corresponding inhibition data.

Therefore, the protonation state should have little effect on the magnitude of inhibition. However, our docking studies suggest another reason for the decrease in binding affinity. The position of the carbonyl oxygen adjacent to the amidine ring is in a less favorable position for hydrogen bonding with the NH of Gly^{66} than the corresponding carbonyl oxygen of the peptidyl analog. No pharmacokinetic data was given.

6. MATRIX METALLOPROTEASES (MMPs)

The matrix metalloproteases are a family of zinc metallo endopeptidases involved in the degradation and remodeling of connective tissues. They degrade nearly all constituents of the major extracellular matrix macromolecules under both physiological and pathological conditions.[43,44] The overexpression and activation of some of these enzymes have been involved in a range of diseases associated with unwanted degradation of connective tissues. Examples include, the loss of collagen from cartilage associated with rheumatoid arthritis,[45,46] and, the loss of collagen from bone associated with osteoporosis.[47,48] MMPs are also currently under investigation for cancer therapy[49] since they are known to be essential for the growth of new blood vessels, a phenomenon necessary for the growth of tumor cells.

As at the time of this writing (June 1998), the crystal structures of five different MMPs are available: human fibroblast collagenase, human fibroblast stromelysin, human neutrophil collagenase, matrilysin, and gelantinase. Reports of the use of the crystal structures of stromelysin, collagenase and matrilysin to design amide bond isosteres to improve pharmacokinetics have appeared in the literature.

At Roche Bioscience, a team of scientists [50] designed a novel series of MMP inhibitors which contain imidazole and benzimidazole groups as replacements for the P_2'–P_3' amide bond of their MMP inhibitor RS 39066 (**26**) (Fig. 12.14(b)). The benzimidazole analog, **27**, most similar to RS 39066 and also the

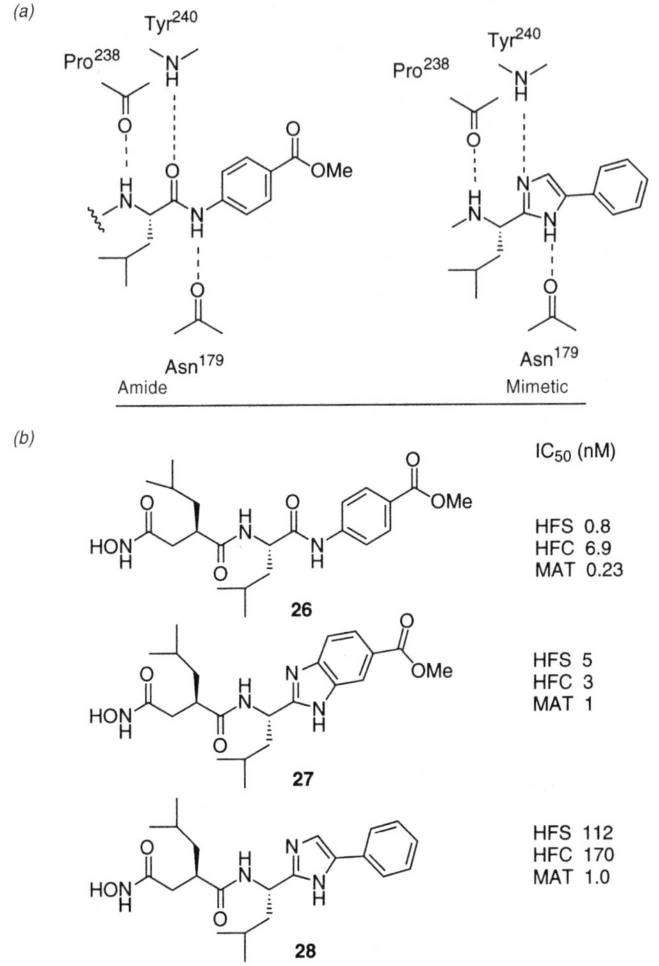

FIGURE 12.14. (*a*) Hydrogen bonding network between the amide (mimetic) and the target protein MMPs; partial structure; (*b*) Comparison of the amide and mimetic structures and their corresponding inhibition data.

most active compound against the three MMPs (fibroblast stromelysin (HFS), fibroblast collagenase (HFC) and matrilysin (MAT)), displayed less than 10-fold decrease in binding affinity relative to RS 39066 in stromelysin and matrilysin and a slight increase in affinity for collagenase. No crystal structure of **27** bound to any of these MMPs is available. Therefore, we docked this compound into the active site of matrilysin (structure 1MMQ) and found that this compound forms the same hydrogen bonds as that of the NH and carbonyl of the P_2'–P_3' amide bond observed in the structure of **31** co-crystallized with matrilysin (Fig. 12.15).[51] However, the distances between the nitrogen proton and the carbonyl oxygen were somewhat greater than those observed in the matrilysin/**31** crystal structure.

The best compound in the imidazole series, **28**, was approximately 100 times less potent than RS 39066. This may be due in part to the lack of the terminal ester group which is present in RS 39066 and **27** but not in the imidazole analog **28**. The authors also suggest that **28** may be protonated at the pH of the assay conditions. In that case, the hydrogen bond acceptor nitrogen will be protonated and will be repulsed by the binding site NH of Tyr.[240]

The effect of replacing an amide bond with an imidazole can also be seen by comparing Galardin, **29**,[52] and Compound **30** (Fig. 12.15). The crystal structure of **30** bound to the MMP matrilysin has been determined;[50] it shows that when the P_2'–P_3' amide bond is replaced by the heterocycle, the positions of the nitrogen atoms of the heterocycle correspond well to those of the oxygen and nitrogen atoms of the amide bond of **31** as observed in the matrilysin/**31** crystal structure.[51] Interestingly, there is a decrease in inhibition with either the incorporation of an imidazole, **30**, (which may be due to the protonated form) as

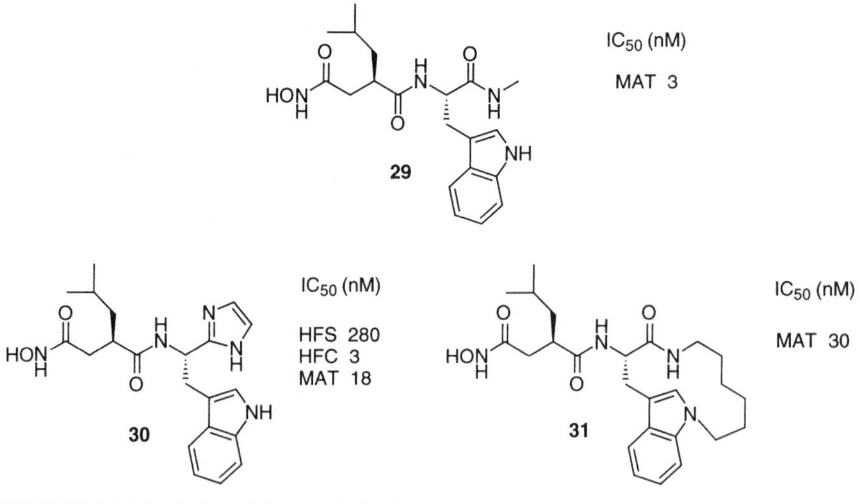

29 — IC_{50} (nM): MAT 3

30 — IC_{50} (nM): HFS 280, HFC 3, MAT 18

31 — IC_{50} (nM): MAT 30

FIGURE 12.15. Series of 3 MMP inhibitors used to compare the effect of incorporating an imidazole as an amide bond mimic. The crystal structures of **30** and **31** bound with matrilysin show that both compounds have similar hydrogen bonding networks.

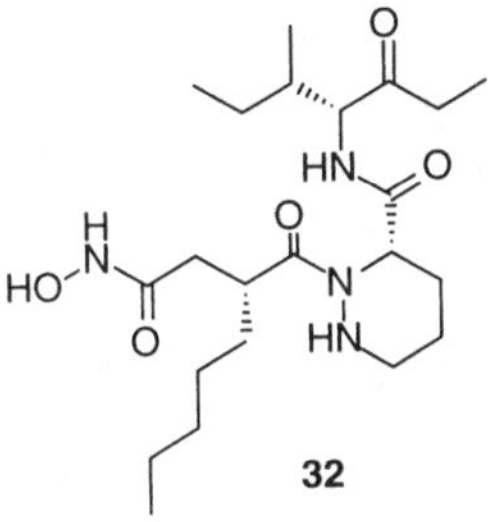

FIGURE 12.16. Matylistatin, a naturally occurring MMP inhibitor.

well as the incorporation of a macrocycle, **31**. Pharmacokinetic studies had not been completed at the time of this publication.

Matylstatin (Fig. 12.16, **32**) is a natural product found to have metalloproteinases inhibition properties. Researchers at DuPont Merck[50] designed a potent stromelysin inhibitor, **34**, which incorporates the piperazic acid moiety of matylstatin (Fig. 12.17). Since no crystal structure of this compound bound to stromelysin is available, we docked the compound into the active site of stromelysin by using the structure 1SLN.[53] Our docking results indicated that the

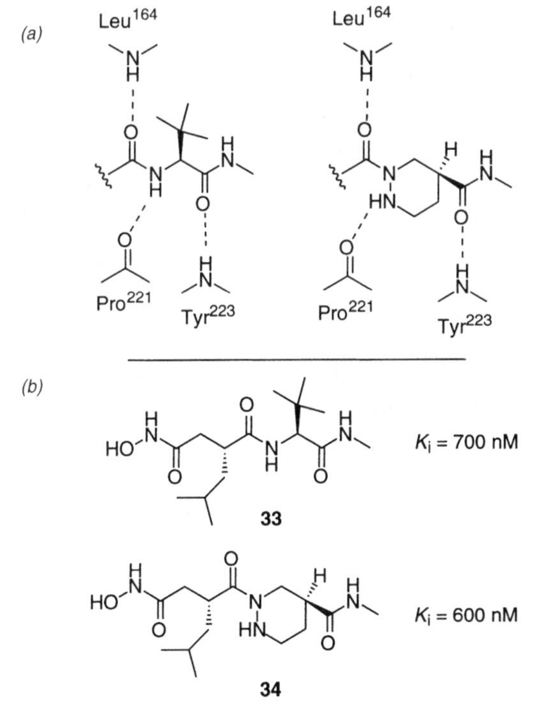

FIGURE 12.17. (*a*) Hydrogen bonding network between the amide (mimetic) and the target protein stromelysin; partial structure; (*b*) Comparison of the amide and mimetic structures and their corresponding inhibition data.

NH of piperazic acid forms a hydrogen bond with the carbonyl of Pro[221] and the carbonyl adjacent to the N of piperazic acid forms a hydrogen bond with Leu[164]. Additionally, the carbon atoms of piperazic acid are in contact with the side chain atoms of Leu[222] forming several hydrophobic interactions. Comparison of the piperazic acid containing Compound, **34**, to its amide counterpart, **33** indicates that incorporation of the piperazic moiety enhances the binding affinity to stromelysin.

The incorporation of piperazic acid greatly improved the solubility of these compounds; however, no pharmacokinetic data was reported.

7. NEP, ECE AND THERMOLYSIN

Neutral endopeptidase (NEP) and endothelin converting enzyme (ECE) are two zinc metalloproteases which play a critical role in the regulation of vasoactive peptides, including atrial natriuretic peptide, endothelin and bradykinin. NEP and ECE, therefore, consitute potential therapeutic targets for the treatment of various cardio-, cerebro- and renovascular diseases.[54,55] The crystal structure of NEP and ECE are not yet available; therefore, the crystal structure of a related enzyme, thermolysin, was used by researchers at Ciba-Geigy to study amide bond mimics as a way to improve the duration of action of one of their series of dual NEP/ECE inhbitors.

At Ciba-Geigy, Compound **35** (Fig. 12.18) was found to inhibit neutral endopeptidase (NEP) with an $IC_{50} =$ of 5 nM.[56] To improve the pharmacokinetic properties of this peptidic inhibitor, it was converted to a non-peptide surrogate by replacing the glycine portion with a tetrazole. This compound, **36**, retained excellent inhibitory activity against NEP ($IC_{50} = 0.9$ nM) and also was found [57] to inhibit ECE ($IC_{50} = 410$ nM).

To understand the binding mode of these compounds and to use this information to aid in the design of improved inhibitors, we turned to molecular modeling. The crystal structure of thermolysin had previously been used successfully as a model for NEP and the design of novel, potent NEP inhibitors.[58,59] However, **35** does not inhibit thermolysin. The biphenyl group

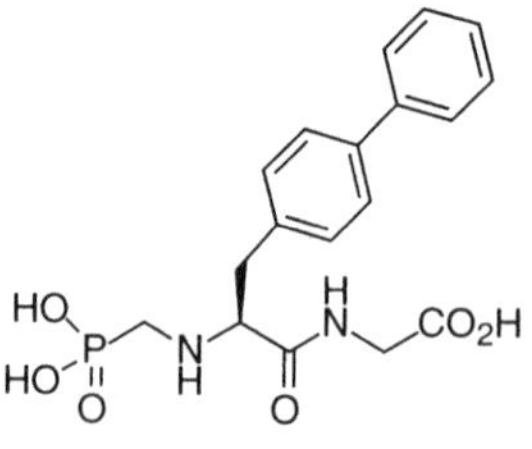

35
IC_{50} NEP 5 nM; ECE inactive

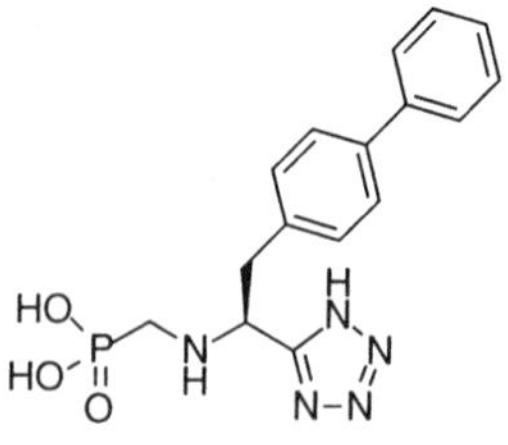

36
IC_{50} NEP 0.9 nM; ECE 410 nM

FIGURE 12.18. Inhibitors of NEP and ECE discovered at Ciba-Geigy.

which confers Compound **35**s selectivity for NEP and ECE, is too large to fit into the thermolysin S_1' pocket. Therefore, in our modeling studies we used a truncated analog with a single phenyl, **38** (Fig. 12.19(b)). This compound has modest activity against thermolysin $IC_{50} = 5.2\ \mu M$), retains good binding to NEP ($IC_{50} = 26\ nM$) but is inactive against ECE. Our modeling studies of **38** predicted that the tetrazole is a beautiful amide bond surrogate, forming two hydrogen bonds with Arg^{203} and one with Asn^{112}. To verify these findings, the crystal structure of this compound bound to thermolysin was determined and found to be in agreement with the modeling predictions.[60]

The hydrogen bonding pattern of these compounds is quite different from that of typical peptidic thermolysin/NEP inhibitors such as thiorphan, **37** (Fig. 12.19(a)). It is known from the crystal structure of thiorphan bound to thermolysin that the carbonyl oxygen of thiorphan forms a bifurcated hydrogen bond with Arg^{203}. Interestingly, as shown in Fig. 12.19(a), each of the two of the

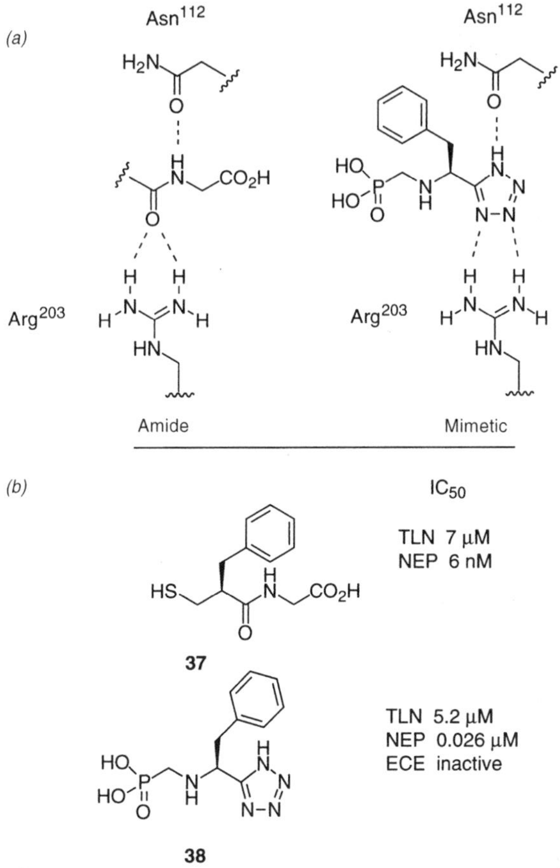

FIGURE 12.19. (*a*) Hydrogen bonding network between the amide (mimetic) and the target protein thermolysin; partial structure; (*b*) Comparison of the amide and mimetic structures and their corresponding inhibition data.

nitrogens of the tetrazole form a hydrogen bond with one of the nitrogens of the guanidinium group of Arg^{203}.

The pharmacokinetic profile of compound **35** was improved by replacing the glycine functionality with a tetrazole. In in vivo models, both **35** and **36** inhibit the conversion of exogenously administered big ET-1 to ET-1. However, **36** has a significantly longer duration of action indicating that replacement of the glycine portion with a tetrazole significantly improves the pharmacokinetic profile of compounds containing this functionality.[57,61]

8. PEPSIN

Unlike the examples discussed above, the pepsin inhibitor described in this section was not based on a peptidic lead but rather was the result of a de novo

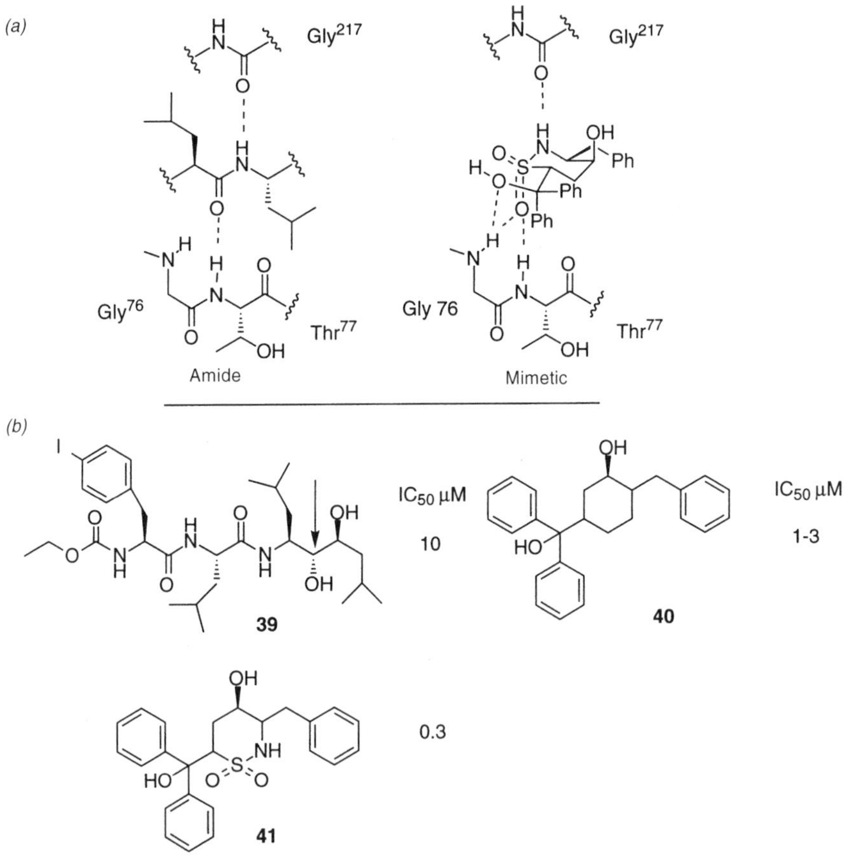

(a) Gly217 Gly76 Thr77 Amide; Gly217 Gly 76 Thr77 Mimetic

(b) 39 IC50 μM 10; 40 IC50 μM 1-3; 41 0.3

FIGURE 12.20. (*a*) Hydrogen bonding network between the amide (mimetic) and the target protein pepsin; partial structure. (*b*) The GrowMol computer program was used to generate novel structures in the pepsin binding site using three atoms from compound **39** as a starting point: the hydroxyl and carbon identified by an arrow. Comparison of **40**, a compound designed using GrowMol, and **41** which incorporates an amide mimic.

design using the computer program GrowMol.[62] This program generates novel structures by using an algorithm which randomly joins atoms and fragments to fill and chemically complement the three dimensional structure of a target binding site.

GrowMol was used in the laboratry of Prof. Daniel Rich at the University of Wisconsin to discover low molecular weight, non-peptidic inhibitors of pepsin. The structures were generated in the binding site of porcine pepsin, using the X-ray structure of A66702, **39**, bound to pepsin (Fig. 12.20(b)).[63] Using only the hydroxyl and carbon of compound **39** (identified by an arrow in Fig. 12.20(b)) as a starting point, GrowMol generated a series of structures containing a cyclohexanol from which side chains extended to fill hydrophobic pockets of the S_2, S_3 subsites of pepsin. The GrowMol-generated structure was simplified to result in **40**. This compound does not contain a nitrogen, nor a carbonyl group. Surprisingly, it nonetheless inhibits pepsin with a K_i between 1 and 3 μM.

Visual inspection of the original GrowMol generated structures suggested that the incorporation of a sulfonamide into the ring would mimic the hydrogen bond interactions of the P_2–P_3 amide bond of **39** (Fig. 12.20(a)). Sulfonamide **40**, was synthesized and shown to increase the binding affinity 10-fold.[64,65] Determination of the crystal structure of this compound bound to pepsin is in progress. This is one of the first examples in which a de novo computer program was used in the successful design of a novel enzyme inhibitor.

9. DISCUSSION

9.1. In Vitro Binding Affinity of Heterocyclic Amide Bond Mimics

Although heterocycles are capable of forming the same hydrogen bonding network as amide bonds, no heterocycle is an exact amide bond isostere. There are distinct steric and chemical differences which distinguish the two. First, the number of atoms in heterocycles commonly employed as amide bond surrogates is greater than that of an amide bond. Second, the direction and flexibility of side chains extending from either side of an amide bond cannot be exactly duplicated by substituents on a heterocycle. Third, the strength of the hydrogen bonding network as exhibited by amides is likely to be different from that formed by heterocycles. And last, depending upon the heterocycle employed, the protonation state can have a profound impact on the effectiveness of the mimetic. These are only some of the potential differences between amide bonds and their heterocyclic replacements which can significantly affect the binding affinity of compounds incorporating these mimics. Not surprisingly, in many of the cases examined in this chapter, substitution of an amide bond with a mimic resulted in a loss of binding affinity relative to the peptidic lead. In those cases in which the affinity is retained or improved, some of the features unique to the heterocycle were used advantageously to form additional interactions that are not possible by the amide bond and its substituents. We will now consider individually the various features of heterocycles as amide bond mimics and discuss the in vitro and in vivo properties of the compounds which contain them (see Table 12.1).

9.2. Position of Hydrogen Bonds

Usually, the first criterion evaluated in the design of any surrogate is how well it can mimic the hydrogen bonding network of the amide bond. In several of the preceding cases, crystal structures of both the mimics and peptidic inhibitors compared favorably, displaying remarkable similarities in the hydrogen bonding networks. Still, the retention of the peptide hydrogen bonding network by the mimic is not necessarily sufficient to retain activity. On the contrary, disruption of the hydrogen bonding network usually results in a significant loss of binding affinity (See for example Compound **2**).

It is also possible for an amide mimetic to form a "better" hydrogen bonding network than that possible for an amide. For example, in the case of NEP/thermolysin inhibitor **38**, the crystal structure reveals that the tetrazole is acting as a hydrogen bond acceptor, forming two hydrogen bonds with the guanidinium group of an Arg (relative to a single, bifurcated hydrogen bond made by the carobnyl of the peptidic inhibitor **37**). Assuming the same interactions take place in NEP, this additional hydrogen bond may partially explain the five fold improvement in activity.

9.3. Additional Ring Atoms

In some binding sites, the amide bond of the substrate or peptidic inhibitor lies in a narrow groove with just enough room to form hydrogen bonds with the binding site atoms. In such situtations, there may simply not be enough room for a heterocycle with its additional atom or atoms. On the other hand, the binding site geometry may be such that the extra atoms not only fit but can form additional favorable interactions not possible with an amide bond.

This is precisely the case with HIV-1 protease inhibitor **4**. The volume occupied by the amide bond in the peptide inhibitor is very narrow. However, the X-ray structure shows that there is enough room for an imidazole which perfectly mimics the hydrogen bonding pattern of the peptidic analog. In addition, one of the imidazole carbons is in van der Waal contact with atoms of the Ile^{50} side chain. These added hydrophobic interactions are likely to be some of the factors which explain why **4** and its peptidic analog **3**, inhibit HIV-1 to the same extent.

Another example is the piperazic acid containing stromelysin inhibitor, **28**. Our modeling studies show that not only does the NH of the piperazic acid form a hydrogen bond but, in addition, several of the carbon atoms of the six-membered ring form hydrophobic interactions with the side chain of Leu^{222}. Here too, the in vitro activity of this inhibitor relative to the peptide is retained.

TABLE 12.1. Summary In vitro and In vivo Data of Peptidic Inhibitors and their Mimetic Analog

	Structures		In Vitro Data			In Vivo Data (mimetics)
Entry	Amides	Mimetics	Amide	Mimetic	Ratio[a]	
1	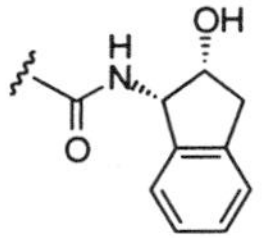	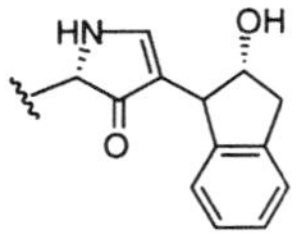	0.03 nM	2 nM	67	• $CIC_{95} = 100$ nM
2	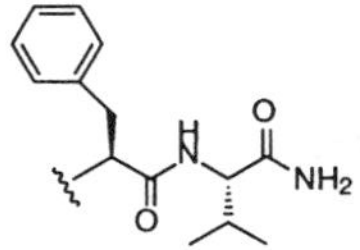	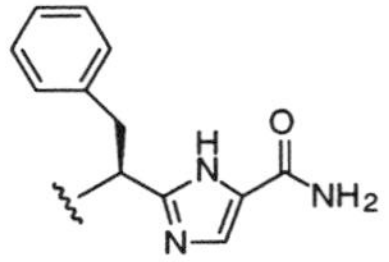	0.7 nM	0.6 nM	0.85	• No cellular data reported
3	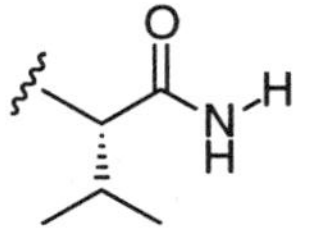	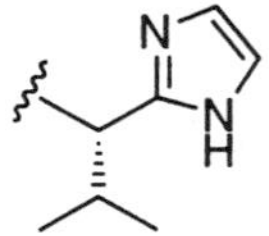	1.4 nM	18 nM	13	• $CIC_{50} = 570$ nM • $t_{1/2\beta} = 84$ min. (monkey) • 30% oral bioavailability
4	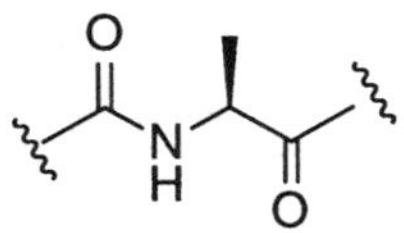	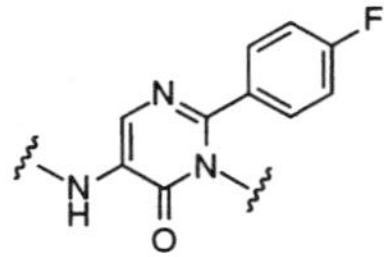	210 nM	101 nM	0.48	• $ED_{50} = 7.5$ mg kg^{-1} • 88% oral bioavailability • Duration of action $>$ 4 hr
5	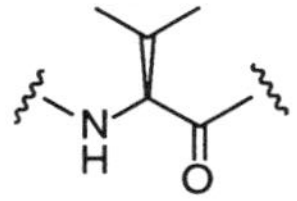	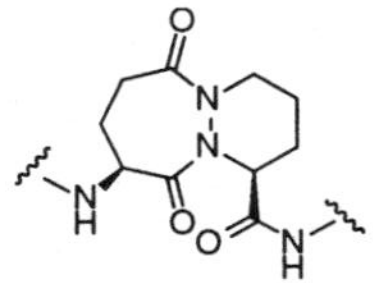	$K_{obs}I^{-1}$ ($M^{-1}s^{-1}$) 432,000	$K_{obs}I^{-1}$ ($M^{-1}s^{-1}$) 572,000	1.3	• 16% oral bioavailability • Stable in ex vivo liver assay

TABLE 12.1. *(continued)*

Entry	Structures: Amides	Structures: Mimetics	In Vitro Data: Amide	In Vitro Data: Mimetic	In Vitro Data: Ratio[a]	In Vivo Data (mimetics)
6	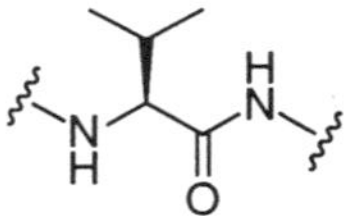	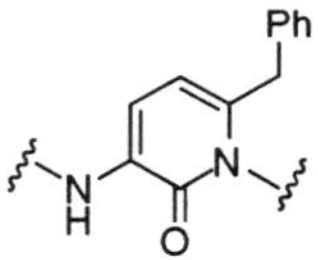 Ph	54 nM	150 nM	3	• No cellular data reported
7	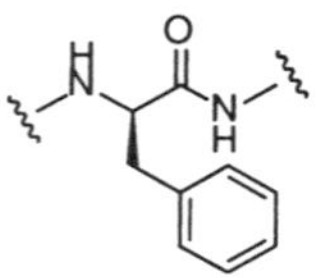	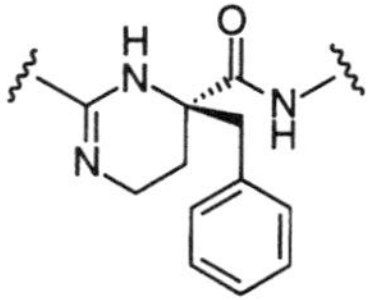	46 nM	790 nM	17	• No cellular data reported
8	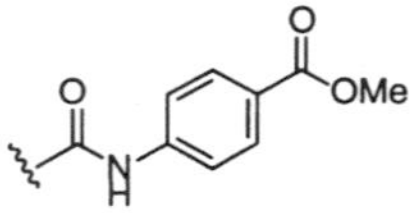 OMe	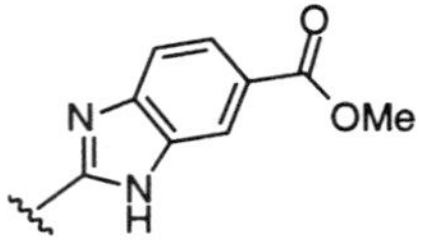 OMe	0.23 nM[c]	1 nM[c]	4.3	• No cellular data reported
9	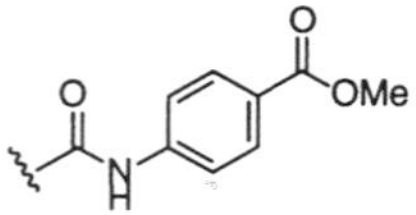 OMe	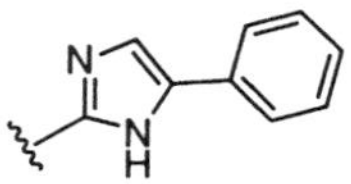	0.23 nM[c]	1 nM[c]	4.3	• No cellular data reported
10	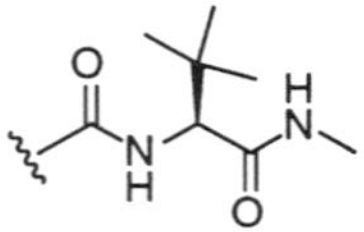	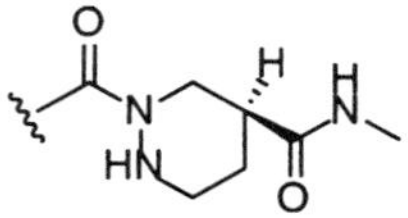	700 nM	600 nM	0.85	• No cellular data reported

11	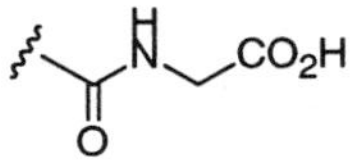	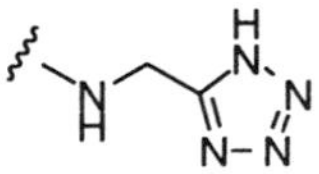	5 nM	0.9 nM	0.18	• Half life > amide
12	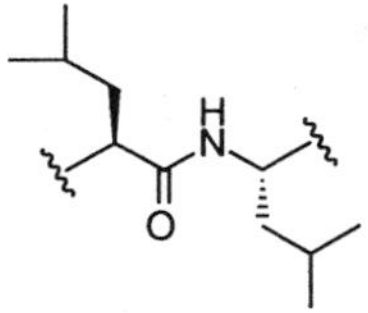	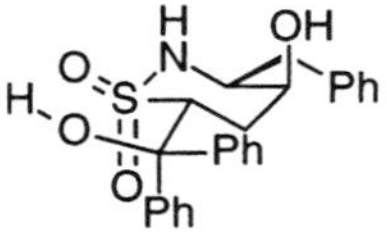	10 μM	0.3 μM	NA[b]	• No cellular data reported

[a]Ratio = mimetic/amide,
[b]No close peptidic analog was available
[c]Data for matrilysin

9.4. Side Chain Directionality

One factor which is likely to influence the activity of an amide bond mimic is the orientation of the side chain on the heterocyclic amide surrogate. At least one of the atoms equivalent to the α-carbon adjacent to an amide bond will now be fixed in the heterocycle. This is in sharp contrast to what is found in normal amide bonds where the bonds between the carbonyl and the α-carbon, and between the α-carbon and the nitrogen, are purely sp^2–sp^3 bonds. Consequently, there is a large degree of rotational freedom about either side of the peptide bond which allows adjustments in side chain orientation such that the maximum interaction with the target protein is obtained. Conversely, the rigidity of the heterocycle may not confer sufficient rotational flexibiltiy to maximize side chain overlap. In particular, small aromatic heterocyles with sp^2 hybridization simply may not present side chains to the protein in the appropriate orientation. Examples where side chain orientations were not ideal and were sited as a reason for a decrease of binding affinity include the pyrolidinone based HIV-1 inhibitor **2** and elastase inhibitor **9**.

One way to alter the direction of the substituents is to use heterocycles which contain more sp^3 hybridization. Although synthetically more challenging, this form of rigidification offers the potential for more precise vectors and ultimately a more perfect overlap. An example of a highly successful application of this strategy is found in the very potent ICE inhibitor **19**.

Although, the directionality of substituents on a heterocycle cannot always mimic that of an amide bond, it can be used advantageously. Compounds can be designed so that a substitutent on a rigid heterocycle is directed into the appropriate binding site pocket. This is nicely exemplified with elastase inhibitors **11** and **12** where substitution onto the pyrimidone ring (position C(6)) allowed aromatic groups to be projected directly into the S_2 pocket above a leucine side chain and stacked against a histidine. Thus, in this particular example, the lack of rotational flexibility was used in a beneficial way to improve potency.

9.5. Protonation State

Another major consideration in the selection of an amide bond mimic is the protonation state of the surrogate. In many of these examples, the hydrogen bond acceptor is an aromatic, unprotonated sp^2 hybridized nitrogen. If, under physiological conditions, this nitrogen becomes protonated, then instead of providing an attractive force, a NH group of the ligand will encounter a NH group of the binding site resulting in repulsion. Therefore, the choice of surrogate must include considerations of its likely ionic state when bound to the target binding site. This is perhaps most relevant for imidazoles and benzimidazole mimetics which, with a pK_a of ~ 7.1, are likely to exist in both forms at physiologically relevant pHs.

Protonation states were implicated at least partially for the observed decrease in in vitro inhibition in several examples, including the HIV-1 protease inhibitor

6, papain inhibitor **25**, and matrix metalloprotease inhibitor **28**. In one example, employing an imidazole in the design of an HIV-1 inhibitor (**4**), the appropriate protonation state was ensured by the addition of an acyl group onto the imidazole. Not only does the acyl carbonyl make a hydrogen bond, it also decreases the basicity of the imidazole ring keeping it unprotonated. Compound **4** inhibits HIV-1 protease with the same affinity as that of the peptidic analog.

Another factor for consideration is the strength of the hydrogen bond. Hydrogen bond donor and acceptor groups are not equal in the strength of the hydrogen bonds they form. Bond energies are a function of both the electronegativity of atoms as well as the directionality of the hydrogen bond. Therefore, it is likely that most amide surrogates do not efficiently replace the hydrogen bonding ability of the original amide bond.

9.6. In Vivo Binding

For those cases where in vivo data was available, the replacement of amide bonds by small heterocyclic templates almost always resulted in compounds with better pharmacokinetic properties than their peptidyl counterparts. Unfortunately, the data set is too limited to draw any firm conclusions regarding whether one template might have a more profound impact than another, and clearly each amide bond replacement has to be approached on an individual basis in accordance with the observations made above. It should be pointed out, however, that in several cases, even though substitution resulted in a compound with decreased in vitro activity, more favorable pharmacokinetic properties ultimately yielded compounds whose in vivo properties were superior.

10. CONCLUSIONS

Several of these examples indicate that incorporation of amide bond mimics is a good strategy for turning a peptide lead into an orally bioavailable drug candidate. They also show, however, that it is difficult to design mimetics that will retain, or at least not lose, too much of in vitro activity. Factors other than the obvious retention of hydrogen bonding patterns must be considered. It is also clear from these examples, that features unique to heterocycles can be exploited to form additional interactions leading to inhibitors as potent or even more potent than their peptidic predecessors.

The choice of heterocycles typically used seems rather limited. In these examples imidazoles, benzimidazoles, tetrazoles, pyridones, pyrimidones, pyrrolidinones and pyridazinodiazapine were selected. The piperazic acid, an interesting functional group which imparted not only a hydrogen bond donor but additional carbon atoms available for hydrophobic interactions, came from a natural product. However, biologically relevant compounds contain a great variety of heterocycles. A search of the MDL Drug Data Report data base revealed 300 different five-membered and 280 different six-membered heterocycles incorporated in the 78,000 compounds listed in this data base.

Because of the large variety of different heterocycles available, a computer program was developed at Ciba which automatically docks these heterocycles into the three dimensional structure of a target binding site in regions originally occupied by a ligand amide bond.[66] The program then identifies those heterocycles which can form the most favorable interactions with binding site atoms. As previously mentioned, some of these scaffolds are likely to be synthetically more challenging, however, access to a wider range of templates is more likely to result in compounds which possess both the desired in vitro activity, physicochemical properties, and appropriate pharmacokinetic profile to become drug candidates.

REFERENCES

1. McMartin, C. *Adv. Drug Res.* 1992, **22**, 39–106.
2. Bernstein, F. C.; Koetzle, T. F.; Williams, G. J. B.; Meyer, E. F.; Brice, M. D.; Rodgers, J. R.; Kennard, T.; Shinamoucchi, T.; Tasumi, M. *J. Mol. Bio.* 1977, **112**, 535–542.
3. McMartin, C.; Bohacek, R. S. *J. Computer-Aided Molecular Design* 1997, **11**, 333–344.
4. Smith, A. B.; Keenan, T. P.; Holcomb, R. C.; Sprengeler, P. A.; Guzman, M. C.; Wood, J. L.; Carroll, P. J.; Hirschmann, R. *J. Am Chem. Soc.* 1992, **114**, 10672–10674.
5. Smith, A. B.; Guzman, M. C.; Sprengeler, P. A.; Keenan, T. P.; Holcomb, R. C.; Wood, J. L.; Carroll, P. J.; Hirschmann, R. *J. Am. Chem. Soc.* 1994, **116**, 9947–9962.
6. Smith, A. B.; Hirschmann, R.; Pasternak, A.; Akaishi, R.; Guzman, M. C.; Jones, D. R.; Keenan, T. P.; Sprengeler, P. A.; Darke, P. L.; Emini, E. A.; Holloway, M. K.; Schleif, W. A. *J. Med. Chem.* 1994, **37**, 215–218.
7. Smith, A. B.; Hirschmann, R.; Pasternak, A.; Guzman, M. C.; Yokoyama, A.; Sprengeler, P. A.; Darke, P. L.; Emini, E. A.; Schleif, W. A. *J. Am. Chem. Soc.* 1995, **117**, 11113–11123.
8. Smith, A. B.; Hirschmann, R.; Pasternak, A.; Yao, W.; Sprengeler, P. A.; Holloway, M. K.; Kuo, L. C.; Chen, Z.; Darke, P. L.; Schleif, W. A. *J. Med. Chem.* 1997, **40**, 2440–2444.
9. Rich, D. H.; Bode, W. *J. Mol. Biol.* 1983, **164**, 283–311.
10. James, M. N. G.; Sielecki, A.; Salituro, F.; Rich, D. H.; Hofmann, T. *Proc. Natl. Acad. Sci. U. S. A.* 1982, **79**, 6137–6141.
11. Foundling, S. I.; Cooper, J.; Watson, F. E.; Cleasby, A.; Pearl, L. H.; Sibana, B. L.; Hemmings, A.; Wood, S. P.; Blundell, T. L.; Valler, M. J.; Norey, C. G.; Kay, J.; Boger, J.; Dunn, B. M.; Leckie, B. J.; Jones, D. M.; Atrash, B.; Hallet, A.; Szelke, M. *Nature* 1987, **327**, 349–352.
12. Miller, M.; Schneider, J.; Sathyanarayana, B. K.; Toth, M. V.; Marshall, G. R.; Clawson, L.; Selk, L. M.; Kent, S. B. H.; Wlodawar, A. *Science* 1989, **246**, 1149–1152.

13. Ghosh, A. K.; Thompson, W. J.; McKee, S. P.; Duong, T. T.; Lyle, T. A.; Chen, J. C.; Darke, P. L.; Zugay, J. A.; Emini, E. A.; Schleif, W. A.; Huff, J. R.; Anderson, P. S. *J. Med. Chem.* 1993, **36**, 292–294.
14. Thompson, S. K.; Murthy, K. H. M.; Zhao, B.; Winborne, E.; Green, D. W.; Fisher, S. M.; DesJarlais, R. L.; Tomaszek, T. A.; Meek, T. D.; Gleason, J. G.; Abel-Meguid, S. S. *J. Med. Chem.* 1994, **37**, 3100–3107.
15. Thompson, S. K.; Eppley, A. M.; Frazee, J. S.; Darcy, M. G.; Lum, R. T.; Tomaszek, T. A.; Ivanoff, L. A.; Morris, J. F.; Sterberg, E. J.; Lambert, D. M.; Fernandez, A. V.; Petteway, S. R.; Meek, T. D.; Metcalf, B.; Gleason, J. G. *Bioorg. Med. Chem. Lett.* 1994, **4**, 2441–2446.
16. Abel-Meguid, S. S.; Metcalf, B. W.; Carr, T. J.; Demarsh, P.; DesJarlais, R. L.; Fisher, S. M.; Green, D. W.; Ivanoff, L.; Lambert, D. M.; Murthy, K. H. M.; Petteway, S. R.; Pitts, W. J.; Tomaszek, T. A.; Winborne, E.; Zhao, B.; Dreyer, G. B.; Meek, T. D. *Biochem.* 1994, **33**, 11671–11677.
17. *Pulmonary Emphysema and Proteolysis: 1986*; Academic Press: New York, 1987.
18. Burchardi, H.; Stokke, T.; Hensel, I.; Koestering, H.; Rahlf, G.; Schlag, G.; Heine, H.; Horl, W. H. *Adv. Exp. Med. Biol.* 1984, **167**, 319–333.
19. Janoff, A. In *Granulocyte Elastase: Role in Arthritis and in Pulmonary Emphysema*; Janoff, A. Ed.; Urban and Schwarzenberg: Baltimore, MD, 1978.
20. Janoff, A. *Annu. Rev. Med.* 1985, **36**, 207–216.
21. Nadel, J. A. *Am. Rev. Respir. Dis.* 1991, **144**, S48–S51.
22. Stein, R. L.; Trainor, D. W.; Wildonger, R. A.*Annu. Rev. Med. Chem.* 1985, **20**, 237–246.
23. Williams, J. C.; Falcone, R. C.; Knee, C.; Stein, R. L.; Strimpler, A. M.; Reaves, B.; Giles, R. E.; Krell, R. D. *Am. Rev. Respir. Dis.* 1991, **144**, 875–883.
24. Williams, J. C.; Stein, R. L.; Giles, R. E.; Krell, R. D. *Ann. N. Y. Acad. Sci.* 1991, **624**, 230–243.
25. Takahashi, L. H.; Radhakrishnan, R.; Rosenfield, R. E.; Meyer, E. F.; Trainor, D. A.; Stein, M. *J. Mol. Biol.* 1988, **201**, 423–428.
26. Brown, F. J.; Andisik, D. W.; Bernstein, P. R.; Bryant, C. B.; Ceccarelli, C.; Damewood, J. R.; Edwards, P. D.; Earley, R. A.; Feeney, S.; Green, R. C.; Gomes, B.; Kosmider, B. J.; Krell, R. D.; Shaw, A.; Steelman, G. B.; Thomas, R. M.; Vacek, E. P.; Veale, C. A.; Tuthill, P. A.; Wolanin, D. J.; Woolson, S. A. *J. Med. Chem.* 1994, **37**, 1259–1261.
27. Warner, P.; Green, R. C.; Gomes, B.; Strimpler, A. M. *J. Med. Chem.* 1994, **37**, 3090–3099.
28. Edwards, P. D.; Andisik, D. W.; Strimpler, A. M.; Gomes, B.; Tuthill, P. A. *J. Med. Chem.* 1996, **39**, 1112–1124.
29. Navia, M. A.; McKeever, B. M.; Springer, J. P.; Lln, T. Y.; Williams, H. R.; Fluder, E. M.; Dorn, C. P.; Hoogsteen, K. *Proc. Natl. Acad. Sci. U. S. A* 1989, **86**, 7.
30. Kostrua, M. J.; Tocci, M. J.; Limjuco, G.; Chin, J.; Cameron, P.; Hillman, A. G.; Chartrain, N. A.; Schmidt, J. A. *Proc. Natl. Acad. Sci. U. S. A.* 1989, **86**, 5227–5231.
31. Black, R. A.; Kronheim, S. R.; Cantrell, M.; Deeley, M. C.; March, J. C.; Prickett, K. S.; Wingnall, J.; Conlon, P. J.; Coseman, D.; Hopp, T. P.; Mochizuki, D. Y. *J. Biol. Chem.* 1988, **263**, 9437.

32. Dinarello, C. A. *Blood* 1991, **77**, 1627.
33. Dinarello, C. A.; Thompson, R. C. *Immunol. Today* 1991, **12**, 404–410.
34. Dinarello, C. A.; Wolff, S. M. *N. Engl. J. Med.* 1993, **328**, 106.
35. Dolle, R. E.; Singh, J.; Rinker, J.; Hoyer, D.; Prasad, C. V. C.; Helaszek, C. T.; Miller, R. E.; Ator, M. A. *J. Med. Chem.* 1994, **37**, 3863–3865.
36. Dolle, R. E.; Prouty, C. P.; Prasad, C. V. C.; Cook, E.; Saha, A.; Ross, T. M.; Salvino, J. M.; Helaszek, C. T.; Ator, M. A. *J. Med. Chem.* 1996, **39**, 2438–2440.
37. Wilson, K. P.; Black, J. F.; Thomson, J. T.; Kim, E. K.; Griffith, J. P.; Navia, M. N.; Murcko, M. A.; Chambers, S. P.; Aldape, R. A.; Raybuck, S. A.; Livingston, D. J. *Nature* 1994, **370**, 270.
38. Dolle, R. E.; Prasad, C. V. C.; Prouty, C. P.; Salvino, J. M.; Awad, M. M. A.; Schmidt, S. J.; Hoyer, D.; Ross, T. M.; Graybill, T. L.; Speir, F. J.; Uhl, J.; Miller, B. E.; Helaszek, C. T.; Ator, M. A. *J. Med. Chem.* 1997, **40**, 1941–1946.
39. Golec, J. M. C.; Mullican, M. D.; Murcko, M. A.; Wilson, K. P.; Kay, D. P.; Jones, S. D.; Murdock, R.; Bemis, G. W.; Raybuck, S. A.; Luong, Y.; Liningston, D. J. *Bioorg. Med. Chem. Lett.* 1997, **7**, 2181–2186.
40. Perspect. *Drug Discovery Des.* 1996, **6**, 1–118.
41. Cheng, H.; Keitz, P.; Jones, J. B. *J. Org. Chem.* 1994, **59**, 7671–7676.
42. Drenth, J.; Kalk, K. H.; Swen, H. M. *Biochemistry* 1976, **15**, 3731.
43. Woessner, J. F. *FASEB* 1991, **5**, 2145–2154.
44. Hagmann, W. K.; Lark, M. W.; Becker, J. W. *Annu. Rep. Med. Chem.* 1996, **31**, 231–240.
45. Andrews, P. R.; Carson, J. M.; Caselli, A.; Spark, M. J.; Woods, R. *J. Med. Chem.* 1985, **28**, 393–399.
46. Blaser, J.; Triebel, S.; Maasjosthusmann, U.; Romisch, J.; Krahlmateblowski, U.; Freudenberg, W.; Fricke, R.; Tschesche, H. *Clin. Chin. Acta* 1996, **244**, 17–33.
47. Ohishi, K.; Fujita, N.; Morinaga, Y.; Tsuruo, T. *Clin. Exp. Metastasis* 1995, **13**, 287–295.
48. Witty, J. P.; Foster, S. A.; Stricklin, G. P.; Matrisian, L. M.; Stern, P. H. *J. Bone Mineral Res.* 1996, **11**, 72–78.
49. Zucker, S.; Lysik, R. M.; Zarrabi, H. M.; Moll, U.; Tickle, S. P.; Stetler-Stevenson, W.; Bakeer, T. S.; Dcoerty, A. J. P. *Ann. New York Acad. Sci.* 1994, **732**, 248–262.
50. Chen, J. J.; Zhang, Y.; Hammond, S.; Dewdney, N.; Ho, T.; Lin, X.; Browner, M. F.; Castelhano, A. L. *Bioorg. Med. Chem. Lett.* 1996, **6**, 1601–1606.
51. Browner, M. , Smith, W. , Castelhano, A. *Biochemistry* 1995, **34**, 6602–6610.
52. Galardy, R. E. *Drugs Future* 1993, **18**, 1109–1111.
53. Becker, J. W.; Marcy, A. I.; Rokosz, L. L.; Axel, M. G.; Burbaum, J. J.; Fitzgerald, P. M. D.; Cameron, P. M.; Esser, C. K.; Hagmann, W. K.; Hermes, J. D.; Springer, J. P. *Protein Science* 1995, **4**, 1966–1976.
54. Ruskoaho, H. *Pharmacol. Rev.* 1992, **44**, 479–602.
55. Rubanyi, G. M.; Polokoff, M. A. *Pharmacol. Rev.* 1992, **46**, 325–415.
56. De Lombaert, S.; Erion, M. D.; Tan, J.; Blanchard, L.; El-Chehabi, L.; Ghai, R.; Sakane, Y.; Berry, C.; Trapani, A. J. *J. Med. Chem.* 1994, **37**, 498–511.

57. Wallace, E. M.; Moliterni, J. A.; Moskal, M. A.; Neubert, A. D.; Marcopulos, N.; Stamford, L. B.; Trapani, A. J.; Savage, P.;. Chou, M.; Jeng, A. Y. *J. Med. Chem.* 1998, **41**, 1513–1523.

58. MacPherson, L. J.; Bayburt, E. K.; Capparelli, M. P.; Bohacek, R. S.; Clarke, F. H.; Ghai, R. D.; Sakane, Y.; Berry, C. J.; Peppard, J. V.; Trapani, A. J. *J. Med. Chem.* 1993, **36**, 3821–3828.

59. Bohacek, R.; DeLombaert, S.; McMartin, C.; Priestle, J.; Gruetter, M. *J. Am. Chem. Soc.* 1996, **118**, 8231–8249.

60. Lombaert, S. D.; Trapani, A.; Priestle, J.; Grutter, M.; Bohacek, R.; Blanchard, L.; Tan, J.; Savage, P.; Jeng, A. Y. *4th International Conference on Endothelin, London, England* 1995.

61. Lombaert, S. D. , Private communication.

62. Bohacek, R. S.; McMartin, C. *J. Am. Chem. Soc.* 1994, **116**, 5560–5571.

63. Abad-Zapatero, C. *Adv. Exp. Med. Biol.* 1991, **306**, 9–21.

64. Rich, D. J.; Bohacek, R. S.; Dales, N. A.; Glunz, P.; Ripka, A. S. In *Combinatorial Design and Combinatorial Synthesis of Enzyme Inhibitors.* ; Rich, D. J.; Bohacek, R. S.; Dales, N. A.; Glunz, P.; Ripka, A. S. Eds.; Elsevier: Amsterdam, 1996, pp. 101–111.

65. Glunz, P.; Bohacek, R. S.; Rich, D. J. 1998.

66. Ding, Y.; Paris, C. G.; Kolb, H.; McMartin, C.; Bohacek, R. S. In: *De Novo Generation of Heterocyclic Systems*; Ding, Y.; Paris, C. G.; Kolb, H.; McMartin, C.; Bohacek, R. S. Eds.; Amer. Chem. Soc.: Orlando, 1996, pp. Comp–166.

CHAPTER 13

AB INITIO CONFORMATIONAL ANALYSIS OF PROTEIN-SUBUNITS: A CASE-STUDY OF THE SERINE DIAMIDE MODEL

ANDRÁS PERCZEL
Department of Organic Chemistry, L. Eötvös University

IMRE G. CSIZMADIA
Department of Chemistry, University of Toronto

1. INTRODUCTION

Protein engineering has changed the way most disciplines of protein chemistry are now being tackled. Gene technology with protein expression has fundamentally changed our views on what protein architecture can achieve. Altering one (or more) amino acid residue(s) is now rather common and widely used to test (and solve) problems in molecular biology. This routinely applied strategy provides more new proteins (and analogs) that have never been dreamt. It looks rather obvious that deletion, insertion or mutation of a residue affect the conformation of the modified "sequential bit." Furthermore, such a change in the primary sequence could also have an influence on the conformational properties at its neighbor. Computational studies along these lines have been performed,[1–11] pinpointing many fascinating aspects of primary-sequence/conformation relationship. The structure of some of these "reshaped" macromolecules with their parent proteins were determined even at an atomic level by X-ray crystallography and/or by NMR-spectroscopy and deposited in databases (e.g., PDB[12–13]).

The conformational consequences of genetic engineering are seldom predictable. The relative orientation of the consecutive amide linkages (peptide bonds) determines the global fold. The exponentially growing number of deposited

The Amide Linkage: Selected Structural Aspects in Chemistry, Biochemistry, and Materials Science,
Edited by Arthur Greenberg, Curt M. Breneman, and Joel F. Liebman
ISBN 0-471-35893-2

conformers, accessible worldwide, has initiated new fields and is an emerging scientific area. New generations of structure prediction algorithms or alternative neural network analysis signal this trend. Although important, compared to these "hot topics" the application of quantum mechanics for the analysis of peptides and protein fragments is less popular. This theoretical approach, although now accepted, still suffers from some skepticism. Undoubtedly, the strategy has its limitation which keeps conservative minds still suspicious. Strictly speaking, ab initio calculations are for gasphase, relatively expensive in terms of CPU time, and still available only for models of moderate sizes. On the other hand, it is independent from parameterization, unlike force-fields, and provides a variety of physico-chemical and spectral properties at a reliable level. In particular, ab initio methods are useful in structure calculations and in the validations of spectral parameters (e.g., vibrational frequencies, chemical shielding anisotropy, etc.). These aspects of the approach make calculations more and more indispensable.

As outlined, some skepticism still holds, questioning the relevance and scientific merit of a theoretical approach worked out for peptides and proteins. One of the most commonly asked questions is: "Why calculate, if one can measure?" The answer is not to be expected in a single sentence. Due to the intrinsic flexibility of fragments of peptides and proteins measuring a single conformer by any spectroscopic method is often not straightforward. From the time scale of NMR, a conformational ensemble is typically expected in a superimposed form. Such a measured averaged structure cannot even be deconvoluted into components. Thus, such results need to be analysed with the background knowledge of all available structures, perhaps determined by computation. In contrast to NMR, other branches of spectroscopy, with faster time-scales, such as CD, UV, and FTIR, reflect the sum of all energetically accessible forms. To understand and to interpret these spectra, they have to be deconvoluted and once again geometrical information could be essential. As pointed out, structural properties (bond length, bond angles, etc.) are required by spectroscopy. To help these analyses, calculated conformers are of great help.

A second inquiry could be stated: Why do expensive ab initio calculation instead of applying only a commonly used force-field? Obviously, if we stick to the state of the art approach, the size and not the type of the molecule will predetermine the method to be used. Needless to say, such a choice depends on the "historic time" when the calculation is to be completed. A third of a century ago, a study of the $HCONH_2$ molecule, which contains only a single amide bond, by ab initio SCF-MO using an exceedingly small basis set was a major achievement[14], even without geometry optimization. In contrast to that, a decade ago a RHF/3-21G ab initio calculation on HCO–Ser–NH_2 with geometry optimization appeared to be close to the upper limit of that time[15]. Today, the same molecule can be tackled, including full geometry optimization, with a triple-zeta quality basis set augmented with diffuse and polarization functions at the *post*-Hartree–Fock level. More important perhaps, is the fact that even a complete library of conformers (e.g., up to dozens of structures in the case of serine) could be determined even by using a sophisticated method. Alternatively, one may compromise on the level of a theory to favor an increasing molecular size. For example, oligopeptides of up to six amino acid residues, HCO–(NH–CHR–CO)$_n$-NH_2, have been computed,[16–18] but with today's super-computers one may possibly go up to octapeptides still performing a calculation of RHF quality. Although polypeptide is rather large for ab initio study, according to the standards of these days, this molecular size is still far from a real protein both in size and complexity. Thus, if proteins are to be computed then one needs to resort to some suitably chosen force-field. This is exactly what is meant above that the size of the molecule predetermines the state of the art method to be used.

The philosophy of a "total-approach," without prejudice, can now be sketched. For a moderate size peptide, where only few conformers can be studied at the RHF level of theory, one would prefer to determine a complete conformational library, perhaps several thousand of structures, by restricting the method at a force-field level of theory. On the other hand, interactions between important functional groups of side-chains within the peptide model may be studied at the *post*-Hartree–Fock level by using a rather extensive basis set. A far more complete understanding could emerge from such a comprehensive analysis, incorporating all three approaches, than anything that may be the outcome of a single theoretical method. This "total approach" can be demonstrated in the case of the -RGD- sequence, part of many important proteins. Using HCO–Arg–Gly–Asp–NH_2 as a model compound one can recognize that even for a single backbone orientation as many as $3^4 \cdot 3^0 \cdot 3^2 = 729$ side-chain conformers are possible. (It will be introduced shortly that a power of 3 is used to estimate the maximum number of conformers.) Furthermore, considering only all *trans* peptide bonds $9 \cdot 9 \cdot 9 = 729$ backbone conformers are to be expected. The frighteningly large number of theoretically allowed -RGD- conformers, more than half of a million ($729 \cdot 729 = 531441$) structures, which could form the structure library, surely cannot be investigated at the RHF/3-21G level of theory. Thus, an appropriately chosen force field is in order. However, it is instructive to study at least a selected few conformations at the modest RHF/3-21G ab initio level of theory. Beyond this, one may like to discover the binding intricacies of the guadinium-carboxylate salt bridge which makes the -RGD- sequence so important. To do so, *post*-Hartree–Fock calculations may well be necessary on the functional groups, which could be simulated by ethyl substituted guanidium cation and acetate anion. The corresponding molecular structure is shown in Fig. 13.1.

The third most frequent objection against a theoretically determined structure library is "why try to calculate all possible conformers, if they are not energetically readily accessible?" First of all when the global minimum is aimed, often a large number of conformers are determined as "byproducts" of the protocol. Even if these "high energy" zero-gradient structures have low probability, they are not useless. They could guide us to understand basic questions like the following one. A protein can occupy its global minimum or one of its local minimum, energetically not much higher than the global one.

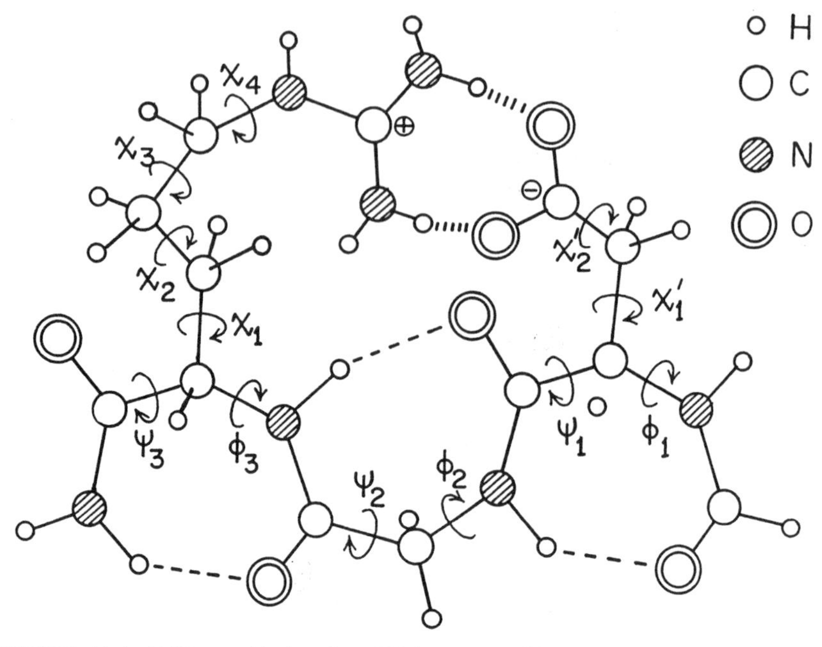

FIGURE 13.1. RGD motif showing HCO–L-Arg–Gly–L-Asp–NH_2 in its γ_L-γ-γ_L backbone conformation. The side-chains are drawn to show an internal salt bridge between guanidium cation and carboxylate anion functional groups.

This problem could be addressed by investigating prion proteins and it will be important to know what is conformationally available and at what energy cost. Therefore, instead of discarding the so-called "high-energy" structures, it is wiser to collect as many of these as possible. An additional rationale for collecting perhaps less probable conformers is related to the understanding of biomolecular complexes. A structure, even the global minimum, regardless of how favorable it might be energetically, may not necessarily have a stereochemically preferred geometrical arrangement for a protein-ligand interaction. If such a binding is thermodynamically exothermic, then the energy released could be used to transform the carrier molecule into one of its energetically less preferred rotamers. The strategy to find the ensemble of otherwise less preferred conformers, stabilized through intermolecular complexation, is one of the fundamental challenges of drug design.

2. AMIDE LINKAGE AND CONFORMERS OF PROTEIN BUILDING UNITS

During the years, a large number of protein structures have been analyzed by diffraction and spectroscopic methods. As a conclusion of these comprehensive analyses, structures of proteins can be subdivided into domains, modules and motifs. The main chain fold of these structural subsets was categorized as secondary structural elements: helices, sheets, turns, etc. In spite of the synthetic, modeling, and computational efforts[19–27] we still do not know enough about the dynamics and folding process of these conformers. On the other hand, the static 3D-structure of peptides and proteins is well characterized and often described by using a simplified rigid model with fixed bond lengths and bond angles. Within this approach, three torsional *angles* (ϕ, ψ and ω) per amino acid residue are sufficient to determine the main chain fold of a protein. Strictly speaking, only ϕ and ψ are true variables, since ω is typically 180° (or seldom 0°). (This is the rational behind the usage of even the simplest models into which that ω, ϕ, ψ, ω' values are to be incorporated. Please see Scheme 1). The result, a two-dimensional potential energy surface, the $E = E(\phi,\psi)$ function, which is also called the Ramachandran map, is used to visualize the energy content of a $[\phi,\psi]$-type conformational change.

From the early sixties, the characterisation of "Ramachandran-type" potential energy surfaces (PES) formed the foundation of systematic computational approaches.[28–30] These two variables, namely ϕ and ψ, are sufficient to describe the backbone flexibility of HCO–Gly–NH_2 or HCO–L-Ala–NH_2. More typical is the case of serine (e.g. HCO–L-Ser–NH_2) where the conformational motion of the residue takes place not on a 2D- but on a 4D-space of $f(\phi,\psi,\chi_1,\chi_2)$. Cys, Asp, Asn, Phe, Tyr, His, and Trp also have two backbone (ϕ,ψ) and two side chain (χ_1,χ_2) torsional variables, like serine representing their conformational status. In all of the above eight natural amino acid residues the side-chain functional group is separated from the backbone by no more than one methylene group. Among these, serine is the smallest amino acid which has a polar side-chain. Its diamide derivative (e.g. the L-enantiomeric form of HCO–NHCH(CH_2OH)CO–NH_2) is a suitable model for a better understanding of the conformational properties of a rather typical peptide building unit. In addition to the determination of the stationary points of this molecule, a closer look at this hyperspace will also be provided within this chapter to demonstrate how to perform an inventive ab initio grid search. The direct investigation of a 4D-space is hard to visualize and even lengthier to determine. There is a possibility to generate selected 2D-cross sections ($f_{\phi,\psi}^{relax}$ $[\chi_1,\chi_2]$) of this 4D-hyperspace at

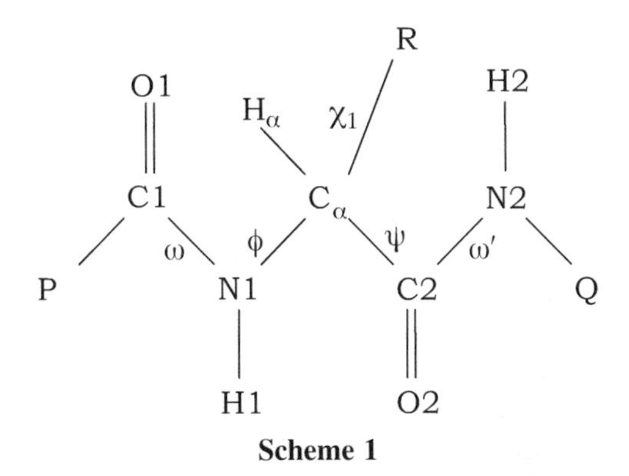

Scheme 1

regions of interest (e.g. at ϕ,ψ values typical for the building unit of an α-helix, β-pleated sheet, inverse γ-turn, etc.). Following certain tracks on these ab initio conformation energy maps, the tracing of some relaxation paths is becoming possible. The monitoring of the various side-chain induced backbone conformational shifts is also feasible by determining relaxation paths on $f_{\phi,\psi}{}^{relax}$ $[\chi_1,\chi_2]$ surfaces.

The energetic and topological results of these (ϕ,ψ) surface-analyses can be generalized. According to the commonly accepted "dipeptide-model," conformational properties of a peptide, made from κ residues, are based on the knowledge of κ number of 2D surfaces. Although not perfect, this strategy is straightforward and thus commonly employed. This simplification is similar to the approach of protein structure prediction based on the "summation" of properties of individual amino acid residues. This model obviously ignores the structure modifying effects of the nearest-neighbors, and also those interactions that originate from far lying molecular fragments. Nevertheless, this model is the starting point of ab initio approaches and provides useful information.

3. BASIC TOPOLOGY OF TWO ADJACENT AMIDE-LINKAGES AND THEIR CONFORMERS

We have outlined the central role and importance of generalized Ramachandran-type surfaces. The key to an effective structure analysis is a quick and reliable strategy to highlight potential locations of stationary points of these hyperspaces. Once localized approximately, the question of fine-tuning becomes a matter of CPU-time and some minimization algorithm. It looks as if the major question is to pinpoint where to look for conformational "candidates," since due to the large number of variables to scan, the strategy of systematic grid-search on hypersurfaces, $E = E(x)$, is not feasible. Note that the components of vector $\underline{x}$ are all the relevant torsional angles (e.g. ϕ, ψ, χ_1, χ_2 etc.). Multidimensional conformational analysis (MDCA) was developed to help to localise various regions containing critical points on a generalised n-dimensional Ramachandran-surface.

The 2D-Ramachandran maps ($E = E(\phi,\psi)$[32]) resulted in a maximum number of nine minima. This is due to the fact that the torsional potential along both ϕ and ψ have three (or sometimes fewer) minima (g^+, a and g^-). Therefore, within the law of conformational analysis, there must be nine *legitimate* conformers donated by g^+g^+; $(+60°, +60°)$, ag^+; $(+180°, +60°)$, g^-g^+; $(+300°, +60°), \ldots, g^-g^-$;$(+300°, +300°)$ (Scheme 2). What is important is the fact that a maximum of nine backbone conformers are expected per residue in peptides[32] and not the way they are labeled. The labeling is optional, but for various reasons these nine conformers were named using Greek letters subscripted by L and D (e.g. α_L, α_D, β_L etc.). By surprise, initial RHF/3-21G calculations carried out on selected peptide models containing a single amino acid resulted in seven minima only. Apparently, two out of the nine legitimate

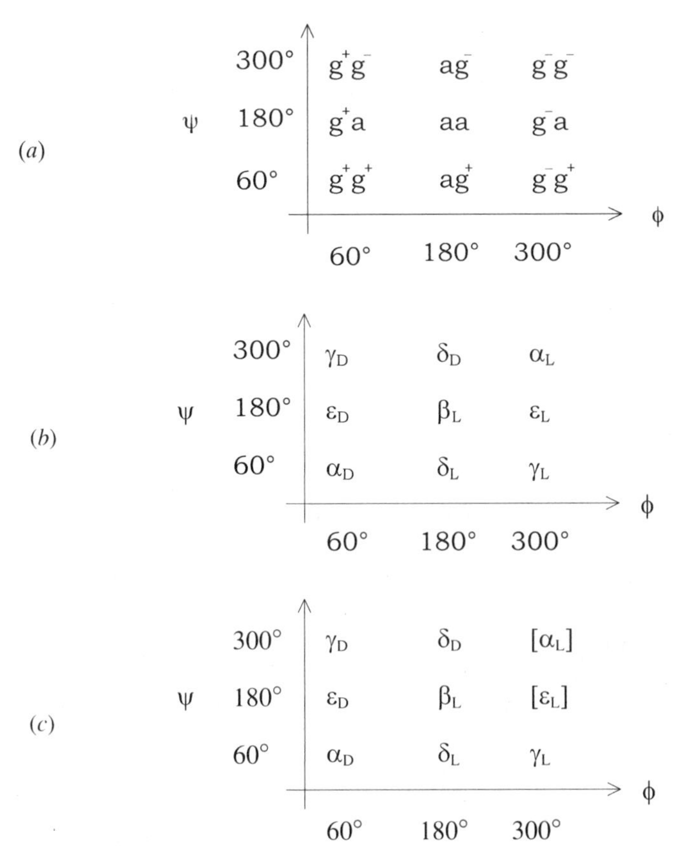

Scheme 2. (*a*) The 9 backbone conformers predicted by MDCA (*b*) These conformers labeled by the conformational symbolism suggested for peptide building units (*c*) Approximate locations of optimized ab initio minima of HCO–Xxx–NH_2 (Xxx = Gly, Ala, Val, Phe). The α_L and ε_L are in brackets since they may vanish.

minima, that of α_L and ε_L have a more complicated nature. (It needs to be emphasized that both of the two missing minima (α_L and ε_L) were found in HCO–L–Ala–L-Ala–NH_2, as sub-conformers of the molecule.[33–35] Consequently, all legitimate conformers have been located by ab initio techniques in the triamide system. Furthermore, all nine minima were found in proteins with known X-ray structure.[35]

Rotation about two bonds connected to a prochiral center produces nine conformers from which eight occurs as four doubly degenerate pairs (α, γ, δ and ε). This is the case for the achiral *N*-formyl-glycineamide: $E_{\gamma L} = E_{\gamma D}$, $E_{\delta L} = E_{\delta D}$ etc.(c.f. Scheme 3).

If the amino acid is chiral (as is the case for 19 out of the 20 natural amino acids) this double degeneracy disappears; $E_{\gamma L} \neq E_{\gamma D}$, $E_{\delta L} \neq E_{\delta D}$ etc. In the case of L-amino acids the four conformers at the right-hand side will be favored (i.e., lowered in energy) with respect to their counterparts at the left side. Therefore,

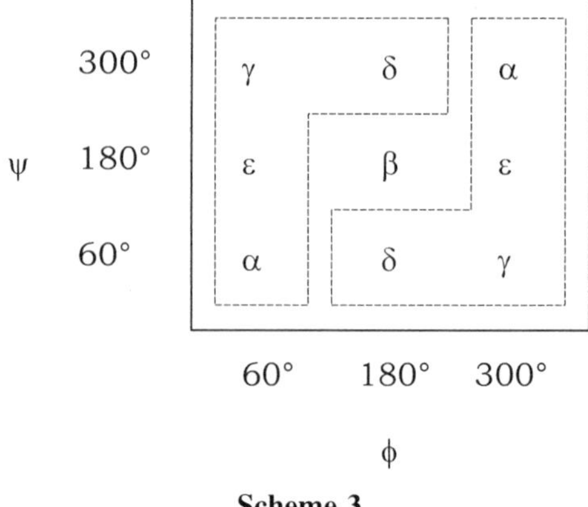

Scheme 3

minima with subscript L have lower energy than their D counterparts, and therefore appear with a higher probability in proteins. Thus, when protein X-ray data are analyzed, all the frequently assigned protein secondary structural elements are to be denoted with symbols having L subscript. For example, the common right-handed α-helix is $(\alpha_L)_n$, while β-sheet is $(\beta_L)_n$. This right-hand side of the Ramachandran map may be referred to as the L-valley. Accordingly, the left-hand side section may be referred to as the D-valley (c.f. Scheme 4). (Of course, in terms of stability the opposite is true for a D-amino acid diamide, where the D-valley is the favored one.).

The notation just introduced is useful in several respects: it reflects the rotation topology of peptide, adheres to peptide traditional convention [α-structures =: $(\alpha_L)_n$, β-structures =: $(\beta_L)_n$, inverse γ-turns =: γ_L etc.], reflects the "symmetry" of the Ramachandran surface and the relative energetic of the appropriate pairs of conformer [L-amino acid residues favor the L-valley etc.]).

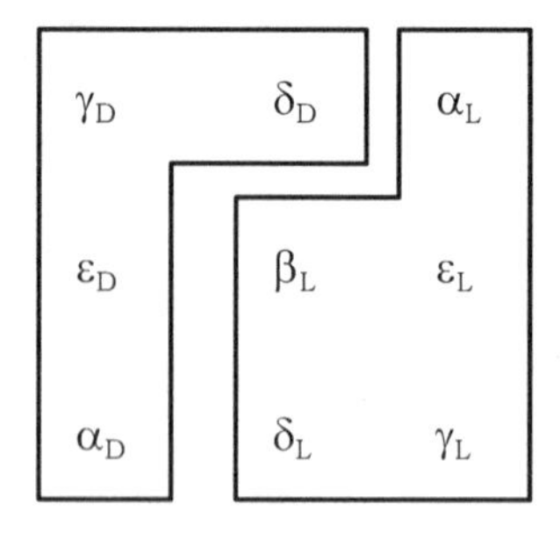

Scheme 4

Furthermore, this convention provides additional features for fine tuning a structure: e.g., $[(\alpha_L)_n]^{3-10}$ or $[(\alpha_L)_n]^{\pi}$ could stand for 3-10- or π-helices, obviously members of the α-helix "family". The only slightly disturbing problem is when we specify right-handed and left-handed helical conformations with this notation:

$$(\alpha\text{-helix})_{\text{right}} = (\alpha_L)_n$$

$$(\alpha\text{-helix})_{\text{left}} = (\alpha_D)_n$$

This could lead to some misunderstanding, but it should be mentioned that the official notations of IUPAC-IUB for the two forms of helical structures are P- or M-, and not R- or L- which are sometimes used to describe right-handed and left-handed helices. It is hoped however, that the practical utility and uniformity of this notation overrides this marginal problem. Similarly, both parallel and anti-parallel β-pleated sheet conformations are regarded as subsets of the $(\beta_L)_n$ conformational homopolymers.

When specifying numerical values of ϕ and ψ, we use the IUPAC-IUB convention (IUPAC-IUB 1970), i.e., $-180° \leq \phi \leq 180°$ and $-180° \leq \psi \leq 180°$. However, occasionally for graphical presentation we also use the traditional cut ($0° \leq \phi \leq 360°$ and $0° \leq \psi \leq 360°$) suggested previously by Ramachandran and Sasisekharan[30] (Scheme 5/C).

Undoubtedly, besides the topology based subdivision, the other alternative is to use the notation of protein crystallography[36]. The suggested subdivision of the $E = E(\phi,\psi)$ surface by Karplus[36] is similar to the one proposed by Efimov[37]. Different researchers in the past subdivided this surface into sub-conformational regions based on a different philosophy. The works of Richardson and Richardson[38] and Rooman et al.[39] have also suggested how to cluster different backbone conformers. We have to mention the pioneering works of Zimmerman et al.[40] and the one of Thornton et al.,[41] which is one of the most recent one, to make the list of studies more complete. For example, in the approach used by Karplus,[36] the Ramachandran surface has 12 distinct regions, labelled as α_L, α_R, β_S, β_P, γ,γ', δ_L, δ_R, ε, ε', ε'' and ζ. Prior to this, Zimmerman and coworkers[40] had suggested a subdivision up to 16 regions (A,B,C,D,E,F,G,H and A*,B*,C*,D*, E*,F*,G*,H*). Both of these latter methods, with several other alternatives, depend on X-ray data. A composition of the topology based notation (**C**) with that used by Karplus[36] (**A**) and that of Zimmerman[40] (**B**) is plotted on Scheme 5.

Ideally, all nomenclatures should pinpoint to centers of conformational clusters. There is no limit to the number of sub-regions (*nonantes, dodecantes* etc.) that the same surface can be dissected into. The question to be addressed is how many different relative backbone orientations are rational to be introduced to simplify the description of the main-chain fold of peptides and proteins. A considerable difference between a peptide-topology based clustering and the one that derives locations from the frequency analysis of observed peptide conformers in proteins is that the former one will not change with the improvement of the database. Although, all notations have some advantages, the

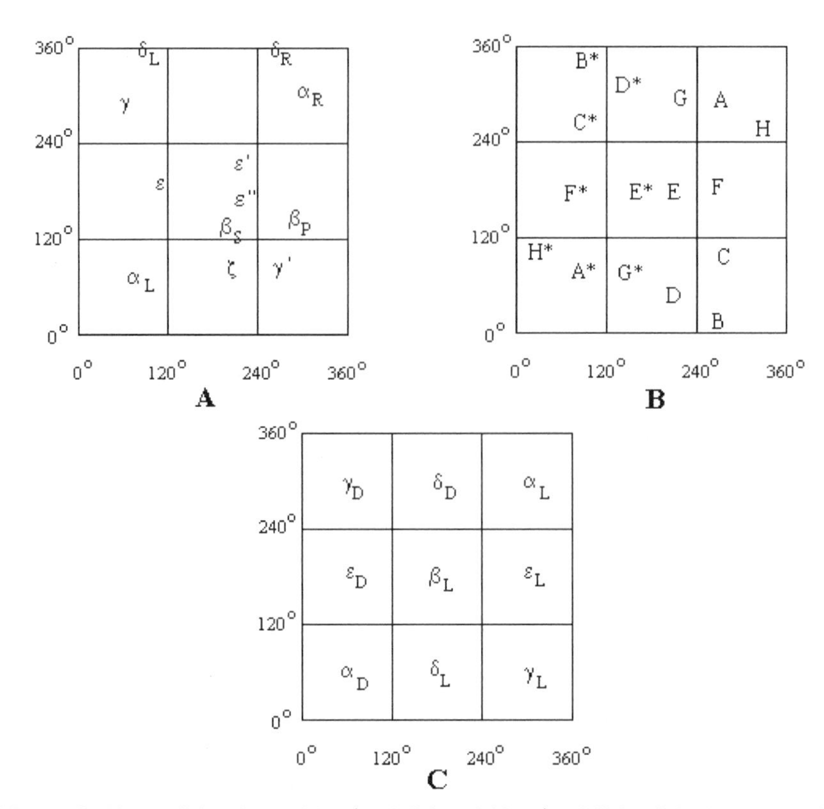

Scheme 5. The traditional cut ($0° \leq \phi \leq 360°$ and $0° \leq \psi \leq 360°$) of the Ramachandran plot subdivided into "*nonantes*" [$0° \leq \phi \leq 120°, 0° \leq \psi \leq 120°$], [$120° \leq \phi \leq 240°, 0° \leq \psi \leq 120°$],...,[$240° \leq \phi \leq 360°, 240° \leq \psi \leq 360°$]. Three alternative methods (**A** to **C**) are reported for a diamide derived from an L-amino acid, all attempting to label the important sub-regions of the Ramachandran-surface. That of Karplus[36] [**A**] and Zimmerman et al.[40] [**B**] are based on the distribution of X-ray data of proteins, while the topology based notation Perczel et al.[31] is reported as **C**.

topology based notation (**C**) will be used in this review since amide-linkage topology is in focus.

4. HOW TO LOCALIZE STATIONARY POINTS ON A HYPER-SURFACE

In the next few lines the strategy and the efficiency of a systematic grid search on an *n*D-surface is compared with the MDCA suggested input determination technique. As pointed out in the introduction, the more conformers that are explicitly determined, the better the picture that emerges for a flexible molecule. This almost trivial goal is rather complicated to achieve in practice. Even the systematic search for stationary points on a 2D-potential energy surface can not only be tedious but also rather expensive. Since peptides are inherently flexible molecules, with multiple relatively low lying stationary points (see Fig. 13.2), a

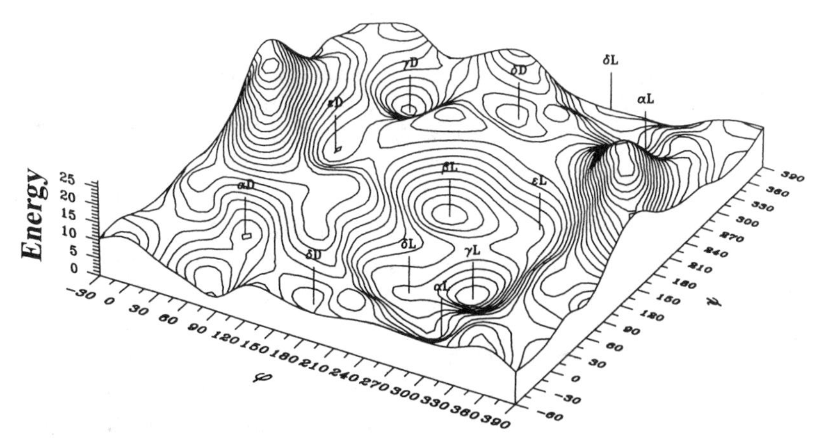

FIGURE 13.2. The Ramachandran surface of HCO–L-Ala–NH_2 determined at RHF/3-21G level of theory.

strategy to cover all available conformational area is necessary. For example, to describe the main chain conformational properties of alanine diamide, the systematic scanning of the $E = E[\phi,\psi]$ surface is needed.[42] The computation of the entire surface ($0° \leq \phi \leq 360°$ and $0° \leq \psi \leq 360°$) requires over a hundred SCF points ($11 \cdot 11 = 121$), if the grid search is executed by 30° increments. Nevertheless, the resulting ab initio Ramachandran-surface is rewarding (Fig. 13.2). Although, extremely visual, such an approach can hardly be used at higher dimension. For example, the same grid search for HCO–L-Val–NH_2 ($E = E[\phi,\psi,\chi_1]$) would require calculations at $11 \cdot 11 \cdot 11 = 1331$ scanning points. Furthermore, the complete structure analysis of HCO–L-Ser–NH_2 or that of HCO–L-Ala–L-Ala–NH_2 requires as many as $11^4 = 14{,}641$ grid points. To fulfill the dream of ab initio structure aided conformational research for peptides and proteins, one must find a radically less laborious and more realistic strategy. Based on previously described theories[44–48] we have suggested the use of MDCA for the locations of minimum-energy structures of peptides. As a test for reliability of MDCA, we have minimised all nine input geometries predicted to be located by MDCA on the Ramachandran surface of HCO–L-Ala–NH_2. This strategy resulted in the same seven stationary points[32] as those from the full grid search carried out by Head-Gordon et al.[42] This could mean that MDCA based minimization requires at least 10 times less CPU! Prompted by the success of MDCA, a systematic grid search was carried out for the serine containing diamide molecule (HCO–L-Ser–NH_2).[49–52] Even though the hydroxymethyl side chain of this molecule can interact strongly with the backbone atoms, the systematic search over a total of $9 \cdot 121 = 1089$ grid points,[51] determined at 3-21G/RHF level of theory, resulted in the same number of minima as the simple minimization of the 81 initial geometries predicted by MDCA. It is probable that the full calculation of the entire 4D-hyperspace by a 30° increment (resulting in a total of 14,641 grid points) would result in the same number of minima. It may be too early to say that MDCA is a foolproof strategy. Naturally, one could fine-tune these approximate locations, if minimization is not achieved at once, by slightly shifting the appropriate input ϕ,ψ or χ values based on previous ab initio results. For example, in the case of the "fragile" δ_L backbone conformer (Fig. 13.2) instead of the use of the MDCA recommended $\phi = -180°$, $\psi = +60°$ input, the $\phi = -150°$, $\psi = +40°$ could be a preferable choice. We have to note that the success of any minimization depends not only on the "correct" input location, but also on the applied minimization algorithm. Nevertheless, we and others have found an MDCA-like strategy rewarding, therefore we strongly encourage its application.

5. THE SUITABLE LEVEL OF THEORY

A large variety of force fields have been introduced in the past for the conformational analyses of peptides and proteins. From time to time, comparisons are published[53–55] to demonstrate their full capacity. Some of these force fields (HYPERCHEM,[56] MM+,[57] AMBER,[58] CHARMM[59] and OPLS[60,61]), along with the common semi-empirical MO approaches (MNDO/3,[62] MNDO,[63] AM1[64] and PM3[65]) have been widely used to investigate molecules too large for ab initio studies. However, these results are too often different from each other and from ab initio data.[53–55] Their inconsistency can be traced to the variation of the relative energy orders for the same conformers with different structural values. Furthermore, even the number of the allowed conformers vary from one method to another. Therefore, the present review does not incorporate force-field or semi-empirical values only in exceptional cases.

Hundreds of ab initio computations have been carried out on various conformations of small amino acids and peptides. It is hard to refer to each and every important mosaic of the complete picture even in a review chapter. Some results were selected when focusing on the important issue of secondary structural building units. Ab initio results of P–CONH–CHR–CONH–Q models, where P = Q = –H or P = Q = $-CH_3$, for glycine R = –H,[32,42,54,66–69] for alanine R = $-CH_3$,[32,42,54,66–69] for valine R = $-CH(Me)_2$,[84] for serine R = $-CH_2OH$[50,51,52,66] and for phenylalanine R = $-CH_2Ph$[85,86] are only some examples, where systematic topological research has been carried out to establish a conformational library as complete as possible. In most of these works, computations were performed by using a relatively small basis set (e.g. 3-21G, 4-21G) although more extensive basis sets with and without the inclusion of electron correlation have also been employed.[70,71]

It may seem to be a mere technicality but to find a basis set which is relatively small in size, yet reliable in quality, remains an intrinsic problem of applied quantum mechanics. The success and the reliability of the peptide ab initio results largely depend on a wise choice of basis set. Undoubtedly, "the larger the basis set, the better the result" principle is expected to hold for peptides too. Furthermore, electron correlation may also have a contribution to structure and stability. However, the true challenge is to find a small basis set without the inclusion of electron correlation, which could yield reasonably good structures and acceptable energy differences. Unless this is achieved, due to CPU limitations, only a handful structures could be optimized. Attempts have been made to monitor the fluctuation of the ϕ,ψ values in small peptides (e.g. HCO–Gly–NH_2, HCO–L-Ala–NH_2) as a function of the applied ab initio method[69,71,72] and pointed out that the 4-21G RHF energies were found to be close to MP2 energies for selected main chain conformers. Surprisingly, the 4-21G RHF relative energies were more similar to the corresponding MP2 energy differences than those obtained by SCF calculations by using larger basis sets. Triggered by this observation, a systematic comparison was started by Endrédi and coworkers[70] for the different conformers of HCO–L-Ala–NH_2. The RHF type geometry optimisations were carried out for a total of 11 basis sets: 3-21G, 4-21G, 4-31G, 6-31G, 6-311G, 6-31++G, 6-31G**, 6-31++G**, 6-311G** 6-311++G and 6-311++G**. The accuracy of the computed results were increased by geometry optimization carried out with the inclusion of electron correlation using a second ordered perturbation method (MP2). Here again the same 11 basis sets were used. Note that the symbol for the basis set 6-311++G**, i.e., 6-311++G(d,p) implies a minimal core representation with the aid of 6s-type Gaussians, and 311 implies that 5s and 5p type Gaussians are contracted in the ratio of 3 : 1 : 1 to a triple-zeta quality representation of the valence electron shell. The ++ sign indicates that diffuse Gaussian functions were added to all heavy atoms (C,N,O) as well as to all hydrogens. The ** superscript specifies that d-type polarization functions for the heavy atoms (C,N,O) and p-type polarization functions for the hydrogen atoms were also included in the basis set. The polarization functions are necessary for the computation of accurate torsional angles. These calculations revealed that the backbone torsional angles vary depending on the level of theory, but the variability is typically smaller than 10°. Variation of relative energies with the accuracy of the calculation was also monitored. Not considering the relative energy differences obtained with an MP2 calculations using small basis sets such as 3-21G and 4-21G, the relative energy differences of the conformers fluctuated by about ≈ 1 or $\approx 2\,\mathrm{kcal\,mol^{-1}}$. For selected backbone conformers the 3-21G RHF relative energy differences were close to similar data obtained from 6-31++G** MP2 and 6-311++G** MP2 calculations. Although, this is the result of a fortuitous cancellation of different "errors," it justifies the selection of the 3-21G and or 4-21G basis set as suitable candidates for studying oligopeptides: (PCO–[NH–CHCR–CO]$_n$–NHQ.

6. THE TWO SIMPLEST DIAMIDE MODELS (HCO–GLY–NH_2 AND HCO–L-ALA–NH_2)

The smallest amino acid diamide system is that of glycine. Glycine itself has as many as eight different structures,[88] while its diamide has only three unique structures out of the 5 expected main chain folds (Table 13.1).[32,42,54] As mentioned earlier, for a chiral amino acid diamide (19 out of the 20 natural amino acids) a maximum of nine relative backbone orientations are expected.

But the achiral Gly has a symmetric Ramachandran surface, so that $\alpha_L = \alpha_D$, β_L, $\gamma_L = \gamma_D$, $\delta_L = \delta_D$ and $\varepsilon_L = \varepsilon_D$, resulting in only five distinct structures.[32,54] Among these three were found.

Alanine is the smallest chiral amino acid (R = $-CH_3$) and has 13 conformers in the gas phase.[89] The appropriate diamide at RHF/3-21G level of theory has only little more than half of the backbone conformers of the free amino acid itself. The alanine diamide model is probably the most intensively investigated system.[32,42,54,68–72]

Further computations were carried out at additional levels of theory for $HCO{-}Gly{-}NH_2$ and $CH_3CO{-}Gly{-}NHCH_3$.[54,66–72] Saddle points were also determined by Head-Gordon and coworkers[42] for the simplest glycine diamide model compound both at the 3-21G RHF and 6-31+G* RHF levels of theory.

Comparing the location and the relative ϕ,ψ shifts as a function of the applied level of theory, it looks obvious that each type of minima remains in its own catchment region (Fig. 13.2). At 6-311++G** MP2 level of theory two additional minima vanish in the case of the Ala model: δ_L and ε_D. The latter one (ε_D) is predicted (MDCA) to be located close to $\phi = 60°$, $\psi = 180°$. Indeed, using the 3-21G RHF approach both Head-Gordon et al.[42] and Perczel et al.[32] determined the actual location at [$\phi = 67.5°$, $\psi = -177.3°$] and at [$\phi = 67.6°$, $\psi = -178.1°$], respectively. (These minima were recalculated by Endrédi et al.[70] [see values in Table 13.1], using different optimisation parameters.) Investigating this region of the Ramachandran surface (Fig. 13.2) ε_D is found in a basin-like extremely shallow valley located on a hill-side (Fig. 13.2). This surface property could explain the minor but significant conformational differences between the calculated structures for ε_D (all using the same 3-21G RHF approach). The local environment of δ_L has a shape rather similar to the one of ε_D. Due to the low energy barriers (saddle points), both the δ_L and ε_D minima look "fragile." In fact neither of these two conformers could have been optimized at the MP2/6-311++G** level of theory (Table 13.1). Comparing the energies (RHF/3-21G and MP2/6-311++G**) relative to γ_L of the residual five minima (α_D, β_L, γ_L, γ_D, and δ_D), the following can be concluded:

- the relative order of the minima is conserved ($E_{\gamma L} < E_{\beta L} < E_{\gamma D} < E_{\alpha D} < E_{\delta D}$),
- the relative energies do change with the applied level of theory,
- the larger the relative energy difference between conformers at RHF/3-21G level of theory, the more the MP2/6-311++G** energy difference increases, e.g.,

$$(\Delta E^{\text{RHF/3-21G}}_{\beta L-\gamma L} < \Delta E^{\text{RHF/3-21G}}_{\alpha D-\gamma L}) \Rightarrow (\Delta E^{\text{RHF/3-21G}}_{\beta L-\gamma L} - \Delta E^{\text{MP2/6-311++G}^{**}}_{\beta L-\gamma L}) < (\Delta E^{\text{RHF/3-21G}}_{\alpha D-\gamma L} - \Delta E^{\text{MP2/6-311++G}^{**}}_{\alpha D-\gamma L}),$$

which in terms of numbers [Table 13.1] is; $1.25 < 5.95 \Rightarrow (1.25 - 1.21) < (5.95 - 3.88)$.

On the basis of this, perhaps the Ramachandran surface at MP2/6-311++G** level of theory would look like that on Fig. 13.2. However, differences expected are outlined as:

TABLE 13.1. Conformers of $HCO-Gly-NH_2$ and $HCO-L-Ala-NH_2$ Molecules and their ϕ and ψ Values with the Relative Energies, Calculated at Different Level of Theory

		$HCO-Gly-NH_2$			$HCO-L-Ala-NH_2$		
bb[a]/method		ϕ	ψ	ΔE	ϕ	ψ	ΔE
γ_L	**A**	−83.3	+64.7	0.00	−84.5	+67.3	0.00
	B	−85.2	+67.4	0.00	−85.8	+78.1	0.00
	C				−86.2	+78.8	0.00
	D				−82.8	+80.6	0.00
β_L	**A**	−180	+180	0.65	−168.3	+170.6	+1.25
	B	−180	+180	−0.58	−155.6	+160.2	+0.19
	C				−155.1	+161.0	+0.11
	D				−157.1	+163.2	+1.21
δ_L	**A**	−121.9	+25.2	3.27	−128.1	+29.8	+3.83
	B		not found		−110.4	+12.0	+2.24
	C				−112.8	+13.2	+2.22
	D					not found	
α_D	**A**		not found		+63.8	+32.7	+5.95
	B		not found		+69.5	+24.9	+4.73
	C				+69.0	+26.9	+4.56
	D				+63.1	+35.5	+3.88
γ_D	**A**	+83.3	−64.7	0.00	+74.0	−57.4	+2.53
	B	+85.2	−67.4	0.00	+75.1	−54.1	+2.56
	C				+75.3	−55.4	+2.54
	D				+74.4	−49.1	+2.19
δ_D	**A**	+121.9	−25.2	3.27	−178.6	−44.1	+7.31
	B		not found		−165.6	−40.7	+5.52
	C				−165.2	−42.1	+5.39
	D				−166.0	−39.9	+5.45
ε_D	**A**		not found		+67.2	−171.9	+8.16
	B		not found			not found	
	C					not found	
	D					not found	

[a] Zero-gradient backbone conformers calculated at four levels of theory (**A**) RHF/3-21G[b], (**B**) RHF/6-31+G*[c], (**C**) RHF/6-311++G**[d] and (**D**) MP2/6-311++G**[e]). (ϕ and ψ in degree, ΔE in $kcal mol^{-1}$) Note that the conformational building unit of the right-handed helix (α_L) and that of polyproline II (ε_L) aren't stationary points.

[b] From Endrédi et al.,[70]

[c] From Head-Gordon et al.,[42]

[d] From Endrédi et al.,[70]

[e] From Endrédi et al.[70]

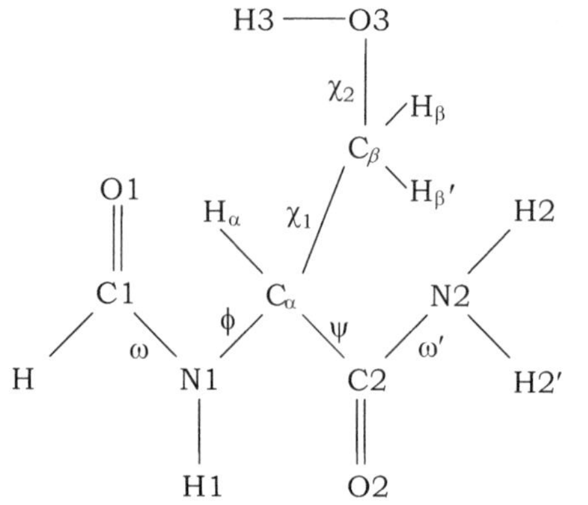

H3 — O3
χ_2 H_β
C_β
O1 H_α χ_1 H_β' H2
C1 C_α N2
ω ϕ ψ ω'
H N1 C2 H2′
H1 O2

Scheme 6

- the z axis (Energy axis) is reduced by some 30%,
- the regions of δ_L and ε_D could be a mountain-side on the MP2/6-311++G** surface, like the α_L and ε_L are located on slopes of the RHF/3-21G Ramachandran map (Fig. 13.2). Saddle points on the Ramachandran surface associated with alanine diamide were also determined by Head-Gordon and co-workers[42] both the 3-21G RHF and the 6-31+G* RHF levels of theory and could be used to estimate the energy requirements of conformational interconversion. This point will be discussed later in connection with Scheme 6.

7. THE TYPICAL DIAMIDE MODEL: HCO–L-SER–NH_2

7.1. Structure and Stability

On the $E = E(\phi, \psi, \chi_1, \chi_2)$ potential energy hypersurface associated with HCO–L-Ser–NH_2 $3^4 = 81$ stationary points are expected: a maximum of 9 different side-chain orientations for each of the 9 backbone conformers. Among these, 44 relaxed structures have been located at the 3-21G RHF levels of theory.[51] Of these 44 stable conformers 19 were located in the L-valley, while 25 were assigned as part of the D-valley of the Ramachandran surface.

The remaining 37 input structures (81–44) have migrated to one of the existing 44 minima. Recently, these conformers were the subject of a more thorough and higher level of ab initio calculations.[73] First the 44 RHF/3-21G structures were reoptimized by using a larger basis set. To avoid misleading terminology the term "migration" is used when an optimized conformer is in a different catchment region then its input was.

When structures were recalculated at RHF/6-311++G** level of theory, using the RHF/3-21G structures as inputs, the following eight migrations were observed:

$$\beta_L(g^-,a) \Rightarrow \gamma_L(g^-,a),$$
$$\delta_L(g^-,a) \Rightarrow \alpha_L(g^-,a),$$
$$\delta_L(a,g^-) \Rightarrow \gamma_L(a,g^-),$$
$$\gamma_L(a,g^+) \Rightarrow \beta_L(a,g^+),$$
$$\varepsilon_D(g^-,g^-) \Rightarrow \gamma_D(g^-,g^-),$$
$$\gamma_D(a,g^-) \Rightarrow \alpha_D(a,g^-),$$
$$\delta_D(g^-,g^+) \Rightarrow \gamma_D(g^-,g^+),$$
$$\delta_D(a,a) \Rightarrow \beta_L(a,a).$$

Note that 4 out of the original 19 L-type conformers (Table 13.2) migrated to other structures all within the same L-valley. Four among the 25 D-type RHF/3-21G backbone structures have migrated, and three ended up in the D-valley, while one $\delta_D(a,a)$ moved into the "L-side" of the map and became $\beta_L(a,a)$. The number of stable structures found so far is 36 (i.e. 44-8) at this higher level of theory. All of these migrations affect only ϕ,ψ and not χ_1 and/or χ_2 values.[73]

It is interesting to compare the new geometrical properties with those obtained with the 3-21G basis set. The direction of the change harmonizes with the general expectation: a larger basis set should result in conformational parameters closer to those derived from X-ray analysis of proteins. This is the case for β_L-, γ_L- and δ_L-type diamide orientations. By surprise for α_L, the building unit of a helical secondary structural element, the ab initio structure at a higher level of theory looks less similar to those observed in proteins than those calculated at 3-21G basis set. A 3_{10}-helix is sharper than a normal one, which is reflected in terms of ϕ,ψ values such as: $\phi \approx -60°, \psi \approx -30°$ for the previous and $\phi \approx -54°, \psi \approx -45°$ for the latter type. While the averages of the three α_L-building units at RHF/3-21G level of theory ($\phi \approx -68.3°$, $\psi \approx -30.5°$) are close to the expected values of a 3_{10}-helix, the same values for the recalculated structures with a 6-311++G** basis set ($\phi \approx -79.1°$, $\psi \approx -22.4°$) suggest a building unit that could provide an even sharper helix-like structure. The α_L- and δ_L-structures are closer to each other in conformational space at higher level of theory than found at 3-21G/RHF. As described, the average values of the conformational shifts as a function of the applied theory are significant but not too large. They can be compared with similar values obtained previously for alanine diamides. The ideal values ($[\phi,\psi]^{ideal}$) are those simply predicted by MDCA. Not surprising, the average shifted values

$$[\phi,\psi]^{\text{ideal}} \Rightarrow [\phi,\psi]^{\text{RHF/3-21G}} \Rightarrow [\phi,\psi]^{\text{RHF/6-311++G}^{**}}$$

are rather similar for serine than observed for alanine (Table 13.3) both in character and magnitude. It looks, therefore, obvious that the above shifts are

TABLE 13.2. Optimised Ab Initio Conformers[a] of For–L-Ser–NH_2 at RHF/6-311++G and RHF/3-21G Level of Theory[b]**

Conf.[a,b]	ω_0	ϕ	ψ	ω_1
$\beta_L(g^-,a) \Rightarrow \gamma_L(g^-,a)$	–(–179.46)	–(–137.86)	–(160.53)	–(177.19)
$\beta_L(g^-,g^+)$	177.65 (177.01)	–174.03 (–179.13)	165.53 (172.54)	179.3 (179.41)
$\beta_L(g^+,g^-)$	171.47(175.01)	–153.30 (–166.61)	173.98 (174.83)	173.21 (176.66)
$\beta_L(g^+,a)$	176.21 (179.41)	–156.76 (–170.14)	173.34 (175.08)	172.55 (177.67)
$\beta_L(a,a)$	172.85 (174.98)	–155.95 (–171.36)	174.89 (–173.38)	174.02 (–179.27)
$\beta_L(a,g^+)$	172.8 (175.23)	–157.29 (–170.31)	–179.1 (–171.5)	174.69 (179.82)
$\delta_L(g^-,g^-)$	–166.39 (–172.21)	–134.06 (–151.98)	17.07 (35.73)	171.86 (176.91)
$\delta_L(g^-,a) \Rightarrow \alpha_L(g^-,a)$	– (–166.62)	– (–130.24)	– (29.98)	– (176.41)
$\delta_L(a,g^-) \Rightarrow \gamma_L(a,g^-)$	– (–173.51)	– (–127.86)	– (33.18)	– (176.41)
$\delta_L(g^+,a)$	–170.57 (–172.71)	–114.38 (–118.21)	11.79 (20)	174.12 (177.4)
$\gamma_L(g^-,g^-)$	–171.19 (–171.15)	–86.96 (–77.26)	73.12 (62.95)	–174.17 (62.95)
$\gamma_L(g^-,a)$	–169.29 (–168.8)	–86.99 (–77.05)	71.14 (61.42)	–174.96 (179.61)
$\gamma_L(g^-,g^+)$	176.42 (–177.95)	–86.66 (–85.29)	82.85 (67.1)	–172.81 (–179.22)
$\gamma_L(a,g^-)$	–177.22 (–173.78)	–86.45 (–83.42)	69.16 (62.5)	–178.11 (179.79)
$\gamma_L(a,g^+) \Rightarrow \beta_L(a,g^+)$	– (–177.86)	– (–86.59)	– (77.83)	– (–176.83)
$\gamma_L(g^+,g^+)$	–176.82 (–176.2)	–85.43 (–83.66)	72.24 (71.42)	–175.09 (–177.55)
$\alpha_L(a,a)$	–170.61 (–174.55)	–73.91 (–62.39)	–37.99 (–42.81)	171.58 (–179.72)
$\alpha_L(g^-,a)$	–164.89 (–169.36)	–80.77 (–70.53)	–14.24 (–24.94)	172.39 (–179.77)
$\alpha_L(g^-,g^-)$	–166.59 (–171.05)	–82.60 (–71.98)	–14.9 (–23.68)	173.44 (–179.17)
$\varepsilon_D(g^-,g^-) \Rightarrow \gamma_D(g^-,g^-)$	– (–173.8)	– (68.93)	– (178.24)	– (179.55)
$\varepsilon_D(g^-,g^+)$	–168.61 (–165.26)	56.02 (64.49)	–136.97 (177.92)	174 (179.68)
$\varepsilon_D(a,a)$	–164.54 (–166.85)	71.28 (68.42)	–170.18 (–172.85)	–176.31 (–179.03)
$\varepsilon_D(a,g^+)$	–164.61 (–167.12)	69.72 (66.89)	–167.07 (–168.95)	–179.73 (179.75)
$\varepsilon_D(g^+,g^-)$	–170.13 (–175.46)	89.91 (99.84)	–141.95 (–116.85)	177.77 (179.85)
$\varepsilon_D(g^+,a)$	–175.18 (178.31)	41.83 (43.02)	–115.63 (–105.53)	171.23 (176.85)

$\gamma_D(g^-,g^-)$	170.61 (170.77)	75.33 (74.66)	−50.48 (−55.23)	−177.54 (−178.23)
$\gamma_D(g^-,a)$	170.18 (170.32)	77.05 (75.42)	−55.49 (−56.12)	179.53 (−178.81)
$\gamma_D(g^-,g^+)$	177.18 (177.67)	76.48 (72.21)	−55.35 (−57.54)	−178.59 (−178.3)
$\gamma_D(a,g^-) \Rightarrow \alpha_D(a,g-)$	− (172.41)	− (67.46)	− (−31.2)	− (−176.13)
$\gamma_D(a,a)$	−179.32 (178.25)	75.33 (73.99)	−74.56 (−64.98)	172.47 (−179.46)
$\gamma_D(a,g^+)$	179.59 (176.57)	74.23 (71.3)	−61.43 (−52.17)	−178.11 (−175.86)
$\gamma_D(g^+,g^-)$	−178.51 (−179.61)	80.21(77.96)	−37.71 (−45.24)	−169.55 (−176.74)
$\gamma_D(g^+,a)$	172.4 (168.47)	58.45 (51.9)	−27.74 (−28.66)	−173.04 (−177.43)
$\gamma_D(g^+,g^+)$	170.05 (171.24)	62.66 (62.93)	−18.59 (−40.34)	−176.01 (−177.71)
$\alpha_D(g^-,a)$	165.47 (170.43)	66.39 (60.25)	33.17 (37.6)	−173.43 (−179.9)
$\alpha_D(g^+,g^+)$	165.89 (167.53)	48.28 (46.39)	50.81 (53.58)	−170.33 (−177.4)
$\alpha_D(a,g^-)$	170.4 (174.86)	64.30 (62.32)	34.72 (34.14)	−174.34 (179.56)
$\alpha_D(a,g^+)$	172.39 (177.17)	66.54 (60.1)	39.84 (43.84)	−172.81 (−179.76)
$\delta_D(g^-,g^-)$	173.68 (173.92)	−148.44 (−157.92)	−50.65 (−51.78)	−171.96 (−178.56)
$\delta_D(a,g^+)$	175.77 (174.52)	−147.48 (−173.29)	−72.03 (−49.35)	−173.3 (−175.69)
$\delta_D(g^+,g^-)$	173.12 (173.74)	−153.88 (−159.3)	−62.82 (−67.57)	−172.52 (−179.08)
$\delta_D(g^+,a)$	−179.98 (179.87)	−154.02 (−163.74)	−58.32 (−63.28)	−173.26 (179.71)
$\delta_D(g^-,g^+) \Rightarrow \gamma_D(g^-,g^+)$	− (174.77)	− (146.26)	− (−33.93)	− (−175.62)
$\delta_D(a,a) \Rightarrow \beta_L(a,a)$	− (174.31)	− (−172.41)	− (−55.07)	− (−177.17)

[a]The backbone conformers are labeled according to the set of abbreviation introduced in the past[30]: α_L, α_D, β_L, γ_L, γ_D, δ_L, δ_D, ε_L and ε_D.

[b]The values in parenthesis are calculated at RHF/3-21G level of theory.

TABLE 13.3. Selected Conformational Parameters of HCO–L-Ala–NH_2 and For–L-Ser–NH_2 at RHF/6-311++G Levels of Theory**

	For–L-Ser–NH_2		Ideal		HCO–L-Ala–NH_2	
Conf.[a] averaged[c]	$\phi^{average}$	$\psi^{average}$	ϕ	ψ	ϕ	ψ
β_L 5 (6)	−159.5 (−165.9)[b]	173.7 (176.4)	−180.0	180.0	−155.1[d](−168.3)	161.0(170.6)
δ_L 2 (4)	−124.2 (−132.1)	14.4 (29.7)	−180.0	60.0	−112.8(−128.1)	13.2(29.8)
γ_L 5 (6)	−86.5 (−82.2)	73.7 (67.2)	−60.0	60.0	−86.2(−84.5)	78.8(67.3)
α_L 3 (3)	−79.1 (−68.3)	−22.4 (−30.5)	−60.0	−60.0	–	–
ε_L 0 (0)			−60.0	180.0	–	–
ε_D 5 (6)	65.8 (68.6)	−146.4 (−154.7)	60.0	−180.0	−(67.2)	−(−171.9)
γ_D 8 (9)	72.5 (69.8)	−47.7 (−47.9)	60.0	−60.0	75.3(74.0)	−55.4(−57.4)
α_D 4 (4)	61.4 (57.3)	39.6 (42.3)	60.0	60.0	69.0(63.8)	26.9(32.7)
δ_D 4 (6)	−151.0 (−173.4)	−61.0 (−53.5)	−180.0	−60.0	−165.2(−178.6)	−42.1(−44.1)

[a]Backbone conformers (ϕ and ψ in degree) were calculated at RHF/6-311++G** level of theory. ($\phi^{average}$ and $\psi^{average}$ are reported for serine diamides).

[b]Values in parentheses are the relevant values calculated at RHF/3-21G levels of theory.

[c]Number of RHF/6-311++G** side-chain conformers of HCO–L-Ser–NH_2 within the same backbone catchment region. Values in parentheses are those of RHF/3-21G.

[d]The HCO–L-Ala–NH_2 parameters are form Endrédi et al.[70]

more characteristic to the applied level of theory than to the polarity of the side chain of the amino acid residue.

Due to its flexible and polar side chain HCO−L-Ser−NH_2 has the potential to establish all the important forms of the intramolecular H-bonds operative in peptides and proteins. The side-chain functional hydroxy methyl group can be a proton donor as well as acceptor. Writing first the symbol of the acceptor followed by that of the donor-group three different interactions are possible: bb/bb, bb/sc and sc/bb. A detailed hydrogen bond analysis, focusing on patterns and stability, offers the possibility to compare the effect of the applied level of theories on these factors. The 3-21G basis set has the well known effect of over-emphasising the energetics of polar interactions. This presumption is strongly supported when analysing H-bond lengths and angles in the two γ-turns (γ_D and γ_L) (see Table 13.4). Comparing the seven-membered bb/bb-type H-bond parameters for both 8γ_D and 5γ_L structures not only the heavy atom- ($d_{O1...N2}$) but also the hydrogen bond- ($d_{O1...H-N2}$) distances are elongated by ≈0,2 Å by using a 6-311++G** basis set. Thus, in average the H-bond angle ($\theta_{O1\ H-N2}$) looks more bent, by an amount of 5° at RHF/6-311++G** level of theory.

In contrast to the strong interaction in γ-turns, β-structures (β_L) have rather weak hydrogen bonds. (This interaction incorporates the NH and the CO group within the same residue.) It is not surprising therefore, that the 6-311++G** H-bond parameters as compared with that of the 3-21G, show only minor shifts: $\Delta d_{O2...N1}$ and $\Delta d_{O2...H-N1}$ is less than ≈0.1 Å. From the comprehensive analysis of these two backbone forms incorporating the two extremes of H-bonds (a strong one and a weak one) it looks as if the H-bond parameters vary within the range of ≈0.2 Å as a function of the applied level of theory.

Beside the above two bb/bb H-bond patterns, a polar side-chain like that of serine diamide can form two additional intramolecular H-bond types: bb/sc and sc/bb. To establish some correlation between total energy changes and H-bond patterns looks more complicated than to compare changes of local geometrical parameters as function of the H-bond type. In general, the total energy shift between two conformers, calculated by using both 3-21G and 6-311++G** basis sets, is smaller when a sc/bb type hydrogen bond is present. Thus the basis set induced relative energy increase is more important in those conformers

TABLE 13.4. The Average Values of Backbone/Backbone Type Hydrogen Bond Parameters Determined in the Case of For−L-Ser−NH_2 by Using the RHF/6-311++G Levels of Theory**

Conf.[a]	$d_{O1...N2}$[b]	$d_{O1...H-N2}$	$\Theta_{O1\ H-N2}$
γ average	3.00 (2.83)[c]	2.16 (1.94)	140.9 (145.9)
β average	2.62 (2.59)	2.17 (2.07)	106.3 (109.8)

[a]For−L-Ser−NH_2 conformers. Distances (d) in Å and angles (Θ) are in degrees.
[b]The numbering of the atoms are according to that of Scheme 6.
[c]Values in parenthesis are those obtained at RHF/3-21G levels of theory.

TABLE 13.5. Backbone Side-chain Type Hydrogen Bond Parameters of For–L-Ser–NH_2 at RHF/6-311++G Levels of Theory**

H-bond Type		Heavy atom dist.	H-bond dist.	H-bond ang.
sc/bb	Atoms involved[a]	$d_{O1...O3}$[b]	$d_{O1...H\text{-}O3}$	$\Theta_{O1.H.O3}$
	Average values	2.9(2.7)[c]	2.2(1.9)	133.4(139.1)
sc/bb	Atoms involved[a]	$d_{O2...O3}$	$d_{O2...H\text{-}O3}$	$\Theta_{O2.H.O3}$
	Average values	2.9(2.8)	2.3(2.1)	122.2(125.7)
bb/sc	Atoms involved[a]	$d_{O3...N1}$	$d_{O3...H\text{-}N1}$	$\Theta_{O3\ H\ N1}$
	Average values	2.8(2.6)	2.5(2.1)	97.1(107.0)
bb/sc	Atoms involved[a]	$d_{O3...N2}$	$d_{O3...H\text{-}N2}$	$\theta_{O3\ H\ N2}$
	Average values	2.9(2.7)	2.2(1.9)	128.6(134.8)

[a]For–L-Ser–NH_2 conformers. Distances (d) in Å and angles (Θ) are in degrees.
[b]The numbering of the atoms are according to that of Scheme 6.
[c]Values in parenthesis are those obtained at RHF/3-21G levels of theory.

where fewer H-bonds are operative. Four backbone-side-chain interactions are now analysed in details at both levels of theory (Table 13.5).

The eight typical amino acid residues with polar side-chains (Ser, Thr, Asp, Asn, etc.) can be donors or acceptors by forming intramolecular H-bonds. In the serine model compound the hydroxy methyl side-chain group can donate its proton (abbreviated as sc/bb in Table 13.5) or accept that of an amide bond (labeled as bb/sc in Table 13.5) during such interaction. As found in the case of backbone–backbone type H-bond formation the pair-wise analysis of the H-bond parameters calculated at both levels of theory could reveal basis set induced effects. The distance and bond angle modifying effect of the basis set in these structures, in average, is within the previously mentioned ≈ 0.2 Å ($5°$) range.

Among the three different types of intramolecular hydrogen bonds (*bb/bb, sc/bb* and *bb/sc*) the third pattern, where one of the two backbone amide protons is donated to the oxygen atom of the side-chain, shows the most significant basis set dependence. A rather significant difference is observed whether the first or the second amide group is involved in the interaction. If the NH of the first (N1) amide group is involved in structures like $\delta_L(g^+, a)$, $\gamma_L(g^-, g-)$, $\gamma_L(g-, a)$, $\alpha_L(g-, a)$, $\alpha_L(g-, g-)$ the H-bond distance is extremely weakened when reoptimized at a larger basis set. The parameters calculated at RHF/6-311++G** level of theory (e.g. $d^{average}_{O3...H-N1} = 2.5$ Å, $\Theta^{average}_{O3...H-N1} = 97.1°$) signal two sings. First, a significant correlation between the applied method and the determined structural parameters, second the non-existence of this H-bond at least according to the classical terminology. Similar to the bb/bb type five-member hydrogen-bonds in β-structures, a polar interaction, although present in the above conformers, may not be hydrogen bonds in a classical term. On the other hand, if the NH of the second (N2) amide group is donated to the side-chain oxygen atom (O3), the interaction could be strong, even when values calculated at the larger basis set level. In the latter case, even at larger basis set,

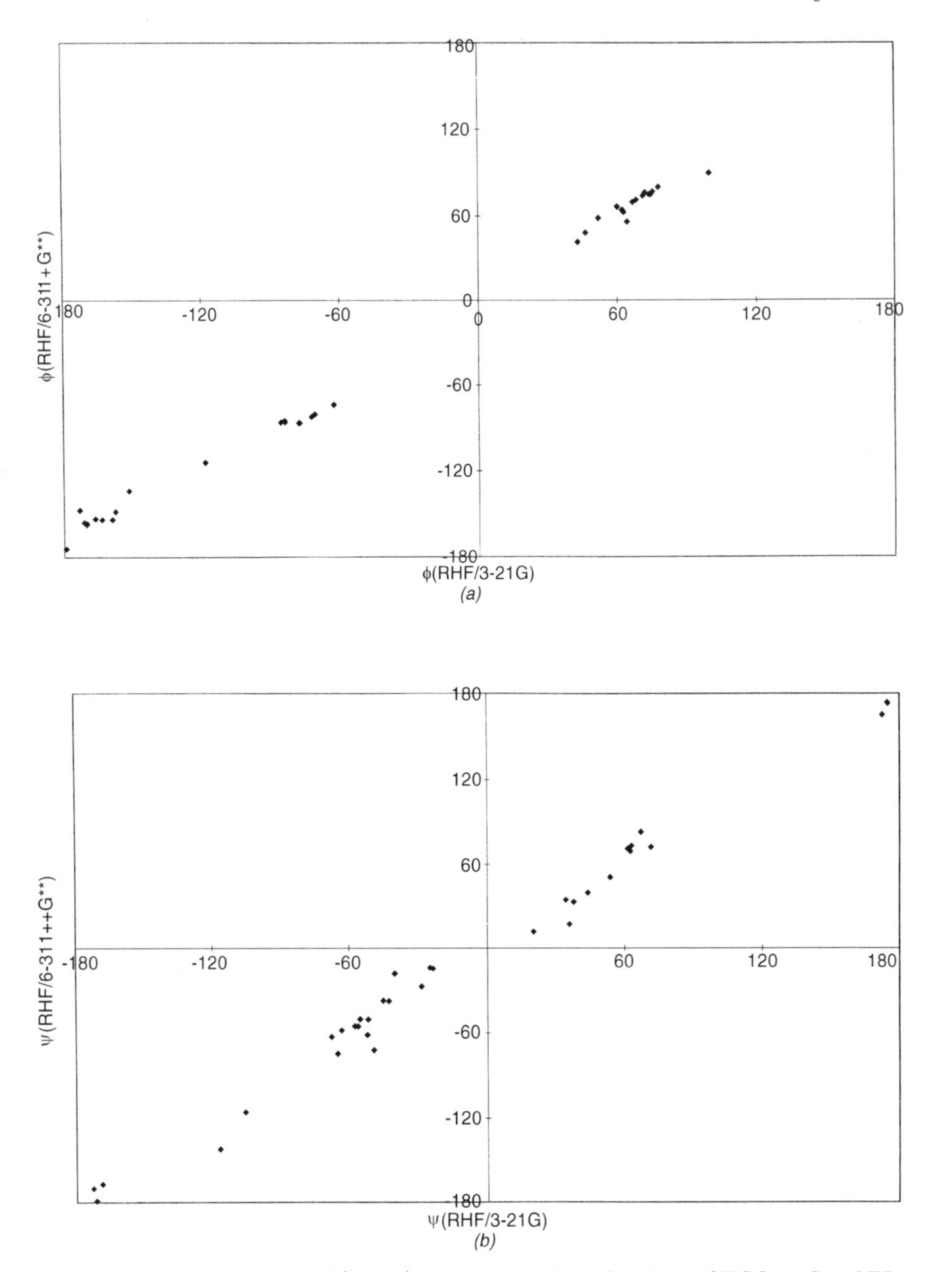

FIGURE 13.3. Correlation of ϕ (*a*), ψ (*b*), χ_1 (*c*) and χ_2 (*d*) values of HCO–L-Ser–NH_2 calculated at RHF/3-21G and RHF/6-311++G** levels of theory.

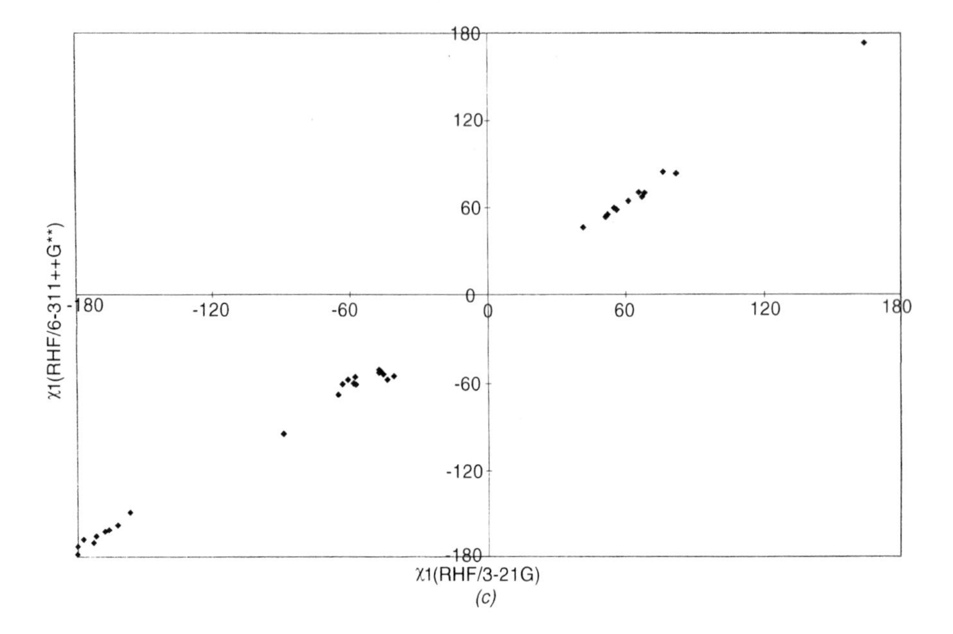

(c)

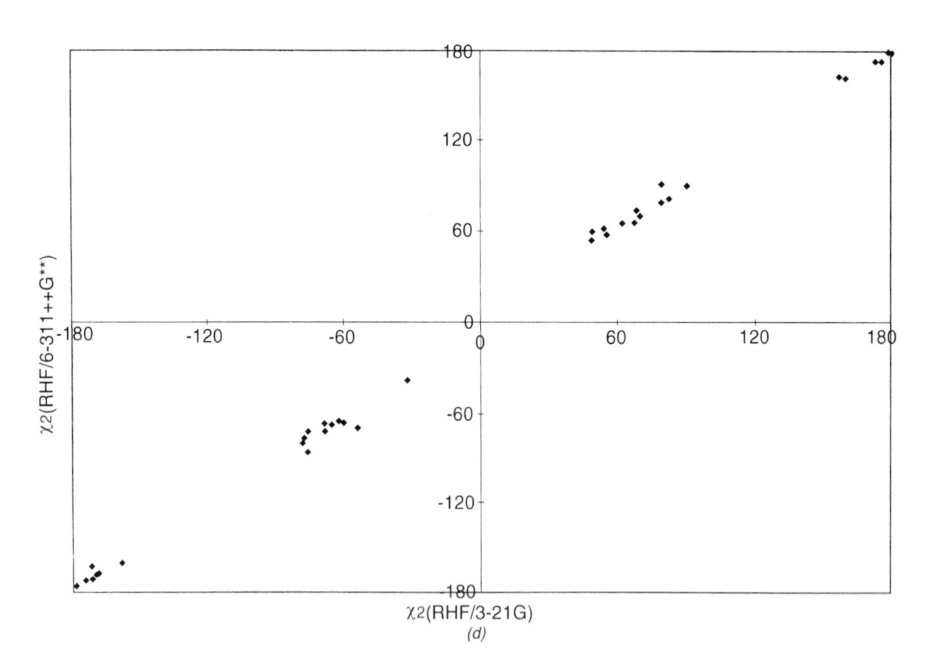

(d)

FIGURE 13.3. *(Continued)*

the average H-bond distance is favorable (e.g. $d_{O3...H-N2} = 2.2$ A). As a closing remark, it looks straightforward to say that the stability and the structural parameters of H-bonds depend on its size; e.g. five-, or seven-member H-bonds. In general it looks as if H-bond parameters are more relaxed and therefore such an interaction is less emphasised when a larger basis set is used.

7.2. Basis Set Dependence of Structure and Stability

The accuracy of any chemical and structural conclusion obtained, depends on the accuracy of the used data set. False conclusion could result from partially incorrect or inaccurate data. This is to be avoided at all costs. But, the possession of entirely perfect data is rather unlikely, so as good a conclusion has to be drawn as possible from the available data set. In ab initio structural investigations this question is more cumbersome than in most other scientific area, since a compromise on size and on the level of the applied theory is rather typical. We are facing this problem as pointed out in the introduction. The RHF/3-21G data cannot be numerically correct, even though qualitatively and perhaps even semi-quantitatively are acceptable. Yet, we wish to draw conclusions that could stand the test of time. For this reason it would make sense to attempt to calibrate our RHF/3-21G results against values computed by using a larger basis set. For structure reliability and transferability we wish to compare all torsional variables (ϕ, ψ, χ_1 and χ_2) of HCO–L-Ser-NH$_2$ obtained by using a larger (RHF/6-311++G**) and a smaller (RHF/3-21G) basis set. For the above torsional variables (Z), like for any other structural parameters, a linear relationship is expected:

$$Z^{\mathrm{RHF/6-311++G^{**}}} = m^* Z^{\mathrm{RHF/3-21G}} + b \tag{13.1}$$

The established correlation between the two different basis set for all four dihedral angles (shown in Fig. 13.3) are convincing. The fitted parameters, as reported in Table 13.6, summarise that the standard error (σ) is between 5° and 10° but not higher.

This result agrees quantitatively with previous conclusions established for example in the case of HCO–L-Ala–NH$_2$ (Table 13.1). First, it is to be mentioned that the established error limit ($< 10°$) for all conformational

TABLE 13.6. The Fitting Parameters of 36 HCO–L-Ser–NH$_2$ Conformers Calculated at RHF/6-311++G** and at RHF/3-21G Levels of Theory

Parameters[a]	ϕ	ψ	χ_1	χ_2
m	1.0	1.0	1.0	1.0
b	−1.7	0.0	−1.6	−0.5
σ	8.3	9.9	5.9	5.3

[a] m and b values are from Eq. 13.1. Standard error (σ) is in degree.

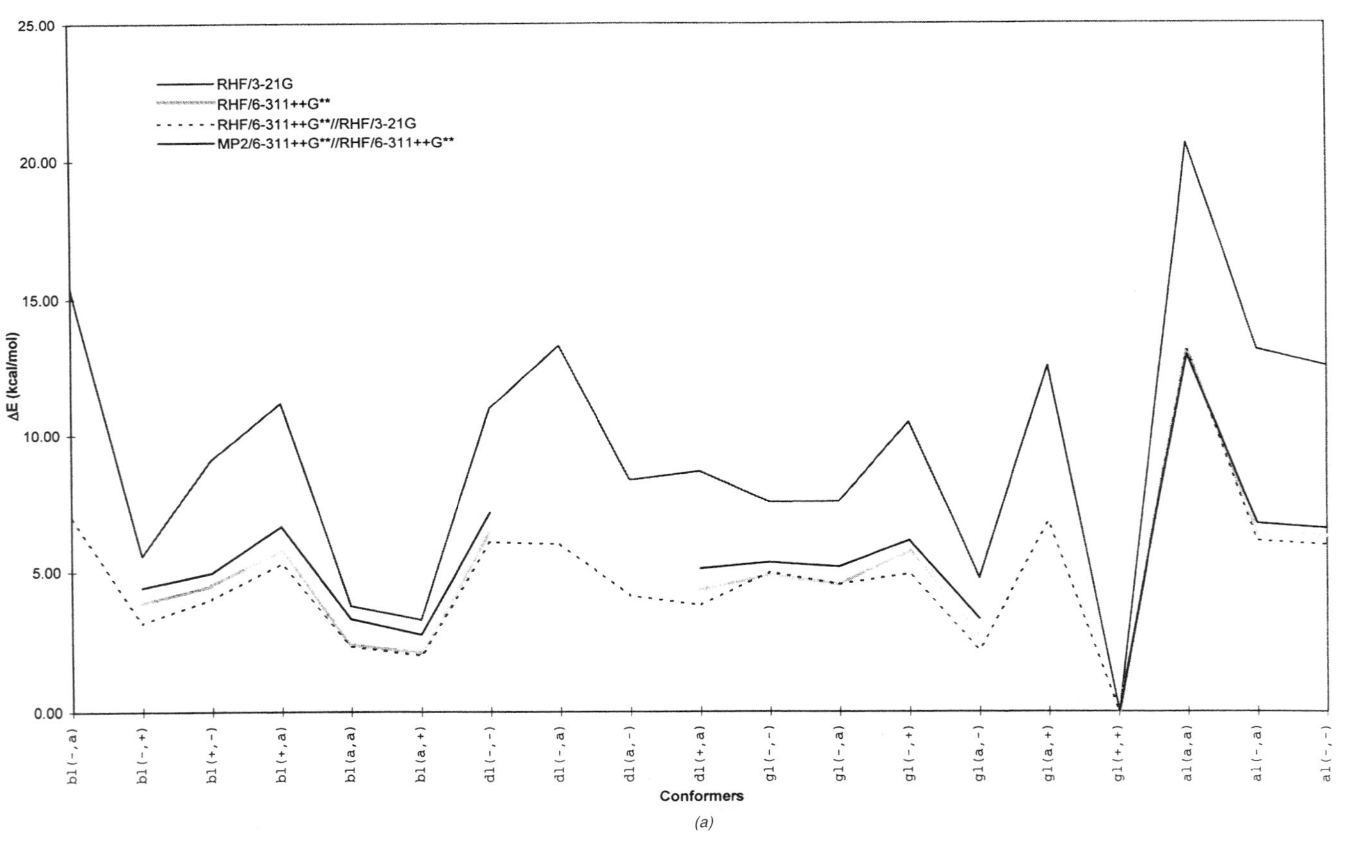

FIGURE 13.4. Relative energy differences of the HCO–Ser–NH_2 conformers calculated at RHF/3-21G, RHF/6-311++G**, RHF/6-311++G**//RHF/3-21G and MP2/6-311++G**//RHF/6-311++G** levels of theory. [(a) for conformers in the L-valley, (b) for conformers in the D-valley.)

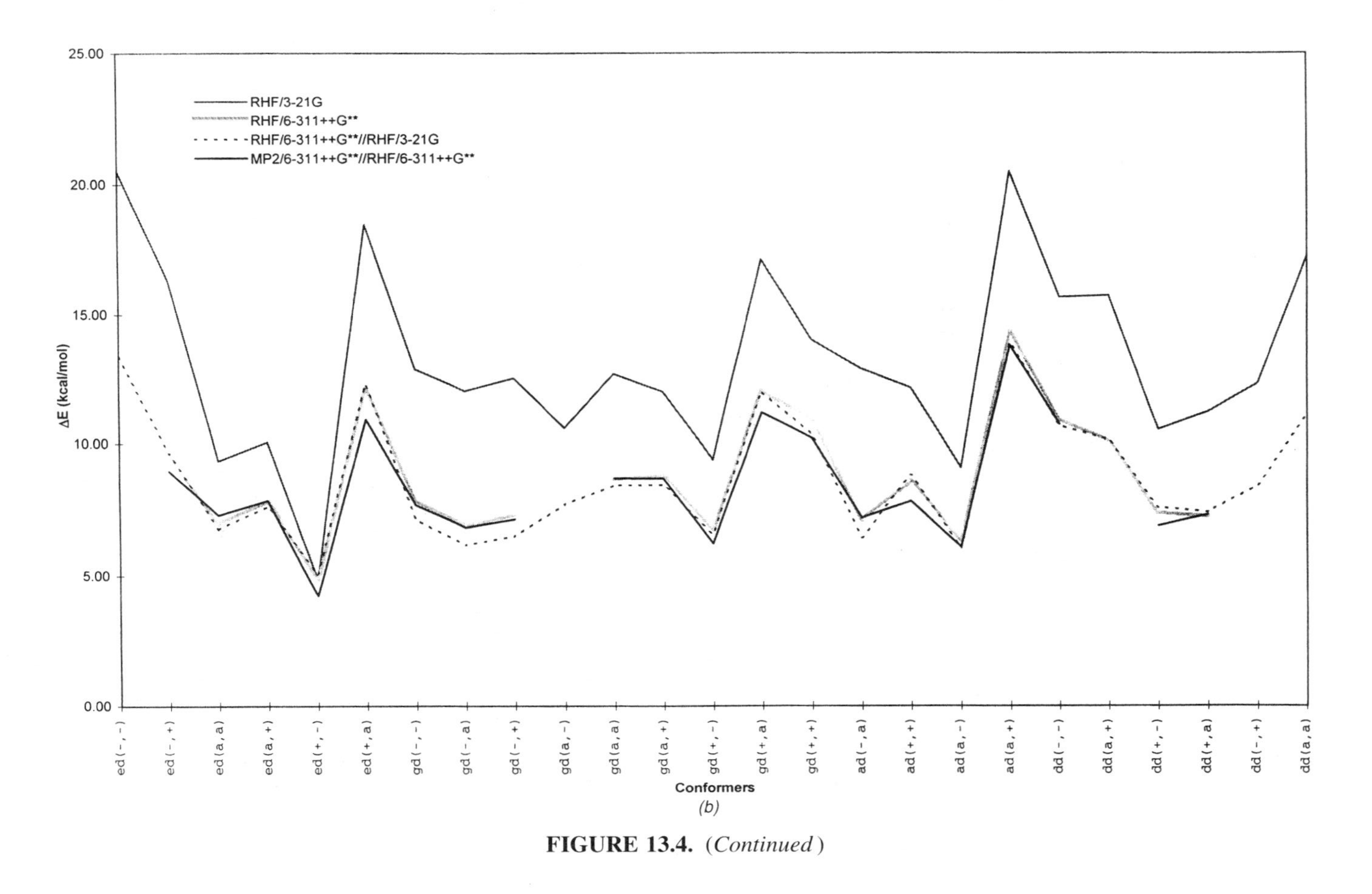

FIGURE 13.4. *(Continued)*

variables confirms that no dramatic change is expected when the level of theory is improved. Second, by using the above linear equation, the scaling factor is now available.

Besides the structure, the relative energy scaling would be of great importance. Comparing the relative energy differences ($\Delta E = E_{x\mathrm{L}} - E_{\gamma\mathrm{L}(g^+,g^+)}$) of the same conformers calculated at four levels of theory ($\Delta E_{\mathrm{RHF/3\text{-}21G}}$, $\Delta E_{\mathrm{RHF/6\text{-}311{+}{+}G^{**}}}$, $\Delta E_{\mathrm{RHF/6\text{-}311{+}{+}G^{**}//RHF/3\text{-}21G}}$, and $\Delta E_{\mathrm{MP2/6\text{-}311{+}{+}G^{**}//RHF/6\text{-}311{+}{+}G^{**}}}$) has been performed (Fig. 13.4). By referencing each and every energy values (E_x) to the appropriate global minimum ($E_{\gamma\mathrm{L}(g^+,g^+)}$), the relative energies (ΔE_x) were plotted as function of the conformation type (Fig. 13.4).

Before correlating the relative energies as a function of the applied level of theory, a quick overview of the plotted trend should be rewarding. All L-type backbone structures (x_{L}) (except $\alpha_{\mathrm{L}}(a,a)$) have lower relative energy than any structure that has D-type backbone orientation (x_{D}). The single exception is that of $\varepsilon_{\mathrm{D}}(g^+,g^-)$. Furthermore, ΔE values are typically smaller than 6 kcal mol^{-1} at RHF/6-311++G** level of theory, with an average value close to 4.9 kcal mol^{-1}. In contrast, for D-type backbone orientation the same value ($E_{xD} - E_{\gamma\mathrm{L}(g^+,g^+)}$) is significantly larger, with an average close to 8.7 kcal mol^{-1}. A similar qualitative picture was established for relative energies obtained at the other three levels of theory. This finding agrees with the results of statistical analysis performed on proteins. By using the X-ray determined structures of the macromolecules and focusing in general on the relative occurrence of the above structural building units of proteins, it is known that D-type backbone structures (found here as high energy conformers) are less frequent as compared to those of L-type backbone orientation. Structures from the latter type, more common in proteins, were found as low relative energy building units according to ab initio calculations. Monitoring the fluctuation of ΔE, determined by using a 3-21G and 6-311++G** basis sets ($\Delta E_{\mathrm{RHF/6\text{-}311{+}{+}G^{**}}}$ and $\Delta E_{\mathrm{RHF/3\text{-}21G}}$ values), a systematic energy decrease was established that favors the result obtained for a larger basis set (Fig. 13.4). By comparing values from single point energy calculations ($\Delta E_{\mathrm{RHF/6\text{-}311{+}{+}G^{**}//RHF/3\text{-}21G}}$) with those determined by geometry optimization using the same basis set ($\Delta E_{\mathrm{RHF/6\text{-}311{+}{+}G^{**}}}$) only a small difference can be noted (Fig. 13.4). This agrees with earlier findings that the conformational parameters do not change intensively when the molecule is reoptimized at a larger basis set. Consequently, the clear decrease in energy when moved from a 3-21G to a 6-311++G** basis set is mainly due to the enlarged basis set and not to the improved geometry. Due to a fortunate trend, in the case of serine one can attempt to condense the effect of basis set into a single value. To perform such a scaling the single point energies ($\Delta E^{\mathrm{RHF/6\text{-}311{+}{+}G^{**}//RHF/3\text{-}21G}}$) are to be compared with those determined by using a 3-21G basis set ($\Delta E^{\mathrm{RHF/3\text{-}21G}}$). Similarly the relative energies calculated by optimizing on a larger basis set ($\Delta E^{\mathrm{RHF/6\text{-}311{+}{+}G^{**}}}$) and those that includes even electron correlation ($\Delta E^{\mathrm{MP2/6\text{-}311{+}{+}G^{**}//RHF/6\text{-}311{+}{+}G^{**}}}$) are also analyzed in conjunction with the RHF/3-21G values. Once again, a linear relationship is expected. The fitted parameters are summarised in Table 13.7 while the correla-

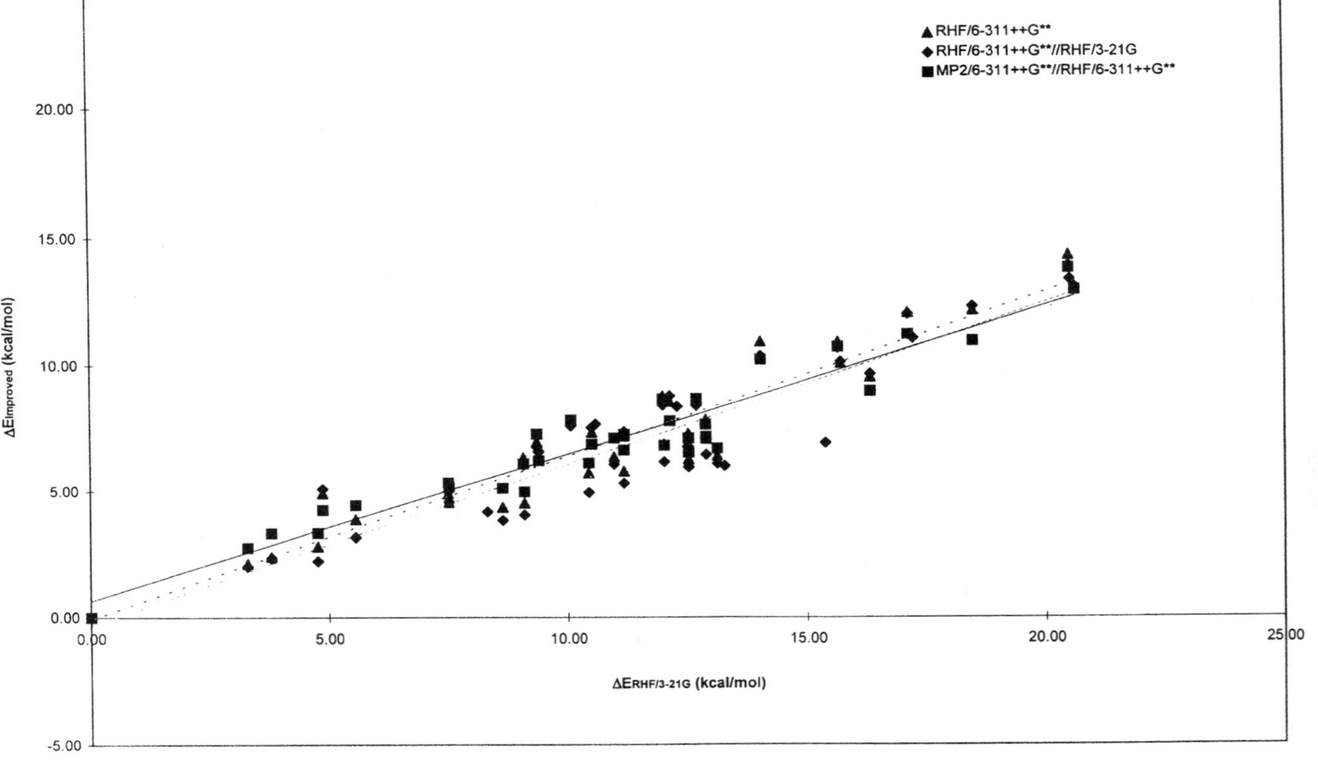

FIGURE 13.5. Correlation of the relative energies (at different levels of theory) in respect to the energy values determined at RHF/3-21G level of theory. (All available conformers of HCO–L-Ser–NH$_2$ are included.) Parameters of the fitted linear (dashed line) between $\Delta E^{RHF/6\text{-}311++G^{**}}$ and $\Delta E^{RHF/3\text{-}21G}$ (data points represented by triangle) are in Table 13.7. Similarly, values associated with the fitting of $\Delta E^{RHF/6\text{-}311++G^{**}//RHF/3\text{-}21G}$ and $\Delta E^{RHF/3\text{-}21G}$ (dotted line, diamond) as well as those of $\Delta E^{MP2/6\text{-}311++G^{**}//RHF/6\text{-}311++G^{**}}$ and $\Delta E^{RHF/3\text{-}21G}$ (solid line, squares) are also in Table 13.7.

TABLE 13.7. Fitting Parameters of HCO–L-Ser–NH_2 when ΔE Values are Obtained at MP2/6-311++G//RHF/6-311++G**, RHF/6-311++G**//RHF/3-21G, RHF/6-311++G** and at RHF/3-21G Levels of Theory. The Improved Method of Energy Calculation is Compared to Values Obtained from the Common RHF/3-21G Approach**

Improved Method	Number of Conformers	m	b	σ
RHF/6-311++G**//RHF/3-21G	44	0.6	0.3	1.2
RHF/6-311++G**	36	0.7	0.1	1.0
MP2/6-311++G**//RHF/6-311++G**	35	0.6	0.6	0.8

tions are shown in Fig. 13.5.

$$\Delta E^{\text{Improved}} = m^* \Delta E^{\text{RHF/3-21G}} + \text{b} \tag{13.2}$$

First comparing the relative energy shift in the case of RHF/6-311++G**// RHF/3-21G the following difference was established (in kcal mol^{-1}):

$$\Delta E^{\text{RHF/6-311++G**//RHF/3-21G}} = \Delta E^{\text{RHF/3-21G}} 4.5 \pm W \tag{13.3}$$

If the threshold ($|W|$) is set to be smaller than ≤ 1.0 kcal mol^{-1}, 11 conformers would be within the range. By increasing the allowed threshold values to $W \leq 2.1$ kcal mol^{-1}, then 34 structures would fall in. Similar type of estimation can also be established for the optimized RHF/6-311++G** results (in kcal mol^{-1}):

$$\Delta E^{\text{RHF/6-311++G**}} = \Delta E^{\text{RHF/3-21G}} - 4.0 \pm W \tag{13.4}$$

where $|W| \leq 1.0$ kcal mol^{-1} for 12 and $|W| \leq 2.1$ kcal mol^{-1} for 25 structures.

It would be too early to generalize these results. However, in the case of HCO–L-Ser–NH_2 by simply knowing the RHF/3-21G energies one can estimate the energy value of the same conformer obtained at a different level of theory (e.g. $E^{\text{RHF/6-311++G**}}$ or $E^{\text{RHF/6-311++G**//RHF/3-21G}}$). Although at this stage, the present finding is theoretical and perhaps has no practical utility at the present time, it envisages the possibility to parametrise at some future date stage a working force-field model for peptides and proteins with rather accurate features.

7.3. Correlation Between Natural Occurrence of Conformers and Computed Stability

Comparison of structural parameters from experimental databases (X-ray and/or NMR) with those obtained from ab initio results is a challenging possibility for computational chemistry. Thus, the comparison of relative energies and the relative probabilities for the same conformers by using a non-homologous database is also a potential technique for the cross-validation of the two approaches. To make the sizes comparable, first the backbone of a protein is truncated into conformational building units: e.g. amino acid diamides. To keep the modeling as simple as possible it will be assumed that the probability (p_x) of conformer x in proteins depends only on its relative energy. (Obviously in this model several well known phenomena are neglected, such as inter-residue interactions, long-range effects, hydration etc.). Adhering to the present limitations, it is possible to correlate, in a simple way, the relative energy of a conformer and its relative probability in an ensemble of proteins. Choosing $\gamma_L(g^+,g^+)$ as the reference conformer for the energy scale ($\Delta E = E_x - E_{\gamma L(g^+,g^+)}$), the relative population is assumed to be related to relative stabilities in a Boltzman type exponential distribution (Eq. (13.5)). Using the logarithmic form (Eq. (13.6)) of this equation, the correlation between relative energies and relative probabilities should be linear. In a rigorous way, as outlined above, (Eqs. (13.5) and (13.6)) are expected to be valid for HCO–L-Ser–NH$_2$ in the gas phase only. Thus, if the relative populations ($p_x/p_{\gamma L(g^+,g^+)}$) could be measured in the gas phase (e.g. by FTIR) and accurate ΔE could be calculated by ab initio methods, including electron correlation, then the above linear equation should hold. However, we are attempting to correlate the solid state occurrence of serine conformers taken from globular proteins with energy values obtained for an isolated HCO–L-Ser–NH$_2$ model compound. In the event of any significant correlation between the above to data sets we could conclude that the diamide model is a useful approach in some sense, and perhaps it can even reproduce selected features of protein folding. A protein data set described elsewhere[74,75] was used in which as many as 9511 Ser residues were found in 1135 proteins. Each and every conformational parameter (ϕ, ψ, χ_1 and χ_2) from RHF/6-311++G** and RHF/3-21G calculations pinpoints the center of a 4D hyper-sphere. The experimentally determined serine conformers were clustered by using the ab initio determined geometries as cluster centres. All clusters had a shape similar to a 4D hypersphere with an actual size determined by preset values of radius ($\Delta\phi$, $\Delta\psi$, $\Delta\chi_1$ and $\Delta\chi_2$). Setting the radius as 60°, 45° and 30° a different number of serine residues (called as sum in Table 13.8) was assigned within the selected conformational area. Since these hyperspheres can overlap, the number of serine residues found in these overlapping regions (called overlap in Table 13.8) was considered.

$$\left(\frac{p_x}{p_{\gamma L(g^+,g^+)}}\right) = e^{+\frac{\Delta E}{m}} \cdot e^{-\frac{b}{m}} \tag{13.5}$$

$$\Delta E = m \cdot \ln\left(\frac{p_x}{p_{\gamma L(g^+,g^+)}}\right) + b \tag{13.6}$$

As outlined, the limitations of the present approximation make correlation problematic, and this is manifested in the relatively large values of standard

TABLE 13.8. The Fitting Parameters Between Relative Energies of Serine Diamide Conformers Compared to $\gamma_L(g^+,g^+)$ and the Relative Probabilities of Similar Backbone Structures in a Set of Proteins with Known X-ray Structure

Size[a]	Method[b]	m^c	b^d	Stand. error[e]	Sum[f]	Overlap[g]
60	RHF/6-311++G**	−0.84	6.67	2.80	10622	4395
60	RHF/6-311++G**//RHF/3-21G	−0.46	6.69	3.09	10374	4364
60	RHF/3-21G	−0.59	11.09	4.53	10374	4364
45	RHF/6-311++G**	−0.66	7.04	2.96	7214	2640
45	RHF/6-311++G**//RHF/3-21G	−0.42	6.90	3.15	7074	2640
45	RHF/3-21G	−0.61	11.49	4.46	7074	2640
30	RHF/6-311++G**	−0.63	7.05	3.00	3637	1314
30	RHF/6-311++G**//RHF/3-21G	−0.37	6.90	3.21	4027	1375
30	RHF/3-21G	−0.28	11.48	4.59	4027	1375

[a]The radius of the hypersphere pinpointed by the ϕ, ψ, χ_1 and χ_2 conformational values calculated with the given method.
[b]The used ab initio level of theory to determine the ϕ, ψ, χ_1 and χ_2 conformational parameters and the relative energies.
[c]The slope of the fitted curve.
[d]The $\Delta E(0)$ value.
[e]Standard error in kcal mol^{-1}.
[f]The sum of serines residues incorporated in the analysis.
[g]The number of overlapping serine residues between clusters.

error. To our surprise, the enlargement of the allowed region (the increase of the volume of the hypersphere) does not improve significantly the standard error of the fitting (Table 13.8). This indicates that even for the smallest data set (obtained with a radius of 30°) the distribution of the 3827 ± 200 serine residues among the conformational clusters is close to that calculated for more than 10000 residues. On the other hand, the standard error varies markedly with the change of the applied levels of theory (RHF/3-21G, RHF/6-311++G**//RHF/3-21G, and RHF/6-311++G**). This implies that the precise locations of the hyper-spheres (ϕ, ψ, χ_1 and χ_2 values centering the conformational regions) or the relative energy differences changes significantly with the applied method. The fact that the 6-311++G** ($E^{\mathrm{RHF/6\text{-}311++G^{**}}}$) and the single point ($E^{\mathrm{RHF/6\text{-}311++G^{**}//RHF/3\text{-}21G}}$) energy values are close to each other offer an explanation. Monitoring the variation of the standard errors of the line-fitting as function of the applied ab initio strategy (Table 13.8) it is convincing that the improved energy and not the improved geometry is responsible for the better fitting. The present finding is in good agreement with previous conclusions:

- RHF/3-21G geometries for peptides, due to fortuitous cancellations, provide good structural data,
- single point calculations furnish inexpensive and rapid ways to get valuable information on molecular energy.

Although promising, this strategy should be further tested on other amino acid residues and on additional model sizes.

8. SURFACE ANALYSES

For small molecules containing a single rotor only (e.g. H_3C–CHO), the conformational analysis requires a 1D (i.e. single variable) function to visualize the energy cost of a conformational change:

$$E = E(\phi).$$

Conformational analysis extended to systems containing several rotors (e.g. peptides with torsion angles ϕ,ψ,ω, etc.) leads to the investigation of potential energy hypersurfaces (PEHS):

$$E = E(\phi, \psi, \omega, \ldots).$$

In terms of mathematical formalism, the quintessence of analyzing a 1D- or an nD-space is not much different and requires only some "expertise" in mathematical analysis of multivariable functions. However, in practice the visualization of conformational motions associated with a hypersurface is rather hopeless. The qualitative concept of a generalized or multidimensional conformational analysis (MDCA), using the HCO–L-Ser–NH_2 molecules as an example, will be used to demonstrate how to overcome the practical problems of not being able to plot and "see" a 4D conformational space in which energy is a function of four variables: $E = E(\phi,\psi,\chi_1,\chi_2)$.

It was shown previously that if three conformers per torsional variables are expected for all rotors within the molecule, then for a 2D-PES $\{E(x_1, x_2)\}$, a maximum of nine, while for n dimension $\{E[x]\}$ a maximum of 3^n zero-gradient minimum energy points are expected. Furthermore, an nD-surface has not only zero gradient minimum energy points but also further types of critical points. The index (λ) of a critical point is the number of negative eigen-values of the Hessian matrix (collecting the second partial derivatives of the energy with respect to $3n$-6 internal coordinates). Thus $\lambda = 0$ is always a minimum, $\lambda = 1$ is a first order saddle point (transition structure) and $\lambda = 2$ is a second order critical point which is in fact a maximum on an ordinary PES with two independent variables. In summary, such an idealized conformational potential energy surface has nine minima ($N_0 = 9$), eighteen saddle points ($N_1 = 18$) and nine maxima ($N_2 = 9$), yielding a grand total of $9 + 18 + 9 = 36$ critical points.

However, for a non-ideal system, some of those conformers may disappear. In other words, some of the *legitimate conformers*, which are expected to be present on an ideal surface are missing from the real (non-ideal) surface. This phenomenon is referred to as the *annihilation of critical points*. In contrast to the above, the appearance of critical points, which are not expected to be present on an ideal surface, is called the *creation of critical points*. Not surprisingly, critical points cannot just be created or annihilated on a conformational PES. These occurrences are governed by strict selection rules[76] which are usually specified in terms of the indices (λ) of the critical points involved.

Since the purpose of this chapter is to demonstrate how to perform and how to visualize conformational analysis some qualitative rules and their consequences must be placed in focus. In terms of critical point changes, a brief overview of moving from an ideal to a non-ideal surface is summarized below. The selection rule "in action" is demonstrated in the case of annihilation and creation of critical points. To make it as simple as possible, a PES of two independent variables are used. Mountain ridges are denoted now by solid lines while valley floors are plotted by using broken lines.

Two minima (with critical point of index 0) separated by a saddle point ($\lambda = 1$), can merge into a single minimum as shown by path A of Scheme 7. The "fusion" of two saddle points, neighboring a common minimum, can result in a single saddle point (Path B of Scheme 7). If the two saddle points surround a maximum then the result can be a single saddle point (Path C of Scheme 7). On Path D of Scheme 7, the way how the two maxima separated from each other by a saddle point may collapse into a single maximum is demonstrated. These selection rules are based on the periodicity of torsional angles, which demand that higher and lower indexed critical points must appear or disappear by pairs. If this condition is not met, the 360° periodicity is not maintained and the PES

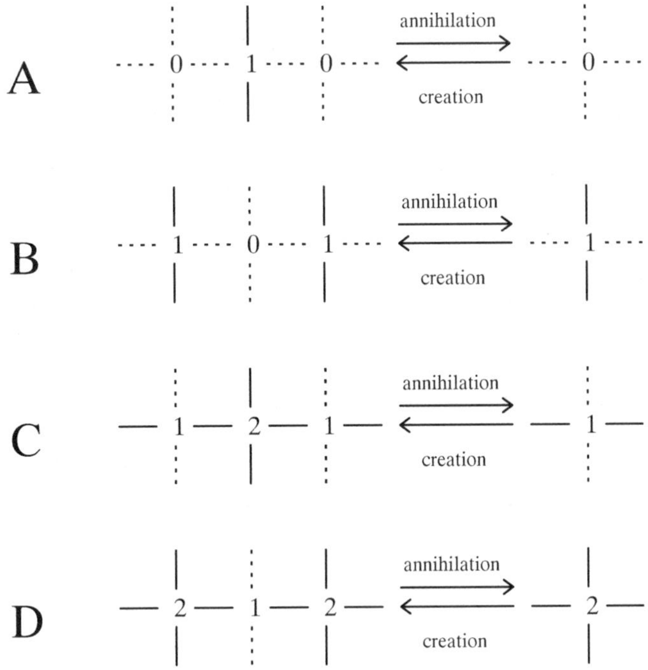

A ···· 0 ···· 1 ···· 0 ···· $\underset{\text{creation}}{\overset{\text{annihilation}}{\rightleftarrows}}$ ···· 0 ····

B ···· 1 ···· 0 ···· 1 ···· $\underset{\text{creation}}{\overset{\text{annihilation}}{\rightleftarrows}}$ ···· 1 ····

C — 1 — 2 — 1 — $\underset{\text{creation}}{\overset{\text{annihilation}}{\rightleftarrows}}$ — 1 —

D — 2 — 1 — 2 — $\underset{\text{creation}}{\overset{\text{annihilation}}{\rightleftarrows}}$ — 2 —

Scheme 7

ceases to describe conformational changes which are, by definition, always periodic.

8.1. The Classical Example

The topological analyses of the $E(\phi,\psi)$ surface of HCO–L-Ala–NH_2 is one of the first step to take. As pointed out (Scheme 2C) for this model system, in the case of ideal behavior, the conformational potential energy surface should have nine minima ($\lambda = 0$), eighteen saddle points ($\lambda = 1$) and nine maxima ($\lambda = 2$), resulting in a total of $9+18+9=36$ critical points. Preliminary ab initio calculations revealed that some of the expected minima may not be present.[67] Upon counting of the seven minima of HCO–L-Ala–NH_2, calculated at the 3-21G/RHF level of theory, the obvious difference becomes striking. Two minima, those of the expected α_L and ε_L, were missing. When Head-Gordon first reported in 1991[42] the 3-21G/RHF surface of alanine diamide, the stationary point calculations were justified; indeed these two expected minima were not present. It appears that in the case of the smaller methyl (alanine), the larger isopropyl (valine) or for the aromatic side-chain containing i.e., benzyl (phenylalanine) diamide models, these two stationary points are missing. It looks as if some unfavorable interactions between the atoms of the two amide moieties destabilize these two main-chain orientations; namely that of α-helix

(α_L) and polyproline II (ε_L) secondary structures. First, focusing on the conformational building unit of the right handed helix, this region of interest (around $\phi = -54°$, $\psi = -45°$) has been reoptimized at higher levels of theory by using a more rigorous grid search, but the results were still the same. These observations initiated further, more complex studies.[28,32–34,42,77–79] Considering the importance of the PES of HCO–L-Ala–NH_2 let's investigate the number of critical points on this ab initio Ramachandran surface ($0° \leq \phi \leq +360°$, $0° \leq \psi \leq +360°$ calculated at a step size of 30°). The idealized as well as the computed surface topology is shown in Scheme 8.

Head-Gordon et al.[42] and Perczel et al.[32] optimized independently at the same time all the existing seven minima. Furthermore Head-Gordon et al.[42] have also calculated twelve transition structures, labeled from **8** to **19** (Scheme 9). Nobody has yet optimized the five maxima which would be the second order saddle points (i.e. critical points with $\lambda = 2$). The topology of the PES, showing seven minima ($\lambda = 0$), twelve first order saddle points ($\lambda = 1$) or transition structures with the conformational reaction coordinates, are also shown by heavy lines in Scheme 9.

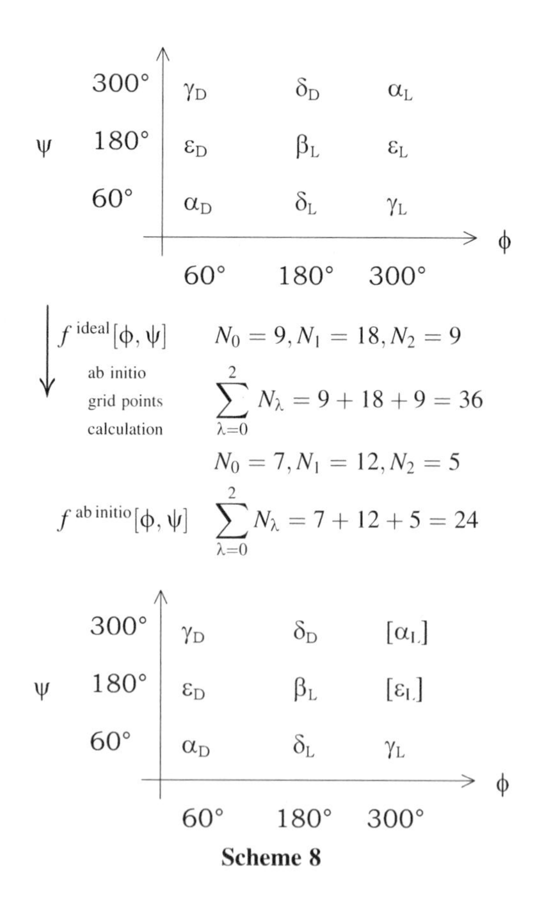

Scheme 8

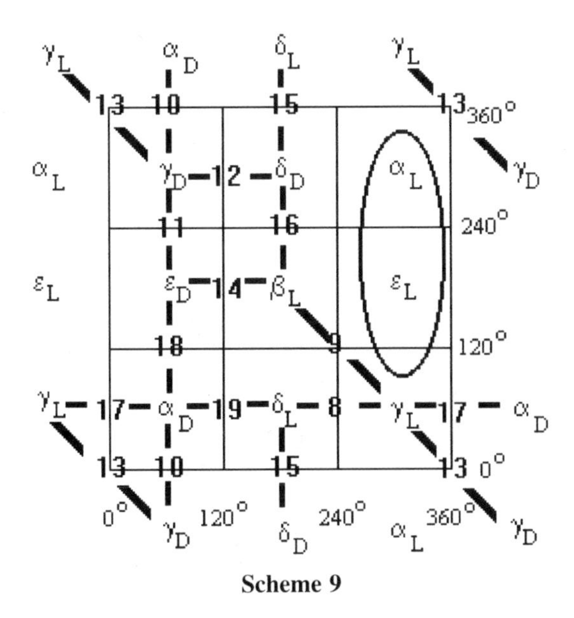

Scheme 9

8.2. Conformational Analysis of PES of Ser, Cys, Asp, Asn, Phe, Tyr, His, Trp as Typical Examples of MDCA

The use of MDCA is equally relevant for the conformational analyses of backbone as well as for side-chain PES(s), especially since in peptides both types of rotors have rather similar multiplicity; 3^n in the case of n independent dihedral variables. The side chains (all $Z{-}C^{\beta}H_{2-}$) of Ser, Cys, Asp, Asn, Phe, Tyr, His and Trp residues contain two side-chain conformational variables (χ_1 and χ_2). These are expected to have a maximum number of nine different conformers. On the one hand, three stable orientations are expected from the rotation about the $C^{\alpha}{-}C^{\beta}$ covalent bond (χ_1 torsional angle) and on the other hand, each and every minimum from this class is coupled with an additional set, perhaps three minima arising from the three possible orientations associated with the rotation about the $C^{\beta}{-}Z$ bond (χ_2 torsional angle). (Note that Gly, Ala and Pro have fewer side-chain conformers, while the other nine natural amino acid residues could have more than expected for the above eight amino acid residues.)

In an idealized case, the locations of the above nine unique minima ($\lambda = 0$) on a side-chain surface are presumed to be: g^+g^+; [60°,60°], g^+a; [60°,180°], ..., g^-g^-[−60°,−60°]. On the same surface, where an idealized topological behavior is expected, nine unique maxima (M_A; [120°,0°], M_B; [120°,120°], M_C; [120°,240°],... M_H; [360°,120°], M_I; [360°,240°]) should be present. The topology of such an f^{ideal} [χ_1,χ_2] is shown in Scheme 10. (The nine unique maxima ($\lambda = 2$) are circled, the uncircled maxima are repetitions of these critical points of index 2.). There are 18 transition structures ($\lambda = 1$) located at the edges

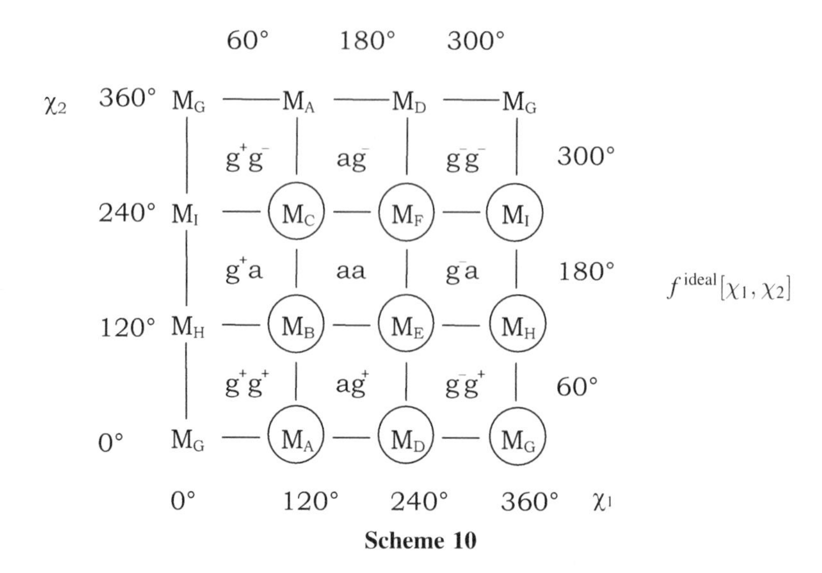

Scheme 10

of the nine squares. These are understood to be present although not shown in Scheme 10. Due to specific interactions between sidechain/sidechain or sidechain/backbone atoms, the annihilation of some sidechain conformers may well be expected.[40,50,52] The annihilation pattern depends on the type of backbone conformation.[32,54]

From the detailed analysis of X-ray data of peptides and proteins we know that the preference of side-chain conformation of an amino acid residue could depend on its backbone conformation.[16,23,24,80–82] The backbone dictated side-chain modifying effects may be due to both intra-residual and inter-residual interactions. However, the preference of inter-residual interactions, typical in the solid states, make it impossible to separate and assess the conformation modifying effect of the intra-molecular type of interactions. On the other hand, the ab initio calculations described here investigate purely intra-residue side-chain/backbone conformation modifying effects. These results are strictly correct for the vacuum state. However, they can model the conformational feature of amino acid residues located at the hydrophobic core region of a globular protein, where a low local dielectric constant is expected and where water molecules are normally excluded.

8.3. Surface Cross-Sections of Hypersurfaces

The direct calculation of a 4D-hyperspace could be lengthy, rather expensive and it may not even be needed. We are about to show in the case of HCO–L-Ser–NH_2 how this tedious grid-search covering the entire 4D-hyperspace can be reduced. Although to explore this hyperspace we use data points calculated at RHF/3-21G level of theory, we think that the level of theory applied has no intrinsic limitation on the overall approach. More sophisticated surfaces, when available, should be analyzed just like this.

The comprehensive structure analyses of the alternative side-chain geometries associated with a common backbone conformation type α_L or β_L or γ_L etc. resulted in characteristic [ϕ,ψ] values at the RHF/3-21G level of theory. The evaluation of these ab initio conformers provided characteristic backbone torsional angles (i.e. ϕ and ψ) used for the systematic grid search of the side-chain conformational subspace:

$$E = E_{[\phi,\psi]}\,[\chi_1, \chi_2]$$

Side-chain maps composed of 144 grid points were generated using a 30° incrementation along both χ_1 and χ_2. Since these maps are associated with a particularly fixed backbone (bb) conformation, they are called *backbone fixed side-chain maps* ($f_{\mathrm{bb}}^{\mathrm{fixed}}\,[\chi_1,\chi_2]$), even though [(3$n$-6)-4] internal coordinates are fully relaxed. The values of the fixed ϕ and ψ values are summarized below:

bb	ϕ	ψ
α_L	−60	−40
β_L	−170	170
γ_L	−84.5	68.7
δ_L	−125	30
ε_L	−95	150

In order to obtain "zero gradient" surfaces, [(3n-6)-2] internal coordinates had to be fully relaxed. These surfaces are called *backbone relaxed side-chain maps* ($f_{\mathrm{bb}}^{\mathrm{relaxed}}\,[\chi_1,\chi_2]$). In order to avoid ambiguities, it perhaps should be emphasized, that the $f_{\mathrm{bb}}^{\mathrm{relaxed}}\,[\chi_1,\chi_2]$ data points were originating from the corresponding backbone fixed side-chain map grid points ($f_{\mathrm{bb}}^{\mathrm{fixed}}\,[\chi_1,\chi_2]$) by relaxing ϕ and ψ for each and every $[\chi_1,\chi_2]$ values. The above technique is "direction independent," since the mapping procedure has been started from a given side-chain conformation. Otherwise the i-th grid point of the map would depend on the $(i-1)$th grid point in term of the ϕ,ψ values.

8.4. The Example of HCO–L-Ser–NH_2 Model Compound

Let us restrict this discussion to those five side-chain maps $f_{\mathrm{bb}}^{\mathrm{fixed}}\,[\chi_1,\chi_2]$ that are associated with typical L-type backbone orientations: α_L, β_L, γ_L, δ_L and ε_L). First, the characteristic ϕ,ψ values are kept constant throughout the calculations of the grid points. After determining the shape of these *fixed* surfaces $f_{\mathrm{bb}}^{\mathrm{fixed}}$ $[\chi_1,\chi_2]$, as a second step the appropriate ϕ,ψ values were relaxed. On going from $f_{\mathrm{bb}}^{\mathrm{fixed}}\,[\chi_1,\chi_2]$ to $f_{\mathrm{bb}}^{\mathrm{relaxed}}[\chi_1,\chi_2]$ the ϕ,ψ values are optimized, thus they may change even qualitatively, leading over to another catchment region. Thus, the corresponding $f_{\rm bb}^{\rm relaxed}$ $[\chi_1,\chi_2]$ surfaces could have a mixed backbone character e.g., "islands" of alternative backbone types may appear on the grid.

$$f_{\gamma L}^{\rm fixed}[\chi_1,\chi_2] \rightarrow f_{\gamma L}^{\rm relaxed}[\chi_1,\chi_2]$$
$$f_{\beta L}^{\rm fixed}[\chi_1,\chi_2] \rightarrow f_{\beta L}^{\rm relaxed}[\chi_1,\chi_2]$$
$$f_{\varepsilon L}^{\rm fixed}[\chi_1,\chi_2] \rightarrow f_{\varepsilon L}^{\rm relaxed}[\chi_1,\chi_2]$$
$$f_{\alpha L}^{\rm fixed}[\chi_1,\chi_2] \rightarrow f_{\alpha L}^{\rm relaxed}[\chi_1,\chi_2]$$
$$f_{\delta L}^{\rm fixed}[\chi_1,\chi_2] \rightarrow f_{\delta L}^{\rm relaxed}[\chi_1,\chi_2]$$

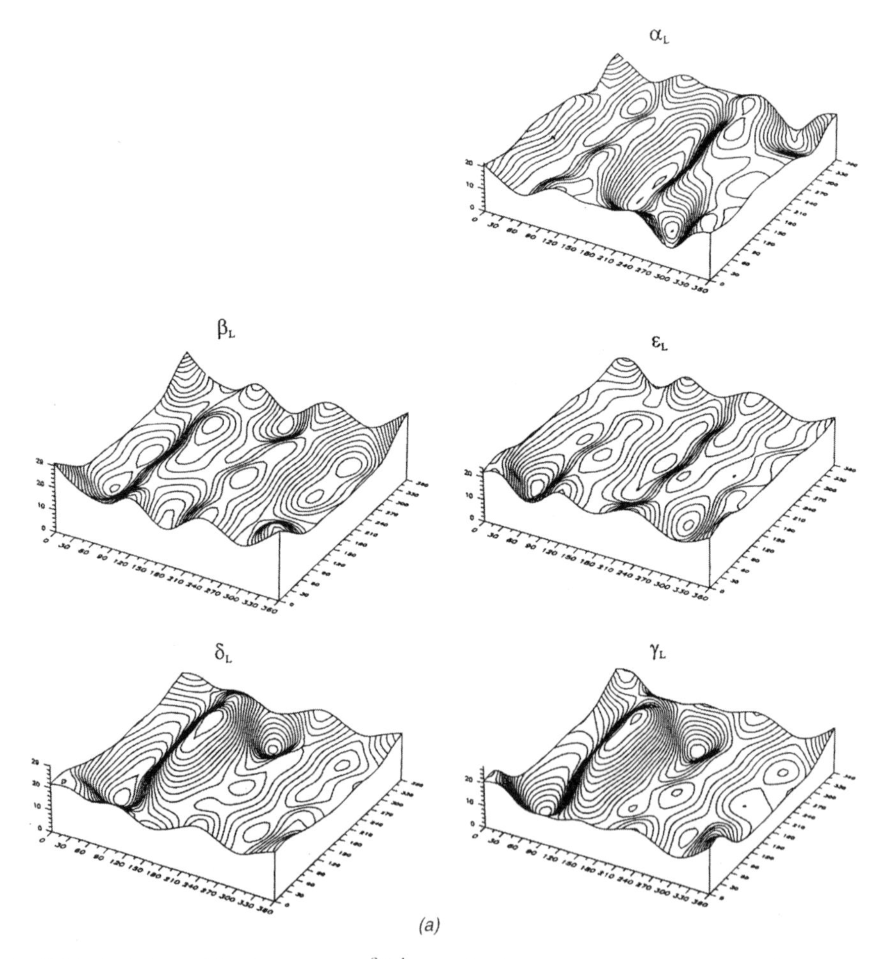

FIGURE 13.6. (*a*) The five fixed ($f_x^{\rm fixed}[\chi_1,\chi_2]$) side chain maps associated with HCO–L-Ser–NH_2 calculated at RHF/3-21G level of theory. (*b*) The five relaxed ($f_x^{\rm relaxed}[\chi_1,\chi_2]$) side chain maps associated with HCO–L-Ser–NH_2 calculated at RHF/3-21G level of theory.

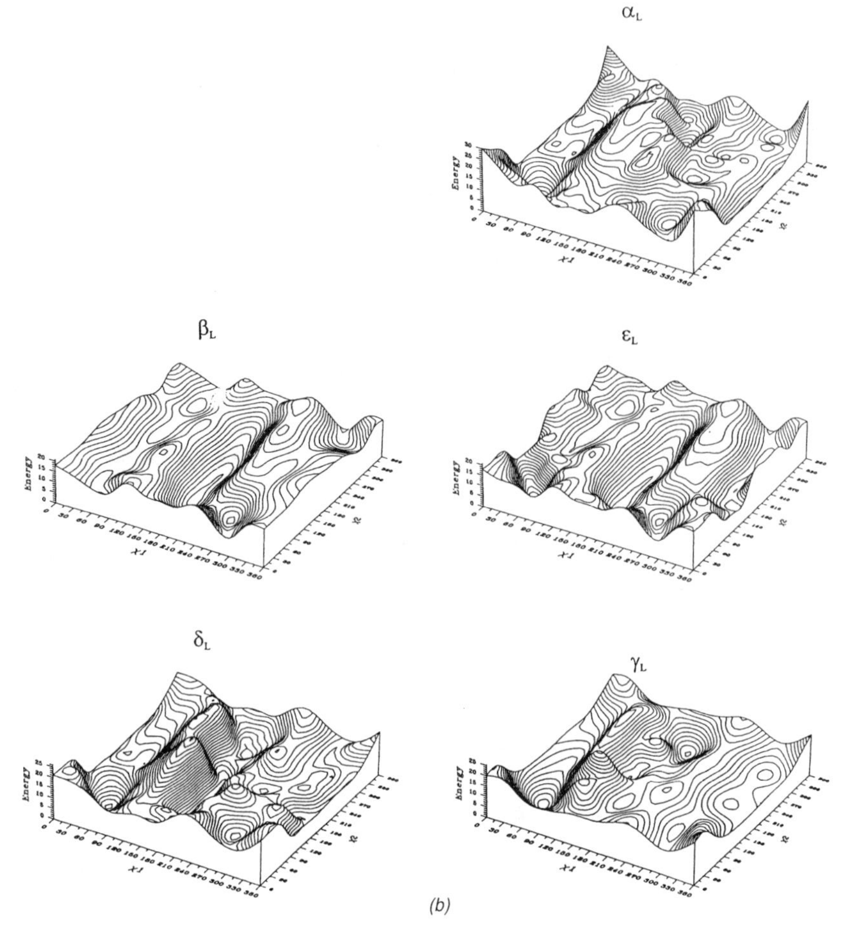

(b)

FIGURE 13.6. (*Continued*)

Through counting the number of the different critical points and *via* the comprehensive analysis of the $f_{\mathrm{bb}}^{\mathrm{fixed}}[\chi_1,\chi_2]$ and $f_{\mathrm{bb}}^{\mathrm{relaxed}}[\chi_1,\chi_2]$ surfaces intrinsic features of the molecule will be revealed.

First, the two ab initio surfaces associated with the most stable backbone orientation, that of γ_L ($f_{\gamma L}^{\mathrm{fixed}}[\chi_1,\chi_2]$, $f_{\gamma L}^{\mathrm{relaxed}}[\chi_1,\chi_2]$) are explored. A grand total of 22 critical points were counted on the $f_{\gamma L}^{\mathrm{fixed}}[\chi_1,\chi_2]$ surface (Fig. 13.6(a)). Six minima ($N_0 = 6$), eleven saddle points ($N_1 = 11$) and five maxima ($N_2 = 5$) were identified (Scheme 11) on the $f_{\gamma L}^{\mathrm{fixed}}[\chi_1,\chi_2]$ surface. In contrast to this, on the $f_{\gamma L}^{\mathrm{relaxed}}[\chi_1,\chi_2]$ side-chain map (Fig. 13.6(b)) which is associated with relaxed backbone torsional angles 32 critical points were found. A net increase of 10 critical points were observed, just by relaxing the ϕ,ψ values. Eight minima ($N_0 = 8$), 16 saddle points ($N_1 = 16$) and eight maxima ($N_2 = 8$) were located on the "relaxed" surface (Scheme 11). The net increase in numbers of critical

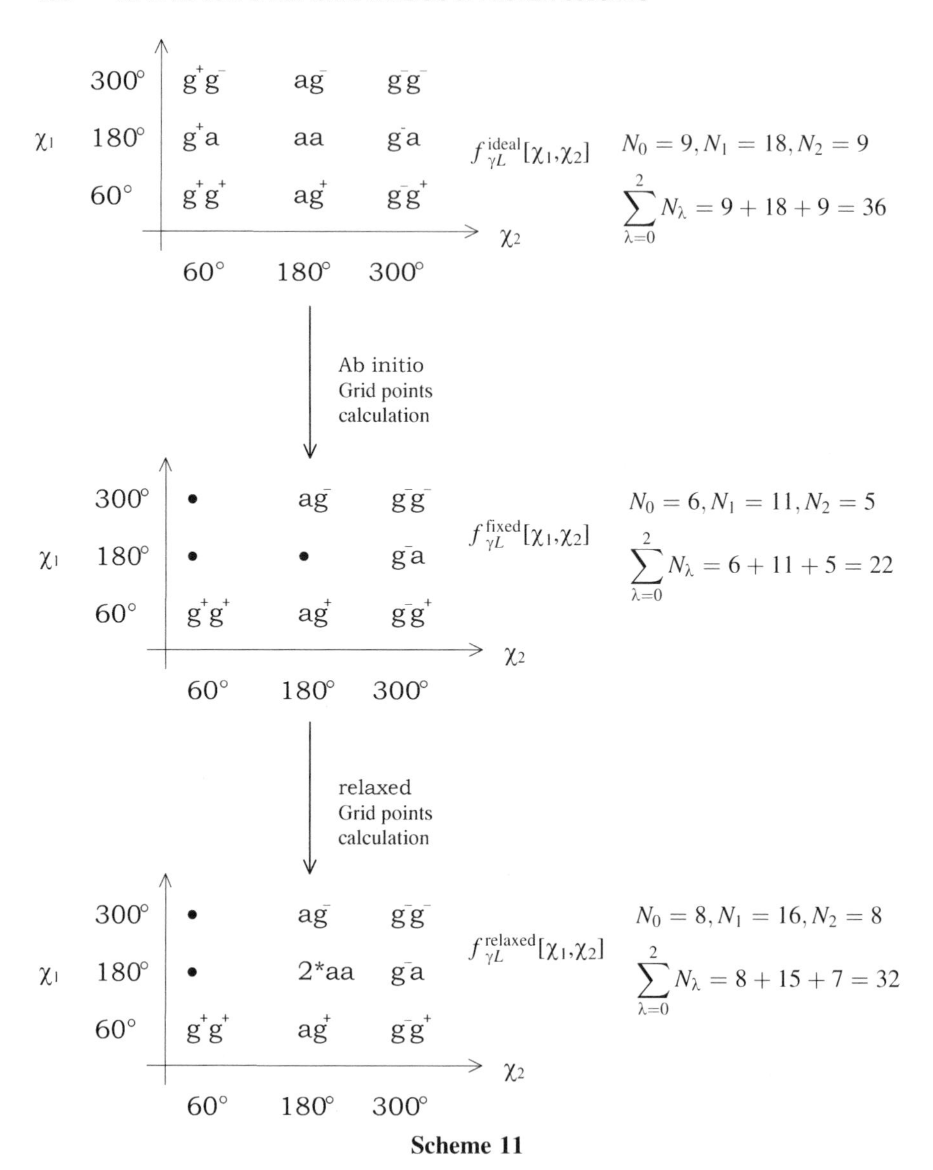

Scheme 11

points indicates that the $f_{\gamma L}^{relaxed}[\chi_1,\chi_2]$ surface is more "crammed" than that of the $f_{\gamma L}^{fixed}[\chi_1,\chi_2]$.

All changes, responsible for the above 45% of increase in the total number of critical points were localized on one side of the PES at a well defined elliptical region close to $[\chi_1 = 120°, \chi_2 = 180°]$. To understand this, first we have to compare the number of critical points in the same elliptical region in the case of the $f_{\gamma L}^{ideal}[\chi_1,\chi_2]$ and that of the $f_{\gamma L}^{fixed}[\chi_1,\chi_2]$ maps. A net decrease is observed. The original surface (Scheme 12(I)), which has an ideal topological behavior, has three maxima (M_A, M_B and M_C) separated from each other by three saddle points denoted by 1. The $f_{\gamma L}^{\text{fixed}}[\chi_1,\chi_2]$ side-chain surface of HCO–L-Ser–NH_2 has only a single large maximum (M_B) (cf. Scheme 12(II)) with one saddle point ($\lambda = 1$). The saddle point at the top and the one at the bottom are the same. On the other hand, in the same elliptical region, the $f_{\gamma L}^{\text{relaxed}}[\chi_1,\chi_2]$ surface has a grand total of twelve critical points (four maxima, six saddle points and two minima). Therefore, keeping the backbone at a fixed conformation will decrease the total number of critical points from the ideal case, while relaxing the backbone it will increase the number of critical points from that of the rigid surface toward the ideal topological character. The number of maxima as well as the number of saddle points and minima are affected. The splitting of the [aa] minimum into two closely spaced minima [2*aa] (Scheme 11) is observed. These two minima are denoted as 0,0 on Scheme 12(III). Such duplicating split-up of a minimum usually occurs on a PES in order to reduce excessive repulsion between moieties which became too close in a particular conformation.

Here also, an unfavorable side-chain conformation is the probable source of this major topological modification (M_A-1-M_B-1-$M_C \rightarrow M_B$) observed in the $\chi_1 = 120°$ and $\chi_2 = 180°$ region of the $f_{\gamma L}^{\text{fixed}}[\chi_1,\chi_2]$ side-chain map. The subsequent relaxation of the ϕ and ψ internal coordinates results in the splitting-up of this "composite" M_B maximum into four maxima (M'_A, M''_A, M'_B and M'_C). Such an analysis shows that the relaxation of the backbone torsional angles affects the side-chain orientation of the HCO–L-Ser–NH_2 molecule. When the backbone is kept fixed ($f_{\gamma L}^{\text{fixed}}[\chi_1,\chi_2]$) fewer relaxation paths are available to avoid unfavorable side-chain conformations. On the other hand, if relaxation is

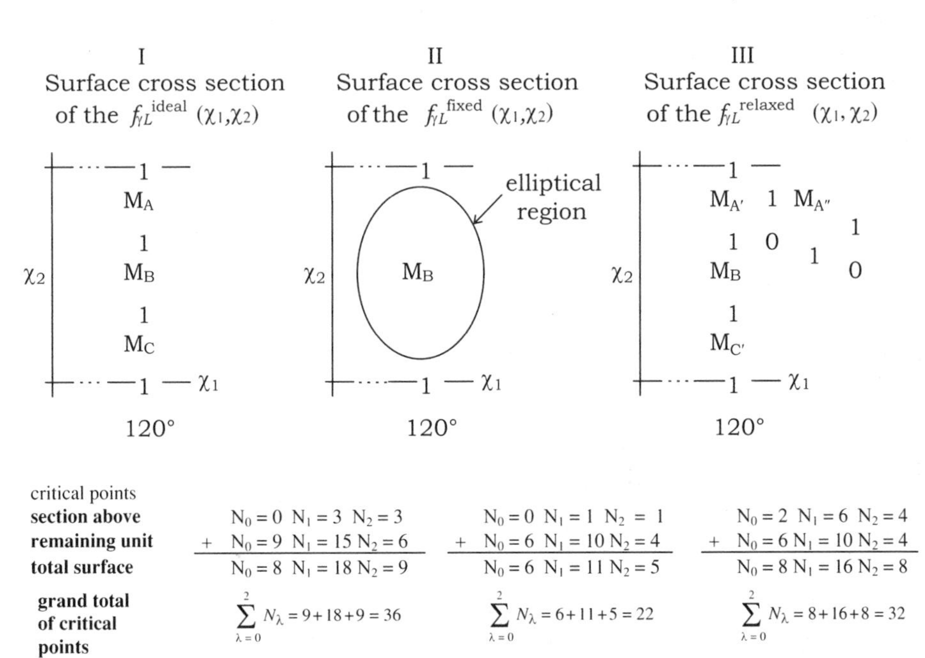

Scheme 12

allowed throughout the backbone, a larger variety of relaxation modes are made available to the molecular system. The observed topological changes are indicators of the significant impact of the side-chain on the orientation of the main-chain atoms, because most regions of the $f^{\text{relaxed}}_{\gamma L}[\chi_1,\chi_2]$ surface the backbone conformation remains an inverse gamma turn (γ_L) structure. This indicates that most of the side-chain conformations do not interfere with a γ_L structure. However, in the above mentioned elliptical region ($90° \leq \chi_1 \leq 150°$, $0° \leq \chi_2 \leq 360°$) due to side-chain induced effect, the backbone orientation is shifted. In the middle of the $f^{\text{relaxed}}_{\gamma L}[\chi_1,\chi_2]$ surface, where the side-chain conformation is close to an *anti,anti* ($120° \leq \chi_1 \leq 150°$, $150° \leq \chi_2 \leq 270°$) orientation, the relaxation has a qualitative effect on the backbone (Scheme 13).

As a conclusion, depending on the value of χ_2, the side-chain conformational twist at $\chi_2 = 120°$ results in the shift from γ_L to β_L- or from γ_L to δ_L-type backbone structure. More plausibly the γ_L surface has a leak both toward the appropriate section of the β_L and that of the δ_L surfaces. The present coupling between these three surfaces is the quantitative description of what is called "side-chain induced backbone effect" by stereo-chemists. This phenomenon is not all that different from the well recognized "stereo-electronic-effect" observed at rather unexpected and structurally unrelated areas.[83]

By analyzing the β_L relaxed surface ($f^{\text{relaxed}}_{\beta L}[\chi_1,\chi_2]$) the immediate question to be answered is the following: is this "leaking" mutual? It is of interest to determine, if the "hyperchanel" driving conformers from γ_L to β_L has its pair to allow a backward-type conformational shift. In the case of the β_L backbone conformer a total of 22 critical points could be counted on the $f^{\text{fixed}}_{\beta L}[\chi_1,\chi_2]$ surface (Fig. 13.6(A)). Five minima ($N_0 = 5$), 11 saddle points ($N_1 = 11$) and six maxima ($N_2 = 6$) were identified (Scheme 14). The same conformational area of the $f^{\text{relaxed}}_{\beta L}[\chi_1,\chi_2]$ side-chain map (Fig. 13.6(B)), which is associated with

χ_2 \ χ_1	0°		60°		120°		180°		240°		300°		360°
360°	γ_L	γ_L	γ_L	γ_L	γ_L	γ_L	γ_L	γ_L	γ_L	γ_L	γ_L	γ_L	γ_L
	γ_L	γ_L	γ_L	γ_L	γ_L	γ_L	γ_L	γ_L	γ_L	γ_L	γ_L	γ_L	γ_L
300°	γ_L	γ_L	γ_L	γ_L	γ_L	γ_L	γ_L	γ_L	γ_L	γ_L	γ_L	γ_L	γ_L
	γ_L	γ_L	γ_L	γ_L	β_L	γ_L	γ_L	γ_L	γ_L	γ_L	γ_L	γ_L	γ_L
240°	γ_L	γ_L	γ_L	γ_L	β_L	γ_L	γ_L	γ_L	γ_L	γ_L	γ_L	γ_L	γ_L
	γ_L	γ_L	γ_L	γ_L	β_L	β_L	γ_L	γ_L	γ_L	γ_L	γ_L	γ_L	γ_L
180°	γ_L	γ_L	γ_L	γ_L	β_L	β_L	γ_L	γ_L	γ_L	γ_L	γ_L	γ_L	γ_L
	γ_L	γ_L	γ_L	γ_L	β_L	β_L	γ_L	γ_L	γ_L	γ_L	γ_L	γ_L	γ_L
120°	γ_L	γ_L	γ_L	γ_L	γ_L	γ_L	γ_L	γ_L	γ_L	γ_L	γ_L	γ_L	γ_L
	γ_L	γ_L	γ_L	γ_L	δ_L	γ_L	γ_L	γ_L	γ_L	γ_L	γ_L	γ_L	γ_L
60°	γ_L	γ_L	γ_L	γ_L	γ_L	γ_L	γ_L	γ_L	γ_L	γ_L	γ_L	γ_L	γ_L
	γ_L	γ_L	γ_L	γ_L	γ_L	γ_L	γ_L	γ_L	γ_L	γ_L	γ_L	γ_L	γ_L
0°	γ_L	γ_L	γ_L	γ_L	γ_L	γ_L	γ_L	γ_L	γ_L	γ_L	γ_L	γ_L	γ_L

Scheme 13

relaxed backbone torsional angles, has 26 critical points. Six minima ($N_0 = 6$), 13 saddle points ($N_1 = 13$) and seven maxima ($N_2 = 7$) were encountered (Scheme 14). By surprise, the relaxation of the β_L backbones ($f_{\beta L}^{\text{fixed}}[\chi_1,\chi_2] \rightarrow f_{\beta L}^{\text{relaxed}}[\chi_1,\chi_2]$) has a less dramatic effect than observed previously for γ_L. In the case of β_L only a minimum and a maximum are split introducing four new critical points (one minimum, two saddle points and one maximum) during the individual relaxation of each and every surface point.

Most important about the $f_{\beta L}^{\text{relaxed}}[\chi_1,\chi_2]$ surface is its "monotony." Almost all grid points (143 out of 144) are associated with the same β_L-type backbone

$f_{\beta L}^{\text{ideal}}[\chi_1,\chi_2]$

χ_1 \ χ_2	60°	180°	300°
300°	g^+g^-	ag^-	g^-g^-
180°	g^+a	aa	g^-a
60°	g^+g^+	ag^+	g^-g^+

$$N_0 = 9, N_1 = 18, N_2 = 9$$
$$\sum_{\lambda=0}^{2} N_\lambda = 9 + 18 + 9 = 36$$

↓ ab initio grid points calculation

$f_{\beta L}^{\text{fixed}}[\chi_1,\chi_2]$

χ_1 \ χ_2	60°	180°	300°
300°	g^+g^-	•	•
180°	g^+a	aa	•
60°	g^+g^+	•	g^-g^+

$$N_0 = 5, N_1 = 11, N_2 = 6$$
$$\sum_{\lambda=0}^{2} N_\lambda = 5 + 11 + 6 = 22$$

↓ relaxed grid points calculation

$f_{\beta L}^{\text{relaxed}}[\chi_1,\chi_2]$

χ_1 \ χ_2	60°	180°	300°
300°	g^+g^-	•	•
180°	g^+a	aa	g^-a
60°	g^+g^+	•	g^-g^+

$$N_0 = 6, N_1 = 13, N_2 = 7$$
$$\sum_{\lambda=0}^{2} N_\lambda = 6 + 13 + 7 = 26$$

Scheme 14

structure. If there were no other backbone conformer type than that of the β_L on the $f^{\text{relaxed}}_{\beta L}[\chi_1,\chi_2]$ surface, the β_L structure would be a "conformational dead-end-street." The $[\chi_1 = 90°, \chi_2 = 0°$ or $360°]$ side chain orientation forms a unique example where a side-chain rotamer is associated with a none β_L-like backbone structure. Thus, at the $[\chi_1 = 90°, \chi_2 = 0°$ or $360°]$ point of the relaxed surface the backbone conformation is γ_L ($\phi = -84.96°$, $\psi = +84.75°$). In such a way not only γ_L can be transformed into β_L, but also β_L conformation can be switched back into γ_L. It is important to note that the two types of backbone transformations occur at two different side-chain orientations, so the two "hyper-channels" are not the same.

In contrast to β_L, the ε_L backbone conformation is not stable for our model system, therefore the relaxation of the $f^{\text{fixed}}_{\varepsilon L}[\chi_1,\chi_2]$ surface is of great importance for a better understanding of this feature of serine diamide. If we count the number of critical points on a side PES associated with the fixed ε_L backbone conformer ($\phi = -95.0°$, $\psi = +150.0°$), $f^{\text{fixed}}_{\varepsilon L}[\chi_1,\chi_2]$, we find eight maxima ($N_0 = 8$), 15 saddle points ($N_1 = 15$) and seven maxima ($N_2 = 7$), according to our ab initio results. Thus, a sum of 30 critical points can be counted on the $f^{\text{fixed}}_{\varepsilon L}[\chi_1,\chi_2]$ surface (Fig. 13.6(a)). On the relaxed $f^{\text{relaxed}}_{\varepsilon L}[\chi_1,\chi_2]$, side-chain map (Fig. 13.6(b)), a total of 36 critical points were found; seven minima ($N_0 = 7$), 18 saddle points ($N_1 = 18$) and 11 maxima ($N_2 = 11$). Thus when the backbone torsional angles (ϕ and ψ) are allowed to relax ($f^{\text{fixed}}_{\varepsilon L}[\chi_1,\chi_2] \rightarrow f^{\text{relaxed}}_{\varepsilon L}[\chi_1,\chi_2]$) a more complex surface is obtained. Investigating the different backbone-conformational shifts, it looks obvious that for the HCO–Ser–NH_2 molecule, the ε_L backbone structure is not a favored orientation (Scheme 16).

Among the 144 grid points less than eight percent remains ε_L, while the vast majority ($\sim 80\%$) is transformed into the neighboring β_L structure. In the case of the remaining 12% a $\varepsilon_L \rightarrow \gamma_L$ conformational-shift is observed. These two

χ_2 \ χ_1	0°		60°		120°		180°		240°		300°		360°
360°	β_L	β_L	β_L	γ_L	β_L	β_L	β_L	β_L	β_L	β_L	β_L	β_L	β_L
	β_L	β_L	β_L	β_L	β_L	β_L	β_L	β_L	β_L	β_L	β_L	β_L	β_L
300°	β_L	β_L	β_L	β_L	β_L	β_L	β_L	β_L	β_L	β_L	β_L	β_L	β_L
	β_L	β_L	β_L	β_L	β_L	β_L	β_L	β_L	β_L	β_L	β_L	β_L	β_L
240°	β_L	β_L	β_L	β_L	β_L	β_L	β_L	β_L	β_L	β_L	β_L	β_L	β_L
	β_L	β_L	β_L	β_L	β_L	β_L	β_L	β_L	β_L	β_L	β_L	β_L	β_L
180°	β_L	β_L	β_L	β_L	β_L	β_L	β_L	β_L	β_L	β_L	β_L	β_L	β_L
	β_L	β_L	β_L	β_L	β_L	β_L	β_L	β_L	β_L	β_L	β_L	β_L	β_L
120°	β_L	β_L	β_L	β_L	β_L	β_L	β_L	β_L	β_L	β_L	β_L	β_L	β_L
	β_L	β_L	β_L	β_L	β_L	β_L	β_L	β_L	β_L	β_L	β_L	β_L	β_L
60°	β_L	β_L	β_L	β_L	β_L	β_L	β_L	β_L	β_L	β_L	β_L	β_L	β_L
	β_L	β_L	β_L	β_L	β_L	β_L	β_L	β_L	β_L	β_L	β_L	β_L	β_L
0°	β_L	β_L	β_L	γ_L	β_L	β_L	β_L	β_L	β_L	β_L	β_L	β_L	β_L

Scheme 15

χ_2 \ χ_1	0°		60°		120°		180°		240°		300°		360°
360°	β_L	β_L	ε_L	γ_L	β_L	β_L	β_L	β_L	β_L	β_L	β_L	β_L/ε_L	β_L
	β_L	β_L	β_L	β_L	γ_L	β_L	β_L	β_L	β_L	β_L	β_L	β_L	β_L
300°	β_L	β_L	β_L	β_L	β_L	β_L	β_L	β_L	β_L	β_L	β_L	γ_L	β_L
	ε_L	β_L	β_L	β_L	β_L	β_L	β_L	β_L	β_L	β_L	β_L	γ_L	ε_L
240°	ε_L	ε_L	β_L	β_L	β_L	β_L	β_L	β_L	β_L	β_L	β_L	γ_L	ε_L
	ε_L	ε_L	β_L	β_L	β_L	β_L	β_L	β_L	β_L	β_L	β_L	γ_L	ε_L
180°	ε_L	ε_L	β_L	β_L	β_L	β_L	β_L	β_L	β_L	β_L	β_L	γ_L	ε_L
	γ_L	ε_L	β_L	β_L	β_L	β_L	β_L	β_L	β_L	β_L	β_L	γ_L	γ_L
120°	γ_L	ε_L	β_L	β_L	β_L	β_L	β_L	β_L	β_L	β_L	β_L	γ_L	γ_L
	γ_L	γ_L	ε_L	β_L	β_L	β_L	β_L	β_L	β_L	β_L	β_L	β_L	γ_L
60°	β_L	γ_L	γ_L	β_L	β_L	β_L	β_L	β_L	β_L	β_L	β_L	β_L	β_L
	β_L	γ_L	γ_L	γ_L	β_L	β_L	β_L	β_L	β_L	β_L	β_L	β_L	β_L
0°	β_L	β_L	ε_L	γ_L	β_L	β_L	β_L	β_L	β_L	β_L	β_L	β_L/ε_L	β_L

Scheme 16

backbone structural interconversions are not as surprising since both the $\varepsilon_L \rightarrow \beta_L$ and the $\varepsilon_L \rightarrow \gamma_L$ conformational modifications affect only one torsional variable.* Thus the shift led to both the β_L and γ_L conformers which are nearest neighbor conformers to ε_L. During the $\varepsilon_L \rightarrow \beta_L$ shift only the ϕ torsional angle is rotated ($\phi_{\varepsilon L} \approx -75° \rightarrow \phi_{\beta L} \approx -160°$). The $\varepsilon_L \rightarrow \gamma_L$ alteration influences only the value of $\psi(\psi_{\varepsilon L} \approx +150° \rightarrow \psi_{\gamma L} \approx +75°)$. As already discussed, the final consequence is that none of the optimized side-chain conformation has an ε_L backbone structure. This is also clear from Scheme 16 and Fig. 13.6 since the 11 or 12 points with ε_L backbone conformation fall into a region where there is no minimum energy conformation on the surface (Fig. 13.6(b) / ε_L).

We have seen already that β_L and γ_L interconvert mutually and that ε_L conformation is a "highway" toward these two types of backbone orientations. We have shown also that *via* a narrow channel the way from $\gamma_L \rightarrow \delta_L$ is also open (Scheme 13, $\phi \approx 120°$ and $\psi \approx 90°$). However, we have not yet found a single allowed transformation toward an α_L structure. The fixed side-chain potential energy surface ($f^{\text{fixed}}_{\alpha L}[\chi_1,\chi_2]$) associated with the right-handed helical building unit (α_L where $\phi = -60°$, $\psi = -40°$) of the For–L-Ser–NH_2 molecule has five minima ($N_0 = 5$), 11 saddle points ($N_1 = 11$) and six maxima ($N_2 = 6$). Thus a grand total of 22 critical points can be encountered on the $f^{\text{fixed}}_{\alpha L}[\chi_1,\chi_2]$ surface (Fig. 13.6(a)). The same region of the $f^{\text{relaxed}}_{\alpha L}[\chi_1,\chi_2]$ side-chain map (Fig. 13.6(b)), associated with relaxed backbone torsional angles has 46 critical points. Among them ten minima ($N_0 = 10$), 23 saddle points ($N_1 = 23$) and 13 maxima ($N_2 = 13$) were located. Again, when the backbone torsional angles,

*Note that the catchmant regions of β_L, ε_L and γ_L form the common extended β region of the Ramachandran surface.

χ_2													
360°	α_L	α_L	α_L	α_L	γ_L	δ_L	α_L	δ_L	δ_L	δ_L	γ_L	γ_L	α_L
	γ_L	α_L	α_L	α_L	α_L	γ_L	δ_L	δ_L	δ_L	δ_L	δ_L	γ_L	γ_L
300°	γ_L	α_L	δ_L	δ_L	α_L	α_L	δ_L	δ_L	δ_L	δ_L	α_L	α_L	γ_L
	γ_L	δ_L	δ_L	δ_L	α_L	α_L	γ_L	δ_L	δ_L	δ_L	α_L	α_L	γ_L
240°	γ_L	δ_L	δ_L	δ_L	α_L	α_L	α_L	γ_L	δ_L	δ_L	α_L	α_L	γ_L
	γ_L	δ_L	δ_L	δ_L	α_L	α_L	α_L	α_L	γ_L	γ_L	α_L	α_L	γ_L
180°	γ_L	δ_L	δ_L	δ_L	α_L	α_L	α_L	α_L	γ_L	γ_L	α_L	α_L	γ_L
	γ_L	δ_L	δ_L	δ_L	α_L	α_L	α_L	α_L	γ_L	γ_L	α_L	α_L	γ_L
120°	γ_L	γ_L	δ_L	δ_L	α_L	α_L	α_L	α_L	δ_L	δ_L	α_L	α_L	γ_L
	γ_L	γ_L	γ_L	δ_L	δ_L	α_L	α_L	α_L	δ_L	δ_L	δ_L	α_L	γ_L
60°	α_L	γ_L	γ_L	δ_L	δ_L	α_L	α_L	γ_L	δ_L	δ_L	δ_L	α_L	α_L
	α_L	α_L	α_L	γ_L	δ_L	α_L	α_L	γ_L	δ_L	δ_L	δ_L	δ_L	α_L
0°	α_L	α_L	α_L	α_L	γ_L	δ_L	α_L	δ_L	δ_L	δ_L	γ_L	γ_L	α_L
	0°		60°		120°		180°		240°		300°		360°

Scheme 17

originally associated with an α_L fixed molecular conformation are relaxed ($f_{\alpha L}^{\text{fixed}}[\chi_1,\chi_2] \rightarrow f_{\alpha L}^{\text{relaxed}}[\chi_1,\chi_2]$) a more complex surface emerges (Scheme 17).

Among the 144 grid more than 40% remains α_L but almost the same quantity (~39%) is transformed into a δ_L-type structure. The remaining ~21% undergoes an $\alpha_L \rightarrow \gamma_L$ type transformation. Thus, $\alpha_L \rightarrow \delta_L$ conformational modification affects both ϕ and ψ variables ($\phi_{\alpha L} \approx -60° \rightarrow \phi_{\delta L} \approx -120°$ and $\psi_{\alpha L} \approx -40° \rightarrow \psi_{\delta L} \approx +30°$). However, in spite of the relatively high ratio (40%) of α_L points only three minimum energy conformers have been found as shown in Table 13.2.

Finally, the δ_L relaxation path is to be described. The $f_{\delta L}^{\text{fixed}}[\chi_1,\chi_2]$ potential energy surface (Fig. 13.6(a)) is the side-chain map of HCO–L-Ser–NH_2 molecule, where the backbone conformation is fixed to be $\phi = -125°$, $\psi = +30°$ (δ_L). This surface, $f_{\alpha L}^{\text{fixed}}[\chi_1,\chi_2]$, has five minima ($N_0 = 5$), ten saddle points ($N_1 = 10$) and five maxima ($N_2 = 5$). The grand total of 20 critical points reflects that this is the smoothest surface among the five fixed side-chain maps investigated in L-valley (Fig. 13.6(a)). At the same region of the $f_{\alpha L}^{\text{relaxed}}[\chi_1,\chi_2]$ side-chain map (Fig. 13.6(b)) where the backbone torsional angles are relaxed, a grand total of 38 critical points can be identified. This is the highest number of critical points among the relaxed side-chain maps. Eight minima ($N_0 = 8$), 19 saddle points ($N_1 = 19$) and eleven maxima ($N_2 = 11$) were located on $f_{\alpha L}^{\text{relaxed}}[\chi_1,\chi_2]$.

Considering that the number of critical points almost doubles with the relaxation of the backbone torsional angles ($f_{\alpha L}^{\text{fixed}}[\chi_1,\chi_2] \rightarrow f_{\alpha L}^{\text{relaxed}}[\chi_1,\chi_2]$), interesting side-chain induced backbone modifications are expected here. A high number (about 60%) of the 144 grid points do have δ_L-type backbone structure. These are approximately equal amount (~7%) of α_L- and β_L-conformers while

χ_2 \ χ_1	0°		60°		120°		80°		240°		300°		360°
360°	δ_L	δ_L	α_L	γ_L	γ_L	δ_L	α_L	δ_L	δ_L	δ_L	δ_L	γ_L	δ_L
	γ_L	δ_L	δ_L	α_L	γ_L	γ_L	δ_L	δ_L	δ_L	δ_L	δ_L	δ_L	γ_L
300°	γ_L	δ_L	δ_L	α_L	δ_L	γ_L	δ_L	δ_L	δ_L	δ_L	δ_L	δ_L	γ_L
	γ_L	δ_L	δ_L	δ_L	δ_L	β_L	γ_L	δ_L	δ_L	δ_L	δ_L	δ_L	γ_L
240°	γ_L	δ_L	δ_L	δ_L	δ_L	β_L	γ_L	γ_L	δ_L	δ_L	δ_L	δ_L	γ_L
	γ_L	δ_L	δ_L	δ_L	δ_L	β_L	β_L	γ_L	γ_L	δ_L	δ_L	δ_L	γ_L
80°	γ_L	δ_L	δ_L	δ_L	δ_L	α_L	β_L	γ_L	γ_L	γ_L	δ_L	δ_L	γ_L
	γ_L	δ_L	δ_L	δ_L	δ_L	α_L	β_L	γ_L	γ_L	γ_L	δ_L	γ_L	γ_L
120°	γ_L	γ_L	δ_L	δ_L	δ_L	α_L	β_L	β_L	δ_L	δ_L	δ_L	δ_L	γ_L
	γ_L	γ_L	γ_L	δ_L	δ_L	α_L	β_L	γ_L	δ_L	δ_L	δ_L	δ_L	γ_L
60°	δ_L	γ_L	γ_L	δ_L	δ_L	α_L	β_L	γ_L	δ_L	δ_L	γ_L	δ_L	δ_L
	δ_L	δ_L	γ_L	γ_L	δ_L	α_L	α_L	γ_L	δ_L	δ_L	γ_L	δ_L	δ_L
0°	δ_L	δ_L	α_L	γ_L	γ_L	δ_L	α_L	δ_L	δ_L	δ_L	δ_L	γ_L	δ_L

Scheme 18

the γ_L-type backbone orientation (27%) makes up the rest of this surface manifold (Scheme 18). In the central region ($120° \leq \chi_1 \leq 240°$ and $0° \leq \chi_2 \leq 360°$) of the map all four backbone conformers (α_L, β_L, γ_L and δ_L) could be present. This means that the *anti* orientation of the χ_1 torsional angle governs a backbone-conformation flexibility, thus even a relatively small twist on the χ_2 could result in a large variation in the folding of the backbone. (i.e. at $\chi_1 = 180°$ we have in the vertical columns of $0° \leq \chi_2 \leq 360°$ in 30° intervals the following backbone conformations: α_L, α_L, β_L, β_L, β_L, β_L, β_L, β_L, γ_L, γ_L, δ_L, δ_L, α_L). Finally, it is to be mentioned that we have not found any exit toward a D-type backbone structure, signaling that no side-chain conformation may induce such a dramatic backbone shift whereby conformer from L-valley ends into the D-valley.

9. CONCLUDING REMARKS

Diamides of single amino acid residues represent a practical model to study protein folding in spite of all their limitations. Clearly, tripeptides could have been better models, since at least the interactions of the nearest neighbors would be incorporated in the analyses. However, there are $20 \cdot 20 = 400$ different compositions even for the HCO–Xxx–Ala–Yyy–NH_2 sequence bit. Furthermore, each and every one of the above 400 sequences could have as many as $9^3 = 729$ backbone conformations, not considering any of the possible side chain orientations increasing exponentially the number of structures. Thus, such an advanced model system is beyond ab initio computability, at least at the end of the 20th century.

ACKNOWLEDGMENT

This research was supported by grants from the Hungarian Scientific Research Foundation (OTKA T017604 and T017192) and the Hungarian Academy of Sciences (AKP 96/2-427 2,4). The continuous financial support of the NSERC of Canada is gratefully acknowledged.

REFERENCES

1. Howard, J. C.; Ali, A.; Scheraga, H. A.; Momany, F. A. *Macromol.* 1975, **8**, 607.
2. Némethy, G.; Scheraga, H. A. *Biopolymers*, 1965, **3**, 155.
3. Némethy, G.; Miller, M. H.; Scheraga, H. A. *Macromol.* 1980, **13**, 914.
4. Paine, G. H.; Scheraga, H. A. *Biopolymers*, 1986, **25**, 1547.
5. Bruccoleri, R. E.; Karplus, M. *Biopolymers*, 1987, **26**, 137.
6. Ripoll, D. R.; Scheraga, H. A. *Biopolymers*, 1988, **27**, 1283.
7. Schultz, G. E. *Ann. Rev. Biophys. Chem.* 1988, **17**, 1.
8. Wolfe, S.; Bruder, S.; Weaver, D. F.; Yang, K. *Can. J. Chem.* 1988, **66**, 2703.
9. Lambert, M. H.; Scheraga, H. A. *J. Comp. Chem.* 1989, **10**, 798.
10. Lambert, M. H.; Scheraga, H. A. *J. Comp. Chem.* 1989, **10**, 770.
11. Lambert, M. H.; Scheraga, H. A. *J. Comp. Chem.* 1989, **10**, 817.
12. Bernstein, F. C.; Koetzle, T. F.; Williams, G. J. B.; Meyer, E. F.; Brice, M. D. Jr.; Rodgers, J. R.; Kennard, O.; Shimanouchi, T.; Tasumi, M.; "The Protein Data Bank: A Computer Based Archival Data File for Macromolecular structures" *J. Mol. Biol.* 1977, **112**, 535.
13. Abola, E. E.; Bernstein, F. G.; Bryant, S. H.; Koetzle, I. F.; Weng, J. Crystallographic database-information content, software system, scientific applications. In: *Data Commission of the International Union of Crystallography*; Allen, F. H.; Bergerhoff, G. Sievers, R. Eds.; Bonn/Cambridge/Chester, 1987, p. 107.
14. Robb, M. A.; Csizmadia, I. G. *Theoretica Chim. Act.* 1968, **10**, 269.
15. Perczel, A.; Daudel, R.; Ángyán, J. G.; Csizmadia, I. G. *Can. J. Chem.* 1990, **68**, 1882.
16. Liegener, C. M.; Endrédi, G.; McAllister, M. A.; Perczel, A.; Ladik, J.; Cszimadia, I. G. *J. Am. Chem. Soc.* 1993, **115**, 8275.
17. Endrédi, G.; Liegener, C. M.; McAllister, M. A.; Perczel, A.; Ladik, J.; Csizmadia, I. G. *J. Mol. Struct. (THEOCHEM)* 1994, **306**, 1.
18. Endrédi, G.; McAllister, M. A.; Perczel, A.; Császár, P.; Ladik, J.; Csizmadia, I. G. *J. Mol. Struct. (THEOCHEM)* 1995, **331**, 5.
19. Mirskz, A. E.; Pauling, L. *Proc. Natl. Acad. Sci*, 1936, **22**, 439.
20. Liquori, A. M. *Quant. Rev. Biophysics*, 1969, **2**, 65.
21. Lattman, E. E.; Rose. G. D. *Proc. Natl. Acad. Sci.* 1969, **90**, 439.
22. Aubry, A.; Marraud, M.; Protas, J.; Neel, J. *Compt. Rend. Acad. Sci. Paris* 1974, **287c**, 163.
23. Aubry, A.; Ghermani, N.; Marraud, M. *Int. J. Peptide, Protein Res.* 1984, **23**, 113.

24. Raj, P. A.; Soni, S. D.; Ramasubbu, N.; Bhandary, K. K.; Levine, M. J. *Biopolymers* 1990, **30**, 73.
25. Stroup, A. N.; Cole, L. B.; Dhingra, M. M.; Gierach, L. M. *Int. J. Peptide, Protein Res.* 1990, **36**, 531.
26. Gething, M. J.; Sambrook, J. *Nature*, 1992, **335**, 33.
27. Perczel, A.; Foxman, B. M.; Fasman, G. D. *Proc. Natl. Acad. Sci. USA* 1992, **89**, 8210.
28. Sasisekharan, V. Stereochemical criteria for polypeptide and protein structures In *Collagen*; Ramanaathan, N. Ed.; Wiley & Sons: Madras, 1962, 39.
29. Ramachandran, G. N.; Ramakrishnan, C.; Sasisekharan, V. *J. Mol. Biol.* 1963, **7**, 95.
30. Ramachandran, G. N.; Sasisekharan, V. *Adv. Protein Chem.* 1968, **23**, 283.
31. Ramachandran, G. N.; Venkatachalam, C. M.; Krimm, S. *Biophys. J.* 1966, **6**, 849.
32. Perczel, A.; Ángyán, J. G.; Kajtár, M.; Vivini, W.; Rivail, J. L.; Marcoccia, J. F.; Csizmadia, I. G. *J. Am. Chem. Soc.* 1991, **113**, 6256.
33. Perczel, A.; McAllister, M. A.; Császár, P.; Csizmadia, I. G. *J. Am. Chem. Soc.* 1993, **115**, 4849.
34. Perczel, A.; McAllister, M. A.; Császár, P.; Csimadia, I. G. *Can. J. Chem.* 1994, **72**, 2050.
35. Perczel, A.; Csizmadia, I. G. *Int. Reviews in Phys. Chem.* 1995, **14**, 127.
36. Karplus, P. A. *Protein Science* 1996, **5**, 1406.
37. Efimov, A. V. *Prog. Biophys. Mol. Biol.* 1993, **60**, 201.
38. Richardson, J. S.; Richardson, D. C. Principles and patterns of protein conformation. In *Prediction of Protein Structure and the Principles of Protein Conformation*; Fasman, G. D. Ed.; Plenum Press: New York, 1989, 1–98.
39. Rooman, M. J.; Kocher, J. P. A.; Widak, S. J. *Biochemistry* 1992, **31**, 10226.
40. Zimmerman, S. S.; Pottle, M. A.; Némethy, G.; Scheraga, H. A. *Macromol.* 1977, **10**, 1.
41. Thornton, J.; Jones, D. T.; MacArthur, M. W.; Orengo, C. M.; Swindells, M. B. *Phil. Trans. Roy. Soc. Lond.* 1995, **B348**, 71.
42. Head-Gordon, T.; Head-Gordon, M.; Frisch, M. J.; Brooks III, C. L.; Pople, J. A. *Int. J. Quantum Chem., Quantum Biol.*, 1989, **16**, 311.
43. Head-Gordon, T.; Head-Gordon, M.; Frisch, M. J.; Brooks III, C. L.; Pople, J. A. *J. Am. Chem. Soc.*, 1991, **113**, 5989.
44. Csizmadia, I. G.; General and theoretical aspects of the thiol Group. In *The Chemistry of the Thiol Group*; Patai, S.; Ed.; John Wiley & Sons, 1974, 1.
45. Peterson, M.R.; Csizmadia, I. G.; *J. Am. Chem. Soc.* 1978, **100**, 6911.
46. Peterson, M. R.; Csizmadia, I. G. Analytic equations for conformational energy surfaces. In *Progress of Theoretical Organic Chemistry*, Peterson, M. R. Csizmadia, I. G. (Eds.; Elsevier, 1982, 3, 190.
47. Mezey, P. G. Potential Energy Hypersurfaces. Elsevier Science Publishers, 1987.
48. Csizmadia, I. G. Multidimensional theoretical stereochemistry and conformational potential energy surface topology. In *New Theoretical Concept for Understanding Organic Reactions*; Csizmadia, I. G., Bertran, J. D. Eds.; Reidel: Publishing Co., 1989, 1.

49. Farkas, Ö.; Perczel, A.; Marcoccia, J. F.; Hollósi, M.; Csizmadia, I. G. *J. Mol. Struc. (THEOCHEM)*, 1995, **331**, 27.
50. Perczel, A.; Farkas, Ö.; Csizmadia, I. G. *J. Comp. Chem.* 1996, **17**, 821.
51. Perczel, A.; Farkas, Ö.; Csizmadia, I. G. *J. Am. Chem. Soc.* 1996, **118**, 7809.
52. Perczel, A.; Farkas, Ö.; Marcoccia, J. F.; Csizmadia, I. G. *International J. Quantum Chem.* 1997, **61**, 797.
53. Roterman, I. K.; Lambert, M. H.; Gibson, K. D.; Scheraga, H. A. *J. Biomolecular Structure and Dynamics* 1989, **7**, 421.
54. McAllister, M. A.; Perczel, A.; Császár, P.; Viviani, W.; Rivail, J. L.; Csizmadia, I. G.; *J. Mol. Struct. (THEOCHEM)* 1993, **288**, 161.
55. Rodriguez, A. M.; Baldoni, H. A.; Suvire, F.; Nieto-Vasquez, R.; Zamarbide, G.; Enriz, R. D.; Farkas, Ö.; Perczel, A.; Csizmadia, I. G. *J. Mol. Struct. (THEOCHEM)* 1998, (in press)
56. HYPERCHEM 4.5, (1994) Hypercube, Inc. 419 Phillip St., Waterloo, Ontario, Canada N2L 3X2.
57. Allinger, N. L. *J. Am. Chem. Soc.*, 1977, **99**, 8127 (The HyperMM+ force field is derived from the public domain code developed in this paper).
58. Wiener, S. J.; Singh, U. C.; O'Donnel, T. J; Kollman, P. *J. Am. Chem. Soc.* 1984, **106**, 6243.
59. Brooks, B. R.; Bruccoleri, R. E.; Olafson, B. D.; States, D. J.; Swaminathan, S.; Karplus, M. *J. Comput. Chem.* 1983, **4**, 187.
60. Jorgensen, W. L.; Tirado-Rives, J. *J. Am. Chem. Soc.* 1988, **110**, 1657.
61. Pranate, J.; Wierschke, S.; Jorgensen, W. L. *J. Am. Chem. Soc.* 1991, **113**, 2810.
62. Bingham, R. C.; Dewar, M. J. S.; Lo, D. H. *J. Am. Chem. Soc.* 1975, **97**, 1302.
63. Dewar, M. S. J.; Thiel, W. *J. Am. Chem. Soc.* 1977, **99**, 4899.
64. Dewar, M. J. S.; Zoebisch, E. G.; Healy, E. F.; Stewart, J. J. P. *J. Am. Chem. Soc.* 1986, **107**, 3902.
65. Stewardt, J. J. P. *J. Comp. Chem.* 1989, **10**, 221.
66. Scarsdale, J. N.; Van Alsenoy, C.; Klimkowski, V. J.; Schäfer, L.; Momany, F. A. *J. Am. Chem. Soc.* 1983, **105**, 3438.
67. Schäfer, L.; Kimkowski, V. J.; Momany, F. A.; Chuman, H.; van Alsenoy, C. *Biopol.* 1984, **23**, 2335.
68. Wiener, S. J.; Kollman, P. A.; Case, D. A.; Singh, U. C.; Ghio, C.; Alagona, G.; Profetajr. S.; Wiener, P. *J. Am. Chem. Soc.* 1984, **106**, 765.
69. Böhn, H. J.; Brode, S. *J. Am. Chem. Soc.* 1991, **113**, 7129.
70. Böhm, H. J.; Brode, S. *J. Comp. Chem.* 1995, **16**, 146.
71. Endrédi, G.; Perczel, A.; Farkas, Ö.; McAllister, M. A.; Csonka, G. I.; Ladik, J.; Csizmadia, I. G. *J. Mol. Struct. (THEOCHEM)* 1997, **391**, 15.
72. Frey, R. F.; Coffin, J.; Newton, S. Q.; Ramek, M.; Cheng, V. K. W.; Momany, F. A.; Schäfer, L. *J. Am. Chem. Soc.* 1992, **114**, 5369.
73. Ramek, M.; Cheng, V. K. W.; Frey, R. F.; Newton, S. Q.; Schäfer, L. *J. Mol. Struct. (THEOCHEM)* 1991, **235**, 1.
74. Jákli, I.; Perczel, A.; Farkas, Ö.; Csizmadia, I. G. *J. Comp. Chem.* 1998, (in preparation)

75. Hobohm, U.; Scharf, M.; Schneider, R.; Sander, C. *Protein Science* 1 1992, **1**, 409.

76. Hobohm, U.; Sander, C. *Protein Science* 1994 **3**, 522.

77. Ángyán, J. G.; Daudel, R.; Kucsman, Á.; Csizmadia, I. G. *Chem. Phys. Letters.* 1987, **136**, 1.

78. Klimkowski, V. J.; Schäfer, L.; Momany, F. A.; van Alsenoy, C. *J. Mol. Struct. (THEOCHEM)* 1985, **124**, 143.

79. Siam, K.; Klimkowski, V. J.; van Alsenoy, C.; Ewbank, J. D.; Schäfer, L. *J. Mol. Struct. (THEOCHEM)* 1987, **152**, 261.

80. Van Alsenoy, C.; Cao, M.; Newton, S. Q.; Teppen, B.; Perczel, A.; Csizmadia, I. G.; Momany, F. A.; Schäfer, L. *J. Mol. Struct.* 1993, **286**, 149.

81. Smith, J. A.; Pease, L. G. *CRC Crit. Reviews in Biochem.* 1980, **8**, 315.

82. Perczel, A.; Hollósi, M.; Foxman, B. M.; Fasman, G. D. *J. Am. Chem. Soc.* 1991, **113**, 9772.

83. Perczel, A.; Foxman, B. M.; Fasman, G. D. *Proc. Natl. Acad. Sci.* 1992, **89**, 8210.

84. Csonka, G. I.; Loos, M.; Kucsman, Á.; Csizmadia, I. G. *J. Mol. Struct. (THEOCHEM)* 1994, **315**, 29.

85. Viviani, W.; Rivial, J. L.; Perczel, A.; Csizmadia, I. G. *J. Am. Chem. Soc.* 1993, **115**, 8321.

86. Farkas, Ö.; McAllister, M. A.; Ma, J. H.; Perczel, A.; Hollósi, M.; Csizmadia, I. G. *J. Mol. Struc. (THEOCHEM)* 1996, **396**, 105.

87. Perczel, A.; Farkas, Ö.; Császár, A. G.; Csizmadia, I. G. *Can. J. Chem.* 1997, **75**, 1120.

88. Császár, A. G. *J. Am. Chem. Soc.* 1992, **114**, 9569.

89. Császár, A. G. *Phys. Chem.* 1996, **100**, 3541.

CHAPTER 14

GAS-PHASE ION CHEMISTRY OF AMIDES, PEPTIDES, AND PROTEINS

CAROLYN J. CASSADY
Department of Chemistry and Biochemistry, University of Alabama
Former address: Department of Chemistry and Biochemistry, Miami University

1. INTRODUCTION

Mass spectrometry is widely used to probe the gas-phase reactive and dissociative chemistry of ions. Information on intrinsic thermodynamic, kinetic, mechanistic, and structural properties can be obtained in the absence of solvents. Initial studies with amides, which began more than three decades ago, focused on the unimolecular dissociation of ions produced by electron ionization (EI)[1] and chemical ionization (CI).[2] Ion generation by EI and CI requires that the sample be in the gas-phase; thus these studies were limited to small, volatile amides. A second era in gas-phase chemistry studies of amides began in 1981, with the development of fast atom bombardment (FAB).[3] In FAB, bombardment of a sample solution with an energetic ion or atom beam leads to vaporization and ionization of the sample. FAB can routinely produce protonated molecular ions, MH^+, for peptides with molecular masses up to about 5000 Da; other adducts such as $[M+Na]^+$ may also form. The unimolecular dissociation of peptide ions generated by FAB became an active area of research and led to the development of mass spectrometry/mass spectrometry (MS/MS) as a tool for obtaining sequence information. In addition, reports on studies of gas-phase reactions of peptide and amide ions began to appear, with most of the research involving proton transfer properties. Proton transfer is of importance to biological activity and also to mass spectrometry because protonation is the dominant pathway for ion production during desorption/ionization events such as FAB. A third era in the ion chemistry of amides, peptides, and proteins began in the late 1980s with the development of two ionization techniques that expanded the mass range to greater than 100,000 Da. One technique

The Amide Linkage: Selected Structural Aspects in Chemistry, Biochemistry, and Materials Science,
Edited by Arthur Greenberg, Curt M. Breneman, and Joel F. Liebman
ISBN 0-471-35893-2

is electrospray ionization (ESI),[4] where multiply charged ions, such as $[M+nH]^{n+}$, are generated by passage of a sample solution through a high electric field, followed by an evaporation process. In the second technique, matrix-assisted laser desorption/ionization (MALDI),[5] the sample is mixed with an organic matrix that absorbs at the wavelength of a bombarding laser; energy absorbed by the matrix is transferred to the sample, leading to the production of singly charged ions, such as MH^+. ESI and MALDI both give facile production of peptide and protein ions. These ions have been subjected to numerous unimolecular dissociation studies that probe their primary structures. Another area of increasing interest is gas-phase bimolecular chemistry involving proton transfer and hydrogen/deuterium (H/D) exchange reactions. In addition to primary amino acid sequence, three-dimensional conformations of large ions can play an important role in these processes.

Research from each of the three eras in the mass spectrometry of amides, peptides, and proteins is discussed below. This is not an exhaustive review. Instead, the intention is to highlight some of the most important aspects of dissociative and reactive ion chemistry involving the amide linkage.

2. BIMOLECULAR REACTION CHEMISTRY

2.1. Proton Transfer Properties of Singly Charged Small Amide Ions

Studies of proton transfer for small amides have centered around measurements of gas-phase basicity (GB) and proton affinity (PA). For proton transfer Reaction 14.1,

$$M + H^+ \rightarrow MH^+ \tag{14.1}$$

GB corresponds to the negative of the Gibbs free energy charge ($-\Delta G$), while PA is the negative of the enthalpy ($-\Delta H$) of protonation. GB and PA are related by Eq. 14.2:

$$GB = PA + T\Delta S \tag{14.2}$$

Several methods can be employed for measuring GB and PA. For volatile compounds, such as small amides, the most common procedure involves measurement of the equilibrium constant for gas-phase proton transfer between the sample, M, and a reference compound B whose GB is known:

$$MH^+ + B \rightleftharpoons M + BH^+ \tag{14.3}$$

The equilibrium constant for Reaction 14.3, which can be obtained from the rate constants of the forward and reverse reactions, yields the difference in GB between the two compounds by the relationship:

$$-RT \ln K_{eq} = GB(M) - GB(B) \tag{14.4}$$

TABLE 14.1. Gas-Phase Basicities and Proton Affinities for Several Small Amides[a]

Amide	PA, $kJ\,mol^{-1}$	GB, $kJ\,mol^{-1}$
Formamide	822.2	791.2
N-Methylformamide	851.3	820.3
N,N-Dimethylformamide	887.5	856.6
Acetamide	863.6	832.6
Propanamide	876.2	845.3
2-Azetidinone	852.6	821.7
N-methylacetamide	888.5	857.6

[a] All values are from Ref. 13.

Small amides have been the subjects of numerous proton transfer studies.[6–12] The values obtained have been included in an evaluated compilation of GBs and PAs by Hunter and Lias.[13] To illustrate the trends in basicity, Table 14.1 gives PA and GB values for several small amides. As the number of groups attached to the amide nitrogen increases—going from primary to secondary to tertiary amide—the basicity increases. This trend, which is depicted in Table 14.1 by formamide, *N*-methylformamide, and *N,N*-dimethylformamide, is analogous to that found for amines.[7] As the number of substituents increases, the charge site is stabilized by electrostatic solvation effects, which may be combined with resonance and inductive effects. A more stable protonated ion results in increased values of GB and PA. In addition, as the chain length of the group attached to the amide carbon increases (i.e., formamide to acetamide to propanamide), the basicity of the amide increases. This is also analogous to an effect seen with amines, where the polarizable alkyl side chain assists in stabilization of the positive charge.[6]

One of the most interesting aspects of the gas-phase protonation of amides is the location of the proton. In solution, it is well-established that the most energetically favorable site of protonation is the amide oxygen. Recently, the protonation of formamide, the simplest amide, was studied by mass spectrometry and high level ab initio quantum chemical calculations[14–16]. It was found that gas-phase O-protonation is more energetically favorable than N-protonation by about $59\,kJ\,mol^{-1}$.[16]

The impact of cyclization on amide basicity is also noteworthy. For example, consider 2-azetidinone, a β-lactam shown in Structure **I**, and its corresponding secondary acyclic amide, *N*-methylacetamide, shown in Structure **II**.

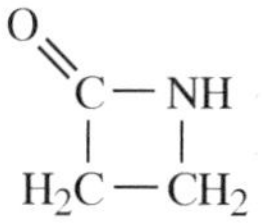

```
O
 \\
  C — NH            O
  |    |            ||
H2C — CH2       CH3CNHCH3

    I               II
```

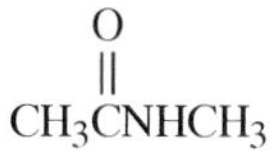

As the data in Table 14.1 indicates, 2-azetidinone is approximately $36\,kJ\,mol^{-1}$ less basic than *N*-methylacetamide. This contrasts results for amines, where cyclization has almost no impact on GBs, and for ketones, where the effect is present but much less pronounced. Abboud et al.[9] have conducted an extensive mass spectral and theoretical study of cyclization effects. They concluded that lactams are generally weaker gas-phase bases than their acyclic amide counterparts. Ab initio calculations indicate that this is due to a negative hyperconjugation effect in amides, which is absent in amines and attenuated in ketones. In other words, the 90° C–C–N bond centered at the amide carbon of lactams dramatically alters the hybridization of this carbon, which leads to a decrease in electronic charge density at the oxygen lone pairs. Therefore, lactams transfer less electronic charge to the incoming proton than their acyclic counterparts and are less basic. As another consequence of the hyperconjugation effect, the difference between oxygen and nitrogen intrinsic basicities is narrower for lactams. For example, O-protonation of 2-azetidinone is approximately 25 kJ mol^{-1} more favorable than N-protonation,[9] which is considerably less than the 59 kJ mol^{-1} difference of formamide.[16] Interestingly, although it has not been studied in detail, this hybridization effect does not appear to exist for less strained cyclic amides with 5-membered rings. For example, the GB of *N*-methylpyrrolidinone is within 10 kJ mol^{-1} of the GB for its acyclic counterpart *N,N*-dimethylpropionamide.[13]

2.2. Proton Transfer Properties of Singly Charged Peptide Ions

With the development of FAB, it became possible to produce protonated ions from low volatility peptides and proteins. As a result, the gas-phase basicities of amino acids and small peptides have recently become an active area of research. The equilibrium method of GB determination is rarely used in these studies because insufficient levels of the neutral biomolecule are present in the gas phase. Instead, GBs of peptides have been determined by deprotonation reaction bracketing and by the kinetic method.[17] In bracketing studies, the protonated ion of interest, MH^+, is allowed to react with a series of neutral compounds (B) of known GB, Reaction 14.5:

$$MH^+ + B \rightarrow M + BH^+ \tag{14.5}$$

Using the deprotonation rate constants as a guide, upper and lower limits for the GB of compound M can be obtained, thus bracketing its GB. In the kinetic method, a proton bound cluster ion is produced between the compound of interest and a reference compound of known GB; this ion is often formed directly in the FAB ionization process by bombarding a mixture of the two compounds. The cluster ion is given excess internal energy by a process such as collision-induced dissociation (CID) and undergoes competitive fragmentation, Reaction 14.6:

$$\begin{aligned}[M\text{–}H\text{–}B]^+ &\rightarrow MH^+ + B \\ &\rightarrow BH^+ + M\end{aligned} \tag{14.6}$$

The most basic compound will preferentially retain the proton and the ratio of MH^+ and BH^+ abundances can be used to obtain the GB of compound M. Although the bracketing and kinetic methods are applied in non-ideal conditions where equilibrium is not achieved, a recent study on propionamide indicated that the results obtained are comparable to those acquired from the proton transfer equilibrium method.[10]

Deprotonation reaction studies for small peptides have been performed by several groups employing Fourier transform ion cyclotron resonance mass spectrometry (FT-ICR or FTMS). To illustrate the spectral data obtained by this technique, Fig. 14.1 shows a series of spectra for MH^+ of the tripeptide GlyProGly reacting with 1-methylpiperidine, B.[18] Product ions corresponding to proton transfer, BH^+, and adduct formation, $M{-}H^+{-}B$, are formed. This cluster ion could be dissociated for kinetic method experiments. For reference, Table 14.2 identifies the twenty common amino acids by their single letter and three letter codes.

In a recent review, Harrison[19] gives a detailed discussion of the methods used to obtain GBs and PAs. In addition, GB and PA values for amino acids and peptides are tabulated and discussed. Therefore, only the major points in this field will be reviewed here. An important point to note prior to this discussion is that, unlike in the solution phase, gas-phase protonated amino acids and peptides do not generally exist as zwitterions.[20] Although the GBs and PAs of amino acids have been well-established,[20–27] little work has dealt with peptides. The earliest peptide reports involve polyglycines, which were independently investigated by three research groups.[28–30] These studies agree that the basicity increases as the number of amino acid residues in the peptide increases. However, the reported GB values differ substantially as the size of the peptide increases. For example, with diglycine, the three studies yielded GBs in the range of 872–884 kJ mol^{-1}. However, for pentaglycine, the values range from 885 kJ mol^{-1} for bracketing experiments of Wu and Lebrilla[29] to 935 kJ mol^{-1} by kinetic method experiments of Wu and Fenselau;[30] values from the laboratory of Cassady[28] were intermediate, at 914 kJ mol^{-1} from bracketing and 920 kJ mol^{-1} from the kinetic method. This trend of greater deviation in GB values between laboratories as the size of the peptide increases also occurs for polyalanines.[31,32] The reason is unclear. Harrison[19] has suggested that the various studies may be sampling different populations of peptide ions. In the mass spectrometer, numerous stable isomers with differences in protonation site and conformation are probably accessible.

Most other works on peptide basicity have dealt with di- and tripeptides. Gorman and Amster[33] used the bracketing method to obtain the GBs of 22 dipeptides containing at least one valine residue. Most GBs were nearly equal to the GB of the amino acid that constitutes the most basic residue in the dipeptide. Several studies from the laboratories of Cassady[34–37] and Fenselau[31,38] have focused on how specific residues and their locations in the peptide sequence affect GBs. To illustrate the major trends, Table 14.3 gives the GBs for di- and tripeptides composed of glycine and proline or histidine, along with the GBs of

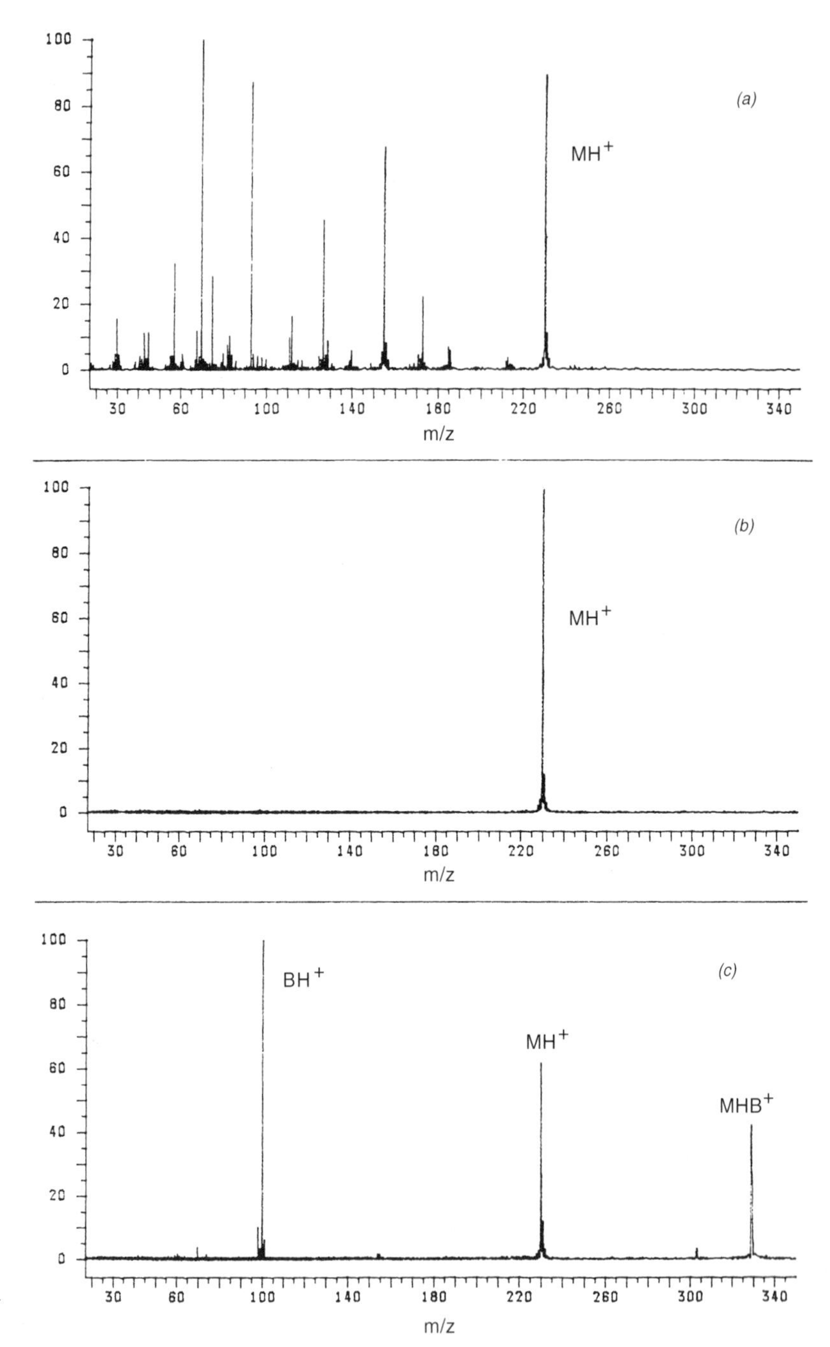

FIGURE 14.1. FT-ICR mass spectra for the reaction of $GlyProGlyH^+$ (MH^+) with 1-methylpiperidine (B) at a static pressure of 7.2×10^{-8} mb. (*a*) All ions produced by FAB on a solution of GlyProGly in glycerol. (*b*) Isolation of MH^+ by swept frequency ejection techniques. (*c*) Reaction of MH^+ and B for 1.0 s.

TABLE 14.2. One and Three Letter Symbols for the Common Amino Acids

One Letter Symbol	Three Letter Symbol	Amino Acid
A	Ala	Alanine
C	Cys	Cysteine
D	Asp	Aspartic Acid
E	Glu	Glutamic Acid
F	Phe	Phenylalanine
G	Gly	Glycine
H	His	Histidine
I	Ile	Isoleucine
K	Lys	Lysine
L	Leu	Leucine
M	Met	Methionine
N	Asn	Asparagine
P	Pro	Proline
Q	Gln	Glutamine
R	Arg	Arginine
S	Ser	Serine
T	Thr	Threonine
V	Val	Valine
W	Trp	Tryptophan
Y	Tyr	Tyrosine

the component amino acids. For the proline-containing peptides, GB increases when the proline residue is located at the N-terminus.[36] This also occurs for small peptides containing glycine residues with alanine[34] or serine.[35] In contrast, there is no notable increase in GB when a histidine residue is placed at the N-terminus; this is also true of lysine-containing peptides.[37] The implication is that for small peptides containing highly basic residues (i.e., histidine, lysine, or arginine) the proton in MH^+ resides on the amino group of the basic residue's side chain. Thus, the position of the residue in the peptide chain has little impact on GB. In contrast, this data suggests that for peptides containing less basic residues, such as the proline–glycine peptides of Table 14.3, the proton resides on the N-terminal amino nitrogen. The basicity of the peptide is enhanced by the ability of the N-terminal residue's nearby side chain to stabilize the positive charge. For example, because it is a secondary amine, the pyrrolidine ring of proline, Structure **III**, is better able to stabilize the charge than the methyl group of glycine; consequently, the GB of ProGly is 24.6 kJ mol^{-1} greater than that of GlyPro.

```
H2C ─ NH      O
 |      \     ||
 |       HC ─ C ─ OH
 |      /
H2C ─ CH2
```

III

TABLE 14.3. Gas-Phase Basicities for Several Amino Acids and Small Peptides[a]

Peptide or Amino Acid	GB, kJ mol^{-1}	Reference
Glycine, Gly	847.2	35
Proline, Pro	898.9	36
GlyPro	905.6	36
ProGly	925.2	36
ProPro	944.8	36
GlyGlyPro	915.5	36
GlyProGly	915.5	36
ProGlyGly	925.2	36
Histidine, His	934.7	37
GlyHis	955.6	37
HisGly	955.6	37
GlyGlyHis	967.7	37
GlyHisGly	955.6	37
HisGlyGly	967.7	37

[a] All values have been adjusted to the GB ladder of Ref. 13.

To gain further insight into protonated peptide structures, several groups have combined mass spectrometry with theoretical calculations. Diglycine, triglycine, and their protonated ions have been studied by ab initio[28,34,39] and semi-empirical[29] techniques. As is consistent with the mass spectral results, the calculations indicate that the most energetically favorable protonation site is the N-terminal amino nitrogen. Figure 14.2 shows several ab initio HF/3-21G minimum-energy structures for protonated triglycine. Structure **IV** is the lowest energy species and involves amino N-protonation. Relative to **IV**, protonation at the N-terminal amide oxygen (**V**) is 21 kJ mol^{-1} higher in energy, O-protonation at the central residue (**VI**) is 51 kJ mol^{-1} higher in energy, and O-protonation at the C-terminal carbonyl (**VII**) is 77 kJ mol^{-1} higher. Structures with amide N-protonation (not shown) are even less favorable, being at least 80 kJ mol^{-1} higher in energy than **IV**. In addition to the structures of Fig. 14.2, many slightly different conformations with similar energies were found.

Although the amide group is not the protonation site for most peptide MH^+, it does have a dramatic impact on the ion's three dimensional structure. A notable feature of the structures of Fig. 14.2 is the presence of extensive intramolecular hydrogen-bonding (H-bonding). Amide carbonyl oxygens and hydrogens attached to amide nitrogens are involved in these interactions; heteroatoms on side chains of amino acid residues may also participate. Several reports indicate that intramolecular H-bonding has a major impact on the gas-phase proton transfer properties of peptides. For example, kinetic method studies of entropy effects for protonated tri- and tetraglycines suggest that H-bonding contributes to the increase in GB as the number of glycine residues increases.[40] In addition, H-bonding involving lysine side chains may be responsible for the greater GB of LysGlyGlyGlyLys (1012 kJ mol^{-1}) over GlyLysLysGlyGLy

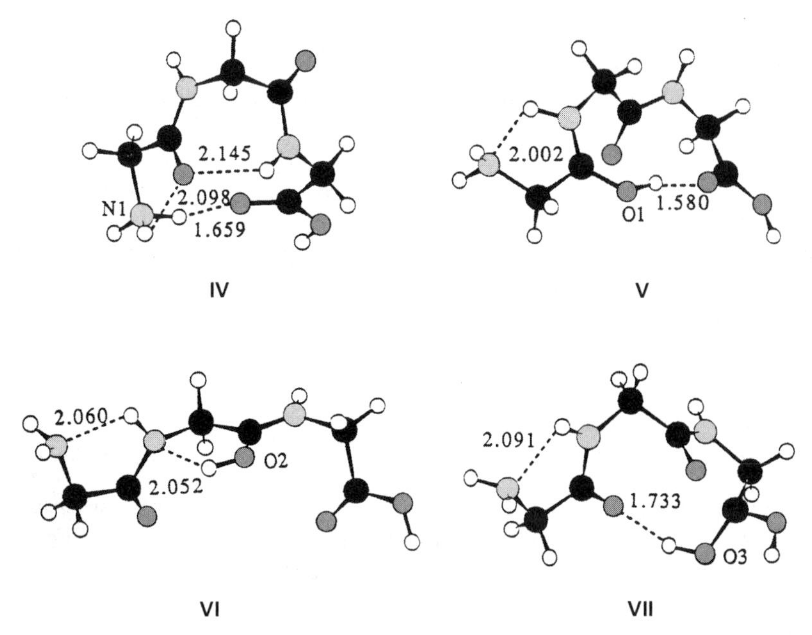

FIGURE 14.2. HF/3-21G minimum-energy structures for protonated triglycines. The site of protonation is denoted by an atom label. Intramolecular H-bonds are shown by dotted lines, with bonds distances given in angstroms. Atoms are identified by shading (H, none; C, dark; N, light; O, medium).

(995 kJ mol^{-1});[31] the terminal lysine residues may have greater flexibility to participate in H-bonding. Along this same line, for the data of Table 14.3, the lower GB of GlyHisGly (relate to HisGlyGly and GlyGlyHis) implies that a central protonated side chain participates less in H-bonding than terminal side chains. In addition, semi-empirical calculations suggest that the relatively high GB of ProPro (as seen in Table 14.3) is the result of a very stable protonated structure with two H-bonds between amino hydrogens and both carbonyl oxygens.[36] This conformation is caused by the unique geometry of the pyrrolidine rings on the two proline residues. It is not found for GlyPro and ProGly, which are 20–40 kJ mol^{-1} less basic.

Kim and Friesner[41] recently published an extensive ab initio study on intramolecular H-bonding involving eleven amino acid side chains. Although this work did not involve mass spectrometry, it should be of great utility in interpreting gas-phase ion chemistry results. The order of H-bond acceptance was found to be $N > O > S$, which is consistent with the trend in basicity. Donor behavior follows the acidity trend, with $OH \gg NH \sim SH > CHO$.

2.3. Proton Transfer Properties of Multiply Charged Peptide and Protein Ions

Electrospray ionization typically produces multiply protonated ions, $[M + nH]^{n+}$, from peptides and proteins. The most basic sites on a peptide are the N-terminal amino group and the side chain functional groups of arginine, lysine, and histidine residues. For many peptides, a good correlation exists between the number of attached protons and the number of these basic sites, thus suggesting that the protons are located at these positions.[42–44] Using a model that estimates proton transfer energetics based on intrinsic reactivity of basic sites and point charge Coulomb interactions, Williams and co-workers[45] have suggested that the higher charge state ions may have a significant fraction of ions protonated at the less basic residues of proline, tryptophan, and glutamine. This would occur to lessen Coulomb repulsion between the various charged sites. Note that Coulomb or electrostatic energy is related to the distance between two charged atoms, r_{ij} $(i \neq j)$, as shown in Eq. 14.7:

$$\text{Coulomb energy} = \sum_{ij} \frac{z_i z_j e^2}{4\pi\varepsilon_0\varepsilon_r r_{ij}} \tag{14.7}$$

(ε_r is the dielectric polarizability, which may have a value near 1.0 for gas-phase processes.[46–48]) However, there is also experimental evidence to suggest that multiple adjacent basic residues can be protonated. Proton transfer reactions on the dodecapeptide K_4G_8 suggest that the four adjacent lysine residues are all protonated by ESI.[49,50] In addition, two independent CID studies of peptides containing highly basic arginine residues suggest protonation occurs at adjacent residues.[51,52]

The study of Williams and co-workers[45] indicates that the solvent influences the charge state of the protein and peptide ions generated by ESI. In ESI, ions are formed from a peptide solution that is sprayed into a high electric field. If the solvent is more basic than a peptide ion, $[M+nH]^{n+}$, the ion may be deprotonated during the ionization process. Thus, the GB of the solvent dictates the maximum charge state that is observed during ESI experiments.

Obtaining thermodynamic information for $[M+nH]^{n+}$ is not as straightforward as it is for MH^+. For the deprotonation of $[M+nH]^{n+}$, Reaction 14.8,

$$[M+nH]^{n+} \rightarrow [M+(n-1)H]^{(n-1)+} + H^+ \tag{14.8}$$

a reverse activation barrier (RAB) exists due to the Coulomb repulsion between the two charged product ions. The Gibbs free energy change for Reaction 14.8 is equal to the gas-phase acidity (GA) of the parent ion, as shown in Eq. 14.9:

$$\text{GA}([M+nH]^{n+}) = \Delta G_{\text{rxn}} \tag{14.9}$$

The gas-phase basicity (GB) of the deprotonated ion is the negative of the Gibbs free energy change for the deprotonation reaction:

$$[M+(n-1)H]^{(n-1)+} + H^+ \rightarrow [M+nH]^{n+} \tag{14.10}$$

$$\text{GB}([M+(n-1)H]^{(n-1)+} = -\Delta G_{\text{rxn}} \tag{14.11}$$

Therefore, the quantity of $GA([M+nH]^{n+})$ is essentially the same as that of $GB([M+(n-1)H]^{(n-1)+})$. It has been suggested[53–55] that GA is a more sensible term to describe ions generated by ESI because it is the quantity determined from deprotonation reactions and indicates the reactivity of $[M+nH]^{n+}$. In contrast, the protonation of $[M+(n-1)H]^{(n-1)+}$, Reaction 14.10, involves bringing together of two like-polarity ions and is barrier-inhibited; thus, Reaction 14.10 is not observed by mass spectrometry. However, both terms, GB and GA, have been used in recent studies to describe the properties of $[M+nH]^{n+}$. In the present report, the term gas-phase acidity, GA, will be used consistently.

To further complicate matters, the GA itself is not directly obtained from a gas-phase deprotonation reaction, such as Process 14.12:

$$[M + nH]^{n+} + B \rightarrow [M + (n-1)H]^{(n-1)+} + BH^{+} \qquad (14.12)$$

The quantity actually measured by mass spectrometry is the sum of the GA and the reverse activation barrier (RAB) present in Reaction 14.12. This is referred to as the apparent gas-phase acidity, GA_{app}. The relationship between GA and GA_{app} is shown in Eq. 14.13:[53–56]

$$GA_{app}([M + nH]^{n+}) = GA([M + nH]^{n+}) + RAB \qquad (14.13)$$

GA_{app} values can be measured by mass spectrometry using deprotonation reaction bracketing[44,47,50,56–59] or the kinetic method.[60–62] In several studies, the magnitude of the RAB has been estimated by using either kinetic energy release measurements coupled with molecular dynamics calculations[60,61] or a computation model that also employs molecular dynamics.[47] The values obtained are system-dependent, suggesting that the locations and numbers of protonation sites as well as ion conformation have a great impact on the RAB. In addition, it should be emphasized that although GA represents the fundamental acidity of a multiply charged ion, GA_{app} is also important because it directly relates to the "observable acidity" of $[M+nH]^{n+}$. Thus, GA_{app} is the value that a mass spectrometrist needs to know in order to experimentally shift the charge state of an ion population.

For small organic ions, the dominating force in proton transfer reactions is the intrinsic acidity/basicity of the site being deprotonated/protonated. For $[M+nH]^{n+}$ produced by ESI, intrinsic basicity is still important but other factors also play prominent roles. As the size of the molecule increases, its conformation often has a greater influence on reactivity. This is particularly true for large biopolymers such as peptides and proteins where intramolecular H-bonding can occur. In addition, reactivity is affected by Coulomb repulsion resulting from the presence of multiple charge sites on the ion. In the past few years, several studies have provided insight into these factors. As an indication of the importance of Coulomb repulsion, initial work on ions from cytochrome c[63] and ubiquitin[56,58] revealed that the rate of deprotonation increased as the number of protons attached to the peptide increased. To illustrate this trend, Fig. 14.3 shows a plot of the GA_{app} values for ubiquitin ions as a function of charge state. Note that a lower GA_{app} value means that an ion more readily donates a proton (i.e., is more acidic) and a higher value means that an ion more readily retains a proton (i.e., is more basic). Ubiquitin is a moderate-size peptide with 76 amino acid residues, a molecular mass of 8565 Da, and 13 basic residues that are protonated by ESI. The sequence of ubiquitin is also shown in Fig. 14.3.

The magnitude of the deprotonation rate constants for $[M+nH]^{n+}$ is significant. For example, if the neutral accepting the proton is sufficiently basic, some of the ubiquitin reactions have rate constants in excess of $10^{-8}\,cm^3\,molecule^{-1}\,s^{-1}$.[58] This is an order of magnitude higher than the rate constants observed for singly charged ions. In general, the rates of ion/molecule reactions increase with increasing reaction exoergicity and eventually reach the collision rate. Endoergic reactions proceed at only a few percent of the collision rate, if they proceed at all. For highly charged ions, faster reactions are driven by a greater need to lessen the Coulomb repulsion experienced by the ion. However, it should also be noted that the theoretical Langevin collision rate[64] is directly proportional to a reactant ion's charge. Thus, an ion with a charge state of 10+ will react approximately 10 times faster than an ion with 1+. Although

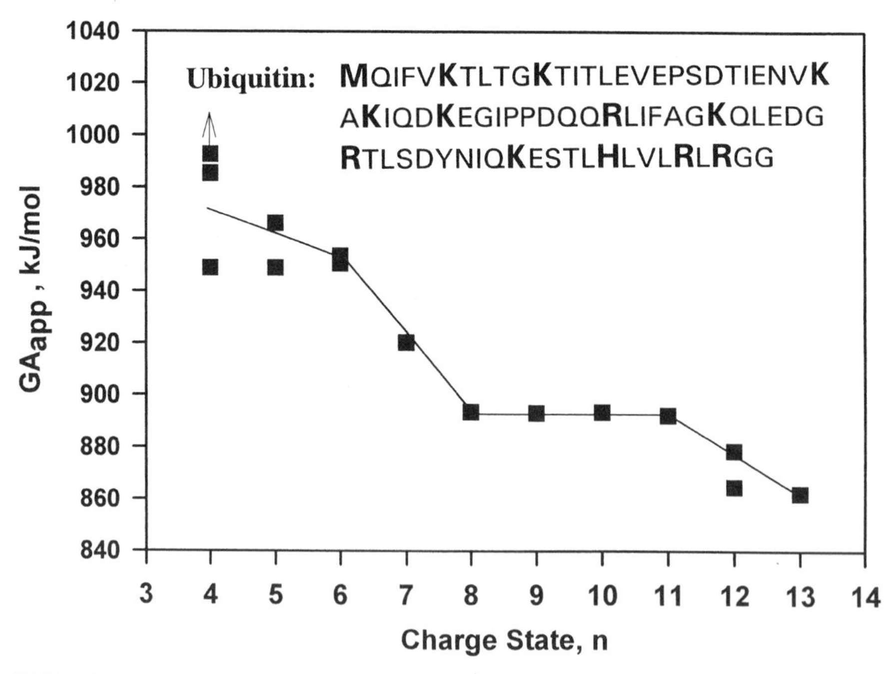

FIGURE 14.3. Plot of GA_{app} for $[M+nH]^{n+}$ from ubiquitin as a function of charge state, n. Basic residues in the ubiquitin sequence are shown in larger, bold type.

Langevin theory and its modifications, the average dipole orientation (ADO) theory[65] and the thermal capture theory,[66,67] have been used on peptide and protein ions, there is currently no proven model for calculating the collision rate constants of these ions. The Langevin-related models consider ions to be spherical point charges, with negligible volumes. The probability that a large, multiply charged ion will undergo a collision should be higher than the probability of a point charge undergoing a collision. Cassady and coworkers[58] have suggested that this is the reason why experimental rate constants for ubiquitin ions are up to 40% higher than the collision rate constants predicted by ADO theory.

The impact of Coulomb repulsion on reactivity was evident in a study[50] of three dodecapeptides containing only high basicity lysine (K) residues and low basicity glycines (G):

$$\mathrm{KGGKGGKGGKGG} \equiv (\mathrm{KGG})_4$$

$$\mathrm{KKGGGGKKGGGG} \equiv (\mathrm{K_2G_4})_2$$

$$\mathrm{KKKKGGGGGGGG} \equiv \mathrm{K_4G_8}$$

These peptides give primarily $4+$ ions under ESI conditions, suggesting that the four lysine residues are protonated. Due to their identical amino acid compositions, the reactive sites should have identical intrinsic basicities. However, the various orderings of residues yield differences in conformations and Coulomb energies. This work also included molecular dynamics calculations to gain information on these factors. It was found that $[(\mathrm{KGG})_4+4\mathrm{H}]^{4+}$ and $[(\mathrm{K_2G_4})_2+4\mathrm{H}]^{4+}$ have virtually identical GAs, as well as very similar Coulomb energies and geometries. In contrast, the presence of four adjacent lysine residues for $[\mathrm{K_4G_8}+4\mathrm{H}]^{4+}$ leads to a calculated Coulomb energy that is 75 kJ mol^{-1} higher than that of the other peptide ions. As a result, this ion readily losses a proton, requiring 65 kJ mol^{-1} less energy to deprotonate than $[(\mathrm{KGG})_4+4\mathrm{H}]^{4+}$ or $[(\mathrm{K_2G_4})_2+4\mathrm{H}]^{4+}$. In addition, the energetic threshold for dissociation of $[\mathrm{K_4G_8}+4\mathrm{H}]^{4+}$ was significantly lower during CID experiments.[49] During a follow-up study[68] of $3+$ ions for these peptides, the removal of a proton from one of the four adjacent lysine residues of $\mathrm{K_4G_8}$ lowered the Coulomb energy, making it comparable to the Coulomb energies of the other model peptide ions. Moreover, the conformations of all $3+$ ions were similar. (In addition, the modeled structures for $3+$ ions show little evidence of a $\mathrm{NH_2}\cdots\mathrm{H^+}\cdots\mathrm{NH_2}$ interaction.) As a result, the GAs for $[\mathrm{M}+3\mathrm{H}]^{3+}$ from the three peptides varied over a relatively narrow range of 10 kJ mol^{-1}.

The impact of conformation was demonstrated by Ogorzalek Loo et al.[69–71] by using comparisons of deprotonation reactivity between disulfide-intact and -reduced protein ions. They found that ions with intact disulfide bonds underwent proton transfer more readily than the analogous ions with cleaved linkages. In contrast, for the protein hen egg-white lysozyme, Williams and coworkers[59] reported that ions with cleaved linkages react more rapidly for most charge states. The explanation of this seeming discrepancy may be that cleavage of disulfide bonds affects the conformations of various peptides in different ways. For some compounds, intact bonds may give more compact structures, greater Coulomb energies, and faster rates of proton transfer. For other compounds, the enhanced flexibility of cleaved bonds may result in more compact structures and faster rates. If a structure is too compact, proton transfer rates may also be lessened because the reactant neutral may have difficulty in accessing the site to be deprotonated.

Studies by Williams and coworkers have provided further information on conformational effects and, consequently, on intramolecular H-bonding for proton transfer in large ion systems. They used ion/molecule reactions and molecular modeling to study $[M+nH]^{n+}$ produced from gramicidin S,[47] cytochrome *c*,[57] and lysozyme.[59] This group has also developed a model for calculating dielectric polarizability (ε_r) of isolated peptide and protein ions using gas-phase proton transfer data.[46,47] Williams has authored a review of the proton transfer reactivity for large, multiply charged ions, which includes detailed summaries of the pioneering work by his group.[72] In addition, Green and Lebrilla[73] have recently reviewed the proton transfer and H/D exchange reactions of peptide and protein ions.

The intrinsic basicity/acidity of the site being deprotonated is also a factor in deprotonation reactions for $[M+nH]^{n+}$. Carr and Cassady[44] studied deprotonation of several small peptide ions with a comparable number of amino acid residues (i.e., 11–14) but varying in composition and sequence. All of the ions were sequentially deprotonated to the 1+ charge state. The GA_{app}s ranged from 809 kJ mol^{-1} (for $[M+4H]^{4+}$ of renin substrate, the ion most readily deprotonated) to >973 kJ mol^{-1} (for $[M+2H]^{2+}$ of ACTH (11–24), the ion most difficult to deprotonate). While Coulomb effects are important, they are sometimes overcome by intrinsic basicity of the protonation site. This can result in some lower charge state ions deprotonating more readily than other peptide ions with higher charges but with more basic protonation sites. However, as a second study[56] indicates, intrinsic basicity becomes less of a factor in proton transfer reactions as the size of the peptide increases. For three peptides with average molecular masses of 1759, 3496, and 8565 Da, the GA_{app}s generally agreed to within 40 kJ mol^{-1} at a given *m*/*z* value despite large differences in molecular mass and charge. The smaller peptide, renin substrate, contained ions that deviated the most from this trend.

There have been several reports[44,56,58,59,74] of proton transfer reactions, revealing the presence of multiple structural isomers for $[M+nH]^{n+}$. Although gas-phase deprotonation reactions, such as Reaction 14.12, are first-order in both the ionic and neutral reactants, the overwhelming excess of neutrals inside of the mass spectrometer (even at pressures of 10^{-7}–10^{-8} mb) results in pseudo-first-order kinetic behavior. Therefore, an ion population reacting at one rate should give a linear plot of $\ln[M+nH]^{n+}$ versus reaction time. For processes such as the reaction of oxidized insulin chain B with 2-fluoropyridine, as shown in Fig. 14.4, non-linear behavior indicates the presence of two ion populations reacting at different rates.[18] In general, evidence for structural isomers is only

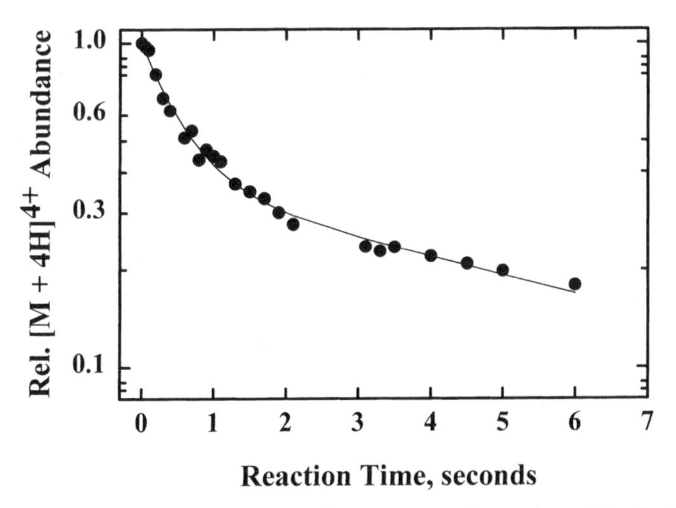

FIGURE 14.4. Reactant loss curve for the reaction of oxidized insulin B $[M+4H]^{4+}$ with 2-fluoropyridine at 1.3×10^{-7} mb. The logarithm of the abundance of $[M+4H]^{4+}$ (relative to the total abundance for $[M+nH]^{n+}$ product and parent ions) is plotted as a function of reaction time. The fitted curve represents a nonlinear regression of the data to the sum of two exponentials.

experimentally observed with near thermoneutral deprotonation reactions. As such, these are very slow processes and usually have reaction efficiencies of 0.1 or less (i.e., $\leq 10\%$ of all collisions result in a reaction). The most pronounced evidence to date for multiple isomers involves $[M+12H]^{12+}$ of ubiquitin. Proton transfer reactions, H/D exchange reactions, and CID—all indicate the existence of two ion populations with distinctly different gas-phase chemistry. These isomers may differ in conformation and/or proton location. Interestingly, the 13+ and 11+ ions of ubiquitin show only one population under the same experimental conditions. The proton transfer and CID data suggests that the isomers may differ in protonation sites. Ubiquitin has 13 basic sites that are readily protonated by ESI. Thus, the two 12+ populations may each be unprotonated at different sites. Based on the H/D exchange data, it is likely that the two populations also have different conformations. However, which came first—different protonation sites or different conformations—is unknown. If the two populations have different conformations in solution, this may have resulted in varying accessibility to the basic sites during protonation in the ESI process. Or, conversely, differing sites of protonation during ion production may lead to different conformations as the ion folds to minimize Coulomb repulsion. How solution phase conformation relates to gas-phase ion conformation is an intriguing question that should be the subject of considerable research in the years to come.

In a 1996 report, Camara et al.[75] considered the issue of chirality in protein ion chemistry. In reactions with (2R)- and (2S)-2-butylamine with cytochrome *c* $[M+nH]^{n+}$, they found that the R-isomer is another order of magnitude more reactive than the S-isomer for $n = 7$–9. Moreover, the 7+ and 8+ ions exhibited kinetic behavior which suggested the presence of isomers.

Proton transfer reactions were used in noteworthy studies of conformational effects for ions generated by ESI. Clemmer and co-workers[76] explored the conformation of ubiquitin ions before and after exposure to proton transfer reagents. Ion mobility mass spectrometry was used to measure cross sectional areas of the ions.[77,78] In this technique, the mass-selected ions flow through a chamber with an inert gas at a known pressure. The time that an ion requires to traverse the flow tube and reach the detector is related to its cross section. More compact conformers have shorter drift times and lower cross sections. In the ubiquitin study,[76] higher charge states for $[M + nH]^{n+}$ were found to favor unfolded conformations. This is consistent with unfolding to increase the distances between charge sites and minimize Coulomb repulsion. Proton transfer reactions were used to shift the charges states of the ubiquitin ions. After reaction, the products favored more compact conformations. The different conformations had a general GA_{app} order of compact < partially folded < elongated; that is, compact conformations more readily donated a proton. This same group also explored conformation and folding/unfolding pathways for disulfide-intact and -reduced lysozyme ions generated by ESI.[79] For both forms of the protein, proton stripping again caused the ions to fold up, yielding more compact conformations.

Proton transfer reactions have not been limited to peptide ions and neutral molecules. McLuckey and co-workers[80–83] reported on ion/ion reactions between peptide ions and oppositely charged organic ions. Proton transfer dominates, Reaction 14.14:

$$[M + nH]^{n+} + A^- \rightarrow [M + (n-1)H]^{(n-1)+} + AH \qquad (14.14)$$

Adduct formation, $[M + nH + A]^{(n-1)+}$, has also been reported but there is little evidence of electron transfer alone. These studies have involved peptide cations and organic anions,[80–82] as well as peptide anions and organic cations.[83] Because of Coulomb attraction between the reactants, these processes are very exoergic and occur rapidly. This makes them virtually insensitive to ion structure. Consequently, ion/ion reactions are an effective method of shifting high charge state ions down to lower charge states. This can increase the sensitivity of the mass spectral analysis because analyte signal will usually become distributed among fewer peaks. In addition, this is a useful method for decreasing the complexity of spectra during mixture analysis.[82]

2.4. Hydrogen/Deuterium Exchange Reactions

In H/D exchange reactions, a labile hydrogen attached to a heteroatom exchanges with a deuterium from a reagent, as shown in Reaction 14.15 for exchange at an oxygen:

$$ROH + R'D \rightarrow ROD + R'H \qquad (14.15)$$

Solution phase H/D exchange studies have been used for over 30 years to investigate structural changes in biomolecules, including peptides and proteins.[84] H/D exchange is particularly useful for probing conformation features. Hydrogens that are involved in intramolecular H-bonding or are shielded by the peptide chain, and thus not readily accessible to the deuterating reagent, may exchange at rates up to 10^8 times slower than exposed hydrogens.[85] Intrinsic basicity of the site undergoing exchange is also a factor in the rate of reaction. In solution, labile hydrogens at the amino acid side-chains and the N- and C-termini usually exchange too rapidly to be measured by techniques such as NMR. Thus, exchange of the amide hydrogens on the peptide backbone is studied as a structural probe. Although the most widely used technique to measure H/D exchange rates is NMR, several recent studies have used mass spectrometry to analyze solutions undergoing H/D exchange.[85–91] Mass spectrometry provides information on the number of exchanges occurring as a function of reaction time. Such studies are particularly amenable to ESI because the solution containing the biomolecule and the exchange reagent can be injected directly into the mass spectrometer.

Gas-phase H/D exchange reactions of small organic ions have been studied for many years.[92–97] The process is considered to be essentially thermoneutral, with the exchanged species being a proton and the exchange proceeding through a H-bonded complex, Process 14.16:

$$MH^+ + RD \rightleftharpoons MH^+ \cdots RD \rightleftharpoons M \cdots H^+RD \rightleftharpoons MD^+ \cdots RH \rightleftharpoons MD^+ + RH \tag{14.16}$$

For exchange to occur, sufficient energy must be available to overcome the barrier to proton transfer within the complex. Therefore, the ability of a system to undergo exchange is related to the PA or GB difference between the two reacting species. In general, the deuterating reagents are relatively low basicity compounds such as D_2O or CD_3OD. In an early study, Ausloos and Lias[92] estimated that, for exchange to occur, the PA of the neutral deuterating species must be within 84 kJ mol^{-1} of the PA of the neutral form of the sample. (This assumes that the deuterating reagent is of lower PA than the sample.)

Peptide and protein ions have recently been the subjects of gas-phase H/D exchange studies. This area was reviewed in 1997 by Green and Lebrilla.[73] Unlike solution-phase studies, gas-phase exchanges are even observed for faster-reacting sites. (Moreover, with mass spectrometry, it is the slow exchanges that prove difficult to study; ion trapping times of hours may be needed for hydrogens in a large molecule to exchange at FT-ICR pressures of ca. 10^{-7} mb.) In addition, unlike gas-phase exchanges with smaller ions, peptide and protein ion exchanges may occur even when the PA difference between the reactants is much greater than 84 kJ mol^{-1}. For example, in a study by Cheng and Fenselau[98] of renin substrate MH^+ reacting with ND_3, the PA difference was at least 200 kJ mol^{-1}. Other H/D exchange processes involving very intrinsically exothermic proton transfers within the reactive complex have also been reported.[99–101] For peptide and protein ions, it is believed that the barrier to proton transfer within the complex is lowered by the presence of multiple H-bonds.[102–105] For example, an intermediate for exchange involving a protonated amino group is shown in Structure **VIII**. Thus, heteroatoms in the amide backbone are responsible for increasing the rate of H/D exchange in peptides and proteins. As recent evidence in support of this intermediate, Cassady and coworkers[50,68] have reported negligible exchange for hydrogens on lysine amino groups that are extended too far from the backbone to form Structure **VIII**. Additional mechanisms for exchange have also been proposed.[104,105] The mechanism invoked may depend on the basicity of the deuterating reagent and the type of hydrogen undergoing exchange. However, the peptide's carbonyl sites remain important to these processes.

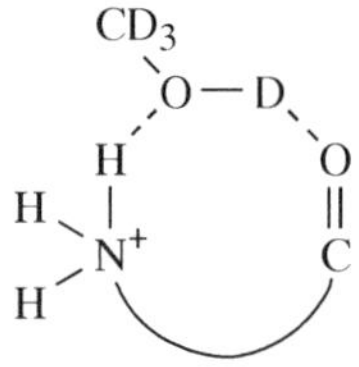

CD_3 O—D O H H N$^+$ C H

VIII

The type of hydrogen can greatly affect the exchange rate. Kinetic studies have been performed on a number of MH^+ from small peptides and amino acids.[101,104,105] The basicity of the site undergoing exchange is important, with hydrogens at more basic sites (e.g., arginine residues) exchanging slowly or not at all.[98,106] However, the results are not always easy to interpret and rates for specific sites may depend on the overall ion structure. For example, with protonated glycine the carboxylic hydrogen exchanges before the amine hydrogens; however, with protonated diglycine, the amine hydrogens exchange first.[101]

Large, multiply protonated peptide and protein ions have also been the subjects of gas-phase H/D exchange reactions. The primary emphasis has been to probe higher-order structures. As noted for solution studies, compact conformations generally exchange more slowly than extended structures. Thus, H/D exchange reactions can be used to distinguish between conformational isomers. As an example of typical H/D exchange data obtained by FT-ICR, Fig. 14.5 shows oxidized insulin chain B $[M+4H]^{4+}$ reacting with CD_3OD.[18] What was initially a single distribution of isotopomers has split into a bimodal distribution. This indicates that there are two populations of 4+ ions undergoing H/D exchange at different rates. (CH_3OD should give similar results because the alcohol deuterium is the transferring atom.)

McLafferty and co-workers[107,108] have authored two notable studies of coexisting stable conformers of large protein ions. For cytochrome *c* $[M+nH]^{n+}$, $n=6$–17, they found six distinct levels of exchange, suggesting six levels of folding/unfolding. Infrared laser heating and collisions were used to induce unfolding, while proton transfer reactions to produce lower charge states

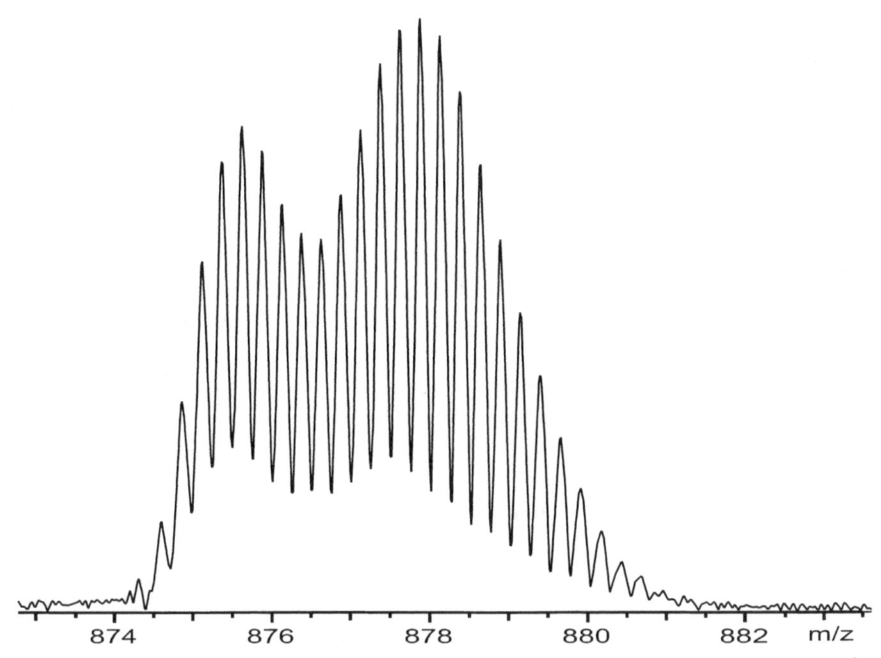

FIGURE 14.5. FT-ICR mass spectra for the reaction of oxidized insulin B $[M+4H]^{4+}$ with CD_3OD at 7.5×10^{-8} mb. The reaction time is 90.0 s.

may fold or unfold the ions.[107] In a study of S-derivatized RNase A ions, this group found that two gaseous conformers were stable in the mass spectrometer for hours.[108]

The study of ubiquitin $[M+12H]^{12+}$ isomers, discussed previously, combined proton transfer reactions and H/D exchange.[74] Both techniques indicated the existence of two ion populations. It was found that the population that underwent the fastest deprotonation also H/D exchanged most readily. Gross and Williams[47] studied singly and doubly protonated ions from the small cyclic peptide gramicidin S by deprotonation, H/D exchange, and molecular modeling. Their modeled structure was consistent with the measured exchange rates. Interestingly, the 2+ ion exchanged at rates 3–4 times more faster than the 1+ ion.

In general, H/D exchange rates decrease as the molecular size increases. For example, Campbell et al.[100,104] found that the first exchange with MH^+ from diglycine occurred with a rate constant of 3.1×10^{-10} cm^3 $molecule^{-1}$ s^{-1}, while the first exchange with pentaglycine was at 2×10^{-12} cm^3 $molecule^{-1}$ s^{-1}. This trend has been attributed to the increased intramolecular H-bonding of large peptides leading to greater basicity at the reactive site. For larger protein ions, the exchange rate is generally very low. In the studies of McLafferty and co-workers,[107,108] it was on the order of 10^{-13} cm^3 $molecule^{-1}$ s^{-1}. Here, the multiple charging should decrease the overall basicity of the ion (but may have little influence on the basicity at the reactive site). In addition, even for exposed reactive sites, the deuterating reagent may have more difficulty accessing the site as the size of the biomolecule increases.

Although the use of mass spectrometry to probe the conformations of biomolecules is a field with great potential, there is evidence to suggest that H/D exchange results must be viewed cautiously. In the initial report of H/D exchange occurring inside of a mass spectrometer for protein ions, Winger et al.[91] studied the disulfide-reduced and -intact forms of proisulin and α-lactabumin. They found that the intact forms, which were believed to be more compact, exchanged more readily. This is the opposite of the commonly accepted trend in H/D exchange of more compact conformers exchanging more slowly. However, this is not the only report that goes against the common interpretation. In an investigation that coupled H/D exchange with ion mobility measurements, low exchange levels were reported for cytochrome *c* ions with diffuse conformations.[109] Also, in a H/D exchange and molecular modeling study[50] of dodecapeptides containing only lysine and glycine residues (whose deprotonation reactions were discussed earlier), the more compact $[K_4G_8 + 4H]^{4+}$ exchanged fewer hydrogens and more slowly than the elongated $[(KGG)_4 + 4H]^{4+}$ and $[(K_2G_4)_2 + 4H]^{4+}$. The implication is that while gas-phase H/D exchange of multiply protonated protein and peptide ions can be slowed by compact structures, there are cases in which extended conformations may limit exchange. A balance may exist between conformational extension that promotes accessibility and contraction that enhances reaction intermediate stabilization by H-bonding to the peptide backbone. Therefore, as several recent studies have concluded, caution must be exercised in using H/D exchange reactions to assign structures for biological ions.[50,68,104,109]

2.5. Adduct Formation Reactions

Cluster ions or adducts may form when smaller molecules attach to peptide and protein ions, Reaction 14.17.

$$[M + nH]^{n+} + B \rightarrow [M + nH + B]^{n+} \qquad (14.17)$$

This process can occur for either singly or multiply charged ions. As an example, Fig. 14.1(c) shows the adduct $[GlyProGlyH \cdots (1\text{-methylpiperidine})]^+$ as a reaction product. In a study of MH^+ from polyglycines,[28] Reaction 14.17 was more pronounced as the size of the molecule increased. This is consistent with larger molecules having more degrees of freedom available to disperse energy and thus limit dissociation of the adduct ion. Reaction 14.17 is very much dependent on energetics and is observed in the greatest abundance when the reaction is close to be thermoneutral. As proton transfer becomes increasingly exoergic, the amount of excess internal energy in the adduct ion (which can be considered as the reaction intermediate for the proton transfer process) increases; thus, its rate of dissociation increases and the observed adduct intensity decreases. In fact, the magnitude of adduct formation can serve as a rough indicator of the energetics of the proton transfer reaction for MH^+.

As studied for singly-protonated polyglycines, Gly_nH^+, adduct formation is also pressure dependent.[28] If proton transfer is slightly endoergic, increasing the pressure of either the reference compound or of an inert gas increases the observed Gly_nHB^+ intensity because of collisional stabilization. In contrast, if the proton transfer reaction is exoergic, increasing the pressure gives a decreased Gly_nHB^+ intensity; here, the combination of excess internal energy and increased pressure facilitates dissociation of the adduct.

For peptides, $[M+nH]^{n+}$ have been observed to attach multiple smaller molecules.[44,63,110] The rate and number of attachments depend on the peptide sequence; however, the exact structural features that influence adduct formation are unknown. In a study of oxidized bovine insulin chain B ions, Morris et al.[110] have suggested that the number of clustering ammonia molecules depend on the number of available acidic sites. It is doubtful that this is a universal conclusion because, for several other peptide ions, adduct attachment does not correlate well with the number of acidic residues.[44]

Stephenson and McLuckey[111] have investigated the attachment of hydroiodic acid HI, to peptide $[M+nH]^{n+}$, Reaction 14.18.

$$[M+nH]^{n+} + xHI \rightarrow [M+nH+xHI]^{n+} \qquad (14.18)$$

For all the 21 compounds studied, the sum of the ion charge n, and the maximum number of attached HI molecules, x, is equal to the total number of basic sites on the ion. (Basic sites being defined as arginine, lysine, and histidine residues, as well as the amino group of the N-terminal residue.) This suggests that HI is attached to unprotonated basic sites on the ions. Thus, HI adduction reactions can be used to count the number of basic sites in a peptide. In a follow-up study,[112] iodide anions, I^-, were also found to react with $[M+nH]^{n+}$ of the peptide ubiquitin. This process competes with proton transfer to the anion. Surprisingly, reaction rates increased inversely with ion charge. The data is not consistent with the formation of proton-bound cluster ions. Instead, it suggests the formation of ion pairs or dipole/dipole bonding involving neutral acid and neutral basic sites on the peptide. The implication is that gas-phase peptide and protein ions are amphoteric, with the nature of the second reactant dictating whether the biomolecule's acidic or basic sites would be involved in reactions.

3. UNIMOLECULAR DISSOCIATION

3.1. Small Amide Ions

For molecular ions formed by loss of an electron, $M^{\cdot+}$, a dominant unimolecular dissociation pathway is cleavage at the (O=C)–C bond involving the amide

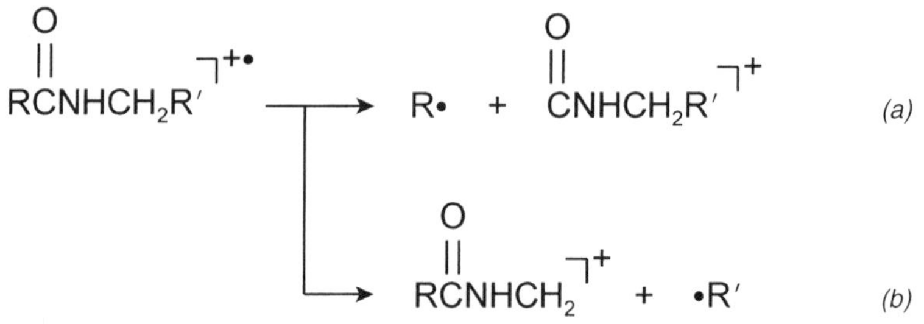

$$\mathrm{RC(=O)NHCH_2R'}^{+\bullet} \longrightarrow \mathrm{R^{\bullet}} + \mathrm{C(=O)NHCH_2R'}^{+} \quad (a)$$

$$\longrightarrow \mathrm{RC(=O)NHCH_2}^{+} + {}^{\bullet}\mathrm{R'} \quad (b)$$

FIGURE 14.6. Common dissociation pathways for a secondary amide molecule ion, $M^{\cdot+}$, under EI conditions.

carbon.[1] This is shown in path (a) of Fig. 14.6 and is analogous to the cleavage that occurs alpha to the carbonyl carbon of ketones, aldehydes, esters, and carboxylic acids. With secondary and tertiary amides, dissociation initiated by the nitrogen also occurs. An example is cleavage of the C–C bond beta to the nitrogen, as shown in path (b) of Fig. 14.6. Ionization energies (IE) associated with the removal of an electron from the amide oxygen are low. For example, the IE of formamide is 10.16 eV, in contrast to an IE of 11.52 eV for ethane.[113] Thus, most ions in the electron ionization mass spectra of amides have the charge on the fragment containing the amide functional group.

Protonated ions, MH^+, from small amides can be produced by chemical ionization. As noted earlier, for amides which do not contain other basic functional groups, the most energetically favorable site of protonation is the carbonyl oxygen. Thus, immediately following ion formation, MH^+ are probably protonated at the amide oxygen. However, upon ion activation in MS/MS processes, the proton may transfer to the amide nitrogen. The result is that, after an additional proton transfer, the amide C–N bond cleaves with retention of the charge on the N-containing fragment. Cleavage of the C–N bond is rare for $M^{\cdot+}$ because, in simple amides, this bond is strengthened by n-π conjugation involving the nitrogen long pair electrons and the electrons of the carbonyl group.[114] However, protonation of the amide nitrogen destroys this conjugation, thus lowering the energy needed to cleave the C–N bond for protonated amides.

Lin et al.[15] studied protonated formamide by ab initio calculations and CID. Their work suggests that dissociation pathways are dependent upon the location of the proton. The more stable O-protonated isomer is the main precursor for pathways involving H_2O loss and NH_3 loss, while elimination of CO occurs from the less stable N-protonated isomer.

Tu and Harrison[114] have recently conducted an extensive study of MH^+ fragmentation for secondary amides under low-energy CID conditions. They found that the nature of secondary functional group attached to the amide nitrogen can play an important role in fragmentation. If this neighboring group will readily accept a proton, the energy barrier to proton transfer is lowered which further facilitates cleavage of the C–N bond.

3.2. Singly Protonated Peptide Ions

In recent years, the dissociation of peptide MH^+ has been one of the most widely studied areas of mass spectrometry. Over a thousand reports have appeared on this topic, some emphasizing fundamental aspects of the dissociation process and others focusing on its analytical utility for peptide sequencing. The process actually involves an activation step to add excess internal energy to the ion, followed by an ion dissociation step. The MS/MS techniques employed have included low-energy CID, high-energy CID, surface-induced dissociation (SID), and metastable decay. Approximately fifty reports have also appeared on the dissociation of deprotonated peptide ions, $[M-H]^-$. Payannopoulos[115] has published an extensive review of the interpretation of CID spectra of peptides. This includes information on how specific residues affect dissociation. Because of the large number of papers in this area, the present report will be limited to a general discussion of peptide dissociation with an emphasis on features of fundamental interest to gas-phase ion chemistry.

Descriptions of peptide fragment ions are based on a nomenclature proposed by Roepstroff and Fohlman.[116] Small letters denote dissociation sites and subscript numbers specify the number of the amino acid residues on the ion. For multiply charged ions, superscripts represent the ion's charge state. The most common cleavages for peptides are shown in Fig. 14.7. Ions retaining the N-terminus have a-, b-, and c-cleavages, while C-terminal ions have x-, y-, and z-cleavages. Ions formed from c- and y-cleavages also involve rearrangements with two added hydrogens. Thus, they may be known as $[c_n + 2H]^+$ and $[y_n + 2H]^+$. Alternatively, two prime symbols can be used as a shorthand for added hydrogens; that is, c_n'' has the same meaning as $[c_n + 2H]^+$. In addition, using a modification by Biemann,[117] the symbols c_n and y_n refer to ions formed by these corresponding cleavages plus the addition of two hydrogens; that is, c_n means $[c_n + 2H]^+$. The present report will use the nomenclature as modified by Biemann. For internal ions formed by cleavages from both ends of the peptide chain, the residues incorporated in the fragment ions are denoted by standard one letter amino acid symbols.

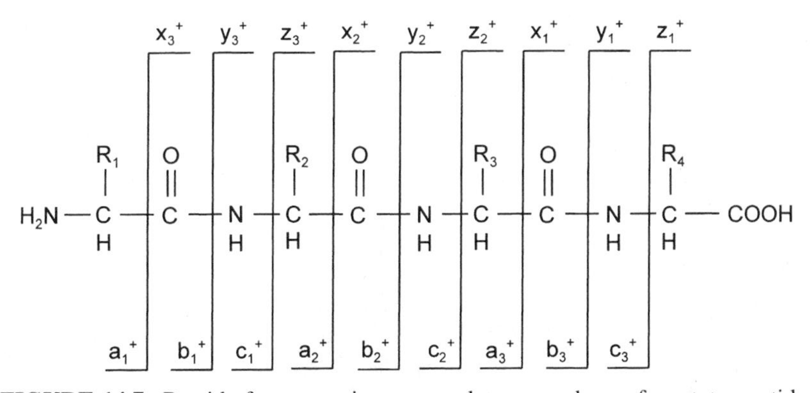

FIGURE 14.7. Peptide fragmentation nomenclature, as shown for a tetrapeptide.

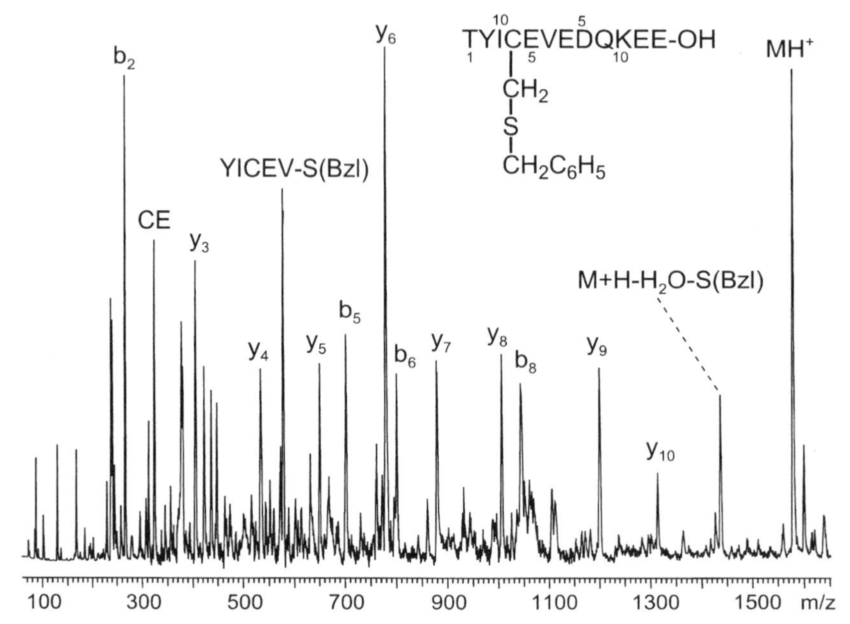

FIGURE 14.8. MALDI-TOF PSD spectrum of MH^+ from benzylated Cys(bzl)[84]-CD_4(81–92).

As an example of a typical peptide MH^+ dissociation, Fig. 14.8 shows the post-source decay (PSD) spectrum of benzylated Cys(Bzl)[84]-CD_4(81–92).[18] This is a fragment from a peptide under study as an HIV inhibitor.[118,119] PSD is a metastable dissociation technique that is performed in a time-of-flight (TOF) mass spectrometer.[120] For Fig. 14.8, the ions were produced by MALDI. A notable feature of the spectrum is the almost complete series of y_n, $n = 3–10$. The y-ions' N-terminal analog, b-ions, are also common in the spectrum. Both series are formed by cleavage at the (O=C)–NH amide linkage, which is known as the "peptide bond". The y- and b-ions dominate in dissociation processes that impart a relatively low level of internal energy to a peptide ion (e.g., low-energy CID and PSD). Losses of small molecules, such as H_2O, NH_3, or CO, from the sequence ions are also common. When higher energies are imparted to an ion (e.g., high energy CID and SID), a-, c-, x-, and z-ions become more of a factor.[117] Higher energies can also lead an increased observation of side chain cleavages, which are known as d- and w-ions.[121,122] However, there are no universal rules and, for various peptides, virtually any type of ion may be observed at any energy regime.

The presence of basic amino acid residues can impact fragmentation pathways.[49,52,122–126] Charge localization on a basic residue produces a homogeneous population of MH^+. Fragmentation that occurs adjacent to the charge site may be charge-directed, while the remainder of the fragmentation ions have charge-remote mechanisms. Charge-remote cleavages are less common (especially for low-energy techniques) and the result is general reduction in the extent of fragmentation. This is particularly true for peptides containing several arginine residues; however, histidine and lysine residues may also limit fragmentation.

In contrast, a lack of highly basic sites, leads to protons located at many sites along the peptide backbone. These include protonation at amide carbonyl oxygens, which (as discussed earlier) is more energetically favored than N-protonation.[16,28,29,34,39] However, semi-empirical calculations[127] suggest that other less stable protonated structures may also form and participate in dissociation. The ion activation process may provide energy for thermodynamically unfavorable proton transfers that promote fragmentation. In addition, not all protonated forms fragment. In support of this "mobile proton model,"[52] a series of deuterium labeling studies[128–130] suggest that rapid intramolecular proton transfer can occur after ion activation. The overall implication is that for less basic peptides, there is a more heterogenous population of charge sites. This results in numerous charge-directed (or charge-initiated) fragmentations and a wider range of fragment ions, which facilitates peptide sequencing.

There are many reports of fragmentations related to the presence of specific amino acid sequences; these have been reviewed by Papayannopoulas.[115] Perhaps the most pronounced effect involves proline residues. For both singly and multiply protonated peptide ions, abundant y-ions arise from cleavage of the amide bond on the N-terminal side of the proline residue.[121,125,131–136] Among the twenty common L-α-amino acids, proline, as shown in Structure **III**, is the only α-imino acid and the only secondary amine. The cause of the "proline effect" on peptide dissociation is probably due to a combination of factors. First, because of the pyrrolidine ring, a b-ion produced by cleavage adjacent to, and incorporating, a proline residue would be a strained [3.3.0] bicyclic system.[132] This accounts for the low intensity of this process. Second, proline has a relatively high GB—only arginine, lysine, and histidine are more basic.[19,21] Thus, some preferential protonation at the proline residue is possible and may enhance the charge-directed fragmentation adjacent to proline.[131] Finally, proline's ring imposes rigid constraints on $N{-}C^{\alpha}$ rotation and may introduce a kink of 20° or more into α-helices. Because of this, proline plays a significant role in protein folding, which is necessary to achieve the physiological active, three-dimensional structure of the protein. The result is that the proline linkages may "jut out" from the peptide backbone and, especially for larger peptides and proteins, be more accessible to the ionizing proton.[136] Again, this would enhance charge-directed fragmentation adjacent to the proline residue.

3.3. Multiply Protonated Peptide and Protein Ions

Multiply protonated ions have the same fragmentation pathways as their singly charged counterparts. In addition, charge-transfer and charge-stripping processes may also occur.[137] Generally, as charge on the parent ion increases, the ion dissociates more readily.[138–141] This is consistent with greater Coulomb energy facilitating dissociation.

Overall, the dissociation spectra of $[M + nH]^{n+}$ are complex because the product ions can have a variety of charge stages. For example, a parent $[M + 10H]^{10+}$ has the potential for yielding products with charges in the range of 10+ to 1+. However, as first noted by Fabris et al.,[142] when residues are cleaved, charges are also lost. Thus, mass-to-charge ratios (m/z) remain relatively constant and there is a marked tendency of product peaks to be close in m/z to the parent peak.

For $[M + nH]^{n+}$, dissociation products usually contain no more protons than the number of basic residues (arginine, lysine, histidine, N-terminal residue) in the fragment. The implication is that, if the charge sites are relatively well-distributed, protons reside on the basic residues.[44] Studies with high-energy CID,[124] low-energy CID,[49] and SID[52] indicate that charge can remain localized on two adjacent residues. Fragmentation is often more prominent at positions adjacent to basic residues.[49,143] However, if a peptide ion contains fewer charges than its maximum number of basic residues (i.e., the ion is "unsaturated" with respect to protons), the trend is for charges to be distributed as far apart as possible to minimize Coulomb repulsion.[142] In addition, for peptide ions with severe Coulomb repulsion, protons may migrate to less basic sites to minimize repulsion.[49,143] For example, low-energy CID on $[K_4G_8 + 4H]^{4+}$, which has four adjacent basic residues at the N-terminus, the major dissociation products are b-ions formed from cleavages near the C-terminus. This suggests that high Coulomb repulsion has resulted in proton transfer to the C-terminal residue. Also, analogous to the proposed dissociation mechanisms for MH^+, $[M + nH]^{n+}$ may have mobile protons that promote charge-directed dissociations.[52,123,144]

4. CONCLUDING REMARKS

Gas-phase amide, peptide, and protein ions exhibit a rich chemistry. Although much work has been performed, the surface has only been scratched. Further expansion should occur in areas such as the sequencing of very large proteins, the use of ion/molecule reactions to obtain structural information on peptides and proteins, and the probing of biomolecular conformation by mass spectrometry.

ACKNOWLEDGMENTS

Support from the National Institutes of Health and the Ohio Academic Challenge Program is gratefully acknowledged. Special thanks are extended to Nigel Ewing and Jaran Jai-nhuknan for their assistance in the preparation of illustrations.

REFERENCES

1. McLafferty, F. W.; Turecek, F. *Interpretation of Mass Spectra*, University Science Books: Mill Valley, CA, 1993.
2. Harrison, A. G. *Chemical Ionization Mass Spectrometry*, CRC Press: Boca Raton, FL, 1992.
3. Barber, M.; Bordoli, R.; Segwick, R. D.; Tyler, A. N. *J. Chem. Soc. Chem. Commun.* 1981, **7**, 325.
4. Whitehouse, C. M.; Dreyer, R. N.; Yamashita, M.; Fenn, J. B. *Anal. Chem.* 1985, **57**, 675.
5. Karas, M.; Hillenkamp, F. *Anal. Chem.* 1988, **60**, 2299.
6. Aue, D. H.; Bowers, M. T. Stabilities of positive ions from equilibrium gas-phase basicity measurements. In *Gas Phase Ion Chemistry*, Bowers, M. T. Ed.; Academic Press: New York, 1979, Vol. 2, pp. 1–51.
7. Aue, D. H.; Betowski, L. D.; Davidson, W. R.; Bowers, M. T.; Beak, P.; Lee, J. *J. Am. Chem. Soc.* 1979, **101**, 1361.
8. Alvarez, E. J.; Brodbelt, J. S. *J. Mass Spectrom.* 1996, **31**, 901.
9. Abboud, J.-L. M.; Canada, T.; Homan, H.; Notario, R.; Cativiela, C.; Diaz de Villegas, M. D.; Bordeje, M. C.; Mo, O.; Yanez, M. *J. Am. Chem. Soc.* 1992, **114**, 4728.
10. Witt, M.; Grutzmacher, H. F. *Int. J. Mass Spectrom. Ion Proc.* 1997, **164**, 93.
11. Kinser, R. D.; Ridge, D. P.; Hvistendahl, G.; Rasmussen, B.; Uggerud, E. *Chem. Eur. J.* 1996, **2**, 1143.
12. Witt, M.; Grutzmacher, H. F. *Int. J. Mass Spectrom. Ion Proc.* 1997, **165**, 49.
13. Hunter, E. P.; Lias, S. G. Proton affinity evaluation. In *NIST Chemistry WebBook, NIST Standard Reference Database Number 69*, Mallard, W. G.; Linstrom, P. J. Eds.; National Institute of Standards and Technology: Gaithersburg, MD, March 1998.
14. Hvistendahl, G.; Uggerud, E. *Org. Mass Spectrom.* 1991, **26**, 67.
15. Lin, H. Y.; Ridge, D. P.; Uggerud, E.; Vulpius, T. *J. Am. Chem. Soc.* 1994, **116**, 2996.
16. Krug, J. P.; Popelier, P. L. A.; Bader, R. F. W. *J. Phys. Chem.* 1992, **96**, 7604.
17. McLuckey, S. A.; Cameron, D.; Cooks, R. G. *J. Am. Chem. Soc.* 1981, **103**, 1313.
18. Cassady, C. J.; Ewing, N. P.; Jai-nhuknan, J. (unpublished results).
19. Harrison, A. G. *Mass Spectrom. Rev.* 1997, **16**, 201.
20. Locke, M. J.; Hunter, R. L.; McIver, R. T. *J. Am. Chem. Soc.* 1979, **101**, 272.
21. Gorman, G. S.; Speir, J. P.; Turner, C. A.; Amster, I. J. *J. Am. Chem. Soc.* 1992, **114**, 3986.
22. Li, X.; Harrison, A. G. *Org. Mass Spectrom.* 1993, **28**, 366.
23. Bojesen, G. *J. Am. Chem. Soc.* 1987, **109**, 5557.
24. Campbell, S.; Beauchamp, J. L.; Rempe, M.; Lichtenberger, D. L. *Int. J. Mass Spectrom. Ion Proc.* 1992, **117**, 83.
25. Isa, K.; Omote, T.; Amaya, M. *Org. Mass Spectrom.* 1990, **25**, 620.
26. Hunter, E. P.; Lias, S. G. *J. Phys. Chem. Ref. Data* 1998, **27**, 413.

27. Wu, Z.; Fenselau, C. *Rapid Commun. Mass Spectrom.* 1992, **6**, 403.
28. Zhang, K.; Zimmerman, D. M.; Chung-Phillips, A.; Cassady, C. J. *J. Am. Chem. Soc.* 1993, **115**, 10812.
29. Wu, J.; Lebrilla, C. B. *J. Am. Chem. Soc.* 1993, **115**, 3270.
30. Wu, Z.; Fenselau, C. *J. Am. Soc. Mass Spectrom.* 1992, **3**, 863.
31. Wu, Z.; Fenselau, C. *Tetrahedron* 1993, **49**, 9197.
32. Wu, J.; Lebrilla, C. B. *J. Am. Soc. Mass Spectrom.* 1995, **6**, 91.
33. Gorman, G. S.; Amster, I. J. *J. Am. Chem. Soc.* 1993, **115**, 5729.
34. Cassady, C. J.; Carr, S. R.; Zhang, K.; Chung-Phillips, A. *J. Org. Chem.* 1995, **60**, 1704.
35. McKiernan, J. W.; Beltrame, C. E. A.; Cassady, C. J. *J. Am. Soc. Mass Spectrom.* 1994, **5**, 718.
36. Ewing, N. P.; Zhang, X.; Cassady, C. J. *J. Mass Spectrom.* 1996, **31**, 1345.
37. Carr, S. R.; Cassady, C. J. *J. Am. Soc. Mass Spectrom.* 1996, **7**, 1203.
38. Wu, Z.; Fenselau, C. *Rapid Commun. Mass Spectrom.* 1994, **8**, 777.
39. Zhang, K.; Cassady, C. J.; Chung-Phillips, A. *J. Am. Chem. Soc.* 1994, **116**, 11512.
40. Cheng, X.; Wu, Z.; Fenselau, C. *J. Am. Chem. Soc.* 1003, **115**, 4844.
41. Kim, K.; Friesner, R. A. *J. Am. Chem. Soc.* 1997, **119**, 12952.
42. Loo, J. A.; Edmonds, C. G.; Udseth, H. R.; Smith, R. D. *Anal. Chem.* 1990, **62**, 693.
43. Guevremont, R.; Siu, K. W. M.; Le Blanc, J. C. Y.; Berman, S. S. *J. Am. Soc. Mass Spectrom.* 1992, **3**, 216.
44. Carr, S. R.; Cassady, C. J. *J. Mass Spectrom.* 1997, **32**, 959.
45. Schnier, P. D.; Gross, D. S.; Williams, E. R. *J. Am. Soc. Mass Spectrom.* 1995, **6**, 1086.
46. Gross, D. S.; Rodriguez-Cruz, S. E.; Bock, S.; Williams, E. R. *J. Phys. Chem.* 1995, **99**, 4034.
47. Gross, D. S.; Williams, E. R. *J. Am. Chem. Soc.* 1995, **117**, 883.
48. Gronert, S. 45th ASMS Conference on Mass Spectrometry and Allied Topics, Palm Springs, CA, June 3, 1997.
49. Zhang, X.; Jai-nhuknan, J.; Cassady, C. J. *Int. J. Mass Spectrom. Ion Proc.* 1997, **171**, 135.
50. Zhang, X.; Ewing, N. P.; Cassady, C. J. *Int. J. Mass Spectrom. Ion Proc.* 1998, **175**, 159.
51. Downard, K. M.; Biemann, K. *Int. J. Mass Spectrom. Ion Proc.* 1995, **148**, 191.
52. Dongre, A. R.; Jones, J. L.; Somogyi, A.; Wysocki, V. H. *J. Am. Chem. Soc.* 1996, **118**, 8365.
53. Gronert, S. *J. Am. Chem. Soc.* 1996, **118**, 3525.
54. Petrie, S.; Javahery, G.; Bohme, D. K. *Int. J. Mass Spectrom. Ion Proc.* 1993, **124**, 145.
55. Petrie, S.; Javahery, G.; Wincel, H.; Bohme, D. K. *J. Am. Chem. Soc.* 1993, **115**, 6290.
56. Zhang, X.; Cassady, C. J. *J. Am. Soc. Mass Spectrom.* 1996, **7**, 1211.
57. Schnier, P. D.; Gross, D. S.; Williams, E. R. *J. Am. Chem. Soc.* 1995, **117**, 6747.

58. Cassady, C. J.; Wronka, J.; Kruppa, G. H.; Laukien, F. H. *Rapid Commun. Mass Spectrom.* 1994, **8**, 394.
59. Gross, D. S.; Schnier, P. D.; Rodriguez-Cruz, S. E.; Fagerquist, C. K.; Williams, E. R. *Proc. Natl. Acad. Sci. USA* 1996, **93**, 3143.
60. Kaltashov, I. A.; Fabris, D.; Fenselau, C. C. *J. Phys. Chem.* 1995, **99**, 10046.
61. Kaltashov, I. A.; Fenselau, C. C. *J. Am. Chem. Soc.* 1995, **117**, 9906.
62. Kaltashov, I. A.; Fenselau, C. C. *Rapid Commun. Mass Spectrom.* 1996, **10**, 857.
63. McLuckey, S. A.; Van Berkel, G. J.; Glish, G. L. *J. Am. Chem. Soc.* 1990, **112**, 5668.
64. Gioumousis, G.; Stevenson, D. P. *J. Chem. Phys.* 1958, **29**, 294.
65. Su, T.; Bowers, M. T. *Int. J. Mass Spectrom. Ion Phys.* 1973, **12**, 347.
66. Su, T.; Chesnavich, W. J. *J. Chem. Phys.* 1982, **76**, 5183.
67. Su, T. *J. Chem. Phys.* 1988, **88**, 5355.
68. Cassady, C. J. *J. Am. Soc. Mass Spectrom.* 1998, **9**, 716.
69. Ogorzalek Loo, R. R.; Loo, J. A.; Udseth, H. R.; Fulton, J. L.; Smith, R. D. *Rapid Commun. Mass Spectrom.* 1992, **6**, 159.
70. Ogorzalek Loo, R. R.; Smith, R. D. *J. Am. Soc. Mass Spectrom.* 1994, **5**, 207.
71. Ogorzalek Loo, R. R.; Winger, B. E.; Smith, R. D. *J. Am. Soc. Mass Spectrom.* 1994, **5**, 1064.
72. Williams, E. R. *J. Mass Spectrom.* 1996, **31**, 831.
73. Green, M. K.; Lebrilla, C. B. *Mass Spectrom. Rev.* 1997, **16**, 53.
74. Cassady, C. J.; Carr, S. R. *J. Mass Spectrom.* 1996, **31**, 247.
75. Camara, E.; Green, M. K.; Penn, S. G.; Lebrilla, C. B. *J. Am. Chem. Soc.* 1996, **118**, 8751.
76. Valentine, S. J.; Couterman, A. E.; Clemmer, D. E. *J. Am. Soc. Mass Spectrom.* 1997, **8**, 954.
77. von Helden, G.; Wyttenbach, T.; Bowers, M. T. *Science* 1995, **267**, 1483.
78. Clemmer, D. E.; Jarrold, M. F. *J. Mass Spectrom.* 1997, **32**, 577.
79. Valentine, S. J.; Anderson, J. G.; Ellington, A. D.; Clemmer, D. E. *J. Phys. Chem.* 1997, **101**, 3891.
80. Stephenson, J. L.; Van Berkel, G. J.; McLuckey, S. A. *J. Am. Soc. Mass Spectrom.* 1997, **8**, 637.
81. Stephenson, J. L.; McLuckey, S. A. *J. Am. Chem. Soc.* 1996, **118**, 7390.
82. Stephenson, J. L.; McLuckey, S. A. *Anal. Chem.* 1996, **68**, 4026.
83. McLuckey, S. A.; Herron, W. J.; Stephenson, J. L.; Goeringer, D. E. *J. Mass Spectrom.* 1996, **31**, 1093.
84. Englander, S. W.; Kallenbach, N. R. *Quart. Rev. Biophys.* 1984, **16**, 521.
85. Winston, R. L.; Firzgerald, M. C. *Mass Spectrom. Rev.* 1997, **16**, 165.
86. Smith, D. L.; Deng, Y.; Zhang, Z. *J. Mass Spectrom.* 1997, **32**, 135.
87. Katta, V.; Chait, B. T. *J. Am. Chem. Soc.* 1993, **115**, 6317.
88. Thevenon-Emeric, G.; Kozlowski, J.; Zhang, Z.; Smith, D. L. *Anal. Chem.* 1992, **64**, 2456.
89. Smith, D. L.; Zhang, Z.; Liu, Y. *Pure Appl. Chem.* 1994, **66**, 89.

90. Robinson, C. V.; Chung, E. W.; Kragelund, B. B.; Knudsen, J.; Aplin, R. T.; Poulsen, F. M.; Dobson, C. M. *J. Am. Chem. Soc.* 1996, **118**, 8646.
91. Winger, B. E.; Light-Wahl, K. J.; Rockwood, A. L.; Smith, R. D. *J. Am. Chem. Soc.* 1992, **114**, 5897.
92. Ausloos, P.; Lias, S. G. *J. Am. Chem. Soc.* 1981, **103**, 3641.
93. Grabowski, J. J.; DePuy, C. H.; Van Doren, J. M.; Bierbaum, V. M. *J. Am. Chem. Soc.* 1985, **107**, 7384.
94. Ranasinghe, A.; Cooks, R. G.; Sethi, S. K. *Org. Mass Spectrom.* 1992, **27**, 77.
95. Hunt, D. F.; Sethi, S. K. *J. Am. Chem. Soc.* 1980, **102**, 6953.
96. Brauman, J. L. In *Kinetics of Ion-Molecule Reactions*; Ausloss, P. Ed.; Plenum Press: New York, 1979, pp. 153–164.
97. Lias, S. G. *J. Phys. Chem.* 1984, **88**, 4401.
98. Cheng, X.; Fenselau, C. C. *Int. J. Mass Spectrom. Ion Proc.* 1992, **122**, 109.
99. Hemling, M. E.; Conboy, J. J.; Bean, M. F.; Mentzer, M.; Carr, S. A. *J. Am. Soc. Mass Spectrom.* 1994, **5**, 434.
100. Campbell, S.; Rodgers, M. T.; Marzluff, E. M.; Beauchamp, J. L. *J. Am. Chem. Soc.* 1994, **116**, 9765.
101. Gard, E.; Willard, D.; Bregar, J.; Green, M. K.; Lebrilla, C. B. *Org. Mass Spectrom.* 1993, **28**, 1632.
102. Gard, E.; Green, M. K.; Bregar, J.; Lebrilla, C. B. *J. Am. Soc. Mass Spectrom.* 1994, **5**, 623.
103. Green, M. K.; Gard, E.; Bregar, J.; Lebrilla, C. B. *J. Mass Spectrom.* 1995, **30**, 1103.
104. Campbell, S.; Rodgers, M. T.; Marzluff, E. M.; Beauchamp, J. L. *J. Am. Chem. Soc.* 1995, **117**, 12840.
105. Gur, E. H.; de Koning, L. J.; Nibbering, N. M. M. *J. Am. Soc. Mass Spectrom.* 1995, **6**, 466.
106. Dookeran, N. N.; Harrison, A. G. *J. Mass Spectrom.* 1995, **30**, 666.
107. Wood, T. D.; Chorush, R. A.; Wampler III, F. M.; Little, D. P.; O'Connor, P. B.; McLafferty, F. W. *Proc. Natl. Acad. Sci., U.S.A.* 1995, **92**, 2451.
108. Suckau, D.; Shi, Y.; Beu, S. C.; Senko, M. W.; Quinn, J. P.; Wampler, F. M.; McLafferty, F. W. *Proc. Natl. Acad. Sci., U.S.A.* 1993, **90**, 790.
109. Valentine, S. J.; Clemmer, D. E. *J. Am. Chem. Soc.* 1997, **119**, 3558.
110. Morris, M.; Thibault, P.; Boyd, R. K. *Rapid Commun. Mass Spectrom.* 1993, **7**, 1136.
111. Stephenson, J. L.; McLuckey, S. A. *Anal. Chem.* 1997, **69**, 281.
112. Stephenson, J. L.; McLuckey, S. A. *J. Am. Chem. Soc.* 1997, **119**, 1688.
113. Lias, S. G. Ionization Energy Evaluation. In *NIST Chemistry WebBook, NIST Standard Reference Database Number 69*; Mallard, W. G.; Linstrom, P. G. Eds.; National Institute of Standards and Technology: Gaithersburg, MD, March 1998; http://webbook.nist.gov
114. Tu, Y.-P.; Harrison, A. G. *J. Am. Soc. Mass Spectrom.* 1998, **9**, 454.
115. Papayannopoulos, I. *Mass Spectrom. Rev.* 1995, **14**, 49.
116. Roepstorff, P.; Fohlman, J. *Biomed. Mass Spectrom.* 1984, **11**, 601.
117. Biemann, K. *Biomed. Environ. Mass Spectrom.* 1988, **16**, 99.

118. Nara, P. L.; Hwang, K. M.; Rausch, D. M.; Lifso, J. D.; Eiden, L. E. *Proc. Natl. Acad. Sci., U.S.A.* 1989, **86**, 7139.
119. Lifson, J. D.; Hwang, K. M.; Nara, P. L.; Fraser, B.; Padgett, M.; Dunlop, N. M.; Eiden, L. E. *Science* 1988, **241**, 712.
120. Spengler, B. *J. Mass Spectrom.* 1997, **32**, 1019.
121. Bean, M. F.; Carr, S. A.; Thorne, G. C.; Reilly, M. H.; Gaskell, S. J. *Anal. Chem.* 1991, **63**, 1473.
122. Johnson, R. S.; Martin, S. A.; Biemann, K. *Int. J. Mass Spectrom. Ion Proc.* 1988, **86**, 137.
123. Tang, X. J.; Thibault, P.; Boyd, R. K. *Anal. Chem.* 1993, **65**, 2824.
124. Downard, K. M.; Biemann, K. *J. Am. Soc. Mass Spectrom.* 1994, **5**, 966.
125. Martin, S. A.; Biemann, K. *Int. J. Mass Spectrom. Ion Proc.* 1987, **78**, 213.
126. Vachet, R. W.; Asam, M. R.; Glish, G. L. *J. Am. Chem. Soc.* 1006, **118**, 6252.
127. Somogyi, A.; Wysocki, V. H.; Mayer, I. *J. Am. Soc. Mass Spectrom.* 1994, **5**, 704.
128. Kenny, P. T. M.; Nomoto, K.; Orland, R. *Rapid Commun. Mass Spectrom.* 1992, **6**, 95.
129. Johnson, R. S.; Krylov, D.; Walsh, K. A. *J. Mass Spectrom.* 1995, **30**, 386.
130. Harrison, A. G.; Yalcin, T. *Int. J. Mass Spectrom. Ion Proc.* 1997, **165**, 339.
131. Schwartz, B. L.; Bursey, M. M. *Biol. Mass Spectrom.* 1992, **21**, 92.
132. Vaisar, T.; Urban, J. *J. Mass Spectrom.* 1996, **31**, 1185.
133. Barinaga, C. J.; Edmonds, C. G.; Udseth, H. R.; Smith, R. D. *Rapid Commun. Mass Spectrom.* 1989, **3**, 160.
134. Hunt, D. F.; Yates, J. R., III; Shabanowitz, J.; Winston, S.; Hauer, C. R. *Proc. Natl. Acad. Sci., USA* 1986, **83**, 6233.
135. Ashton, D. S.; Beddell, C. R.; Cooper, D. J.; Green, B. N.; Oliver, R. W. A. *Org. Mass Spectrom.* 1994, **28**, 672.
136. Loo, J. A.; Edmonds, C. G.; Smith, R. D. *Anal. Chem.* 1993, **65**, 425.
137. Smith, R. D.; Barinaga, C. J.; Udseth, H. R. *J. Phys. Chem.* 1989, **93**, 5019.
138. Rockwood, A. L.; Busman, M.; Smith, R. D. *Int. J. Mass Spectrom. Ion Proc.* 1991, **111**, 103.
139. Busman, M.; Rockwood, A. L.; Smith, R. D. *J. Phys. Chem.* 1992, **96**, 2397.
140. Senko, M. W.; Speir, J. P.; McLafferty, F. W. *Anal. Chem.* 1994, **66**, 2801.
141. Summerfield, S. G.; Gaskell, S. J. *Int. J. Mass Spectrom. Ion Proc.* 1997, **165**, 509.
142. Fabris, D.; Kelly, M.; Murphy, C.; Wu, Z.; Fenselau, C. C. *J. Am. Soc. Mass Spectrom.* 1993, **4**, 652.
143. Ishikawa, K.; Nishimura, T.; Koga, Y.; Niwa, Y. *Rapid Commun. Mass Spectrom.* 1994, **8**, 933.
144. Burlet, O.; Orkiszewski, R. S.; Ballard, K. D.; Gaskell, S. D. *Rapid Commun. Mass Spectrom.* 1992, **6**, 658.

CHAPTER 15

β-SHEET INTERACTIONS BETWEEN PROTEINS

SANTANU MAITRA and JAMES S. NOWICK
Department of Chemistry, University of California Irvine

1. INTRODUCTION

Proteins consist of 20 natural amino acids that are interconnected in innumerable combinations by amide bonds. Their diverse biological functions arise not only from the amino acids and their connectivity, but also from their folded three-dimensional structures. Although the overall folded structures of proteins are complex, they can be described in terms of simpler structural elements, such as α-helices, β-sheets, turns, and loops. Many non-covalent interactions contribute to these structures, including hydrophobic effects, electrostatic interactions, and hydrogen bonds. Hydrogen bonds between the amide groups of the polypeptide backbone are arguably the most distinctive of these interactions, forming regular and repeating patterns in both α-helices and β-sheets. In parallel β-sheets these patterns consist of a series of hydrogen-bonded 12-membered rings; in antiparallel β-sheets they consist of an alternating array of ten- and 14-membered rings (Fig. 15.1).

β-Sheet formation between proteins constitutes an important form of molecular recognition between amide groups and represents one general mode of protein–protein interactions. The following three sections describe three categories of β-sheet interactions between proteins: the formation of dimers and higher oligomers, β-sheet formation between different proteins, and protein aggregation. The final section explores the development of strategies that interrupt β-sheet interactions and block β-sheet formation.

The Amide Linkage: Selected Structural Aspects in Chemistry, Biochemistry, and Materials Science,
Edited by Arthur Greenberg, Curt M. Breneman, and Joel F. Liebman
ISBN 0-471-35893-2

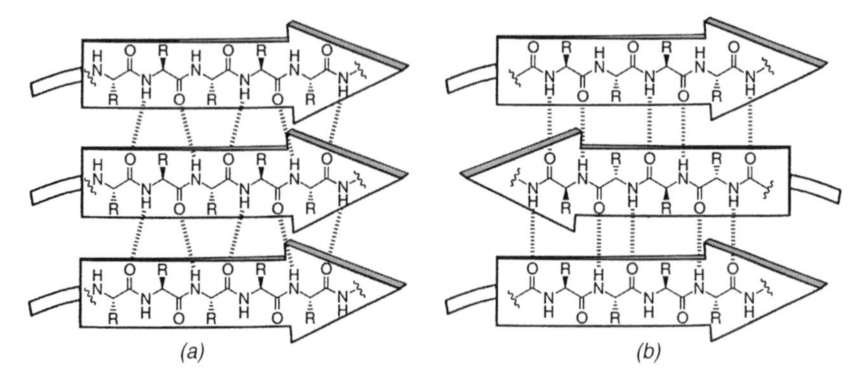

FIGURE 15.1. Ribbon diagrams and molecular representations of (*a*) parallel and (*b*) antiparallel β-sheets.

2. β-SHEET INTERACTIONS WITHIN PROTEIN DIMERS AND HIGHER OLIGOMERS

Many proteins function as dimers that are held together in part by the formation of a β-sheet. HIV-1 protease, an enzyme that plays an integral role in the reproduction of the AIDS virus, forms a dimer composed of two interdigitating β-sheets.[1] Figure 15.2 illustrates the structure of the dimer, as a protein ribbon diagram.[2]* Each monomer consists of 99 amino acids, and the N- and C-termini of the monomers form salt bridges in the β-sheets. Figure 15.3 provides a molecular representation of the dimerization interface. The Rous sarcoma virus (RSV) protease has a related dimeric β-sheet structure (PDB reference 2rsp).[3,4]

The *met* repressor protein, which regulates the biosynthesis of methionine and related biomolecules by binding to the corresponding genes, is also a dimer (Fig. 15.4).[5,6] The dimerization interface that binds to the major groove of DNA is a two-stranded antiparallel β-sheet. The amino acid side-chains projecting from the β-sheet help bind the DNA and help account for the specificity of the *met* repressor. The side-chains of Lys-23 and 23′ and Thr-25 and 25′, which project from the upper face of the β-sheet, make noteworthy contacts with the DNA bases. Figure 15.5 provides a molecular representation of this β-sheet.

A number of other DNA binding proteins function as oligomers with β-sheet interfaces between their monomeric units. The β-subunit (processivity factor) of *E. coli* DNA polymerase III forms a ring-shaped dimer that encircles DNA (Fig. 15.6).[7] The interior of the ring is lined with α-helices, and the outside of the ring is made up of β-sheets. The ring forms a sort of sliding clamp that helps

*Protein structures are available from the Research Collaboratory for Structural Bioinformatics (http://www.rcsb.org) and can be viewed on almost any computer with RasMol (http://www.umass.edu/microbio/rasmol/).

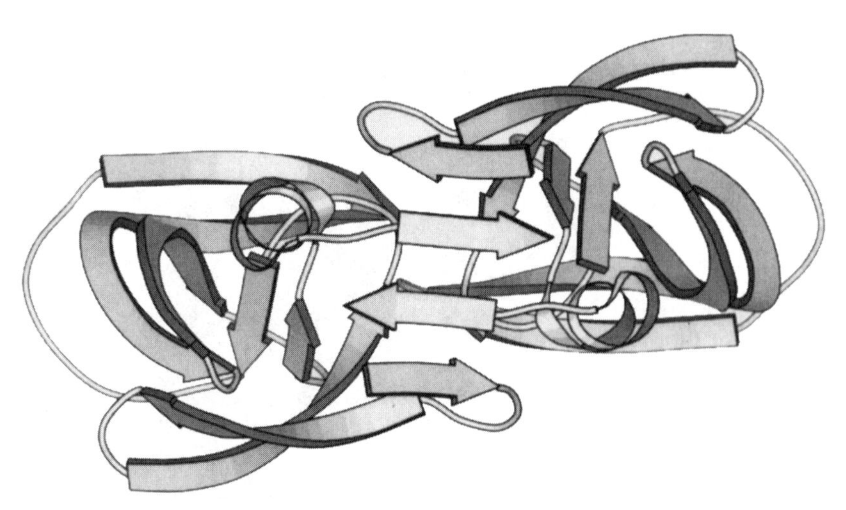

FIGURE 15.2. Ribbon diagram of HIV-1 protease illustrating four-stranded antiparallel β-sheet dimerization interface. (PDB reference 9hvp; see also 3hvp and many related HIV-1 protease structures.)

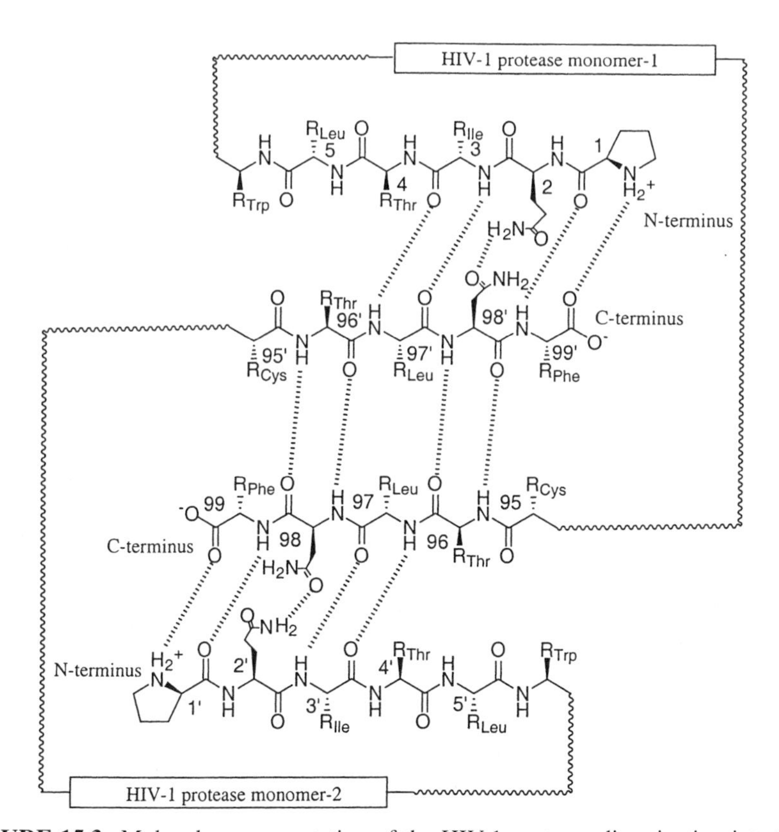

FIGURE 15.3. Molecular representation of the HIV-1 protease dimerization interface.

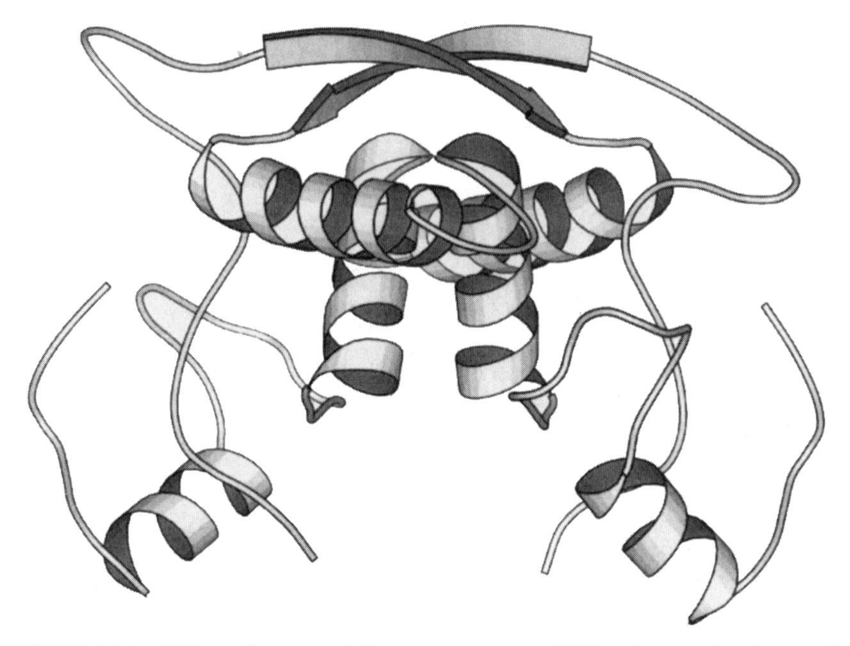

FIGURE 15.4. Ribbon diagram of the *met* repressor. (PDB reference 1cmb; see also 1cma and 1 cmc.)

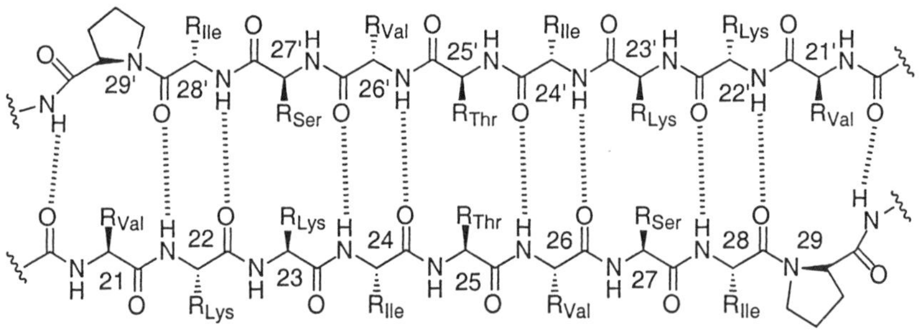

FIGURE 15.5. Molecular representation of the β-sheet dimerization interface of the *met* repressor.

keep the polymerase attached to the DNA during replication. β-Sheet interactions between its two halves help hold the clamp together. The eukaryotic DNA processivity factor has a similar ring shape but is a trimer (PDB reference 1plq and others).[8]

The lectins function as dimers, trimers, and tetramers. Lectins bind to carbohydrates and are involved in a wide variety of cellular adhesion processes. Their oligomeric structures make the lectins multivalent, endowing them with multimeric quaternary structures with multiple carbohydrate binding sites. In

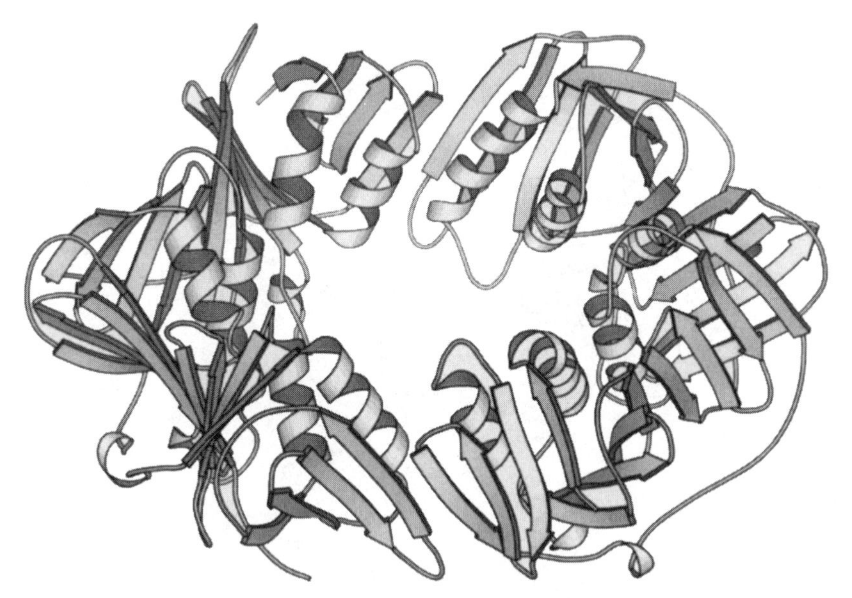

FIGURE 15.6. Ribbon diagram of the β-subunit of *E. coli* DNA polymerase III. The two halves of the dimer are on the left and right. (PDB reference 2pol.)

some of the lectins, β-sheet formation is key in holding the monomers together. Figure 15.7 illustrates the dimeric structure of a β-galactosyl-binding lectin (galectin) from toad (*Bufo arenarum*) ovary.[9] In this structure, each monomer is a sandwich comprising two antiparallel β-sheets; one edge of the sandwich is self-complementary and forms the dimer. Concanavalin A, a widely studied lectin, forms a tetrameric structure consisting of two dimeric units that resemble those shown in Fig. 15.7 (PDB references 2cna, 3cna, and many others).[10,11] Transthyretin, a non-lectin protein that is involved in the binding and transport of thyroxine and retinol, forms a similar tetrameric structure (PDB reference 1tta and many others).[12–14]

The defensins also form and function as β-sheet dimers. These small amphiphilic proteins are produced by the immune system and help destroy pathogens by disrupting their plasma membranes. Figure 15.8 illustrates the β-sheet dimeric structure of defensin HNP-3.[15] Many snake venom toxins, such as cardiotoxin V_4^{II}, κ-bungarotoxin, and erabutoxin b, form similar β-sheet dimers (PDB references 1cdt, 1kba, 6ebx, and others).[16–19]

β-Sheet interactions are involved in many other aspects of immune system function. A number of immunoregulatory proteins (cytokines) form β-sheet dimers or higher oligomers. Interleukin-8 (IL-8), which promotes the accumulation and activation of neutrophil leukocytes and is involved in inflammation, forms a β-sheet dimer (Fig. 15.9).[20–22] Monocyte chemoattractant protein-1 (MCP-1) is also involved in recruiting leukocytes and forms a β-sheet

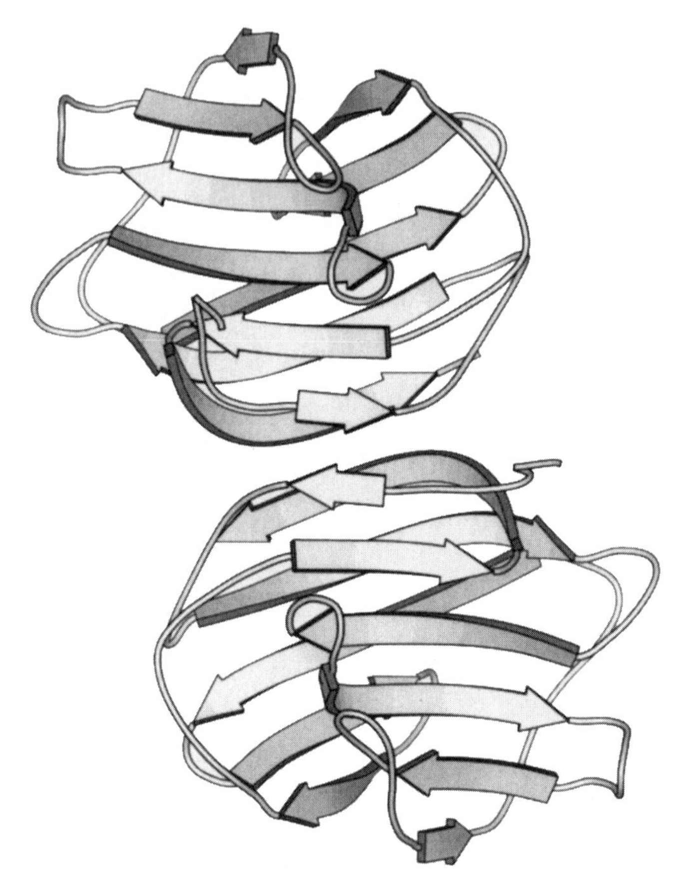

FIGURE 15.7. Ribbon diagram of the β-galactosyl-binding lectin from toad ovary. (PDB reference 1gan; see also 5cna.)

dimer (PDB references 1dol, 1dom).[23,24] Glycosylation-inhibiting factor (GIF), a cytokine that helps regulate immunoglobulin E (IgE) synthesis, forms a trimeric barrel consisting of three six-stranded β-sheets (PDB reference 1gif).[25]

The major histocompatibility complex (MHC) molecules present peptide and protein antigens to T cells to elicit a response from the immune system. Class II MHC molecules are heterodimers, which resemble the IL-8 dimer and consist of a dimeric β-sheet with two α-helices lying along one face. Peptide antigens are bound in the region between the α-helices. Figure 15.10 illustrates the structure of a class II I-A^k MHC molecule bound to a peptide from hen egg lysozyme.[26] Many similar structures of class II MHC molecules are known (PDB references 1dlh, 1seb, and others).[27–30] Interleukin-1β converting enzyme is also a β-sheet heterodimer, consisting of a β-sheet with α-helices lying on both faces (PDB references 1ice and others).[31,32]

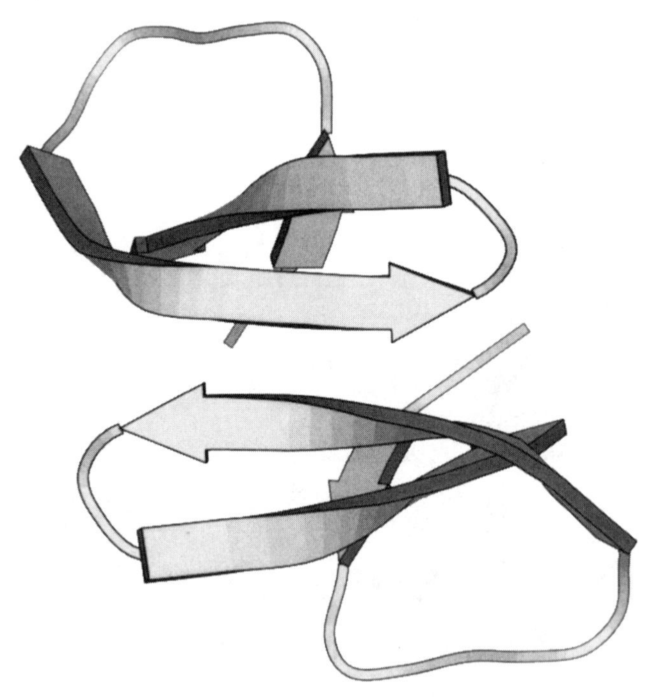

FIGURE 15.8. Ribbon diagram of defensin HNP-3. (PDB reference 1dfn.)

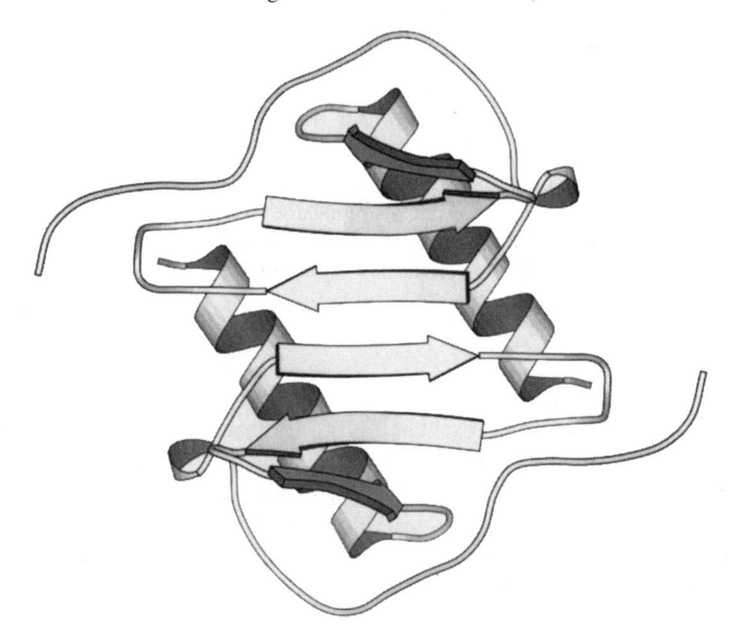

FIGURE 15.9. Ribbon diagram of the interleukin-8 dimer. The upper three β-strands and right-hand helix comprise one half of the dimer, and the lower three β-strands and left-hand helix comprise the other half. (PDB reference 1il8; see also 3il8.)

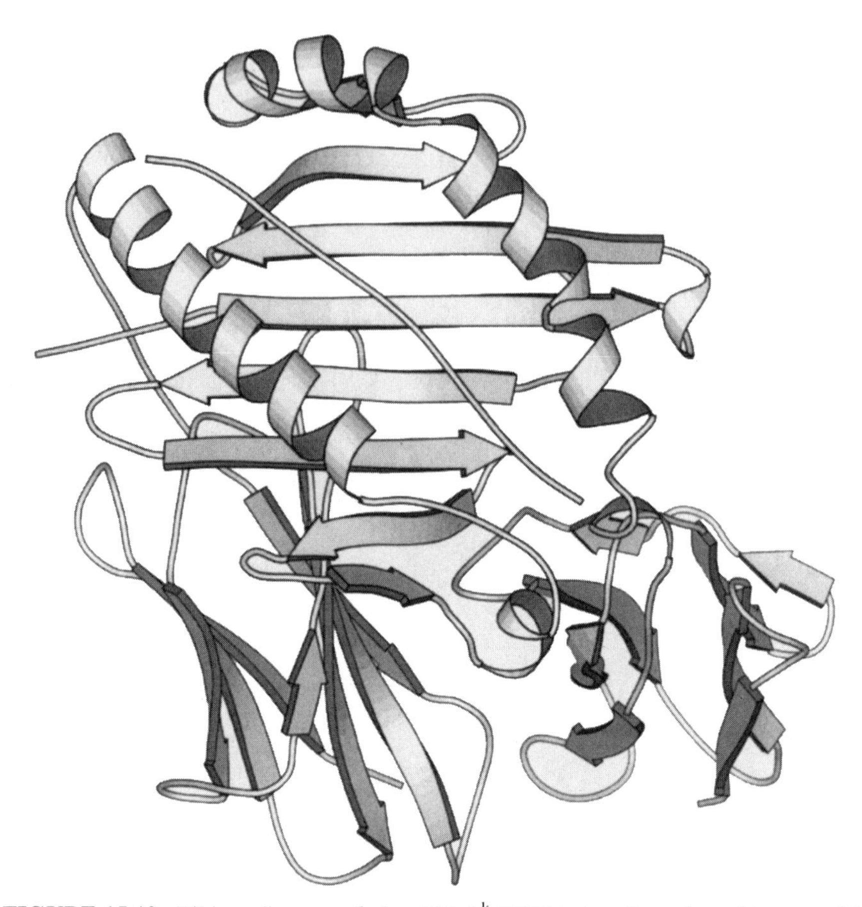

FIGURE 15.10. Ribbon diagram of class II I-A^k MHC heterodimer bound to a peptide from hen egg lysozyme. The upper four β-strands, the β-strands in the lower right-hand corner, and the helices on the top and right belong to the β-chain of the dimer; the β-strands and α-helix toward the lower and left-hand region belong to the α-chain of the dimer. (PDB reference liak.)

In the eye lens, β-sheet packing between proteins play a key structural role.[33] The γ-crystallin proteins are present at high concentrations in the densely packed regions of the lens, which have a high refractive index. Two closely homologous proteins, γ_B-crystallin (γII-crystallin) and γ_F-crystallin (γIVa-crystallin), exhibit different crystal packing structures. Although both proteins have edges that consist of exposed β-sheets, only in γ_F-crystallin do the edges hydrogen-bond together to form extended β-sheets. The difference between the packing modes of these two γ-crystallins is thought to result from subtle structural differences in the proteins and to contribute to their different propensities to form cataracts. Figure 15.11 illustrates the β-sheet interactions in the packing of γ_F-crystallin.

FIGURE 15.11. Ribbon diagram of bovine γ-crystallin. (PDB referene 1a45.)

3. β-SHEET INTERACTIONS BETWEEN DIFFERENT PROTEINS

β-Sheet formation is involved in interactions between different proteins, as well as in protein dimerization. A particularly noteworthy example of this sort of interaction involves the binding of the Ras oncoproteins to their kinase receptors. Ras oncoproteins act as molecular switches that activate the serine/threonine

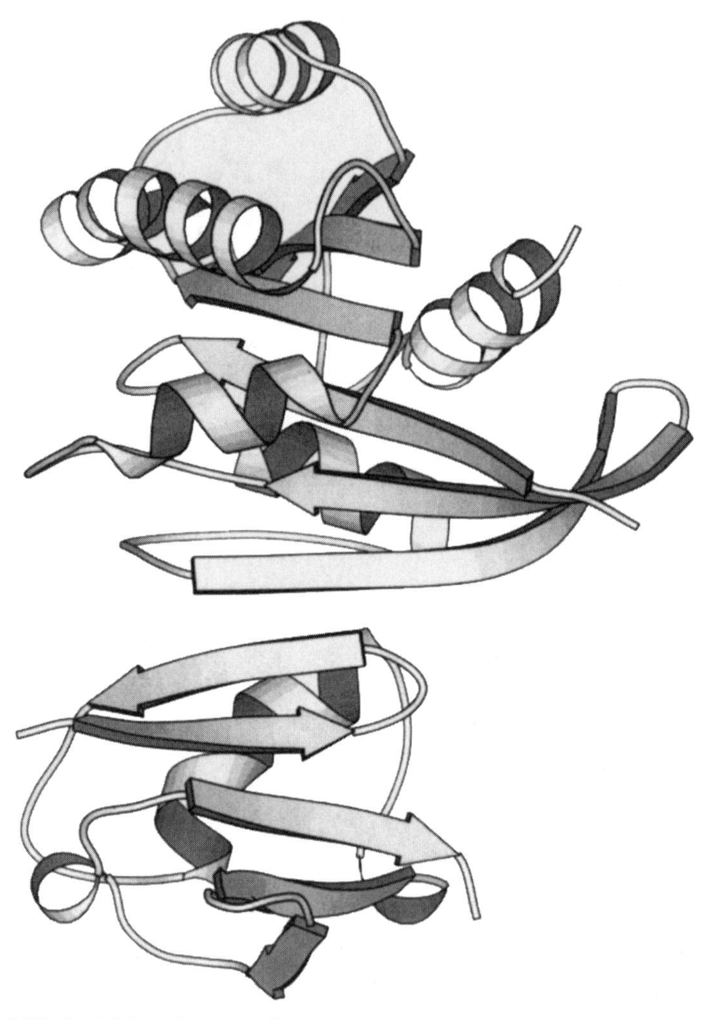

FIGURE 15.12. Ribbon diagram of the complex between the Ras-binding domain of the c-Raf1 kinase (upper) and the Ras analog, Rap1A (lower). (PDB reference 1gua.)

kinase c-Raf1 (Raf) by binding to its Ras-binding domain (RBD).[34–36] This activation is part of the cell-signaling pathway that leads to cell growth and plays a key role in cancer. The Ras-related protein Rap1A (Rap) is very similar in structure to Ras and has an identical effector region, but it is not membrane bound. Rap also binds to the RBD of Raf, acting as an antagonist, rather than an agonist. The X-ray crystallographic structure of the complex between Raf and Rap shows that the main-chains of the two proteins form an antiparallel β-sheet and their side-chains form a rich array of polar contacts (Fig. 15.12).[37] Figure 15.13 illustrates some of these contacts. Particularly prominent is a salt bridge

c-Raf1

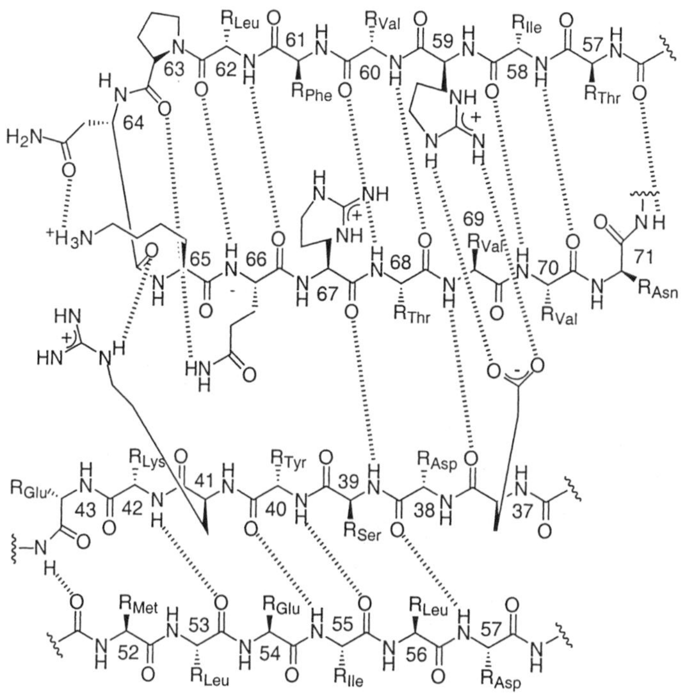

Rap1A

FIGURE 15.13. Molecular representation of the complex between the Ras-binding domain of the c-Raf1 kinase and the Ras analog, Rap1A. The upper two β-strands are from c-Raf1; the lower two are from Rap1A.

between Arg-59 (upper peptide strand in Fig. 15.13) and Glu-37 (third peptide strand in Fig. 15.13).

β-Sheet interactions also occur between immune system proteins and other proteins. Immunoglobulin G (IgG) and related proteins are composed primarily of β-sheets, and interactions between these proteins and a number of other proteins involve β-sheet formation. Protein G, a cell surface-associated protein from *Streptococcus*, binds IgG through β-sheet interactions with its three binding domains. Figure 15.14 illustrates the interaction between protein G domain III and the Fab portion of IgG.[38,39] Interactions between the side-chains of the proteins, as well as the main-chains, help stabilize the complex (Fig. 15.15). β-Sheet interactions have also been noted in the binding of an antibody to a peptide derived from HIV-1 protease (PDB reference 2hrp).[40]

HIV-1 protease and a variety of other proteolytic enzymes bind their substrates through β-sheet interactions.[41] Although it is not possible to directly observe the enzyme-substrate complexes, the structures of complexes between

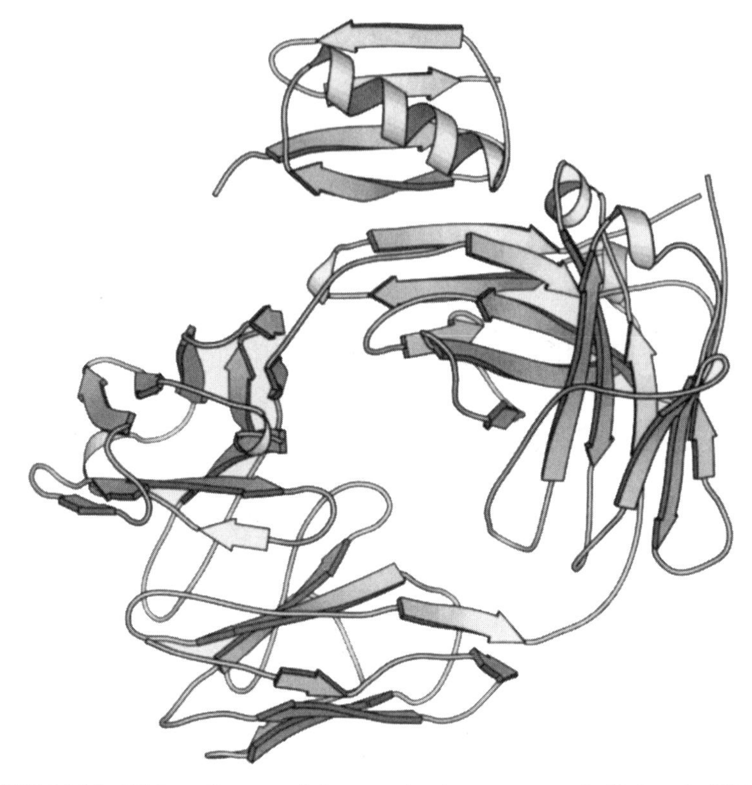

FIGURE 15.14. Ribbon diagram of the complex between protein G domain III (upper) and the Fab portion of IgG (lower). (PDB reference 1igc.)

the enzyme and various inhibitors are readily determined by X-ray crystallography. Figure 15.16 illustrates the structure of the complex between the HIV-1 protease dimer and a C_2 symmetric inhibitor.[42] In this structure, hydrogen bonding between the inhibitor, the enzyme, and a molecule of water generates a β-sheetlike structure. Many other β-sheetlike complexes between HIV-1 protease and inhibitors are known.[43]

β-Sheet formation is critical in forming cell–cell junctions and clustering of ion channels and receptors. In these processes, membrane-bound proteins that contain PDZ domains bind the C-termini of the ion-channel and receptor proteins through β-sheet interactions. Figure 15.17 illustrates the structure of the complex between the third PDZ domain of the synaptic protein PSD-95 and a peptide corresponding to the C-terminal residues of a protein that it binds.[44] The C-terminal peptide makes many contacts with the PDZ domain, in addition to forming a hydrogen-bonded β-sheet. Figure 15.18 illustrates some of these contacts. A network of hydrogen bonds forms between the C-terminal peptide carboxylate group and the Gly-Leu-Gly-Phe

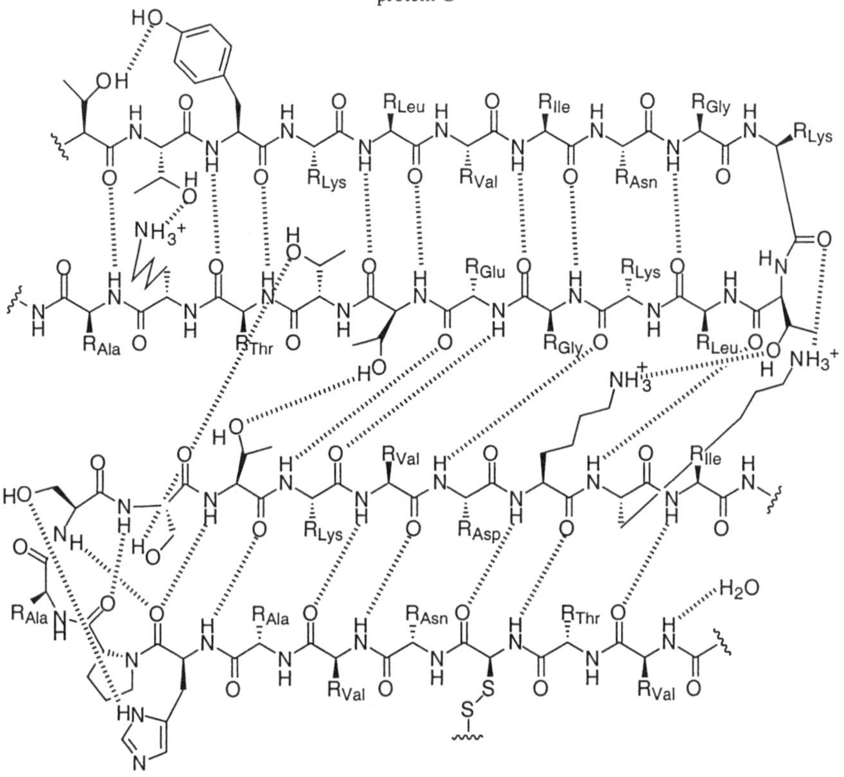

protein G

Fab portion of IgG

FIGURE 15.15. Molecular representation of the complex between domain III of protein G and the Fab portion of IgG. The upper two β-strands are from protein G; the lower two are from IgG.

loop of the protein (residues 322–325) and Arg-318. Additional hydrogen bonds and hydrophobic contacts form between the peptide side-chains and the protein.

Vancomycin, ristocetin, and related antibiotics act by forming similar types of β-sheetlike interactions with the C-terminus of a polypeptide (Fig. 15.19). These antibiotics are glycopeptides with crosslinked aromatic side-chains and main-chains that adopt β-strandlike conformations. The main-chains bind the C-terminal D-Ala-D-Ala groups of the bacterial cell-wall peptidoglycan precursor through β-sheetlike hydrogen bonds with its main-chain and additional hydrogen bonds to its carboxylate group (Fig. 15.20).[45–47] The vancomycin family antibiotics dimerize through β-sheetlike interactions between their peptide chains; this dimerization cooperatively contributes to the binding of their ligands.

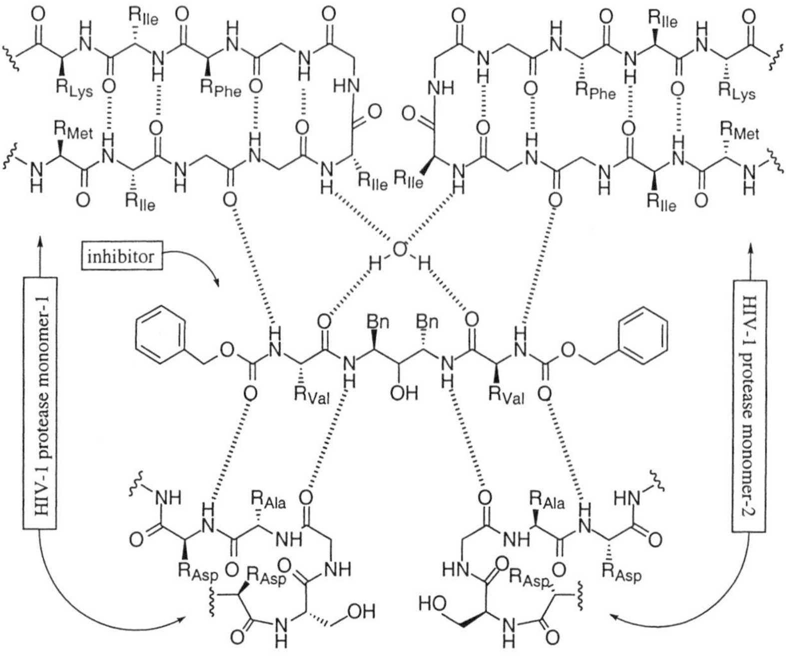

FIGURE. 15.16. Molecular representation of the complex between HIV-1 protease dimer and an inhibitor. (From PDB reference 9hvp; see also 4hvp and many related structures of HIV-1 protease complexed with inhibitors.)

4. β-SHEET INTERACTIONS IN PROTEIN AGGREGATION

Protein aggregation occurs in a variety of diseases and is a common feature of many neurodegenerative diseases, including Alzheimer's disease, Creutzfeldt–Jakob disease, and Huntington's disease. In each of these diseases, soluble proteins form insoluble or aggregated structures that contain β-sheets. Although the precise structures of the aggregates are much less understood and well-defined than those of the protein dimers and complexes described in the preceding sections, β-sheet formation appears to be critical in the aggregation processes. Also not fully understood are the precise roles that formation of the aggregates plays in the progression of these diseases.

In Alzheimer's disease, β-sheet containing aggregates, called β-amyloid, deposit as fibrils in the brain.[48–52] Although the role of β-amyloid deposition in the progression of the disease is still the subject of investigation, it is known to be neurotoxic and its deposition is generally thought to be important in the progression of the disease. The aggregates are composed of polypeptides, called Aβ or amyloid β-peptide, which are generally 40 or 42 amino acids in length; the

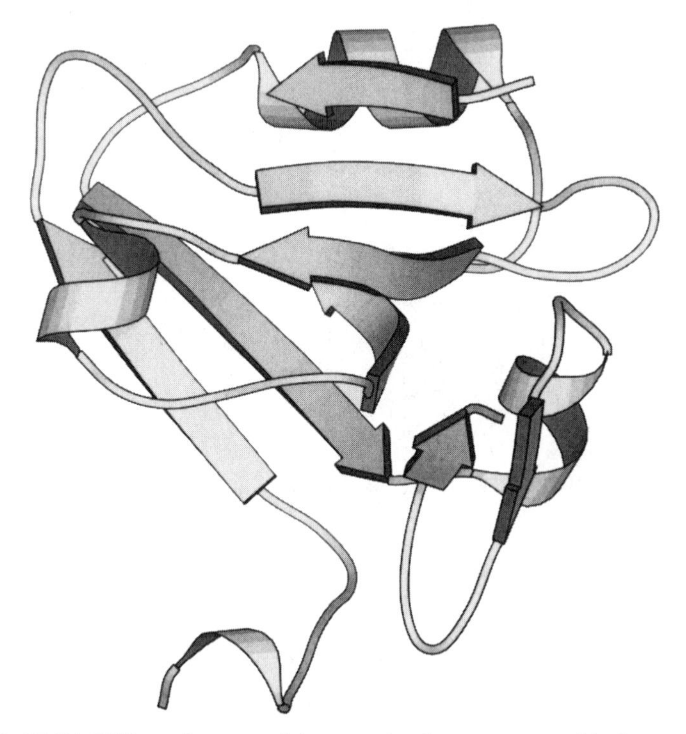

FIGURE. 15.17. Ribbon diagram of the complex between a peptide (upper β-strand) and the third PDZ domain of the synaptic protein PSD-95. (PDB reference 1be9; see also 1kwa.)

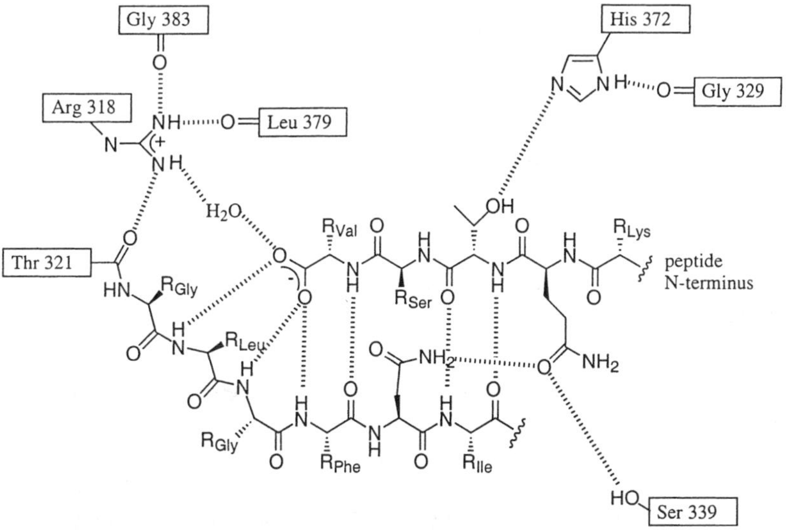

FIGURE. 15.18. Molecular representation of the complex between a peptide and the third PDZ domain of the synaptic protein PSD-95.

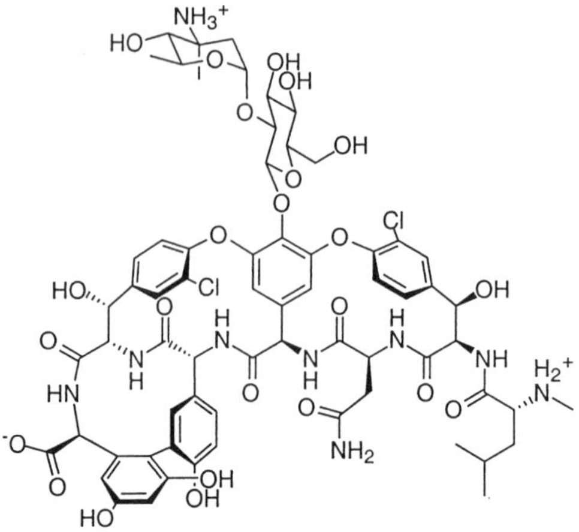

FIGURE 15.19. Molecular representation of vancomycin.

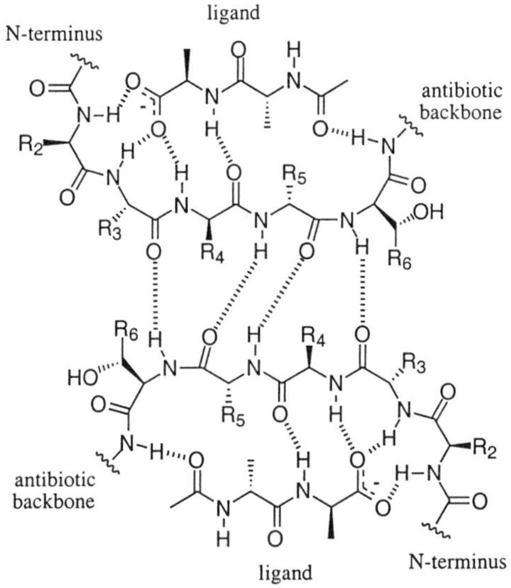

FIGURE 15.20. Molecular representation of the complex between vancomycin and related glycopeptide antibiotics and the C-terminal D-Ala–D-Ala groups of bacterial cell wall peptidoglycan precursors.

42 amino acid variant, Aβ1-42, is especially fibrillogenic. β-Amyloid has long been known to have a high degree of β-sheet structure. Recent synchrotron X-ray diffraction studies of β-amyloid fibrils and of transthyretin amyloid fibrils suggest that these β-sheets aggregate in a helical array.[53,54] Atomic force microscopy studies have established that Aβ first aggregates to form small protofibrils, which then self-assemble to form β-amyloid fibrils.[55,56] Many other solution phase and solid phase studies of the structures of β-amyloid and

fragments of β-amyloid have been performed, and the determination of β-amyloid structure remains an active area of research.

The formation of β-sheets and fibrils is also involved in the prion diseases.[57,58] These diseases include Creutzfeldt–Jakob disease (CJD), bovine spongiform encephalopathy (BSE, mad cow disease), scrapie, kuru, and fatal familial insomnia (FFI) and were the subject of the 1997 Nobel Prize in medicine.[59] Unlike Alzheimer's disease, the prion diseases are infectious. The infectious agent is a protein, called PrP^{Sc}, which replicates and forms fibrils. This protein is present in healthy cells as a form called PrP^{C}. In replicating, the protein undergoes a conformational conversion from a structure that is largely α-helical (42% α-helix and 3% β-sheet) to a structure that has a high β-sheet content (43% β-sheet and 30% α-helix). The precise mechanism for the conversion of PrP^{C} to PrP^{Sc} is a subject of controversy, and two different models have been proposed. In one model, the PrP^{Sc} form acts as a template to promote the refolding of a partially unfolded monomer of PrP^{C} into the PrP^{Sc} form.[60] In the other, PrP^{Sc} is an aggregate that acts as a seed to promote the aggregation of the PrP^{C} form.[61]

β-Sheet formation and protein aggregation also occur in Huntington's disease, various spinocerebellar ataxias, and a number of other progressive neurodegenerative genetic disorders associated with trinucleotide (CAG) repeats, which code for polyglutamine.[62,63] Proteins coded for by the Huntington's disease gene and related genes contain long (e.g. $\geq$40-mer) polyglutamine regions, and individuals with longer repeats are affected at earlier ages and with greater severity. The repeats make the proteins aggregate. An attractive model for the aggregation process is that the glutamine repeats act as "polar zippers" that make the proteins aggregate by forming β-sheets through both main-chain and side-chain interactions.[64–66] The role of protein aggregation in these diseases is, however, still controversial.[67,68] Although the proteins form amyloid-like aggregates and neuronal intranuclear inclusions, recent studies suggest that formation of the aggregates is not required for pathogenesis.[69–72]

5. CONCLUSION AND OUTLOOK

As the preceding sections have shown, β-sheet interactions between proteins are important in a variety of biological processes. Many proteins form β-sheet dimers and higher oligomers, and often these dimers and oligomers are the functional forms of the proteins. β-Sheet formation also plays a key role in recognition interactions between different proteins, providing a fundamental form of molecular recognition between proteins. β-Sheet formation is involved in protein aggregation processes that occur in a variety of devastating neurodegenerative diseases.

All of the examples of β-sheet interactions in protein–protein recognition and in the formation of dimers and oligomers that were described in Sections 2 and 3 of this chapter are based upon crystallographic and NMR data from the Protein Data Bank. The number of structures in the Protein Data Bank is growing rapidly; it currently contains 9000 structures and is projected to grow to 48,000 by the year 2004.[73] As these structures are determined, many more β-sheet interactions between proteins will be discovered. Additional β-sheet interactions will likely be identified from genetic sequence data from the Human Genome Project and other gene-sequencing efforts.

The discoveries of β-sheet interactions between proteins offer intriguing possibilities for developing new drugs based on compounds that can block, modulate, or mediate β-sheet interactions between proteins. A number of researchers have begun to achieve this goal in systems involving protein dimerization, protein–protein interactions, and protein aggregation. Several teams of researchers have used peptides from the β-sheet dimerization interface of HIV-1 protease to interrupt dimerization of the protease and inhibit its activity.[74–77] Chmielewski and coworkers have employed cross-linked peptides from the N- and C-termini of HIV-1 protease, in a particularly elegant example of this strategy.[78–82] Smith, Hirschmann, and co-workers have invented pyrrolinone-based β-strand mimics as inhibitors of HIV protease and renin.[83–85] Unlike the aforementioned dimerization inhibitors, these traditional inhibitors block the β-sheet interactions between the enzyme and its substrate. Michne and Schroeder have developed a β-sheet mimic designed to block a putative β-sheet interaction between lymphocyte function-associated antigen-1 (LFA-1) and the intercellular adhesion molecule-1 (ICAM-1).[86] Rebek, Pallai, and coworkers have developed a cyclic vinylamide β-strand mimic as a potential inhibitor of a postulated β-sheet interaction between gp120 and the CD4 receptor.[87,88] Kirsten and Schrader have developed an aminopyrazole that binds to peptides in a β-sheet conformation through β-sheetlike hydrogen-bonding interactions.[89,90]

Many compounds disrupt the aggregation of Aβ into β-amyloid fibrils and inhibit its neurotoxicity, and the discovery of compounds that block Aβ aggregation is an active area for drug development.[91–93] Many of the compounds that have been discovered thus far are aromatic molecules, such as Congo Red, that bind β-amyloid. Others are peptides containing sequences from Aβ. In 1996, Tjernberg and coworkers reported that a pentapeptide (Lys-Leu-Val-Phe-Phe) matching the Aβ amino acid sequence from residues 16 to 20 inhibits fibril formation by Aβ.[94] This result is particularly promising, because it may allow the development of peptidomimetic inhibitors of Aβ aggregation using standard medicinal chemistry techniques. As a step in this direction, these researchers have reported the development of protease-resistant pentapeptides containing D-amino acids that inhibit Aβ aggregation.[95] Concurrently, Sahasrabudhe and coworkers reported that an octapeptide, related to the Aβ sequence from 17 to 24, also inhibits fibril formation.[96] Shortly thereafter, Kiessling, Murphy, and co-workers described a longer peptide, containing Aβ residues 15 to 25 linked to a polylysine unit, that retards Aβ aggregation.[97] Further efforts to develop drugs from both non-peptide and peptide lead compounds are underway.

Kelly and co-workers are pursuing a complementary approach to the problem of amyloid fibril formation; rather than directly blocking formation of the aggregates, these researchers are interested in stabilizing the unaggregated form of the protein to prevent its formation of amyloid fibrils.[98,99] In these studies, the Kelly group is using an amyloidogenic form of transthyretin, which exists as a soluble tetramer that can dissociate, change conformation, and aggregate to form amyloid fibrils. Thus far, the researchers have identified several classes of small compounds that stabilize the tetramer and prevent amyloid formation.

Another clever approach to blocking β-sheet interactions between proteins involves covalently modifying their main-chains to prevent formation of a hydrogen-bonded β-sheet. A number of researchers have shown that this modification can be achieved by *N*-methylating one or more of the amide groups on the exposed edge of a β-sheet. Thus, Clark-Lewis and co-workers interrupted the dimerization of IL-8 by *N*-methylating one residue of its β-sheet dimerization interface.[100] Ghadiri and coworkers and Kelly and coworkers have noted that selective *N*-methylation of one edge of a peptide β-strand can prevent β-sheet aggregation.[101–104] Doig has used this strategy to prevent aggregation of a peptide that folds into a three-stranded β-sheet.[105] These diverse efforts to block β-sheet interactions between proteins may eventually lead to new therapies for diseases ranging from AIDS to cancer to Alzheimer's disease.

NOTES ADDED IN PROOF

The *Arc* repressor forms a β-sheet dimer that resembles the *met* repressor dimer and binds DNA through similar interactions (PDB references 1aar, 1arq, 1par, and others).[106,107] Neuronal nitric oxide synthase inhibitory protein (PIN), also known as the dynein light chain (LC8), forms a β-sheet dimer that binds a peptide from neuronal nitric oxide synthase (nNOS) through β-sheet interactions (PDB references 1b1w and 1cmi).[108]

ACKNOWLEDGMENTS

The authors thank Dr. Roman Laskowski (Department of Biochemistry, University College London) for identifying proteins containing β-sheet interactions using the PDBsum database. We also thank Prof. Jon Clardy (Department of Chemistry and Chemical Biology, Cornell University), Prof. Andrew Hamilton (Department of Chemistry, Yale University), Dr. Kim Henrick (European Bioinformatics Institute), Prof. John Hess (Department of Biochemistry and Anaerobic Microbiology, Virginia Polytechnic Institute), Prof. Leslie Thompson (Department of Biological Chemistry, University of California, Irvine), Prof. Douglas Tobias (Department of Chemistry, University of California, Irvine), and Prof. David Van Vranken (Department of Chemistry, University of California, Irvine) for bringing to our attention examples of β-sheet interactions between proteins. J.S.N. thanks the following agencies for support in the form of awards: the National Science Foundation (Presidential Faculty Fellow Award), the Camille and Henry Dreyfus Foundation (Teacher-Scholar Award), the Alfred P. Sloan Foundation (Alfred P. Sloan Research Fellowship), and the American Chemical Society (Arthur C. Cope Scholar Award). Additional financial support was provided by NIH Grant GM-49076 and NSF Grant CHE-9813105.

REFERENCES

1. Wlodawer, A.; Miller, M.; Jaskólski, M.; Sathyanarayana, B. K.; Baldwin, E.; Weber, I. T.; Selk, L. M.; Clawson, L.; Schneider, J.; Kent, S. B. H. *Science* 1989, **245**, 616–621.
2. Kraulis, P. J. *J. Appl. Crystallogr.* 1991, **24**, 946–950. See, also: http://www. avatar. se/molscript/.
3. Miller, M.; Jaskólski, M.; Rao, J. K.; Leis, J.; Wlodawer, A. *Nature* 1989, **337**, 576–579.
4. Jaskólski, M.; Miller, M.; Rao, J. K. M.; Leis, J.; Wlodawer, A. *Biochemistry* 1990, **29**, 5889–5898.
5. Rafferty, J. B.; Somers, W. S.; Saint-Girons, I.; Phillips, S. E. V. *Nature* 1989, **341**, 705–710.
6. Somers, W. S.; Phillips, S. E. V. *Nature* 1992, **359**, 387–393.
7. Kong, X. -P.; Onrust, R.; O'Donnell, M.; Kuriyan, J. *Cell* 1992, **69**, 425–437.
8. Krishna, T. S. R.; Kong X. -P.; Gary, S.; Burgers, P. M.; Kuriyan, J. *Cell* 1994, **79**, 1233–1243.
9. Ahmed, H.; Pohl, J.; Fink, N. E.; Strobel, F.; Vasta, G. R. *J. Biol. Chem.* 1996, **271**, 33083–33094.
10. Hardman, K. D.; Ainsworth, C. F. *Biochemistry* 1972, **11**, 4910–4919.
11. Reeke, G. N., Jr.; Becker, J. W.; Edelman, G. M. *J. Biol. Chem.* 1975, **250**, 1525–1547.
12. Blake, C. C. F.; Swan, I. D. A.; Rerat, C.; Berthou, J.; Laurent, A.; Rerat, B. *J. Mol. Biol.* 1971, **61**, 217–224.
13. Blake, C. C. F.; Geisow, M. J.; Swan, I. D. A.; Rerat, C.; Rerat, B. *J. Mol. Biol.* 1974, **88**, 1–12.
14. Hamilton, J. A.; Steinrauf, L. K.; Braden, B. C.; Liepnieks, J. L.; Benson, M. D.; Holmgren, G.; Sandgren, O.; Steen, L. *J. Biol. Chem.* 1993, **268**, 2416–2424.
15. Hill, C. P.; Yee, J.; Selsted, M. E.; Eisenberg, D. *Science* 1991, **251**, 1481–1485.
16. Rees, B; Samama, J. P.; Thierry, J. C.; Gilibert, M.; Fischer, J.; Schweitz, H.; Lazdunski, M.; Moras, D. *Proc. Natl. Acad. Sci. U. S. A.* 1987, **84**, 3132–3136.
17. Rees, B.; Bilwes, A.; Samama, J. P.; Moras, D. *J. Mol. Biol.* 1990, **214**, 281–297.
18. Dewan, J. C.; Grant, G. A.; Sacchettini, J. C. *Biochemistry* 1994, **33**, 13147–13154.
19. Saludjian, P.; Prangé, T.; Navaza, J.; Ménez, R.; Guilloteau, J. P.; Riès-Kautt, M.; Ducruix, A. *Acta Crystallogr., Sect. B* 1992, **48**, 520–531.
20. Clore, G. M.; Appella, E.; Yamada, M.; Matsushima, K.; Gronenborn, A. M. *J. Biol. Chem.* 1989, **264**, 18907–18911.

21. Clore, G. M.; Appella, E.; Yamada, M.; Matsushima, K.; Gronenborn, A. M. *Biochemistry* 1990, **29**, 1689–1696.
22. Baldwin, E. T.; Weber, I. T.; St. Charles, R.; Xuan, J. -C.; Appella, E.; Yamada, M.; Matsushima, K.; Edwards, B. F. P.; Clore, G. M.; Gronenborn, A. M.; Wlodawer, A. *Proc. Natl. Acad. Sci. U.S.A.* 1991, **88**, 502–506.
23. Handel, T. M.; Domaille, P. J. *Biochemistry* 1996, **35**, 6569–6584.
24. Lubkowski, J.; Bujacz, G.; Boqué, L.; Domaille, P. J.; Handel, T. M.; Wlodawer, A. *Nat. Struct. Biol.* 1997, **4**, 64–69.
25. Kato, Y.; Muto, T.; Tomura, T.; Tsumura, H.; Watarai, H.; Mikayama, T.; Ishizaka, K.; Kuroki, R. *Proc. Natl. Acad. Sci. U. S. A.* 1996, **93**, 3007–3010.
26. Fremont, D. H.; Monnaie, D.; Nelson, C. A.; Hendrickson, W. A.; Unanue, E. R. *Immunity* 1998, **8**, 305–317.
27. Brown, J. H.; Jardetzky, T. S.; Gorga, J. C.; Stern, L. J.; Urban, R. G.; Strominger, J. L.; Wiley, D. C. *Nature* 1993, **364**, 33–39.
28. Stern, L. J.; Brown, J. H.; Jardetzky, T. S.; Gorga, J. C.; Urban, R. G.; Strominger, J. L.; Wiley, D. C. *Nature* 1994, **368**, 215–221.
29. Jardetzky, T. S.; Brown, J. H.; Gorga, J. C.; Stern, L. J.; Urban, R. G.; Chi, Y. I.; Stauffacher, C.; Strominger, J. L.; Wiley, D. C. *Nature* 1994, **368**, 711–718.
30. Jardetzky, T. S.; Brown, J. H.; Gorga, J. C.; Stern, L. J.; Urban, R. G.; Strominger, J. L.; Wiley, D. C. *Proc. Natl. Acad. Sci. U. S. A.* 1996, **93**, 734–738.
31. Thornberry, N. A. *Nature* 1994, **370**, 251–252.
32. Wilson, K. P.; Black, J. -A. F.; Thomson, J. A.; Kim, E. E.; Griffith, J. P.; Navia, M. A.; Murcko, M. A.; Chambers, S. P.; Aldape, R. A.; Raybuck, S. A.; Livingston, D. J. *Nature* 1994, **370**, 270–275.
33. White, H. E.; Driessen, H. P. C.; Slingsby, C.; Moss, D. S.; Lindley, P. F. *J. Mol. Biol.* 1989, **207**, 217–235.
34. Avruch, J.; Zhang, X. -F.; Kyriakis, J. M. *Trends Biochem. Sci.* 1994, **19**, 279–283.
35. Marshall, M. *Molec. Reprod. Dev.* 1995, **42**, 493–499.
36. Sprang, S. R. *Structure* 1995, **3**, 641–643.
37. Nassar, N.; Horn, G.; Herrmann, C.; Scherer, A.; McCormick, F.; Wittinghofer, A. *Nature* 1995, **375**, 554–560.
38. Derrick, J. P.; Wigley, D. B. *Nature* 1992, **359**, 752–754.
39. Derrick, J. P.; Wigley, D. B. *J. Mol. Biol.* 1994, **243**, 906–918.
40. Lescar, J.; Stouracova, R.; Riottot, M. -M.; Chitarra, V.; Brynda, J.; Fabry, M.; Horejsi, M.; Sedlacek, J.; Bentley, G. A. *J. Mol. Biol.* 1997, **267**, 1207–1222.
41. Smith, A. B., III; Hirschmann, R.; Pasternak, A.; Akaishi, R.; Guzman, M. C.; Jones, D. R.; Keenan, T. P.; Sprengeler, P. A. *J. Med. Chem.* 1994, **37**, 215–218, and references contained therein.
42. Erickson, J.; Neidhart, D. J.; Vandrie, J.; Kempf, D. J.; Wang, X. C.; Norbeck, D. W.; Plattner, J. J.; Rittenhouse, J. W.; Turon, M.; Wideburg, N.; Kohlbrenner, W. E.; Simmer, R.; Helfrich, R.; Paul, D. A.; Knigge, M. *Science* 1990, **249**, 527–533.
43. Miller, M.; Schneider, J.; Sathyanarayana, B. K.; Toth, M. V.; Marshall, G. R.; Clawson, L.; Selk, L.; Kent, S. B. H.; Wlodawer, A. *Science* 1989, **246**, 1149–1152.

44. Doyle, D. A.; Lee, A.; Lewis, J.; Kim, E.; Sheng, M.; MacKinnon, R. *Cell* 1996, **85**, 1067–1076.
45. Mackay, J. P.; Gerhard, U.; Beauregard, D. A.; Westwell, M. S.; Searle, M. S.; Williams, D. H. *J. Am. Chem. Soc.* 1994, **116**, 4581–4590.
46. Schäfer, M.; Schneider, T. R.; Sheldrick, G. M. *Structure* 1996, **4**, 1509–1515.
47. Williams, D. H.; Maguire, A. J.; Tsuzuki, W.; Westwell, M. S. *Science* 1998, **280**, 711–714.
48. Harper, J. D.; Lansbury, P. T., Jr. *Annu. Rev. Biochem.* 1997, **66**, 385–407.
49. Lansbury, P. T., Jr. *Acc. Chem. Res.* 1996, **29**, 317–321.
50. Selkoe, D. J. *J. Neuropath. Exp. Neurol.* 1994, **53**, 438–447.
51. Verbeek, M. M.; Ruiter, D. J.; de Waal, R. M. W. *Biol. Chem.* 1997, **378**, 937–950.
52. Cotman, C. W. *Annals New York Acad. Sci.* 1997, **814**, 1–16.
53. Blake, C.; Serpell, L. *Structure* 1996, **4**, 989–998.
54. Sunde, M.; Serpell, L. C.; Bartlam, M.; Fraser, P. E.; Pepys, M. B.; Blake, C. C. F. *J. Mol. Biol.* 1997, **273**, 729–739.
55. Harper, J. D.; Wong, S. S.; Lieber, C. M.; Lansbury, P. T., Jr. *Chem. Biol.* 1997, **4**, 119–125.
56. Harper, J. D.; Lieber, C. M.; Lansbury, P. T., Jr. *Chem. Biol.* 1997, **4**, 951–959.
57. Edenhofer, F.; Weiss, S.; Winnacker, E. -L.; Famulok, M. *Angew. Chem., Int. Ed. Engl.* 1997, **36**, 1674–1694.
58. Ng, S. B. L.; Doig, A. J. *Chem. Soc. Rev.* 1997, **26**, 425–432.
59. Prusiner, S. B. *Proc. Natl. Acad. Sci. U. S. A.* 1998, **95**, 13363–13383.
60. Cohen, F. E.; Pan, K. -M.; Huang, Z.; Baldwin, M.; Fletterick, R. J.; Prusiner, S. B. *Science* 1994, **264**, 530–531.
61. Lansbury, P. T., Jr. *Chem. Biol.* 1995, **2**, 1–5.
62. The Huntington's Disease Collaborative Research Group *Cell* 1993, **72**, 971–983.
63. Paulson, H. L.; Fischbeck, K. H. *Annu. Rev. Neurosci.* 1996, **19**, 79–107.
64. Perutz, M. F.; Johnson, T.; Suzuki, M.; Finch, J. T. *Proc. Natl. Acad. Sci. U.S.A.* 1994, **91**, 5355–5358.
65. Stott, K.; Blackburn, J. M.; Butler, P. J. G.; Perutz, M. *Proc. Natl. Acad. Sci. U.S.A.* 1995, **92**, 6509–6513.
66. Perutz, M. F. *Curr. Opin. Struc. Biol.* 1996, **6**, 848–858.
67. Ross, C. A. *Neuron* 1997, **19**, 1147–1150.
68. Sisodia, S. S. *Cell* 1998, **95**, 1–4.
69. Davies, S. W.; Turmaine, M.; Cozens, B. A.; DiFiglia, M.; Sharp, A. H.; Ross C. A.; Scherzinger, E.; Wanker, E. E.; Mangiarini, L.; Bates, G. P. *Cell* 1997, **90**, 537–548.
70. Scherzinger, E.; Lurz, R.; Turmaine, M.; Mangiarini, L.; Hollenbach, B.; Hasenbank, R.; Bates, G. P.; Davies, S. W.; Lehrach, H.; Wanker, E. E. *Cell* 1997, **90**, 549–558.
71. Klement, I. A.; Skinner, P. J.; Kaytor, M. D.; Yi, H.; Hersch, S. M.; Clark, H. B.; Zoghbi, H. Y.; Orr, H. T. *Cell* 1998, **95**, 41–53.
72. Saudou, F.; Finkbeiner, S.; Devys, D.; Greenberg, M. E. *Cell* 1998, **95**, 55–66.

73. Sussman, J. L.; Abola, E.; Ritter, O. "The Protein Data Bank. Proposal for the Macromolecular Structure Database". The Protein Data Bank, May 27, 1998.

74. (a) Schramm, H. J.; Nakashima, H.; Schramm, W.; Wakayama, H.; Yamamoto, N. *Biochem. Biophys. Res. Commun.* 1991, **179**, 847–851. (b) Schramm, H. J.; Breipohl, G.; Hansen, J.; Henke, S.; Jaeger, E.; Meichsner, C.; Riess, G.; Ruppert, D.; Rücknagel, K. -P.; Schäfer, W.; Schramm, W. *Biochem. Biophys. Res. Commun.* 1992, **184**, 980–985.

75. Schramm, H. J.; Billich, A.; Jaeger, E.; Rücknagel, K. -P.; Arnold, G.; Schramm, W. *Biochem. Biophys. Res. Commun.* 1993, **194**, 595–600.

76. Zhang, Z. -Y.; Poorman, R. A.; Maggiora, L. L.; Heinrikson, R. L.; Kézdy, F. J. *J. Biol. Chem.* 1991, **266**, 15591–15594.

77. Babe, L. M.; Rose, J.; Craik, C. S. *Protein Sci.* 1992, **1**, 1244–1253.

78. Franciskovich, J.; Houseman, K.; Mueller, R.; Chmielewski, J. *Bioorg. Med. Chem. Lett.* 1993, **3**, 765–768.

79. Zutshi, R.; Franciskovich, J.; Shultz, M.; Schweitzer, B.; Bishop, P.; Wilson, M.; Chmielewski, J. *J. Am. Chem. Soc.* 1997, **119**, 4841–4845.

80. Shultz, M. D.; Chmielewski, J. *Tetrahedron: Asymmetry* 1997, **8**, 3881–3886.

81. Zutshi, R.; Brickner, M.; Chmielewski, J. *Curr. Opin. Chem. Biol.* 1998, **2**, 62–66.

82. Zutshi, R.; Shultz, M. D.; Ulysse, L.; Lutgring, R.; Bishop, P.; Schweitzer, B.; Vogel, K.; Franciskovich, J.; Wilson, M.; Chmielewski, J. *Synlett* 1998, 1040–1044.

83. Smith, A. B., III; Keenan, T. P.; Holcomb, R. C.; Sprengeler, P. A.; Guzman, M. C.; Wood, J. L.; Carroll, P. J.; Hirschmann, R. *J. Am. Chem. Soc.* 1992, **114**, 10672–10674.

84. Smith, A. B., III; Guzman, M. C.; Sprengeler, P. A.; Keenan, T. P.; Holcomb, R. C.; Wood, J. L.; Carroll, P. J.; Hirschmann, R. *J. Am. Chem. Soc.* 1994, **116**, 9947–9962.

85. Smith, A. B., III; Hirschmann, R.; Pasternak, A.; Guzman, M. C.; Yokoyama, A.; Sprengeler, P. A.; Darke, P. L.; Emini, E. A.; Schleif, W. A. *J. Am. Chem. Soc.* 1995, **117**, 11113–11123.

86. Michne, W. F.; Schroeder, J. D. *Int. J. Pept. Protein Res.* 1996, **47**, 2–8.

87. Roberts, J. C.; Pallai, P. V.; Rebek, J., Jr. *Tetrahedron Lett.* 1995, **36**, 691–694.

88. Boumendjel, A.; Roberts, J. C.; Hu, E.; Pallai, P. V.; Rebek, J., Jr. *J. Org. Chem.* 1996, **61**, 4434–4438.

89. Schrader, T.; Kirsten, C. *Chem. Commun.* 1996, 2089–2090.

90. Kirsten, C. N.; Schrader, T. H. *J. Am. Chem. Soc.* 1997, **119**, 12061–12068.

91. Schenk, D. B.; Rydel, R. E.; May, P.; Little, S.; Panetta, J.; Lieberburg, I.; Sinha, S. *J. Med. Chem.* 1995, **38**, 4141–4154.

92. Bandiera, T.; Lansen, J.; Post, C.; Varasi, M. *Curr. Med. Chem.* 1997, **4**, 159–170.

93. Lansbury, P. T., Jr. *Curr. Opin. Chem. Biol.* 1997, **1**, 260–267.

94. Tjernberg, L. O.; Näslund, J.; Lindqvis, F.; Johansson, J.; Karlström, A. R.; Thyberg, J.; Terenius, L.; Nordstedt, C. *J. Biol. Chem.* 1996, **271**, 8545–8548.

95. Tjernberg, L. O.; Lilliehöök, C.; Callaway, D. J. E.; Näslund, J.; Hahne, S.; Thyberg, J.; Terenius, L.; Nordstedt, C. *J. Biol. Chem.* 1997, **272**, 12601–12605.

96. Hughes, S. R.; Goyal, S.; Sun, J. E.; Gonzalez-DeWhitt, P.; Fortes, M.; Riedel, N. G.; Sahasrabudhe, S. R. *Proc. Natl. Acad. Sci. U. S. A.* 1996, **93**, 2065–2070.
97. Ghanta, J.; Shen, C. -L.; Kiessling, L. L.; Murphy, R. M. *J. Biol. Chem.* 1996, **271**, 29525–29528.
98. Baures, P. W.; Peterson, S. A.; Kelly, J. W. *Bioorg. Med. Chem.* 1998, **6**, 1389–1401.
99. Peterson, S. A.; Klabunde, T.; Lashuel, H. A.; Purkey, H.; Sacchettini, J. C.; Kelly, J. W. *Proc. Natl. Acad. Sci. U. S. A.* 1998, **95**, 12956–12960.
100. Rajarathnam, K.; Sykes, B. D.; Kay, C. M.; Dewald, B.; Geiser, T.; Baggiolini, M.; Clark-Lewis, I. *Science* 1994, **264**, 90–92.
101. Ghadiri, M. R.; Kobayashi, K.; Granja, J. R.; Chadha, R. K.; McReem, D. E. *Angew. Chem., Int. Ed. Eng.* 1995, **34**, 93–95.
102. Kobayashi, K.; Granja, J. R.; Ghadiri, M. R. *Angew. Chem., Int. Ed. Eng.* 1995, **34**, 95–98.
103. Nesloney, C. L.; Kelly, J. W. *J. Am. Chem. Soc.* 1996, **118**, 5836–5845.
104. Choo, D. W.; Schneider, J. P.; Graciani, N. R.; Kelly, J. W. *Macromolecules* 1996, **29**, 355–366.
105. Doig, A. J. *Chem. Commun.* 1997, 2153–2154.
106. Breg, J. N.; van Opheusden, J. H. J.; Burgering, M. J. M.; Boelens, R.; Kaptein, R. *Nature* 1994, **367**, 754–757.
107. Raumann, B. E.; Rould, M. A.; Pabo, C. O.; Sauer, R. T. *Nature* 1990, **346**, 586–589.
108. Liang, J.; Jaffrey, S. R.; Guo, W.; Snyder, S. H.; Clardy, J. *Nature Struct. Biol.* 1999, **6**, 735–740.

CHAPTER 16

HEAD-TO-TAIL CYCLIC PEPTIDES AND CYCLIC PEPTIDE LIBRARIES

ARNO F. SPATOLA and PETERIS ROMANOVSKIS
Department of Chemistry, University of Louisville

1. INTRODUCTION

Naturally occurring peptides often exist in cyclic, bicyclic, or polycyclic forms.[1] In most cases, the ring structures are formed through the oxidation of cysteine residues to form cystine disulfide bridges. But peptides can form alternative ring structures, through formation of amide bonds or through aryl–aryl linkages, and a variety of other functional group combinations. Most of the latter groups have been found in lower animals and plants. Examples of a few such ring structures are shown in Fig. 16.1.

A common method of classification of amide ring types is by the location of the ring juncture. Thus the link may involve two side-chains, one side-chain and the backbone, or two backbone elements (Fig. 16.2). Head-to-tail cyclic peptides are the most common form of the last group and typically form a ring via amide bond formation through the juncture of the N-terminal amine and the C-terminal carboxylic acid. This type of head-to-tail cyclic peptide is the primary class that will be discussed in this chapter.

1.1. Examples of Synthetic Head-to-Tail Cyclic Peptides

To date, there are no known examples of head-to-tail cyclic peptides in mammals. But such structures have often been discovered in bacteria, fungi, and other lower organisms, and many have unusual biological activities. There is currently great interest in these compounds as potential pharmaceutical agents. In contrast to linear peptides, head-to-tail cyclic peptides are not susceptible to attack by the exopeptidases (amino- and carboxy-peptidases). Furthermore, their constrained nature reduces the entropic penalty and offers the potential for

The Amide Linkage: Selected Structural Aspects in Chemistry, Biochemistry, and Materials Science, Edited by Arthur Greenberg, Curt M. Breneman, and Joel F. Liebman
ISBN 0-471-35893-2

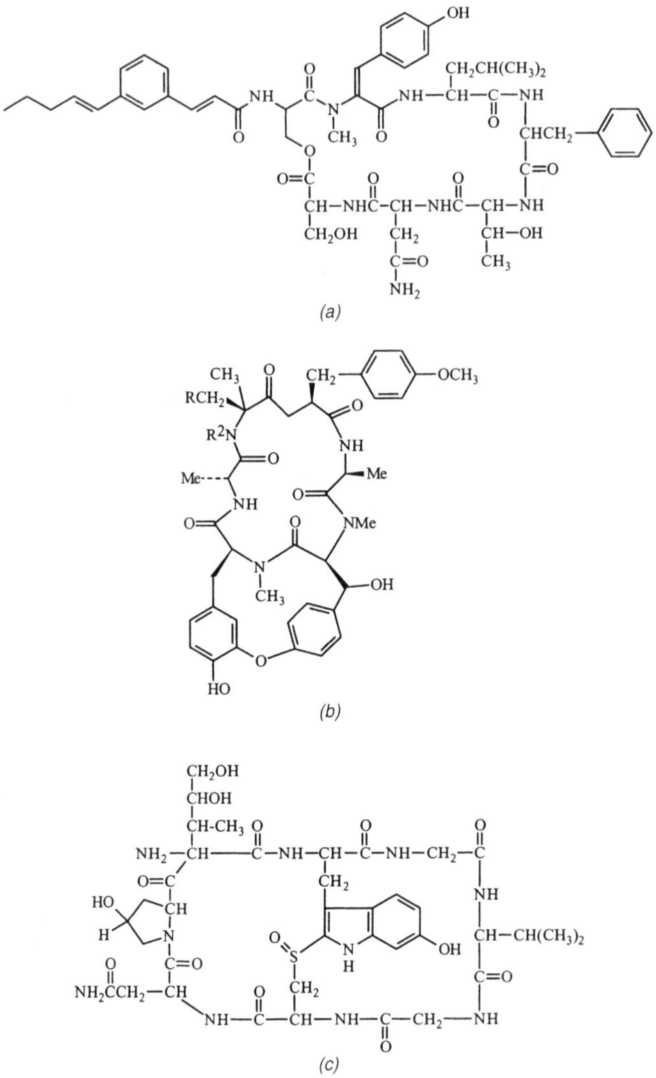

(a)

(b)

(c)

FIGURE 16.1. Examples of naturally occurring cyclic peptides with amide and nonamide groups in the ring. (*a*) Tachykinin antagonist from *Streptomyces violceonigen*; (*b*) structure of Bouvardins; (*c*) structure of the mushroom-derived toxin α-amantin.

improved binding, and often improved selectivity, toward a given biological receptor. These properties render cyclic peptides more like peptidomimetics in the continuum between peptides and nonpeptides in the perpetual search for potent, orally active, and selective pharmaceuticals.

Natural products have served as a rich source of head-to-tail cyclic peptides. Some of these, such as stylostatin, cyclo(Ala–Ile–Pro–Phe–Asp–Ser–Leu) have been isolated from sea sponges and have reasonably potent activity in an

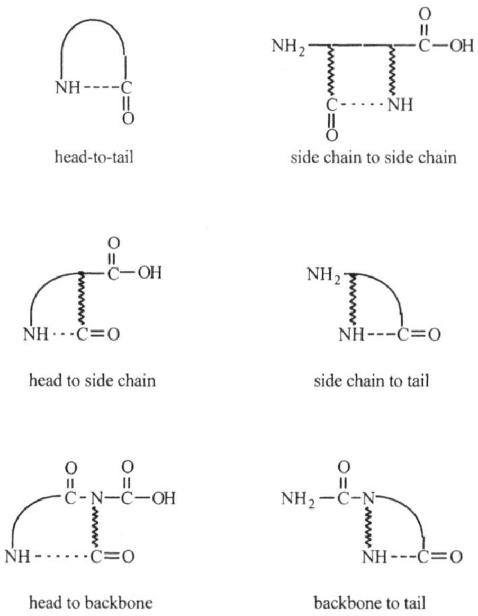
NH----C
O
head-to-tail
NH2
C—OH
C·····NH
side chain to side chain
C—OH
NH···C=O
head to side chain
NH2
NH---C=O
side chain to tail
C-N—C—OH
NH······C=O
head to backbone
NH2—C-N
NH---C=O
backbone to tail

FIGURE 16.2. Various types of peptide monocycle closures that can feature new amide (- - -) formation.

anti-leukemia assay.[2] Lower organisms have yielded compounds with other important biological actions, such as Gramicidin S, (Fig. 16.3(a)) a potent antimicrobial compound,[3] cyclosporin (Fig. 16.3(b)), the clinically important immune suppressant,[4] and the cyclic pentapeptide, BQ-123 (Fig. 16.3(c)), an antagonist to the endothelin peptides with potential utility as an antihypertensive agent.[5] Gramicidin S was originally isolated from a strain of *Bacillus brevis*,[3] while BQ-123 is actually a modified version of the natural product found in Streptomyces broth.[5]

Cyclosporin is a cyclic undecapeptide containing no less than seven *N*-methylated amino acid residues. In one conformer, it contains no amide hydrogens that are not intramolecularly hydrogen bonded.[6] Cyclosporin is a relatively rare example of a fairly large peptide that retains significant oral activity in humans. One possible explanation for this is that the lack of free polar amide functions precludes extensive solvation by water molecules during biological transport and this hydrophobic character facilitates its passage through a variety of nonpolar cell membranes.

Figure 16.3(d) depicts the structure of a somatostatin analog which retains only five of the original amino acids in the 14-residue hypothalamic peptide hormone, somatostatin 14. While somatostatin contains a disulfide bridged ring, the synthetic analog, designed by Merck chemists and referred to as "mini-somatostatin," possesses a head-to-tail all-amide structure.[7] This compound is comparable in activity to its parent and had a brief clinical history before being withdrawn due to unacceptable side effects.

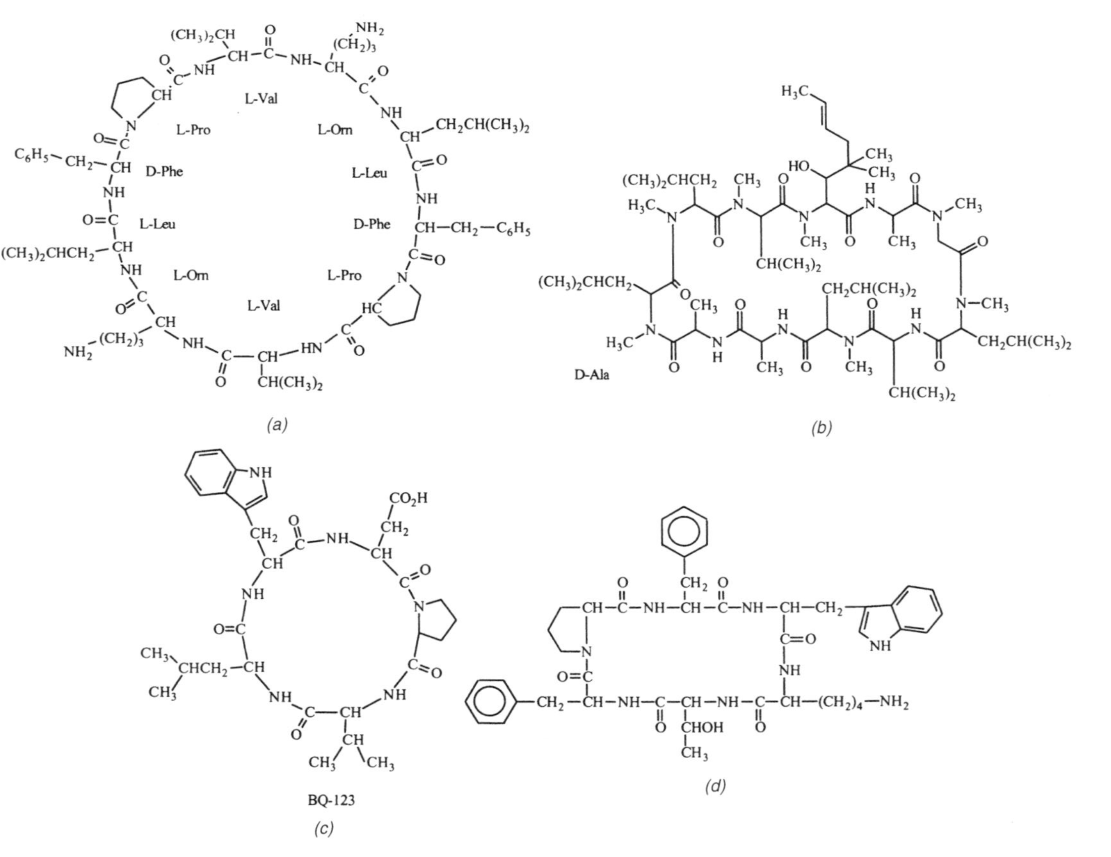

FIGURE 16.3. Examples of potent biologically active cyclic peptides: (*a*) the broad spectrum cyclic peptide anitbiotic, Gramicidin; (*b*) cyclosporin, an immune suppressant; (*c*) the cyclic pentapeptide, BQ-123, an endothelin antagonist; and (*d*) mini-somatostatin, a cyclic hexapeptide inhibitor of growth hormone release.

Current research with cyclic peptides tends to emphasize their potential as pharmaceuticals or to find leads for constrained peptide mimetics. But head-to-tail cyclic peptides provide an imaginative approach for producing new biomaterials and for studying aggregation tendencies. For example, Ghadiri and colleagues have shown that by alternating D and L residues in a cyclic peptide, these compounds can exist in an extended stacked structure referred to as nanotubes.[8] The exploration of these molecules and their derivatives is one of the more exciting areas in a fascinating arena of structural diversity.

2. METHODS OF SYNTHESIS

2.1. General Approaches

Historically, cyclic peptides were synthesized by classical methods of peptide chemistry in solution.[9] Often linear peptides were intramolecularly cyclized using azide methods or by heating linear active esters of peptides (e.g., *p*-nitrophenyl esters) for several hours until cyclization was judged complete. But yields were often low and dimerization and epimerization were significant problems.

Several strategies exist for synthesizing cyclic peptides including side chain-to-side chain, side chain-to-terminal group, and terminal group-to-terminal group (head-to-tail) cyclizations. Head-to-tail cyclizations can be accomplished either in solution[9] or while attached to a solid phase resin using carbodiimide or some other form of chemical coupling agent to form C $\rightarrow$ N terminal amides (Fig. 16.4). Only at the end of 1960s did reports start to appear on the successful synthesis of head-to-tail cyclized peptides on the solid phase.[10,11] An excellent review article[12] covers the present state-of-the-art for synthesizing cyclic peptides on solid phase carriers.

In this chapter, we emphasize new developments on the synthesis of cyclic peptides, primarily head-to-tail cyclic peptides, in which all steps have been carried out in the solid-phase mode and emphasizing work from the authors' laboratory. Nevertheless, when dealing with the synthesis of other types of cyclopeptides (head-to-side chain, side chain-to-side chain, side chain-to-tail), similar tactical considerations involving choice of protecting groups, condensing reagents, and problems of chirality are usually considered.

On-resin head-to-tail cyclization presumes anchoring of the first amino acid to the solid phase carrier through its side chain and features the following steps:

1. side chain anchoring of an initially protected amino acid residue to a polymeric support;
2. stepwise solid-phase assembly of the linear sequence;
3. orthogonal deprotection to liberate selectively a free C-α-carboxylic group and the N-terminal amine for the subsequent cyclization step;
4. efficient activation of the C-α-carboxylic group and its condensation with free N-α-amino group to close the desired head-to-tail ring, taking

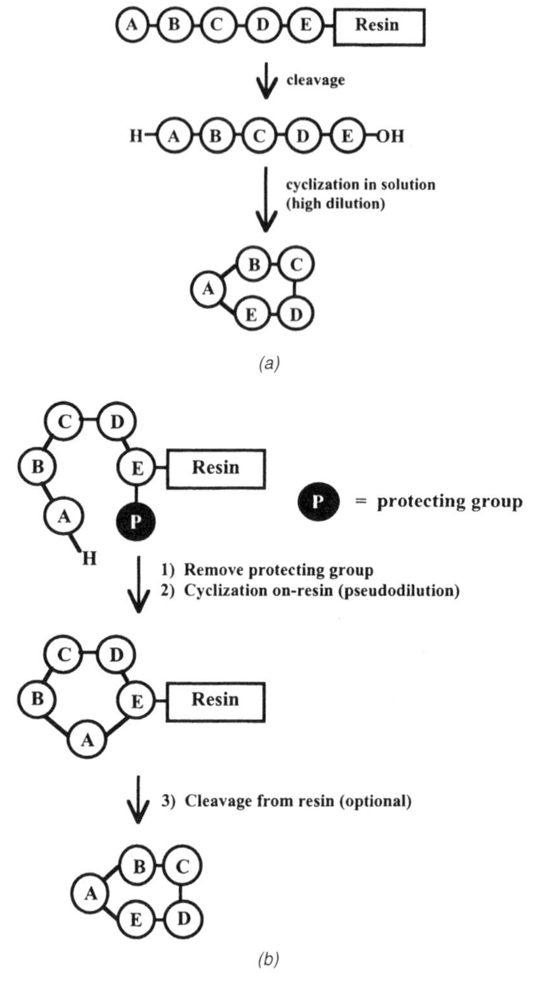

FIGURE 16.4. Representation of head-to-tail cyclic peptides prepared either by (*a*) solution cyclization or via (*b*) on-resin cyclization.

advantage of the pseudo-dilution phenomenon which favors intramolecular resin-bound reactions;[13] and

5. final deprotection and cleavage to release the required free cyclic peptide into solution. Side chain attachment also allows the synthesis of other peptide derivatives that would be otherwise inaccessible by standard solid phase synthesis procedures, e.g., peptide *p*-nitroanilides.

2.2. Examples of Synthetic Strategies Utilizing Orthogonality

Depending on the synthetic goal the anchor bond should be compatible with at least one of the two main peptide synthesis strategies, based on Boc- and/or

Fmoc chemistries. For various trifunctional amino acids, the anchor bond can be represented by an ester bond (Asp, Glu, Ser, Thr), amide bond (Asn, Gln), urethane bond (Lys, Orn [ornithine], Dab [α,γ-diamionbutyric acid]), mixed carbonate (Ser, Tyr, Thr), ether bond (Tyr), thioether bond (Cys), C–N bond in a special case (His), or a sulfamide-based linker (Arg). Simultaneously the anchor bond is expected to provide the protection for the side chain functional group during synthesis. The anchor group must also be orthogonal to the α-carboxyl protecting group: the latter is to be cleaved off selectively under mild conditions at the end of the peptide chain assembly before the cyclization step.

A system for peptide synthesis is defined as orthogonal when two or more kinds of protecting groups are used and each protecting group is removable separately and selectively based on different chemical mechanisms without affecting the other group's stability.[14] In contrast to the synthesis of a linear peptide (demanding two levels of orthogonality—one for the N-terminal amino group, another for the C-terminal carboxyl group and side chains of trifunctional amino acids), design of a cyclopeptide synthesis must include one additional (third) level of orthogonality. The three-dimensional orthogonal strategy is achieved by combining protecting groups that are, for example, cleavable by acid (Boc–), strong acid (benzyl-based), base (Fmoc–), palladium (Aloc,–OAl), fluoride ion (trimethylsilylethyl-based), or with nucleophiles (Dde). Examples of these combinations are collected in Table 16.1 and selected structures are provided in Table 16.2. These demands in turn suggest that a trifunctional amino acid can be attached directly to polystyrene resin (Merrifield resin, hydroxymethyl resin, aminomethyl resin), to the resin through a linker, or to another suitable support such as polyethylene glycol modified polystyrene (TentaGel or PEG-PS), polyamide, or a membrane-based solid. The exact chemistries may be expected to influence the anchor bond stability, chemistry, and the ease of cleavage of the final product from the resin. Usually the most critical decision involves the choice of the C-terminal carboxylic acid protecting group and the nature of the linkage of the side chain functional group to the solid support.

2.2.1. Carboxyl Protecting Groups

2.2.1.1. OFm Esters. A general method for the SPPS of head-to-tail cyclic peptides containing an aspartyl residue in their sequence was proposed by Rovero et al.[15] Synthesis was performed starting from the Asp residue linked to PAM-resin through the β-carboxyl function and protected as its fluorenylmethyl ester on the α-carboxylic acid group. Once the linear precursor had been synthesized by the Boc/benzyl strategy and the Boc protecting group on the N-terminal amino function was removed with TFA, the C-terminal carboxyl group of the Asp residue was selectively deprotected with piperidine and head-to-tail cyclization was easily accomplished by the BOP method. Final HF deprotection of the side chains of trifunctional residues, with concomitant cleavage from the resin, gave the cyclic peptide as a single product of good purity in reasonable yield.

TABLE 16.1. Selected Examples of Orthogonality of Protecting Groups for SPPS of Head-to-Tail Cyclic Peptides

C^{α}–COOH Protection	N^{α}–NH_2 Protection	Side Chain Protection	Anchoring Bond/Cleavages	Resins Used	Amino Acids	References
–OFm	Boc–*	benzyl-based or related	ester, urethane, amide/HF	hydroxymethyl-PAM	Asp, Glu, Lys, Orn, Dab	15, 19
				MBHA	Asn, Gln	19
–ODmb	Fmoc–*	*t*-Bu-based	ester/TFA	4-alkoxybenzyl	Asp, Glu	23
–OAR	Boc–	benzyl-based or related	mixed carbonate/HF	hydroxymethyl-	Ser, Tyr	47
	Fmoc–	*t*-Bu-based	ester, amide, BAL, urethane/TFA	4-alkoxybenzyl	Asp, Glu, Lys	27
				PAL-aminomethyl	Asn, Gln	27
–ONB	Boc–	benzyl-based or related	ester, ether, urethane/HF	hydroxymethyl	Asp, Glu, Tyr, Lys	51
			ester (hemi-succinate linker)/NH_3	aminomethyl-	Ser	51
–ODmab	Fmoc–	*t*-Bu-based	ester/TFA	2-chlorotrityl	Glu, Asp	39

*Boc = *tert*-butyloxycarbonyl; Fmoc – fluorenylmethyloxycarbonyl.

This approach was further developed in our laboratory for the synthesis of head-to-tail cyclic peptide libraries.[16] However, as suggested by Bednarek et al.,[17] the C-terminal-OFm ester is not completely stable under the conditions used for Boc-based solid phase synthesis. The neutralization cycle required before coupling, when Boc- is used on the α-amino group, exposes the –OFm ester protecting the C-terminal α-carboxylic group to the tertiary base diisopropylethylamine (DIPEA) which can result in a small amount of deprotection.[18] Detailed studies revealed that epimerization accompanied chain growth and was therefore most likely caused by premature removal of the C-terminal protecting group during and after the neutralization step.[19] This unintended exposure of the acid function can lead to inappropriate activation by the condensing agent during subsequent stepwise N-terminal elongation and accounts for ever-increasing amounts of epimerization as synthesis proceeds. Nevertheless, we found that in situ neutralization could avert many of these problems; this orthogonality was successfully used for the synthesis of Lys-, Orn-, and Dab-containing cyclic peptides through a similar side chain attachment mode.[20] This side chain attachment approach was successfully applied for the synthesis of a small library of stylostatin analogs[21] as well as a large library of cyclic pentapeptides which included the endothelin antagonist analog BQ-123[22] (Fig. 16.3(c)), discussed later in this chapter.

2.2.1.2. ODmb Esters. In an on-resin cyclization technique developed by McMurray,[23] the first amino acid attached to a 4-alkoxybenzyl-resin was either Fmoc–Asp-α-2,4-dimethoxybenzyl (Dmb) ester or Fmoc–Glu–ODmb ester. The resin-bound Dmb esters are compatible with hydrolysis using 1% TFA in methylene chloride. (See Table 16.2 for descriptions.) The *t*-butyl- and sulfonyl-based side chain protecting groups widely used in solid phase Fmoc chemistry are mostly stable to these conditions. If Asn or Gln residues are desired these amino acid esters can be attached to the acid-labile amide handle such as peptide amide linker (PAL)[24] or the methoxy-substituted benzhydrylamine handles.[25] A comparable Fmoc/ODmb protection strategy was also used by Brugghe et al.[26] to prepare cyclic peptides containing 7–14 amino acid residues as potential synthetic vaccines for bacterial meningitis.

2.2.1.3. OAl Esters. Several authors have applied allyl chemistry to provide the third dimension of orthogonality to the initially anchored Asp or Glu via their α-carboxyl-protected allyl esters.[27,28] The allyl group is stable to the conditions used for removal of most acid and base labile protecting groups. Thus, it is compatible for both Boc and Fmoc methods of peptide synthesis. At the appropriate stage the allyl esters can be selectively deblocked with palladium(0) under nearly neutral conditions. While this is a potentially attractive approach to orthogonality, the homogeneous palladium reagents are air sensitive and can sometimes give rise to greater amounts of side products.[29]

TABLE 16.2. Structures of Common Acid Protecting Groups and Methods for their Removal

	Removed by	Stable to
–OFm = –OCH_2–(9-fluorenyl)	20% piperidine/ DMF	TFA
–ODmb = –OCH_2–(2-MeO, 4-OMe-phenyl)	1% TFA/ CH_2Cl_2	20% piperidine/ DMF
OAl = $-OCH_2CH=CH_2$	Pd^0/morpholine as allyl acceptor	TFA; 20% piperidine/DMF
–ONB = –OCH_2–(C6H4)–NO_2	$SnCl_2$/DMF	TFA
–ODmab = –OCH_2–(C6H4)–NH–(ivDde: O, Me, Me, O, Me, Me)	2% hydrazine/ DMF	TFA; 20% piperidine/DMF

2.2.1.4. ONB Esters. *p*-Nitrobenzyl (–ONB) esters are stable to most conditions of peptide synthesis; they can be selectively cleaved in the presence of *t*-butyl and benzyl based protecting groups by reducing agents (Zn in HOAc, Na_2S, $Na_2S_2O_4$ or $SnCl_2$ in aqueous organic solvents).[30] These esters have been used to provide orthogonality for the solution phase synthesis of cyclic analogs of some ACTH/α-MSH linear fragments.[31] However, upon reductive cleavage of ONB esters, quinonimine methide is formed, and this leads to intractable contaminants. Recently, a new approach was suggested[32] for the removal of the reactive intermediate (quinonimine methide) from the reaction mixture based on its reaction with benzene sulfonic acid.[33]

Recently, it has been reported that a new *p*-nitrobenzyl-based linker, 5-hydroxymethyl-2-nitrophenoxy-acetyl (2-NPA), is HF-resistant. By using this group for resin attachment, this provides for the orthogonal cleavage of protected and unprotected peptides from resins in organic and aqueous media using sodium dithionite under very mild reductive conditions.[34] This represents a swap between the groups used for the side chain versus main chain functionalities and suggests that other protecting groups are often similarly reversible as appropriate. This and other examples in which the side chain acid groups can be linked to solid supports by a variety of cleavable linkers are contained in Table 16.3.

TABLE 16.3. Selected Examples of Linkers Attached to Solid Supports by their Carboxylic Acid Function

No.	Linker	Structure (Linker-Resin)	Peptide Released from Linker by	Reference
1	5-hydroxymethyl-2-nitrobenzoic acid	—OCH_2— (benzene ring) —NO_2; CO-Resin	$Na_2S_2O_4$	34
2	5-hydroxymethyl-2-nitrophenoxyacetic acid	—OCH_2— (benzene ring) —NO_2; OCH_2—CO-Resin	$Na_2S_2O_4$	35
3	4-oxocrotonic acid	–OCH_2–CH=CH–CO-Resin	Pd^0/morpholine as allyl acceptor	36
4	N-[(9-hydroxymethyl)-2-fluorenyl]-succinamic acid	—OCH_2— (fluorene); HN-CO-CH_2CH_2-CO-Resin	secondary amine	46
5	4-hydroxymethyl-3-nitrobenzoic acid	—OCH_2— (benzene ring) —CO-Resin; NO_2	photolysis at 350 nm in TFE-CH_2Cl_2 for 16 h	47

2.2.1.5. ODmab Esters. The Dde group, (1-(4,4-dimethyl-2,6-dioxocyclo-hexylidene)ethyl), typically has been used for side chain amine protection.[37] The Dde group is stable to the acidic reagents used in Boc-based solid phase synthesis. It is mostly stable to 20% piperidine/DMF, but can be removed by treatment with 2% hydrazine in DMF. The compatibility of Dde- and allyl-chemistries has been exploited to provide two "mild" extra orthogonal steps for an otherwise Fmoc/*t*-butyl-based SPPS of an intrachain-branched Lys-based cyclic peptide.[38]

By combining the desirable properties of the N-Dde group with the known lability of 4-aminobenzyl esters, this led to the development of a new carboxyl-protecting group (–ODmab) 4-(N-[1-(4,4-dimethyl-2,6-dioxocyclohexylidene)-3-methylbutyl]amino) benzyl ester (Table 16.2) that retains all the desired chemical properties for masking either the α- or the $\beta(\gamma$-) carboxylic acid groups of Asp/Glu. This group is stable to acidic reagents, and it is also reported to be completely stable to Fmoc-deprotection conditions.[39] Removal can be effected by using 2% hydrazine in DMF, thus providing another approach to orthogonality.

Applications of orthogonality of new systems can lead to further complications, thus removal of the Dde group by hydrazine results in side reactions in peptides containing the allyloxycarbonyl (Aloc) protecting group. The addition of allyl alcohol as scavenger prevents the hydrogenation of the Aloc group. Under these conditions, Dde and Aloc groups can also be used as fully orthogonal protection techniques in solution and SPPS.[40] These precautions may also be required when using the ODmab group together with allyl chemistries.

2.2.2. Side Chain Linkage Methods

2.2.2.1. Ester or Amide Linkages. Most amino acid side chain attachment strategies have involved the use of aspartic acid and glutamic acid residues since the functional group involves the same chemistries as normal α-carboxylic acid solid phase peptide synthesis. If the side chain acid is attached to the resin through a benzyl ester-type linkage, acid cleavage provides the Asp or Glu residues. If ammonolysis is used as the cleavage method or if initial attachment is to a benzhydrylamine or *p*-methylbenzhydrylamine resin, the cleavage will yield the corresponding asparagine or glutamine derivatives (Fig. 16.5).

Choices for the α-carboxylic acid protecting group will be based on the nature of the amine protection (Boc or Fmoc) and on the approach towards orthogonality that is desired. A latter section will describe some of the specific derivatives used for head-to-tail cyclic peptide synthesis.

2.2.2.2. Urethane-based Linkages. Trifunctional amino acids with amino acid side chains such as lysine and ornithine have primarily been attached to a solid support using a benzyloxycarbonyl-type urethane bond. The urethane or carbamate group is compatible with both Boc and Fmoc strategies and can be formed from hydroxymethyl resins with amines. Such resins were first prepared

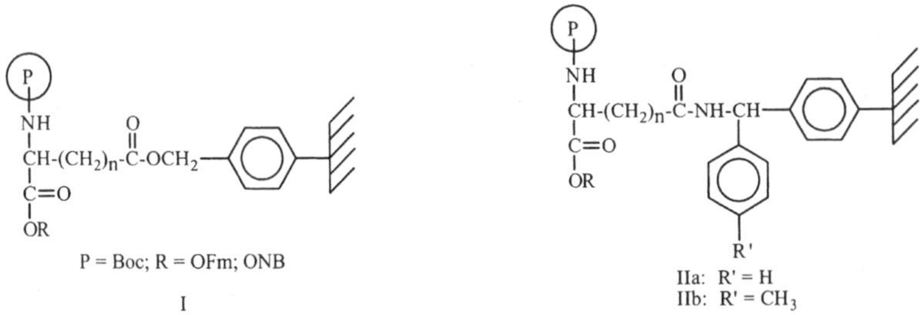

P = Boc; R = OFm; ONB

I

IIa: R' = H
IIb: R' = CH_3

FIGURE 16.5. Side chain attachment of aspartic acid ($n = 1$) or glutamic acid ($n = 2$) to polystyrene resin (I) or to benzhydrylamine; (IIa) $R' = H$ or *p*-methylbenzhydrylamine resin; (IIb) $R' = CH_3$ for formation of Asp, Glu(I) or Asn, Gln(II) containing cyclic peptides.

from the hydroxymethyl derivative of the classical Merrifield resin by reaction with phosgene, to furnish the corresponding chloroformate derivative, which was followed by displacement of halogen by the amine component.[41] Thus, Sklyarov and Shashkova attached a preformed linear tetrapeptide via the δ-amino group of an ornithine residue to a chloroformate derivatized polystyrene support (Scheme 1). After the addition of six more residues, the peptide product, an analog of Gramicidin S, was deprotected at both ends, cyclized on the resin and detached from the resin.[11]

However, the use of phosgene is dangerous and not compatible with more acid labile resins required for orthogonal schemes. Other synthetic pathways useful for preparing the reactive species leading to alkoxycarbonylation of amines involve the preparation of various activated derivatives of alcohols (activated mixed carbonates or chloroformates). Several of these methods have also been used for the preparation of polyethylene glycol derivatives.[42]

Thus, hydroxymethyl resin can be easily converted to activate mixed carbonate in nearly quantitative yield by treatment with *p*-nitrophenyl

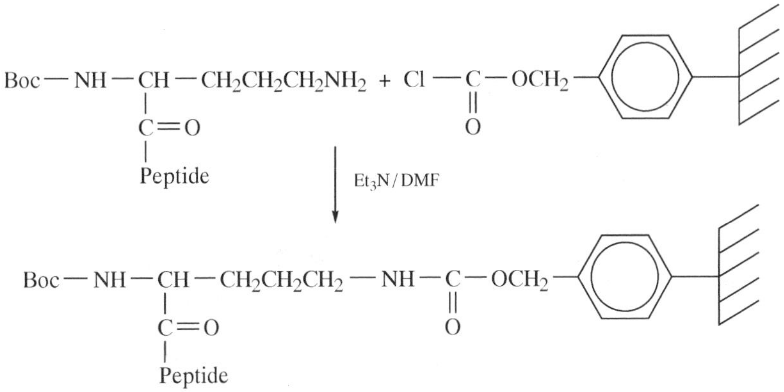

Boc—NH—CH(C=O–Peptide)—$CH_2CH_2CH_2NH_2$ + Cl—C(=O)—OCH_2—(resin)

Et_3N/DMF

Boc—NH—CH(C=O–Peptide)—$CH_2CH_2CH_2$—NH—C(=O)—OCH_2—(resin)

Scheme 1. Attachment of the ornithine side chain to activated polystyrene resin to furnish a urethane linkage.

chloroformate and *N*-methyl morpholine.[43] This activated mixed carbonate is claimed to be a robust polymer-bound reagent well suited for urethane formation as well as for other modifications that can be used for combinatorial library production.

N,N′-disuccinimidyl carbonate (DSC) has recently been proposed as a mild agent for alkyoxycarbonylation of amines from the corresponding alcohols in solution.[44] This method can also be used to prepare the resin-bound urethanes. In a recent comparison of some of the various derivatization agents (Fig. 16.6) used to form activated carbonates from hydroxymethyl resin (*p*-nitrophenyl chloroformate, carbonyldiimidazole, DSC, or di(pentafluorophenyl)carbonate)[45]), DSC and di(pentafluorophenyl)carbonate were the most effective, providing high levels of substitution, reasonable stability and ease of handling. Both showed good reactivity in the subsequent urethane formation processes. Urethane-based side chain attachment has been used with the Fmoc–/–OAl strategy,[46] Boc–/–OFm strategy,[20] and it is also compatible with the Boc–/–ONB strategy.

2.2.2.3. Mixed Carbonate Links. Activated carbonates such as those used for the preparation of urethanes from amines can also be used to form stable links with the hydroxyl side chains of amino acids. For example, the active mixed carbonate resins formed from disuccinimidyl carbonate and hydroxymethyl resins were reported to undergo transesterification with alcohols in the presence of dimethylaminopyridine as catalyst.[47] The resulting asymmetric carbonates are reported to be stable for Boc-based synthesis. Using this chemistry, several model cyclic peptides attached to solid polystyrene-based supports via their hydroxyl (serine) or phenolic (tyrosine) side chains could be prepared with the Boc strategy and with the allyl group used for α-carboxylic acid protection. Cleavage of the products from resin was achieved using either anhydrous HF or with trifluoromethane sulfonic acid.

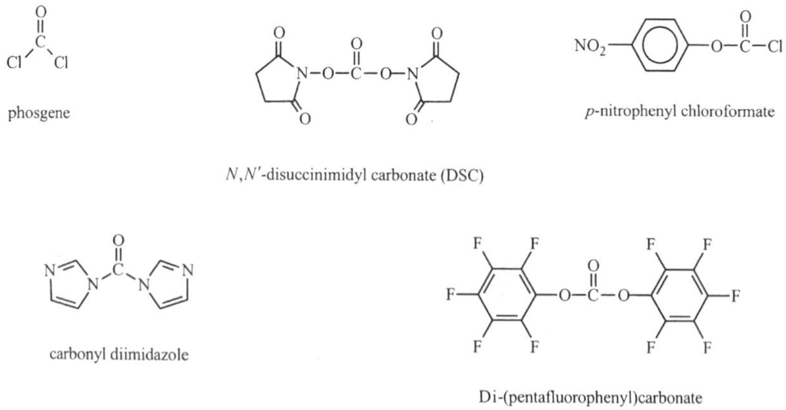

phosgene

N,N′-disuccinimidyl carbonate (DSC)

p-nitrophenyl chloroformate

carbonyl diimidazole

Di-(pentafluorophenyl)carbonate

FIGURE 16.6. Structures of reagents used for activation of resin bound alcohols as a first step toward urethane formation.

2.2.3. Examples of Amino Acid Derivatives Used for Cyclic Peptides. Once a suitable orthogonal protection strategy is selected, the head-to-tail cyclic peptides can be conveniently prepared by linking the desired trifunctional amino acid to a solid support via its side chain. In the sections below, we describe methods that have been used for preparation of cyclic peptides using several of the coded trifunctional amino acids and their analogues.

2.2.3.1. Asp, Glu, Asn, Gln. A convenient method for preparing a cyclic peptide with on-resin cyclization is shown in Fig. 16.7. Orthogonality is provided by use of a base-labile fluorenylmethyl ester. Boc–Asp–OFm can be obtained by reacting Boc–Asp–anhydride with 1 equivalent of fluorenylmethanol in the presence of base followed by fractional crystallization from ethyl acetate and hexane to obtain the α-ester. Boc–Glu–OFm could be obtained by reacting commercially-available Boc–Glu(OBzl)–OH with fluorenylmethanol in ethyl acetate in the presence of Boc anhydride and pyridine[48] and removing the γ-benzyl ester by catalytic hydrogenolysis.

Esterification of the β-carboxylic group of Asp to the hydroxymethyl–polystyrene support is achieved through a Mitsunobu reaction[49] mediated

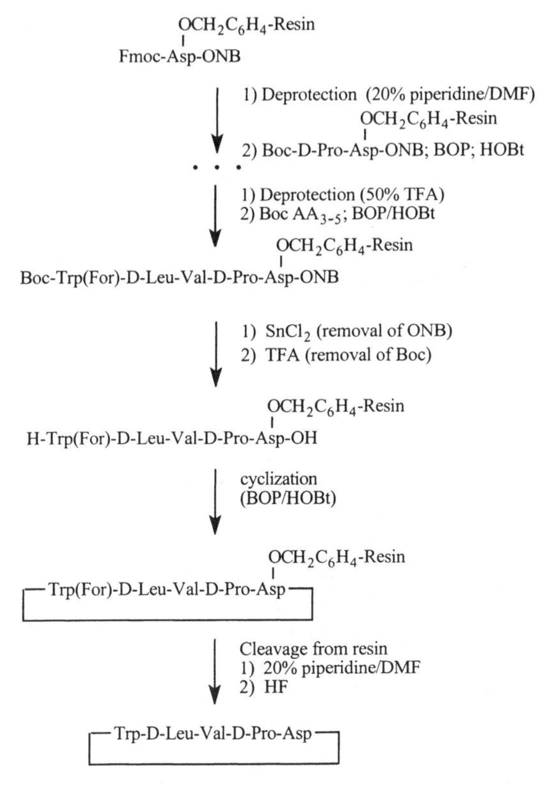

FIGURE 16.7. Synthesis of a model cyclopentapeptide using Asp–ONB for orthogonality.

through triphenylphosphine and diethyl azodicarboxylate in THF or by using Boc_2O/pyridine in methylene chloride. Esterification of the γ-carboxylic group of Glu onto the support can also be accomplished with Boc_2O/pyridine. By coupling the same ω-carboxylic groups with BOP to para-methylbenzhydrylamine resin, Asn and Gln peptides are produced upon cleavage (Table 16.2).

Allyl esters provide a separate approach for orthogonal protection of the α-carboxylic acid. Fmoc–Asp(OH)–OAl and Fmoc–Glu(OH)–OAl were prepared by reacting commercially-available Fmoc–Asp/Glu(OtBu)–OH in neat allyl bromide in the presence of diisopropylethylamine (2 equivalents). Esterification of ω-carboxyls of Fmoc–Asp–OAl and Fmoc–Glu–OAl to *p*-alkoxybenzyl supports was achieved by *N,N'*-diisopropylcarbodiimide (DIC) or TBTU in the presence of 4-dimethylaminopyridine. Coupling of the same ω-carboxyls on amide resin ultimately generates Asn or Gln peptides upon cleavage.[46,50]

Another approach to orthogonality involves the *p*-nitrobenzyl ester as a carboxylic acid protecting group. Fmoc–Asp–ONB could be prepared by reacting Fmoc–Asp internal anhydride with 1 equivalent of *p*-nitrobenzyl alcohol and base.[51] This was followed by fractional crystallization to remove the undesired β-ester. The analogous glutamic acid derivative was prepared by reacting Fmoc–Glu(OtBu)–OH with *p*-nitrobenzyl bromide in the presence of DIPEA with selective OtBu removal by TFA to produce Fmoc–Glu–ONB.

2.2.3.2. Lys, Orn, Dab, Dap. The diamino acids lysine, ornithine, diaminobutyric acid, and diaminopropionic acid can similarly be differentially protected using the corresponding allyl or *p*-nitrobenzyl esters. Fmoc–Lys–OAl can be obtained by refluxing commercially-available Fmoc–Lys(Boc)–OH in neat allyl bromide in the presence of DIPEA. The final product is obtained by removal of the Boc-amino protecting group after which it can be coupled to the solid support.[46] The ONB ester is also obtained by starting with Fmoc–Lys(Boc)OH with removal of the Boc group after esterification.[51] The resulting product, Fmoc–Lys–ONB, can be attached to an activated mixed carbonate resin or other suitably functionalized support to provide a linkage useful for the preparation of head-to-tail cyclic peptides (Fig. 16.8).

Another approach toward the preparation of orthogonally protected diamino acids involves the use of a suitably protected glutamine or asparagine precursor which is subjected to the Hofmann rearrangement.[52,53] Thus, Boc–Gln–OFm can be treated with *bis*(trifluoroacetoxy)iodobenzene to afford the corresponding diamino butyric analog, Boc–Dab–OFm. This derivative is then attached to a solid support via its side chain amine group by using a suitable linker.

2.2.3.3. Ser, Tyr, Thr. One method for preparing cyclic peptides by using the side chain hydroxyl function for resin attachment also involves the allyl group for carboxyl protection (Fig. 16.8(b)). The reaction of DSC with 4-hydroxymethyl-polystyrene resins affords the corresponding active carbonates, which in the presence of base, can react smoothly with compounds containing hydroxyl

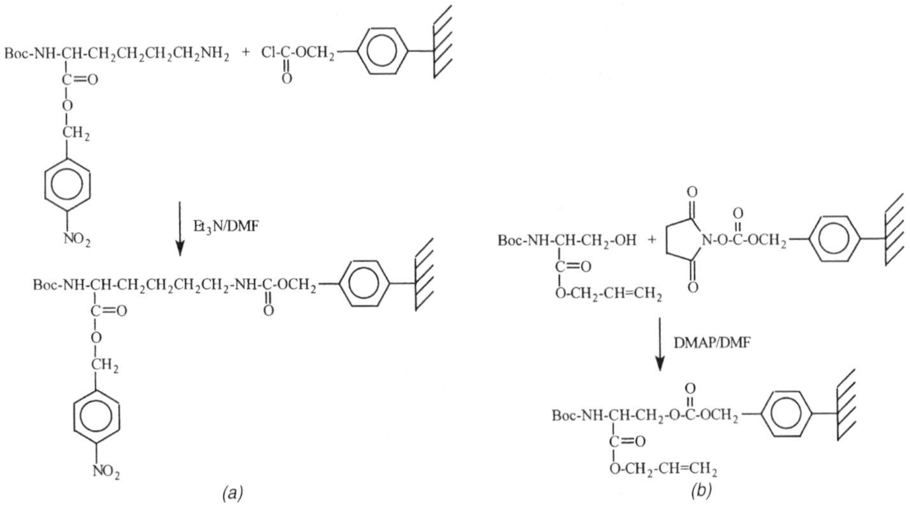

FIGURE 16.8. The *p*-nitrobenzyl and allyl esters can be used for carboxyl protection when attaching lysine (*a*) or serine (*b*) to resin supports.

functions (Boc–Ser–OAllyl, Boc–Tyr–OAllyl and others) to afford the corresponding mixed carbonates.[47]

Another comparable approach developed in our laboratory uses the *p*-nitrobenzyl group (ONB) for acid protection. Thus, Boc–Ser–ONB is prepared from Boc–Ser–OH and the product attached to the support as the Boc–Ser(Suc)–ONB derivative. Using this approach, a serine analog of the BQ-123 endothelin antagonist was prepared in good yield.[20]

The use of an asymmetric carbonate bond for tyrosine side chain attachment has been considered in analogy with serine attachment. We chose an aryl benzyl ether bond to attach Boc–Tyr–ONB to the hydroxymethyl-PS resin through a Mitsunobu reaction[49] (Scheme 2). This reaction has been used successfully to form aryl ethers.[54] More recently Mitsunobu etherification of polymer-supported phenols with alcohols in solution has been studied[55] and conditions for attachment of protected tyrosines to the hydroxymethyl resin using this approach have been developed.[56] We found the Mitsunobu conditions quite useful for etherification: the triphenyl phosphonium activated hydroxymethyl resin, when treated with Boc–Tyr–ONB, in the presence of tertiary amine, formed an aryloxide which, provided a good yield of the side chain attached Tyr derivative.

2.2.3.4. His. Orthogonality for His has been achieved by attaching the amino acid to polystyrene resin through a 2,4-dinitrophenyl-type side chain protecting group. For that purpose, the aminomethyl resin was reacted with excess 1,5-fluoro-2,4-dinitrobenzene to give polymer-bound 1-fluoro-2,4-dinitrobenzene. Reaction of the latter with Boc–His gave Boc–His(DNP) attached to the resin through its side chain (Fig. 16.9(a)). The His-derivatized resin was used for the synthesis of a cyclic hexapeptide with on-resin cyclization. After removal from

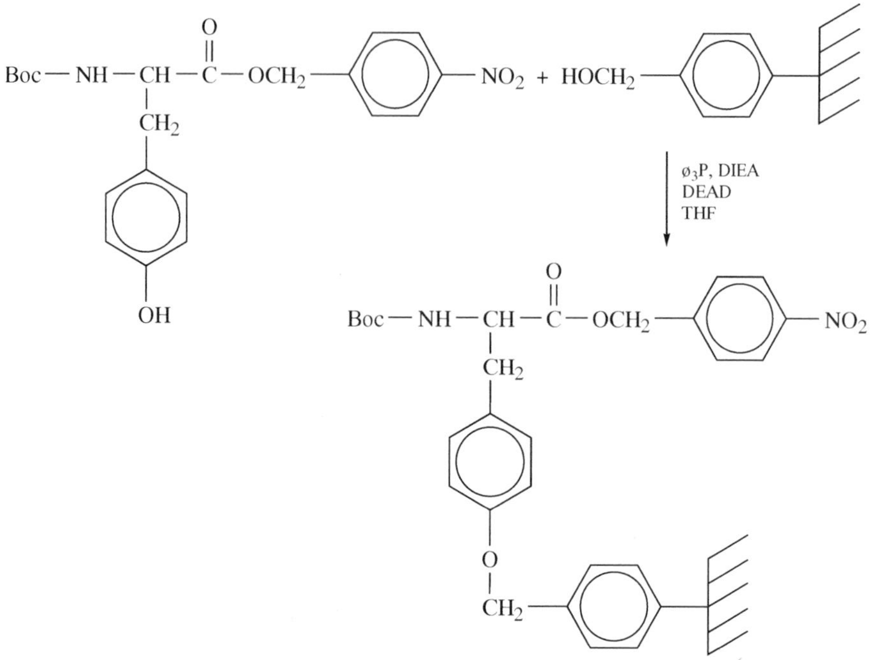

Boc—NH—CH—C(=O)—OCH_2—C_6H_4—NO_2 + $HOCH_2$—C_6H_4—resin (Tyr side chain CH_2—C_6H_4—OH)

ø$_3$P, DIEA
DEAD
THF

Boc—NH—CH—C(=O)—OCH_2—C_6H_4—NO_2 (side chain CH_2—C_6H_4—O—CH_2—C_6H_4—resin)

Scheme 2. Use of Mitsunobu reaction conditions to attach tyrosine through its side chain to a hydroxymethyl resin.

the resin by thiophenol/DMF treatment, cyclic (Gly–His)$_3$ was obtained in good yield and purity.[57]

2.2.3.5. Arg. The guanidyl side chain of arginine is typically protected by an aryl sulfonyl derivative such as the 4-methoxy-2,3,6-trimethylbenzenesulfonyl (Mtr) group. Arginine can be attached to a solid support by using a modified form of the Mtr group in which the 4-methoxy group methyl has been replaced by a 4-carboxymethyl moiety. This new linker-protecting group, CMtr, (Fig. 16.9(b)) allows synthesis of Arg-peptides and Arg-peptide derivatives that are inaccessible by current procedures. While the more traditional Mtr group may lead to incomplete cleavage, the use of the modified linkage should result in a purer product.[58] But this approach has not yet been applied to the synthesis of head-to-tail cyclic peptides.

2.2.3.6. Cys. The cysteine residue provides an interesting challenge for side chain attachment strategies. Sulfur is a good leaving group and there is always a danger of a β-elimination reaction. One approach is to use an indirect attachment to a solid support via an intervening linker or handle. Using this approach, C-terminal cysteines with an Fmoc-group for N-α-protection and *t*-butyl or allyl groups for C-α-protection have been attached to solid supports using an S-xanthenyl (2-Xal) handle[59] (Fig. 16.9(c)). By using this anchoring handle, a

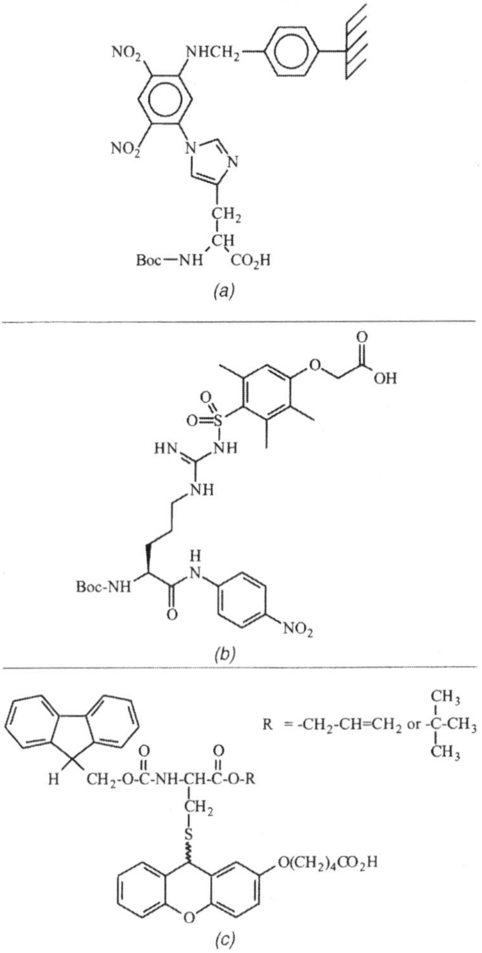

FIGURE 16.9. Amino acid derivatives used for side chain attachment of glutamic acid; (a) histidine; (b) arginine; (c) and cysteine (d).

common side reaction during the base deprotection step, involving formation of a dehydroalanine intermediate followed by piperidine addition,[60] is circumvented. Cleavage of the S-(2-Xal) by an acidolytic or oxidative mechanism in concert with appropriate protecting groups for the C-α-carboxyl acid provides options for the preparation of free or protected disulfide containing peptides.[59]

Attachment of cysteine through its side chain can be exploited to provide the elimination products. It has been shown that a polymer-bound cysteine, protected at its N-terminus, can be readily esterified or coupled with an amine or amino acid ester. Oxidation of the corresponding sulfides with m-chloroper-benzoic acid furnishes the desired sulfone derivative attached to the resin. (Scheme 3) In this case, dehydroalanine derivatives are obtained by treating the

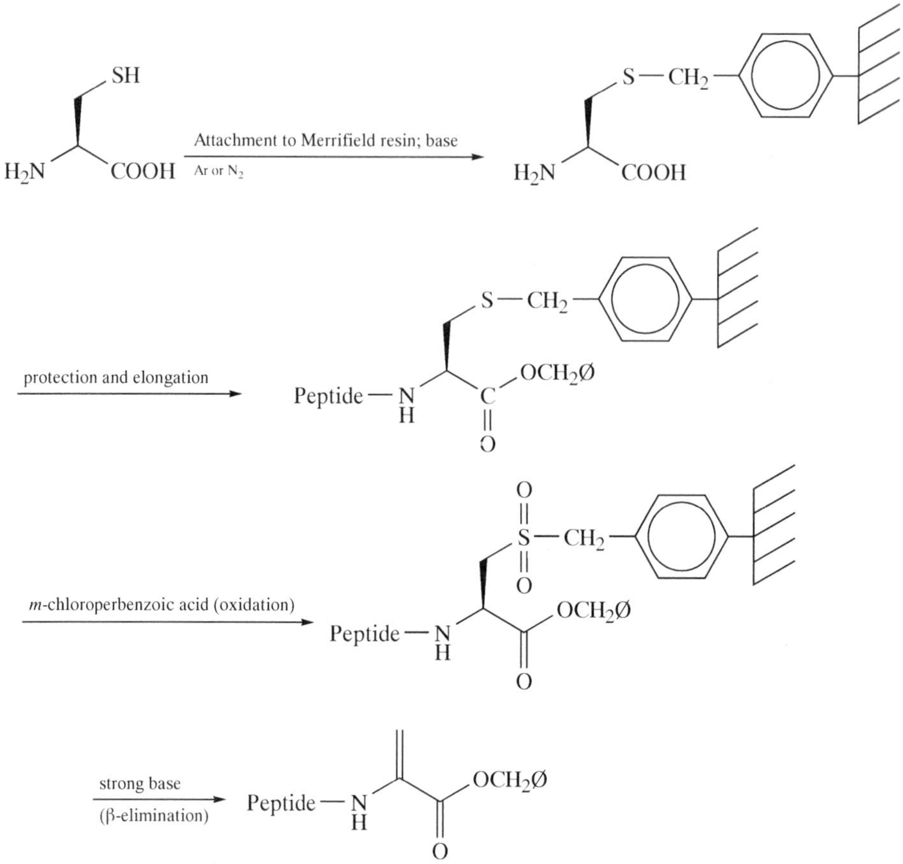

SH

H_2N COOH

Attachment to Merrifield resin; base

Ar or N_2

S—CH_2

H_2N COOH

protection and elongation

S—CH_2

Peptide—N(H)—C(=O)—$OCH_2Ø$

m-chloroperbenzoic acid (oxidation)

O=S(=O)—CH_2

Peptide—N(H)—C(=O)—$OCH_2Ø$

strong base

(β-elimination)

Peptide—N(H)—C(=O)—$OCH_2Ø$

Scheme 3. Synthesis of dehydroalanine peptides based on cysteine side chain attachment.

sulfone derivatized resin with an equimolar amount of the strong base DBU (1,8-diazabicyclo[5.4.0]undec-7-ene) in CH_2Cl_2.[61]

2.2.4. Cyclic Peptides from Amide (NH) Attachment to Resin. The attachment of amino acids to solid supports through their side chains leads to the consideration of alternative approaches. Alkoxybenzaldehyde-based linkers have been used for the temporary protection of amide nitrogens.[62] These same derivatives allow anchoring of the first amino acid residue for SPPS through the amide nitrogen of the peptide backbone (Fig. 16.10). Initially an aldehyde precursor to PAL was coupled through a reductive amination procedure to the α-amine of the prospective C-terminal amino acid, which was protected as a *tert*-butyl, methyl, or allyl ester, or modified to a dimethyl acetal.[63] The resultant intermediates, all secondary amines, were treated with Fmoc–Cl or Fmoc–succinimide to give the corresponding protected amino acid preformed handle

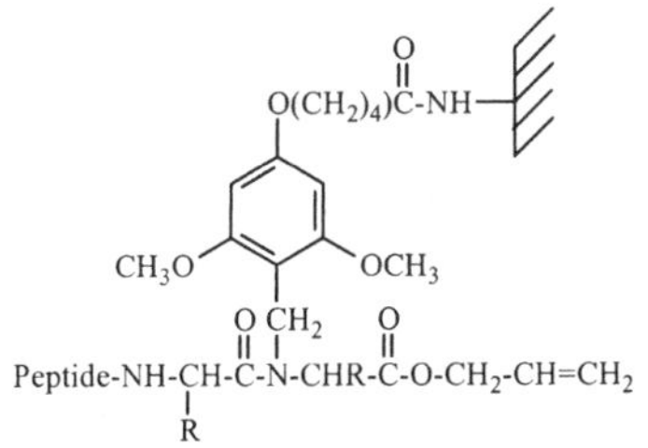

O(CH$_2$)$_4$C(=O)-NH-
CH$_3$O OCH$_3$
Peptide-NH-CH(R)-C(=O)-N(CH$_2$)-CHR-C(=O)-O-CH$_2$-CH=CH$_2$

FIGURE 16.10. Structure of a peptide attached to a solid support through a backbone amide linker (BAL) and with an allyl ester for C-terminal protection.

derivatives in 40–70% yields. These preformed handles are then ready to attach to PEG-PS supports through a BOP/HOBt mediated coupling.

The Fmoc-preformed BAL handle derivative of Gly–OAllyl has been coupled onto polystyrene resin and extended (including the modified protocol to avoid diketopiperazine formation) to generate the protected peptide-resin Fmoc–Pro–Tyr–Leu–Ala–(Bal–PS)Gly–OAl. Cleavage with 5% trifluoroacetic acid provided the protected peptide Fmoc–Pro–Tyr–Leu–Ala–Gly–OAl in greater than 85% purity and 90% yield. In a preliminary experiment, the allyl ester was removed selectively from the BAL-anchored peptide with Pd(0), and the Fmoc group was then removed with piperidine/DMF (1 : 4). On-resin cyclization using BOP/HOBt/DIPEA gave the expected five-residue head-to-tail cyclic peptide as the main product.

More recent work has focused on improved handles which can be prepared more readily and which provide a wider array of amide derivatives.[64,65]

In conclusion, the backbone amide linker (BAL) anchoring has been applied to the solid phase synthesis of linear peptides with a considerable range of C-terminal modifications, and appears to provide an alternative general approach for the synthesis of head-to-tail cyclic peptides.

2.3. Coupling Agents

The azide method has often been used for the solution cyclization of linear peptides, primarily due to its relatively low danger of racemization as compared to other methods. Azide formation and cyclization are carried out in separate stages following removal of the N-α-protecting group of the linear precursor (Scheme 4). Azide formation proceeds so fast that the α-amino group is not affected by deamination. Cyclization is generally conducted at high dilution (for a review, see Ref. 66). A related compound that generates an azide intermediate, diphenylphosphoryl azide, has been the reagent of choice for linear peptide cyclization in solution since its introduction for that purpose in 1979,[67] and has been combined with dimethylaminopyridine and hydroxybenzotriazole to accelerate the rate of cyclization.[68] 2-Ethoxy-1-ethoxycarbonyl-1,2-dihydroquinoline (EEDQ)[69] has been used for solid phase hexapeptide cyclization (presumably through the formation of an intermediate mixed anhydride).[57]

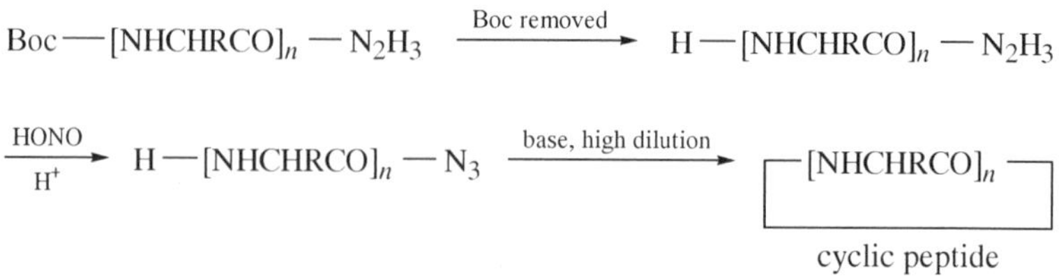

$$\text{Boc}-[\text{NHCHRCO}]_n-\text{N}_2\text{H}_3 \xrightarrow{\text{Boc removed}} \text{H}-[\text{NHCHRCO}]_n-\text{N}_2\text{H}_3$$

$$\xrightarrow[\text{H}^+]{\text{HONO}} \text{H}-[\text{NHCHRCO}]_n-\text{N}_3 \xrightarrow{\text{base, high dilution}} \text{cyclo}[\text{NHCHRCO}]_n$$

cyclic peptide

Scheme 4. Intramolecular cyclization of a peptide in solution using azide generation from a linear precursor.

Dicyclohexylcarbodiimide (DCC), first described in the 1950s,[70] is still one of the most popular activating agents in peptide cyclizations; it is used either in conjunction with hydroxybenzotriazole (HOBt) or with other activated hydroxyl compounds for active ester generation.[71] Addition of HOBt to carbodiimide-mediated activations is beneficial: intermediate hydroxybenzotriazole esters are generated. They need not be isolated, the reaction is extremely fast, the method counteracts racemization, and other side reactions are few.[72] Another variant, diisopropylcarbodiimide (DIC), is considered advantageous as it forms a derivative, diisopropylurea, that is more soluble than dicyclohexylurea (DCU).[73] Recent results indicate that HOAt, a 4-nitrogen containing variant, is a very effective coupling additive, more efficient than HOBt for solution or solid-phase synthesis.[74] Both HOBt and HOAt effectively catalyze the active ester aminolysis.

Phosphonium and uronium (a urea-derived carbocation)-based reagents (BOP, PyBOP, PyBroP, HBTU, TBTU, HAPyU, and others) have become popular in recent years.[74–77] Structures of these analogues are shown in Fig. 16.11. They couple quickly and smoothly even with sterically hindered amino acids.[78] The manufacture of BOP involves hexamethylphosphoric triamide that is highly carcinogenic. However, in most cases, BOP can be substituted by PyBOP without loss of performance.

On investigating coupling efficiencies and rates for several reagents and esters they were arranged in the following order BOP/HOBt > DIC/HOBt > DIC/HOPfp.[79] Uronium and phosphonium derivatives of HOAt, namely, HATU and PyAOP are even better coupling agents.[80] X-ray structure analysis of HATU and HBTU has revealed that the solid phase structures of these compounds are not the *N,N,N′,N′*-tetramethyluronium salts as commonly presented in the literature, but rather the guanidinium N-oxides[81] (Fig. 16.12). Also, the existence of HAPyU in the form of its guanidinium N-oxide in the solid state has been confirmed by X-ray analysis, as well as by ^{13}C NMR spectroscopy in solution.[82] It has been noted that uronium-based coupling reagents can sometimes interact with the amino group leading to linear guanidino-derivatives instead of the expected cyclic products.[83–85] Therefore, PyAOP has been recommended for the synthesis of cyclic peptides. This reagent precludes guanidinium formation that can occur when uronium salts such as HATU or HBTU react with the amino groups during a slow coupling reaction.

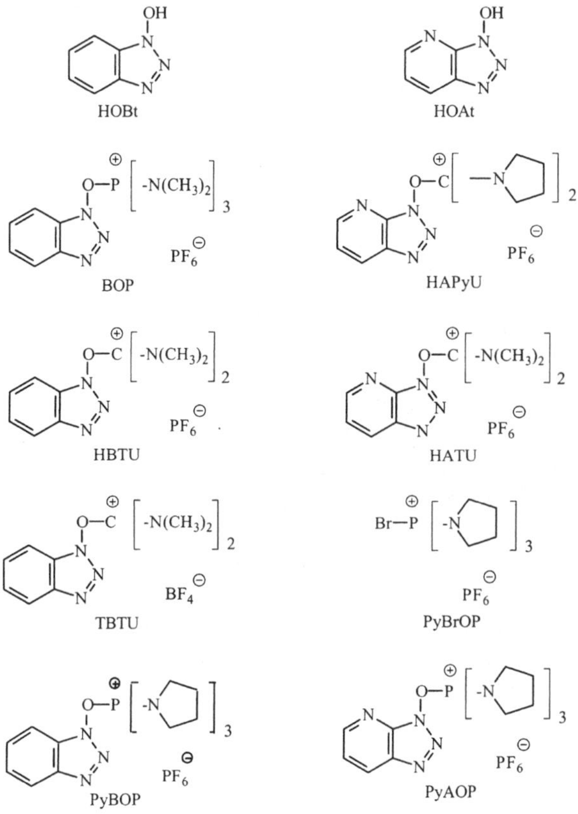

FIGURE 16.11. Common coupling agents used for the peptide cyclization step and their abbreviations.

2.4. Cyclization Techniques

Once a peptide has been synthesized on a solid support via side chain attachment, it can be intramolecularly condensed to form the desired head-to-tail cyclic product. Several approaches have been developed to carry out the cyclization reaction.

2.4.1. Cyclization in Solution. In this method of preparing head-to-tail cyclic peptides, the linear precursors are assembled either by classical methods in solution or cleaved after SPPS from the resin, purified and the cyclization reactions are carried out in solution. Most reactions are carried out under conditions of high dilution (10^{-5} M or less) and this can complicate subsequent reaction work-up.

2.4.2. Resin-bound Active Esters as Peptide Anchoring Groups for End-to-End Cyclization. In this approach, the C-terminal attachment point on the resin is turned into an active ester and upon cyclization, the product is released into

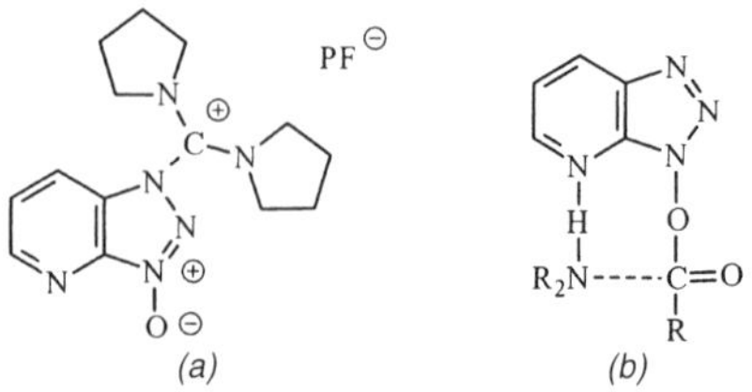

FIGURE 16.12. Structure of the condensing agent HAPyU as the guanidinium N-oxide in the solid state (*a*) as contrasted to the postulated reactive intermediate in solution during coupling where the adjacent 7-azabenzotriazole additive (*b*) can assist in amide bond formation via intramolecular base catalysis.

solution. Various groups including resin-bound *o*-nitrophenol,[10] *p*-sulfonylphenol[86] or *p*-nitrobenzophenone oxime[87] have been used as anchoring groups; their esterification with N-protected peptides produced polymeric active esters. After cleavage of the amino protecting group the resin-bound peptide active ester spontaneously cyclizes in the presence of tertiary amine, simply in suspension in highly swelling solvents (CH_2Cl_2, DMF, or pyridine).

2.4.3. On-resin Cyclization. There are two approaches currently available for the on-resin preparation of the head-to-tail cyclic peptides: the first involves selective deprotection of the C-terminal carboxylic group, activation of the liberated carboxyl function by converting it into an active ester[88]—either by carbodiimide/phenol or transesterification reagent, e.g., pentafluorophenyl trifluoroacetate[89] or *p*-nitrophenyl trifluoroacetate[90] and subsequent liberation of the amino function with cyclization initiated upon neutralization by organic base. The other approach involves simultaneous or concomitant liberation of both the C-terminal carboxylic and N-terminal amino groups and creating the lactam bridge by use of the highly effective phosphonium or uronium-based coupling reagents.[91] Both these methods retain the advantage of the pseudodilution effect provided by the resin carrier which concomitantly simplifies the reaction work-up step.

2.4.4. Enzyme-assisted Cyclization. Enzyme assisted cyclization of linear peptide esters is a relatively new but promising approach for the preparation of head-to-tail cyclic peptides.[92] Due to the entropic barriers for cyclization reactions and competing intermolecular oligomerization, chemical methods are considered largely inefficient for cyclization of longer peptides. Several head-to-tail cyclic peptides (12–25 residues in length) have been obtained via enzymatic cyclization of linear peptide C-terminal glycolate phenylalanylamide using the enzyme, subtiligase. In this particular case, cyclization efficiency appeared to depend primarily on peptide sequence and not concentration. No side chain protection was necessary because the subtiligase enzyme does not accept the ε-amino group of Lys as a substrate and thus this ensures the desired head-to-tail ligation.

2.5. Analytical Techniques

Cyclization reactions in solution can be conveniently monitored by analytical reversed-phase C_{18} HPLC typically using the familiar 0.05% TFA/acetonitrile gradients. Alternatively, on-resin cyclization reactions can be monitored by the qualitative and quantitative ninhydrin reaction (Kaiser test)[93] or the bromophenol blue test for support-bound amines.[94,95] Following the cyclization step, the reaction products are typically isolated by preparative reversed phase HPLC. The identity and purity of all isolated products can be confirmed by various mass spectrometry techniques (FABMS, ESMS, MALDI-TOF and others).[84,85,96–98] Cyclic peptides are identified as those products having a mass of 18 mass units less than their linear counterparts due to the loss of water upon cyclization. The UV-extinction coefficients of starting peptides and cyclic products usually are identical, while dimerization products have extinction coefficients twice those of the corresponding monomers.

2.6. Side Reactions in Cyclization

Oligomerization and C-terminal amino acid epimerization are two major side reactions that affect on-resin cyclization. The first variant occurs from the intermolecular condensation of neighboring peptide molecules to form dimers or oligomers, a phenomenon observed during side chain-to-side chain as well as head-to-tail cyclizations.[99] This is caused by the condensation of peptides adjacent to each other on the same polymer strand or by condensation of peptides on different strands of polymer that are within bonding distance. Racemization is not an unexpected phenomenon since it is also observed as a result of activation of the carboxyl termini of peptides in fragment condensations.

Side reactions are naturally more prevalent when cyclizations are slow. Thus, during the cyclization of an all L-hexapeptide in DMF solution, significant amounts of by-products were observed.[100] These included higher molecular weight masses such as + 60 amu (*N*-α-acetylated linear peptides; acetic acid is sometimes found as an impurity in acetonitrile); + 87 amu (*N*-α-acetylated linear peptide dimethylamides from DMF); and an additional impurity at + 166 amu was seen but not identified. Linear piperidine amides (+ 85 amu) are often observed if piperidine is used during synthesis, e.g., for Fmoc deprotection. Small amounts of piperidine can be difficult to remove through normal washes and can couple to the activated C-terminus when intramolecular cyclizations are slow. Another common impurity observed when uronium condensing reagents are used is the C-terminal guanidinium derivatized by-product (+ 98 amu). This was observed during the synthesis of small ring cyclic opioid peptide analogues.[83]

Aspartic acid and glutamic acid groups in peptides are often the cause of common side reactions. Loss of 18 mass units can be due to imide formation, but dehydration of asparagine and glutamine residues, while less common, can also be a problem. When aspartic acid is linked to a resin through its side chain, the possibility of an α-aspartyl to β-aspartyl rearrangement is often a possibility.

Among major side products observed while cyclizing glutamic acid-containing peptides in side chain to tail or side chain to side chain derivatives in 5-mer to 14-mer epitopes were the linear peptides modified at their Glu side chain carboxylic groups. Thus a +27 amu peak was attributed to a linear peptide adduct formed with dimethylamine, a trace impurity found in DMF. And a +69 amu peak resulted from an adduct formed with morpholine, a possible trace component in *N*-methylmorpholine, a base often used with difficult cyclization sequences.[101]

Sometimes, side reactions can be misinterpreted as evidence of difficult cyclization sequences. Unusually long cyclization times using BOP/HOBt/DIPEA in DMF, as monitored by the ninhydrin test appeared to be due to a reluctance to form the ring. But upon FABMS analysis of the reaction products, a major side product was seen that turned out to be a linear piperidide (+85 amu). This common derivative,[16] reported also by others[27] would result in a continuous positive ninhydrin test, since there would no longer be a carboxylic acid available for amide formation. This difficulty was overcome by more extensive washes in order to effect complete removal of residual piperidine salts that form following N-Fmoc- or C-OFm cleavage using 20% piperidine/DMF before cyclization.

3. EXAMPLES OF CYCLIC PEPTIDES OF VARYING RING SIZE

3.1. Lactams

While peptide cyclization normally implies the presence of at least two amino acids, lactam rings can often form from side chain to backbone processes that may also be in competition with oligomerization or with the formation of larger ring sizes.

The γ-, δ-, and ε-amino acids ($NH_2(CH_2)_nCO_2H$, where $n = 3–5$), bound as active esters to the resin support, provided only the 5-, 6-, and 7-membered lactam rings on cyclization with no or only slight oligomerization.[102] Resin-bound β-alanine gave mainly the cyclic tri- and tetrapeptides, less cyclopentapeptide and very little eight-membered cyclodipeptide as well as some macrocyclic hexa- and heptapeptides. ε-Aminoundecanoic acid and ω-aminododecanoic acid yield mainly cyclic diamides (24 or 26 ring atoms, respectively) and decreasing amounts of higher rings up to hexamers. Lactams with unfavorable ring sizes (4-, 12-, and 13-membered rings) are not formed at all during ring-closure of the polymer-bound ω-amino acids.[102] It seems that steric isolation of the grafted polymer chains leading to monomeric rings only occurs to a minor degree. Instead very large rings (up to about 300–400 ring atoms) are formed.[102]

Polystyrene-based resins have been suggested to act chemically as a rigid lattice,[103] separating the loci of polypeptide growth. When insoluble cross-linked poly-4-hydroxy-3-nitro-styrene was used as a phenol component for a peptide active ester, condensation between the active peptide moieties was markedly reduced and internal aminolysis led to the formation of the desired cyclic peptides[10] which were released from the insoluble carrier. It has been suggested that the enhanced preference for cyclization over open-chain coupling of peptides bound to high molecular-weight polystyrene—2% divinylbenzene based resins takes place due to separation of the specific reaction sites in the rigid lattice, an effect equivalent to high dilution (pseudodilution factor). However, convincing evidence has appeared that complete intra-resin site separation is not achieved: in fact, it has been unambiguously demonstrated that even moderately cross-linked polymers (up to 8% divinylbenzene) must be regarded as rather flexible entities, and that steric hindrance due to the polymer backbone can be discounted as a significant factor in preventing intermolecular reactions.[104]

3.2. Cyclodipeptides

Diketopiperazines (DKP) or dioxopiperazine analogs of amino acids are the simplest form of head-to-tail cyclic peptide. Diketopiperazines are frequently observed as unwanted side products, arising from intramolecular aminolysis at the dipeptide level in both solution and SPPS. DKP formation is favored by the presence in either the first or second positions of amino acids (Gly, Pro, or *N*-alkyl) that can easily adopt a cis-configuration in the resulting amide bond. Another favorable combination is to have one L- and one D-amino acid in the dipeptide, due to the minimal steric interference between the two side chains. Furthermore, rates of DKP formation differ considerably depending on the nature of the peptide–resin linkages or the structure of the C-terminal carboxyl protecting group. For example, when solid phase syntheses were performed with BAL-anchored amino acid allyl or *n*-alkyl esters, the authors[63] observed an almost quantitative side reaction due to DKP formation during Fmoc group removal at the dipeptide level. Interestingly, DKP products formed in this fashion have been exploited as scaffolds for combinatorial chemistry, with simultaneous ring formation and release from the resin. The structures themselves provided a large variety of DKP derivatives since they provided at least three different points of diversity: both amino acid side chains and one (of the two) amide bonds.[105]

3.3. Cyclic Tripeptides and Tetrapeptides

Triproline can be cyclized on a polymeric support to give the nine-membered cyclotriprolyl ring.[102] With other amino acids, cyclic tri- and tetrapeptides are formed only to a small or even negligible extent when polymer-bound *p*-sulfophenyl esters are used as the anchoring groups for end-to-end cyclization.[102] All the tripeptide sequences studied yielded oligomeric products featuring a homologous series of cyclopeptides containing up to eight tripeptide residues, i.e., 72 ring atoms. In the case of resin-bound trisarcosine, a sequential set of cyclic peptides with up to 27 sarcosine residues could be identified by gel chromatography.[102]

While investigating the tetrapeptide [4-alanine]-chlamydocin cyclo(Aib-Phe–D–Pro–Ala) precursor cyclization in solution, it was reported[106] that one of the four sequences, namely, H–Ala–Aib–Phe–D–Pro–OH (activated as its ONSu ester), produced the desired product in a much higher (at least an order of magnitude) yield than what was obtained from cyclization of the other linear tetrapeptides. This result again supports a strong sequence dependence on the nature of products formed during a cyclization step.

In cyclic diprolines and cyclic triprolines, both the smaller ring size and the proline side chains contribute to the relative rigidity of the backbone, restricting the peptide to an all-*cis* conformation.[107] Cyclic tetrapeptides serve as useful model compounds to study the relatively slow *cis–trans* isomerization equilibria, *cis/trans* isomerization determined conformations, energy barriers between them, and their interconversions as well as the relative tendencies of the linear precursors to cyclize.[107]

In another study examining the effect of ring size, on-resin cyclization was studied with variable length side-chain amino and carboxyl groups on the first and fourth residues in a model tetrapeptide[108] (Fig. 16.13). BOP reagent was used as the condensing agent to generate lactams containing 14–18 atoms. After cleavage from the resin, the desired cyclic tetrapeptide and a major side product were obtained. It was found that the degree of cyclization difficulty increased significantly as the ring size of the lactam decreased from 18 to 14 atoms. Complete lactam formation required repeated condensations by using 3 equiv of BOP reagent. Thus, an 18-membered ring was formed after four couplings during a period of nearly two days; in contrast, the 14-membered ring required 14 couplings of 12 h each to reach completion as judged by the ninhydrin test. The major side product was identified as a cyclic dimer, obtained as a consequence of interchain cyclization on the resin.

When there is no turn-inducing structural feature (*N*-methyl amino acid, proline, or D-amino acid) present in the peptide, the linear precursor of a cyclopeptide may be unfavorably disposed toward cyclization. In the latter case, a reversible chemical modification of the peptide main chain favoring the cisoid conformation can help. Tetra(phenylalanine) whose preferred linear and rigid conformation is unfavorable for cyclization was used as a test compound for this approach. Three *t*-Boc groups were introduced as substituents on the main chain nitrogen atoms and this altered the molecule enough to allow induction of a

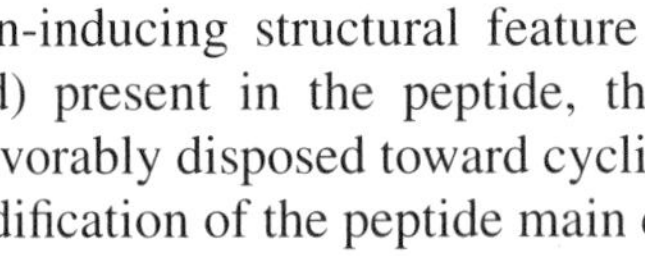

O O
|| ||
CH_3C-NH-CH-CO-Pro-Gly-NH-CH-CNH_2
$(CH_2)m$ $(CH_2)n$
NH——————CO

FIGURE 16.13. Structure of a tetrapeptide ($m = 4–1$; $n = 2,1$) with ring sizes from 14 to 18 subjected to cyclization. Cyclization times range from two days (18-membered ring) to one week (14-membered ring).

conformation with extremities close enough together for cyclization. Cyclization yields increased from less than 1–27% after this chemical modification.[109] On the other hand, cyclic tri-, tetra-, penta-, and hexacarbamates (with correspondingly, 13, 18, 23, and 28 ring atoms) were obtained in high yield and purity by a solid phase synthesis of side chain protected linear derivatives followed by a cyclization reaction by TBTU/HOBt in CH_2Cl_2 solution.[110]

3.4. Cyclizations of Medium Sized Rings

Medium sized rings typically contain 5–8 amino acid residues, and most studies have thus far focused at the lower range of cyclic penta- and hexapeptides. The monocyclization of peptides is often controversial and can still be regarded as an unsettled problem. Theoretical studies predict that intramolecular head-to-tail cyclizations are dependent on peptide length and can also be heavily influenced by side chain configurations. Thus, a DLDLDL hexapeptide is predicted to cyclize intramolecularly while the all-L sequence should resist head-to-tail ring closure, according to these predictions. It has been suggested that cyclization reactions of linear penta- or hexapeptides that do not contain turn-inducing amino acids such as Gly, Pro, or a D-amino acid may be extremely slow and give rise to side reactions such as epimerization and cyclodimerization even at high dilutions (10^{-3}–10^{-4} M).[111]

Significant variations of cyclization yield with sequence were observed in the DPPA mediated condensation of three different linear hexapeptides that were designed to produce the same cyclic product.[112] Cyclo-(Phe–Pro–Phe–Pro–Phe–Pro) was isolated in only 2% yield from the linear all-L-peptide H-(Phe–Pro)$_3$-OH, and low yields were also obtained from two linear diastereomers (Table 16.4). In contrast, the cyclization yield from the diastereomeric linear precursor, H–D–Phe–Pro–Phe–Pro–Phe–Pro–OH, was 57%. From their observations the authors concluded that the best precursor of a cyclic hexapeptide containing a D-amino acid residue is the sequence with the D-residue at the amino terminus.

In another instructive example involving the thymopentin pentapeptide H–Tyr–Arg–Lys–Glu–Val–Tyr–OH, cyclization in solution was performed starting from H–Tyr–Arg–Lys–Glu–Val–OH and using carbodiimide/DMAP. This resulted in a product of high yield and purity, better than obtained using the azide method. However, the strong activation conditions led to complete

TABLE 16.4. Cyclic Hexapeptide Yields Vary Greatly Depending on Configurations Present

Linear Sequence	Yield of Cyclic Peptide (%)
1. H–D-Phe–Pro–Phe–Pro–Phe–Pro–OH	57
2. H–Phe–Pro–D-Phe–Pro–Phe–Pro–OH	2
3. H–Phe–Pro–Phe–Pro–D-Phe–Pro–OH	1
4. H–Phe–Pro–Phe–Pro–Phe–Pro–OH	2

inversion of the C-terminal valine in those cases. This problem was not seen when the linear precursors contained a C- or N-terminal D-amino acid. It was suggested that the presence of a D-amino acid in one of these positions favored a conformation in the linear precursor that would lead to rapid cyclization in good yield.[113] In the absence of such a favorable conformation, the strong and long-lasting activation would lead to racemization at the C-terminus, in which case the now LLLLD linear peptide could undergo cyclization readily.[113]

On the other hand, synthesis of the same cyclic thymopentin pentapeptide from a different linear precursor, H–Arg–Lys–Asp–Val–Tyr–OH, using DPPA with in situ azide formation and a different point of cyclization (Tyr) provided only one product: the all-L-amino acid cyclic pentapeptide. Determination of the chirality of all five amino acids confirmed the absence of racemization.[114] Surprisingly, when the highly effective cyclization agent HAPyU was used, no intermolecular reactions were observed, even at high peptide concentrations (0.1–0.2 M), indicating that at least for the head-to-tail and side chain cyclizations studied, application of the principle of dilution is not required.[115] Attempts to cyclize a thymopentin-related hexapeptide analogue, H–Val–Arg–Lys(Ac)–Ala–Val–Tyr–OH (0.01 M in solution) using BOP and its HOBt-based analogues yielded 5–25% of the desired cyclic peptide. However, the process was accompanied by extensive racemization of the C-terminal tyrosine residue. In contrast, using HAPyU, the all-L-cyclohexapeptide was formed in 55% yield within 30 minutes; and less than 0.5% of the D-Tyr isomer was detected in the reaction mixture. But even HAPyU was not able to effect cyclization of the corresponding all-L pentapeptide without racemization.

In another instructive example, a set of six diastereomeric cyclic hexapeptides of the sequence H–Leu–Tyr–Leu–Gln–Ser–Leu–OH, in which each residue was serially replaced with the D-amino acid, was synthesized (classical methods in solution) and cyclized via their azides.[116] The authors found no significant difference in the cyclization yields with respect to the position of the D-residue, and they could not confirm that a β-turn inducing subunit (a DL pair) needed to be placed only at the center or only at the ends of the peptide chain in order to induce mono-cyclization. Nevertheless, the corresponding all-L-hexapeptide, H–Leu–Tyr–Leu–Gln–Ser–Leu–OH, failed to cyclize when using the azide method or with water-soluble carbodiimide (EDCI).

When two isomeric linear hexapeptides, H–Leu–Met–Gln–Trp–Phe–Gly–OH and H–Met–Gln–Trp–Phe–Gly–Leu–OH (the two linear sequences did not coelute in RP-HPLC) were submitted to cocyclization (PyBOP, DIPEA, DMF), only a single cyclic product was obtained, indicating that both peptides cyclized independently in a head-to-tail fashion to form the same structure. No dimers or polymeric sequences were detected by HPLC or FABMS.[117]

Often the imino acid proline is included to help induce β-turn structure and facilitate cyclization. However, whereas the pentapeptide, H–Arg–Lys(Ac)–Ala–Val–Tyr–OH, underwent cyclization at room temperature (HAPyU, DIPEA, 0.001 M in DMF) to give a mixture of 33% cyclomonomer and 38% cyclodimer in only 10 min, the presence of Pro, regardless of its position in the linear sequence, dramatically hindered formation of cyclomonomers.[118] Similarly, the cyclization probability of hexapeptides was nearly abolished by the presence of proline.[118] In contrast to proline, a single D-amino acid, glycine or *N*-methylamino acid, or just the backbone-modified 2-hydroxy-4-methoxybenzyl (Hmb) -derivatized Ala residues dramatically increased the formation of cyclic pentapeptide monomer.[118]

Oligomer formation during cyclizations of peptides attached to a solid support has been investigated by numerous groups. Again, results have been mixed. In a detailed study of longer peptide sequences by McMurray, the amount of dimer content was uniformly high, and this was not significantly affected by changes in the resin type, resin load, solvent, or coupling agents.[119] In contrast to these results, our group's experience with smaller cyclic pentapeptides has been more positive. Only minimal amounts of dimer content could be found in several studies when using a side-chain attachment, on-resin head-to-tail cyclization approach.[21] A possible explanation for this dichotomy may relate to β-sheet formation.[21,104,120] Longer peptides may be more likely to form anti-parallel β-sheet dimers while still attached to the solid support and this may facilitate intermolecular dimer formation. In contrast, with shorter linear sequences, this type of β-sheet structure may be less prevalent, comparable to the behavior of linear peptides in solution, and thus intermolecular reactions may be entropically disfavored. In any case, the tendencies of compounds to form dimers and higher oligomers is almost certainly sequence specific. The cyclization rules governing product monomer/dimer ratios are likely to resist a facile discovery for some time to come.

3.5. Larger Cyclic Peptides

An interesting study regarding the relationship between primary structure and cyclization efficiency was conducted on a 14-residue cyclic peptide that, when correctly cyclized takes on an amphipathic β-sheet form. A set of purified linear peptides, prepared by solid phase methods, were head-to-tail cyclized in DMF with BOP or HBTU reagents. HPLC analysis of the cyclization products convincingly demonstrated that:

1. Low yields and significant racemization (up to 40%) can occur during cyclization reactions, especially when the carboxyl terminus is not proline;
2. peptides which cyclized more slowly showed greater amounts of racemization;
3. the reaction conditions used had a major influence on the amount of racemization.[121]

The synthesis of a series of 25 11-residue side chain-to-side chain lactam cyclized peptides of related sequences with ring sizes ranging from 15 to 39-

atoms was also investigated.[88] The highest yields of the target cyclic peptides were observed with ring sizes of 21–30 atoms, corresponding to ring sizes of seven to ten amino acid residues. Good yields were also reported for some of the smaller ring sizes. However, the largest cyclic lactams were obtained with only poor yields of the desired monocyclic peptides, with cyclodimers being the major contaminants.

Side chain-to-side chain and side chain-to-tail lactam formation on a TentaGel resin was studied using peptides of varying length and based on a 12-mer sequence, H–Glu–Gly–Val–Gln–Gln–Glu–Gly–Ala–Gln–Gln–Pro–Ala–OH. Among the cyclic peptides studied, the 8-mer sequence gave the highest yield. When yields were low, the major side products observed included the linear peptides modified at their Glu side chain carboxyl groups, and intermolecularly cross-linked dimers. When ring sizes were larger than 11 amino acids or smaller than six amino acids, dimers and linear peptide adducts were more common.

Since the ring-size dependence of cyclization efficiency may be specific to the epitope sequence and Pro has been reported to assist peptide cyclizations, a series of oligoalanyl peptides with the sequence H–$(Ala)_n$–Pro–Ala–Glu(OAl)–β–Ala–β–Ala–Lys–Met–TentaGel ($n = 2$–11) were synthesized and subsequently cyclized in a head-to-side chain (Glu) fashion on TentaGel resin. The results indicated that the cyclization of the oligoalanyl peptides larger than eight amino acids was poor. Peptides with deletion sequences were detected and became problematic with increasing length, even though double couplings were employed during the synthesis.[101]

Enzymes represent a potentially attractive way to produce cyclic peptides through an efficient catalytic condensation process. Enzyme assisted cyclization of peptide C-terminal glycolate phenylalanylamide esters using the enzyme, subtiligase, revealed a minimum length requirement of 12 residues. Peptides shorter than 12 residues hydrolyzed or dimerized but did not cyclize, presumably because the enzyme could not accommodate both ends in a productive binding geometry. As the length of the peptides increased, the cyclization efficiency also increased while dimerization and hydrolysis decreased. No dimerization was observed for peptides greater than 14 residues in length, indicating that intramolecular cyclization is much faster than the corresponding intermolecular dimerization.[92]

4. PROPERTIES OF HEAD-TO-TAIL CYCLIC PEPTIDES

By forming a new amide bond between the N-terminal amine and the C-terminal carboxylate, the resulting cyclic peptide acquires new physical characteristics. The elimination of two ionizable groups renders the cyclic peptide more hydrophobic with commensurably longer retention times on most reversed phase-high performance liquid chromatography columns. The cyclic peptides are naturally resistant to exopeptidases and in general can be expected to be more stable to most proteolytic enzymes although this is not necessarily an absolute tendency.[122]

When assayed for their biological activities, cyclic peptides are often more potent than their linear counterparts. This behavior has been exploited through the preparation of cyclic analogs of the Arg–Gly–Asp cell-adhesion motif, cyclo(Arg–Gly–Asp–Phe–Pro–),[123] by the synthesis of a cyclic hexapeptide analogue of somatostatin, cyclo(Phe–Pro–Phe–D-Trp–Lys–Thr–),[7] and by a constrained analog of leucine enkephalin, Tyr-cyclo(D-Lys–Gly–Phe–Leu–), known as the DiMaio–Schiller compound.[124] The latter case involved ring formation from the side chain of D-lysine to the C-terminal acid, since the N-terminal amine group is known to be part of the essential pharmacophore for this opioid analog. Cyclic compounds often possess the added bonus of receptor selectivity. For example, the DiMaio–Schiller analog was found to be much more potent towards the μ versus the δ class of opioid receptors. Interestingly, an even more constrained analog with a 13-atom ring, Tyr-cyclo(D-Lys–Phe–Ala–), was even more potent, with subnanomolar IC_{50}s towards both receptor classes, but was therefore surprisingly nonselective.[125]

While cyclic peptides might appear to be ideal drug lead candidates because of their stabilized structure, they tend to exhibit limited bioavailability.[122] Through the study of transport properties of peptides, several workers have concluded that the problem may devolve to the nature of the amide bond itself. Because of its hydrogen bonding potential, the amide groups accumulate waters of hydration that inhibit transport through membrane barriers.[126] The resulting "desolvation penalty" is a significant energy cost which is a reasonable rationale for the lack of oral activity of even relatively constrained, small ring cyclic peptides.

The cyclic undecapeptide immune-suppressant cyclosporin provides a spectacular exception to this limited bioavailability. Cyclosporin actually retains significant oral activity and has been a mainstay of postsurgical transplant regimens. But cyclosporin's structure contains seven *N*-methyl amides which precludes any role as hydrogen bond donors. And in one of its known multiple low-energy conformers, cyclosporin's remaining four amides are each involved in a pair of *intra*molecular hydrogen bonds.[127]

The lessons from cyclosporin could serve as models for future structure–function studies. The first steps would involve a strategy to use cyclic peptides first to find biologically active lead candidates for a given biological response. Then, in analogy with cyclosporin, the parent structure might be modified with amide bond replacements, and/or other conformational constraints designed to enhance membrane transport while inhibiting enzymatic degradation. This general approach has led to considerable research for suitable amide bond surrogates that go beyond *N*-methylation. These efforts are supplemented by studies with other backbone modifications including β-turns and γ-turn mimics.[128]

5. CYCLIC PSEUDOPEPTIDES

Pseudopeptides have been defined as peptides with one or more amide bond replacements (referred to as amide bond surrogates).[129] Examples of these replacements include ketomethylene ($COCH_2$), reduced carbonyl (CH_2NH), retroamide (NHCO), thioamide (CSNH), alkene (CH=CH), thiomethylene (CH_2S), and hydroxyethylene ($CHOHCH_2$).[129,130] Several of these can be readily prepared using solid phase methods of synthesis as shown in Scheme 5. Some of the earliest examples of pseudopeptides were intended as inhibitors of such enzymes as collagenase[131] and renin[132] and have more recently found applications for inhibition of HIV aspartyl protease.[130] Reviews have appeared which cover the "psi-bracket" nomenclature of pseudopeptides and the breadth of backbone changes found in nature and synthetic analogs of peptides.[129,133,134]

Cyclic pseudopeptides represent a natural extension of backbone-modified linear peptides; they combine the inherently greater resistance of cyclic peptides toward enzyme attack with one or more enzyme resistant amide replacements. In those cases where pseudopeptide backbone changes introduce unwanted flexibility in linear peptides, small cyclic peptides represent better hosts by providing increased conformational constraint.[135,136]

While the conformational properties of the amide bond have been extensively studied, there is less known about the ability of well-known structural features such as helices, loops, β-turns and β-sheets to coexist with a plethora of possible backbone modifications. One way to study this question is by introducing an amide surrogate as a "guest" in one or more positions of a cyclic peptide "host."[137] For example, the well-studied cyclic pentapeptides, cyclo(Gly–Pro–Gly–D-Xxx–Pro), were shown to form a relatively common stabilized structure, consisting of two intramolecular hydrogen bonds (Fig. 16.14(a)), using four of the five amide bonds, participating either as H-bond donors or H-bond acceptors.[139,140] This cyclic peptide was thus recognized as a suitable model system to ask whether various amide bond replacements could coexist with the

$$\text{Boc—NH—}\underset{\text{R}}{\text{CH}}\text{—}\overset{\text{O}}{\overset{\|}{\text{C}}}\text{—H} + \text{NH}_2\text{—}\underset{\text{R}'}{\text{CH}}\text{—}\overset{\text{O}}{\overset{\|}{\text{C}}}\text{—Peptide–Resin} \longrightarrow$$

$$\text{Boc—NH—}\underset{\text{R}}{\text{CH}}\text{—CH=N—}\underset{\text{R}'}{\text{CH}}\text{—}\overset{\text{O}}{\overset{\|}{\text{C}}}\text{—Peptide–Resin} \xrightarrow{\text{NaBH}_3\text{CN}}$$

$$\text{Boc—NH—}\underset{\text{R}}{\text{CH}}\text{—CH}_2\text{NH—}\underset{\text{R}'}{\text{CH}}\text{—}\overset{\text{O}}{\overset{\|}{\text{C}}}\text{—Peptide–Resin}$$

Scheme 5. Solid phase synthesis of a ψ[CH_2NH] pseudopeptide.

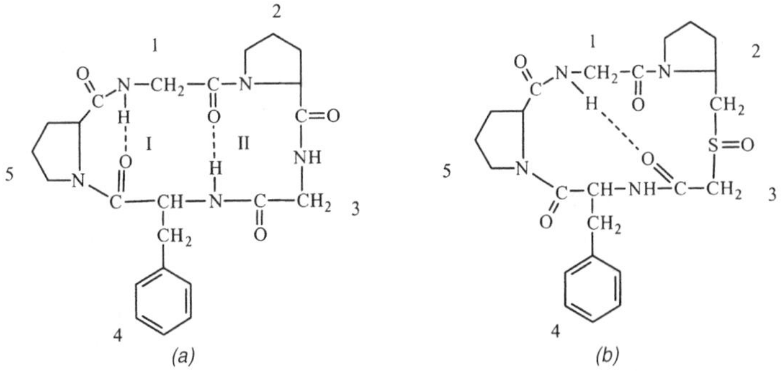

FIGURE 16.14. Structure of a cyclic pentapeptide (*a*) and a cyclic ψ[CH_2S] pseudopentapeptide (*b*) showing shift of intramolecular β-turn.

intramolecular β- and γ-turns found in the parent molecule.[137] Remarkably, the sulfoxide group, normally an excellent hydrogen bond acceptor, does not form any inter- or intramolecular hydrogen bond in the solid state.

As shown in Table 16.5, a variety of amide bond surrogates were introduced into the model pentapeptides at several positions. NMR spectrometry was used to assess the presence of intramolecular hydrogen bonding using chemical shift, temperature dependence and solvent dependence data. While analysis was often complicated by the presence of two or more rapidly interconverting conformers, especially in the solvent dimethylsulfoxide, it is clear that the different amide

TABLE 16.5. Amide Bond Replacements Within the Cyclic Peptide cyclo(Gly–Pro ψ[XX]Gly–D-Phe–Pro) and cyclo(Pro ψ[YY]Gly–Pro–Gly–D-Phe)

	Amide Bond Replacement	$CDCl_3$	DMSO	Reference
XX =				
ψ[CONH]	Amide	+	+	139
ψ[CH_2S]	Thiomethylene ether	+	+	137, 138
ψ[CH_2SO]	Thiomethylene ether sulfoxide (R or S)	±	±	142
ψ[CH_2NH]	Reduced carbonyl	±	–	143
ψ[CH_2NH] · TFA	Reduced carbonyl (protonated)	–	–	143
ψ[CSNH]	Thioamide	+	+	141
YY =				
ψ[CONH]	Amide	+	+	139
ψ[CH_2S]	Thiomethylene ether	+	+	137, 138
ψ[CH_2SO]	Thiomethylene ether sulfoxide (R or S)	–	+	143
ψ[CH_2NH]	Reduced carbonyl	±	±	143
ψ[CH_2NH] · TFA	Reduced carbonyl (protonated)	±	±	143

+ = compatible with turns; – = not compatible with turns.
± = compatible with only one turn or alters structure.

replacements can exhibit a wide range of conformational disruptions. For example, with a single ψ[CH_2SO] replacement in the original β-turn position, an intramolecular hydrogen bond is observed in the solid state but it is now in an altered relative position (Fig. 16.14(b)). Remarkably, the sulfoxide group, normally an excellent hydrogen bond acceptor, does not form any inter- or intramolecular hydrogen bond in the solid state. It will be of interest to investigate the effects of amide surrogates in rings both larger and smaller than cyclic pentapeptides (15 atoms) and to establish the sequence dependence of these structural modifications.

The thioamide replacement, ψ[CSNH], represents a particularly subtle amide bond surrogate. It was first introduced into a head-to-tail cyclic peptide in the model cyclopentapeptides, cyclo[Proψ[CSNH]Gly–Pro–Gly–D-Phe] and cyclo[Pro–Gly–Proψ[CSNH]Gly–D-Phe][141] and soon thereafter in a biologically active cyclic enkephalin (but in a side chain-to-tail analogue).[144] The reduced electronegativity of sulfur places higher electron density on the thioamide nitrogen, resulting in its potentially improved H-bond donor ability. In fact, when the thioamide was introduced into a cyclic hexapeptide model system, cyclo(Pro–Phe–Val–Phe–Phe–Gly) at the Phe–Phe position, the molecule altered its parent conformation, presumably to favor the donor ability of the thioamide[145] (Fig. 16.15). In contrast, within a cyclic pentapeptide, cyclo-(Proψ[CSNH]Gly–Pro–Gly–D-Phe), where the thioamide NH would have been the donor contributor to an intramolecular γ turn, this feature was *not* observed. This was attributed to a steric clash between the bulkier sulfur of the thioamide and the adjacent proline side chain.[141]

Pseudopeptides can also be prepared in which two or more amide bonds are replaced: this can give rise to unique forms of macrocycles, especially if the building blocks incorporate chiral amino acids with defined side chains. One of the first examples of such pseudopeptide macrocycles was the cyclic pseudohexapeptide, (Glyψ[CH_2S]Phe)$_3$.[146] Similar head-to-tail cyclic analogues containing ψ[CH_2NH] amide bond replacements were also prepared using solid phase methods of synthesis, and some were shown to bind metal ions, presumably via the basic amine backbone moieties.[147]

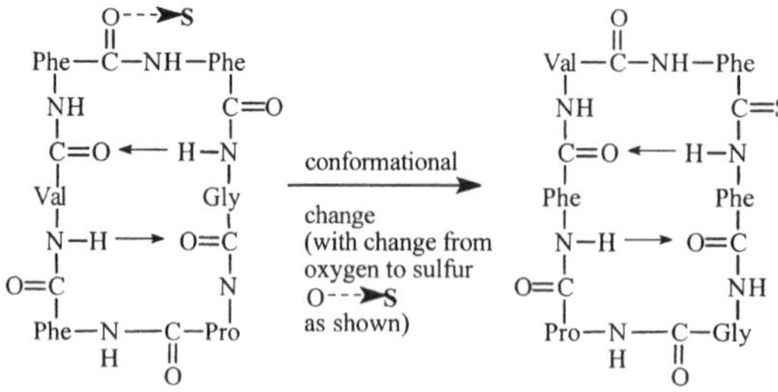

conformational change (with change from oxygen to sulfur O- - -➤S as shown)

FIGURE 16.15. Conformational change accompanies thionation of a more sterically accessible Phe–Phe amide to form the Pheψ[CSNH]Phe-containing pseudopeptide. This leads to a new hydrogen bonded structure with thioamide participation as a donor. Nevertheless all structures displayed significant conformational heterogeneity.

The combination of flexible amide replacements with rigid amino acids represents further evidence of the evolution of cyclic peptide analogs. One such structure is represented by the compound, D-Phe[1]–Cys–Tyrψ[CH_2N]D-Tca–Lys–Thr–Pen–Thr–NH_2, where Tca represents a constrained β-carboline-related tryptophan analog.[148] This analog of somatostatin is a significant departure from the original hypothalamic-derived tetradecapeptide. Unlike mini-somatostatin,[7] the head-to-tail cyclic hexapeptides developed by the Merck group, this structure retains the disulfide bridge of somatostatin, but in a much smaller, 18-atom ring. Furthermore, the ring contains both a ψ[CH_2NH] amide replacement as well as the highly constrained amino acid derivative, tetrahydro-β-carboline. Remarkably, many of these somatostatin analogues possess opioid activity although the reduced potency and lack of selectivity seen with the pseudopeptides compared to their all-amide parents were attributed to the flexible ψ[CH_2N] unit. It is significant that such flexibility appears to persist in spite of the presence of the constrained Tca residue within the ring portion of the molecule.

6. LIBRARIES OF HEAD-TO-TAIL CYCLIC PEPTIDES

In 1991, a number of groups revealed their detailed research involving methods to generate and test mixtures or arrays of linear peptides known as peptide libraries.[149–153] Soon thereafter, the field became known as combinatorial chemistry. Thus, the pioneering work of Mario Geysen with linear peptides[154] was expanded to include a much wider group of organic molecules including carbohydrates, nucleic acids, and rigid peptidomimetic heterocycles such as the benzodiazepines. Numerous reviews of this approach have recently appeared,[155–161] as well as several texts,[162–165] and a useful on-line listing of references to the field is maintained.[166]

In 1993, our group reported one of the initial efforts to extend this field to include head-to-tail cyclic peptides.[16] A year earlier, DeGrado and colleagues had impressively demonstrated, using disulfide bridged cyclic peptides prepared on bacterial viruses known as phages, that cyclic peptides can provide analogues that are 2–3 orders of magnitude more potent than their linear counterparts in their bioassays.[167] This finding accelerated efforts by several groups to prepare more constrained peptides, pseudopeptides and peptidomimetic structures, both individually and in library (mixture) formats. For example, Houghten's group reported that while linear peptides tended to yield a higher percentage of active structures or "hits," their cyclic counterparts yielded fewer but higher quality candidates, overcoming the entropic penalty of the more flexible linear chains.[28]

The same San Diego group reported an imaginative approach toward molecular diversity in a cyclic scaffold by introducing diversity in the side chains of a cyclic oligolysine precursor.[168] Thus, the central ring can be composed of cyclo(Gly–Lys)$_n$ and the ε-amino groups of the three lysines can be used to attach additional functional groups in a combinatorial fashion (Fig. 16.16(a)).

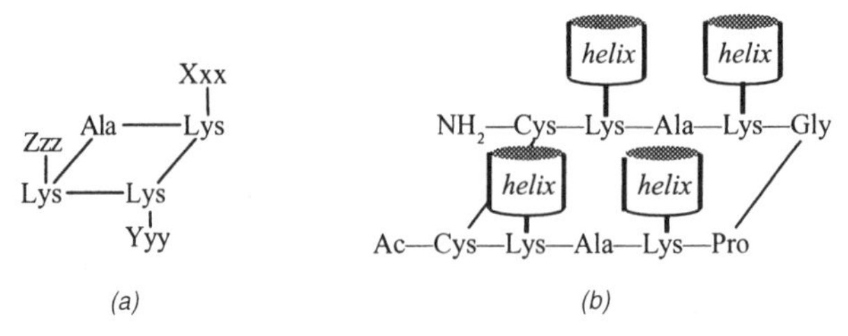

FIGURE 16.16. Structure of a cyclic peptide library template and comparison with the TASP template for the synthesis of protein mimics.

This structure is reminiscent of a relatively rigid "TASP" template introduced by Mutter and co-workers[169] to prepare protein mimics with relatively constrained linear peptides tethered to the oligolysine core (Fig. 16.16(b)).

In our approach, we exploited known chemistries for preparing cyclic peptides using a side chain resin attachment Fmoc strategy as described earlier.[15] While this tactic fixes the linkage point, the remaining positions may be substituted by small or large mixtures of amino acids or their analogues. By adopting the positional scan approach,[170] we were able to prepare an 80,000-component library (with about 20,000 nonredundant cyclic peptides) in 48 separate vials with two fixed positions represented in each vial (the D-aspartic acid attachment link and one other). (Fig. 16.17.) Once these mixtures were assayed for binding against an endothelin receptor, the most active vials corresponded exactly to the four residues of a potent endothelin antagonist,

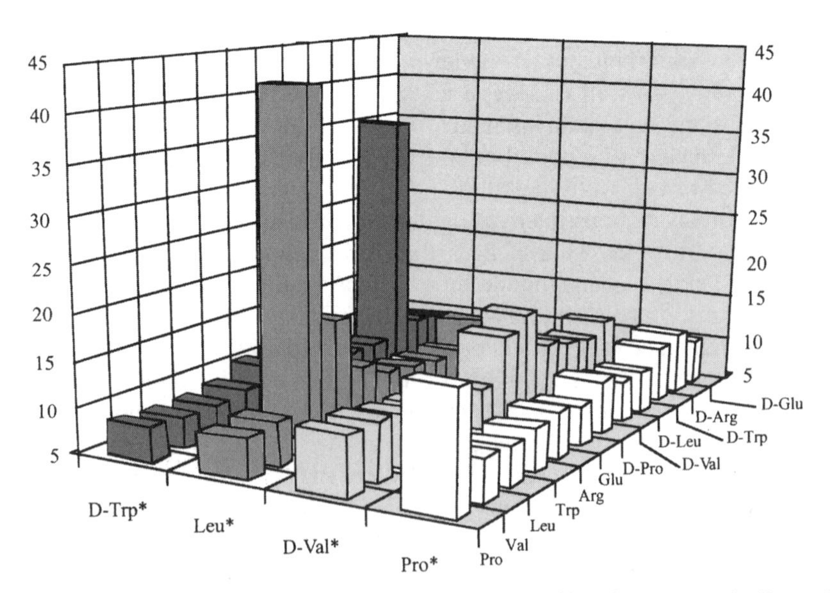

FIGURE 16.17. Endothelin antagonist bioactivity profiles for a set of 48 cyclic pseudopeptide "library" mixtures incorporating the positional scan approach.

cyclo(Phe–D-Val–Leu–D-Trp–D-Asp–).[22] This experiment supported the utility of mixtures for lead discovery by providing a positive control. It also demonstrated the effectiveness of the positional scan approach to find rather quickly a few lead compounds after a single set of assays.

A solid phase strategy with concomitant cyclization and cleavage using *p*-nitrobenzophenone oxime resin has been utilized to obtain a cyclic pentapeptide library cyclo(Arg–Gly–Asp–Xxx–Aca), where Xxx = Ser, Lys, Gln, Ala, Glu, Pro, Val, Tyr, Phe, Trp(CHO), and Aca = ω-aminocaproic acid.[171] The oxime resin procedure was effective at producing nearly equivalent peptide mixtures when using either a separate coupling protocol for the individual amino acids or by a mixed coupling in which the 10 amino acids were represented by equivalent molar quantities.

Using a solution cyclization approach, a set of more than 400 head-to-tail cyclic hexapeptides were prepared having the sequence RGDSPXxx, where Xxx = all usual L- and D-amino acids and Gly and RZzzDSZzzY, where Zzz = all possible combinations of the 20 L-amino acids, analogues were studied.[100] It was found that L- and D-amino acids in the C-terminal position have no influence on the cyclization of the hexapeptide analogues. Ile, Val, Thr (D- or L-) in the C-terminal position reduced cyclization yields. This finding is in contrast to prior reports[13] that suggested that the presence of either Gly, Pro or a D-configuration amino acid residue in the sequence is an essential prerequisite for ring closure of linear hexapeptides.

These methods for cyclic library production should also be amenable for the synthesis of related peptidomimetics. In fact, libraries of cyclopseudopeptides have been prepared in which dipeptide units in the ring were substituted with either Proψ[CH_2S]Gly pseudodipeptides or by Xxxψ[CH_2NH]Yyy pseudopeptide units.[172] In the latter case, the cyclic analogs were prepared on the solid support by coupling an amino aldehyde to the resin bound amine followed by sodium cyanoborohydride reduction of the Schiff base intermediate (Scheme 5). Following cleavage, the pseudopeptides containing one or more ψ[CH_2NH] units were cyclized in solution to furnish several head-to-tail cyclic peptides including cyclo(D-Pheψ[CH_2NH]Xxx–Arg–Gly–Asp),[172] cyclo(Glyψ[CH_2NH]Phe)$_3$ and cyclo(Glyψ[CH_2NH]Phe)$_4$.[147] Alternatively, using side chain attachment, this approach can be adapted for the synthesis of cyclic ψ[CH_2NH] pseudopeptides via on-resin cyclization.

Future applications of cyclic peptides can be envisaged in which they can be used for drug discovery, for the production of novel biomaterials, and even for use in affinity separations. In many cases, their use may not require cleavage from the solid support in which case more options are available for permanent attachment of side chain functional groups to a solid support.[173] Resin bound cyclic peptide libraries have been described and cyclic peptides attached to photolabile linkers provide an additional approach to orthogonality. It may be anticipated that additional innovations will involve cyclic peptides as important contributors to the quest for practical applications of molecular diversity.

REFERENCES

1. Ovchinnikov, Y. A.; Ivanov, V. T. In *The Proteins*, Neurath, H.; Hill, R. L. Eds.; Academic Press; New York, 1982, p. 307.
2. Pettit, G. R.; Srirangam, J. K.; Herald, D. L.; Erickson, K. L.; Doubek, D. L.; Schmidt, J. M.; Tackett, L. P.; Bakus, G. J. *J. Org. Chem.* 1992, **57**, 7217.
3. Gause, G. F.; Brazhnikova, M. G. *Lancet* 1994, **247**, 715.
4. Hassan, M. A.; Al-Yahya, M. A. In *Analytical Profiles of Drugs Substances, 16*, Florey, K. Ed.; Academic Press: New York, 1987, p. 145.
5. Ihara, M.; Noguchi, K.; Saeki, T.; Fukuroda, T.; Tsuchida, S.; Kimura, S.; Fukami, T.; Ishikawa, K.; Nishibe, M.; Yano, M. *Life Sciences* 1992, **50**, 247.
6. Fesik, S. W.; Gampe, Jr., R. T.; Holzman, T. F.; Egan, D. A.; Edalji, R.; Luly, J. R.; Simmer, R.; Helfrich, R.; Kishore, V.; Rich, D. H. *Science* 1990, **250**, 1400.
7. Veber, D. F.; Freidinger, R. M.; Perlow, D. S.; Paleveda, W. J.; Holly, F. W.; Strachen, R. G.; Nutt, R. F.; Arison, B. H.; Hommick, C.; Randall, W. C.; Giltzer, M. S.; Saperstein, R.; Hirschmann, R. *Nature* 1981, **292**, 55.
8. Ghadiri, M. R.; Granja, J. R.; Buehler, L. K. *Nature* 1994, **369**, 301.
9. Schroder, E.; Lubke, K. In *The Peptides, Vol. II. Synthesis, Occurrence and Action of Biologically Active Polypeptides*, Academic Press: New York, 1966.
10. Fridkin, M.; Patchornik, A.; Katchalski, E. *J. Am. Chem. Soc.* 1965, **87**, 4646.
11. Sklyarov, L. Y.; Shashkova, I. V. *J. Gen. Chem.* 1969, **39**, 2778.
12. Kates, S. A.; Solé, N. A.; Albericio, F.; Barany, G. In *Peptides: Design, Synthesis and Biological Activity*; Basava, C.; Anantharamaiah, G. M. Eds.; Birkhauser: Boston, 1994, p. 39.
13. Mazur, S.; Jayalekshmy, P. *J. Am. Chem. Soc.* 1978, **101**, 677.
14. Barany, G.; Merrifield, R. B. In *The Peptides*, Vol. II, Gross, E.; Meienhofer, J. Eds.; Academic Press: New York, 1979, p. 1.
15. Rovero, P.; Quartara, L.; Fabbri, G. *Tetrahedron Lett.* 1991, **32**, 2639.
16. Darlak, K.; Romanovskis, P.; Spatola, A. In *Peptides: Chemistry, Structure, and Biology*; Hodges, R.; Smith, J. Eds.; ESCOM: Leiden, 1994, p. 981.
17. Bednarek, M.; Bodanszky, M. *Int. J. Peptide Protein Res.* 1983, **21**, 196.
18. Bodanszky, M.; Deshmane, S. S.; Martinez, J. *J. Org. Chem.* 1979, **49**, 1622.
19. Spatola, A. F.; Darlak, K.; Romanovskis, P. *Tetrahedron Lett.* 1996, **37**, 591.
20. Romanovskis, P.; Spatola, A. F. In *Peptides 1996*; Ramage, R.; Epton, R. Eds.; 1998, p. 761.
21. Spatola, A. F.; Romanovskis, P. In *Combinatorial Peptide and Nonpeptide Libraries*; Jung, J. Ed.; VCH: Weinheim, 1996, p. 327.
22. Spatola, A. F.; Crozet, Y.; deWit, D.; Yanagisawa, M. *J. Med. Chem.* 1996, **39**, 3842.
23. McMurray, J. S. *Tetrahedron Lett.* 1991, **32**, 7679.
24. Albericio, F.; Kneib-Cordonier, W.; Biancalana, S.; Gero, L.; Madada, R. I.; Hudson, D.; Barany, G. *J. Org. Chem.* 1990, **55**, 3730.
25. Matsueda, G.; Stewart, J. *Peptides* 1991, **2**, 45.
26. Brugghe, H. F.; Timmermans, H. A. M.; Vanunen, L. M. A.; Jan Ten Hove, G.; van De Werken, G.; Poolman, J. T.; Hoogerhout, P. *Int. J. Peptide Protein Res.* 1994, **43**, 166.

27. Kates, S. A., Solé, N. A.; Johnson, C. R.; Hudson, D.; Barany, G.; Albericio, F. *Tetrahedron Lett.* 1993, **34**, 1549.
28. Eichler, J.; Lucka, A. W.; Pinilla, C.; Houghten, R. A. *Molecular Diversity* 1996, **1**, 233.
29. Valero, M. L.; Giralt, E.; Andreu, D. *Tetrahedron Lett.* 1996, **37**, 4229.
30. Guibe-Jampel, E.; Wakselman, M. *Synthetic Commun.* 1982, **12**, 219.
31. Romanovskis, P. J.; Syskov, I. V.; Liepkaula, I. K.; Porunkevich, J. A.; Ratkevich, M. P. in *Peptides: Synthesis, Structure, Function*; Gross, E.; Rich, D. Eds.; Pierce Chem. Co.: Rockford, 1981, p. 229.
32. Hocker, M. D.; Caldwell, C. G.; Macsata, R. W.; Lyttle, M. H. *Peptide Res.* 1995, **8**, 310.
33. Hinsberg, O.; Himmelschein, A. *Chem. Ber.* 1896, **29**, 2019.
34. Pande, C. S.; Gupta, S. K.; Glass, J. D. *Indian J. Chem.* 1987, **26B**, 957.
35. Kulikov, S. V.; Selivanov, R. S.; Ginak, A. I. In *Peptides: Chemistry, Structure and Biology*; Kaumaya, P. T. P.; Hodges, R. S. Eds.; Mayflower Scientific Ltd.: 1966, p. 48.
36. Birr, C. In *Innovation and Perspectives in Solid Phase Synthesis*; Epton, R. Ed.; Oxford, 1994, p. 83.
37. Bycroft, B. W.; Chan, W. C.; Chhabra, S. R.; Hone, N. D. *J. Chem. Soc. Chem. Commun.* 1993, 778.
38. Bloomberg, G. B.; Askin, D.; Gargaro, A. R.; Tanner, M. J. A. *Tetrahedron Lett.* 1993, **34**, 4709.
39. Chan, W. C.; Bycroft, B. W.; Evan, D. J.; White, P. *J. Chem. Soc. Chem. Commun.* 1995, 2209.
40. Rohwedder, B.; Mutti, Y.; Dumy, P.; Mutter, M. *Tetrahedron Lett.* 1998, **39**, 1175.
41. Letsinger, R. L.; Kornet, M. J. *J. Am. Chem. Soc.* 1963, **85**, 3045.
42. Zalipsky, S. *Bioconjugate Chem.* 1995, **6**, 150.
43. Dressman, B. A.; Spangle, L. A.; Kaldor, S. W. *Tetrahedron Lett.* 1996, **37**, 937.
44. Ghosh, A. K.; Duong, T. T.; McKee, S. P.; Thompson, W. J. *Tetrahedron Lett.* 1992, **33**, 1775.
45. Medvedkin, V. N.; Mitin, Y. V.; Klimenko, L. V.; Podgornova, N. N.; Bystrychenko, A. I.; Zabolotskikh, V. F.; Korobeinikova, L. I.; Pozdeeva, V. V. *Bioorg. Khim (in Russian)* 1989, **15**, 460.
46. Alsina, J.; Rabanal, F.; Giralt, E.; Albericio, F. *Tetrahedron Lett.* 1994, **35**, 9633.
47. Alsina, J.; Chiva, C.; Ortiz, M.; Rabanal, F.; Giralt, E.; Albericio, F. *Tetrahedron Lett.* 1997, **38**, 883.
48. Pozdnev, V. F. *Int. J. Peptide Protein Res.* 1992, **40**, 407.
49. Mitsunobu, O. *Synthesis* 1981, **1**, 1.
50. Trzeciak, A.; Bannwarth, W. *Tetrahedron Lett.* 1992, **33**, 4557.
51. Romanovskis, P.; Spatola, A. F. *J. Peptide Res.* 1998, **52**, 356.
52. Waki, M.; Kitajima, Y.; Izumiya, N. *Synthesis* 1981, 266.
53. Loudon, G. M.; Radhakrishna, A. S.; Almond, M. R.; Blodgett, J. K.; Boutin, R. H. *J. Org. Chem.* 1984, **49**, 4272.
54. Bittner, S.; Assaf, Y. *Chem. and Ind.* 1975, 281.

55. Krchňák, V.; Flegelova, Z.; Weichsel, A. S.; Lebl, M. *Tetrahedron Lett.* 1995, **36**, 6193.
56. Richter, L. S.; Gadek, T. R. *Tetrahedron Lett.* 1994, **35**, 4705.
57. Isied, S. S.; Kuehn, C. G.; Lyon, J. M.; Merrifield, R. B. *J. Am. Chem. Soc.* 1982, **104**, 2632.
58. Urban, J.; Dattilo, J. W. In *15th APS (American Peptide Symposium), Abstracts*; June 14–19, 1997, Nashville, TN, p. 131.
59. Han, Y.; Barany, G. *J. Org. Chem.* 1997, *62*, 3841.
60. Lukszo, J.; Patterson, D.; Albericio, F.; Kates, S. A. *Letters in Peptide Sci.* 1996, **3**, 157.
61. Yamada, M.; Miyajima, T.; Horikawa, H. *Tetrahedron Lett.* 1998, **39**, 289.
62. Johnson, T.; Quibell, M.; Owen, D.; Sheppard, R. C. *J. Chem. Soc. Chem. Commun.* 1993, 369.
63. Jensen, K. J.; Songster, M. F.; Vagner, V.; Alsina, J.; Albericio, F.; Barany, G. *J. Am. Chem. Soc.* 1998, **120**, 5441.
64. Fivush, A. M.; Wilson, T. M. *Tetrahedron Lett.* 1997, **38**, 7151.
65. Sarantakis, D.; Bicksler, J. J. *Tetrahedron Lett.* 1997, **38**, 7325.
66. Meienhofer, J. In *The Peptides*; Gross, E.; Meienhofer, J. Eds.; Academic Press: New York, 1979, Vol. 1, p. 197.
67. Shioiri, T.; Ninomiya, K.; Yamada, S. *J. Am. Chem. Soc.* 1972, **94**, 6203.
68. Spatola, A. F.; Anwer, M. K.; Rockwell, A. L.; Gierasch, L. M. *J. Am. Chem. Soc.* 1986, **108**, 825.
69. Belleau, B.; Malek, G. *J. Am. Chem. Soc.* 1968, **90**, 1651.
70. Sheehan, J. C.; Hess, G. P. *J. Am. Chem. Soc.* 1955, **77**, 1067.
71. Rich, D. H.; Singh, J. In *The Peptides* Gross, E.; Meienhofer, J. Eds.; Academic Press: New York, 1979, Vol. 1, p. 241.
72. Koenig, W.; Geiger, R. *Chem. Ber.* 1970, **105**, 788.
73. Sarantakis, D.; Teichman, J.; Lien, E. L.; Fenichel, R. L. *Biochem. Biophys. Res. Commun.* 1976, **73**, 336.
74. Carpino, L. A. *J. Am. Chem. Soc.* 1993, **115**, 4397.
75. Castro, B.; Dormoy, J. R.; Evin, G.; Selve, C. *Tetrahedron Lett.* 1975, 1219.
76. Henklein, P.; Beyermann, M.; Bienert, M.; Knorr, R. In *Peptides 1990*; Giralt E.; Andreu, D. Eds.; ESCOM, 1991, p. 67.
77. Carpino, L. A.; El-Faham, A.; Minor, C. A.; Albericio, F. *J. Chem. Soc. Chem. Commun.* 1994, 201.
78. Coste, J.; Frérot, E.; Dufour, M. N.; Pantaloni, A.; Jouin, P. In *Peptides 1990*; Giralt, E.; Andreu, D. Eds.; ESCOM, 1991, p. 76.
79. Hudson, D. *J. Org. Chem.* 1988, **53**, 617.
80. Carpino, L. A.; El-Faham, A.; Truran, G. A.; Minor, C. A.; Kates, S. A.; Griffin, G. W.; Shroff, H.; Triolo, S. A.; Albericio, F. In *Innovations and Perspectives in Solid Phase Synthesis*; Epton, R. Ed.; Mayflower: Birmingham, 1994, p. 95.
81. Abelmoty, I.; Albericio, F.; Carpino, L. A.; Foxman, B. M.; Kates, S. A. *Lett. Peptide Sci.* 1994, **1**, 57.

82. Henklein, P.; Costisella, B.; Wray, V.; Domke, T.; Carpino, L.; El-Faham, A.; Kates, S. A.; Abdelmoty, I.; Foxman, B. M. In *15th American Peptide Symposium, Abstracts*; June 14–19, 1997, Nashville, TN, p. 173.
83. Story, S. C.; Aldrich, J. V. *Int. J. Peptide Protein Res.* 1994, **43**, 292.
84. Delforge, D.; Dieu, M.; Delaive, E.; Art, M.; Gillon, B.; Devreese, B.; Raes, M.; Van Beeumen, J.; Remacle, J. *Lett. Peptide Sci.* 1996, **3**, 89.
85. Artmangkuls, S.; Arbogast, B.; Barofsky, D.; Aldrich, J. V. *Lett. Peptide Sci.* 1996, **3**, 357.
86. Flanigan, E.; Marshall, G. R. *Tetrahedron Lett.* 1970, 506.
87. DeGrado, W. F.; Kaiser, E. T. *J. Org. Chem.* 1980, **45**, 1295.
88. Tumelty, D.; Needels, M. C.; Antonenko, V. V.; Bovy, P. R. In *Peptides: Chemistry, Structure and Biology*; Kaumaya, P. T. P.; Hodges, R. S. Eds.; Mayflower, 1966, p. 121.
89. Green, M.; Berman, J. *Tetrahedron Lett.* 1990, **41**, 5851.
90. Sakakibara, S.; Inukai, N. *Bull. Chem. Soc. Japan* 1964, **37**, 1231.
91. Hruby, V.; Meyer, J. -P. In *Bioorganic Chemistry: Peptides and Proteins*, Hecht, S. M. Ed.; Oxford University Press: New York, 1998, p. 27.
92. Jackson, D. Y.; Burnier, J. P.; Wells, J. A. *J. Am. Chem. Soc.* 1995, **117**, 819.
93. Kaiser, E.; Colescott, R.; Bossinger, C.; Cook, P. *Anal. Biochem.* 1970, **34**, 595.
94. Krchnak, V.; Vagner, J.; Lebl, M. *Int. J. Peptide Protein Res.* 1988, **32**, 415.
95. Krchnak, V.; Vagner, J.; Safar, P.; Lebl, M. *Coll. Czech. Chem. Commun.* 1988, **53**, 2542.
96. Schnölzer, M.; Jones, A.; Alewood, P. F.; Kent, S. B. H. *Anal. Biochem.* 1992, **204**, 335.
97. Metzger, J. W.; Kempter, C.; Weismüller, K. H.; Fung, G. *Anal. Biochem.* 1994, **219**, 261.
98. Eckart, K. *Mass. Spectrometry Rev.* 1994, **13**, 23.
99. McMurrary, J. S.; Lewis, C. A.; Obeyesekere, N. U. *Peptide Res.* 1994, **7**, 195.
100. Feiertag, S.; Weismüller, K. H.; Nicholson, G. J.; Jung, G. In *Peptides: Chemistry, Structure and Biology*; Kaumaya, P. T. P.; Hodges, R. S. Eds.; Mayflower, 1996, p. 693.
101. Yu, H.; Yu, C.; Chu, Y. *Tetrahedron Lett.* 1998, **39**, 1.
102. Rothe, M.; Lohmüller, Fischer, W.; Taibe, W.; Breuksch, U. In *Solid Phase Synthesis* Epton, R. Ed.; SPCC (UK) Ltd.: Birmingham, 1990, p. 551.
103. Regen, S.; Lee, D. P. *J. Am. Chem. Soc.* 1974, **96**, 294.
104. Rothe, M.; Alt, A.; Taiber, W.; Berginski, R.; Bernard, H. In *Solid Phase Synthesis*; Epton, R. Ed.; Mayflower: Birmingham, 1994, p. 111.
105. del Fresno, M.; Alsina, J.; Royo, M.; Barany, G.; Albericio, F. *Tetrahedron Lett.* 1998, **39**, 2639.
106. Pastuszak, J.; Gardner, J. H.; Singh, J.; Rich, D. H. *J. Org. Chem.* 1982, **47**, 2982.
107. Link, U.; Mästle, W.; Rother, M. *Int. J. Peptide Protein Res.* 1993, **42**, 475.
108. Rao, M. H.; Yang, W.; Joshua, H.; Becker, J. M.; Naider, F. *Int. J. Peptide Protein Res.* 1995, **45**, 418.

109. Cavelier-Frontin, F.; Achmad, S.; Verducci, J.; Jacquier, R.; Pepe, G. *J. Mol. Structure (Theochem)* 1993, **286**, 125.
110. Warrass, R.; Wiesmüller, K. H.; Jung, G. *Tetrahedron Lett.* 1998, **39**, 2715.
111. Mutter, M. *J. Am. Chem. Soc.* 1977, **99**, 8307.
112. Brady, S. F.; Varga, S. L.; Freidinger, R. M.; Schwenk, D. A.; Mendlowski, M.; Holly, F. W.; Veber, D. F. *J. Org. Chem.* 1979, **44**, 3101.
113. Kessler, H.; Kutscher, B. *Liebigs Ann. Chem.* 1986, 869.
114. Heavner, G.; Audhya, T.; Doyle, D.; Tjoeng, S.; Goldstein, G. *Int. J. Peptide Protein Res.* 1991, 37, 198.
115. Ehrlich, A.; Rothemund, S.; Brudel, M.; Beyermann, M.; Carpino, L. A.; Bienert, M. *Tetrahedron Lett.* 1993, **34**, 4781.
116. Kessler, H.; Hasse, B. *Int. J. Peptide Protein Res.* 1992, **39**, 36.
117. Sowemimo, V.; Scanlon, D.; Jones, P.; Craik, D. J. *J. Protein Chem.* 1994, **13**, 339.
118. Ehrlich, A.; Klose, J.; Heyne, H. U.; Beyermann, M.; Carpino, L. A.; Bienert, M. In *Peptides: Chemistry, Structure and Biology*, Kaumaya, P. T. P.; Hodges, R. S. Eds.; Mayflower, 1996, p. 75.
119. McMurray, J. S.; Lewis, C. A.; Obeyesekere, N. U. *Peptide Res.* 1994, **7**, 195.
120. Plaue, S. *Int. J. Peptide Protein Res.* 1990, **35**, 510.
121. Kondejewski, L. H.; Semchuk, P. D.; Daniels, L.; Wilson, I.; Hodges, R. S. In *15th APS Abstracts*; June 14–19, 1997, Nashville, TN, p. 193.
122. Hirschmann, R. *Angew. Chem. Int. Ed. Engl.* 1991, **30**, 1278.
123. Kessler, H.; Diefenbach, B.; Finsinger, D.; Geyer, A.; Gurrath, M.; Goodman, S. L.; Holzemann, G.; Haubner, R.; Jonczyk, A.; Muller, G.; Graf von Roedern, E.; Wermuth, J. *Lett. Peptide Sci.* 1995, **2**, 155.
124. Schiller, P. W.; DiMaio, J. *Nature* 1982, **297**, 79.
125. Ro, S.; Zhu, Q.; Lee, C. W.; Goodman, M.; Darlak, K.; Spatola, A. F.; Chung, N. N.; Schiller, P. W.; Malmberg, A. B.; Yakksh, T. L.; Burks, T. F. *J. Peptide Sci.* 1995, **1**, 157.
126. Conrad, R. A.; Hilgers, A. R.; Ho, N. F. H.; Burton, P. S. *Pharmaceutical Res.* 1992, **9**, 435.
127. Wenger, R. M. *Angew Chem. Int. Ed. Engl.* 1985, **24**, 77.
128. Gillespie, P.; Cicariello, J.; Olson, G. L. *Biopolymers (Peptide Sci.)* 1997, **43**, 191.
129. Spatola, A. F. In *Chemistry and Biochemistry of Amino Acids, Peptides and Proteins*; Vol. VII, Weinstein, B. Ed.; Marcel Dekker: New York, 1983, p. 267.
130. Sawyer, T. K. In *Structure-Based Drug Design: Diseases, Targets, Techniques and Developments*; Veerpandian, P. Ed.; Marcel Dekker: New York, 1997, p. 559.
131. Yankeelov, Jr., J. A.; Fok, K. F.; Carothers, D. J. *J. Org. Chem.* 1978, **43**, 1623.
132. Szelke, M.; Leckie, B.; Hallett, A.; Jones, D. M.; Sueiras, J.; Atrash, B.; Lever, A. F. *Nature* 1982, **299**, 533.
133. Gante, J. *Angew. Chem. Int. Ed. Engl.* 1994, **33**, 1699.
134. Goodman, M.; Ro, S. In *Burger's Medicinal Chemistry and Drug Discovery*; Vol. I, Wolff, M. E. Ed.; Wiley: New York, 1995, p. 803.
135. Gero, T. W.; Spatola, A. F.; Torres-Aleman, I.; Schally, A. V. *Biochem. Biophys. Res. Commun.* 1984, **120**, 840.

136. Spatola, A. F.; Edwards, J. V. *Biopolymers* 1986, **25**, S229.
137. Spatola, A. F.; Anwer, M. K.; Rockwell, A.; Gierasch, L. *J. Am. Chem. Soc.* 1986, **108**, 825.
138. Spatola, A. F.; Gierasch, L.; Rockwell, A. *Biopolymers* 1983, **22**, 147.
139. Pease, L. G.; Watson, C. *J. Am. Chem. Soc.* 1978, **100**, 1279.
140. Karle, I. *J. Am. Chem. Soc.* 1978, **100**, 1286.
141. Sherman, D. B.; Spatola, A. F. *J. Am. Chem. Soc.* 1990, **112**, 433.
142. Ma, S.; Richardson, J.; Spatola, A. F. *J. Am. Chem. Soc.* 1991, **113**, 8529.
143. Ma, S.; Spatola, A. F. *Int. J. Peptide Protein Res.* 1992, **41**, 204.
144. Sherman, D. B.; Spatola, A. F.; Wire, W. S.; Burks, T. F.; Nguyen, T. M. D.; Schiller, P. W. *Biochem. Biophys. Res. Commun.* 1989, **162**, 1126.
145. Kessler, H.; Geyer, A.; Matter, H.; Köck, M. *Int. J. Peptide Protein Res.* 1992, **40**, 25.
146. Spatola, A. F.; Darlak, K. *Tetrahedron* 1988, **44**, 821.
147. Chen, J. J.; Teesch, L. M.; Spatola, A. F. *Lett. Peptide Sci.* 1996, **3**, 17.
148. Kazmierski, W. M.; Ferguson, R. D.; Knapp, R. J.; Lui, G. K.; Yamamura, H. I.; Hruby, V. J. *Int. J. Peptide Protein Res. I* 1992, **39**, 401.
149. Geysen, H. M.; Meloen, R. H.; Barteling, S. J. *Proc. Natl. Acad. Sci., U.S.A.* 1984, **81**, 3998.
150. Furka, A.; Sebestyen, F.; Asgedom, M.; Dibo, G. *Int. J. Peptide Protein Res.* 1991, **37**, 487.
151. Houghten, R. A.; Pinilla, C.; Blondelle, S. E.; Appel, J. R.; Dooley, C. T.; Cuervo, J. H. *Nature* 1979, **354**, 84.
152. Nikolaiev, V.; Stierandova, A.; Krchnak, V.; Seligmann, B.; Lam, K. S.; Salmon, S. E.; Lebl, M. *Peptide Research* 1993, **6**, 161.
153. Fodor, S.; Read, J. L.; Pirrung, M. C.; Stryer, L.; Lu, A. T.; Solus, D. *Science* 1991, **251**, 767.
154. Geysen, A. M.; Roddo, S. J.; Mason, T. J. *Molecular Immunology* 1986, **23**, 709.
155. Bunin, B. A.; Plunkett, M. J.; Ellman, J. A. *Proc. Natl. Acad. Sci., U.S.A.* 1994, **91**, 4708.
156. Hobbs-DeWitt, S.; Kiely, J. S.; Stankovic, C. J.; Schroeder, M. C.; Cody, D. M. R.; Pavia, M. R. *Proc. Natl. Acad. Sci., U.S.A.* 1993, **90**, 6909.
157. Gallop, M. A.; Barrett, R. W.; Dower, W. J.; Fodor, S. P. A.; Gallop, M. A. *J. Med. Chem.* 1994, **37**, 1233.
158. Gordon, E. M.; Barrett, R.; Dower, W. J.; Fodor, S. P. A.; Gallop, M. A. *J. Med. Chem.* 1994, **37**, 1385.
159. Terrett, N. K.; Gardner, M.; Gordon, D. W.; Kobylecki, R. J.; Steele, J. *Tetrahedron* 1995, **51**, 8135.
160. Thompson, L. A.; Ellman, J. A. *Chem. Rev.* 1996, **96**, 555.
161. Lam, K. S.; Lebl, M.; Krchnak, V. *Chem. Rev.* 1997, **97**, 411.
162. Jung, G. *Combinatorial Peptides and Nonpeptide Libraries* VCH: Weinheim; 1996.
163. Bunin, B. *The Combinatorial Index*, Academic Press, 1998.

164. *Combinatorial Chemistry: Synthesis and Applications*; Wilson, S. R.; Czarnik, A. W. Eds.; John Wiley, 1998.
165. *Combinatorial Peptide Library Protocols. Methods in Molecular Biology*; Vol. 87, Cabily, S. H. Ed.; Academic Press, 1997.
166. Lebl, M. *Molecular Diversity References* http//:www.5z.com.
167. O'Neil, K. T.; Hoess, P. H.; Jackson, S. A.; Ramachandran, N.; Mousa, S. A.; DeGrado, W. F. *Proteins* 1992, **14**, 509.
168. Eichler, J.; Lucka, A. W.; Houghten, R. A. *Peptide Res.* 1994, **7**, 300.
169. Sila, U.; Mutter, M. *J. Mol. Recog.* 1995, **8**, 29.
170. Dooley, C. T.; Houghten, R. A. *Life Sci.* 1993, **52**, 1509.
171. Mihara, H.; Yamabe, S.; Niidome, T.; Aoyagi, H.; Kumagai, H. *Tetrahedron Lett.* 1995, **36**, 4837.
172. Wen, J. J.; Spatola, A. F. *J. Peptide Res.* 1997, **49**, 3.
173. Winkler, D.; Schuster, A.; Hoffmann, B.; Schneider-Mergener, J. In *Peptides 1994*; Maia, H. L. S. Ed.; Escom: Leiden, 1994, 485.

CHAPTER 17

FROM CRYSTAL STRUCTURES OF OLIGOPEPTIDES TO PROTEIN FOLDING: THE IMPORTANCE OF PEPTIDE BOND-SIDE CHAIN HYPERCONJUGATION

ANDRZEJ S. CIEPLAK
Department of Chemistry, Fordham University; Laboratory of Crystallography, University of Berne; Department of Chemistry, Yale University; Department of Chemistry, State University of New York at Stony Brook; Department of Chemistry, Bilkent University

1. INTRODUCTION

The question that inspired the work described in this chapter goes back nearly a century. In the mid 1920s, McKenzie et al., pursuing for two decades the puzzle of asymmetric induction in the atrolactic synthesis, have discovered mutarotation of phenylglyoxalates, lactates, and related esters of chiral alcohols.[1,2] For instance, a solution of (−)-menthyl (+)-phenylchloroacetate in EtOH was found to quickly reverse the sense of optical rotation when treated with a single drop of a potassium ethoxide solution and continue to change its specific rotation for ~90 min. The reversal was shown to result from the base-catalyzed epimerization of the optically pure ester to an equilibrium mixture of the (−)- and (+)-phenylchloroacetates (57 : 43). A transfer of stereogenic information has occurred from the carbinol C to the acyl C^{α}. Thus, the mutarotation poses what may be called the vinylogous threo–erythro problem: how do these two tetrahedral C centers, separated by two trigonal centers of the carboxylate moiety, "know" of each other's configuration and conformation?

The commonly accepted tenet, based on the success of Prelog's rule,[3,4] attributes the outcome of these mutarotations and related processes (e.g. the atrolactic synthesis) exclusively to the differences in steric strain between the diastereoisomers or the diastereoisomeric transition states. It is not difficult,

The Amide Linkage: Selected Structural Aspects in Chemistry, Biochemistry, and Materials Science,
Edited by Arthur Greenberg, Curt M. Breneman, and Joel F. Liebman
ISBN 0-471-35893-2

however, to quote instances of high asymmetric induction obtained in the absence of any steric bias in the fragments with the inducing and the incipient stereogenic centers separated by one, two or more conjugated trigonal centers; the reactions of Meyers' 2-acyl, 2-alkyl and 2-vinyloxazolines offer a classic example of such a situation.[5] Apparently, the mechanism of asymmetric induction in these reactions involves not only the steric strain effect but also a stereoelectronic effect: the π-electron density itself seems to mediate the transfer of stereogenic information through a system of conjugated trigonal centers.

This concept was in fact used to explain asymmetric induction in nucleophilic additions to vinyl sulfoxides (mediated by the antiperiplanarity of the nascent carbanion and the σ_{SO} bond),[6] asymmetric induction in cycloadditions and phosphite addition to *N*-(1′-furanosyl)nitrones (mediated by the antiperiplanarity of the nascent amine and the C1′–O4′ bond),[7–9] and the stereospecific hydride transfer in reactions of pyridine coenzyme-dependent alcohol dehydrogenases (mediated by the antiperiplanarity of the NADH enamine group and the C1′–O4′ as well as the scissile C–H bonds).[10–12] It was also explored in the study of the N and C pyramidalization as the basis of asymmetric induction in C-acylations and C-alkylations of enamines.[13]

Taken together, these hypotheses imply that the relay of chirality through a system of conjugated trigonal centers involves a cascade of stereoelectronic effects: (1) a blend of interactions with a single bond at the stereogenic center stabilizes σ-antiperiplanar pyramidalization at the ajacent trigonal center; (2) pyramidalization at one trigonal center induces anti pyramidalization at the neighboring trigonal center; (3) hyperconjugation of a pyramidalized trigonal center stabilizes development of the antiperiplanar vicinal incipient bond. The underpinning of these postulates is the notion that the preferred sense of pyramidalization is always the one that results in pseudostaggering about the fragment's σ framework. Such an assumption is certainly in agreement with the scarcely available experimental and computational evidence and the qualitative PMO arguments used in the analysis of the secondary bonding and nonbonding interactions in the allyl group.[14–19] Is this then the solution to the vinylogous threo-erythro problem?

To address this question, we have examined crystal structures of the aliphatic amides and linear oligopeptides.[20,21] A simple model to demonstrate operation of the implied cascade of stereoelectronic effects is the Ala*–Ala* dipeptide (Ala* are the residues other than Gly and Pro) where the two stereogenic centers, C_i^α and C_{i+1}^α are attached to two conjugated trigonal centers, C′ and N: if the proposed transfer of stereogenic information does occur, the ψ_i and ϕ_{i+1} angles should correlate with each other in a predictable manner. However, our investigation has shown that only one of the three expected molecular geometry correlations is found in this model situation. The ψ_i angle, that is the conformation about the C_i^α–C′ bond, does indeed correlate quite well with the out-of-plane distortion at C′ defined by the $\theta_{C'}$ parameter introduced by Winkler and Dunitz.[20–22] In contrast, the correlation between the pyramidalization at N and the ϕ_{i+1} angle (i.e., the conformation about the N–C_{i+1}^α bond), which should be even better according to the premise of the generalized anomeric effect, is nonexistent.[20] Furthermore, the expected antiperiplanar pyramidalization at C′ and the adjacent N is not necessarily the preferred way of out-of-plane bending of peptide bonds, at least not in the solid state, for the synperiplanar pyramidalization occurs in the crystal structures just as often. In short, either there is no stereoelectronic control of the peptide conformations, or its understanding, as presented above, is too naive to capture even its most basic features.

Thus, our attention has turned to the stereoelectronic (hyperconjugative) contributions to the barriers to rotation about the peptide C_i^{α}–C′ and N–C_{i+1}^{α} bonds as the key to understanding of the transfer of stereogenic information through the π-electron systems. To gain a better qualitative appreciation of such contributions, several other structure correlation studies of bonding in aliphatic amides, oligopeptides and lactams have been undertaken.[23–27] The results of that effort are now reviewed and a new theory of the stereoelectronic cascade responsible for the transfer of stereogenic information in the peptide chain is proposed. Its implications for the theory of the secondary structure propensities of the Ala* amino acids are then formulated and the available evidence from single-site mutation studies, recently found to confirm the importance of the peptide bond-side chain hyperconjugation,[28] is reviewed.

2. DIVERSITY OF AMIDE BONDING AND CONFORMATIONAL PREFERENCES IN THE CRYSTAL STRUCTURES OF OLIGOPEPTIDES

2.1. Diversity of Amide Bonding in Primary Carboxamides, Oligopeptides, and Lactams

The structure correlation approach assumes that the changes in the amide bond geometry in the crystal structures are essentially deformations along the path of *cis–trans* isomerization.[29] Pauling's resonance model predicts that the shift along this path involves (1) a negative (inverse) correlation of the C=O and C–N bond distances (e.g., C=O increases when C–N decreases) and (2) a positive correlation of the C–N bond distance and the out-of-plane bending at the amide N (e.g., C–N increases when N pyramidalization increases).[30] Surprisingly, however, the correlations found in the fragments retrieved from the Cambridge Structural Database[31] are consistent with Pauling's model only in the case of the tertiary and the more extensively substituted secondary amides and common-ring lactams.[23]

Contrary to the predictions of the resonance model, the correlations between the bond distances in the less substituted primary amides are positive (e.g., C=O increases when C–N increases); the unexpected positive correlation of the bond distances (C–N and C=O) is also found in the least substituted methyl carboxylates.[23] Furthermore, for the majority of the less substituted amides and lactams, an increase in pyramidalization at the N atom is associated with the shortening of the C–N bond instead of the expected lengthening. Thus, the r(C=O), r(C–N), $|\theta_N|$ coordinates are coupled in more than one way.

In principle, there are four patterns of coupling of three structural parameters. The factor analyses for 42 subclasses of amides and lactams reveal that one of the eigenvector patterns: [r(C=O), r(C–N), $|\theta_N|$], is never dominant. The dominance of the other three patterns appears to be related to the major types of the amide embedding, i.e., the extent of the alkyl substitution and the lactam-ring size. Thus, the structural variation in the primary and secondary amides and the small-ring lactams is largely described by the patterns [r(C=O), r(C–N), $-|\theta_N|$] and [r(C=O), $-r$(C–N), $|\theta_N|$], while the pattern [$-r$(C=O), r(C–N), $|\theta_N|$] (the resonance model) becomes more important in the tertiary amides and common-ring lactams. Therefore, the observed diversity of the patterns of structural variation ought to originate in the diversity of the amide bond properties that are affected in similar ways by alkyl substitution and the size of the embedding ring. Two such properties are the energy and symmetry (the s–p character) of the N lone pair.

The hypothesis proposed to explain these findings is shown in Fig. 17.1. It is assumed, as has been done previously,[29] that the changes in the fragment geometry are deformations along the path of rehybridization and rotation. This path begins with the saddle-point structure P, where N is nonhybridized (N lone pair resides on the 2s orbital). The rotation to the structure Q initiates rehybridization

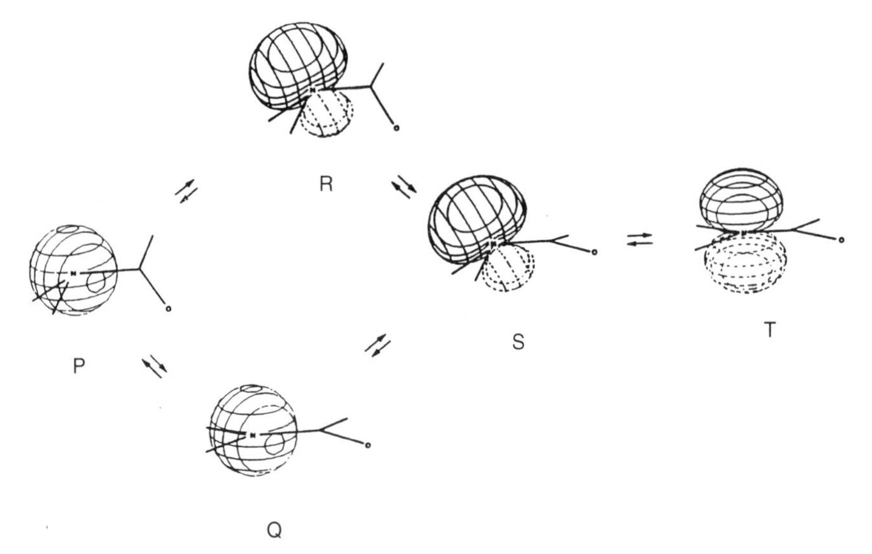

FIGURE 17.1. Proposed path for the internal rotation and rehybridization of the amide bonds. Positions of the minimum and the saddle point for a given class of amides depend on the alkyl substitution and on the size of the ring incorporating N. Structural variation observed in the crystal structures of the primary carboxamides corresponds to a deformation along the Q–S path; structural variation found in the crystals of the extensively substituted secondary amides and tertiary lactams corresponds to a deformation along the S–T path. Schematic representation of the hybridization and geometry of the structures Q and S is exaggerated to accentuate the direction of the change.

of N which occurs gradually, the lone pair being put first on an sp^3 orbital, the structure S, and subsequently on the $2p_z$ orbital, the structure T.

In the early stages of the Q–S–T path, the variation in the C=O and C–N bond distances is governed by the change in electronegativity of N associated with rehybridization. Along the Q–S path, the C–N bond shortens mainly as a consequence of the shift of the s character to the σ bond. The increasing electronegativity of N in the C–N bond causes in turn a shortening of the C=O bond (hence the unexpected positive correlation) which is not yet offset by an increase in the relatively poor π-bonding. The subsequent improvement of the π-bonding along the S–T path, which overrides now the smaller change in the C–N bond s character, results in the inverse correlation of the bond distances.

Given the change in the s character of the N lp, the N pyramidalization is expected to have two maxima along the entire path, the structures P and S, and two minima, the structures Q and T. This is suggested by the trends in the activation energies for the nitrogen inversion.[32–35] As the N n orbital gradually loses the s character, stabilization of the pyramidal geometry initially decreases; when the n orbital subsequently turns into an sp^3 orbital, the stabilization reaches the second maximum. The activation energies (in kcal mol^{-1}) are NH_3 5.8, CH_3NH_2 4.8, $(CH_3)_2NH$ 4.4, $(CH_3)_3N$ 7.5–10.0; for *N*-methyl and *N*-chloro azetidines 10.2 and 13.4, pyrrolidines 8.3 and 10.3, piperidines 8.7–9.6 (eq) and 13.7, azepines 6.6 and 8.4 kcal mol^{-1}. Consequently, an increase in the N pyramidalization results in the shortening of the C–N bond along the Q–S path and in the lengthening of that bond along the T–S path.

The alternative path via the saddle-point structure R, cf. Fig. 17.1, involves rehybridization of N prior to the rotation to the structure S which is likely to require a more extensive alkyl substitution. In general, positions of the saddle point and the minimum for a given class of amides depend on the alkyl substitution and on the size of the ring incorporating N.

2.2. Out-of-Plane Bending at C′ and N in Oligopeptides and Penicillins: Dependence on the C′–N Bond Order

Pyramidalization at the amide N may affect the geometry of the carbonyl group, i.e., produce pyramidalization at the carbonyl C. As was mentioned earlier, such transfer of stereogenic information from a chiral trigonal nitrogen has in fact been implicated in the mechanism of asymmetric induction in reactions of enamines etc.[13] It seems reasonable to assume that the out-of-plane distortion at C will result from either (1) the destabilizing N lp interaction with the C=O π orbital, or (2) the stabilizing N lp interaction with the C=O π^* orbital. The *anti* pyramidalization would occur to minimize the lp, π overlap and thereby the repulsion, the *syn* pyramidalization would maximize the lp, π^* overlap to improve the bonding.

Examination of the geometry of the lactam bonds in the crystal structures of penicillins and cephalosporins suggests that the balance between the two overlap demands, i.e., the relative mode of pyramidalization, depends on the C′–N bond distance and the energy of the N lp.[24] The results are shown in Fig. 17.2; the positive values of θ_C signify the *anti* pyramidalization with respect to the N pyramid. As the energy level of the N lp decreases in penams, penam 1-oxides and penam 1,1-dioxides, Figs. 17.2(b–d), the improvement of the lp, π^* overlap (bonding) apparently becomes more important than the minimization of the lp, π overlap (repulsion) and mainly the *syn* pyramidalization is observed in the two latter fragments, while the maximum due to the repulsion is shifted to the shorter C′–N bond distances.

The peptide carbonyl group is often pyramidalized as well.[36–39] To determine whether the mode of pyramidalization at C′ and N is related to the C′–N bond order, the patterns of coupling of the r(C=O), r(C–N), and $|\theta_N|$ coordinates were examined for the samples of the *anti* and *syn* bent peptide bonds. The results suggest that the *syn* pyramidalized bonds deform along the S–T path and the *anti* pyramidalized bonds along the Q–S path, i.e., the former bonds are more polarized than the latter.[23] Thus, the relative mode of pyramidalization at C′ and N is an additional factor that may contribute to the shift of a peptide bond along the rehybridization/polarization path.

2.3. Correlations of the ψ_i and ϕ_{i+1} Angles in Oligopeptides

The question whether the ψ_i and ϕ_{i+1} angles are correlated and whether the changes in electronic configuration of the peptide bonds play any role in controlling these angles can be addressed by using the ψ_i, ϕ_{i+1} maps instead of the normal Ramachandran maps with the ψ_i, ϕ_i coordinates. Two such maps in Fig. 17.3 show conformational preferences about the endo and C-terminal Ala*–Ala* peptide bonds, i.e., with the CHRC(=O)NH– and CHRC(=O)O– groups, respectively, attached to the peptide N. These groups have similar steric demand but different electronic properties, the carboxylate group being, inter alia, a better inductive electron acceptor (Charton's constants[40] σ_I are CH_2CONH_2 0.06 and CH_2COOEt 0.019; Swain–Lupton constants[41] F are CH_2CONH_2 0.08 and CH_2COO^- 0.19).

The ψ_i, ϕ_{i+1} distribution indicates that there are two preferred loci for the endo Ala*–Ala* bonds which correspond, interestingly, to the α-helix and β-sheet minima of the Ramachandran map, Fig. 17.3a. The detailed picture of the distribution of the *anti* and *syn* pyramidalized bonds might change in the future since a few structures with the very small $\theta_{C'}$ and θ_N values or with the H positional parameters that are not refined are included in the plot for the want of data; at present, the *anti* and *syn* pyramidalized bonds appear to occur in both regions.[25]

In contrast, these two loci are nearly empty in the case of the C-terminal Ala*–Ala* bonds, Fig. 17.3b. Instead, a considerable fraction of the data points are found at the ψ_i values characteristic of the I, II, I′ and II′ type β-turn residues (e.g., 0°, +30°, −120° and +120°),[42–44] while the most populated region is that of the polyproline helix. A majority of the *syn* pyramidalized bonds adopts

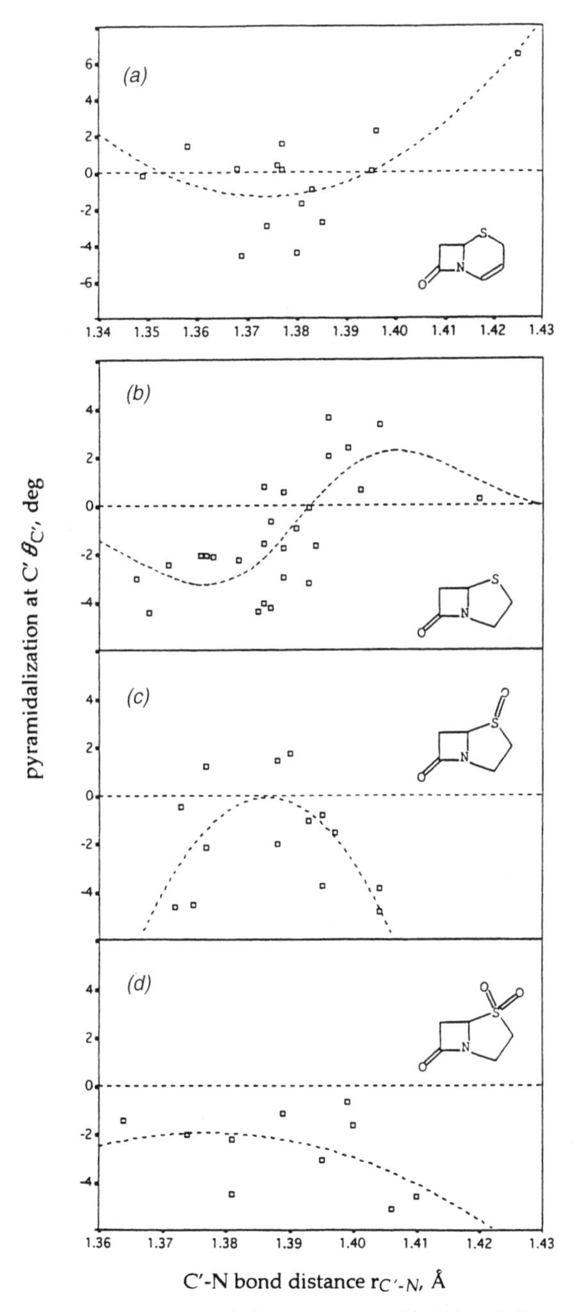

FIGURE 17.2. Pyramidalization at C′ ($\theta_{C'}$) versus the C′–N bond distance in the crystal structures of penicillins and cephalosporins. Positive $\theta_{C'}$ corresponds to *anti* pyramidalization at C′ and N: (*a*) cephems; (*b*) penams; (*c*) penam 1-oxides; (*d*) penam 1,1-dioxides. The curves in panels A, C and D represent the cubic fit, the curve in panel B is obtained by fitting the function $(r_{C'N} - r^0_{C'N})/\exp(r_{C'N} - r^0_{C'N})^2$.

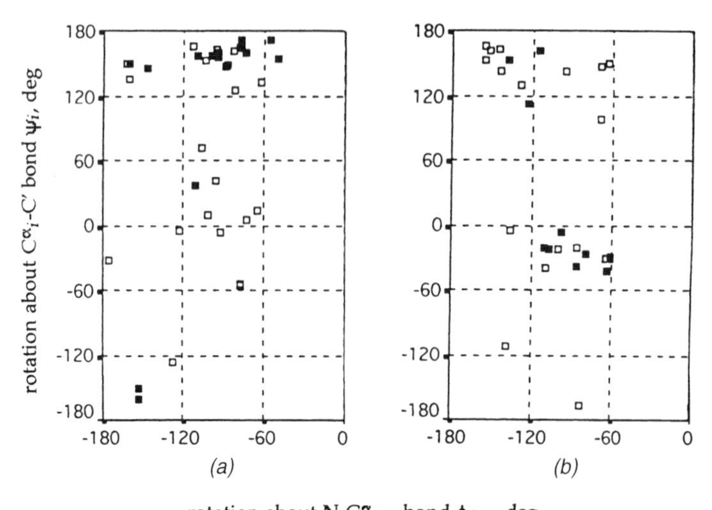

FIGURE 17.3. The ψ_i, ϕ_{i+1} conformational maps for the C-terminal (*a*) and endo (*b*) Ala*–Ala* peptide bonds in the crystal structures of oligopeptides. The empty markers represent the peptide bonds pyramidalized *anti* at C′ and N, the filled markers represent the *syn* pyramidalized peptide bonds, see text.

the latter conformation, the *anti* pyramidalized bonds tend to adopt the turn conformations.

The stereoelectronic origin of these differences is suggested by the manner in which the ψ and ϕ angles correlate with the out-of-plane bending at C′ and N. The C′ pyramidalization in the crystal structures of oligopeptides is described by the function $A \sin 3\psi$ whose amplitude is invariably positive ($A > 0$).[20] This corresponds to pseudostaggering about the peptide C^{α}_{i}–C′ bonds: when one of the bonds is approximately normal to the peptide plane, the apex of the C′ pyramide points in the opposite direction. On the other hand, the N pyramidalization is described by the function $A' \sin 3\phi$ whose amplitude depends on the relative mode of pyramidalization and the N substitution of the peptide bonds. In the case of the C-terminal Gly–Ala* and Ala*–Ala* bonds, A′ is always positive, Fig. 17.4. However, in the case of the endo Gly–Ala* and Ala*–Ala* bonds, A′ is positive only in the *anti* pyramidalized bonds, as shown in Fig. 17.5a. In the *syn* pyramidalized bonds, the amplitude is negative, that is when one of the bonds is approximately normal to the peptide plane, the apex of the N pyramide points in the same direction which is referred to as pseudoeclipsing about the peptide N–C^{α}_{i+1} bonds.[26,27]

To explain the observed preferences, it seems reasonable to propose that the inductive effect of the CHRC(=O)X groups and the improvement of π-bonding upon the *syn* C′ and N pyramidalization combine to place the Ala*–Ala* bonds along the amide rehybridization/polarization path in the following order: *anti*-C-terminal < *anti*-endo < *syn*-C-terminal < *syn*-endo. Thus, the N lp has the highest s character in the *anti*-pyramidalized C-terminal bonds and, in spite of the largest

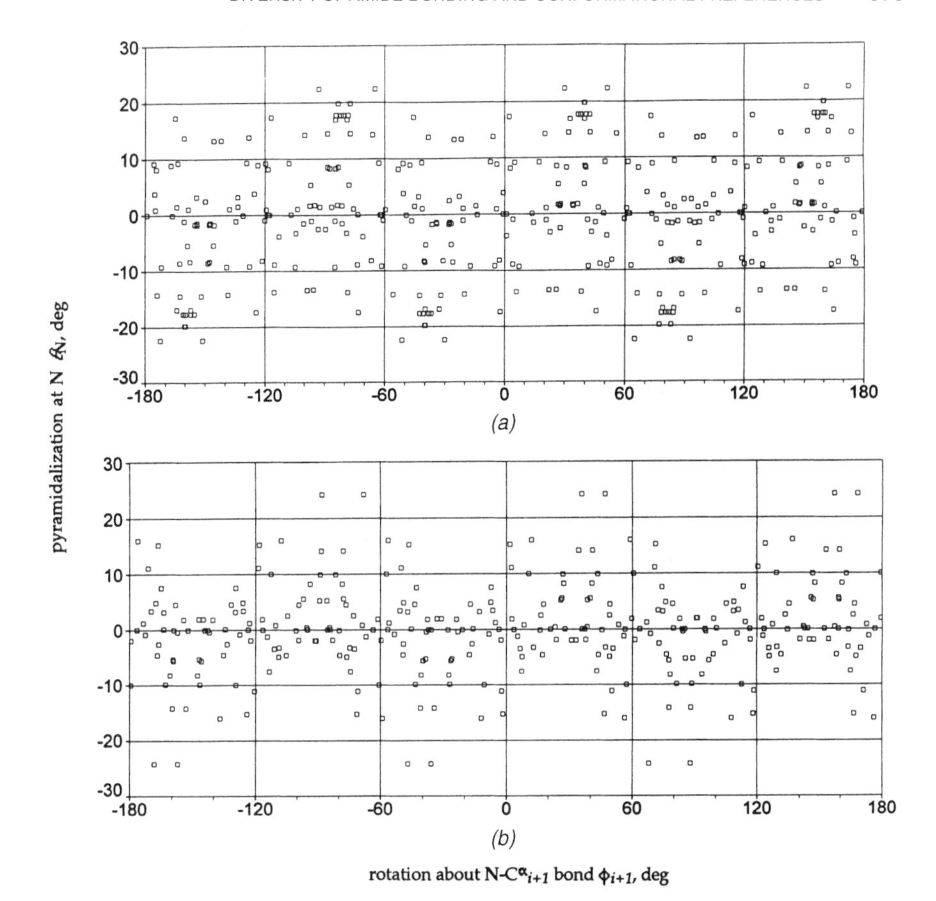

FIGURE 17.4. Pyramidalization at N (θ_N) versus rotation about the N_{i+1}–C^{α}_{i+1} bond measured by the ϕ_{i+1} angle in the *anti* (*a*) and *syn* (*b*) bent C-terminal Gly–Ala* and Ala*–Ala* peptide bonds in the crystal structures of oligopeptides (fragment definition: N–CHRC(=O)NH–CHR′COO, R = H, C; R′ = C); symmetry expanded data sample. Positive amplitude corresponds to pseudostaggering about the N_{i+1}–C^{α}_{i+1} bond (cf. Fig. 17.7b).

amplitude, the N pyramidalization has a small effect on the geometry of the carbonyl group and depends relatively little on the ϕ_{i+1} angle, cf. Fig. 17.4a. Consequently, the ψ_i and ϕ_{i+1} angles are not correlated and a variety of conformations can be adopted where one of the C^{α}_i bonds tends to be in the amide plane.

In the *anti*-endo bonds, the N pyramidalization correlates quite well and in the expected fashion with the ϕ_{i+1} angle, as shown in Fig. 17.5a, which suggests that it is stabilized by the N lp, σ^*_{C-C} hyperconjugation. Its control of the C′ pyramidalization is here more effective and C^{α}_i tends to adopt the pseudo-staggered conformations with the largest substituent, the *i* side chain, out of the amide plane, which results in populating the α-helix and β-sheet minima since

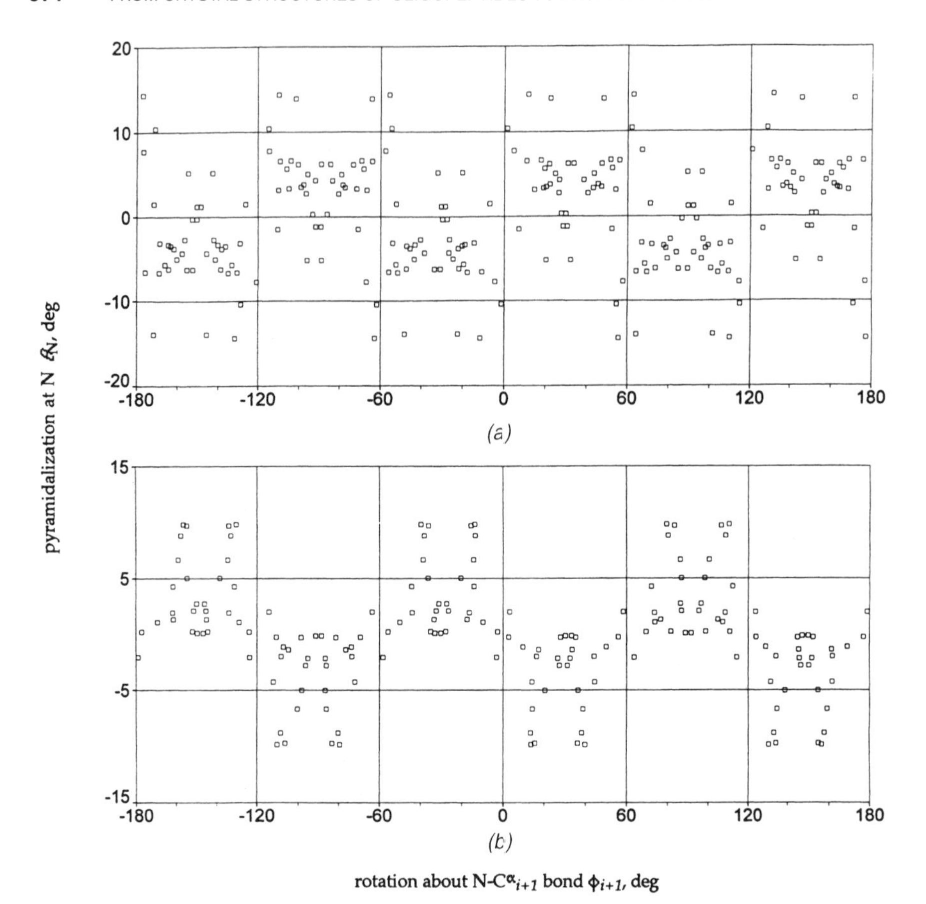

FIGURE 17.5. Pyramidalization at N (θ_N) versus rotation about the N_{i+1}–C^{α}_{i+1} bond measured by the ϕ_{i+1} angle in the *anti* (*a*) and *syn* (*b*) bent endo Gly–Ala* and Ala*–Ala* peptide bonds in the crystal structures of oligopeptides (fragment definition: N–CHRC(=O)NHCHR′CON, R = H, C; R′ = C); symmetry expanded data sample. Positive amplitude corresponds to pseudostaggering about the N_{i+1}–C^{α}_{i+1} bond (cf. Fig. 17.7b), negative amplitude corresponds to pseudoeclipsing about the said bond (e.g. the apex of the N pyramid eclipses the $C^{\alpha}_{i+1}R_{i+1}$ bond when $\phi_{i+1} = -150°$ as in Fig. 17.7c).

either the $i+1$ side chain ($\phi_{i+1} \sim -150°$) or the carboxamide ($\phi_{i+1} \sim -90°$) can be the out-of-plane group on the other end of the peptide bond.

Similarly, the N pyramidalization in the *syn*-C-terminal Ala*–Ala* bonds appears stabilized by the N lp, σ^{*}_{C-C} hyperconjugation, cf. Fig. 17.4b, provided that the C–C bond is a sufficiently good electron acceptor (the peptide bond is now a poorer N lp donor due to polarization); this is certainly the case for the C^{α}_{i+1}–C(=O)O bonds but not always for the C^{α}_{i+1}–C^{β} bonds. The control of the C′ pyramidalization by the N lp is also effective here and the preference of the carboxylate group for the ϕ region $-120° < \phi_{i+1} < -60°$ results in populating the polyproline-helix region.

In the *syn*-pyramidalized endo Ala*–Ala* bonds, the N lp orbital is no longer an electron donor in the hyperconjugative interactions with the C–C bonds because of the polarization of the amide bond. The N pyramidalization is considerably smaller and apparently stabilized now by the σ_{C-C} donation into the electron-deficient N $2p_z$ orbital which results in pseudoeclipsing about the N–C^{α}_{i+1} bond. However, its control of the C′ pyramidalization is still effective and produces the already seen preferences for the α-helix minimum (when the N $2p_z$ orbital is stabilized by the carboxamide C^{α}_{i+1}–C which is a good donor as long as the assistance of the Z lone pair of the carbonyl O is possible) and β-sheet minimum (when the N $2p_z$ is stabilized by the side-chain C^{α}_{i+1}–C^{β}).

2.4. Transfer of Stereogenic Information in Peptide Chains

The emerging picture of the cascade of stereoelectronic effects that produces the correlations of the ψ_i and ϕ_{i+1} angles in oligopetides can now be generalized as follows:

1. the nature of the interactions of the single bonds at C^{α}_{i+1} with the N lp depends on the polarization of the peptide bond: in the least polarized bonds, when the N lp has a relatively high s character, the N pyramidalization tends to be antiperiplanar to the σ bonds at C^{α}_{i+1} but the correlation is very poor; in the moderately polarized bonds, the C^{α}_{i+1} σ bond aligned with the N $2p_z$ axis stabilizes the σ-antiperiplanar pyramidalization at N which leads to pseudostaggering about the N–C^{α}_{i+1} bond; in the most polarized bonds, the $C^{\alpha}{}_{i+1}$σ bond aligned with the N $2p_z$ axis stabilizes the σ-synperiplanar pyramidalization at N (as long as the aligned C^{α}_{i+1} σ bond is a good electron donor) which leads to pseudoeclipsing about the N–C^{α}_{i+1} bond;
2. the effect of the N pyramidalization on the geometry of the carbonyl group also depends on the polarization of the peptide bond: when the C′–N bond order is relatively low and the N lp has a relatively high s character, this effect is small; in contrast, the anti pyramidalization at C′ is stabilized when the C′–N bond order has an intermediate value, and the *syn* pyramidalization at C′ is stabilized when the peptide bond is most polarized;
3. the C′ pyramidalization always stabilizes pseudostaggering about the C^{α}_i–C′ bond; in the case of the Ala* residues, the C^{α}_i side chain is most likely to be out-of-plane as the largest group. The C′ pyramidalization increases with the increase in the C′ $2p_z$ occupancy. In the absence of the C′ pyramidalization, the preferred conformation about the C^{α}_i–C′ bond places one of the C^{α}_iσ bonds in the peptide plane; for the Ala* residues, it is either the C^{α}_i–N bond (the poorest donor), or the C^{α}_i–H bond (the smallest group).

3. ELECTRONIC CONFIGURATION OF PEPTIDE BONDS AND STABILITY OF PROTEIN FOLDS

One way to evaluate the significance of the proposed stereoelectronic control of peptide conformations is to test whether one can predict the effect of amino acid residues on the thermodynamic stability of protein folds by invoking primarily the inductive and resonance effects of the side chains. The following sections outline the hypothesis which translates our picture of the cascade of stereoelectronic effects in peptides into the corresponding theory of the secondary structure propensities of the Ala* amino acids.

3.1. Amide Rehybridization/Polarization Path and Electronic Configuration of Peptide Bonds in Protein Folds

The structural and spectroscopic data indicate that the isomerization transitions of the polypeptide chain are accompanied by changes in electronic configuration of the peptide bonds. This evidence suggests that the peptide bonds in β-sheet are more polarized than those in α-helix, and the latter bonds more polarized than the ones in 3_{10}-helices and β-turns. For instance, the mean C=O···H distance and the β angle, the out-of-plane component of the C=O···H approach angle, are 2.17(16) Å and ~60° in 3_{10}-helix and III β-turn; 2.05(15) Å and 27(8)° in α-helix; 1.96(16) Å and 15° and 11° (median |β|) in antiparallel and parallel β-sheets.[45] Those average geometries were used in an ab initio study of the formamide–formaldehyde complex employed as a model for the hydrogen bonds found in the common protein folds.[46] The calculations reveal that the interaction energy attributable to the hydrogen bond is indeed greater for the β-sheets. Such conclusion is also supported by the IR data: the amide I frequencies in 3_{10}-helices and turns are 1665–1694 cm^{-1}; in α-helix 1651–1659 cm^{-1}; in β-sheet 1620–1640 cm^{-1}.[47–49] Similar results are obtained in the studies of model peptides, e.g., amide I bands in $K_2(LA)_n$ peptides are 1659 cm^{-1} (α-helix) and 1626 cm^{-1} (antiparallel β-sheet).[50–53] Furthermore, the chemical shifts in ^{13}C and ^{15}N NMR spectra of proteins support the above interpretation: the C′ signal is on average 3.1 ppm upfield and the N signal 2.9 ppm downfield in β-sheet compared to α-helix,[54] as implied by the canonical resonance structures in Fig. 17.6. Finally, it seems reasonable to expect that the ESCA spectroscopy, which was shown to provide useful information about the charge distribution in the amide group,[55] will also indicate the change in electronic configuration of peptide bonds that occurs upon folding.

Thus, it is proposed that there are three electronic configurations available to the peptide bonds in proteins: *turn* (required e.g. in 3_{10}-helices, I, II, I′, and II′ β-turns, C–1–C–cap bonds), *helix* (required e.g. in α-helix), and *strand* (required, for example, in antiparallel and parallel β-sheet, polyproline helix, and possibly in N–cap–N + 1 bonds). The three configurations can be considered as the contiguous mid-region segments of the amide rehybridization and polarization path, Fig. 17.6. They differ with respect to the N lp energy and symmetry, and

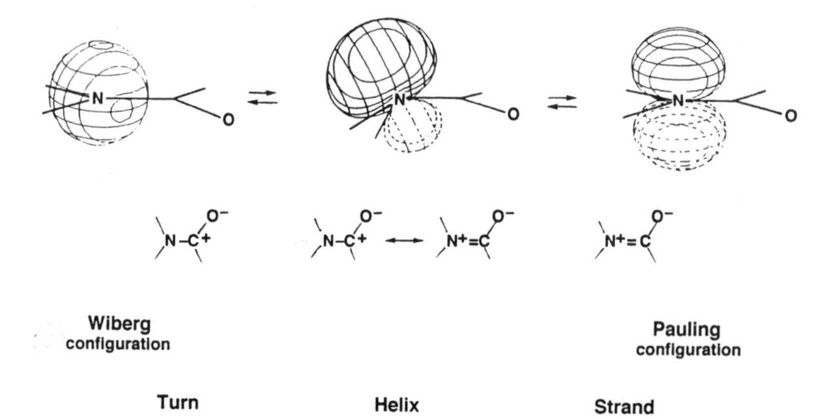

FIGURE 17.6. Location of the *turn*, *helix* and *strand* peptide bonds along the proposed amide rehybridization and polarization path. Electron-releasing substitution at N shifts peptide bonds to the right, electron-releasing substitution at C′ shifts peptide bonds to the left. Schematic representation of the hybridization and geometry of the structures defining the path is exaggerated to accentuate the direction of the change. Only the polarized canonical Lewis structures are shown for the sake of clarity.

hence the degree of charge polarization; the latter is illustrated in Fig. 17.6 using the polarized canonic Lewis structures (the configurations with the lowest and highest C′–N bond order are represented by the canonical structures introduced by Wiberg and Pauling, respectively[30,56]). These differences translate into the increasing strength of H bonding (increasing basicity of the carbonyl O and acidity of the amide N–H), and the narrowing and subsequent reversal of conformational preferences about the C_i^α–C′ and N–C_{i+1}^α bonds:

1. in the least polarized *turn* bond (like in the *anti*-C-terminal peptide bonds, vide supra), the N lp has a relatively high s character and its effect on the geometry of the carbonyl group is small. In the absence of the C′ pyramidalization, the preferred conformation about the C_i^α–C′ bond places one of the C_i^α σ bonds in the peptide plane;
2. the *helix* bond is moderately polarized (like the *anti*-endo and *syn*-C-terminal peptide bonds); in the α-helix, the C_{i+1}^α conformation is stabilized by the N lp, $\sigma^*_{C-Csp^2}$ donation and the resulting σ-antiperiplanar N pyramidalization stabilizes a small negative ψ_i angle via the anti C′ pyramidalization (a preference for pseudostaggering about the N–C_{i+1}^α, C′–N and C_i^α–C′ bonds), as shown in Fig. 17.7b;
3. the *strand* bond is most polarized (like the *syn*-endo peptide bonds). In the antiparallel β-sheet, the C_{i+1}^α conformation stabilized by the σ_{C-C} donation into the N $2p_z$ orbital and the resulting σ-synperiplanar N pyramidalization stabilizes a large positive ψ_i angle via the *syn*-C′ pyramidalization (a preference for pseudoeclipsing about the N–C_{i+1}^α and C′–N bonds, and for pseudostaggering about the C_i^α–C′ bond), Fig. 17.7c.

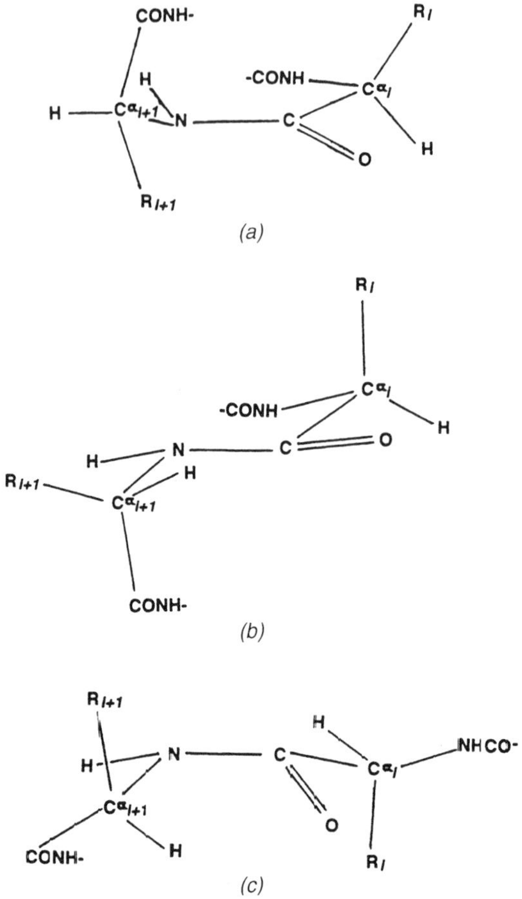

FIGURE 17.7. (*a*) The *turn* bond in the conformation of the C–1–C–cap bond in Rop protein from plasmid ColE1 (29L-30D); (*b*) The *helix* bond in the α-helix conformation; (*c*) The *strand* bond in the antiparallel β-sheet conformation. Schematic representation of nonplanarity of the peptide bonds is exaggerated to accentuate the direction of the change.

As the C′–N bond order increases along the amide rehybridization/polarization path, the stability of a protein fold changes due to a combination of the changes in the H-bonding strength and in the conformational preferences: the β-turn stability decreases, the α-helix stability first increases but reaches a maximum and decreases, and the β-sheet stability increases. Location of a peptide bond along the amide rehybridization and polarization path will depend on the extent of H-bonding, N substitution (Gly, Ala*, Pro), and the inductive and hyperconjugative interactions of the side chains with the peptide bonds' C′ and N. For instance, increasing the C_i^{α} electronegativity shifts the $(i, i+1)$ bond from *helix* to *strand* region and the $(i-1, i)$ bond from *helix* to *turn* region, cf. the canonical Lewis structures in Fig. 17.6; increasing the hyperconjugative C_i^{α}–C^{β} electron donation into the C′ $2p_z$ orbital shifts the $(i, i+1)$ bond from *helix* to *turn* region, while the donation into the N $2p_z$ orbital shifts the $(i-1, i)$ bond from *helix* to *strand* region. Consequently, such inductive and hyperconjugative interactions give rise to the thermodynamic secondary structure propensities of the Ala* amino acids as described below. In this model, it is essential that the backbone chain is moderately electron deficient due to electronegativity of the carbonyl O and it is likely to be stabilized by electron donation from the side-chain C^{α}–C^{β} bonds. It should be noted here that the concept of hyperconjugation between a moderately electron-deficient reaction site and the vicinal bonds is successfully employed to explain the effect of nucleophile basicity and the effect of induction and resonance of substituents on π-face selection in nucleophilic addition to carbonyls.[28,57–60]

3.2. Origin of the Thermodynamic Secondary Structure Propensities of Ala* Amino-Acids

3.2.1. Turn. An example of the *turn* bond is shown in Fig. 17.7a in the conformation of the C–1–C–cap as well as I'_1–I'_2 and $II'1$–II'_2 β-turn bonds; the II_2–II_3 and I'_2–I'_3 β-turn bonds adopt closely related conformations.[42–44] The *turn* configuration is stabilized by the electron-withdrawing effect of the $i+1$ side-chain, the N lp delocalization into the C^{α}_{i+1}–C bonds, and the $\sigma_{C^{\alpha}i-H}$ and $\sigma_{C^{\alpha}i-C}$ donation into the C'_i $2p_z$ orbital. In addition, the steric strain of the C^{α}_{i+1} conformation ($60° < \phi_{i+1} < 90°$, both the C^{α}_{i+1} side chain and the carboxamide group are placed near the $i+1$ carbonyl O) can be offset by the σ_{N-H}, σ^{*}_{C-C} hyperconjugation.

The inductive effect of the Ala* side-chain is then expected to play a major role in stabilization of the conformations of the preceding Ala* residue that place one of the $C^{\alpha}{}_i$ bonds in the plane of the $i, i+1$ peptide, either C_i^{α}–N since it is the poorest hyperconjugative electron donor among those bonds, or C_i^{α}–H since it is the smallest ligand. Such conformations should invariably be associated with, for example, protonation of H or metal cation–sulfide complexation of M. Let us also note that most $i, i+1$ peptide bonds with G as the $i+1$ residue will tend to maintain the *turn* configuration.

3.2.2. Helix. The *helix* bond is shown in Fig. 17.7b. The major σ hyperconjugation of this bond is the C_i^{α}–C^{β} electron donation into the C'_i $2p_z$ orbital which decreases the C′–N bond order and pushes the *strand* bond toward the *helix* region (increasing α-helix stability) and the *helix* bond toward the *turn* region (decreasing α-helix stability). The resulting $\Delta\Delta G^0$ minimum along the hyperconjugation coordinate is expected to depend on charge polarization of the $i, i+1$ bond and to shift in the direction of stronger donation when the polarization increases.

The inductive electron-withdrawing effect of C^{α}_{i+1} also shifts this bond toward the *turn* configuration. The shift toward the *strand* configuration is promoted by the electron-acceptor C_i^α, due to both resonance and inductive interactions with C′.

3.2.3. Strand. The *strand* bond in the antiparallel β-sheet conformation is shown in Fig. 17.7c. Two hyperconjugative C^α–C^β interactions are possible in this case. Since this is the most polarized configuration, the dominant interaction is the stabilization of the electron deficient N_{i+1} $2p_z$ by the electron-donor C_{i+1}^α–C^β bonds. However, good electron donors at C_i^α can destabilize the *strand* configuration. The electron-acceptor side-chains at C_i^α might stabilize the $(i, i+1)$ *strand* bond due to both inductive effect and the C′ $2p_z$ donation into the C_i^α–C^β bond, in particular when the peptide group is relatively less polarized as for instance an edge-strand bond that is H-nonbonded. It should also be added that most i, $i+1$ peptide bonds with P as the $i+1$ residue will tend to maintain the *strand* configuration.

4. SIDE-CHAIN HAMMETT CONSTANTS σ_{ALA^*} AND THERMODYNAMIC STABILITY OF PROTEIN FOLDS

The currently available secondary structure propensities allow to establish whether the inductive and hyperconjugative interactions of the Ala* side-chains with peptide bonds are important for the stability of the protein folds.[61–89] The experimental data come from the host–guest studies of monomeric peptides and single-domain proteins measuring the change in free energy of folding caused by a single-site mutation.[90,91] The method compares the stability to unfolding of a standard protein or peptide with those of mutants in which other amino-acids are individually substituted into the guest site. Using for instance Gly as the reference residue, the secondary structure propensity is given as the change in protein stability relative to the Gly mutant: $\Delta\Delta G^0 = \Delta G^0$ (mutant X) $- \Delta G^0$ (mutant G). The analysis of data from the thermal or urea denaturation of proteins is based on the simple two-state approximation (two-state helix–coil transition). The two-state model is not applicable to peptide helices and the data analysis requires in this case that the multistate models of the helix–coil transition be used. By measuring helix content in solutions of a series of mutants of a reference peptide, the absolute helix propensities are determined, expressed as the helix propagation parameters s of Zimm–Bragg theory or w of Lifson–Roig theory. The relative propensity can be then calculated from the ratio of the propagation parameters:

$$\Delta G^0 \,(\text{mutant X}) - \Delta G^0 \,(\text{mutant G}) = RT \ln (s_X / s_G)$$

Our hypothesis implies that such relative secondary structure propensities depend on the inductive and resonance donor-acceptor abilities of the C^α–C^β bonds which, in turn, depend on the inductive and resonance effects of the C^β substituents X_j. The question therefore is whether the fold-forming tendencies of the Ala* amino-acids can be adequately described by the functions of the σ_I and σ_R constants of the substituents X_j.

There are several difficulties that this approach encounters. Ideally, one would wish to proceed with a set of isosteric substituents designed to minimize any non-local interactions and to ensure that the relevant variables are independent. The set of the coded amino-acids does not meet all these conditions. It is encouraging, however, to find, vide infra, similarity in behavior of residues such as C, E^-, H, V and Y that are not at all similar by the standard criteria of protein chemistry[92,93] but are all expected to be good hyperconjugative C^α–C^β electron donors. Furthermore, the condition that the two variables of our model, the σ_I and σ_R constants, are independent, is met by the set of the coded amino-acids. Thus, the more difficult problem is the availability and adequacy of the σ_R constants which measure the effect of the $2p_z$ electron donors and acceptors interacting with the benzene ring.[94,95] The interaction with the C^α–C^β bond has different steric constraints and to account for that some values need to be adjusted.[94,95] This interaction is also weaker and does not lower basicity of a donor such as OH. The OH groups of S and T are expected therefore to be H-bonded in aqueous solution and be poorer electron donors than predicted by the σ_R constant. As shown below, the behavior of these residues is usually qualitatively consistent with our hypothesis but they are not included in the regression analysis. The same argument is valid for E^-. Finally, some of the σ_R constants are not available and Swaine–Lupton R constants are used as their estimates when possible. Thus, Charton's σ_I and σ_R constants are used to describe the inductive and resonance effect of the side-chains of A: 0.00, 0.00; C: 0.27, -0.34; D^0: 0.30, 0.11; D: -0.19, 0.23; E^0: 0.19, -0.12; F: 0.12, -0.22; I: -0.02, -0.30; L: -0.01, -0.16; M: 0.12, -0.16; N: 0.28, 0.08; Q: 0.06, -0.12; S: 0.24, -0.62; T: 0.23, -0.78; V: -0.02, -0.32; and Y: 0.11, -0.28 (SE's are in the range 0.01–0.04);[40] the calculated modified Swain–Lupton F and R constants are tabulated for E^- (0.19, -0.35) and can be readily estimated for K^+ (0.12, -0.13) and R^+ (0.29, -0.15).[41] No simple estimates of the σ_R or R constants are available for H^0, H^+, and W. Based on distribution of the side-chain χ_2 angle in proteins, two σ_R constants (F, Y) are adjusted (increased by the factor of 2 and 1.5, respectively) to correct for the differences in the $2p_z/2p_z$ overlap of X_j with the peptide C^α–C^β bond[78] ($\theta \sim 90°$) and with the phenyl ring (e.g., biphenyl in gas phase[94] $\theta = 44°$, and in solution[95] $\theta = 32°$).

Given these limitations, a number of conclusions of our regression analysis can be only tentative. Nonetheless, the available information seems sufficient to correctly appraise the major trends in the data. These trends are best shown using the plots of the relative secondary structure propensities $\Delta\Delta G^0$ against σ_{Ala^*}, the side-chain Hammett constants:

$$\sigma_{Ala^*} = \lambda\sigma_I(X_j) + \sigma_R(X_j)$$

The adjusted λ values vary somewhat within the ranges that seem to be related to the nature of the mutation site but the differences are small and we use a few approximate, rounded λ values. As was shown elsewhere,[28] such plots reveal three major patterns of dependence of the fold stability on the side-chain inductive and resonance effects which are consistent with our theory of the secondary structure propensities.

4.1. Helix

The six representative sets of helix propensities, Figs. 17.8 and 17.9, display the first such pattern. The relative stability of the α-helix is determined mainly by the resonance effect of the side-chains and reaches a maximum ($\Delta\Delta G^0$ minimum), usually located at $\sigma_{\mathrm{Ala^*}} \sim 0$. Thus, α-helix stability is decreased by both electron-donor and electron-acceptor *i* side-chains: electron donation into the C_i' $2p_z$ orbital pushes the *helix* bond toward the *turn* configuration, electron withdrawal pushes the *helix* bond toward the *strand* configuration.

The S and T data sets are not included in the plots except in two cases shown in Fig. 17.10. The propensities of T are always among the lowest and qualitatively in agreement with expectations. However, the linearity of the $\Delta\Delta G^0$ distribution for $\sigma_{\mathrm{Ala^*}} < 0$ is lost when T is included in the plots, Fig. 17.10. It is not clear at present whether this is a real effect or, as was argued earlier, the very large σ_R constant is not realistic. The resonance effect of S is expected to be even more influenced by the H-bonding and therefore it is not surprizing within the framework of our theory to find that the S propensities are most variable in terms of the rank order.

On the other hand, the consistent behavior of H, H^+ and W has encouraged us to adopt reasonable estimates of the side-chain Hammett constants and place these data sets in the plots in Figs. 17.8 and 17.9 without including them in the regression analysis. Thus, σ_I is available for W (0.01); the most accurate regressions on $\sigma_{\mathrm{Ala^*}}$, for the rates of base-catalyzed hydrogen exchange, Fig. 17.15, give by extrapolation σ_{R} of -0.21. Similarly, σ_I is available for H (0.12) and σ_R of -0.40 gives a good fit to the five sets of helix data, Figs. 17.8 and 17.9 (the range for 2-furyl groups is between -0.19, σ_R of 2-furyl,[40] and -0.39, *R* of 5-acetyl-2-furyl[41]). The set (0.30, 0.15) is adopted for H^+ as a reasonable estimate. The only major failure to follow the helix pattern is encountered at the position 44 of lysozyme T4 where the standard error of determination of $\Delta\Delta G^0$ is unusually high;[75,78] the scatter, presumably due to non-local interactions, is also increased when the range of $\Delta\Delta G^0$ is very small, as for poly(hydroxyalkyl)glutamine peptides, Fig. 17.8(c), or in 40% TFE.[89] The salient points that emerge from the helix data can be summarized as follows:

1. the $\Delta\Delta G^0$ minimum in Figs. 17.8 and 17.9 can be shifted from the usual location at $\sigma_{\mathrm{Ala^*}} \sim 0$. As was mentioned earlier, Section 3.2.2. the $\Delta\Delta G^0$ minimum is expected to shift toward the more negative $\sigma_{\mathrm{Ala^*}}$ values of the *i* residue when the charge polarization of the *i*, *i+1* peptide bond increases. The available data suggest that such a shift occurs when the site is either insulated from the solvent, cf. the shift toward the more negative

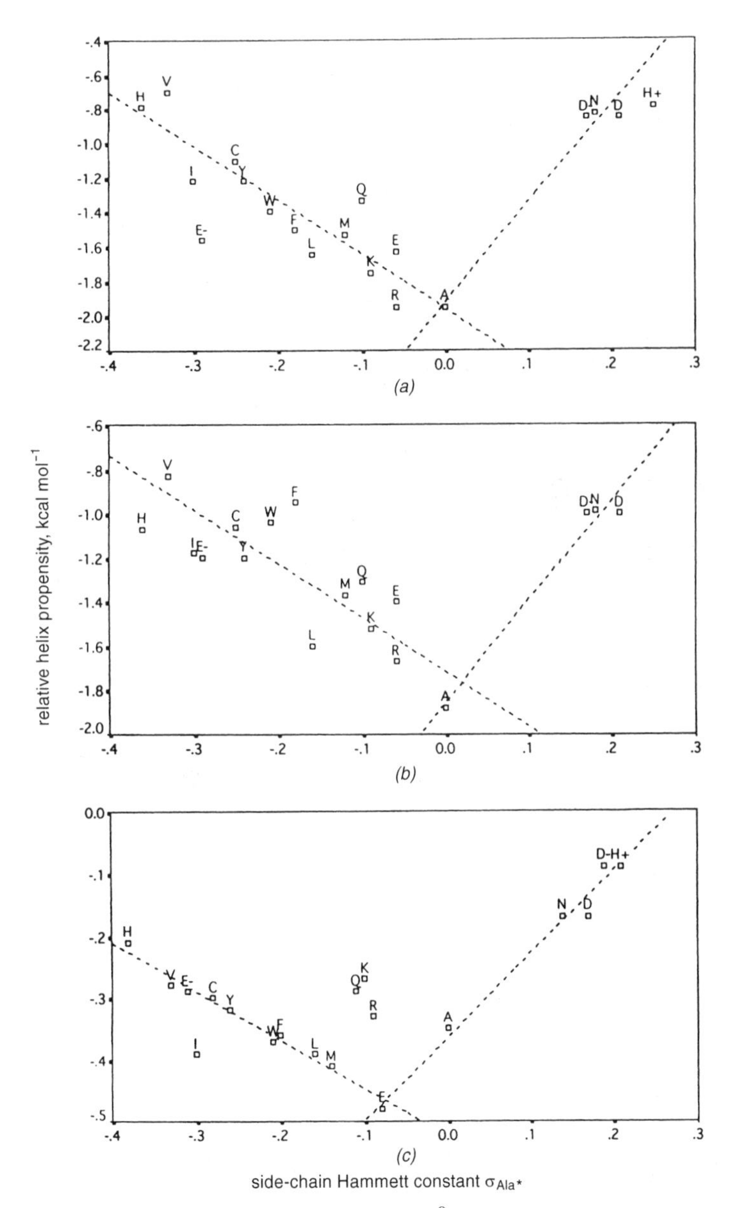

FIGURE 17.8. Relative α-helix propensities $\Delta\Delta G^0_{Gly}$ for the central sites in monomeric peptide systems versus σ_{Ala^*}, r^2 for $\sigma_{Ala^*} < 0$ (E^-, H and W not included in regression, see Section 4.1): (*a*) Ac–YSEEEEKKKKXXXEEEEKKKK–NH_2 (Zimm–Bragg modified to account for *i*, *i* + *4* side-chain interactions), SE = 0.01–0.03 kcal mol^{-1}, 277 K, 10 mM KF, pH 7, $\lambda = 0.33$, $r^2 = 0.795$;[64,66] (*b*) $(AAKAA)_n$ (Lifson–Roig modified to account for helix-end interactions), 273 K, 1.0 M NaCl, pH 7, $(AAQAA)_n$ for D^-, R^+, E^-, H^+, 273 K, 10 mM NaCl, $\lambda = 0.33$, $r^2 = 0.719$;[70–72,77] (*c*) Random copolymers with (hydroxyalkyl) glutamines (Lifson–Allegra–Poland–Scheraga), SE = 0.1 kcal mol^{-1}, water, pH 7, $\lambda = 0.2$, $r^2 = 0.976$ (I, K, Q, R not included in regression).[62]

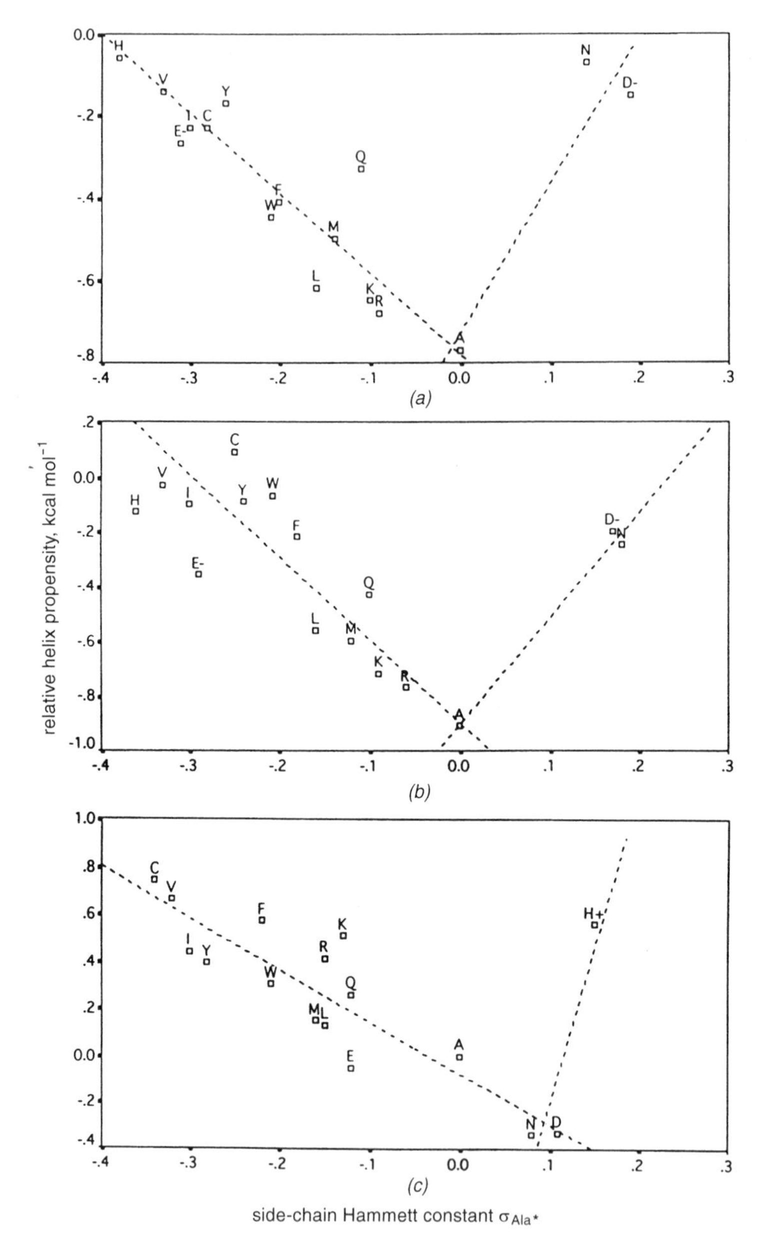

FIGURE 17.9. Relative α-helix propensities $\Delta\Delta G^0{}_{\mathrm{Gly}}$ ($\Delta\Delta G^0{}_{\mathrm{Ala}}$) for the central sites in proteins versus $\sigma_{\mathrm{Ala^*}}$, r^2 for $\sigma_{\mathrm{Ala^*}} < 0$ (E^-, H and W not included in regression, see Section 4.1): (*a*) Leucine zipper Ac–EWEALEKKLAALEXKLQALEKKLEALEHG, SE = 0.1 kcal mol^{-1}, 296 K, 5 M urea, 1 M NaCl, pH 7.5 (MOPS buffer), $\lambda = 0.2$, $r^2 = 0.802$;[63] (*b*) Barnase (site 32), SE = 0.03 kcal mol^{-1}, 298 K, urea denaturation, pH 6.3, 50 mM MES, $\lambda = 0.33$, $r^2 = 0.852$;[69] (*c*) Ribonuclease T_1 (site 21), SE = 0.10 kcal mol^{-1}, 298 K, urea denaturation, 30 mM glycine buffer, pH 2.5, $\lambda = 0.0$, $r^2 = 0.801$.[86]

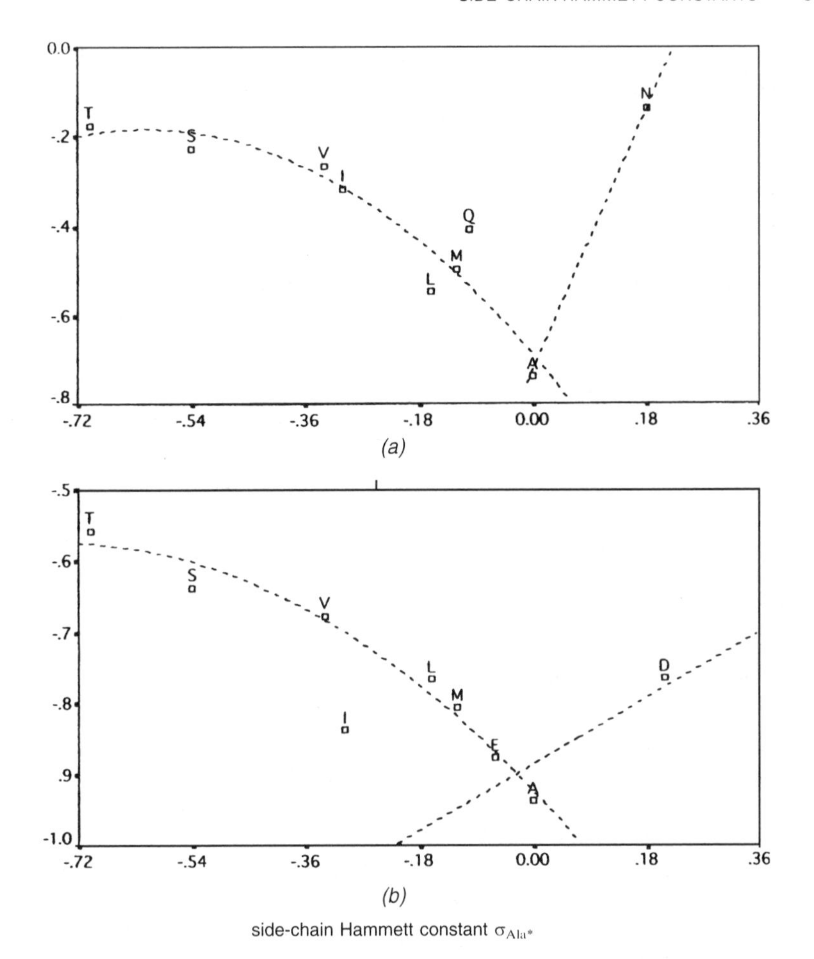

FIGURE 17.10. Relative α-helix propensities $\Delta\Delta G^0_{Gly}$ versus σ_{Ala^*}, $\lambda = 0.33$: (*a*) Ac–YEAAAKEAXAKEAAAKA–NH_2 (Lifson–Roig), SE = 0.01–0.04 kcal mol^{-1}, 273 K (extrapolated to 0 M KCl);[73,74] (*b*) phage T4 lysozyme (site 131), SE = 0.05 kcal mol^{-1}, 326 K, 0.25 M KCl, 3 mM H_3PO_4, 17 mM KH_2PO_4, pH 3.01.[75,78]

Hammett constants in poly(hydroxyalkyl)glutamine peptides, see Fig. 17.8c (possibly also at the buried site of the VBP leucine zipper[88]), or when the H-bonding capacity of the solvent is lowered as in 40% TFE.[89] On the other hand, a shift toward the more positive Hammett constant occurs at the position 21 of RNase T_1 at pH = 2.5, Fig. 17.8e, but not in the corresponding peptide[86];

2. in terms of the $\Delta\Delta G^0$ distribution pattern and the range of relative propensities, there is no difference in behavior of the peptide and protein systems. The range of the $\Delta\Delta G^0$ values varies significantly, from a few tenths of a kilocalorie to several kilocalories per mol. Interestingly, the lowest sensitivity of the fold to the side-chain effects appears to be

associated with the hydrophobic environment as in the 40% TFE solution,[89] or in the poly(hydroxyalkyl)glutamines, Fig. 17.8c, where the α-helix is wrapped in a layer of the fatty acid-like chains;

3. systematic deviations from the helix pattern suggest long-range interactions of the side-chains or a modification of the side-chain resonance effect by the helix environment. For instance, E^- deviations are negative, cf. Figs. 17.8a and 17.9b, which is expected when the resonance effect is diminished by H-bonding. The deviations of K, Q and R, which can be long-range H-bond donors, are positive, Figs. 17.8a, 17.8c and 17.9a;
4. destabilization of α-helix by the strong electron acceptors might involve the inductive electron-withdrawing effect which destabilizes the *helix* configuration. This is the probable reason for the often large effect of protonation of H, consistent with the dramatic effect of phosphorylation of S and T.[96]

4.2. Strand

The plots of the β-sheet propensities show the second pattern of dependence on the inductive and resonance effect of the side-chains. As previously, the relative stability also appears to depend mostly on the resonance effect and reaches a maximum. However, the $\Delta\Delta G^0$ minimum is shifted so far toward the negative σ_{Ala^*} values ($\sigma_{Ala^*} \sim -0.3$ in Figs. 17.11a–c, and apparently off the chart in Fig. 17.11d) that the fold can be said to be stabilized, as expected, by the electron-donor side-chains.

The stabilizing interaction is electron donation from the C^{α}_{i+1} side-chain into the electron-deficient N_{i+1} $2p_z$ orbital of the *strand* bond. Destabilization of the β-sheet around $\sigma_{Ala^*} \sim -0.3$ occurs because of the competing C^{α}_i–C^{β}_i donation into the C'_i $2p_z$ orbital. This destabilization seems to be offset in the case of the strongest donors, S and T, whose β-sheet propensities are very high.

As for the anticipated stabilizing effect of the electron-acceptor side-chains at C^{α}_i, the data in Figs. 17.11c and 17.11d suggest that it might be relevant at the edge-strand sites (in addition, the site shown in Fig. 17.11c is H-nonbonded) since the β-sheet stability reaches a minimum in these plots at $\sigma_{Ala^*} \sim 0.0$ ($\Delta\Delta G^0$ maximum). This effect could become more important in the case of those N–cap–N + 1 bonds that adopt the *strand* configuration to terminate the α-helix in the N-end direction. Such a mechanism of the N-end termination is suggested by the high preference for P at the N + 1 position.[97]

4.3. Turn

The data on the relative turn propensities are obtained for the bonds that share the basic features of the conformation depicted in Fig. 17.7a: the C–1–C–cap bonds, Figs. 17.12a and 17.12b, and the B1–L1 bond, Fig. 17.12d, which all

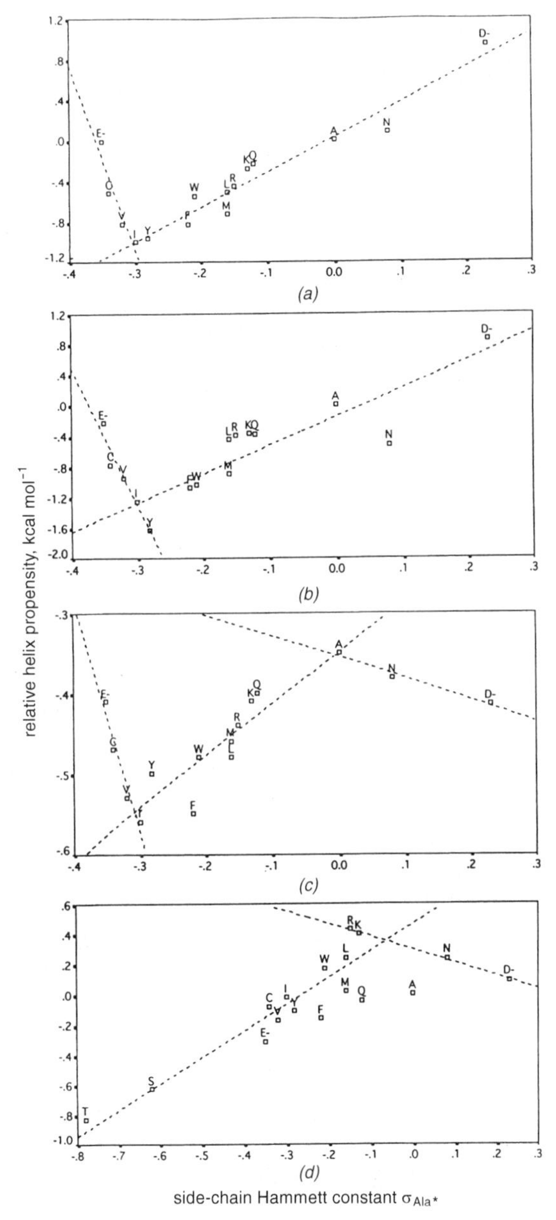

FIGURE 17.11. Relative β-sheet propensities $\Delta\Delta G^0_{\mathrm{Ala}}$ ($\Delta\Delta G^0_{\mathrm{Gly}}$) versus $\sigma_{\mathrm{Ala^*}}$, $\lambda = 0$ (W not included in regression, see Section 4.1): (*a*) IgG binding B1 domain of streptococcal protein G (I6A/T44A/T51S/T55S, site 53, central strand), SE = 0.06 kcal mol^{-1}, 321 K, 50 mM AcONa, 150 mM NaCl, pH 5.4, $r^2 = 0.950$ for $\sigma_{\mathrm{Ala^*}} > -0.3$;[81] (*b*) IgG binding B1 domain of streptococcal protein G (I6A/T44A, site 53, central strand), SE < 0.05 kcal mol^{-1}, 333 K, 50 mM AcONa, pH 5.2, $r^2 = 0.770$ for $\sigma_{\mathrm{Ala^*}} > -0.3$;[79] (*c*) Consensus zinc finger peptide (site 3, edge strand), SE = 0.01–0.05 kcal mol^{-1}, 293 K, 100 mM HEPES, 50 mM NaCl, pH 6.8, $r^2 = 0.801$ for $-0.3 < \sigma_{\mathrm{Ala^*}} < 0.3$[76] (*d*). IgG binding B1 domain of streptococcal protein G (E42A/D46A/T53A, site 44, edge strand), SE = 0.06 kcal mol^{-1}, 321 K, 50 mM AcONa, 150 mM NaCl, pH 5.4, $r^2 = 0.861$ for $\sigma_{\mathrm{Ala^*}} < -0.1$ (Q omitted from regression).[82]

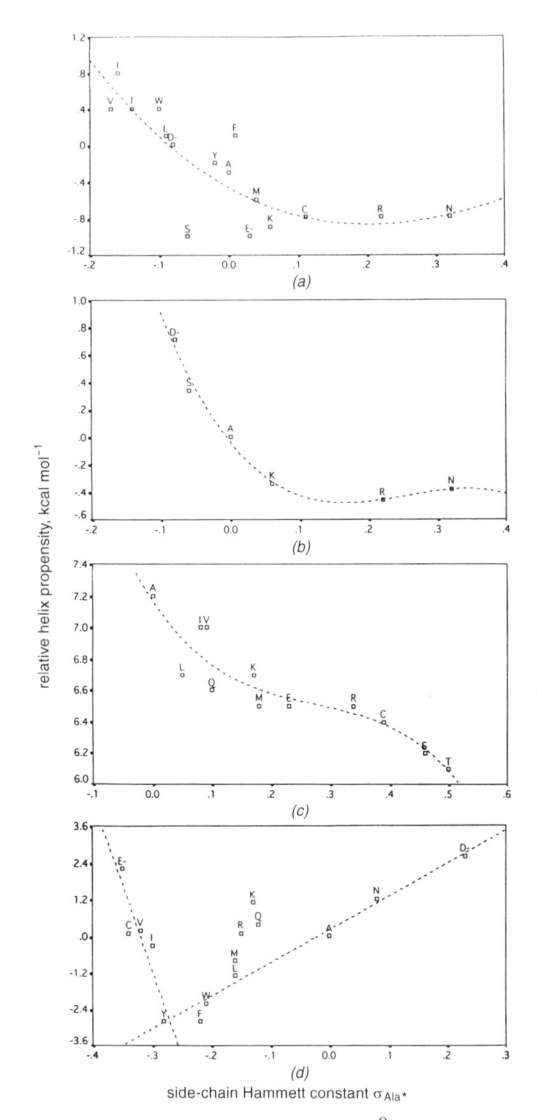

FIGURE 17.12. (*a*) Relative C-cap propensities $\Delta\Delta G^0{}_{Asp}$ in Rop protein from plasmid ColE1 (site 30) versus σ_{Ala^*}, $\lambda' = 0.5$, SE = 0.15 kcal mol^{-1}, 10 mM sodium phosphate, 350 mM NaCl, pH = 7.0, $r^2 = 0.712$ (Q data set is omitted from the plot);[84] (*b*) Relative C-cap propensities $\Delta\Delta G^0{}_{Ala}$ in barnase (site 18, average values for the Trp94 and Leu94 data) versus σ_{Ala^*}, $\lambda' = 0.5$, SE = 0.03 kcal mol^{-1}, 298 K, urea denaturation, 50 mM MES, pH 6.3, $r^2 = 0.998$;[68] (*c*) II_3 β-turn propensities expressed as temperature coefficients for the resonance of the Asp4 amide proton, $\Delta\delta/\Delta T \times 10^3$ (ppm K^{-1}), for the series YPXDV, versus σ_{Ala^*}, $\lambda' = -0.33$, H_2O, pH 4.1, $r^2 = 0.848$ (D, F, N, Y data sets omitted from the plot).[61] (*d*) Relative β-hairpin propensities in staphylococcal nuclease β-barrel (site 27) versus σ_{Ala^*}, $\lambda = 0$, SE < 0.10 kcal mol^{-1}, 293 K, 100 mM NaCl, 25 mM sodium phosphate, pH 7.0, $r^2 = 0.960$ for $\sigma_{Ala^*} > -0.3$ (K, Q, R not included in regression).[87]

have the I'_1–I'_2 conformation, and the reverse turn bond, Fig. 17.12c, which has the II_2–II_3 conformation.[42–44] Propensities at both $i+1$, Figs. 17.12a–c, and i, Fig. 17.12d, sites were examined, and the results at both sites are in agreement with the expectation.

The plots in Figs. 17.12a–c for the $i+1$ site reveal the third pattern of the $\Delta\Delta G^0$ distribution: the relative stability of these turns is controlled, as expected, mainly by the inductive effect of the $i+1$ side-chain substituents. The resonance contribution is small and its sign varies.

In the case of the C-cap residues, Figs. 17.12a and b, the resonance contribution is destabilizing. The C-cap residue adopts the I'_2 conformation where the stabilizing interaction of the $i+1$ side-chain with the $C'_{i+1}\,2p_z$ orbital is not effective ($\psi_{i+1} \sim 30°$). On the other hand, the steric strain of the I'_2 conformation, cf. Section 3.2.1., is offset by the σ_{N-H}, σ^*_{C-C} hyperconjugation which would be diminished by that resonance.

In the case of the II_3 residue of the reverse turn formed by the YPXDV pentapeptides, Fig. 17.12c, the resonance contribution is stabilizing. The side-chain of the examined $i+1$ residue can now interact with the $C'_{i+1}\,2p_z$ orbital (II_3 conformation: $\psi_{i+1} \sim 0°$).

The plot for the i site, Fig. 17.12d, resembles the plots for the β-sheet sites, Figs. 17.11a–c. This set of data is obtained at the position 27 of the staphylococcal nuclease β-barrel, the first position of a Type I′ β-turn linking two antiparallel strands. It is the –B1 (I'_1) residue, preceding the L1 and L2 (I'_2 and I'_3) residues of the characteristic two-residue β-hairpin.[42–44] Thus, the similarity is superficial because the side-chain in this position cannot effectively interact with the N $2p_z$ orbital ($\phi_i \sim -90°$). The major interaction here is the donation into the $C'_i\,2p_z$ orbital that stabilizes the *turn* configuration of the –B1–L1 bond. As the electron-donor ability of the i side chain increases, the fold stability first increases and reaches a maximum at $\sigma_{Ala^*} \sim -0.25$. However, further increase in the electron-donor ability of the i side-chain apparently improves the ineffective interaction with the $N_i\,2p_z$ orbital which stabilizes the *strand* configuration of the preceding –B2–B1 bond. The change in ϕ_i and ψ_i that is required for the latter donation destabilizes the hairpin. The plot also shows positive deviations of K, Q and R, found earlier in the α-helix, Fig. 17.8c.

5. REDISTRIBUTION OF CHARGE IN THE PEPTIDE BACKBONE AND C=O ^{13}C CHEMICAL SHIFTS IN PROTEIN FOLDS

Since the peptide bond-side-chain hyperconjugation would affect the charge distribution in the backbone chain, the chemical shifts in proteins might also be functions of the $\sigma_I(X_j)$ and $\sigma_R(X_j)$ constants and correlate with the thermodynamic secondary structure propensities of the Ala* amino acids. In particular, the dominant paramagnetic term of the chemical shift for the carbonyl carbon C′ is expected to depend on partial occupancy of the $2p_z$ orbital and thus the C′ chemical shifts could be a sensitive probe of the peptide bond-side-chain hyperconjugation.[98]

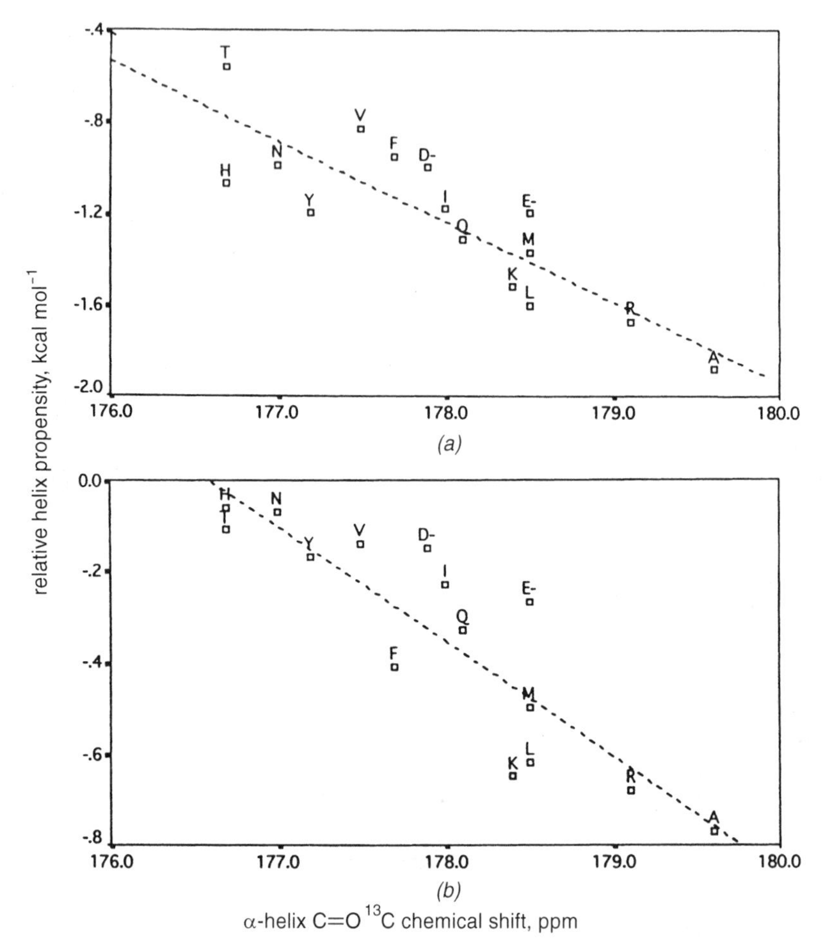

FIGURE 17.13. Relative α-helix propensities $\Delta\Delta G^0_{\mathrm{Gly}}$ versus α-helix C=O ^{13}C chemical shifts (W not included for paucity of the NMR data; C, S are omitted as the two most upfield data sets that require a nonlinear fit): (*a*) $(\mathrm{AAKAA})_n$ peptides, $r^2 = 0.736$;[70–72,77] (*b*) leucine zipper Ac–EWEALEKKLAALEXKLQALEKKLEALEHG, $r^2 = 0.773$.[63]

The electron-donor side-chains which destabilize the α-helix should increase electron density at C′ by increasing occupancy of its $2p_z$ orbital. Consequently, the α-helix propensities are expected to correlate with the C′ chemical shifts. The average ^{13}C C=O shifts of the Ala* residues in the α-helices are available,[54] and the plots shown in Fig. 17.13 confirm that this is true. In principle, a correlation with the opposite sense should exist between the β-sheet propensities and the C′ chemical shifts. In this fold, however, the electron-donor side-chains interact primarily with the electron-deficient N $2p_z$ orbital and the C $2p_z$ occupancy is affected only indirectly. Therefore, the correlation might be very poor or even lost altogether. This prediction is also correct (data not shown).

6. REDISTRIBUTION OF CHARGE, ^{15}N CHEMICAL SHIFTS AND RATES OF BASE-CATALYZED HYDROGEN EXCHANGE IN MODEL DIPEPTIDES

The range of the peptide ^{15}N chemical shifts in proteins is too large to assign average fold-characteristic value to each residue.[54] However, the importance of hyperconjugative donation into the N $2p_z$ orbital is confirmed by the upfield shift caused by the electron-donor side-chains in ^{15}N NMR of *N*-acetyl amino-acids in DMSO, Fig. 17.14.[99] The upfield shift caused by the electron-acceptor side chains can be explained as a result of the inductive effect that destabilizes the charge polarization of the amide group.[28] Furthermore, the rates of base-catalyzed hydrogen exchange in the Ac-Ala*–NHMe dipeptides[100] are expected to depend on the transition-state stabilization of the nascent peptide anions by the side-chain inductive and resonance effects. The plots of log $k_{ex}(\text{Ala}^*)/k_{ex}(\text{Ala})$ versus σ_{Ala^*} suggest that indeed both effects contribute to the relative rates of exchange, as shown in Fig. 17.15.[28]

7. SUMMARY AND CONCLUSIONS

The transfer of stereogenic information through the system of conjugated trigonal centers has been in the past implicated in the asymmetric induction in reactions of chiral enamines, nitrones, vinyl sulfoxides etc. The mechanism of such a transfer would involve a cascade of stereoelectronic effects that stabilize pyramidalization at the trigonal N or C in a locally asymmetric environment. According to the qualitative PMO arguments, the preferred sense of pyramidalization is the one that results in pseudostaggering about the fragment's σ framework. However, the solid-state conformations and out-of-plane

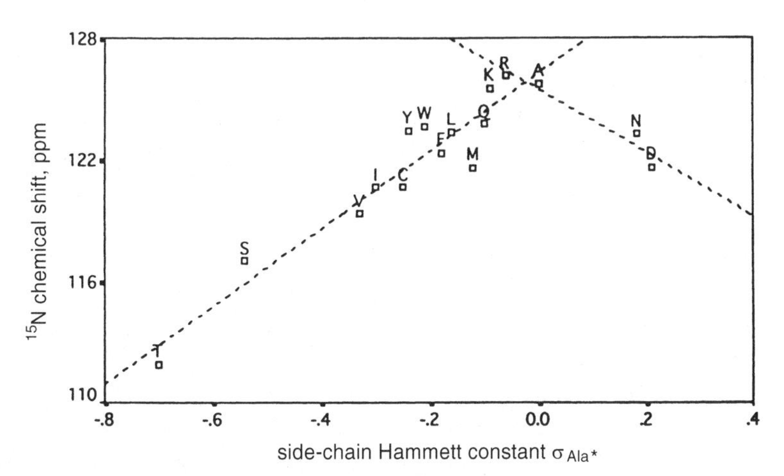

FIGURE 17.14. ^{15}N chemical shifts in *N*-acetyl amino acids in DMSO versus σ_{Ala^*}, $\lambda = 0.33$, $r^2 = 0.922$ for $\sigma_{\text{Ala}^*} < 0$ (E is omitted since its σ_{Ala^*} is highly dependent on the ionization state which is unknown).

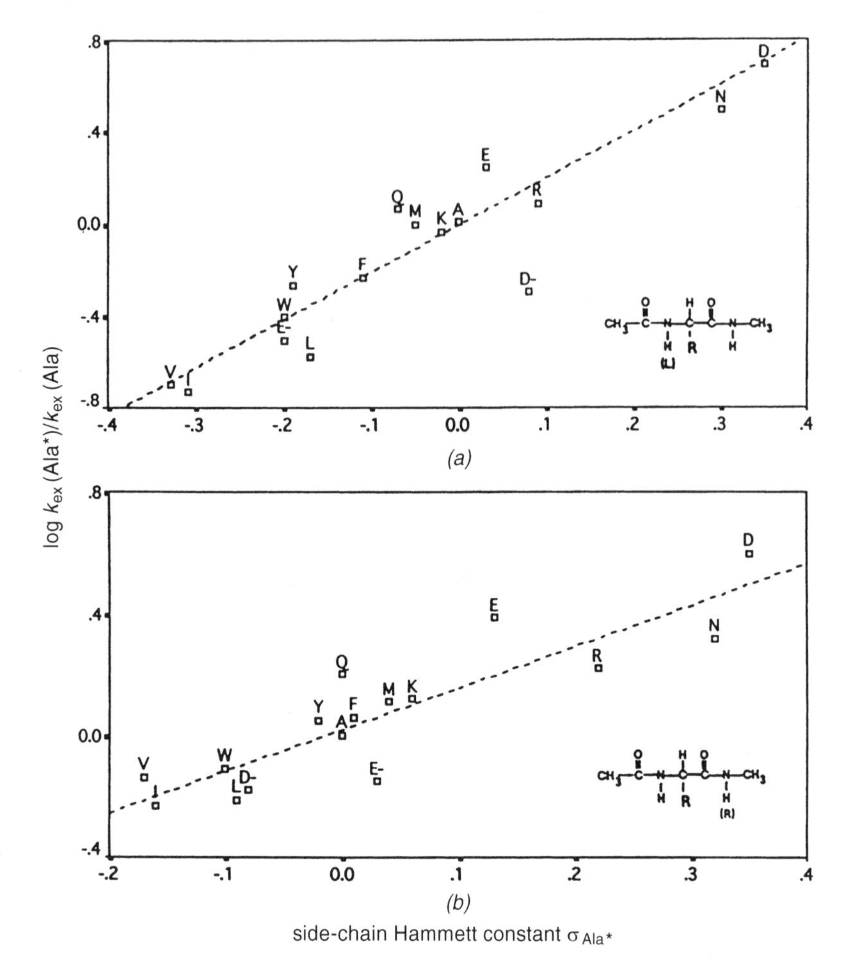

FIGURE 17.15. Relative rates of base-catalyzed hydrogen exchange in Ac–Ala*–NHMe dipeptides log $k_{ex}(\text{Ala}^*)/k_{ex}(\text{Ala})$ versus σ_{Ala^*}: (*a*) Left N–H (see the diagram in the panel), $\lambda = 0.8$, $r^2 = 0.918$ (D^- is plotted but not included in regression); (*b*) Right N–H, $\lambda' = 0.5$, $r^2 = 0.774$. C, S and T are omitted from the plots since it is not certain that the σ_R constants adequately describe electron-donor ability of these residues in the acidic aqueous solution (pD < 6).

distortions of linear oligopeptides are not consistent with this model. Our examinations of (1) the coupling of the internal coordinates in the crystal structures of amides and lactams, (2) the correlations of the C′ and N pyramidalization and the C′–N bond order in penicillins and oligopeptides, and (3) the correlations of the ψ_i and ϕ_{i+1} angles and the C′ and N pyramidalization in the endo and C-terminal Ala*–Ala* peptide bonds, suggest a different picture of the hypothetical stereoelectronic cascade. The interactions of the C_i^α and C_{i+1}^α σ bonds and the pyramidalized C′ and N depend on the C′–N bond order, i.e., polarization of the peptide bond, the energy and symmetry of the N lp, and the donor and acceptor properties of the C_i^α and C_{i+1}^α σ bonds: they may stabilize either pseudostaggering or pseudoeclipsing about the peptide σ framework, or actually fail to correlate the sense of distortion and orientation of the vicinal bonds.

This hypothesis, combined with the structural and spectroscopic evidence of the differences in electronic configuration of peptide bonds in different protein folds, leads to a new theory of the thermodynamic secondary structure propensities of the Ala* amino-acids and can thus be tested using the recently available propensities from single-site mutation studies. The results of the regression analysis show that a good description of the experimental data is obtained with the functions of the side-chain Hammett constants $\sigma_{\text{Ala}^*} = \lambda\sigma_I(X_j) + \sigma_R(X_j)$. The resonance effect of the side-chain unexpectedly emerges as a major factor to destabilize α-helix and stabilize β-sheet, while the inductive effect appears to be a major contributor in formation of turns. This implies that the stability of protein folds is related to charge distribution in the backbone chain and it is indeed observed that secondary structure propensities correlate with the C=O ^{13}C chemical shifts in α-helix but not in β-sheet. Thus, the variations in the plots of relative propensities $\Delta\Delta G^0$ versus σ_{Ala^*} that characterize the site and its environment (range of $\Delta\Delta G^0$, $\Delta\Delta G^0$ extremums, residue deviations etc.) can perhaps be anticipated on the basis of the ^{13}C and ^{15}N NMR. It is noteworthy that no other theory of fold stabilization predicts correlations between the secondary structure propensities and the chemical shifts in peptides and proteins.[75,78,101–108]

These findings support the notion that conformational transitions of the polypeptide chain are accompanied by changes in electronic configuration of the peptide bonds. It is argued that it is convenient to think of such changes in terms of a shift of a peptide bond along the amide rehybridization/polarization path. As the C′–N bond order increases along this path, the concomitant changes in Lewis basicity and conformational preferences of the peptide bond combine to affect the fold stability: the β-turn stability decreases, the α-helix stability first increases but reaches a maximum and decreases, and the β-sheet stability increases. Secondary structure propensities of the Ala* amino-acids are largely due to hyperconjugation and inductive effects of the side-chains because the resulting electron-density shifts stabilize or destabilize electronic configurations of peptide bonds, i.e., transport them from one region to another along the amide rehybridization/polarization path.

ACKNOWLEDGMENTS

The author thanks Professor Ken Wiberg and Dr. Robert Hesse for the invaluable comments and suggestions, and Viktoria Petrova for the enthusiastic, thorough and insightful effort in the investigation of H bonding in the crystal structures of linear oligopeptides.

REFERENCES

1. McKenzie, A.; Smith, I. A. *J. Chem. Soc.* 1924, **125**, 1582.
2. McKenzie, A.; Smith, I. A. *Chem. Ber.* 1925, **58**, 894.
3. Prelog, V. *Helv. Chim. Acta* 1953, **36**, 308.
4. Prelog, V. *Bull. Soc. Chim. Fr.* 1956, 987.
5. Meyers, A. I. *Acc. Chem. Res.* 1978, **11**, 375.
6. Tsuchihashi, G.; Mitamura, S.; Ogura, K. *Tetrahedron Lett.* 1976, 855.
7. Vasella, A. *Helv. Chim. Acta* 1977, **60**, 426, 1273.
8. Vasella, A.; Voeffray, R. *J. Chem. Soc., Chem. Comm.* 1981, **97**.
9. Bernet, B.; Krawczyk, E.; Vasella, A. *Helv. Chim. Acta* 1985, **68**, 2299.
10. Benner, S. A. *Experientia* 1982, **38**, 633.
11. Nambiar, K. P.; Stauffer, D. M.; Kolodziej, P. A.; Benner, S. A. *J. Am. Chem. Soc.* 1983, **105**, 5886.
12. Benner, S. A.; Nambiar, K. P.; Chambers, G. K. *J. Am. Chem. Soc.* 1985, **107**, 5513.
13. Brown, K. L.; Damm, L.; Dunitz, J. D.; Eschenmoser, A.; Hobi, R.; Kratky, C. *Helv. Chim. Acta* 1978, **61**, 3108.
14. Glasfeld, A.; Zbinden, P.; Dobler, M.; Benner, S. A.; Dunitz, J. D. *J. Am. Chem. Soc.* 1988, **110**, 5152.
15. Schweizer, W. B.; Procter, G.; Kaftory, M.; Dunitz, J. D. *Helv. Chim. Acta* 1978, **61**, 2783.
16. Radom, L.; Pople, J. A.; Mock, W. *Tetrahedron Lett.* 1972, 479.
17. Volland, W. V.; Davidson, E. R.; Borden, W. T. *J. Am. Chem. Soc.* 1979, **101**, 533.
18. Houk, K. N.; Rondan, N. G.; Brown, F. K. *Isr. J. Chem.* 1983, **23**, 3.
19. Jeffrey, G. A.; Houk, K. N.; Paddon-Row, M. N.; Rondan, N. G.; Mitra, J. *J. Am. Chem. Soc.* 1985, **107**, 321.
20. Cieplak, A. S. *J. Am. Chem. Soc.* 1985, **107**, 271.
21. Cieplak, A. S. In *Structure Correlation*; Vol. 1, Dunitz, J. D., Bürgi, H.-B. Eds.; Verlag Chemie: Weinheim, 1994; p. 207.
22. Winkler, F. K.; Dunitz, J. D. *J. Mol. Biol.* 1971, **59**, 1969.
23. Cieplak, A. S. *Struct. Chem.* 1994, **5**, 85.
24. Cieplak, A. S.; Shoja, M. (in preparation).
25. Cieplak, A. S. (unpublished results).
26. Cieplak, A. S. European Conference on Stereochemistry, Bürgenstock, 1988, Poster Session.
27. Cieplak, A. S.; Petrova, V. (unpublished results).
28. Cieplak, A. S. *Chem. Rev.* 1999, **99**, 1265.
29. Gilli, G.; Bertolasi, V.; Bellucci, F.; Ferretti, V. *J. Am. Chem. Soc.* 1986, **108**, 2420.
30. Pauling, L. *The Nature of the Chemical Bond*; 3rd edn., Cornell University Press: Ithaca, NY, 1960; pp. 281–282.
31. Allen, F. H.; Kennard, O.; Taylor, R. *Acc. Chem. Res.* 1983, **16**, 146.
32. Sorriso, S. In *The Chemistry of Functional Groups*; Patai, S. Ed.; Wiley: New York; 1982; suppl. F, Part 1, p. 1.

33. Lambert, J. B. *Top. Stereochemistry* 1971, **6**, 19.
34. Katritzky, A. R.; Patel, R. C.; Riddel, F. G. *J. Chem. Soc., Chem. Comm.* 1979, 674.
35. Eades, R. A.; Well, D. A.; Dixon, D. A.; Douglas, Jr., C. H. *J. Phys. Chem.* 1981, **85**, 976.
36. Ramachandran, G. N. *Biopolymers* 1968, **6**, 1494.
37. Ramachandran, G. N.; Lakshminarayanan, A. V.; Kolaskar, A. S. *Biochim. Biophys. Acta* 1973, **303**, 8.
38. Dunitz, J. D.; Winkler, F. K. *Acta Crystallogr., Sect. B* 1975, **B31**, 251.
39. Benedetti, E. *Chem. Biochem. Amino Acids, Pept., Proteins* 1982, **6**, 105.
40. Charton, M. *Progr. Phys. Org. Chem.* 1981, **13**, 119.
41. Hansch, C.; Leo, A.; Taft, R. W. *Chem. Rev.* 1991, **91**, 165.
42. Hutchinson, E. G.; Thornton, J. M. *Protein Sci.* 1994, **3**, 2207.
43. Sibanda, B. L.; Thornton, J. M. *Methods Enzymol.* 1991, **202**, 59.
44. Sibanda, B. L.; Thornton, J. M. *Nature* 1985, **316**, 170.
45. Baker, E. N.; Hubbard, R. E. *Prog. Biophys. Mol. Biol.* 1984, **44**, 97.
46. Mitchell, J. B. O.; Price, S. L. *J. Comp. Chem.* 1990, **11**, 1217.
47. Byler, D. M.; Susi, H. *Biopolymers* 1986, **25**, 469.
48. Surewicz, W. K.; Mantsch, H. H. *Biochim. Biophys. Acta* 1988, **952**, 115.
49. Van Stokkum, I. H. M.; Linsdell, H.; Hadden, J. M.; Haris, P. I.; Chapman, D.; Bloemendahl, M. *Biochemistry* 1995, **34**, 10508.
50. Dieudonne, D.; Gericke, A.; Flach, C. R.; Jiang, X.; Farid, R. S.; Mendelsohn, R. *J. Am. Chem. Soc.* 1998, **120**, 792.
51. Reisdorf, Jr., W. C.; Krimm, S. *Biochemistry* 1996, **35**, 1383.
52. Yoder, G.; Pancoska, P.; Keiderling, T. A. *Biochemistry* 1997, **36**, 15123.
53. Overman, S. A.; Thomas, Jr., G. J. *Biochemistry* 1998, **37**, 5654.
54. Wishart, D. S.; Sykes, B. D. *Methods Enzymol.* 1994, **239**, 363.
55. Greenberg, A.; Moore, D. T.; DuBois, T. D. *J Am. Chem. Soc.* 1996, **118**, 8658.
56. Wiberg, K. B.; Laidig, K. E. *J. Am. Chem. Soc.* 1987, **109**, 5935.
57. Cieplak, A. S. *J. Am. Chem. Soc.* 1981, **103**, 4540.
58. Cieplak, A. S.; Tait, B. D.; Johnson, C. R. *J. Am. Chem. Soc.* 1989, **111**, 8447.
59. Cieplak, A. S.; Wiberg, K. B. *J. Am. Chem. Soc.* 1992, **114**, 9226.
60. Cieplak, A. S. *J. Org. Chem.* 1998, **63**, 521.
61. Dyson, H. J.; Rance, M.; Houghten, R. A.; Lerner, R. A.; Wright, P. E. *J. Mol. Biol.* 1988, **201**, 161.
62. Wojcik, J.; Altmann, K.-H.; Scheraga, H. A. *Biopolymers* 1990, **30**, 121.
63. O'Neil, K.T.; DeGrado, W. F. *Science* 1990, **250**, 646.
64. Lyu, P. C.; Liff, M. I.; Marky, L. A.; Kallenbach, N. R. *Science* 1990, **250**, 669.
65. Merutka, G.; Lipton, W.; Shalongo, W.; Park, S. H.; Stellwagen, E. *Biochemistry* 1990, **29**, 7511.
66. Gans, P. J.; Lyu, P. C.; Manning, M. C.; Woody, R. W.; Kallenbach, N. R. *Biopolymers* 1991, **31**, 1605.

67. Bell, J. A.; Becktel, W. J.; Sauer, U.; Baase, W. A.; Matthews, B. W. *Biochemistry* 1992, **31**, 3590.
68. Serrano, L.; Sancho, J.; Hirshberg, M.; Fersht, A. R. *J. Mol. Biol.* 1992, **227**, 544.
69. Horovitz, A.; Matthews, J. M.; Fersht, A. R. *J. Mol. Biol.* 1992, **227**, 560.
70. Armstrong, K. M.; Baldwin, R. I. *Proc. Natl. Acad. Sci. U.S.A.* 1993, **90**, 11337.
71. Huyghues-Despointes, B. M. P.; Scholtz, J. M.; Baldwin, R. L. *Protein Sci.* 1993, **2**, 1604.
72. Sholtz, J. M.; Qian, H.; Robbins, V. H.; Baldwin, R. L. *Biochemistry* 1993, **32**, 9668.
73. Park, S. H.; Shalongo, W.; Stellwagen, E. *Biochemistry* 1993, **32**, 7048.
74. Park, S. H.; Shalongo, W.; Stellwagen, E. *Biochemistry* 1993, **32**, 12901.
75. Blaber, M.; Zhang, X. J.; Matthews, B. W. *Science* 1993, **260**, 1637.
76. Kim, C. A.; Berg, J. M. *Nature* 1993, **362**, 267.
77. Chakrabartty, A.; Kortemme, T.; Baldwin, R. L. *Protein Sci.* 1994, **3**, 843.
78. Blaber, M.; Zhang, X. J.; Lindstrom, J. D.; Pepiot, S. D.; Baase, W.; Matthews, B. W. *J. Mol. Biol.* 1994, **235**, 600.
79. Smith, C. K.; Withka, J. M.; Regan, L. *Biochemistry* 1994, **33**, 5510.
80. Regan, L. *Curr. Biol.* 1994, **4**, 656.
81. Minor, D. L. Jr; Kim, P. S. *Nature* 1994, **367**, 660.
82. Minor, D. L. Jr; Kim, P. S. *Nature* 1994, **371**, 264.
83. Padmanabhan, S.; York, E. J.; Gera, L.; Stewart, J. M.; Baldwin, R. L. *Biochemistry* 1994, **33**, 8604.
84. Predki, P. F.; Agrawal, V.; Brunger, A. T.; Regan, L. *Nature Struct. Biol.* 1996, **3**, 54.
85. Rohl, C.; Chakrabartty, A.; Baldwin, R. L. *Protein Sci.* 1996, **5**, 2623.
86. Myers, J. K.; Pace, C. N.; Scholtz, J. M. *Biochemistry* 1997, **36**, 10923.
87. Bhat, M. G.; Ganley, L. M.; Ledman, D. W.; Goodman, M. A.; Fox, R. O. *Biochemistry* 1997, **36**, 12167.
88. Moitra, J.; Szilak, L.; Krylov, D.; Vinson, C. *Biochemistry* 1997, **36**, 12567.
89. Myers, J. K.; Pace, C. N.; Scholtz, J. M. *Protein Sci.* 1998, **7**, 383.
90. Chakrabartty, A.; Baldwin, R. L. *Adv. Protein Chem.* 1995, **46**, 141.
91. Smith, C. K.; Regan, L. *Acc. Chem. Res.* 1997, **30**, 153.
92. Broger, C.; Muller, K. In *Structure Correlation*, Vol 2, Dunitz, J. D., Bürgi, H.-B. Eds.; Verlag Chemie: Weinheim, 1994; p. 685.
93. Ladunga, I.; Smith, R. F. *Protein Eng.* 1997, **10**, 187.
94. Almenningen, A.; Bastiansen, O.; Fernholt, L.; Cyvin, B. N.; Cyvin, S. J.; Samdal, S. *J. Mol. Struct.* 1985, **128**, 59.
95. Eaton, V. J.; Steele, D. *J. Chem. Soc., Faraday Trans. 2* 1973, 1601.
96. Szilak, L.; Moitra, J.; Krylov, D.; Vinson, C. *Nature Struct. Biol.* 1997, **4**, 112.
97. Richardson, J. S.; Richardson, D. C. *Science* 1988, 1648.
98. Wiberg, K. B.; Hammer, J. D.; Keith, T. A.; Zilm, K. *Tetrahedron Lett.* 1997, **38**, 323.
99. Glushka, J.; Lee, M.; Coffin, S.; Cowburn, D. *J. Am. Chem. Soc.* 1990, **112**, 2843.
100. Bai, Y.; Milne, J. S.; Mayne, L.; Englander, S. W. *Proteins* 1993, **17**, 75.

101. Creamer, T. P.; Rose, G. D. *Proc. Natl. Acad. Sci. U.S.A* 1992, **89**, 5937.
102. Creamer, T. P.; Rose, G. D. *Proteins* 1994, **19**, 85.
103. Aurora, R.; Creamer, T. P.; Srinivasan, R.; Rose, G. D. *J. Biol. Chem.* 1997, **272**, 1413.
104. Matthews, B. W. *Adv. Prot.* Chem. 1995, **46**, 249.
105. Bai, Y.; Englander, S. W. *Proteins* 1994, **18**, 262.
106. Avbelj, F.; Moult, J. *Biochemistry* 1995, **34**, 755.
107. Qian, H.; Chan, S. I. *J. Mol. Biol.* 1996, **261**, 279.
108. Luque, I.; Mayorga, O. L.; Freire, E. *Biochemistry* 1996, **35**, 13681.

CHAPTER 18

ROLE OF THE PEPTIDE BOND IN PROTEIN STRUCTURE AND FOLDING

NEVILLE R. KALLENBACH, ANTHONY J. BELL, JR., and ERIK J. SPEK
Department of Chemistry, New York University

1. INTRODUCTION

Proteins acquire their ability to function by folding to form specific native structures. Native structure has traditionally been described in terms of four "levels". The primary structure of a protein is determined by its amino acid sequence, including modifications to this sequence such as glycosylation, phosphorylation and/or formation of disulfide bonds. The secondary structure is defined as recurrent local conformations of the peptide backbone, forming α-helices, β-sheets and different types of turns. The tertiary structure describes the conformation of the side chains and reflects the overall protein structure, formed from interactions among helices, sheets and turns. Quaternary structure refers to the state of oligomerization of the protein, the number of chains that form the final structure.

The native structure of proteins turns out to be a delicate balance among large energetic contributions, each arising from summation of a great number of individually weak interactions. For instance, the native (N) and unfolded (U) forms of a protein molecule with 100 amino acids differ by less than 15 kcal mol^{-1}, while the chain entropy, hydrogen bonds, van der Waals forces, hydrophobic interactions and electrostatic forces individually can amount to hundreds of kcal mol^{-1}.[1] This makes it precarious to predict the native structure of a protein from its primary sequence simply by calculating its free energy of stabilization: small errors in calculating the energy or free energy of any of the component interactions can easily exceed the final $N \rightleftharpoons U$ free energy difference. At the same time, there is a pressing need to be able to predict folded structures from sequence alone. This would make it possible to translate the mass of

The Amide Linkage: Selected Structural Aspects in Chemistry, Biochemistry, and Materials Science, Edited by Arthur Greenberg, Curt M. Breneman, and Joel F. Liebman
ISBN 0-471-35893-2

genomic sequence data that is rapidly accumulating directly into protein structures. The situation at present is that sequence information alone is not sufficient to predict structure accurately. If highly accurate prediction of the overall structure of a protein is unattainable, can one predict the secondary structure in a protein? How important are individual propensities, the natural tendency of each amino acid to adopt one or another secondary structure, in folding of proteins? What other factors determine the extent of secondary structures in a protein? How are folding pathways and native structure related? Are certain stable structures unattainable because of non-existent folding pathways? In this review we will address each of these questions.

2. STRUCTURE

The primary structure of a protein is determined by its polypeptide chain and amino acid sequence. In the 1940s Pauling's group solved several dipeptide crystal structures and made the discovery that the peptide bond is planar because of the partial double-bond character of the carbon–nitrogen bond. The bonds between the peptide and the C_α can rotate more freely and this flexibility endows the peptide or protein with a significant conformational space. This space is restricted by the van der Waals radii of the backbone and pendant side chains. A polypeptide can then be approximated as a linked chain of rigid planar peptide groups. The dihedral angles between these amide planes are designated ϕ (C_α–N) and ψ (C_α–C) as depicted in Fig. 18.1. As in the case of many simple organic molecules, the set of conformations accessible is severely restricted, both by the properties of individual side chains and polymer effects such as the excluded volume. Certain peptide conformations are energetically favored over others. A staggered conformation of the peptide planes is favored over an eclipsed one. Except in rare cases, proteins do not form extended chains with a staggered conformation; this is because solvation and other interactions come into play to determine the final structure of a protein. The conformational space accessible to a dipeptide unit can be represented in a Ramachandran plot, named for its inventor (Fig. 18.2). The shaded area represents the allowed conformations based on calculations using the van der Waals radii of the polypeptide (glycine is excluded from this picture because its conformation space is much more extensive and not representative). As can be seen from the diagram in Fig. 18.2, two major conformational regions are generally accessible in a protein. These

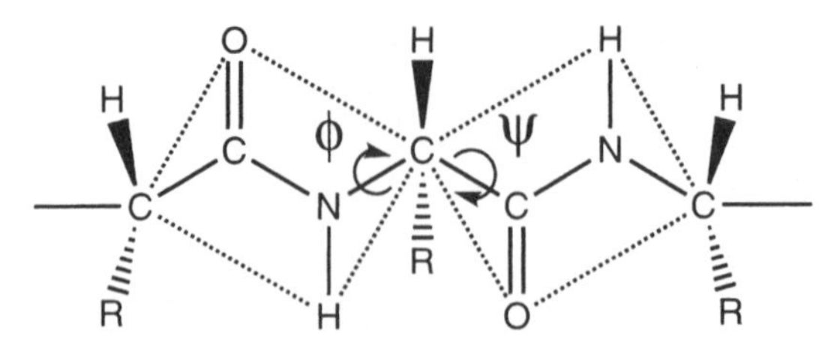

FIGURE 18.1. Conformation of a peptide bond showing ϕ and ψ rotation angles.

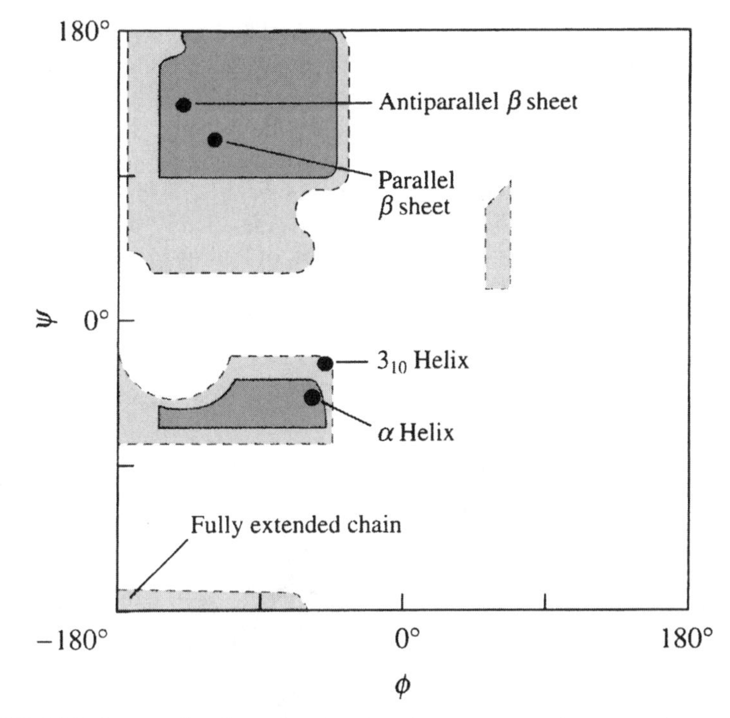

FIGURE 18.2. Ramanchandran plot showing the most occurring secondary structures. (After: *Moran and Scrimgeour: Biochemistry*, Prentice Hall, Englewood Cliffs, 1994.)

areas correspond to the standard secondary structures and are occupied by about half the residues in proteins of known structure.

The general organization of globular proteins consists of secondary structured regions connected by turns, very necessary for compact folding. Secondary structure formation ensures H-bonding of the polar groups in the backbone; sheets and helices both mandate networks of H-bonds. Polar side chains lie at the surface, while non-polar side chains can reside in the interior or at the surface. However, a significant fraction of protein surface is composed of nonpolar side chains. Non globular proteins tend to assume specific, often repetitive structures. Collagen for instance is built up from three strands tightly wrapped around each other. Each strand has a general Gly-X-Y repeated sequence. In this review we will focus on the more general case of globular proteins.

2.1. α-Helix

Given that helices, sheets and turns account for many if not most of the amino acids in globular proteins of known structure, one can ask whether these structures stabilize proteins and whether they exist outside native proteins. Polypeptide chains naturally adopt conformations that correspond to the lowest free energy, the minimal energy conformation compatible with folding constraints. Secondary structure is highly ordered and its formation is enthalpy driven; secondary structural signatures such as the circular dichroism and characteristic IR bands disappear as proteins unfold. The most frequently occurring element of secondary structure is the α-helix, first described by Pauling and his colleagues in 1951.[2] In the α-helix, the peptide chain forms a tight right handed helix with a pitch of 5.4 Å, 3.6 residues per turn and $\phi = -57°$, $\psi = -47°$, and the amino acid side chains directed outwards (Fig. 18.3). The helix is stabilized by a network of hydrogen bonds between the

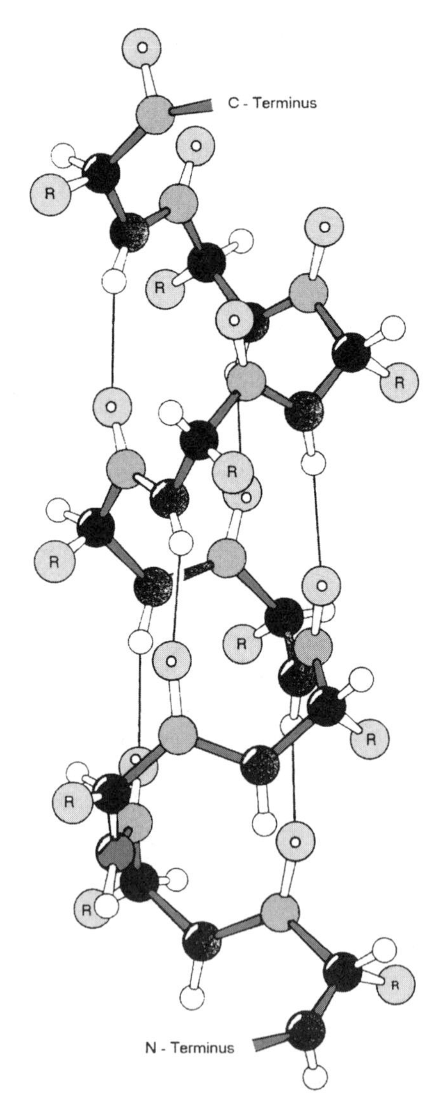

FIGURE 18.3. Right handed α-helix (After: *Moran and Scrimgeour: Biochemistry*, Prentice Hall, Englewood Cliffs, 1994.)

N–H and the C=O of the amide groups spaced four residues apart. Pauling himself emphasized the major contribution from H-bonding, which he anticipated would contribute more than $8\,\mathrm{kcal\,mol^{-1}}$ to helix stability.[3] Subsequent measurements have not borne this out, for reasons that need to be understood. Titration studies of the helix coil transition in pH dependent helical polypeptides such as poly-Glu and poly-Lys show that the enthalpy of helix formation is small, $1\,\mathrm{kcal\,mol^{-1}}$ or less.[4] Calorimetry of Ala rich oligomers containing Glu residues for solubility confirms this value.[5] Pauling's estimate was based on evaluation of the energy of H bonds between NH and O groups in liquids, a reasonable model. One has to bear in mind that if H-bonding contributed $8\,\mathrm{kcal\,mol^{-1}}$ to helix stability, protein structure might be so stable that its conformational flexibility might be lost, impairing its functional capability. Why is helical H-bonding so much weaker than the predicted value? One possibility is strain: helical structure conceivably allows only imperfect H-bond formation, such that strain energy cancels most of the intrinsic bonding free energy available in small molecular systems.

Additional helix stabilization comes from van der Waals interactions in the core, which partly account for the favorable regions in Ramachandran plots. Alternative helical secondary structures are possible too, and occur in native proteins, but are rare. The 3_{10} helix incorporates an $(i, i+3)$ hydrogen bonding pattern while the π-helix has an $(i, i+5)$ system. The conformation angles in both structures are less favorable than those of the α-helix, and there is severe steric clashing in propagating extended 3_{10} structure. The flattened π-helix gives rise to a cavity in the center that is probably destabilizing. Both of these alternative helical structures are much less abundant in native proteins than α-helix or β-sheets.

Each amino acid can be considered to have an intrinsic tendency to adopt an α-helix conformation. The quantity referred to as the α-helix propensity of an amino acid is defined as the equilibrium constant for extending helix with the added residue in helical conformation. This constant includes side chain–backbone interactions that occur when the residue assumes helical conformation. Glycine and proline have the lowest propensities among natural amino acids for reasons that seem clear: glycine, because it has extensive regions of Φ, Ψ space accessible, hence is less likely to adopt an ordered helical conformation. Proline, because its rigidity precludes adopting an unstrained α-helix conformation. Alanine has been identified as a common amino acid in α helical protein sequences as shown in a statistical survey by Chou and Fasman[6] and later surveys.[7] A variety of experiments based on short peptides with natural side chains implicate Ala as a highly helix stabilizing natural residue.[8–11] The methyl side chain of Ala blocks access of solvent to the backbone and has a favorable van der Waals radius allowing it to adopt helix conformation, yet lacks the unfavorable free energy that penalizes larger nonpolar side chains in the helix conformation.[12]

The first systematic analysis of helix propensities arrived at the conclusion that alanine is, however, a helix indifferent side chain. The systematic determination of helix propensities was undertaken initially by Scheraga's group.[13] They copolymerized each of the natural amino acids with host side chains consisting of hydroxy-propylated (HPG) or hydroxy-butylated (HBG) glutamine, which alone form soluble stable helix structure. Poly-Ala for example is not soluble in water and its propensity has therefore to be measured indirectly. The propensities for each amino acid were determined by analyzing the melting temperature and the transition profiles of the circular dichroism (CD) signal of copolymers with varying composition. Helical structure is conventionally dissected into two processes: helix nucleation, and helix propagation.[14,15] In order to nucleate a helix from disordered chain, six dihedral angles need to lie within the helix region of the Ramachandran plot; once a helix is nucleated, propagating it requires restricting only two additional dihedral angles (see Fig. 18.2). In each case, values of the helix propensity (s) and nucleation constant (σ) for the guest side chain could be fit to the transition data, once values of s and σ for the host species were determined from homopolymer experiments. The results are summarized in Table 18.1. Most amino acids are helix indifferent or destabilizing as guests in HPG or HBG polypeptides, while bulky hydrophobic side chains such as Leu and Met are stabilizing. Alanine in particular is not helix stabilizing in this background, although it is one of the most frequently occurring amino acid in helices.[6] One explanation for the discrepancy between the helix stabilizing effect and frequency of side chains might be simply that the determination of helical structure in proteins is unrelated to that in polypeptides. Alternatively, the host side chains HPG and HBG used in these experiments interact with neighboring guest side chains and distort the scale of helix stability. Analysis of short helical peptides containing HPG and HBG indicates that these show strong side chain-side chain interactions that influence the apparent propensity of neighboring side chains.[16] One concrete prediction from the polypeptide host-guest studies is that helix formation in short peptides (for example $n < 30$ AA's) containing Ala side chains is improbable.

Subsequently several new model systems were introduced to determine helix propensities: short alanine rich peptides by the group of Baldwin[8,11] or Glu, Lys block peptides.[9] Both were synthesized by solid phase peptide methods. The models were used as test-beds for determining helix propensities, principally by deconvoluting CD spectral data[17,18] and they both contradicted predictions from HBG/HPG polypeptides. Baldwin's models consist of a set of short alanine based peptides in which particular sites in the middle or end of the helix could be substituted by 20 different amino acids. Solubility is provided for by introducing glutamine or charged amino acids such as Glu or Lys. As can be seen in Table 18.1, alanine and arginine are helix stabilizing in these peptides, and as in most scales the other residues are neutral or destabilizing. An independent study of helix propensity in coiled-coil models was carried out by O'Neil and DeGrado[19] who designed a dimeric helical peptide which is in equilibrium with the random coil monomers. The propensity scale derived from this system agrees reasonably with data obtained from other peptide studies and with the statistically derived scale of Chou and Fasman. The scale does not correlate with the scale originally

TABLE 18.1. Free Energy of α-Helix Formation*

	Fasman[6]	Scheraga[13]	Baldwin[11]	Kallenbach[37]	Stellwagen[10]	DeGrado[19]	Matthews[22]
Ala	1.59	−0.04	−0.27	−0.36	−0.42	−0.77	−0.96
Arg	0.67	−0.02	−0.05	0.05	−0.39	−0.68	−0.77
Asn	0.53	0.15	0.69	0.49	0.24	−0.07	−0.39
Asp	0.53	0.23	0.54	0.42	0.18	−0.15	−0.42
Cys	0.33	0.01	0.64	0.81	0.06	−0.43	−0.42
Gln	0.98	0.01	0.28	−0.01	−0.17	−0.33	−0.80
Glu	1.45	0.02	0.35	−0.15	−0.22	−0.27	−0.53
Gly	0.53	0.31	1.7	0.69	0.59	0	0
His	0.87	0.22	0.84	0.18	0.22	−0.06	−0.57
Ile	1.22	−0.08	0.44	0.02	−0.02	−0.23	−0.84
Leu	1.91	−0.08	0.10	0.00	−0.27	−0.62	−0.92
Lys	1.13	0.04	0.02	0.23	−0.24	−0.65	−0.73
Met	1.25	−0.11	0.25	0.17	−0.21	−0.50	−0.86
Phe	1.14	−0.05	0.73	0.75	−0.24	−0.41	−0.59
Pro	0	1.10	>3.8	2.2			
Ser	0.7	0.16	0.52	0.40	0.10	−0.35	−0.53
Thr	0.75	0.12	0.95	0.53	0.24	−0.11	−0.54
Trp	1.33	−0.06	0.69	0.75	−0.12	−0.45	−0.58
Tyr	0.58	−0.01	0.42	1.59	0.07	−0.49	−0.72
Val	1.42	0.03	0.77	0.37	0.24	−0.14	−0.63

*All values in ΔG (kcal mol^{-1}) except Fasman (statistical probability); ΔG is related to the helix propensity according to $\Delta G = -RT \ln s$.

proposed by Scheraga's group. More recently a third approach has been developed by the group of Kemp.[20] They synthesized a template group that can nucleate helical structure in short peptide chains in one of its conformations, but not in the other(s). Propensities are evaluated by monitoring the relative population of the two conformations of the template in the presence of short stretches of appropriate amino acid sequence. A multi-state helix-coil analysis is applied to deconvolute the data. Interestingly, the propensity derived for alanine in Kemp's system proves to be essentially identical to that obtained by Scheraga, $s_{Ala} = 1.07$, seeming to confirm the conclusion that alanine is intrinsically indifferent to helical structure. Different explanations for the discrepancy between different models have been offered: according to calculations by Scheraga, the presence of charged or polar side chains such as Gln or Glu in Ala rich peptides makes neighboring Ala side chains helix stabilizing.[21] According to Kemp, long side chains such as that of Lys interact in a special way with the helix backbone and enhance the apparent helicity of proximal Ala groups. The charged NH_3^+ group associates with the CO of the main chain, while the methylenes interact with the CH_3 groups of vicinal Ala's, in what they refer to as a helix–barrel interaction. Whether or not helix barrel interactions occur, the issue remains whether Ala "intrinsically" stabilizes α-helical structure or not.

Several studies have taken up the question of helix propensities in proteins, analyzing the stability of proteins by amino acid substitutions in helical sequences. The most detailed such study has been produced by the group of Matthews.[22] They substituted two exposed helical sites on the surface of phage T4 lysozyme with each of the 20 natural amino acids. Stability was monitored by thermal unfolding, assuming two state unfolding of the proteins. Structure was monitored by X-ray crystallography of each mutant protein. Both positions produced consistent propensity scales, with alanine as the most stabilizing residue. The scales correlate well with results from peptide studies, other protein substitutions, and even with the Chou–Fasman scale, but do not correlate with propensities determined by Scheraga, Kemp and their coworkers. Thus, there remain fundamental discrepancies among different scales of helix propensity and the debate is ongoing. Because each scale is based on different model systems, they cannot be compared easily. A global approach presented by Munoz and Serrano[23] does not reconcile the discrepancies we have discussed relative to the role of Ala for example. More detailed information on the various model systems and algorithms that have been used to determine propensities of natural as well as unnatural amino acids can be found in the review by Kallenbach et al.[24] Many unnatural amino acids have been investigated in terms of helix propensity or their ability to interact with natural side chains.

Apart from intrinsic propensities several other factors significantly influence helix stabilization: packing interactions between side chains spaced appropriately, capping of the ends of helices, and electrostatic interactions between side chains, and between side chains and the helix dipole. Electrostatic interactions affect helicity in at least two ways: charge–dipole interactions, whereby charged residues interact with the dipole moment of the helix itself, and inter-side chain salt bridges between pairs or multiples of charged residues. The former interaction reflects the fact that in an α helix, the dipole moments of each of the NH and CO groups sum up to give an effective "macrodipole" with a positive partial charge at the N-terminus and a negative partial charge at the C-terminus. It was shown in studies of short helical fragments of RNase A that α-helix is stabilized by a negatively charged side chain placed at the N-terminus or a positively charged side chain at the C-terminus, and destabilized by the opposite arrangement.[25] Interactions that enhance the dipole moment destabilize helix structure, and vice versa. Due to the form of the dipolar potential, the effect is most significant if the charged amino acid residues are located at the ultimate or penultimate residue of the helix and less significant otherwise.[26,27] The second interaction involves formation of ion pairs between appropriately spaced charged side chains which influence the helix through mutual interactions. Helices are stabilized by salt bridges formed between pairs of opposite charged side chains, including Glu–Lys and Asp–Arg, located at sites at spacings (i, $i+3$ or i, $i+4$) in a helix. On the other hand, charges of the same sign at sites spaced (i, $i+1$) along the helix may destabilize helix structure. The stabilizing salt bridge interaction includes a neutral hydrogen bond and an ionic interaction component, so that titration of the acidic side chain does not completely eliminate pairing.[28,29] The strength of a salt bridge depends on the number and nature of the amino acids involved, their spacing and their overall position within the helix, as a consequence of the dipole interaction. Surface salt bridges can contribute roughly $0.5\,\mathrm{kcal\,mol^{-1}}$ to helical structure.[28,29] Evidence from thermophilic proteins suggests that complex salt bridges may be more stabilizing than simple ion-pairing.[30] More complex salt bridge arrangements can indeed involve greater free energy contributions. A third side chain-side chain interaction involves long nonpolar groups spaced so that they interact along an α-helix.[31,32] Strong effects between rigid aromatic side chains are seen, for example, Trp and His.

In general, helical structure is cooperative, with nucleation constants in the range of 10^{-3} or so. Bonds in the interior of a helix form a cage, and are harder to break than those at an end. Helix structure frays at the ends: in short helices (the mean length of helices in globular proteins is only about 12 amino acids) four NH groups and four CO groups cannot be satisfied by the main chain H-bonding pattern of Pauling's model. Fraying can be counteracted by interactions that restrict the conformational freedom at the ends of helices, often involving specific structures called "capping boxes". One strong capping "box" motif at the N-terminus of helices in proteins consists of the sequence Ser (Thr)-X-X-Glu (Asp).[33,34] In this motif, the serine side chain folds back to form a hydrogen bond with the main chain amide group of the acidic side chain, while the acidic side chain hydrogen bonds with the main chain NH of the serine or threonine. The net effect is to form a bend or turn at the N-terminus, a structure that can be seen in helices in a number of proteins including myoglobin. A common C terminal motif occurs when glycine is the penultimate residue, because the conformational flexibility of Gly allows the final residue to fold backwards and interact with backbone groups of the helix.[35,36]

The thrust of the above studies is that helix stability is a complex but determinable function of the intrinsic propensity of a side chain together with its interactions with side chains up to four positions in either direction along the helix, including capping positions. A satisfactory algorithm needs to include a large library of interaction free energies as well as propensities; a necessary test of such a program is that it will be able to predict helicity of any peptide from the sequence alone. No such program is currently available; a measure of the deficiency can be taken by the apparent context dependence of propensity values.[37] Nevertheless, the frequent use of helix prediction programs, such as AGADIR[23] or by the Baldwin group[38] indicates the ongoing interest in this problem.

2.2. β-Sheets

The second major group of secondary structures consists of β-sheets. In contrast to α-helices, hydrogen bonds are formed between different strands capable of spanning distant regions of a chain. Sheets occur in both parallel and antiparallel fashion as depicted in Fig. 18.4. Sheets in principle can build up indefinitely as in silk; a variety of sizes is found in proteins from simple two-stranded turns to complex networks within β-barrels. The sheets within protein tend to be twisted rather than flat; the flattened sheets seen in silk contain the compact Gly side chain. As in the case of helices, some residues are more likely to adopt a β-sheet conformation than others.[6] However, β-sheet propensities have proven more difficult to assess than helix propensities because the interactions among side chains in sheets are stronger than in helices.[39,40] Threonine for example has a high probability of β-sheet occurrence, but its ability to stabilize sheet structure depends on side chain–side chain interactions at neighboring sites. A β-sheet propensity scale was proposed by Kim and Berg.[41] They employed a solvent exposed (edge) site in a β-sheet zinc finger peptide. The peptide has a metal binding site so that the thermodynamics of metal binding reflect the peptide folding energies, allowing construction of a β-sheet propensity scale. In Kim and Berg's scale, Ile was the most stabilizing and Ala the least (excluding Pro and Gly). The range of stability differences between Ile and Ala that they find is only 0.2 kcal mol^{-1}, significantly smaller than the range for α-helices. There is a reasonable correlation between this scale and the statistically derived Chou–Fasman scale. A different scale was derived using the IgG binding domain of protein G as a model.[39] Each of the amino acids was substituted on a solvent exposed sheet site to obtain a thermodynamic scale of β-sheet propensities. In this model the substituted site lies in the middle of a three stranded sheet structure in which a larger number of side chain–side chain interactions are possible. To minimize these interactions the residues proximal to the substitution site were changed into small neutral residues such as alanines and serines. Thermodynamic data were obtained by thermal unfolding

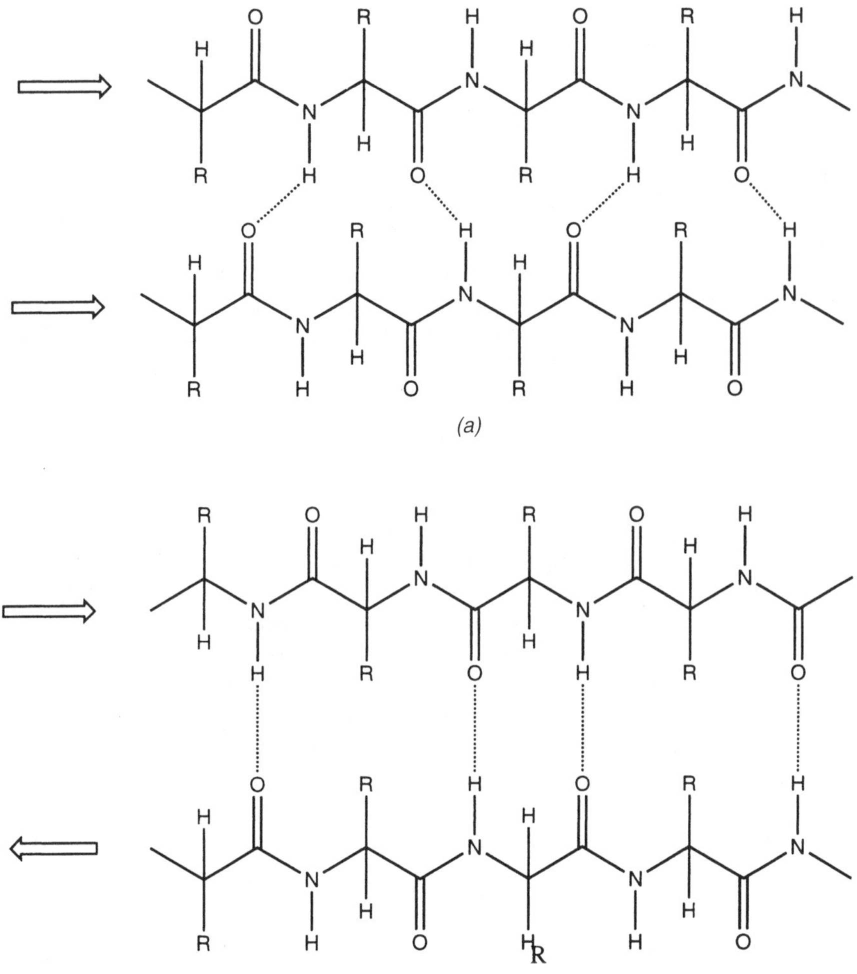

(a)

(b)

FIGURE 18.4. Parallel (*a*) and Antiparallel (*b*) β-sheets.

monitored by CD. The scale has a much larger range in stabilities (almost 2 kcal mol^{-1}) than the scale determined by Kim and Berg. There is only a modest correlation between this scale, the scales of Kim and Berg and the statistically derived Chou–Fasman scale. As in the case of α-helical models, it is not straightforward to compare various model systems, lacking a complete library of side chain–side chain interactions. Smith and Regan[40] investigated the role of side chain–side chain interactions in β-sheet formation using the same IgG binding subdomain of protein G as a model. They constructed pairwise mutations in the 4 stranded β-sheet. Pairs with Thr ranked lower in stability than expected. While the intrinsic propensity for Thr is high, its side chain–side chain interactions are unfavorable. Aliphatic pairs such as Phe–Phe and Ile–Tyr show significantly stabilizing side chain–side chain interactions, as do charged pairs such as Glu–Arg and Glu–Lys that have a low β-sheet propensity. The strongest overall stabilization is observed by Phe–Tyr and Phe–Phe pairs. These sets combine favorable β-sheet propensities with favorable side chain–side chain interactions. The strength of the side chain interactions makes it clear that interactions other than intrinsic β-sheet propensities play a crucial role in β-sheet formation. This is true of helices as well. However, a propensity scale can only be useful if the effect of modulating side chain–side chain and other interactions is roughly comparable in free energy to the propensity values themselves. If the modulating interactions are ten times stronger than intrinsic propensities, for example, a propensity scale would have limited significance. Finally, it should be pointed out that the ends of sheets can show capping effects, just as those of helices do.

2.3. Turns

Globular protein structure requires frequent turns, allowing the polypeptide to assume a compact overall shape despite the extended forms of helices or sheets. In principle, turns are the most obvious features in a protein sequence: they contain polar side chains, since they occur at the surface, frequently with Gly or Pro residues.[1] One question of interest is whether turns are active or passive: Do they stabilize native protein structure or not? This issue has been addressed by mutational studies of turn motifs in four helix bundle type proteins[42] and by analysis of turns using a phage display system.[43] The answer depends on context, unfortunately. Four helix bundle proteins contain four sequential helical domains, linked by turns. A detailed mutational study reveals that the turns that span the helices in these proteins show no evidence for selectivity in their turn sequences. Any of a wide range of side chains can fulfill the requirement to form a turn. This is not the case for turns spanning a β-sheet region: specific side chains stabilize turns more than others and the differences are significant. Thus some turns are active, seemingly, while others are passive. However, the experiments on four helix bundles did not evaluate the stability of different turns directly: the screen used simply selected folded proteins over unfolded ones. The phage display experiment evaluated the relative stability of individual turn sequences, and so the two systems may not be so far apart in their conclusions.

2.4. Implications for Protein Structure

Despite progress in identifying helix stabilizing side chains and many interactions that mediate helix stability, it is still not possible to predict α-helices, β-sheets and turns by inspection of primary amino acid sequences. Several algorithms have been developed to predict secondary structure in proteins with limited success.[23,38,44,45] An alanine rich stretch may or may not form an α-helix, and a threonine rich stretch may or may not form a β-sheet. The difficulty is that secondary structure is only marginally stable. Analysis of substitutions in helical proteins shows that simple patterns of hydrophobicity or hydrophilicity influence secondary structure formation, at least at a crude level.[46] Thus, it is important to underline the conditional nature of helix propensities. A run of five Ala residues should signal helix from the results described above; in fact, in the context of spider drag-line silk proteins, this number of adjacent Ala side chains forms a β-sheet.[47] Many of the helical residues in membranes, or the interior of globular proteins, are fully or partially buried. These helices are stabilized by hydrophobic and van der Waals interactions that cannot be predicted using propensity scales derived from short model helices in water.[48] The determinants of secondary structure in membrane environments reflect several differences from the behavior of peptides or proteins in water: for example, proline is not rare in trans-membrane helices, many side chains appear to be helix stabilizing in addition to Ala, and polar side chains are rare, as expected. Contrary to scales derived from experiments in water, Ile and Val are helix stabilizing, and Ala much less so.[48] On the other hand, capping interactions can be detected at the boundaries of membrane spanning helices, for example in glycophorin, suggesting that these motifs are preserved. The role of H-bonding in principle might be enhanced within a membrane. If one can extrapolate from studies of helical peptides in the helix stabilizing solvent TFE (trifluoroethanol), the H-bonding contribution increases by 0.2 kcal mol^{-1}.[49]

We have pointed out that the native state of proteins is only slightly more stable than unfolded states. Since the seminal review by Kauzmann in 1959,[50] the major reason for the stabilization of native proteins has been attributed to the hydrophobic effect in which burial of nonpolar side chains from water in the interior of a protein drives formation of native structure. In one extreme view, secondary structure in fact is supposed to arise entirely as a consequence of hydrophobicity,[51] driven purely by compaction of the folding chain. This view is contradicted by a wealth of experimental studies on substitutions in proteins which show consistency with scales of helix propensity from peptide models.[52] The classical hydrophobic effect postulates that a thermodynamically unfavorable entropy results upon transfer of nonpolar groups into water, giving rise to a high heat capacity for protein folding. Folding a protein is certainly accompanied by a major loss in cratic entropy for the polypeptide chain itself. This is compensated by an entropy gain upon desolvating the hydrophobic side chains from their partially aqueous environment. Along with the hydrophobic side chains, parts of the peptide backbone are also transferred from water to the nonpolar environment. This transfer is unfavorable since the peptide backbone itself is polar and opposes burial in a hydrophobic milieu.[53] This effect is thought to be enthalpically driven, and enthalpic effects arguably dominate in folding a protein despite the evident hydrophobic contribution to heat capacity. The temperature dependence of the hydrophobic effect gives rise to the general phenomenon of cold unfolding, in which native structure is destabilized by low temperatures as well as high temperatures. The explanation is that below room temperature there is a reversal of the usual unfavorable free energy of transfer of nonpolar side chains to water.

Why should a polypeptide chain depart from its potentially more stable staggered conformation in order to adopt what seems to be a sterically and energetically less favored alternative? One major player in the stabilization of backbone conformation is the hydrogen bond. As mentioned before H-bonding is a major feature of secondary structure. Secondary structural elements become more stable when a protein is transferred from water to ethanol which cannot compete as effectively as water for hydrogen bonds, favoring the hydrogen bonds within the protein. However, high concentrations of alcohols and TFE destabilize the native state in proteins because of their effect on water structure, regardless of the stabilization of helices or sheets. Other hydrogen bonds can occur between any hydrogen acceptor and donor pair in the protein either on side chains or on the backbone and define the tertiary structure of the protein.

It is now appreciated that a protein fold also depends on details such as van der Waals interactions rather than hydrophobicity per se. One reason is that scales of hydrophobicity have evolved since the first estimates based on octanol–water partition coefficients. Mutational experiments and sequence comparisons show that the identity of hydrophobic side chains in the interior of proteins is not conserved. Repacking the core of proteins is generally possible without disruption of the native structure;[54] proteins are remarkably tolerant of multiple substitutions among bulky side chains in their interior. One view of folding is that the simple patterning or distribution of polar versus nonpolar side chains alone establishes the native fold.[46] However, this view is oversimplified. Selective van der Waals interactions have been shown to exist most clearly in forming the more extended hydrophobic core of α helical coiled-coil proteins, referred to as leucine zippers.[55] These are chains containing repeating sets of seven amino acids, $(\mathbf{abcdefg})_n$, in which the **a** and **d** residues are nonpolar groups, frequently leucine, that associate to form dimers. Replacing Leu by Ile can switch the stoichiometry of a coiled coil from dimers to trimers or even tetramers. Why? The effect resides in fine details of the packing between the side chains in the core of the protein, not simple hydrophobicity or patterning of hydrophobic side chains. It is possible that the interior of a globular protein may be a more fluid environment than that of a coiled-coil which demands extended and sterically stringent knob-into-hole packing of Leu or its replacement side chains.

3. PROTEIN FOLDING

3.1. Introduction

How is the native structure (N) of a protein formed? *In vivo*, folding accompanies the biosynthesis of a protein, on the surface of a ribosome. This reaction bears little resemblance to refolding experiments *in vitro*, and would be difficult to study experimentally. *In vivo* protein chaperones are available to divert misfolded chains from aggregating.[56–58] In experimental studies of folding, a solution of unfolded protein, at a pH, temperature or in a solvent favoring the unfolded state, is mixed with solvent favoring native structure to initiate the folding reaction. Rapid mixing devices make it possible to accomplish this in milliseconds, and more recently devices have been built to mix solutions in microseconds. The progress of the reaction monitored by spectroscopic signals allows determination of the bulk folding rate detected by the signal that is followed: the CD signal at 222 nm is often used to follow helix formation, fluorescence at 340 nm reports on the burial of Trp side chains, etc. In general, the folding rates detected by different probes are not identical, indicating the presence of intermediates in the reaction. For very small proteins, the rates are nearly identical and there is no substantial population of intermediates. Other strategies have proved more informative than simple spectroscopic monitoring of folding. Trapping transient intermediates by using sulfhydryl reagents in the case of SH containing proteins,[59–61] or labeling amide NH groups within an intermediate with H or D isotope by exchange with solvent presents a higher resolution view of folding.[62–64] The strategy is to label the unfolded chain in D_2O for example, rapidly mix the solution with H_2O in conditions that allow folding and exchange, then quench the reaction in a stopped-flow device, slowing exchange by a pH shift. The particular sites which have folded are labeled with H isotope and then can be identified by NMR spectroscopy. This procedure allows definition of the extent of folding at individual residues in a protein, and provides the most detailed picture that is presently available.

The traditional view of the refolding reaction is that early in the process, the extended chain collapses into a compact state, driven by unfavorable interactions similar to those experienced by any polymer chain in a poor solvent. Subsequently, the compact intermediate is thought to anneal in order to maximize van der Waals and favorable electrostatic interactions. The first step can be extremely rapid.[65] Depending on the protein, the second phase can also be fast, and complete folding of small proteins can be achieved within milliseconds.[66] In other cases, folding requires a time scale of hours, as has been reported for some β-sheet proteins.[67] A complete understanding of folding entails specification of the order of intermediates in a pathway, a detailed description of each intermediate, as well as the population of molecules following a pathway if several parallel reaction channels are available. For a reaction in which hundreds or thousands of atoms organize on a fast time scale, elucidating a folding mechanism presents a formidable challenge despite the power of the methodology available.

3.2. Early Protein Folding Proposals

An early and simple model of folding postulated that nucleation occurs by forming specific scaffold-like intermediates, which assemble to form the final folded state.[68] This framework model accounts for many features of folding: the presence of intermediates that are partially ordered, the assembly of a core, and the possibility of trapping misfolded states by slow *cis–trans* isomerization of prolines or other interactions. Proceeding along a pathway involving a defined series of progressively more and more ordered intermediate states obviates Levinthal's paradox as well. Levinthal pointed out that any extensive search through the vast conformational manifold accessible to even a small random coiled unfolded protein would take so long that no protein could fold in a reasonable time scale.[69] His point was that intermediates are essential. Moreover, even in denaturing solvents unfolded proteins do not lose all identity with the folded state; in 7 M urea for example, a repressor protein is found to retain a globular shape and even rudiments of a hydrophobic core.[70] Thus the unfolded polypeptide chain does not have to explore more than a tiny fraction of the phase space accessible to the random coil and even this is close to that occupied by the native state. Proceeding through a series of sequential intermediates further simplifies the problem.

However, there are challenges to this simplistic view. First, the smallest known proteins fold in two state fashion, with no evidence for intermediates.[71] Plots of the rate of folding or unfolding as a function of denaturant concentration yield characteristic V shaped "chevron" curves, consistent with the absence of stable intermediates. Larger proteins such as the ribonuclease barnase show distinct intermediates. In both cases, the nature of the transition state can be defined by measuring the kinetic effects of mutations.[71] It is perhaps simplistic to imagine a traditional small molecule transition state as applicable to the folding reaction of a protein. Surprisingly most folding reactions that have been studied to date show simple exponential rate behavior, implying some large energy barrier in the rate limiting step, however complex this step might be. In a classical series of experiments in which the heme ligand of myoglobin was transiently dissociated by a flash of light over a very wide range of temperatures, Frauenfelder's group repeatedly detected non-exponential rate processes, revealing extensive conformational mobility within the native state "manifold".[72] Their term for the most extreme processes that occur in the native state was "quakes"—fluctuations of the native state so massive that their kinetics are sensitive to the viscosity of the solvent. Above about 200 K, the discrete conformation changes are not detected, although similar processes must still occur. Thus the bounds on the native state and its conformational repertory are not well understood. Preparing a protein in an unfolded state leads to a much higher diversity in conformations, and it would be natural to expect non-exponential behavior in some cases at least.

Second, are all folding intermediates native-like or not? Early seminal disulfide trapping experiments by Creighton's group indicated that the pancreatic trypsin inhibitor BPTI folds via obligatory formation of an intermediate that contains a combination of disulfide bonds that is not present in the final native structure.[73] This conclusion was weakened by subsequent analysis which showed that in fact only a small fraction of chains form this non-native like intermediate.[74] So far, there is no convincing evidence that non-native states mediate folding, although the possibility exists in principle. Uncovering even a small population of such intermediates raises the possibility that different molecules in an ensemble of folding proteins follow different paths in folding. In a sense this rephrases the question discussed above concerning the bounds of the native set of conformations. Are the barriers between intermediates so great that the order is invariant? This seems unlikely given the complexity of real proteins, and there is experimental evidence for heterogeneous paths in refolding of lysozyme for example.[75] A peptide model with some features of an intermediate in BPTI folding was studied by Oas and Kim.[76] Their 30 residue analogue (PαPβ) was designed to mimic a highly populated intermediate found along the BPTI folding pathway [30–51]. The intermediate, [30–51], contains a disulfide linkage between residues 30 and 51. After initial collapse, sequential intermediates are thought to form until the native structure emerges. This relationship is depicted in Fig. 18.5. In both PαPβ and [30–51] a disulfide bond connects the C-terminal α-helical region to a strand of the central antiparallel β-sheet. CD and NMR spectroscopy indicate that PαPβ adopts a conformation containing definite secondary structure with native-like contacts yet lacking specific long-range tertiary structure. This result suggests that the BPTI pathway conforms to a frame-work model, in which the most populated intermediates are native-like.

3.3. Molten Globule State

Ptitsyn and his colleagues have argued cogently that intermediates in folding share common features: they contain secondary structure but not rigid tertiary packing, they are more expanded in volume than the native conformation, and they expose more hydrophobic surface than the native state.[77] This led them to propose that molten globules serve as a universal intermediate state in protein folding. Molten globules are postulated to arise early in the protein folding pathway, and to be stable under mild denaturing conditions. Stable molten globules can populate for example under mild acidic or basic conditions, or in

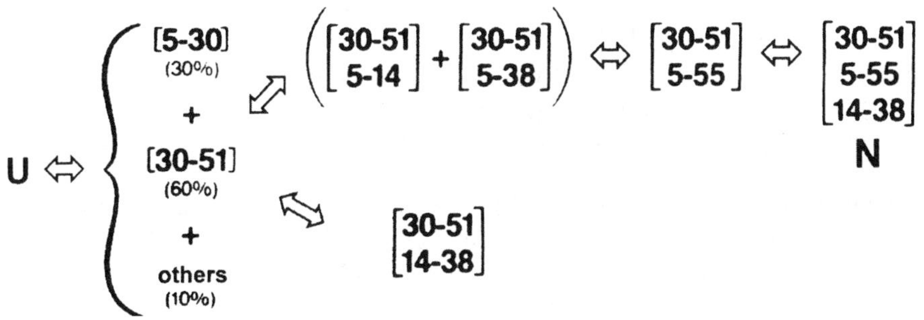

FIGURE 18.5. Folding pathway of BPTI intermediate [30–51]. U depicts the unfolded, reduced protein. N represents the native, folded protein. (After: Oas, T. G.; Kim, P. S. *Nature* 1988, **336**, 42.)

the presence of concentrations of urea or guanidinium chloride that do not produce complete unfolding at neutral pH, or in the absence of a ligand.[78]

There are several identifying characteristics of the molten globule state, including:

1. *Significant secondary structure.* For example, β-Lactamase (neutral pH, 2 M urea) displays far UV ellipticity that is 85% analogous to native form;[79]
2. *Absence of well-defined tertiary packing.* In particular aromatic residues have a larger degree of solvent accessibility than in the native state.[80] NMR studies of the acid stable state of α-lactalbumin indicate an appreciable secondary structure that is stable in comparison to isolated peptides, while there is mobility in the hydrophobic core and significant disorder in other regions of the protein;[81]
3. *Compactness relative to unfolded states.* The dimensions of molten globules have been investigated by a number of experimental methods including X-ray scattering,[82] gel exclusion chromatography,[77] viscosity,[83] urea gel electrophoresis,[84] and light scattering.[85] In each case the MG is smaller than U, and larger than N.
4. *Limited solubility.* Aggregation of globules is expected because of the potential intermolecular interactions of exposed hydrophobic residues.

One difficulty with the concept of molten globules is that it seems to lack consistency: for example substitution experiments on the molten globule intermediates in α-lactalbumin show that the molten globule state folds noncooperatively,[86] while similar experiments in apomyoglobin show cooperative folding behavior.[87] The helical apomyoglobin molecule was thought to represent a molten globule[88] but in fact possesses a well defined folded structure.[89] In this case there is clear evidence favoring a sequential assembly pathway, in which the H and G helices fold to form an intermediate that anchors subsequent packing of the A helix and the remaining helices in the protein.

The folding pathway of myoglobin has been analyzed by mutation[87] and by use of synthetic peptide intermediates. Myoglobin contains a well-defined helical hairpin near its carboxy terminus. This region involves interactions between the G (residues 100–118) and H (residues 124–150) helices. Theoretical and experimental studies implicate this region as a folding initiation site.[90] A series of peptide models have been employed to investigate the initiation site,[91] reverse turn,[92] and long-range interactions of myoglobin.[93]

In order to monitor early forming myoglobin intermediates, Waltho et al.[91] constructed peptide fragments corresponding to the G and H helices. CD and NMR data indicate that the H fragment contains a considerable amount of helicity, while the G fragment does not. Monomeric H peptide acts as a fast forming highly populated intermediate with a defined secondary structure, a likely candidate for a framework intermediate in myoglobin folding. Shin et al.[91]

investigated the long range interactions involved in myoglobin folding by constructing a 51-residue fragment spanning the entire G–H helical hairpin including the turn sequence between the two helices. The G–H helical hairpin was modeled in two ways: in the absence (Mb-GssH) and in the presence of the helical turn region (Mb-GH51). In Mb-GssH, a disulfide bond was employed to connect the G and H helices. CD and NMR data indicate stabilization of intermediate structures via long range interactions only in the disulfide linked peptide. This is again consistent with a framework model, in which tertiary interactions are essential to promote further folding in the pathway. Detailed mutational analysis by Kay and Baldwin shows that apomyoglobin folds in a cooperative process.[87]

In their mutational analysis of the molten globule state of α-lactalbumin, Schulman and Kim[86] introduced single and multiple prolines into the sequence in order to destabilize particular helical regions in the protein. Proline substitutions could be shown to knock out individual helices within the globule. However, the surprising result they obtained was that knocking out individual helices had little influence on the residual structure in the molten globule. Knock-outs of more than one helix produced an additive effect on the overall structure present. The conclusion is that molten globule structure must form in a noncooperative manner. How can the same state be formed cooperatively in one protein and noncooperatively in another? The issue is easily resolved if the molten globule is simply identified as the ensemble of intermediates lying between U and N. However, if different proteins exhibit intermediates with distinct properties, little is gained by identifying them as "molten globules", a state of proteins with universal properties. In the same way, the vexed question of whether "molten globule" refers to intermediates in folding in proteins where the disulfides are broken or intact becomes irrelevant.

3.4. Folding via Trapped Intermediates

If the framework model accounts adequately for many features of folding, what can be said about the rate limiting step in folding? Present evidence does not support a sequential framework model in all cases: for example, helices that should be present early in folding have been seen to be added late in folding. Studies on cytochrome c have clarified features of the transition state in this protein, which folds by virtue of a covalently attached heme group.[94] The chain lacking heme does not fold to a native structure. Sosnick et al. employed a stopped-flow apparatus to determine the relationship between kinetically trapped intermediates and folding/refolding conditions of cytochrome c.[94] The pH of the reaction plays a key role in the heme ligation system. At pH values below the pK of two histidines (His-26 and His-33), cytochrome c folds rapidly. At higher pH, misfolding is likely to occur because the His groups ligate heme and impede folding as depicted in Fig. 18.6. The lower curve in the figure corresponds to the molecular collapse from random coil to an intermediate form which occurs early in folding. The middle curve corresponds to the time course of ligation of the

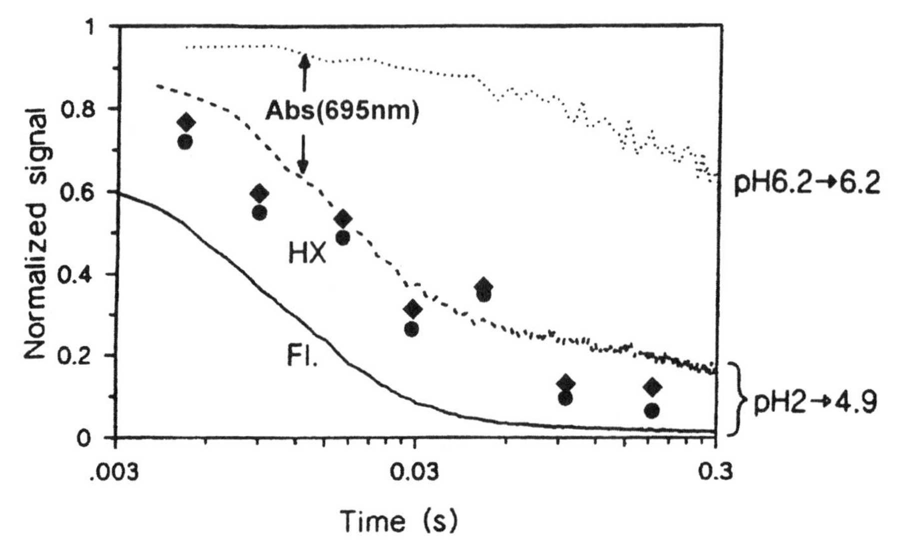

FIGURE 18.6. Fast cytochrome c folding from the unfolded state at pD 2 to native state at pH 4.9. Folding conditions were monitored in terms of initial heme ligations (Abs 695, dashed line), and formation of secondary (solid circles) and tertiary structure (solid diamonds). (After: Sosnick, T. R.; Mayne, L.; Hiller, R.; Englander, S. W. *Nature. Struct. Biol.* 1994, **1**, 149.)

heme iron by native Met-80 which is regarded as the last step in cytochrome c folding. The significance of this experiment is that at lower pH values the protein folds to the native state on the same time scale as molecular collapse, bypassing the effectively artificial kinetic barriers imposed at higher pH values.

Hydrogen exchange experiments were conducted in parallel with the stopped flow kinetics. Labeling at pH 6.2 displayed increased protection values for amide NH which correspond to kinetically trapped intermediates. However, labeling at pH 4.9 shows that kinetic intermediates are not significantly populated. It appears that the barriers which block folding and favor trapped intermediates at higher pH are not intrinsic to folding. In cytochrome c, intrinsic barriers appear to be associated only with chain condensation itself, which occurs early in folding. During condensation, any mis-folded intermediates can become populated, causing a reduction in the folding rate. The transition state is thought to be a compact collapsed state of the protein, containing little secondary structure, which forms only after a relatively slow search, ($\tau \sim 1$ ms), of the large conformational space of the chain. This is then the rate limiting step in folding-once populated, folding is energetically favored and rapid, showing single exponential rate behavior. The latter implies that the transition state is common to all molecules present and is defined by barriers that are substantial. Intermediates can be detected during the second stage of folding provided the protein is large enough: a study of the HD exchange from cytochrome c as a function of denaturant concentration shows that discrete intermediate states exist, each consisting of distinct subdomains of the native protein. For example, one intermediate involves unfolding of a segment of the protein that forms the floor of the heme, creating a loop. A similar intermediate can also be detected using limited proteolytic digestion of cytochrome c.[95] Questions remain concerning whether or not these post-transition state intermediates form sequentially or in parallel.[64] However, the detailed picture suggests that chain collapse is a crucial part of the folding process, even in a framework model. The role of secondary structure in the early transition state is not clear; the authors[64] believe it is minor.

Another result that is not easy to reconcile with a simple framework model comes from a study of the folding kinetics of a coiled-coil GCN4 dimer. As described above, coiled-coils represent the simplest tertiary fold possible in certain respects: they possess a rudimentary hydrophobic core, based on the 3, 4 pattern of nonpolar side chains at a and d positions; their stability depends also on interstrand salt bridges.[96] In addition, the determinants of antiparallel versus parallel strand association[97] and the number of strands participating in the structure have been worked out.[98] Investigation of the folding kinetics reveals simple two-state chevron behavior as a function of denaturant concentration, with no evidence for intermediates.[99] The surprise is that replacement of side chains by Ala or Gly hardly seem to influence the folding rate. Substitutions were made at various sites in the protein, with essentially no discernible effect on the folding branch. The implication is that the transition state contains little or no helix structure, hence bears no resemblance to N, which of course is almost fully helical. The helix stabilizing solvent TFE both enhances the folding rate and stabilizes the protein, as it would if the transition state contains helical structure. However, this result is considered to reflect destabilization by the solvent of U,[100] not direct stabilization of the transition state. Single Ala$\rightarrow$Gly substitutions show no such effect, although multiple substitutions do indicate fractional helix content in the transition state.[101] There is thus a puzzle here: any framework model would postulate a transition state consisting of a short helical nucleus that propagates to maximize helical structure in the dimer, a process not unlike the formation of short duplex DNA structures in a hybridization reaction. This is not observed, nor has the nonexponential rate behavior that would characterize later stages of dimerization, in which the chains anneal, been reported. One possibility then is that the transition state occurs before formation of secondary structures and the framework itself.

Several recent studies extend the resolution of folding experiments to the submicrosecond time scale. Improved stopped flow spectrometers allow mixing on a time scale shorter than 1 ms,[102] for example, while optical techniques using IR make it possible to monitor secondary structure changes on still faster time scales. In several cases, folding or unfolding processes can be detected on time scales down to nanoseconds,[103,104] implying that some steps in folding are remarkably fast. A critique of these experiments has been offered however: in fragments of cytochrome c that lack segments essential for folding to a native-like state, rapid processes are still detected as the unfolded fragments are rapidly "jumped" into native conditions.[105] The implication is that the most rapid kinetic processes might have little bearing on the folding reaction per se, and perhaps arise from changes in the state of the unfolded chain driven by the different solvent conditions. Detection of ultra-fast processes in folding does not necessarily correlate with acquisition of native structure.

3.5. Energetic Landscape of Protein Folding

The simplicity of framework models has not persuaded theoreticians working on protein folding either. Minimalistic theoretical folding models which describe proteins as a necklace of beads, or beads on a lattice, have been studied.[106] According to theoretical models, a protein traverses an energy "landscape" in which numerous possible paths lead to the native state. The topography of a folding energy landscape can be generated via computer models based on a self-interacting necklace of beads that is coupled with dynamic simulations. The density of states is initially very great, but decreases during the search until the native state is attained-a picture resembling a funnel.[107] A hypothetical folding funnel for a 60 residue protein is displayed in Fig. 18.7.[108] The funnel shows rapid folding of a string of beads through what are imagined to be collapsed molten globule states. The bottle neck region of the folding map corresponds to a variety of structures, since multiple pathways are assumed to be available during folding. "Trapping" an intermediate in one of these states can result in a reduced folding rate until a different path is selected to reach the native state. Intermediates trapped in valleys are supposed to represent misfolded intermediates.

While insights into possible folding mechanisms can be obtained by crude simulations, hard conclusions for protein folding must be taken with caution. Lattice models oversimplify the process a great deal for one thing, poorly representing secondary structure and severely under-estimating the chain entropy. For another, the role of the polar backbone cannot be determined unless solvent is included and this is best done explicitly. It should be noted that the form of the funnel in Fig. 18.7 misrepresents a series of intermediates on a pathway. The diagram would need to be modified to contain progressive narrowings corresponding to population of successive I states, deforming the funnel severely. The resulting figure would resemble a series of sequentially smaller funnels, terminating in a single N state.

As noted before, there are contrary views about the role of secondary structure in protein structure: on the one hand, it is argued that formation of secondary structure is an essential component of early intermediates in folding, hence that secondary structure is crucial to folding. On the other, secondary structure is pictured to be a by-product of compaction of the chain as it condenses to the native state. According to the latter idea, secondary structure is not a property of proteins that merits study. In fact it can be shown that local sequence is insufficient to dictate absolute folding preference: an 11 residue sequence can fold to form either α-helix or β-sheet depending on context.[109] Despite this, the results of many experimental studies of simple peptide models show that

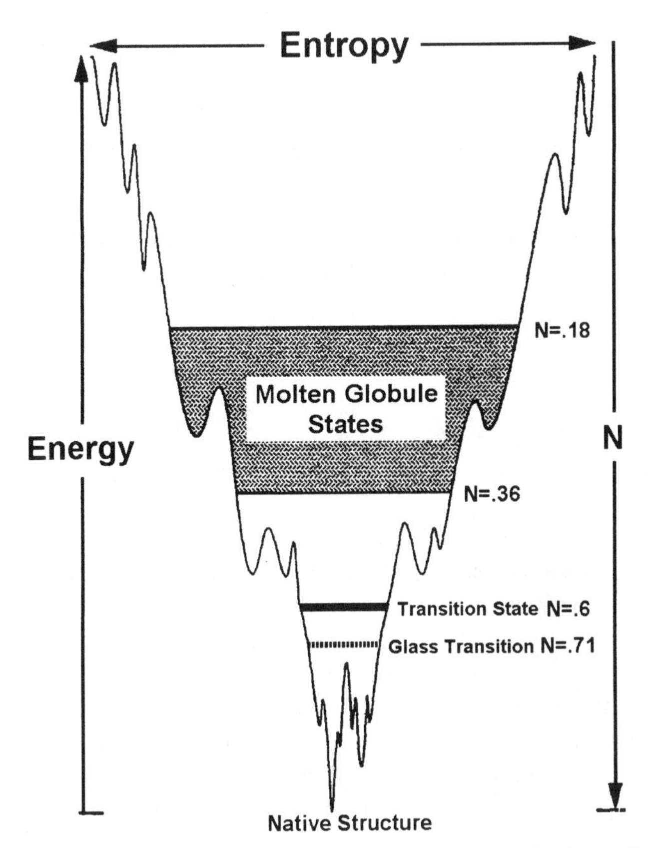

FIGURE 18.7. Folding funnel for a rapid-folding 60 residue helical protein. The width of the funnel represents entropy, and depth, the energy, N represents the number of native contacts correctly made by the protein as is traverses the folding funnel in route to the native state. It is not clear what the valleys in walls means. (After: Onuchic, J. N.; Wolynes, P. G.; Luthey-Schulten, Z.; Socci, N. D. *Proc. Natl. Acad. Sci. U.S.A.* 1995, **92**, 3626.)

secondary structure is important per se, in contrast to theories based on simplistic models of folded proteins such as beads on a lattice.

To summarize this discussion, it may be more useful to identify questions that are outstanding rather than review principles of protein folding. These include:

1. Why are folding rates simple exponentials if the underlying processes can in principle be more complex?
2. What is the role of secondary structure in stabilizing intermediates in folding? Are there universal folding units such as omega loops that mediate any or all folding reactions?
3. Do different molecules follow different pathways to the folded state?

4. What are the energy barriers responsible for the transition state in protein folding?
5. Can the native folded state of a protein be predicted from sequence without information about the pathway, or given information about the pathway?

Answers to these questions will require further experimentation and analysis. The problem is of fundamental importance and interest, and warrants additional effort.

REFERENCES

1. Creighton, T. E. *Proteins: Structures and Molecular Principles.* W. H. Freeman: New York, 1983.
2. Pauling, L.; Corey, R. B.; Branson, H. B. *Proc. Natl. Acad. Sci.* 1951, **37**, 205.
3. Pauling, L. *General Chemistry.* Dover Publications Inc.: New York, 1970.
4. Hermans, J. Jr. *J. Phys. Chem.* 1966, **70**, 510.
5. Scholtz, J. M.; Marqusee, S.; Baldwin, R. L.; Stewart, J. M.; Santoro, M.; Bolen, D. W. *Proc. Natl. Acad. Sci. U.S.A.* 1991, **88**, 2854.
6. Chou, P. Y.; Fasman, G. D. *Biochemistry* 1974, **13**, 211.
7. Richardson, J. S.; Richardson, D. C. *Science* 1988, **240**, 1648.
8. Marqusee, S.; Baldwin, R. L. *Proc. Natl. Acad. Sci. U.S.A.* 1989, **86**, 5286.
9. Lyu, P. C.; Liff, M. I.; Marky, L. A.; Kallenbach, N. R. *Science* 1990, **250**, 669.
10. Park, S. H.; Shalongo, W.; Stellwagen, E. *Biochemistry* 1993, **32**, 7048.
11. Rohl, C. A.; Chakrabartty, A.; Baldwin, R. L. *Protein Sci.* 1996, **5**, 2623.
12. Creamer, P. T.; Rose, G. D. *Proc. Natl. Acad. Sci. U.S.A.* 1992, **89**, 5937.
13. Wojcik, J.; Altmann, K. H.; Scheraga, H. A. *Biopolymers* 1990, **30**, 121.
14. Zimm, B. H.; Bragg, J. K. *J. Chem. Phys.* 1959, **31**, 526.
15. Lifson, S.; Roig, A. *J. Chem. Phys.* 1961, **34**, 1963.
16. Padmanabhan, S.; York, E. J.; Gera, L.; Stewart, J. M.; Baldwin, R. L. *Biochemistry* 1994, **33**, 8604.
17. Gans, P. J.; Lyu, P. C.; Manning, M. C.; Woody, R. W.; Kallenbach, N. R. *Biopolymers* 1991, **31**, 1605.
18. Rohl, C. A.; Scholtz, J. M.; York, E. J.; Stewart, J. M.; Baldwin, R. L. *Biochemistry* 1992, **31**, 1263.
19. O'Neil, K. T.; DeGrado, W. F. *Science* 1990, **250**, 646.
20. Groebke, K.; Renold, P.; Tsang, K. Y.; Allen, T. J.; McClure, F. F.; Kemp, D. S. *Proc. Natl. Acad. Sci. U.S.A.* 1996, **93**, 4025.
21. Vasquez, M.; Scheraga, H. A. *Biopolymers* 1988, **32**, 41.
22. Blaber, M.; Zhang, X. J.; Matthews, B. W. *Science* 1993, **260**, 1637.
23. Munoz, V.; Serrano, L. *Nature Struct. Biol.* 1994, **1**, 399.

24. Kallenbach, N. R.; Lyu, P. C.; Zhou, H. X. In *Circular Dichroism and the Conformational Analysis of Biomolecules*; Fasman, G. D. Ed.; Plenum Press: New York, 1996; pp. 201–259.
25. Bierzynski, A.; Kim, P. S.; Baldwin, R. L. *Proc. Natl. Acad. Sci. U.S.A.* 1982, **79**, 2470.
26. Lockhart, D. J.; Kim, P. S. *Science* 1993, **260**, 198.
27. Huyghues-Despointes, B. M. P.; Scholtz, J. M.; Baldwin, R. L. *Protein Sci.* 1993, **2**, 1604.
28. Lyu, P. C.; Gans, P. J.; Kallenbach, N. R. *J. Mol. Biol.* 1992, **223**, 343.
29. Scholtz, J. M.; Qian, H.; Robbins, V. H.; Baldwin, R. L. *Biochemistry* 1993, **32**, 9668.
30. Yip, K. S. P.; Stillman, T. J.; Britton, K. L.; Artymiuk, P. J.; Baker, P. J.; Sedelnikova, S. E.; Engel, P. C.; Pasquo, A.; Chiaraluce, R.; Consalvi, V.; Scandurra, R.; Rice, D. W. *Structure* 1995, **3**, 1147.
31. Padmanabhan, S.; Baldwin, R. L. *Protein Sci.* 1994, **3**, 1992.
32. Creamer, T. P.; Rose, G. D. *Protein Sci.* 1995, **4**, 1305.
33. Harper, E.; Rose, G. D. *Biochemistry* 1993, **32**, 7605.
34. Gong, Y.; Zhou, H. X.; Guo, M.; Kallenbach, N. R. *Protein Sci.* 1995, **4**, 1446.
35. Schellman, C. In *Protein Folding*; Jaenicke, R. Ed.; Elsevier/North Holland Biomedical Press: Amsterdam, 1980, pp. 53–61.
36. Aurora, R.; Srinivasan, R.; Rose, G. D. *Science* 1994, **264**, 1126.
37. Yang, J.; Spek, E. J.; Gong, Y.; Zhou, H.; Kallenbach, N. R. *Protein Sci.* 1996, **6**, 1264.
38. Doig, A. J.; Chakrabartty, A.; Klinger, T. M.; Baldwin, R. L. *Biochemistry* 1994, **33**, 3396.
39. Minor, D. L. Jr.; Kim, P. S. *Nature* 1994, **367**, 660.
40. Smith, C. K.; Regan, L. *Science* 1995, **270**, 980.
41. Kim, C. A.; Berg, J. M. *Nature* 1993, **362**, 267.
42. Roy, S.; Helmer, K. J.; Hecht, M. H. *Fold. Des.* 1997, **2**, 89.
43. Zhou, H. X.; Hoess, R. H.; DeGrado, W. F. *Nat. Struct. Biol.* 1996, **3**, 446.
44. Finkelstein, A. V.; Ptitsyn, O. B. *J. Mol. Biol.* 1976, **103**, 15.
45. Srinivasan, R.; Rose, G. D. *Proteins* 1995, **22**, 81.
46. West, M. W.; Hecht, M. H. *Protein Sci.* 1995, **4**, 2032.
47. Spek, E. J.; Wu, H. C.; Kallenbach, N. R. *J. Amer. Chem. Soc.* 1997, **119**, 5053.
48. Li, C. S.; Deber, C. M. *Nature Stuct. Biol.* 1994, **1**, 368.
49. Luo, P.; Baldwin, R. L. *Biochemistry* 1997, **36**, 8413.
50. Kauzmann, W. *Adv. Protein Chem.* 1959, **14**, 1.
51. Dill, K. *Biochemistry* 1990, **29**, 7133.
52. Myers, J. K.; Pace, C. N.; Scholtz, J. M. *Proc. Natl. Acad. Sci. U.S.A.* 1997, **94**, 2833.
53. Makhatadze, G. I.; Privalov, P. L. *Adv. Protein Chem.* 1995, **47**, 307.
54. Behe, M. J.; Lattman, E. E.; Rose, G. D. *Proc. Natl. Acad. Sci. U.S.A.* 1991, **88**, 4195.

55. Harbury, P. B.; Zhang, T.; Kim, P. S.; Alber, T. *Science* 1993, **262**, 1401.
56. Guoling, T.; Vainberg, I. E.; Tap, W. D.; Lewis, S. A.; Cowan, J. *Nature* 1995, **375**, 250.
57. Mayhew, M.; Hartl, F. H. *Science* 1996, **271**, 161.
58. Fenton, W. A.; Horwich, A. L. *Protein Sci.* 1997, **6**, 743.
59. Creighton, T. E. *Prog. Biophys. Mol. Biol.* 1978, **33**, 231.
60. Creighton, T. E. *J. Phys. Chem.* 1985, **89**, 2452.
61. Schmid, F. *Biochemistry* 1983, **22**, 4690.
62. Englander, S. W.; Mayne, L. *Annu. Rev. Biophys. Biomol. Struct.* 1992, **21**, 243.
63. Baldwin, R. L. *BioEssays* 1993, **16**, 207.
64. Englander, S. W.; Sosnick, T. R.; Englander, J. J.; Mayne, L. *Curr. Opin. Struct. Biol.* 1996, **6**, 18.
65. Rodor, H.; Colon, W. *Curr. Opin. Struct. Biol.* 1997, **7**, 15.
66. Burton, R. E.; Huang, G. S.; Dougherty, M. A.; Fullbright, P. W.; Oas, T. G. *J. Mol. Biol.* 1996, **263**, 311.
67. Clore, G. M.; Gronenborn, A. M. *Science* 1993, **260**, 1110.
68. Kim, P. S.; Baldwin, R. L. *Annu. Rev. Biochem.* 1990, **59**, 631.
69. Levinthal, C. J. *J. Chim. Physique* 1968, **65**, 44.
70. Neri, D.; Billeter, M.; Wider, G.; Wuthrich, K. *Science* 1992, **257**, 1559.
71. Fersth, A. R. *Curr. Opin. Struct. Biol.* 1997, **7**, 3.
72. Ansari, A.; Berendzen, J.; Bowne, S. F.; Frauenfelder, H.; Iben, I. E.; Sauke, T. B.; Shyamsunder, E.; Young, R. D. *Proc. Natl. Acad. Sci. U.S.A.* 1985, **82**, 5000.
73. Creighton, T. E. *J. Mol. Biol.* 1977, **113**, 275.
74. Wiessman, J. S.; Kim, P. S. *Nature Struct. Biol.* 1995, **2**, 1123.
75. Itshaki, L. S.; Evans, P. A.; Dobson, C. M.; Radford, S. E. *Biochemistry* 1994, **33**, 5212.
76. Oas, T. G.; Kim, P. S. *Nature* 1988, **336**, 42.
77. Ptitsyn, O. B.; Pain, R. H.; Semisotnov, G. V.; Zerovnik, E.; Razgulyaev, O. J. *FEBS Letters* 1990, **262**, 20.
78. Christensen, H.; Pain, R. H. *Eur Biophys J.* 1991, **19**, 221.
79. Carrey, E. A.; Pain, R. H. *Biochem. Biophys. Acta* 1978, **553**, 12.
80. Mitchinson, C.; Pain, R. H. *J. Mol. Biol.* 1985, **184**, 331.
81. Baum, J.; Dobson, C. M.; Evans, P. A.; Hanley, C. *Biochemistry* 1989, **28**, 7.
82. Damaschun, G.; Germat, Ch.; Damaschun, H.; Bychkova, V. E.; Ptitsyn, O. B. *Int. J. Biol. Macromol.* 1986, **8**, 226.
83. Dolgikh, D. A.; Gilmanshin, R. I.; Brazhnikov, E. V.; Bychkova, V. E.; Semisotnov, G. V.; Venyaminov, S. Y.; Ptitsyn, O. B. *FEBS Letters* 1981, **136**, 311.
84. Creighton, T. E.; Pain, R. H. *J. Mol. Biol.* 1980, **137**, 431.
85. Gast, K.; Ziwer, D.; Welfle, H.; Bychkova, V. E.; Ptitsyn, O. B. *Int. J. Biol. Macromol.* 1986, **8**, 231.
86. Schulman, B. A.; Kim, P. S. *Nature Struct. Biol.* 1996, **8**, 682.
87. Kay, M. S.; Baldwin, R. L. *Nature Struct. Biol.* 1996, **3**, 439.

88. Lin, L.; Pinker, R. J.; Forde, K.; Rose, G. D.; Kallenbach, N. R. *Nature Struct. Biol.* 1994, **1**, 447.
89. Eliezer, D.; Wright, P. E. *J. Mol. Biol.* 1996, **263**, 531.
90. Matheson, R. R.; Scheraga, H. A. *Macromolecules* 1978, **11**, 819.
91. Waltho, J. P.; Feher, V. A.; Merutka, G.; Dyson, H. J.; Wright, P. E. *Biochemistry* 1993, **32**, 6337.
92. Shin, H. C.; Merutka, G.; Waltho, J. P.; Wright, P. E.; Dyson, H. J. *Biochemistry* 1993, **32**, 6348.
93. Shin, H. C.; Merutka, G.; Waltho, J. P.; Tennant, L. L.; Dyson, H. J.; Wright, P. E. *Biochemistry* 1993, **32**, 6356.
94. Sosnick, T. R.; Mayne, L.; Hiller, R.; Englander, S. W. *Nature. Struct. Biol.* 1994, **1**, 149.
95. Wang, L.; Chen, R. X.; Kallenbach, N. R. *Proteins* 1998, **30**, 435.
96. Lavigne, P.; Sonnichsen, F. D.; Kay, C. M.; Hodges, R. S. *Science* 1996, **271**, 1136.
97. Lumb, K. J.; Kim, P. S. *Biochemistry* 1995, **34**, 8642.
98. Harbury, P. B.; Kim, P. S.; Alber, T. *Science* 1993, **261**, 879.
99. Sosnick, T. R.; Jackson, S.; Wilk, R. M.; Englander, S. W.; De Grado, W. F. *Proteins Struct. Funct. Genet.* 1996, **24**, 427.
100. Kentsis, A.; Sosnick, T. R. *Biochemistry* 1998, in press.
101. Matthews, B. W. private correspondance.
102. Pascher, T.; Chesick, J. P.; Winkler, J. R.; Gray, H. *Science* 1996, **271**, 1558.
103. Jones, C. M.; Henry, E. R.; Hu, Y.; Chan, C. K.; Luck, S. D.; Bhuyan, A.; Roder, H.; Hofrichter, J.; Eaton, W. A. *Proc. Natl. Acad. Sci. U.S.A.* 1993, **90**, 11860.
104. Phillips, C. M.; Mizutani, Y.; Hochstrasser, R. M. *Proc. Natl. Acad. Sci. U.S.A.* 1995, **92**, 7292.
105. Sosnick, T. R.; Shtilerman, M. D.; Englander, S. W. *Proc. Natl. Acad. Sci. U.S.A.* 1997, **94**, 8545.
106. Hinds, D. A.; Levitt, M. *J. Mol. Biol.* 1996, **258**, 201.
107. Wolynes, P. G.; Onuchic, J. N.; Thirualai, D. *Science* 1995, **267**, 1619.
108. Onuchic, J. N.; Wolynes, P. G.; Luthey-Schulten, Z.; Socci, N. D. *Proc. Natl. Acad. Sci. U.S.A.* 1995, **92**, 3626.
109. Minor Jr., D. L.; Kim, P. S. *Nature* 1996, **380**, 730.

INDEX